Prokaryotic Development

Prokaryotic Development

Edited by

Yves V. Brun
Department of Biology
Indiana University
Bloomington, IN 47405-6801

Lawrence J. Shimkets
Department of Microbiology
University of Georgia
Athens, GA 30602

ASM
PRESS

Washington, DC

Copyright © 2000
American Society for Microbiology
1325 Massachusetts Avenue, N.W.
Washington, DC 20005-4171

Library of Congress Cataloging-in-Publication Data

Prokaryotic development / edited by Yves V. Brun, Lawrence J. Shimkets.
    p.   cm.
    Includes bibliographical references and index.
    ISBN 1-55581-158-2
      1. Prokaryotes—Development. 2. Microbial cell cycle. I. Brun, Yves V.
    II. Shimkets, Lawrence J.

   QR73.4.P76 1999
   571.8′293—dc21   99-044864

CONTENTS

CONTRIBUTORS

David G. Adams
Division of Microbiology, School of Biochemistry and Molecular Biology, University of Leeds, Leeds, LS2 9JT, United Kingdom

Yves V. Brun
Department of Biology, Jordan Hall 142, Indiana University, 1001 E. 3rd St., Bloomington, IN 47405-3700

William F. Burkholder
Department of Biology, Building 68-530, Massachusetts Institute of Technology, Cambridge, MA 02139

Wendy Champness
Department of Microbiology, Michigan State University, East Lansing, MI 48824-1101

Keith F. Chater
Department of Genetics, John Innes Centre, Norwich Research Park, Colney, Norwich, NR4 7UH, United Kingdom

Adam Driks
Department of Microbiology and Immunology, Loyola University Medical Center, 2160 South First Ave., Maywood, IL 60153

Martin Dworkin
Department of Microbiology, University of Minnesota, Minneapolis, MN 55455-0312

Jennifer C. England
Department of Chemistry and Biochemistry and Molecular Biology Institute, University of California, Los Angeles, CA 90095-1569

Gillian M. Fraser
Department of Pathology, University of Cambridge, Tennis Court Road, Cambridge, CB2 1QP, United Kingdom

Richard B. Furness
Department of Pathology, University of Cambridge, Tennis Court Road, Cambridge, CB2 1QP, United Kingdom

James W. Gober
Department of Chemistry and Biochemistry and Molecular Biology Institute, University of California, Los Angeles, CA 90095-1569

Thorsten W. Grebe
Department of Molecular Biology, Princeton University, Princeton, NJ 08544-1014

Alan D. Grossman
Department of Biology, Building 68-530, Massachusetts Institute of Technology, Cambridge, MA 02139

Colin Hughes
Department of Pathology, University of Cambridge, Tennis Court Road, Cambridge, CB2 1QP, United Kingdom

Dean Hung
Department of Genetics, Beckman Center B300, Stanford University School of Medicine, Stanford, CA 94305

Raji Janakiraman
Department of Biology, University of Michigan, 830 North University, Ann Arbor, MI 48109-1048

Dale Kaiser
Department of Biochemistry and Department of Developmental Biology, Stanford University School of Medicine, Beckman Center, B300, Stanford, CA 94305-5329

Petra Anne Levin
Department of Biology, Building 68-530, Massachusetts Institute of Technology, Cambridge, MA 02139

Richard Losick
Department of Molecular and Cellular Biology, The Biological Laboratories, Harvard University, 16 Divinity Ave., Cambridge, MA 02138

William Margolin
Department of Microbiology and Molecular Genetics, University of Texas Medical School, 6431 Fannin, Houston, TX 77030

Akira Matsumoto
Department of Microbiology, Kawasaki Medical School, 577 Matsushima, Kurashiki City, Okayama 701-01, Japan

Harley McAdams
Department of Developmental Biology, Beckman Center B300, Stanford University School of Medicine, Stanford, CA 94305

Austin Newton
Department of Molecular Biology, Princeton University, Princeton, NJ 08544-1014

Noriko Ohta
Department of Molecular Biology, Princeton University, Princeton, NJ 08544-1014

Daniel D. Rockey
Department of Microbiology, Oregon State University, Corvallis, OR 97331-3804

James E. Samuel
Department of Medical Microbiology and Immunology, Texas A&M University System Health Science Center, College Station, TX 77843-1114

Hans Ulrich Schairer
Zentrum für Molekulare Biologie, Universität Heidelberg, Im Neuenheimer Feld 282, D-69120 Heidelberg, Germany

J. William Schopf
IGPP Center for the Study of Evolution and the Origin of Life, Department of Earth and Space Sciences, and Molecular Biology Institute, University of California, Los Angeles, CA 90095-1567

Peter Setlow
Department of Biochemistry, University of Connecticut Health Center, Farmington, CT 06032

Lucy Shapiro
Department of Developmental Biology, Beckman Center B300, Stanford University School of Medicine, Stanford, CA 94305

Lawrence J. Shimkets
Department of Microbiology, 527 Biological Sciences Building, University of Georgia, Athens, GA 30602

Abraham L. Sonenshein
Department of Molecular Biology and Microbiology, Tufts University School of Medicine, Boston, MA 02111-1800

Mandy J. Ward
Department of Molecular and Cell Biology, University of California at Berkeley, Berkeley, CA 94720-3204

David White
Department of Biology, Indiana University, Bloomington, IN 47401

C. Peter Wolk
MSU-DOE Plant Research Laboratory, Michigan State University, East Lansing, MI 48824

David R. Zusman
Department of Molecular and Cell Biology, University of California at Berkeley, Berkeley, CA 94720-3204

PREFACE

Bacteria are capable of amazing feats. They generated the oxidizing environment we inhabit today, and they can eke out a living in environments that are hostile to most forms of life. The focus of this book centers on those select species that perform another amazing feat by undergoing complex developmental transformations. This book follows by 10 or more years three excellent books acknowledging the diversity of prokaryotic developmental systems, including the 1984 book entitled *Microbial Development*, edited by Richard Losick and Lucy Shapiro; *Developmental Biology of the Bacteria* (1985) by Martin Dworkin; and *Genetics of Bacterial Diversity*, edited by David Hopwood and Keith Chater, published in 1989. At that time there was already a great deal of excitement in the field of bacterial development. Many developmental systems were being examined by genetic and biochemical approaches. Molecular analysis was beginning to yield important insight into the function of genes involved in the control of development. Dependence patterns of developmental gene expression were being elucidated, setting the stage for the understanding of the regulatory networks that control development. Around 1995, Greg Payne from ASM Press realized that, with the many exciting new discoveries in prokaryotic development, the time was ripe to synthesize a decade's worth of progress in understanding prokaryotic development.

Research on bacterial development has progressed at an impressive pace in the past 10 years. The goal of this book is to describe the exciting advances that have occurred as a result of the transition from the almost purely genetic to the molecular analysis of bacterial development. We are at the beginning of another important transition in the methods of analysis used to study bacterial development. In the past few years, the application of cytological tools to bacteria has allowed the determination of the subcellular location of many proteins important for development. Knowing where a protein is in the cell, in addition to its sequence and the timing of its expression, can provide major insight into its function. This transition in technology is especially important since the initial impetus to study many of the experimental systems outlined in this book was to understand processes that yield morphological changes at a specific site within the cell. Thus, while most of the chapters describe the regulatory networks involved in the control of development, positional information has a place of choice in many of them. It is also the case that genome sequencing and associated technologies will forever change approaches to studying these regulatory networks. The genome-sequencing projects are so new, however, that they have not yet been exploited.

We chose to focus this book on the experimental systems in which a great deal is known about the molecular mechanisms of development. In contrast to all previous books on prokaryotic development, this book devotes more than one chapter to the organisms that have yielded the most mechanistic information. Thus, most experimental systems are described in more than one chapter, and there are four or five chapters each on *Bacillus*, *Myxobacteria*, and *Caulobacter*. This has allowed a more in-depth coverage of developmental processes than previously available in one place. Introductory chapters describe the biology of each group of organisms in order to place the molecular analysis in a biological and phylogenetic perspective.

This book should be accessible to advanced undergraduate and graduate students and should be especially helpful in courses dealing with microbial diversity and microbial development. It will provide an excellent introduction to current research on prokaryotic development to new investigators, and will be a useful reference for more experienced researchers. We hope that by learning more about other experimental systems, researchers in the field will derive insights for their own work.

We thank Greg Payne from ASM Press for the push at the beginning, for help in focusing an overly ambitious outline, and for help at every stage of the project. We thank Ellie Tupper at ASM Press for producing the book in such an efficient manner that it made us feel like there was nothing left for us to do. We thank the authors of the chapters for their excellent contributions. We appreciate the willingness of many colleagues to review chapters for scientific content, including David Adams, Keith Chater, Wendy Champness, Gene Crawford, Adam Driks, Marty Dworkin, Jim Gober, Jim Golden, Ted Hackstadt, Dale Kaiser, Dan Kearns, Lee Kroos, Sharon Long, Austin Newton, Patrick Piggot, Dan Rockey, Peter Setlow, Jim Shapiro, Lucy Shapiro, David White, Peter Wolk, and David Zusman. Finally, we thank the members of our laboratories and our families for their understanding when our attention was distracted by this project.

Yves V. Brun
Lawrence J. Shimkets

PROKARYOTIC DEVELOPMENT: STRATEGIES TO ENHANCE SURVIVAL

Lawrence J. Shimkets and Yves V. Brun

What constitutes development? In the 1960s it was suggested that "two cells are differentiated with respect to each other if, while they harbour the same genome, the pattern of proteins which they synthesize is different" (Jacob and Monod, 1963). While changes in gene expression are certainly a central feature of developmental systems, it is now clear that such changes are insufficient to constitute development. Insight into global regulation in the 1980s and 1990s has demonstrated that changes in growth temperature or nutrient availability have dramatic consequences for gene expression. In the 1980s a more rigorous definition proposed that "development is a series of stable or metastable changes in the form or function of a cell, where those changes are a part of the normal life cycle of the cell" (Dworkin, 1985). Dworkin (1985) went on to state that "a substantial change in form as well as function defines the developmental process." Following his lead, we suggest that prokaryotic development involves changes in form and function that play a prominent role in the life cycle of the organism. This definition applies equally well to prokaryotic and eukaryotic development.

The purpose of this book is to bring greater visibility to prokaryotic development by acknowledging the insightful contributions of many laboratories to the physiology of development. We hope this book will help bring prokaryotic development out from under the shadow of animal development. It is now clear that prokaryotic development is the product of different organizational strategies, is subject to different selective and evolutionary pressures, and contains a much higher level of genetic and biological diversity. In outlining the composition of this book, we chose to highlight experimental systems that are amenable to genetic analysis and for which the molecular mechanisms of developmental control are understood in some detail. For the major model systems, an introductory chapter will present the biology of the model organism and related bacteria and outline the important problems they pose.

EUKARYOTIC DEVELOPMENT SCHEMES

Eukaryotic development has goals that differ in fundamental ways from those of prokaryotic development. One result of eukaryotic development, specifically in plants and animals, is to build morphological complexity through the differentiation of tissues with specialized physi-

L. J. Shimkets, Department of Microbiology, University of Georgia, Athens, GA 30602. *Y. V. Brun*, Department of Biology, Indiana University, Bloomington, IN 47405.

Prokaryotic Development, edited by Y. V. Brun and L. J. Shimkets,
© 2000 American Society for Microbiology, Washington, DC 20005-4171

ological functions. Complexity, and to a lesser extent size, are roughly proportional to the number of cell types involved. An expanding pattern of cell division and differentiation eventually yields a mature organism. In sharp contrast, multicellularity is barely visible to the naked eye in even the most complex prokaryotic species, like the actinomycetes and the myxobacteria. Morphological complexity is certainly not the modus operandi in prokaryotic development.

A second result of eukaryotic development is sexual reproduction to generate genetic diversity in progeny. In marked contrast, prokaryotic developmental cycles are completely asexual. Apparently the evolutionary need for genetic diversification among prokaryotes is satisfied at the genomic level by the accumulation of mutations and occasional import of new genes and is enhanced at the community level by their large population sizes and rapid replication times. This may be one reason prokaryotic evolution proceeds much more slowly at the morphological level than eukaryotic evolution (see chapter 5).

Finally, animal development is mostly uninformed by the natural environment whereas in many prokaryotic systems development is induced in direct response to environmental changes. The usual consequence of prokaryotic development is the differentiation of a cell type that is more highly adapted to the current environment rather than requiring the organism to deal less optimally with a wider range of environmental situations. Spore-forming cells may remain in the growth phase indefinitely provided nutrients are plentiful, and spores may remain dormant for many years if suitable growth conditions are not encountered. Environmental initiation of development has been the subject of investigation in several systems (see chapters 4, 7, 12, and 13). Interestingly, the uncoupling of growth from development has been exploited in the genetic analysis of prokaryotic development, since many types of developmental mutations do not interfere with growth and hence are not lethal.

PROKARYOTIC DEVELOPMENT SCHEMES

There are several dozen examples of prokaryotic developmental cycles. These can be divided into four categories based on the function of the product of the developmental cycle (Table 1). The most common scheme is the induction of resting cells by nutritional stress. Resting cells vary in their degree of residual metabolic activity but are uniformly more dormant than their vegetative counterparts. They are also more durable to physical and chemical stress than the vegetative cell due to their unique molecular architecture (see chapter 9). This scheme is exemplified by the endospore, the most resistant cell in the world. Endospore formation begins with a partial asymmetric cell division and subsequent engulfment of the prespore by the mother cell to compartmentalize the prespore inside the mother cell (see chapter 8). The genome of the mother cell provides the components for constructing the spore exterior, and the genome of the forespore provides the components for constructing the spore interior. This remarkable division of labor is due to the presence of different sigma factors in each compartment, which ensures that each genome gives rise to a different set of products. An interesting relative of *Bacillus*, *Metabacterium*, produces multiple endospores as a means of proliferation, since binary fission is rare (see chapter 6). Resting cells appear to have evolved many times, since there is an enormous variety in the mechanisms by which they form. Many types of spores are not formed inside a mother cell like the endospore. In *Anabaena* (see chapter 3), *Myxococcus* (see chapter 10), *Azotobacter*, and many other genera the entire cell becomes a spore. In *Methylosinus*, the spores are produced by complete asymmetric division of a progenitor cell, resulting in a mother cell with an attached exospore (Whittenbury et al., 1970). In *Streptomyces* there is extensive fission of a long, multinucleate progenitor cell to form a series of spores (see chapter 2). In the other actinomycetes, spores are formed from a multinucleate cell mass by a series of cell divisions in different planes (see chapter 1).

TABLE 1 Examples of prokaryotic development

Type	Cell	Representative genus or interaction	Group	Function
Resting cells	Endospore	*Bacillus*	Gram positive	
		Metabacterium	Gram positive	
		Thermoactinomyces	Gram positive	
	Aerial spore	*Streptomyces*	Gram positive	
	Zoospore	*Dermatophilus*	Gram positive	
	Cyst	*Azotobacter*	Proteobacteria	
		Methylomonas	Proteobacteria	
		Bdellovibrio	Proteobacteria	
	Myxospore	*Myxococcus*	Proteobacteria	
		Stigmatella	Proteobacteria	
	Exospore	*Methylosinus*	Proteobacteria	
	Small dense cell	*Coxiella*	Proteobacteria	
	Elementary body	*Chlamydia*	Chlamydia	
	Akinete	*Anabaena*	Cyanobacteria	
Complementary cell types	Heterocyst			N_2 fixation
	Vegetative cell	*Anabaena*	Cyanobacteria	Oxygenic photosynthesis
Dispersal cells	Baeocyte	*Pleurocapsa*	Cyanobacteria	
	Elementary body	*Chlamydia*	Chlamydia	
	Gonidium	*Leucothrix*	Proteobacteria	
	Hormogonium	*Oscillatoria*	Cyanobacteria	
	Swarm cell	*Proteus*	Proteobacteria	
	Swarmer cell	*Caulobacter*	Proteobacteria	
	Zoospore	*Dermatophilus*	Gram positive	
Symbiotic development	Bacteroid	*Rhizobium*-legume	Proteobacteria	N_2 fixation
		Frankia-alder	Gram positive	N_2 fixation

A second developmental scheme is the differentiation of cell types whose physiologies are complementary, for example, the *Anabaena* vegetative cell and heterocyst. The vegetative cell provides the heterocyst with organic carbon derived from photosynthesis, while the heterocyst provides the vegetative cell with nitrogen. Sequestering nitrogenase in a specialized cell type allows more efficient nitrogen fixation, since O_2 generated by photosynthesis is a strong inhibitor of nitrogenase. While this example is superficially similar to the differentiation of specialized tissues in eukaryotic organisms, there are important differences. Heterocyst differentiation is not programmed into the cell division cycle but is induced by environmental stress, namely nitrogen limitation (see chapters 3 and 4).

The third scheme is the differentiation of cells for the purpose of dispersal. Such cells may be motile derivatives of a sessile counterpart or they may be nonmotile but released under conditions where they are easily dispersed by wind or water. The gonidium of *Leucothrix*, the hormogonium of *Oscillatoria*, and the swarm cells of *Caulobacter* and *Proteus* are motile cells that provide different means to this end. Differentiation of motile cells can be either induced by environmental stress or coupled to the cell division cycle in a cyclic manner. For a portion of the *Leucothrix* life cycle the cells grow as long nonmotile filaments attached to the substrate by a basal holdfast. In response to conditions that limit growth, ovoid gonidia are differentiated at the tips of the filaments. The gonidia colonize surfaces, where they move by gliding for a time and then produce a holdfast and generate a new filament through growth and cell division (Harold and Stanier, 1955). The filamentous cyanobacteria fragment hormogonia by a septation process that is different from the normal division of cells. Hormogonia in contact with a solid surface move by gliding motility (see chapter 3). The swarmer cells of *Caulobacter* are the progeny of stalked-cell division. Since they grow in oligotrophic environments, there is great value from the vantage point of the stalked cell in making sure the progeny do not become competitors. The swarmer cells swim for an obligatory dispersal period before they develop stalks and holdfasts and colonize a surface (see chapter 15). The swarm cells of *Proteus* differ from the vegetative cells in that they are much larger, are multinucleate, and contain many lateral flagella. Presumably they are more adept at colonizing the human host, since a variety of virulence factors are coordinately expressed with swarm cell differentiation (see chapter 19). Not all dispersal processes involve the production of motile cells. Baeocytes produced by *Pleurocapsa* cyanobacteria may or may not be motile. The reproductive cell undergoes multiple fission without intervening cell growth to produce as many as 1,000 baeocytes (see chapter 3).

The final scheme is symbiotic development, in which a pair of organisms develop a relationship that supercedes the normal free-living state of either organism. Nodulation of legume roots by *Rhizobium* species is the most extensively studied example. Here the free-living *Rhizobium* cells undergo extensive morphological and physiological changes to provide nitrogen fixation for the plant (see chapter 22). Similarly, *Frankia* nodulates *Alnus* and several other plant genera, though the process has not been studied in detail (see chapter 1).

Finally, it should be noted that some types of development fall into multiple categories. For example, *Dermatophilus* zoospores have flagellar motility and are used both for dispersal and as resting cells. The elementary body of the obligate intracellular pathogen *Chlamydia* is both a resting cell and the dispersal cell that transmits the infection to a new host (see chapter 20). The purpose of the fruiting body of the myxobacteria may be to make sure that a large population of spores germinates together because groups are more efficient feeders than isolated cells (see chapter 10).

ORGANIZATION OF PROKARYOTIC DEVELOPMENT CYCLES

Developmental cycles may be divided into four groups based on the manner in which they are organized.

1. Unicellular and cell cycle independent
    A. *Bacillus* endospore formation and germination (cyclic)
    B. *Anabaena* heterocyst differentiation (noncyclic)
    C. *Rhizobium* bacteroid differentiation (noncyclic)
2. Unicellular and cell cycle dependent
    A. *Caulobacter* swarmer cell differentiation (generational)
    B. *Proteus* swarm cell differentiation (multigenerational)
3. Multicellular via directed movement (cell cycle independent)
    A. *Myxococcus* and *Stigmatella* fruiting body formation
4. Multicellular via directed growth (cell cycle dependent)
    A. *Streptomyces* sporulation in aerial mycelia

The book is divided into sections dealing with phylogenetic groups that contain an organism that has been the focus of developmental studies. Each organizational strategy is represented by at least one well-studied group.

Unicellular and cell cycle-independent development is induced by an environmental signal, not as a stage in the cell division cycle. It may be divided into two groups based on whether the differentiated cell type is able to revert to the vegetative cell. The *Bacillus* endospore is induced in response to starvation and can germinate when nutrients are restored, creating a cyclic pattern (see chapter 6). On the other hand, the differentiation of the *Anabaena* heterocyst results in a terminally differentiated cell type (see chapter 4). The vegetative cells can either reproduce by binary fission or differentiate into heterocysts and thus resemble eukaryotic stem cells. In contrast, heterocysts neither reproduce nor dedifferentiate into vegetative cells. Much has been learned about heterocyst differentiation, which involves cell-cell signaling and rearrangement of genetic information, but virtually nothing is known about the mechanism by which the cell division machinery is silenced. This is also the case in *Rhizobium* bacteroid differentiation, where the mature bacteroid is metabolically active but incapable of vegetative growth (see chapter 22).

In unicellular and cell cycle-dependent development, differentiation is the product of the normal cell division process. The sessile *Caulobacter* stalked cell produces a motile swarmer cell at every cell division (see chapter 15). The asymmetric predivisional cell has a flagellum at one pole and a stalk and holdfast at the opposite pole. The establishment of cellular asymmetry prior to cell division is tightly coupled to progression through the cell cycle by various developmental checkpoints (see chapter 17). For example, flagellar biosynthesis is governed both by cell cycle events and by the progression of flagellar assembly (see chapter 16). The subsequent development of the flagellated pole into a stalked pole requires the initiation of cell division. In turn, DNA replication (see chapter 18) and cell division (see chapter 15) are initiated at specific stages during the developmental program. *Proteus* adds an additional element in that differentiation of swarm cells and subsequent dedifferentiation to vegetative cells does not occur in every genome replication. Whether these processes are a response to a replication-counting mechanism or cell-cell signals has not been determined (see chapter 19).

Multicellularity can be achieved by either directed cell movement or directed cell growth. Both processes are essential in mammalian embryogenesis but can be conveniently separated in bacterial systems. Myxobacterial fruiting body development results from the organized movement of tens of thousands of cells (see chapter 10). The *Myxococcus* fruiting body is little more than a raised mound of cells that differentiate into myxospores. The *Stigmatella* fruiting body contains spore-filled sporangioles atop a long stalk (see chapter 14). The mechanism by which the cell moves to the aggregation center remains unclear but may involve chemotaxis. Two different *Myxococcus* sensory transduction systems, homologous to the enteric chemotaxis system, have been shown to be essential for fruiting body development (see chapter 11). Furthermore, certain phosphatidylethanolamine derivatives have been shown to be chemoattractants for *Myxococcus* (Kearns and Shimkets, 1998).

The aerial mycelium of *Streptomyces* forms by directed cell growth and differentiates into a series of spores (see chapter 2). The vegetative mycelium grows in the nutrient substratum by the linear growth of cell wall close to the hyphal tip. Branching of the vegetative mycelium allows close-to-exponential increase of the mycelial mass. Septation is infrequent in the vegetative mycelium, and the vegetative septa do not allow cell separation. With time, the vegetative mycelium becomes more dense, producing aerial hyphae that grow quickly at the expense of nutrients derived from the substrate mycelium and emerge from the surfaces of colonies. Antibiotics are produced, presumably to prevent the growth of other bacteria on nutrients released from lysing substrate mycelium (see chapter 1). Growth of the aerial hyphae eventually stops, and regularly spaced sporulation septa are formed synchronously. Thus, the cell separation required for dispersal occurs by sporulation at the surfaces of colonies.

EMERGING SYSTEMS

The introductory chapters in each section describe other developmental systems within a phylogenetic group that have been described but have not been the subject of substantial investigation. The actinomycetes and cyanobacteria in particular have a rich variety of interesting but virtually unexplored systems. Within the actinomycetes, the genus *Frankia* is interesting in two respects. It undergoes symbiosis with nonlegume plant cells to form nitrogen-fixing nodules and would provide an interesting contrast to the *Rhizobium*-legume system, which is phylogenetically unrelated. *Frankia* also forms spores in terminal swellings of the mycelium by a process that is morphologically distinct from *Streptomyces* development (see chapter 1). Within the cyanobacteria, the *Pleurocapsa* group is interesting for the wide variety of ways in which baeocytes are produced (see chapter 3). Within the endospore-forming bacteria, the closely related genera *Epulopiscium* and *Metabacterium* provide contrasting yet related life styles. *Epulopiscium* produces multiple live young through a series of asymmetric cell divisions within the mother cell. In *Metabacterium* asymmetric cell division gives rise to multiple endospores (see chapter 6). In some stalked bacteria, the stalk is involved in reproduction (see chapter 15). *Rhodomicrobium vanniellii* and *Hyphomicrobium* reproduce by budding at the tip of the stalk, producing a flagellated swarmer cell. An interesting consequence of this mode of growth is that the chromosome must traverse the stalk to be inherited by the budding swarmer cell. How are chromosome replication and segregation coupled to polar growth?

Another developmental system that poses particularly interesting problems but has not been investigated extensively at the genetic and molecular levels is *Bdellovibrio* (Gray and Ruby, 1991; Thomashow and Cotter, 1992). These bacteria are predators of gram-negative bacteria. The *Bdellovibrio* life cycle includes two fundamentally different stages: an obligatory intraperiplasmic growth phase, during which the bacteria use the prey's cytoplasmic contents as their growth substrate, and an attack phase, during which they search for new prey. Growth occurs in the periplasm of the prey bacteria. No DNA replication occurs during the attack phase, but RNA and protein are synthesized. *Bdellovibrio* attack cells attach to prey cells and enter the prey 5 to 10 min after attachment. Once inside the prey, *Bdellovibrio* begins the systematic degradation of host macromolecules, which is complete in about 60 min. DNA replication begins during the intraperi-

plasmic growth phase and occurs without cell division to produce a multinucleate filament. Once growth becomes limited by the depletion of nutrients, elongation ceases and cell division is initiated simultaneously among the nucleoids. Flagella are synthesized de novo, and the bdelloplast is lysed, releasing the attack phase cells. Although the biochemistry and physiology of *Bdellovibrio* development have been well described, many questions remain, especially concerning the regulation of growth and development. How is the transition from attack phase to growth phase regulated? How is DNA replication kept in check in attack cells, even when nutrients are plentiful? The inhibition of cell division during intraperiplasmic growth while DNA replication is occurring presents an interesting contrast to the usual coupling of replication and division in many bacteria. Swarming bacteria like *Proteus vulgaris* also inhibit cell division during growth as part of swarm cell differentiation (see chapter 19).

DEVELOPMENT IN *ARCHAEA*

Quite surprisingly, there are no developing bacteria among the *Archaea* (Table 1). *Methanosarcina* is perhaps the most thoroughly studied candidate. *Methanosarcina mazei* S-6 exhibits three morphological types, single cells, packets of multiple cells, and laminae, in which many packets and cells are embedded (Mayerhofer et al., 1992). Interconversion of these cell types is observed in defined medium and is correlated with differences in protein profiles and antigenicity (Yao et al., 1992). The primary morphological difference is the production of intercellular connective material, which holds the cells together. Thus, there are alterations in form. But it is not yet clear whether each cell type has a different function and whether this interconversion plays an important role in the *Methanosarcina* life cycle. True developmental cycles tend to be stable within phylogenetic groups and provide reliable morphological traits used in taxonomic classification. It is curious that other *Methanosarcina* species cannot form the lamina structure under the conditions tested for *M. mazei* S-6 (Mayerhofer et al., 1992).

The absence of developmental systems among the *Archaea* may simply be due to the fact that isolation and characterization of *Archaea* species are in their infancies. *Archaea* have a preference for extreme habitats, and technical skill is required to isolate them in pure culture. It is also possible that *Archaea* cannot afford to evolve a developmental system because of bioenergetic constraints. *Archaea* tend to use inefficient respiratory pathways, relative to aerobic respiration, and consequently have small, specialized genomes (range, 1.5 to 4.1 Mbp; mean, 2.7 Mbp) (Shimkets, 1998). Development requires extra genes! Most of the developing *Bacteria* discussed in this book have much larger genomes supported by robust means of obtaining energy, generally aerobic respiration (range, 1.5 to 9.9 Mbp; mean, 6.2 Mbp) (Shimkets, 1998). The two notable exceptions are *Chlamydia* and *Coxiella*, which are obligate intracellular pathogens that have more compact genomes because they have dispensed with many biosynthetic pathways. If genome expansion is limited by the efficiency of the electron transport chain, and a larger genome capacity is required for development, then *Archaea* might face bioenergetic constraints that do not apply to the *Bacteria*.

ACKNOWLEDGMENT

We are grateful to Marty Dworkin for reviewing this chapter.

REFERENCES

Dworkin, M. 1985. *Developmental Biology of the Bacteria.* The Benjamin/Cummings Publishing Company, Inc., Menlo Park, Calif.

Gray, K. M., and E. G. Ruby. 1991. Intracellular signalling in the *Bdellovibrio* developmental cycle, p. 333–366. *In* M. Dworkin (ed.), *Microbial Cell-Cell Interactions.* ASM Press, Washington, D.C.

Harold, R., and R. Y. Stanier. 1955. The genera *Leucothrix* and *Thiothrix. Bacteriol. Rev.* **19:**49–64.

Jacob, F., and J. Monod. 1963. Genetic repression, allosteric inhibition and cellular differentiation, p. 30. *In* M. Locke (ed.), *Cytodifferentiation and Macromolecular Synthesis.* Academic Press, New York, N.Y.

Kearns, D. E., and L. J. Shimkets. 1998. Chemotaxis in a gliding bacterium. *Proc. Natl. Acad. Sci. USA* **95:**11957–11962.

Mayerhofer, L. E., A. J. L. Macario, and E. C. de Macario. 1992. Lamina, a novel multicellular form of *Methanosarcina mazei* S-6. *J. Bacteriol.* **174:** 309–314.

Shimkets, L. J. 1998. Structure and sizes of the genomes of Archaea and Bacteria, p. 5–11. *In* F. J. deBruijn, J. R. Lupski, and G. Weinstock (ed.), *Bacterial Genomes: Physical Structure and Analysis.* Chapman & Hall, New York, N.Y.

Thomashow, M. F., and T. W. Cotter. 1992. *Bdellovibrio* host dependence: the search for signal molecules and genes that regulate the intraperiplasmic growth cycle. *J. Bacteriol.* **174:**5767–5771.

Whittenbury, R., S. L. Davies, and S. L. Davey. 1970. Exospores and cysts formed by methane-utilizing bacteria. *J. Gen. Microbiol.* **61:**219–232.

Yao, R., A. J. L. Macario, and E. C. de Macario. 1992. Immunochemical differences among *Methanosarcina mazei* S-6 morphologic forms. *J. Bacteriol.* **174:**4683–4688.

ACTINOMYCETES

ACTINOMYCETE DEVELOPMENT, ANTIBIOTIC PRODUCTION, AND PHYLOGENY: QUESTIONS AND CHALLENGES

Wendy Champness

I

A first-time observer of *Streptomyces coelicolor* colonies as they develop over a several-day period might be impatient at first when nothing is visible on a plate 1 day after streaking and then might grow bored on the second day, seeing only typically bacterial beige colonies. But the third day's observations would be sure to delight: the colonies will have become "heavenly" blue, richly coated with light-gray velvet.

The dramatic changes in the *S. coelicolor* colony's appearance reflect the differentiation processes taking place in the colony as it produces complex sporulation structures on its surface and, underneath, an array of antibiotic compounds, some of which are brightly colored. Experimental analysis of these fascinating aspects of streptomycete biology is yielding intriguing glimpses of the underlying mechanisms regulating development. Much of the genetic and physiological experimentation on developmental regulation focuses on the two streptomycete species *S. coelicolor* and *Streptomyces griseus* because they are relatively tractable as experimental organisms, but the diversity in morphological forms and bioactive compounds

found among streptomycetes and their relatives in the genus *Actinobacteria* is enormous.

ACTINOMYCETE PHYLOGENY AND MORPHOLOGICAL DIVERSITY

The actinomycetes (Holt, 1989) form a major line of descent in the domain *Bacteria*. While actinomycetes are often thought of as "mycelial, branching, and/or spore-forming" bacteria, phylogenetic trees constructed from 16S rRNA sequences (Fig. 1) demonstrate the relatedness of numerous morphologically simple genera, such as *Micrococcus* and *Propionibacterium*, to the more familiar mycelial *Actinomyces* and *Streptomyces*. And not all mycelial bacteria are actinomycetes. As one example, an "actinomycete" look-alike, *Thermoactinomyces*, actually falls into the endospore-forming *Bacillus* group. What essentially distinguishes actinomycetes from other gram-positive bacteria is the fact that the actinomycetes are the high-G + C (>55%) gram-positive group, whereas the low-G + C (<50%) gram-positive group forms the other major gram-positive line of descent, *Bacillales-Clostridiales*. An additional significant criterion that defines actinomycetes' relatedness is a shared insertion of 100 nucleotides located in the 23S rRNA gene. Ten actinomycete suborders (Stackebrandt et al., 1997) and more than 100 genera have been classified

W. Champness, Department of Microbiology, Michigan State University, East Lansing, MI 48824-1101.

Prokaryotic Development, edited by Y. V. Brun and L. J. Shimkets,
© 2000 American Society for Microbiology, Washington, DC 20005-4171

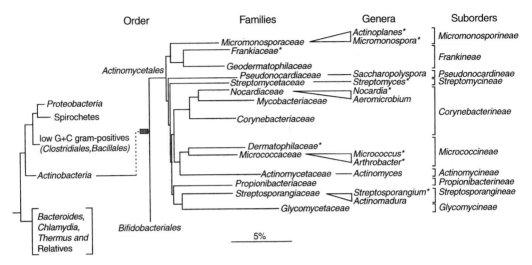

FIGURE 1 Phylogenetic relationships among actinomycetes, based on 16S ribosomal DNA and rRNA sequence comparisons. The 10 suborders of the order *Actinomycetales* are shown, but only representative families and genera (more than 100 genera are classified). Also, the class *Actinobacteria* contains several other minor suborders. The scale bar represents approximately five nucleotide substitutions per 100 nucleotides among the families of the order *Actinomycetales* (Stackebrandt et al., 1997). The hatched bar indicates the presence of a shared insertion in the 23S rRNA. Groups marked with asterisks are represented in Fig. 2 to 10. (Based on Embly and Stackebrandt [1994] and Stackebrandt et al. [1997].)

(Embly and Stackebrandt, 1994). Representatives of these, only a few examples of actinomycete morphological types, are described below.

The type suborder *Actinomycineae* contains the mycelicum-forming genus *Actinomyces*, which includes one of the few mycelium-forming human pathogens, *Actinomyces israelii*, a causative agent of oral and abdominal abscesses. In contrast to most other actinomycetes, which are aerobic, *Actinomyces* spp. are facultative anaerobes.

By far the most extensively studied of the actinomycete suborders is *Streptomycineae*, mostly because the genus *Streptomyces* has been the source of vast numbers of useful antibiotics. Streptomycetes are widespread in soil and are the most frequently isolated actinomycetes. Hundreds of *Streptomyces* spp. have been described, and more than 70 have been phylogenetically sorted.

The spore-to-spore streptomycete life cycle progresses through several stages of differentiation: germ tubes emerge from spores and grow into filamentous hyphae that branch to form a dense "substrate mycelium"; certain hyphae grow away from the substrate to form an "aerial mycelium"; and, finally, the aerial hyphae septate into compartments that become spores. The various *Streptomyces* species' aerial hyphae differ in their spore chain morphologies (e.g., straight, looped, or coiled) and spore surfaces (e.g., smooth, spiny, hairy, or warty). Figure 2 illustrates one example: the aerial hyphae of *Streptomyces violascens*. Chapter 2 details the morphogenesis of *S. coelicolor*, the species for which the most extensive genetic analyses have been done.

Another suborder in which aerial-mycelium formation is common is *Pseudonocardineae*. However, unlike in *Streptomycineae*, in many pseudonocardia genera the substrate mycelium undergoes fragmentation. This correlates with formation of two-layered hyphal septa rather than the single-layered septa that form in streptomycete substrates. Remarkably, the *S. coelicolor* hyphal septae are not essential for vegetative growth (McCormick et al., 1994), as an *ftsZ*

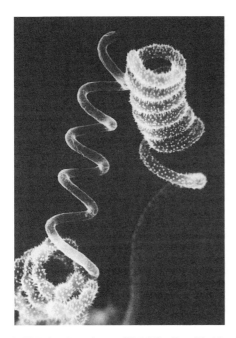

FIGURE 2 *S. violascens* SF 2425. (Provided by S. Amano, S. Miyadoh, and T. Shomura. Reprinted with permission from Miyadoh [1997].)

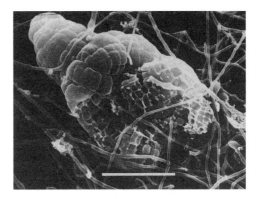

FIGURE 3 *Frankia* sp. strain Brunchorst 1886. (Provided by D. O. Baker and H. A. Lechevalier. Reprinted with permission from Miyadoh [1997].)

null mutant makes no septa yet is viable and forms substrate and aerial hyphae (but not spores). It would be interesting to know if pseudonocardiae's more complex septation is nonessential as well.

Even more elaborate hyphal septation occurs in the suborder *Frankineae*. Besides transverse septa, formed as in the streptomycetes, longitudinal septa also form in hyphae, creating so-called sporangia that are composed of many compartments (Fig. 3); these compartments differentiate into spores. *Frankia* spp. can further differentiate to form nitrogen-fixing nodules, which are laminated by multilayered membranes, and so form symbioses with nonleguminous woody plants.

Although the morphologies of the organisms discussed so far superficially assort with the phylogenetic distinctions shown in Fig. 1 (e.g., the sporangium-forming *Frankineae* diverge from the aerial-mycelium-forming *Streptomycineae* and *Pseudonocardineae*), phylogeny-morphology correlations break down with further exploration of the actinomycete suborders. For example, the family *Geodermatophilaceae* morphologies—involving thalli composed of cuboidal cells, which are produced by transverse and longitudinal septation and which often become motile spores—are very similar to those of *Dermatophilaceae* (Fig. 4), a family in the relatively distantly related suborder *Micrococcineae*. Indeed, this suborder is the most morphologically diverse of all, also encompassing the rela-

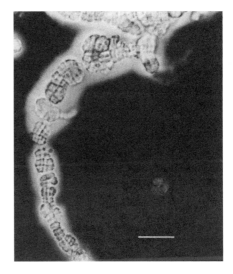

FIGURE 4 *Dermatophilus congolensis*. Bar, 10 μm. (Provided by A. Masters and J. M. Carson. Reprinted with permission from Miyadoh [1997].)

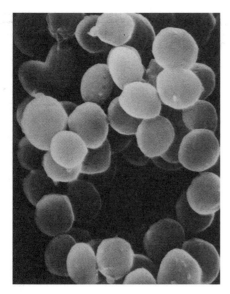

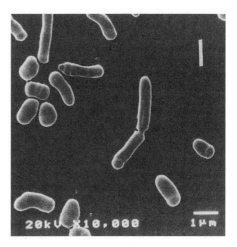

FIGURE 6 *Arthrobacter globiformis*. Bar, 1.0 μm. (Provided by T. Tamura, T. Nishii, and K. Hatano. Reprinted with permission from Miyadoh [1997].)

FIGURE 5 *M. luteus*. Bar, 1.0 μm. (Provided by K. Ochiai, S. Kinoshita, and K. Ando. Reprinted with permission from Miyadoh [1997].)

tively simple, coccoid, nonsporulating bacterium *Micrococcus luteus* (Fig. 5). In addition, the family *Micrococcaceae* includes the genus *Arthrobacter*, which is also nonsporulating, although the vegetative cells themselves are exceptionally desiccation resistant. These bacteria display an interesting growth cycle that alternates between rod-shaped and spherical cells; they also use a curious "snapping" division process, in which cross wall formation occurs in only one of two cell wall layers so that daughter cells remain attached after septum formation, but then one-sided outer-layer rupture produces bent cell groups (Fig. 6).

Propionibacterineae are another nonsporulating group of actinomycete-related bacteria. Although of less interest developmentally, they are gastronomically important: some species are responsible for the holes in Swiss cheese.

Two aerial-mycelium-forming phylogenetic neighbors of *Propionibacterineae* are *Streptosporangineae* and *Glycomycineae*. *Streptosporangium* spp. typically form sac-like sporangia, each produced by septation of a single coiled hypha (Fig. 7).

Sporangia are also produced by members of the suborder *Micromonosporineae*, for example, by *Actinoplanes* spp., which produce motile spores with a polar tuft of flagella (Fig. 8). In these bacteria, the sporangia form on the sub-

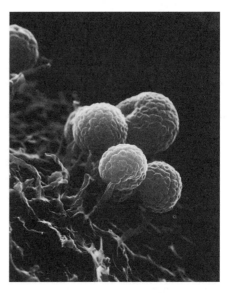

FIGURE 7 *Streptosporangium nondiastaticum*. Bar, 5 μm. (Provided by S. Amano, J. Yoshida, and T. Shomura. Reprinted with permission from Miyadoh [1997].)

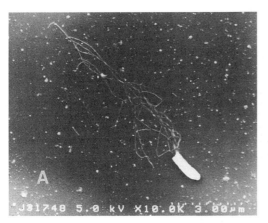
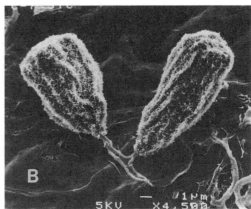

FIGURE 8 *Actinoplanes* sp. Left, spore; right, sporangium. (Provided by S. Amano and S. Miyadoh, and H. Suzuki and A. Seino, respectively. Reprinted with permission from Miyadoh [1997].)

strate mycelium, not the aerial mycelium. The substrate mycelium is also the site of formation of the single spores of *Micromonospora* spp. (Fig. 9). However, aerial-mycelium-forming genera also exist in this suborder: *Catenuloplanes* spp. produce peritrichously flagellated spores by fragmentation of the aerial hyphae.

Last—but not of least importance—is the *Corynebacterineae* suborder, which includes the genera *Corynebacteriacea*, *Mycobacteriacea*, and *Nocardiacea*. This suborder is notable for containing the most seriously pathogenic actinomycetes. *Corynebacteria diphtheriae*, *Mycobacterium tuberculosis*, and *Mycobacterium leprae* are all human pathogens that are related in their mycolic acid-containing cell wall structures. The mycobacteria and corynebacteria generally do not obviously differentiate. However, the name *Corynebacterium* derives from the Greek *koryne*, for "club," referring to the cells' club-shaped appearance due to swelling at one end.

The genetic differences responsible for these diverse actinomycete morphologies are completely undetermined. However, it is amusing to note that morphologies apparent in the phylogenetically distinct *Nocardia* and *Streptosporangium* genera (Fig. 7 and 10) superficially mimic certain developmental-mutant phenotypes of *S. coelicolor*. *S. coelicolor whiA* mutants form exceptionally long coiled serial hyphae like *Nocar-*

FIGURE 9 *Micromonospora* sp. strain SF2259. Bar, 1.0 μm. (Provided by S. Amano, J. Yoshida, and T. Shomura. Reprinted with permission from Miyadoh [1997].)

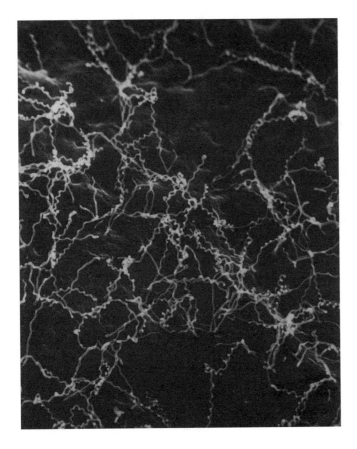

FIGURE 10 *N. brasiliensis* SK 2457. Bar, 1.0 μm. (Provided by S. Miyadoh, S. Amano, and T. Shomura. Reprinted with permission from Miyadoh [1997].)

dia brasiliensis, and some *whiI* mutants form globular structures at aerial hyphal tips (J. Ryding, 1999), like *Streptosporangium*. Perhaps small genetic distinctions can specify morphological outcomes that appear very different!

ANTIBIOTIC DIVERSITY AND GENETICS

The actinomycetes produce an enormous variety of bioactive molecules. The earliest-known examples of these were discovered because of their antibacterial, or antibiotic, activity. One of the first antibiotics used was streptomycin, produced by *S. griseus*. The last 55 years have seen the discovery of more than 12,000 antibiotics. The actinomycetes produce about 70% of these, and the remaining 30% are products of filamentous fungi and nonactinomycete bacteria. The biological activities of many natural products are now known to range well beyond the antibacterial, as illustrated by Table 1 (Strohl, 1997).

Of the plethora of known bioactive compounds, approximately 160 are currently used in clinical practice. *Streptomyces* spp. produce about 55% of these. The disproportionate representation of streptomycetes among the producing strains may have more to do with the relative ease of isolating and screening them than with a lack of biosynthetic capability in other actinomycetes. For example, *Saccharopolyspora erythraea* produces erythromycin (Table 1) and the distant genera *Micromonospora* and *Aeromicrobium* (Fig. 1) produce structurally and genetically related macrolide-type antibiotics. Moreover, many other evolutionarily distant groups of bacteria, including the myxobacteria, are known to produce diverse bioactive compounds, but these organisms are relatively difficult to culture for screening programs.

TABLE 1 Natural actinomycete bioactive products

Antibiological activity	Example of natural product	Example of producing organism
Antibacterial	Erythromycin	*S. erythraea*
Anticoccidial	Monensin	*Streptomyces cinnamonensis*
Antifungal	Amphotericin B	*Streptomyces nodusus*
Antihelminthic	Avermectin (ivermectin)	*Streptomyces avermitilis*
Antitumor	Doxorubicin (adriamycin)	*S. peucetius*
Antiviral	Viderabine (Ara-A)	*S. antibioticus*
Herbicidal	Bialaphos	*Streptomyces hygroscopicus*
Immunosuppressive	Rapamycin	*Streptomyces hygroscopicus*
Insecticidal	Spinosad	*Streptomyces spinosus*

Most of the bioactive compounds from actinomycetes sort into several major structural classes: aminoglycosides (e.g., streptomycin and kanamycin), ansamycins (rifampin), anthracyclines (doxorubicin), β-lactams (cephalosporins), macrolides (erythromycin), peptides and glycopeptides (vancomycin), and tetracyclines. Among these, examples of antibiotics that have been especially well characterized genetically are streptomycin and the antitumor drug doxorubicin (Fig. 11).

Doxorubicin synthesis involves a mechanism related to that employed in fatty acid biosynthesis, since the doxorubicin structure is based on a long carbon chain similar to that of a fatty acid. The enzyme activity responsible for doxorubicin chain synthesis, classified as a polyketide synthase, is evolutionarily related to bacterial fatty acid synthases (Hopwood, 1997). Both types of enzymes assemble the carbon backbones of their respective biopolymer products by successive condensations of small acyl units. The term polyketide describes a general structure consisting of a long carbon chain in which every alternate carbon contains a methylene, keto, enoyl, or hydroxyl function. The polyketide synthase uses coenzyme A-acyl thioesters as building blocks, loading each acyl group onto the phosphopantetheine thiol associated with an acyl carrier protein. Decarboxylative condensations produce β-carbonyl groups, which then undergo variable reduction reactions. Additional cycles of elongation and reduction produce a polyketide chain of defined length: each particular pathway precisely controls chain length. The polyketide chain then cyclizes and, lastly, additional pathway enzymes "tailor" the final product with variable reductions, oxidations, glycosylations, and methylations. Polyketide structures form the cores of many natural products, including not only the anthracycline and macrolide antibiotics but also toxins and other metabolites produced by plants and fungi.

Antibiotic Genes Occur in Clusters

The genes encoding the doxorubicin polyketide synthase, hydroxylase, and other tailoring enzymes are all clustered in the producing organism's genome (Strohl et al., 1997). In fact, the entire set of 37 open reading frames (ORFs) that are required for doxorubicin synthase is linked in a contiguous gene cluster (Fig. 11). Indeed, it is a general observation that the genes for synthesis of a given antibiotic are clustered. Usually, antibiotic genes are chromosomal, although in one case—methylenomycin—the genes are on a plasmid.

The organization of transcripts expressed from the doxorubicin gene cluster is complex, some being monocistronic and some polycistronic, with both convergent and divergent directionality. Transcriptional regulators specific to the doxorubicin genes are also associated with the gene cluster, as are genes that encode doxorubicin resistance functions. Linkage of

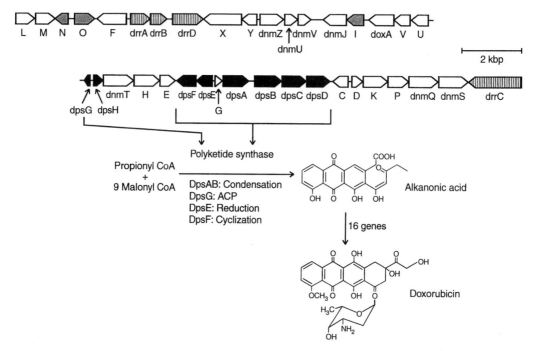

FIGURE 11 Elements of the doxorubicin-daunorubicin gene cluster (Strohl et al., 1997). Doxorubicin is an anthracycline product of a mutant *S. peucetius*; the parent makes the closely related compound daunorubicin. The pointed bars indicate ORFs; those with single letters are *dnr* genes. *dnm*, *dox*, and *dps* indicate various biosynthetic genes; the polyketide synthase includes the solid gene symbols. Gray shading indicates regulatory genes, which are *dnrN*, *-O*, and *-I* (*dnrN* regulates expression of *dnrI*), and *drr* indicates resistance genes. The genes all lie in a single large cluster, which is broken into two lines in the illustration. CoA, coenzyme A.

dedicated regulatory and resistance genes to an antibiotic's biosynthetic genes is another typical feature of antibiotic gene clusters.

The gene cluster for streptomycin synthesis (Pipersberg, 1997) that is found in *S. griseus* repeats the theme established by the doxorubicin cluster: linked biosynthetic, regulatory, and resistance genes arranged in a complex pattern of transcripts. Figure 12 further illustrates that another species, *Streptomyces glaucescens*, contains a 5′-hydroxy-streptomycin gene cluster related to the streptomycin cluster of *S. griseus*, with homologous gene products showing identities ranging from 58 to 86%. The combinations of genes that are cotranscribed are very similar in the two clusters, but the relative positions of the transcription units are quite different.

The nonconserved gene organization of the two streptomycin clusters raises a number of questions about the evolution and dissemination of antibiotic genes. Do the orders and combinations of genes in the transcription units contribute to optimal antibiotic biosynthesis? Are the inversions and rearrangements apparent in the clusters chance occurrences, or do they reflect the existence of especially recombinogenic regions in the clusters? At present there are no definitive answers to these questions, but observations made about the numerous available characterized clusters include the following. First, the positions of genes in a cluster do not necessarily reflect the order of enzyme activities in biosynthetic pathways (Pipersberg, 1997). For example, the *strDELM* subcluster encodes what would be the 12th through 15th steps in streptomycin biosynthesis, but these genes are arranged so that one

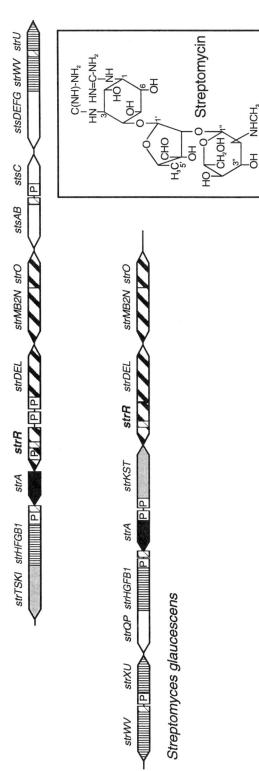

FIGURE 12 Elements of the related streptomycin and 5'-hydroxy-streptomycin gene clusters of the producing organisms (Piepersberg, 1997). The pointed bars indicate transcription units. *sts* and *str* indicate the various biosynthetic genes. *strR* is the streptomycin-specific regulator. *strA* encodes streptomycin phosphotransferase, a resistance enzyme. Heavy diagonal hatching indicates conserved gene organization. Vertical hatching and light-gray shading indicate partially conserved and inverted gene orders, respectively. P indicates promoter regions. Light diagonal hatching indicates defined binding sites for StrR.

transcript includes *strD* (step 12), *strE* (step 13), and then *strL* (step 15), with *strM* (step 14) expressed in a separate, convergent transcript. Further, the second through sixth biosynthetic steps are encoded in monocistronic units that are scattered in the gene clusters (*str-I, -C, -E,* and *-B1* and an unknown gene).

Not only do gene orders often differ between related clusters found in different strains, but the intergenic regions found within the clusters may be quite different as well (Pipersberg, 1997).

Dissemination of Antibiotic Gene Clusters

The distribution of homologous antibiotic gene clusters among actinomycetes does not correlate with phylogenetic relationships. There are many cases in which taxonomically divergent organisms produce structurally and genetically similar antibiotics. For example, streptomycin-like aminoglycosides are produced not only by *Streptomyces* spp. but also in the genera *Streptoverticillium, Micromonospora, Nocardia,* and *Corynebacterium,* and anthracyclines related to streptomycete antibiotics, such as doxorubicin, are produced by *Actinomadura* spp.

It is also typical that a given strain will produce more than one antibiotic, and often several. *S. glaucescens* strains produce not only streptomycin but also an anthracycline, tetracenomycin. The gene cluster for tetracenomycin is genetically completely distinct from the streptomycin cluster. The genetic independence of the multiple antibiotics produced by a single strain is especially well illustrated (Champness and Chater, 1994; Chater and Bibb, 1997) in *S. coelicolor* A3(2), a strain which contains widely separated chromosomal clusters for three structurally unrelated antibiotics as well as one of the very few known plasmid-linked antibiotic clusters.

Besides the characterized antibiotic repertoire of any strain, genes for additional antibiotics may be unexpressed (cryptic) in the usual culture conditions, and so genome sequencing will likely reveal yet more diversity within

strains and sharing of homologous antibiotics among strains. Thus, the antibiotic gene clusters are distributed among bacteria as though they had been passed around like playing cards. Each organism must then make the best of the hand it is dealt.

Is there a genetic mechanism that promotes lateral transfer of the antibiotic clusters? Are the mechanisms for regulation of antibiotics sufficiently conserved throughout the actinomycetes so that a transferred antibiotic cluster could easily "plug in" to the regulatory circuitry of a new host? Or is each antibiotic cluster regulated completely independently? There is very little evidence that bears on the first of these questions, although a recent report that a 241-bp direct repeat flanks the *Streptomyces angillaceus* methramycin gene cluster may be an indicator of lateral gene transfer (Lombó et al., 1999). The last two questions are explored in the following sections.

REGULATION OF ANTIBIOTIC PRODUCTION IN THE COURSE OF DEVELOPMENT

In plate-grown differentiating cultures, synthesis of antibiotics is tied to the developmental cycle, with significant production usually coinciding temporally with sporulation (Bibb, 1996; Champness and Chater, 1994; Chater and Bibb, 1997). Antibiotic gene expression is also spatially localized, by an unknown mechanism, and so antibiotic genes are expressed in cell compartments different from those undergoing sporulation (Hopwood, 1997). In liquid-grown cultures—in which most actinomycetes do not sporulate—the mycelial nature of actinomycete growth complicates distinctions between the exponential and stationary growth phases of a culture, but it is a general observation that antibiotic production occurs only after an extensive period of vegetative growth. Hence, the diverse bioactive antibiotic compounds produced by actinomycetes are often referred to as "secondary metabolites" (Demain, 1992).

Antibiotic-Specific Regulatory Genes

Current evidence suggests that growth phase regulation of antibiotics results from growth

1. ACTINOMYCETES ■ 21

phase regulation of so-called "pathway-specific regulators" (Bibb, 1996). These regulators are cluster-linked genes that regulate transcription of antibiotic biosynthetic genes.

Most antibiotic gene clusters contain pathway-specific regulatory genes—sometimes one, and sometimes two or more. The prototype for such regulators is *actII-ORF4*, which regulates the blue antibiotic actinorhodin in *S. coelicolor* A3(2). The discovery of *actII-ORF4* stemmed from cosynthesis tests of a collection of non-actinorhodin-producing mutants (*act* mutants). Some mutants failed to cosynthesize with any other *act* mutant; these were later found to define the *actII-ORF4* gene. An activator function has been ascribed to the ActII-ORF4 protein because *actII-ORF4* mutants fail to transcribe *act* biosynthetic genes and cloned extra copies of *actII-ORF4* cause actinorhodin overproduction. Most of the antibiotic-specific regulators found so far are also activators.

Two significant observations have suggested that temporal regulation of *actII-ORF4* transcription is largely responsible for growth phase-dependent antibiotic production in defined media (Bibb, 1996). First, accumulation of the *actII-ORF4* transcript is limited to the postexponential growth period. Second, introduction of extra plasmid-borne copies of *actII-ORF4* causes exponential-phase actinorhodin production.

Another *S. coelicolor* pathway-specific activator is *redD*, which regulates the red antibiotic undecylprodigiosin. *redD* is very similar in its genetic and regulatory properties to *actII-ORF4* (Takano et al., 1992).

In sporulating plate-grown *S. coelicolor* cultures, actinorhodin and undecylprodigiosin both appear in the substrate mycelium at about the time when aerial hyphae grow on the colony surface. A dramatic increase in abundance of the *actII-ORF4* and *redD* transcripts correlates with these visible developmental changes (Aceti and Champness, 1998), suggesting that temporal regulation of these genes, in the context of developmental regulation in the differentiating colony, is crucial to the observed

coupling of antibiotic production and sporulation.

The pathway-specific regulators for the streptomycin and doxorubicin clusters discussed above are also well characterized. These are *strR* (Retzlaff and Distler, 1995) and *dnrI* (Strohl et al., 1997), respectively. Both encode proteins which bind specific promoter regions within their clusters (Fig. 11 and 12).

The list of defined pathway-specific regulators is rapidly growing. Several of these define a novel family of regulatory proteins named SARPs (for *Streptomyces* antibiotic regulatory proteins) (Wietzorrek and Bibb, 1997). Examples of better-characterized SARP family members include ActII-ORF4 and RedD, which regulate actinorhodin and undecylprodigiosin, respectively, in *S. coelicolor*; DnrI, which regulates doxorubicin-daunorubicin in *Streptomyces peucetius* (Tang et al., 1996); and CcaR, which regulates cephamycin and clavulanic acid in *Streptomyces clavuligerus* (Paradkar et al., 1998; Pérez-Llarena et al., 1997). Reports of new SARP-like proteins occur frequently. The SARPs are predicted to contain N-terminal OmpR-like DNA-binding domains which bind promoter regions at heptameric direct repeats.

Just as growth phase regulation of the pathway-specific regulators is important in *S. coelicolor*, growth phase regulation of *strR* is known to be significant for regulation of streptomycin in *S. griseus* (Neuman et al., 1996). These observations pose the logical question, what regulates the antibiotic-specific regulators? Answers to this question are only just beginning to be found, as the next section shows.

Pleiotropic Regulators of Antibiotic Synthesis and Morphogenesis

An important genetic aspect of streptomycete antibiotic regulation involves genes identified by *bld* mutations, so named because the mutants fail to develop the "hairy" aerial hyphae characteristic of mature, sporulating colonies and instead remain relatively smooth surfaced, or "bald." Most mutants that have been isolated for their Bld− phenotypes are also blocked for antibiotic production; for example, in *S. coelicolor*, mutations in many *bld* genes affect all four of the organism's known antibiotics whereas some *bld* mutations affect only one or two of the antibiotics (Champness and Chater, 1994; Chater, 1998; Chater and Bibb, 1997). The mechanisms by which most *bld* genes regulate either morphogenesis or antibiotic production are not understood in detail, but recent molecular characterizations are beginning to provide clues.

AUTOREGULATOR SIGNALING SYSTEMS

An actinomycete signaling system that regulates an antibiotic pathway-specific regulator—and also morphogenesis—has been relatively well characterized in *S. griseus* (Horinouchi and Beppu, 1994). A-factor, a hormone-like autoregulator with a structure that permits diffusion through membranes, was discovered more than 30 years ago on the basis of its ability to restore sporulation, as well as streptomycin biosynthesis and resistance, to an *S. griseus* Bld − mutant. A-factor is a γ-butyrolactone. Structurally related γ-butyrolactone molecules regulate differentiation in other streptomycetes as well. These hormones differ primarily in their side chains: Fig. 13A shows a few of the better-studied examples.

The γ-butyrolactone autoregulators bind to specific receptor proteins to modulate their ef-

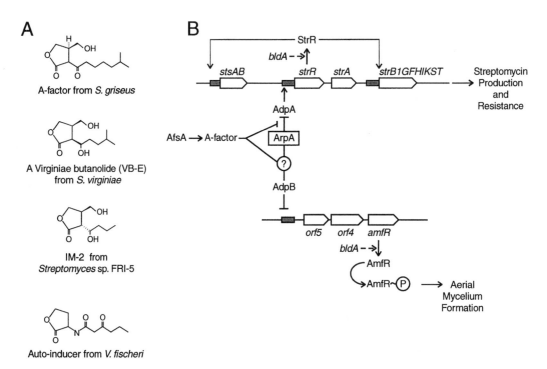

FIGURE 13 (A) A-factor (*S. griseus*) contains a 6-keto group, IM-2 (*Streptomyces* sp. strain FRI-5) contains a 6-β-hydroxy group, and virginiae butanolides (the VB group of *S. virginiae*) contain a 6-α-hydroxy group. The *Vibrio fischeri* autoinducer of luminescence (AHL) is also shown. (B) A-factor regulation of aerial-mycelium formation, streptomycin production, and streptomycin resistance in *S. griseus*. ArpA regulates AdpA, which regulates the streptomycin-specific activator, StrR. StrR translation requires the *bldA*-encoded tRNA^{leu}. StrR binds promoter regions shown as shaded boxes, and so regulates biosynthetic genes and *strA*, the streptomycin phosphotransferase gene. Only part of the streptomycin gene cluster is shown. A-factor also requires the *bldA*-encoded tRNA for translation. AmfR is a response regulator; its cognate kinase is not known, nor are its targets. Orf4 and Orf5 functions are unknown.

fects on transcription. Ligand-binding activity has been a useful criterion for purifying several binding proteins. Each binding protein is highly ligand specific, binding one autoregulator with more than 100-fold-greater affinity than any other.

The A-factor-triggered regulatory cascade regulating antibiotic synthesis in *S. griseus* (Fig. 13B) starts with A-factor binding to its receptor protein, ArpA, which is a cytoplasmic DNA-binding protein (Onaka and Horinouchi, 1997). A-factor binding relieves ArpA repression of the gene *adpA* (for A-factor dependent protein). AdpA is a regulator of *strR*, the pathway-specific activator for streptomycin biosynthesis.

The role of A-factor in linking regulation of morphogenesis to streptomycin biosynthesis is due in part to its effects on the expression of *amfR* (for aerial mycelium formation). AmfR is a response regulator-like protein required for normal aerial-mycelium development. Similarly to its role in derepressing antibiotic synthesis, A-factor also derepresses *amfR* transcription (Fig. 13B). It does this by relieving repression of *amfR* by the protein AdpB, which binds just upstream of the *amfR* promoter (Ueda et al., 1998). Whether *adpB* expression is regulated by ArpA or, if not, how its repressor-like function is otherwise regulated by A-factor is not yet clear.

The following observations have contributed to a model for how A-factor exerts temporal regulation. *arpA* mutations suppress A-factor deficiencies so that the double mutants produce streptomycin and aerial mycelia even earlier than wild-type *S. griseus*, a phenotype consistent with ArpA function as a repressor. ArpA synthesis begins at a very early growth stage. A-factor is produced in a growth-dependent manner; early addition of exogenous A-factor accelerates aerial-mycelium and streptomycin production. Thus, the model proposes that ArpA represses antibiotic and sporulation regulators early in growth before A-factor concentration reaches a threshold at a specific stage of growth, at which point A-factor binds ArpA, thereby releasing it from its DNA targets.

A-factor–ArpA-related signaling systems that regulate development are also being studied in other streptomycetes, including *Streptomyces virginiae* and *S. coelicolor*. In *S. virginiae*, virginiamycin production requires that the VB autoinducer mentioned above bind to its receptor protein, BarA, to induce antibiotic production (Nakano et al., 1998).

S. coelicolor produces at least four small diffusible signaling molecules that can cause precocious antibiotic production; at least one of these, ScbI, is a γ-butyrolactone. An ArpA homologue likely binds ScbI; ScbI antagonizes this protein's specific DNA-binding activity (Takano et al., 1999). Two additional ArpA homologues have been found to regulate development: CprA is a positive regulator, and CprB is a negative regulator. Mutations in *cprA* or *cprB* alter both antibiotic production and morphogenesis (Onaka et al., 1998). The cognate autoinducers for these proteins are not yet known.

The synthetic pathway for γ-butyrolactone synthesis, as established for VB synthesis, involves coupling of a β-keto acid acyl chain with a glycerol-derived three-carbon compound. In the case of *S. griseus* A-factor synthesis, the *afsA* gene likely encodes the coupling enzyme (Horinouchi and Beppu, 1994).

The streptomycete autoregulator chemical structures are somewhat similar to those of the autoinducer acyl homoserine lactones (AHL) that many gram-negative bacteria use to regulate various phenomena, such as luminescence (Fig. 13A), virulence, conjugal transfer, and antibiotic production (Dunlap, 1997). These bacteria use the AHL to signal cell density. An autoinducer synthase, LuxI, directs constitutive synthesis of AHL. The AHL diffuses freely across the cell membranes, and so the extracellular and hence intracellular AHL concentrations increase with population density. AHL binds to a transcriptional activator, LuxR, regulating its ability to influence expression of the luminescence genes or other genes. More than a dozen AHL-, LuxI-, or LuxR-homologous systems are known in gram-negative bacteria. Specificity in these autoinduction systems involves structural variations in the AHL acyl group and response specificity of the dedicated LuxR homologue.

From current evidence, it is not clear if streptomycete γ-butyrolactones signal cell density per se. Rather, experiments such as the following suggest a more complex role, at least for A-factor.

Several recent investigations of the kinetics of antibiotic production in liquid cultures (e.g., Neuman et al., 1996) have noted that it is possible to distinguish two phases of exponential growth; these are separated by a transient lag phase of reduced growth. Interestingly, the transient lag phase is evidently not caused by nutrient limitation. Streptomycin production is first detectable after the transient lag phase, but maximal production, at 100-fold-higher levels, occurs only after the culture has entered stationary phase. In these cultures, A-factor accumulation increases sharply just before production of streptomycin. In a non-A-factor-producing mutant, streptomycin production could be stimulated by added A-factor, but only if A-factor was added very soon after culture inoculation. Surprisingly, A-factor additions at later times, corresponding to when endogenous A-factor is usually detected in the wild type, failed to induce any streptomycin. A proposed explanation for these observations is that a very early A-factor-sensitive "decision phase" can prepare the culture for later antibiotic production (Neuman et al., 1996).

In *Escherichia coli*, homoserine lactone (HSL)-mediated regulation may involve a starvation response pathway in which the RspA protein affects HSL-dependent induction of the stationary-phase-specific sigma factor σ^S by degrading intracellular HSL (Zambrano and Kolter, 1996). *S. coelicolor* contains a homologue of *rspA*, *spaA*. A *spaA* disruption mutant showed a phenotype that suggested a signaling role for *spaA*: actinorhodin and undecylprodigiosin production was reduced and delayed on nutritionally poor medium at low colony density, whereas at high colony density actinorhodin was overproduced (Schneider et al., 1996). The reported failure to detect extracellular interactions between *spaA*+ and *spaA* mutant cultures could be consistent with *spaA* involvement in an intracellular starvation-sensing pathway, such as that proposed for HSL in *E. coli* (Zambrano and Kolter, 1996).

DIVERSE CELL-CELL SIGNALING PHENOMENA

In many of the better-understood cell-cell signaling phenomena of gram-positive bacteria, the signal molecule is a peptide (Dunny and Leonard, 1997). Often, gram-positive bacteria use these systems to signal cell density (Kleerebezem et al., 1997). A dedicated exporter protein complex secretes the peptide signal, and when it reaches a critical extracellular concentration, the peptide triggers a membrane-bound sensor of a two-component-type regulatory system to modulate expression of the appropriate target genes. Peptide-regulated development of *Bacillus subtilis* and *Streptococcus pneumoniae* genetic competence are examples of this behavior. Peptide production evidently occurs at a low constitutive level throughout exponential growth, reaching a concentration sufficient to trigger the signal transduction pathway at the end of the growth phase. The *B. subtilis* competence regulation process also utilizes a second mechanism for peptide-mediated regulation. In this case, the peptide signal's effect on gene expression requires that the cell internalize it via an oligopeptide permease (Spo0K).

The recent discovery that mutation of an *S. coelicolor* oligopeptide permease-encoding locus (*bldK*) blocks aerial-mycelium development implicates peptide-mediated signaling in streptomycete development (Nodwell et al., 1996). A presumptive peptide signal can diffuse from a normally developing culture growing next to a *bldK* mutant. Conditioned culture medium contains a signal candidate substance that has been partially characterized: it contains defined amino acid residues, but a presumptive modification blocks N-terminal sequencing (Nodwell and Losick, 1998). If produced and exported constitutively, the peptide signal might be imported by the BldK oligopeptide permease once it reaches a threshold concentration and then trigger later steps of differentiation.

The "extracellular complementation" of the *bldK* mutant, as described above, is one example of a more extensive array of complementation phenomena observable with *bld* mutants. When certain pairs of *bld* mutants are grown close to one another, one mutant restores aerial-mycelium formation to the other (Willey et al., 1993). Representatives of most of the known *bld* genes sort into a series of extracellular complementation groups. These groups fall into a hierarchical pattern: *bldJ* < *bldK* group < *bldA/H* < *bldG* < *bldC* group < *bldD* group (Nodwell et al., 1999). Mutants of the *bldD* group can complement *bldC*, *bldG*, *bldA*, etc.; *bldC* group mutants can complement *bldG*, *bldA*, etc., but only *bldD* group mutants can complement them; and so forth. An intriguing interpretation of this observation is that many of the *S. coelicolor bld* genes are directly or indirectly involved in production of substances that cells exchange during morphological differentiation. It is not known whether any of these putative "signals" are related to the γ-butyrolactones discussed above or whether antibiotic production follows the same hierarchical pattern.

THE Bld DEVELOPMENTAL PHENOTYPE

As mentioned above, a visual phenotype that characterizes many of the known developmental mutants is Bld. Several mutant hunts have used this phenotype as the primary criterion for identification of developmental genes. The *bld* designation identifies such genes. This chapter discusses only the best-understood Bld mutants; other reviews describe some of the less-characterized mutants (Champness and Chater, 1994; Chater, 1998; Chater and Bibb, 1997; Hodgson, 1992).

The most extensively studied *bld* gene is *bldA*. First defined by mutations that caused the aerial-mycelium-minus phenotype, the *bldA* gene is also important for antibiotic synthesis. *bldA* encodes the only tRNA that translates a leucine codon (TTA), which occurs rarely in *Streptomyces* genes. The *S. coelicolor actII-ORF4* (actinorhodin-specific activator) gene includes a TTA codon, and so actinorhodin production requires *bldA* function. Similarly, the undecyl-prodigiosin-specific activator *redZ* (White and Bibb, 1997; Guthrie et al., 1998), which is an activator of *redD* transcription, contains a TTA codon. Production of two additional antibiotics, methylenomycin and a calcium-dependent antibiotic (CDA), also requires *bldA*, although much less is known about the role of *bldA* in these cases. The *bldA* gene is also important to *S. griseus* morphological and physiological development: *S. griseus bldA* mutants exhibit a pleiotropic, sporulation-minus, antibiotic-minus phenotype (Kwak et al., 1996), owing, at least in part, to the presence of the TTA codon in the regulators *amfR* and *strR* (Fig. 13).

Several additional *bld* genes that were identified in Bld mutant hunts have been cloned by complementation, sequenced, and partially characterized. These include the *bldB* and *bldD* genes: neither of their sequences resemble any known proteins, but both are predicted to encode DNA-binding proteins (Elliot et al., 1998; Pope et al., 1998). *bldG* encodes a complex locus that shows a high degree of similarity (Warawa et al., 1998) to *B. subtilis* anti-sigma and anti-anti-sigma genes (see chapter 8).

The *brgA* gene is another gene that mutates to a Bld phenotype, but it was discovered as the site of a mutation conferring resistance to an inhibitor of ADP-ribosyltransferase (3-aminobenzamide). The *brgA* mutant, and various other *bld* mutants as well, show defects in protein ADP ribosylation (Shima et al., 1996), but the connection of this phenotype to differentiation has not been established.

Disruption mutations in certain other genes identified for a physiological attribute also produce a Bld phenotype, indicating that Bld mutant hunts have not saturated the phenotype. The examples discussed below include mutations generated in *relA* (guanosine 3′-diphosphate 5-diphosphate [ppGpp] synthesis) and *cya* (adenylate cyclase). Another gene that can mutate to give the Bld phenotype, *amfR* (discussed above), was discovered as a cloned sequence that suppressed baldness in an *S. griseus* mutant lacking A-factor (Ueda et al., 1998).

One defect common to many *S. coelicolor* Bld mutants is an inability to produce a morphogenetic proteinaceous substance named SapB (Willey et al., 1993). SapB seems to function in reducing surface tension at the air-water interface of a growing colony, thereby allowing growth of hyphae away from the aqueous colony surface. Application of purified SapB to Bld colonies allows their surface hyphae to grow into the air. The SapB surfactant function is not species specific, and even fungal hydrophobins can substitute for SapB (Tillotson et al., 1998). The first hint of SapB's existence was an observation that diffusible substances from a normally developing colony restored aerial hyphae to *bld* mutants. But purified SapB does not fully rescue the developmental defects in *bld* mutants (Tillotson et al., 1998): the colonies' aerial hyphae do not actually form spores, and they do not produce antibiotics. Hence, SapB production is only one developmental event that the *bld* mutations block.

Global Regulation of Streptomycete Secondary Metabolites

The known antibiotics of *S. coelicolor* are subject to global regulation that is, in part, separate from sporulation regulation. The genetic elements responsible for global regulation are now being discovered through a variety of experimental approaches, primarily utilizing *S. coelicolor* and the closely related *Streptomyces lividans*. These species, especially *S. coelicolor*, offer substantial advantages over other actinomycetes for the study of global regulation. *S. coelicolor* strains produce three characterized antibiotics—actinorhodin, undecylprodigiosin, and CDA—and those strains that carry the SCP1 plasmid produce a fourth antibiotic, methylenomycin. Not only are these antibiotics' gene clusters relatively well characterized, but the actinorhodin and undecylprodigiosin antibiotics are dramatic pigments, as mentioned above. In addition, CDA and methylenomycin can be evaluated with simple, bacterium-killing assays. These attributes make genetic study of co-ordinate regulation relatively easy, compared to other actinomycetes. Moreover, the absence of proprietary interest in *S. coelicolor* antibiotics greatly facilitates information exchange.

SIGNAL TRANSDUCTION

Coordinate regulation of the two *S. coelicolor* pigments, actinorhodin and undecylprodigiosin, was first observed in the course of an early study of the possible role of A-factor in *S. coelicolor* antibiotic production. One outcome of this work was the discovery of the *afsR* gene as a plasmid-cloned sequence that caused actinorhodin and undecylprodigiosin overproduction.

AfsR is related to Ser-Thr-Tyr phosphoproteins, which are common to eukaryotic signal transduction pathways (Matsumoto et al., 1994). AfsR is phosphorylated on Ser and Thr residues by the adjacently encoded AfsK kinase, which autophosphorylates on Ser and Tyr residues. The signal to which the AfsK-AfsR system responds is not known, but the mutant phenotype of Δ*afsR* suggests a role in triggering antibiotic production under high-phosphate conditions (Floriano and Bibb, 1996).

Antibiotic regulation by the two-component-type signal transduction systems that are more common to prokaryotes is also important. Three such *S. coelicolor* two-component systems are known to influence antibiotic production: *absA1-absA2* (Brian et al., 1996), *cutR-cutS* (Chang et al., 1996), and *afsQ1-afsQ2* (Ishizuka et al., 1992). The last gene pair exerts a positive effect on antibiotic production when cloned in a low-copy-number plasmid, but neither *afsQ1* nor *afsQ2* is required for antibiotic synthesis, at least in the growth conditions under which disruption mutants were studied. The first two gene systems function as negative regulators, and disruption mutations result in antibiotic overproduction (Brian et al., 1996; Chang et al., 1996). The signals sensed by these systems' sensor kinases, AbsA1, CutS, and AfsQ2, have not been defined, nor is it known whether these genes function in related or independent signaling pathways.

Other poorly understood regulatory loci,

some of which may also involve signal transduction (e.g., *mia* [Champness et al., 1992; Ryding, 1999]), have been described in several reviews (Champness et al., 1992; Champness and Chater, 1994; Chater and Bibb, 1997).

IS SIGMA FACTOR HETEROGENEITY IMPORTANT?

There is little indication as yet that actinomycetes use alternative sigma factors to regulate growth phase-dependent antibiotic gene expression. In *S. coelicolor*, the actinorhodin and undecylprodigiosin pathway-specific activators are presumed to be transcribed in vivo by a holoenzyme containing the major vegetative sigma factor, HrdB. A second sigma factor, HrdD, can direct transcription from these genes' promoters in vitro but has no obligate in vivo transcriptional role, since disruption of *hrdD* causes no defect in actinorhodin or undecylprodigiosin production (Fujii et al., 1996). The only observed case of involvement of an alternative sigma factor in streptomycete antibiotic synthesis concerns sigma E, a member of the ECF (for extracytoplasmic function) sigma factor subfamily. In *Streptomyces antibioticus*, a *sigE* null mutation prevents actinomycin D synthesis (Jones et al., 1997). At present, this effect is only poorly understood, as the in vivo sigma E targets responsible for the actinomycin-minus phenotype are not yet known.

NUTRITIONAL AND PHYSIOLOGICAL FACTORS AFFECTING DEVELOPMENT

Among the numerous complex nutritional, physiological, and environmental effects on antibiotic synthesis and morphogenesis that have been documented (reviewed in Bibb, 1996; Chater and Bibb, 1997; and Demain, 1992), the following *S. coelicolor* phenomena have been genetically analyzed.

Carbon Metabolism

A shared trait of most *S. coelicolor bld* mutants is the fact that their differentiation defects are carbon source dependent. They are bald on glucose-minimal media but form sporulating aerial hyphae on carbon sources, such as mannitol, galactose, or glycerol. Only *bldB* and *brgA* mutants, among the many *bld* mutants collected, do not exhibit this poorly understood carbon source-conditional aspect of the Bld phenotype.

Another observation suggesting a connection between carbon metabolism and differentiation is the fact that many *S. coelicolor bld* mutants are defective in regulation of carbon source utilization. *bldB* mutants are especially pleiotropically affected, abnormally expressing a variety of metabolic operons in the absence of their normal inducers (Pope et al., 1998).

cAMP

Until recently, evidence for cyclic AMP (cAMP) involvement in actinomycete metabolism was lacking (Chater and Bibb, 1997). However, construction of an *S. coelicolor* adenylate cyclase disruption mutant has allowed a reinvestigation of cAMP's importance, especially in development. The *cya* mutant's pleiotropic phenotype includes defects in spore germination, aerial-mycelium formation, and antibiotic production (Süsstrunk et al., 1998). The observation that the last two defects are pH dependent, since they are reversible at alkaline pH, led the investigators to suggest an interpretation that cAMP mediates a metabolic shift from an acid-generating phase to a differentiation-associated acid-metabolizing phase.

GTP Levels

A drop in the intracellular GTP concentration, occurring as a reflection of nutritional limitation, could be a signal triggering differentiation. In *B. subtilis*, Obg is an essential GTP-binding protein that participates in sporulation initiation and is required for full activation of the phosphorelay (see chapter 7). Similar GTP-binding proteins are present in *S. griseus* and *S. coelicolor*. The *S. coelicolor obg* product has an essential vegetative function, and the observed enhancement of aerial-mycelium formation and antibiotic production by multicopy *obg*[+], but not by mutant alleles with their GTP-binding abilities altered, suggests that Obg is also important to *S. coelicolor* differentiation (Oka-moto and Ochi, 1998). A model for Obg function suggests that the GTP-bound form of Obg stimulates growth but prevents differentiation.

ppGpp

Antibiotic biosynthesis has been correlated with ppGpp accumulation in a variety of studies of various *Streptomyces* species. These observations prompted isolation and characterization of mutants defective in ppGpp metabolism. Two classes of *S. coelicolor* mutants have been studied: *relC* mutants, defective in the L11 protein that activates ribosome-associated ppGpp synthetase under amino acid starvation conditions, and *relA* mutants, defective in the ribosome-dependent ppGpp synthetase RelA. Mutants of both classes are deficient in antibiotic production and morphogenesis (Chakraburtty and Bibb, 1997; Martinez-Costa et al., 1996). For example, a null allele of the *relA* gene affects *S. coelicolor* production of actinorhodin and undecylprodigiosin, as well as aerial-mycelium formation, on some media but not others. Specifically, the defects occur under conditions of nitrogen limitation (Chakraburtty and Bibb, 1997).

PERSPECTIVES

The capacity of actinomycetes to produce a vast array of medically important natural products has profoundly influenced the areas of emphasis chosen by actinomycete researchers, focusing attention on pharmaceutical product discovery and yield enhancement. Enormous research efforts on countless diverse actinomycetes—primarily with the goal of optimizing fermentation conditions for a particular antibiotic—have produced a plethora of data, much of it proprietary, regarding the genetic, nutritional, and environmental factors that influence antibiotic production. In contrast, research on model actinomycetes has been relatively limited.

Our understanding of the development of streptomycetes far exceeds that regarding any other of the actinomycetes, and, of the strep-tomycetes, we know most about *S. coelicolor* A3(2). But even in *S. coelicolor*, which has the most advanced genetics, the lack of an efficient transposon mutagenesis system has slowed the tasks of identification and cloning of regulatory genes. Only one developmental locus, *bldK*, has been cloned from a transposon insertion. Most other regulatory mutants—*whi*, *bld*, and *abs* strains—were identified by mutations induced with UV or chemical mutagenesis, and some regulatory genes were discovered because of their multicopy effects. Fortunately, steady improvements in genetic and cloning techniques and the growing number of available cloned genes are making the recognition and cloning of further genes more straightforward. Moreover, the *S. coelicolor* A3(2) genome sequence will be completed soon (http://www.sanger.ac.uk/pub/s_coelicolor/sequence).

A major challenge for the near future will be to sort out the pathway relationships of the known regulatory genes and then to determine the connections and linkages among various pathways. Gene fusions have been a powerful tool in facilitating systematic analyses of expression dependency relations in bacteria, such as *B. subtilis* and *Mycococcus xanthus*, but in streptomycetes, the common reporter genes have been relatively less useful, slowing such studies. Fortunately, recent results with green fluorescent protein (GFP) have been more promising: development of GFP as a reporter for temporal and spatial gene expression has allowed visualization of predicted patterns of expression of both an inducible and a sporulation-specific promoter (Sun et al., 1999).

Some streptomycetes have a tendency to genetic instability, particularly affecting morphogenesis and antibiotic production (Volff and Altenbuchner, 1998). The genetic basis for high-frequency phenotypic instability is often deletion, usually but not always occurring at the ends of the streptomycete linear chromosome. The deletions may be as large as 2 Mb (25% of the genome), and they are often accompanied by DNA amplification and deletion of regions termed AUD, for amplifiable unit of DNA. The relationships between the DNA sequences involved in the amplifications and deletions and the loci involved in developmental regulation have not been established for any of the phenomena reported. These phenotypic instabilities, which can occur at a frequency of 0.1 to 1% of spores, can seriously complicate genetic analyses in some species. Fortunately, *S. coelicolor* is relatively stable in its antibiotic production and morphological phenotypes, although it does undergo some high-frequency phenotypic variations.

The soon-to-be-completed genome sequence of *S. coelicolor* A3(2), some 8,000 genes, will present a daunting challenge! But the rewards of understanding how the fascinating biology of streptomycetes is genetically programmed should tempt the curious, and a soon-to-be-published comprehensive handbook (Hopwood, 1999) of genetic manipulations will guide their adventures.

ACKNOWLEDGMENTS

I thank M. Bibb, P. Brian, K. Chater, J. Distler, J. Ensign, G. Garrity, S. Horinouchi, L. Katz, B. Leskiw, J. Nodwell, M. Paget, J. Ryding, B. Strohl, K. Ueda, J. Westpheling, and J. Willey for helpful discussions, for comments on the manuscript, and for communicating unpublished results, and S. Miyadoh for generously providing the micrographs. Because of space limitations, I have cited mostly reviews and recent publications that will contain additional references for the interested reader. I apologize to those colleagues whose work has therefore been omitted. I also thank Suzy Peacock and Marlene Cameron for their assistance in preparing the manuscript and illustrations.

Work in my laboratory is supported by the National Science Foundation (grant MCB9604055).

REFERENCES

Aceti, D., and W. Champness. 1998. Transcriptional regulation of *Streptomyces coelicolor* pathway-specific antibiotic regulators by the *absA* and *absB* loci. *J. Bacteriol.* **180:**300–307.

Bibb, M. 1996. The regulation of antibiotic production in *Streptomyces coelicolor* A3(2). *Microbiology* **142:** 1335–1344.

Brian, P., P. J. Riggle, R. A. Santos, and W. C. Champness. 1996. Global negative regulation of *Streptomyces coelicolor* antibiotic synthesis mediated by an *absA*-encoded putative signal transduction system. *J. Bacteriol.* **178:**3221–3231.

Chakraburtty, R., and M. Bibb. 1997. The ppGpp synthetase gene (*relA*) of *Streptomyces coelicolor* A3(2) plays a conditional role in antibiotic production and morphological differentiation. *J. Bacteriol.* **179:** 5854–5861.

Champness, W. C., and K. F. Chater. 1994. Regulation and integration of antibiotic production and morphological differentiation in *Streptomyces* spp., p. 61–94. *In* P. Piggot, C. Moran, and P. Youngman, ed., *Regulation of Bacterial Differentiation.* American Society for Microbiology, Washington, D.C.

Champness, W. C., P. Riggle, T. Adamitis, and P. Van der Vere. 1992. Identification of *Streptomyces coelicolor* genes involved in regulation of antibiotic synthesis. *Gene* **115:**55–60.

Chang, H.-M., J. Y. Chen, Y. T. Shieh, M. J. Bibb, and C. W. Chen. 1996. The *cutRS* signal transduction system of *Streptomyces lividans* represses the biosynthesis of the polyketide antibiotic actinorhodin. *Mol. Microbiol.* **21:**1075–1085.

Chater, K. F. 1998. Taking a genetic scalpel to the *Streptomyces* colony. *Microbiology* **144:**1465–1478.

Chater, K. F., and M. J. Bibb. 1997. Regulation of bacterial antibiotic production, p. 57–105. *In* H. Kleinkauf and H. von Döhren (ed.), *Biotechnology.* vol. 6. *Products of Secondary Metabolism.* VCH, Weinheim, Germany.

Chater, K. F., and R. Losick. 1997. Mycelial life style of *Streptomyces coelicolor* A3(2) and its relatives, p. 149–182. *In* J. A. Shapiro and M. Dworkin (ed.), *Bacteria as Multicellular Organisms.* Oxford University Press, New York, N.Y.

Demain, A. L. 1992. Microbial secondary metabolism: a new theoretical frontier for academia, a new opportunity for industry. *Ciba Found. Symp.* **171:** 3–23.

Dunlap, P. V. 1997. *N*-Acyl-L-homoserine lactone autoinducers in bacteria, p. 69–106. *In* J. A. Shapiro and M. Dworkin (ed.), *Bacteria as Multicellular Organisms.* Oxford University Press, New York, N.Y.

Dunny, G. M., and B. A. B. Leonard. 1997. Cell-cell communication in gram-positive bacteria. *Annu. Rev. Microbiol.* **51:**527–564.

Elliot, M., F. Damji, R. Passantino, K. Chater, and B. Leskiw. 1998. The *bldD* gene of *Streptomyces coelicolor* A3(2): a regulatory gene involved in morphogenesis and antibiotic production. *J. Bacteriol.* **180:**1549–1555.

Embly, T. M., and E. Stackebrandt. 1994. The molecular phylogeny and systematics of the Actinomycetes. *Annu. Rev. Microbiol.* **48:**257–289.

Floriano, B., and M. J. Bibb. 1996. *afsR* is a pleiotropic but conditionally required regulatory gene for antibiotic production in *Streptomyces coelicolor* A3(2). *Mol. Microbiol.* **21:**385–396.

Fujii, T., H. C. Gramajo, E. Takano, and M. J. Bibb. 1996. *redD* and *actII-ORF4*, pathway-specific regulatory genes for antibiotic production in *Streptomyces coelicolor* A3(2), are transcribed in vitro by an RNA polymerase holoenzyme containing σ^{hrdD}. *J. Bacteriol.* **178:**3402–3405.

Guthrie, E., C. Flaxman, J. White, D. A. Hodgson, M. J. Bibb, and K. F. Chater. 1998. A response-regulator-like activator of antibiotic synthesis from *Streptomyces coelicolor* A3(2) with an amino-terminal domain that lacks a phosphorylation pocket. *Microbiology* **144:**727–738.

Hodgson, D. A. 1992. Differentiation in actinomycetes, p. 407–440. *In* S. Mohan, C. Dow, and J. A. Cole (ed.), *Prokaryotic Structure and Function: a New Perspective.* Cambridge University Press, Cambridge, United Kingdom.

Holt, J. G. (ed.). 1989. *Bergey's Manual of Systematic Bacteriology,* vol. 4. Williams and Wilkins, Baltimore, Md.

Hopwood, D. A. 1997. Genetic contributions to understanding polyketide synthases. *Chem. Rev.* **97:**2465–2495.

Hopwood, D. A. 1999. Personal communication.

Horinouchi, S., and T. Beppu. 1994. A-factor as a microbial hormone that controls cellular differentiation and secondary metabolism in *Streptomyces griseus. Mol. Microbiol.* **12:**859–864.

Ishizuka, H., S. Horinouchi, H. M. Kieser, D. A. Hopwood, and T. Beppu. 1992. A putative two-component regulatory system involved in secondary metabolism in *Streptomyces* spp. *J. Bacteriol.* **174:**7585–7594.

Jones, G. H., M. S. B. Paget, L. Chamberlin, and M. J. Buttner. 1997. Sigma-E is required for the production of the antibiotic actinomycin in *Streptomyces coelicolor. Mol. Microbiol.* **23:**169–178.

Kleerebezem, M., L. E. N. Quadri, O. P. Kuipers, and W. M. de Vos. 1997. Quorum sensing by peptide pheromones and two-component signal-transduction systems in Gram-positive bacteria. *Mol. Microbiol.* **24:**895–904.

Kwak, J., L. A. McCue, and K. E. Kendrick. 1996. Identification of *bldA* mutants of *Streptomyces griseus. Gene* **171:**75–78.

Leskiw, B. Personal communication.

Lombó, F., A. F. Braña, C. Méndez, and J. A. Salas. 1999. The mithramycin gene cluster of *Streptomyces argillaceus* contains a positive regulatory gene and two repeated DNA sequences that are located at both ends of the cluster. *J. Bacteriol.* **181:**642–647.

Martinez-Costa, O. H., P. Arias, N. M. Romero, V. Parro, R. P. Mellado, and F. Malpartida. 1996. A *relA/spoT* homologous gene from *Streptomyces coelicolor* A3(2) controls antibiotic biosynthetic genes. *J. Biol. Chem.* **271:**10627–10634.

Matsumoto, A., S.-K. Hong, H. Ishizuka, S. Horinouchi, and T. Beppu. 1994. Phosphorylation of the AfsR protein involved in secondary metabolism in *Streptomyces* species by a eukaryotic-type protein kinase. *Gene* **146:**47–56.

McCormick, J. R., E. P. Su, A. Driks, and R. Losick. 1994. Growth and viability of *Streptomyces coelicolor* mutant for the cell division gene *ftsZ. Mol. Microbiol.* **14:**243–254.

Miyadoh, S. (ed.). 1997. *Atlas of Actinomycetes.* The Society for Actinomycetes, Tokyo, Japan.

Nakano, H., E. Takehara, T. Nihira, and Y. Yamada. 1998. Gene replacement analysis of the *Streptomyces virginiae barA* gene encoding the butyrolactone autoregulator receptor reveals that BarA acts as a repressor in virginiamycin biosynthesis. *J. Bacteriol.* **180:**3317–3322.

Neuman, T., W. Pipersberg, and J. Distler. 1996. Decision phase regulation in *Streptomyces griseus. Microbiology* **142:**1953–1963.

Nodwell, J. R., and R. Losick. 1998. Purification of an extracellular signaling molecule involved in production of aerial mycelium by *Streptomyces coelicolor. J. Bacteriol.* **180:**1334–1337.

Nodwell, J. R., K. McGovern, and R. Losick. 1996. An oligopeptide permease responsible for the import of an extracellular signal governing aerial mycelium formation in *Streptomyces coelicolor. Mol. Microbiol.* **22:**881–893.

Nodwell, J. R., M. Yang, D. Kua, and R. Losick. 1999. Extracellular complementation and the identification of additional genes involved in aerial mycelium formation in *Streptomyces coelicolor. Genetics* **151:**569–584.

Okamoto, S., and K. Ochi. 1998. An essential GTP-binding protein functions as a regulator for differentiation in *Streptomyces coelicolor. Mol. Microbiol.* **30:**107–119.

Onaka, H., and S. Horinouchi. 1997. DNA-binding activity of the A-factor receptor protein and its recognition DNA sequences. *Mol. Microbiol.* **24:**991–1000.

Onaka, H., T. Nakagawa, and S. Horinouchi. 1998. Involvement of two A-factor receptor homologs in *Streptomyces coelicolor* A3(2) in the regulation of secondary metabolism and morphogenesis. *Mol. Microbiol.* **28:**743–753.

Paradkar, A. S., K. A. Aidoo, and S. E. Jensen. 1998. A pathway-specific transcriptional activator regulates late steps of clavulanic acid biosynthesis in *Streptomyces clavuligerus. Mol. Microbiol.* **27:**831–844.

Pérez-Llarena, F. J., P. Linas, A. Rodríguez-García, and J. F. Martín. 1997. A regulatory gene (*ccaR*) required for cephamycin and clavulanic acid production in *Streptomyces clavuligerus*: amplification results in overproduction of both β-lactam compounds. *J. Bacteriol.* **179:**2053–2059.

Pipersberg, W. 1997. Molecular biology, biochemistry and fermentation of aminoglycoside antibiotics, p. 81–163. *In* W. R. Strohl (ed.), *Biotechnology of Antibiotics*, 2nd ed. Marcel Dekker, Inc., New York, N.Y.

Pope, M. K., B. Green, and J. Westpheling. 1998. The *bldB* gene encodes a small protein required for morphogenesis, antibiotic production, and catabolite control in *Streptomyces coelicolor. J. Bacteriol.* **180:** 1556–1562.

Retzlaff, L., and J. Distler. 1995. The regulator of streptomycin gene expression, StrR, of *Streptomyces griseus* is a DNA binding activator protein with multiple recognition sites. *Mol. Microbiol.* **18:**151–162.

Revill, W. P., M. J. Bibb, and D. A. Hopwood. 1996. Relationships between fatty acid and polyketide synthases from *Streptomyces coelicolor* A3(2): characterization of the fatty acid synthase acyl carrier protein. *J. Bacteriol.* **178:**5660.

Ryding, J. 1999. Personal communication.

Schneider, D., C. J. Brunton, and K. F. Chater. 1996. Characterization of *spaA*, a *Streptomyces coelicolor* gene homologous to a gene involved in sensing starvation in *Escherichia coli. Gene* **177:**243–251.

Shima, J., A. Penyige, and K. Ochi. 1996. Changes in patterns of ADP-ribosylated proteins during differentiation of *Streptomyces coelicolor* A3(2) and its developmental mutants. *J. Bacteriol.* **178:** 3785–3790.

Stackebrandt, E., F. A. Rainey, and N. L. Ward-Rainey. 1997. Proposal for a new hierarchic classification system, *Actinobacteria* classis nov. *Int. J. Syst. Bacteriol.* **47:**479–496.

Strohl, W. R. (ed.) 1997. *Biotechnology of Antibiotics*, 2nd ed. Marcel Dekker, Inc., New York, N.Y.

Strohl, W. R., M. L. Dickens, U. B. Rajgarhia, A. J. Woo, and N. D. Priestley. 1997. Anthrocyclines, p. 577–657. *In* W. R. Strohl (ed.), *Biotechnology of Antibiotics*, 2nd ed. Marcel Dekker, Inc., New York, N.Y.

Süsstrunk, U., J. Pidoux, S. Taubert, A. Ullmann, and C. J. Thompson. 1998. Pleiotropic effects of cAMP on germination, antibiotic biosynthesis and morphological development in *Streptomyces coelicolor. Mol. Microbiol.* **30:**33–46.

Sun, J., G. Kelemen, M. Buttner, and M. Bibb. 1999. Personal communication.

Takano, E., H. C. Gramajo, E. Strauch, N. Andres, J. White, and M. J. Bibb. 1992. Transcriptional regulation of the *redD* transcriptional activator gene accounts for growth-phase-dependent production of the antibiotic undecylprodigiosin in *Streptomyces coelicolor* A3(2). *Mol. Microbiol.* **6:** 2797–2804.

Takano, E., R. Chakraburtty, M. Bibb, T. Nihira, and Y. Yamada. 1999. Personal communication.

Tang, L., A. Grimm, Y.-X. Zhang, and C. R. Hutchinson. 1996. Purification and characterization of the DNA-binding protein DnrI, a transcriptional factor of daunorubicin biosynthesis in *Streptomyces peucetius. Mol. Microbiol.* **22:**801–813.

Tillotson, R. D., H. A. B. Wösten, M. Richter, and J. M. Willey. 1998. A surface active protein involved in aerial hyphae formation in the filamentous fungus *Schizophillum commune* restores the capacity of a bald mutant of the filamentous bacterium *Streptomyces coelicolor* to erect aerial structures. *Mol. Microbiol.* **30:**595–602.

Ueda, K., C.-W. Hsheh, T. Tosaki, H. Shinkawa, T. Beppu, and S. Horinouchi. 1998. Characterization of an A-factor-responsive repressor for *amfR* essential for onset of aerial mycelium formation in *Streptomyces griseus. J. Bacteriol.* **180:**5085–5093.

Volff, J.-N., and J. Altenbuchner. 1998. Genetic instability of the *Streptomyces* chromosome. *Mol. Microbiol.* **27:**239–246.

Warawa, J., K. Bilida, and B. Leskiw. 1998. Personal communication.

White, J., and M. Bibb. 1997. *bldA* dependence of undecylprodigiosin production in *Streptomyces coelicolor* A3(2) involves a pathway-specific regulatory cascade. *J. Bacteriol.* **179:**627–633.

Wietzorrek, A., and M. Bibb. 1997. A novel family of proteins that regulates antibiotic production in Streptomyces appears to contain an OmpR-like DNA-binding fold. *Mol. Microbiol.* **25:**1183–1184.

Willey, J., J. Schwedock, and R. Losick. 1993. Multiple extracellular signals govern the production of a morphogenetic protein involved in aerial mycelium formation by *Streptomyces coelicolor. Genes Dev.* **7:**895–903.

Zambrano, M. M., and R. Kolter. 1996. GASPing for life in stationary phase. *Cell* **86:**181–184.

DEVELOPMENTAL DECISIONS DURING SPORULATION IN THE AERIAL MYCELIUM IN *STREPTOMYCES*

Keith F. Chater

2

GENERAL DESCRIPTION OF SPORULATION IN *STREPTOMYCES*

In converting environmental resources into biomass, living organisms generally grow and/or move towards nutrients or sources of energy. However, during reproduction, dispersal, and long-term survival, they behave differently. Examples of such specialized behavior in eukaryotes include movement towards potential mates in animals, coordinated movement of individual cells in a "slug" of a slime mold to produce a fruiting body, growth of the flower-bearing parts of plants into the air (for pollination and seed dispersal), and aerial growth of mycelial fungi for spore formation and dispersal. This chapter deals with an unusual bacterial example involving growth into the air: the formation and metamorphosis of reproductive aerial hyphae in *Streptomyces* spp.

Although most bacteria are essentially unicellular, *Streptomyces* spp. are different. They grow as a mycelium of branching filaments (hyphae). Following the growth of a vegetative mycelium (effectively the conversion of available nutrients into biomass), dispersal to permit colonization of new sites must be achieved. As in mycelial fungi, the *Streptomyces* solution to

this problem is a phase of aerial reproductive growth away from the nutrient substratum into the air. Aerial hyphae grow at least partially parasitically on the substrate mycelium and then metamorphose into long chains of unigenomic spores (Fig. 1).

In the most studied species, *Streptomyces coelicolor* A3(2), the formation of aerial hyphae, at least under some growth conditions, takes place after a rather complex cascade of extracellular signals, which culminates in the production of a surfactant protein, SapB (Willey et al., 1991, 1993; Nodwell et al., 1996; see Kelemen and Buttner, 1998, for a review). SapB appears to coat the emerging aerial hyphae with a hydrophobic outer surface, permitting growth through the surface tension barrier at the air-colony surface interface (see chapter 1). This function resembles that of fungal hydrophobins (Chater, 1991), and indeed, a fungal hydrophobin appears to be able to facilitate aerial growth of a *Streptomyces* mutant defective in this process (Tillotson et al., 1998).

Vegetative hyphae elongate into and over agar medium by the linear growth of the cell wall close to the hyphal tips, with mycelial mass increasing quasi-exponentially by the formation of new tips (i.e., branching [Fig. 2 and Chater and Losick, 1997]). Do aerial hyphae likewise extend by tip growth of cell walls?

K. F. *Chater*, Department of Genetics, John Innes Centre, Norwich Research Park, Colney, Norwich, NR4 7UH, United Kingdom.

Prokaryotic Development, edited by Y. V. Brun and L. J. Shimkets,
© 2000 American Society for Microbiology, Washington, DC 20005-4171

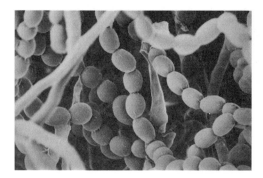

FIGURE 1 Spore chains and hyphal filaments on the surface of a *Streptomyces* colony. (Reproduced from Chater, 1998, with permission.)

Evidence from autoradiography does seem to rule out basal growth in aerial hyphae (Miguélez et al., 1994), but it remains uncertain whether there is a significant degree of intercalary growth, as opposed to tip growth. In many species, including *S. coelicolor*, the aerial hyphae consist of a more or less straight stem surmounted by a long coiled tip, the coiled part going on to form spores. The coiling may be a visible manifestation of an aerial-mycelium-specific change in the architecture of the cell wall, which in turn may reflect a difference in the mode of cell wall synthesis. One idea is that this may be a switch to a more diffuse zone of growth, perhaps involving a reduction in the extent of peptidoglycan cross-linking. Such a change could permit more rapid cell wall extension as the number of genomes in the hy-

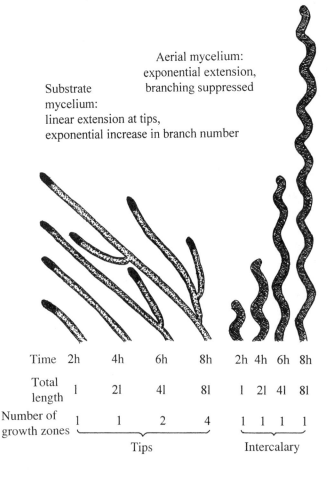

Substrate mycelium: linear extension at tips, exponential increase in branch number

Aerial mycelium: exponential extension, branching suppressed

Time	2h	4h	6h	8h	2h	4h	6h	8h
Total length	1	2l	4l	8l	1	2l	4l	8l
Number of growth zones	1	1	2	4	1	1	1	1
		Tips				Intercalary		

FIGURE 2 Two alternative possible hyphal growth modes. Heavy shading indicates cell wall growth zones. To the left are diagrammed successive stages in rapid growth of a typical segment of vegetative mycelium, in which an exponential increase in total length is maintained by combining linear cell wall growth at the tips with exponentially increasing numbers of branches (and hence, of tips). To the right is diagrammed the situation postulated for a rapidly growing, but unbranched, aerial hypha, in which an exponential increase in total length is achieved by intercalary growth.

phal compartment increases, an important counterbalance to the reduced branching in aerial hyphae compared with substrate hyphae (Fig. 2).

In 10 to 20 h, the number of genomes in an aerial hypha increases by about 2 orders of magnitude. This strongly implies that most or all of the genomes in a hypha replicate repeatedly, to give something close to exponential growth. Since rounds of replication during vegetative growth are apparently non-overlapping (Shahab et al., 1996) and take about 90 to 120 min (Miguélez et al., 1992), the six or seven rounds of doubling needed for aerial hyphae to reach their typical length should take at least 9 to 14 h, which coincides quite well with observations.

At some point, the apical regions of aerial hyphae switch to a different state, in which growth stops and the hyphae undergo multiple cell division (sporulation septation) to give what is often many tens of unigenomic compartments, each destined to become a spore. Sporulation septa differ morphologically from vegetative septa (Wildermuth and Hopwood, 1970) and also show considerable variation among species, particularly in thickness (Hardisson and Manzanal, 1976). One of the main themes of this chapter is the question of how sporulation septation is triggered. Other questions touched on are the extent to which sporulation-associated cell division and metabolic processes involve dedicated genes and how the hyphae are supplied with sufficient water to maintain turgor for growth.

BASIC CELL DIVISION MACHINERY IN VEGETATIVE AND SPORULATION SEPTATION

In simple bacteria that divide by binary fission, many of the essential cell division proteins are typically encoded by a single large gene cluster (termed *dcw*, for division and cell wall), and mutations in most of these genes prevent colony formation (Donachie, 1993). Most often, therefore, the genes are studied by making their proper functions conditional (e.g., by using temperature-sensitive mutations or by the introduction of regulatable promoters). Inactivating gene function typically results in the formation of filaments, which could be regarded as a partial phenocopy of the growth habit of *Streptomyces*. However, these simple bacteria apparently have not evolved to be filamentous, and the mutant cells do not continue to grow for more than a few cell cycles. The most extensively studied gene of this cluster is *ftsZ*. Typically, the FtsZ protein is particularly abundant in comparison to other *dcw* gene products. In *Escherichia coli*, this is partially a consequence of its transcription from multiple promoters scattered through the *dcw* cluster or operon (Flärdh et al., 1998). These may be important in ensuring that the amount of FtsZ protein available during the cell cycle is consistent with requirements: cell division in *E. coli*, and probably other bacteria, is highly sensitive to alterations in the level of FtsZ expression (Palacios et al., 1996; Ward and Lutkenhaus, 1985).

The role of FtsZ is to form a cytoplasmically located, but membrane-coupled, ring at incipient sites of septation to act as a cytoskeletal element in guiding the ingrowth of the septum by contracting, finally disassembling to leave the (FtsZ-free) completed septum (Rothfield and Justice, 1997). This mechanism is universal among bacteria, so it is not surprising that an *ftsZ* gene is present in *Streptomyces* spp. Indeed, immunofluorescence analysis has shown that both vegetative- and sporulation septum formation involves FtsZ ring formation. Such rings exist only transiently and are therefore quite hard to find, so the finding of ladders of as many as 50 FtsZ rings in sporulating aerial hyphae (Schwedock et al., 1997) was a striking demonstration of the simultaneity of multiple sporulation septation in any one hypha. Moreover, since the rings, and the resulting septa (Fig. 3), extend virtually to the tips of hyphae in which they are seen, it seems that growth of the hyphae has stopped even at the earliest stage of septation.

S. coelicolor has a single *ftsZ* gene, which is needed for both kinds of septation. Thus, an *ftsZ* null mutant contained no detectable septa, yet it could form mycelial colonies, albeit

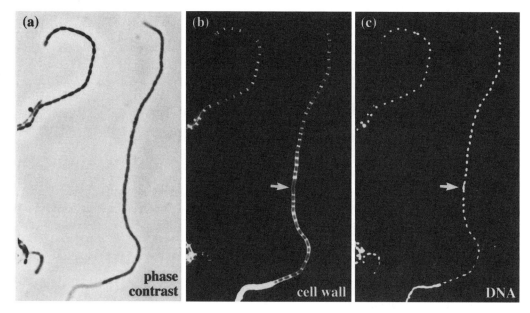

FIGURE 3 Developing spore chains. (a) External morphologies of two spore chains, the left-hand one being more mature. (b) The regular ladders of sporulation septa (revealed by cell wall staining with fluorescein-linked wheat germ agglutinin after lysozyme treatment) extend to the hyphal tips; note that the lateral cell wall stains strongly in aerial-hyphal stalks, weakly in the least mature part of the long spore chain closest to the stalk, and not at all in the more mature parts. (c) Nuclear segregation (revealed by staining with 7-amino actinomycin D) has not occurred in the absence of septation (arrow in panels b and c). (Modified from Flärdh et al., in press, with permission.)

poorly growing, with aerial hyphae (McCormick et al., 1994). (A similar but less extreme phenotype resulted from disrupting the adjacent *ftsQ* gene [McCormick and Losick, 1996].) Transcription analysis of *Streptomyces griseus* showed that *ftsZ* is transcribed from more than one promoter, the strongest one being up-regulated during sporulation in submerged culture (Kendrick, personal communication). Similar observations have been made with sporulating surface cultures of *S. coelicolor* (Flärdh, personal communication). However, at least one of the components involved in cell division appears to be specific to sporulation septation, since the use of a fluorescent derivative of the β-lactam antibiotic 6-amino penicillanic acid showed that a protein with unique β-lactam-binding properties was confined to sporulation septa in *S. griseus* (Hao and Kendrick, 1998).

Sporulation-associated cell division also requires accurate partitioning of DNA to either

side of the new septum. This seems to involve a pair of *par* genes located near the origin of replication (Kim et al., in preparation). These genes encode homologues of ParA and ParB proteins that are important in the partitioning of various plasmids and the chromosomes of many bacteria (e.g., in *Caulobacter crescentus* and, under the aliases of Soj and Spo0J, in *Bacillus subtilis*) (Wheeler and Shapiro, 1997). Deletions of *parB* or of both genes result in imperfect chromosome partitioning during sporulation of *S. coelicolor* (about 13% of the spores in the spore chains of these mutants have atypical amounts of DNA, and some are anucleate). Just as with *ftsZ*, a promoter of the *parAB* operon is up-regulated during sporulation (Kim et al., in preparation).

GENES ESSENTIAL FOR, AND APPARENTLY SPECIFIC TO, THE EARLY STAGES OF SPORULATION

Although very large numbers of genes can be expected to contribute to sporulation, many of them are also important during vegetative growth (for example, genes for macromolecular synthesis and primary metabolism) and others may fulfill very subtle functions that are not readily discernible by investigators. However, some genes function significantly only during sporulation and play such major roles that sporulation is disrupted when they are mutated. Such *Streptomyces* developmental mutants deficient in sporulation were first sought by Hopwood et al. (1970), who took advantage of the observation that the aerial mycelium is white when first formed but becomes gray-brown (in the case of *S. coelicolor*) as spores form, because of a spore wall-associated pigment. Thus, Hopwood and coworkers isolated so-called *whi* mutants, which retained a white aerial mycelium even on prolonged incubation. By excluding mutants with small colonies, mutations in genes that are also important in vegetative growth were avoided.

This approach yielded two broad classes of mutants, which have subsequently been taken to identify "early" and "late" sporulation genes. Mutants with defects in early genes produce few or no spore-like bodies, whereas late mutants produce recognizable spore chains. The existence of early *whi* mutants showed that some early sporulation events are epistatic to the formation of spore pigment and implied that the early *whi* genes are involved, directly or indirectly, in regulating some later genes. Transmission electron microscopy, recently supplemented by sensitive fluorescence microscopy, showed that all of a set of representative early mutants examined lacked the regular multiple sporulation septa typical of the wild type but retained the normal pattern of vegetative septation (McVittie, 1974; Schwedock et al., 1997).

Classical genetic linkage analysis of the early mutations identified just six early *whi* loci: *whiA*, -*B*, -*G*, -*H*, -*I*, and -*J* (Chater, 1972, 1998). Phase-contrast microscopy of aerial-hyphal material showed that mutations mapping to any one locus nearly all gave rise to similar phenotypes, the distinction between loci being based on the extent of coiling and fragmentation of aerial hyphae (Chater, 1972).

The *whi* mutant collection contained multiple representatives of each of the six early loci, suggesting that few if any other genes were specifically needed to modify normal cellular processes for the purposes of sporulation-associated cell division. (An important caveat to this hypothesis is that other genes might be specifically necessary for sporulation but not for gray-pigment formation and may therefore have been missed.) With the availability of cloned *whi* genes (see below) it has become possible to screen large collections of *whi* mutants for any that are not complemented and that may therefore contain unidentified genes. By this route, a few new *whi* loci have been discovered (Ryding et al., 1999). Further analysis, using constructed null mutants, is needed to clarify whether the new *whi* mutations represent special alleles of genes with other functions, as opposed to dedicated sporulation genes. In addition, the use of different culture conditions may perhaps reveal other genes (especially in light of the conditional phenotypes of some *bld* mutants [see chapter 1]).

PRODUCTS OF EARLY *whi* GENES

With the application of recombinant DNA procedures to *S. coelicolor* A3(2), it became possible to identify cloned *whi* genes by virtue of their ability to restore gray pigmentation to *whi* mutants. The earlier genetic-linkage mapping permitted the choice of representative mutants for each locus. All six known early *whi* genes have been cloned and sequenced (Chater, 1998). Three encode members of known families of regulatory genes, and three encode "orphan" proteins that do not closely resemble any functionally characterized protein families.

One of the six proteins appears to be functionally interchangeable with homologues in diverse other bacteria, irrespective of their ability to sporulate. This is the *whiG* gene product, which is an RNA polymerase sigma factor (Chater et al., 1989; Tan et al., 1998). Its homologues include σ^F of *E. coli* and *Salmonella typhimurium* and σ^D of *B. subtilis*, which are known as "motility" sigma factors because they are needed for the formation of the flagella and the chemotaxis systems in these bacteria. These sigma factors, when associated with RNA polymerase, all recognize promoters with a consensus -35, -10 sequence of TAAA (N_{16}) GCCGTAAA (compare this with the consensus promoter sequence of housekeeping genes in *E. coli*: TTGACA (N_{17}) TATAAT). σ^{WhiG} can substitute for σ^F in vivo in *S. typhimurium* (Nodwell and Losick, personal communication) and is capable of in vitro association with *E. coli* RNA polymerase core enzyme, allowing it to transcribe from a *B. subtilis* motility gene promoter (Tan et al., 1998).

Several lines of evidence strongly suggest that σ^{WhiG} is the crucial element in committing hyphae to sporulation. In the absence of *whiG*, aerial hyphae are long, straight, and featureless. Conversely, extra copies of *whiG* enhance sporulation: one extra copy causes colonies to become darker gray, and multiple copies cause ectopic sporulation in the older parts of the substrate mycelium (Chater et al., 1989).

The *whiI* gene also specifies a protein of a well-known family: WhiI resembles the response regulators of so-called two-component sensor-regulator systems (Aínsa et al., in press), which are typically (but not exclusively) transcription factors with an α-helical DNA-binding region near their C-termini. Their ability to function as transcription factors usually depends on phosphorylation of an aspartyl residue in the N-terminal domain by the action of a cognate protein kinase in response to an environmental signal. However, WhiI lacks certain highly conserved and functionally important residues usually needed for phosphorylation to take place, and there is no recognizable kinase gene next to *whiI*. WhiI protein may therefore function in a phosphorylation-independent manner. Although the C-terminal domain of WhiI aligns quite well with those of other response regulators, it gives only a moderate score in a program to detect the DNA-binding helix-turn-helix regions of such proteins, and so further analysis will be needed before we

can be confident that WhiI interacts with DNA at all.

The *whiH* gene also encodes a specialized member of a widespread and large family of DNA-binding transcriptional regulators (Ryding et al., 1998), the most well-known members of which are GntR of *B. subtilis* and FadR of *E. coli*. GntR is a gluconate-sensitive repressor of gluconate metabolism (Miwa and Fujita, 1988), and FadR regulates fatty acid synthesis and degradation in response to certain fatty acid coenzyme A (CoA) esters (DiRusso and Nyström, 1998). Unlike GntR, FadR represses some genes and activates others, and it is implicated in the adjustment of *E. coli* membrane growth to stationary phase.

It is noteworthy that close homologues of *whiG*, *whiH*, and *whiI* are absent from the genome of *Mycobacterium tuberculosis* (Cole et al., 1998), an actinomycete cousin of *Streptomyces* which shows many examples of close conservation of gene sequences with streptomycetes. Thus, the specialized roles of these three genes in *S. coelicolor* sporulation are mirrored by their absence from a nonsporulating actinomycete.

The products of the other three characterized early sporulation genes are unrelated to proteins of well-defined function. The *whiA* gene product has close homologues in all completely sequenced gram-positive bacteria, whether in the low- or high-G + C divisions (Ryding et al., unpublished), and the *whiB* gene product, which has a C-terminal region with features suggestive of DNA binding (Davis and Chater, 1992), is a member of a novel family of proteins found only in high-G + C gram-positive bacteria (Soliveri et al., unpublished). The sixth locus, *whiJ*, is not yet completely defined: DNA complementing *whiJ* mutants contains five genes, paralogues of several of which also occur together in other locations in the genome (Ryding et al., unpublished).

POSSIBLE ROLES OF *whi* GENES AT DEVELOPMENTAL CHECKPOINTS

A detailed phenotypic analysis of early *whi* mutants and of the epistatic interactions revealed in double *whi* mutants has given rise to the idea that several of the *whi* genes act as gatekeepers to consecutive stages on the journey from aerial-mycelium initiation to sporulation septation (Flärdh et al., in press). If a gate is closed by a mutation in one of the genes, a hypha is condemned to continue to grow in the juvenile state, thereby producing an exaggerated version of that state. This kind of analysis has become more reliable and penetrating with the use of the relevant cloned genes to create null mutants in a common genetic background and with the availability of sensitive fluorescence microscopy for cytological analysis. Four early *whi* genes have so far been studied with the aid of these techniques. In the emerging model (Fig. 4), a young growing aerial hypha starts out with an identity essentially indistinguishable from any vegetative branch in the substrate mycelium, except that it is growing inside a tube of self-generated surface material (SapB

or some other hydrophobin-like material [see above and chapter 1]) instead of tunneling into a substrate. At some point in this growth it becomes possible for the hypha to "choose" between continuing in this growth mode or switching to a mode that commits it to eventual sporulation. The switch might, for example, be set off by a decrease in some signal molecule from the ever more remote substrate mycelium. (This change in supply could be exacerbated if septation were to separate the cytoplasm of the tip from direct continuity with the cytoplasm of the substrate mycelium.) The critical determinant of the outcome of this choice is the amount of active σ^{WhiG}. In a *whiG* mutant there is no σ^{WhiG}, removing the element of choice and giving rise to the straight, apparently undifferentiated hyphae typical of such mutants. However, in *whiG*[+] strains the changed signal level would, by an unknown

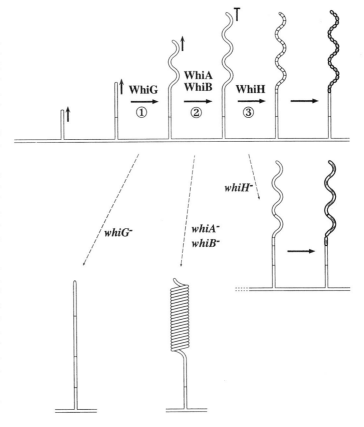

FIGURE 4 Speculative model implicating Whi proteins in developmental decisions. It is proposed that initially σ^{WhiG} is held in an inactive state, but a signal produced while young aerial hyphae are early in growth releases active σ^{WhiG} (1). Among the σ^{WhiG} targets would be genes involved in cell wall biosynthesis, leading to spiral growth. After a moderate number of cell doublings, WhiA and WhiB combine to bring growth to an orderly stop (2), probably in response to one or more further signals. Growth cessation (bold T) would release a further signal, activating WhiH (3). WhiH somehow activates the full initiation of sporulation septation but is less stringently required for certain other late events, such as DNA condensation, changes in the spore wall, and gray-pigment production. The diagrams in the lower part of the figure represent the relevant mutant phenotypes. Bold arrows indicate continuing growth. (Reproduced from Flärdh et al., in press, with permission.)

mechanism, cause an increase in the effective level of σ^{WhiG}, permitting the expression of *whiG*-dependent sporulation genes. The similarity of σ^{WhiG} to σ^{F} of *Salmonella* has prompted speculation that σ^{WhiG} might, like σ^{F}, be regulated by an anti-sigma factor responsive to some morphological cue, such as septum formation (Losick and Shapiro, 1993). One effect of the postulated developmental switch (passage through the first gateway) would be to bring about a change in cell wall growth mode to permit rapid extension, to keep pace with the exponentially increasing number of genomes in the apical compartment (Fig. 2). A visible manifestation of this might be the coiling of aerial hyphae, which is absent from *whiG* mutants. (The untested logical corollary of this is that in *whiG* mutants aerial growth should be confined to the tip region.) Vegetative-type cross wall formation is suppressed in the extending coiled apical compartment of *whiG*+ strains (Schwedock et al., 1997; Flärdh et al., in press). As the compartment extends ever more rapidly, it may become limited for further growth by the rate of supply of nutrients from the substrate mycelium. At this point, a coordinated cessation of growth and completion of ongoing rounds of DNA replication would be advantageous. Perhaps WhiA and WhiB perceive the growth limitation and, after permitting the ongoing round of replication and associated growth to finish, prevent initiation of the next (Flärdh et al., in press). The inability of *whiA* and *whiB* mutants to prevent reinitiation may exclude them from passage through the second gateway, and as a result, they have coiled, unbranched, aseptate tips of exceptional length, probably containing 100 to 200 genomes (Fig. 5).

How might cessation of elongation be cou-

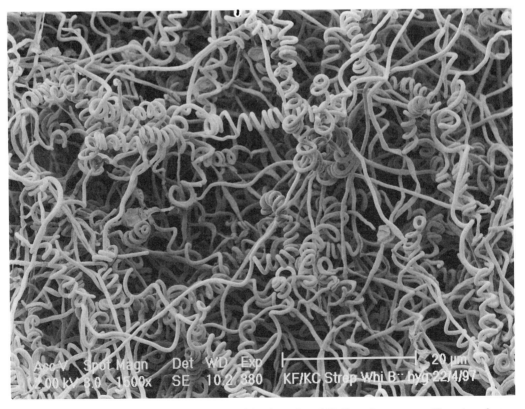

FIGURE 5 The long, often tightly coiled aerial hyphae of a *whiB* disruption mutant. (Scanning electron micrograph kindly provided by K. Findlay and K. Flärdh.)

pled to initiation of sporulation septation? Part of this coupling may involve WhiH and WhI, which both seem to be needed for full sporulation septation but not for cessation of growth. Thus, in the absence of the respective proteins in *whiH* and *whiI* null mutants, aerial hyphae form coils that are more or less similar to those of the wild type in length but contain only a few, widely spaced sporulation septa (Chater, 1972; Flärdh et al., in press). In some respects, WhiI appears to function earlier than WhiH, since WhiH mutants (alone among the early *whi* mutants) show condensation of DNA in aerial hyphae to give extra-brightly staining nucleoid bodies such as usually occur when sporulation septation has taken place. Without such condensation, it is difficult to observe the accuracy of DNA partitioning, but in *whiH* mutants it is easy to see that partitioning is aberrant (Flärdh et al., in press). WhiH may play a role in coordinating sporulation septation with DNA partitioning. *whiH* mutants (including null mutants) not only make a few sporulation septa, but they also retain some capability of performing later events: their aerial hyphae make a certain amount of gray spore pigment; exhibit some expression of the late sigma factor gene, *sigF*; and, like developing spore chains in the wild type, become more opaque in phase-contrast microscopy and less susceptible to staining by fluorescein-coupled wheat-germ agglutinin after lysozyme treatment (Ryding et al., 1998; Kelemen et al., 1998; Flärdh et al., in press). This may indicate that WhiH plays an accessory, rather than integral, role in sporulation. Alternatively, another (WhiH-like?) protein might substitute partially for WhiH.

Is the early *whi* mutant block in sporulation septation—whether partial or complete—an effect manifested before or after FtsZ ring formation? For those mutations examined (*whiA*, *-B*, *-G*, *-H*, and *-I*), no FtaZ ladders could be detected (Schwedock et al., 1997; Flärdh, personal communication).

REGULATION OF EXPRESSION OF THE EARLY *whi* GENES

In order to determine the regulatory interdependences of the early *whi* genes, RNA extracted from surface-grown cultures of the wild type and various *whi* mutants at different developmental stages has been subjected to S1 nuclease protection analysis. This has revealed that both *whiH* and *whiI* are completely dependent on *whiG*, a dependence shown to be direct by in vitro transcription experiments (Fig. 6) (Ryding et al., 1998; Aínsa et al., in press). However, perhaps surprisingly, there is a substantial difference between the timing of *whiG* expression and that of its dependent genes. Transcription of *whiG* is more or less constant throughout the life cycle (with the caveat that very early stages have not been tested) (Kelemen et al., 1996), but expression of *whiH* and *whiI* (and of another *whiG*-dependent gene [see below]) is very strongly up-regulated when aerial growth takes place. What is responsible for this lag? One of several possible explanations would be the hypothetical sequestering of σ^{WhiG} by an anti-sigma factor, as suggested earlier. Autorepression of *whiH* may also contribute to the delay in *whiH* expression, since a *whiH* mutant showed somewhat earlier as well as greatly increased *whiH* transcription (Ryding et al., 1998).

Although transcription of *whiA* and *whiB* is also strongly up-regulated during aerial growth, neither gene has a *whiG*-dependent promoter (Soliveri et al., 1992; Ryding et al., unpublished). Thus, at least two independent transcriptional regulatory pathways are necessary for sporulation septation. There is no clear information on whether there are any mechanisms to coordinate these pathways, but they do converge, inasmuch as late gene expression depends on all the early *whi* genes.

LATE SPORULATION GENES INCLUDE THE DETERMINANT OF ANOTHER SPORULATION-SPECIFIC SIGMA FACTOR

The first late sporulation genes to be cloned were those of the *whiE* spore pigment cluster. Eight genes have so far been identified in this cluster, most of them encoding homologues of proteins involved in the biosynthesis of aromatic polyketide antibiotics (see chapter 1). Further analysis of the *whiE* genes suggests that

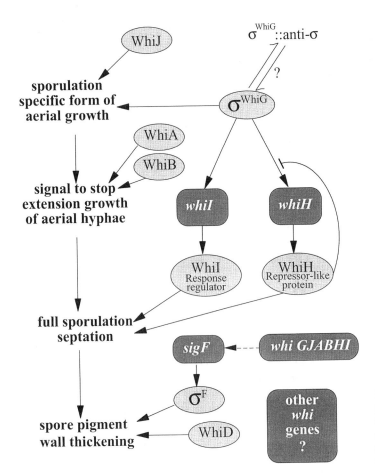

FIGURE 6 Transcriptional dependencies of *whi* genes. This scheme combines results from Kelemen et al. (1996, 1998), Ryding et al. (1998), and Aínsa et al. (in press). (Drawing provided by J. Aínsa.)

the spore pigment is a 24-carbon polyaromatic compound assembled by iterative condensations of malonyl CoA-derived 2-carbon units with an acetyl-CoA starter unit (Yu et al., 1998). It is interesting to note that aromatic polyketides also cause pigmentation of conidia and ascospores in *Aspergillus* spp. (Brown et al., 1993, 1994). Since the early *whi* mutants were identified by their lack of spore pigment, it was not surprising to find that the *whiE* promoters are inactive, or nearly so, in all such mutants. Thus, the two (or more) transcriptional regulatory pathways defined by the early *whi* mutants converge at or before transcription of *whiE* (Kelemen et al., 1998).

An important further link in this scheme is another sigma factor, σ^F. This is one of a large number (now approaching 20 [Kelemen, Paget, and Buttner, personal communication]) of *Streptomyces* sigma factors mostly discovered in surveys with oligonucleotide probes or PCR methods based on conserved features of known sigma factors. σ^F is a member of a subfamily of sigma factors which have been found only in gram-positive bacteria and whose characterized members are associated with stress responses or sporulation (e.g., σ^B, σ^F, and σ^G of *B. subtilis*: see chapters 6 and 8). Disruption of σ^F in *S. coelicolor* alters spore pigmentation, giving greenish or pale-gray spores (Kelemen et al., 1998), as well as interfering with spore wall thickening and possibly other aspects of spore maturation (Potúčková et al., 1995). Transcription of one of the *whiE* promoters is σ^F dependent (though it is not certain that this is a direct dependence) (Kelemen et al., 1998).

This promoter (*whiEP2*) drives expression of a hydroxylase that carries out an ill-defined late modification of the *whiE* polyketide and hence causes a change in its color (Blanco et al., 1993).

The observed dependence of *whiEP2* on early *whi* genes is at least partially explained by the dependence of *sigF* on these genes: *sigF* mRNA, which is detectable in surface cultures only when sporulation septation is taking place, is absent from most of the early *whi* mutants (weak expression was found in a *whiH* and a *whiJ* mutant) (Kelemen et al., 1996).

AN OVERALL REGULATORY MODEL

Figure 5 summarizes the transcriptional dependence relationships of the early and late *whi* genes. A cascade of at least four steps, involving at least two specific sigma factors, converges with another regulatory tributary from the *whiA-whiB* system. Perhaps this complex arrangement has evolved to permit the integration of a number of signal inputs, such as were described earlier. The number of these probably reflects the involvement of both growth and cellular differentiation in sporulation. What might such signals consist of? This has already been discussed for σ^{WhiG}. In the case of WhiH, it is tempting to extrapolate from the finding that its closest homologues all respond to carboxylate-containing compounds and to speculate that the cessation of tip growth of aerial hyphae may cause a burst of production of such a substance (perhaps connected with fatty acid metabolism, as discussed for FadR of *E. coli* by DiRusso and Nyström [1998]) (Ryding et al., 1998). WhiI, with its atypical residues in what is usually a conserved phosphorylation pocket in typical members of the response regulator family, might also bind reversibly to some unidentified signal metabolite and thereby become changed in its regulatory activity for target genes (Aínsa et al., in press).

The WhiA and WhiB proteins may also be responsive to signals. Four cysteine residues are completely conserved among the growing family of WhiB-like proteins from actinomy-

cetes, encouraging speculation that these proteins may be sensitive to a hyperoxidant state (Soliveri et al., 1993, unpublished). Studies of *E. coli* indicate that endogenously generated cytoplasmic oxidative stress can cause an increase in cytoplasmic protein oxidation, including disulfide bond formation, during stationary phase (Dukan and Nyström, 1998), and reversible disulfide bond formation of the *E. coli* cytoplasmic transcription factor OxyR leads to its activation (Storz and Imlay, 1999), providing a potential precedent for suggesting that a significant change in the regulatory properties of WhiB could also be mediated by the oxidation-reduction state of its thiols consequent on growth cessation. Alternatively, the thiols of WhiB may interact with a redox-sensitive metal ion, as in SoxR (Storz and Imlay, 1999). WhiA does not have intrinsic features that suggest particular functions, and there are no clues to guide further speculation about its biochemical role (Ryding et al., unpublished).

DOES THE *S. COELICOLOR* PARADIGM APPLY TO OTHER *STREPTOMYCES* SPP.?

Early *whi* genes have been sequenced and further studied in streptomycetes that are phylogenetically fairly distant from *S. coelicolor*, most notably *Streptomyces aureofaciens*. The *whiG* homologue of *S. aureofaciens* (called *rpoZ*) is known to transcribe the *S. aureofaciens whiH* homologue (Kormanec et al., 1999), and mutations of *rpoZ* generate a phenotype like that of *S. coelicolor whiG* mutants (Kormanec et al., 1994). Differences have been observed between *rpoZ* and *whiG* in the details of their transcription: Kormanec et al. (1996) detected two *rpoZ* promoters, one up-regulated and the other down-regulated when aerial hyphae begin to form and with an apparent net increase in overall expression at this stage. This departure from the single, constitutive promoter of *whiG* in *S. coelicolor* may partially account for the more rapid sporulation of *S. aureofaciens*. The *whiB* homologue of *S. aureofaciens* has regulation very similar to that of *S. coelicolor*, and the aerial hyphae of a disruption mutant showed no sporulation or fragmentation, but the aerial hyphae remained straight, unlike the coils typical of *S. coelicolor whiB* mutants (the spore chains of wild-type *S. aureofaciens* are also straight, unlike those of *S. coelicolor*) (Kormanec et al., 1998).

At least some late sporulation events in *S. aureofaciens*, like those in *S. coelicolor*, depend on σ^F. In *S. aureofaciens*, a *sigF* mutant had green spores, in contrast to the dark gray of the wild type, indicating that, as in *S. coelicolor*, a late step of spore pigment biosynthesis, but not the polyketide synthase itself, was σ^F dependent (Řežuchova et al., 1997). As in *S. coelicolor*, *sigF* transcription approximately coincides with sporulation septation and is eliminated in an *rpoZ* (equivalent to *whiG*) mutant (Kormanec et al., 1996).

These and other *whi* gene sequence comparisons indicate extensive conservation of the sporulation regulatory cascade among streptomycetes, consistent with a monophyletic evolution of the process, such as we might intuitively predict; but the details of *whiG* activation, which sets the cascade in motion, and of the cell wall structure of aerial hyphae (to give coiled or straight aerial hyphae in *whiB* mutants) show species-specific variation. The timing of entry into sporulation and the detailed external architecture of aerial hyphae (e.g., coiling) and spores (e.g., smooth or ornamented with "hairs" or spines) are both aspects of the sporulation process that could affect adaptive interactions of streptomycetes with the environment and so might be expected to evolve in a species-specific manner (and, conversely, to contribute to speciation).

HOW ARE GROWTH REQUIREMENTS SUPPLIED TO AERIAL HYPHAE?

The growth of aerial hyphae, and the metamorphosis of the hyphal tips into spore chains, must involve some degree of specialized metabolic provision. This aspect of metabolism is very difficult to study because of the heterogeneity of aerial-mycelium development and the comparatively small amounts of material available from surface cultures. Two developments will probably begin to clarify this area: the further analysis of the early *whi* genes, permitting an understanding of the signals to which Whi proteins respond and of the DNA sequences to which they bind, and the completion of the DNA sequencing of the *S. coelicolor* A3(2) chromosome (http://www.sanger.ac.uk/Projects/S_coelicolor/), followed by global transcriptome and proteome analysis. Once assembled, this information can be expected to indicate the biochemical functions controlled by each regulatory protein and how they are up- or downregulated during the developmental process.

Some aspects of developmental metabolism are relatively amenable to analysis. For example, the study of the flow of carbon through the developing colony is helped because localized carbon storage reservoirs are cytologically detectable by transmission electron microscopy (whether as electron-transparent poly-β-hydroxyalkanoate or oil deposits or as the electron-opaque granules that result from silver staining of glycogen or other polysaccharides) (Braña et al., 1986; Plaskitt and Chater, 1995). Some (perhaps all) of the genes typically involved in bacterial glycogen metabolism are present in two copies in *S. coelicolor* (Bruton et al., 1995; Martín et al., 1997) and *S. aureofaciens* (Homerová et al., 1996). Each gene is implicated in the synthesis or subsequent reuse of glycogen in only one of two locations: in the substrate mycelium at its junction with the aerial mycelium, or in aerial-hyphal tips as sporulation septation takes place and individual compartments mature into spores. Although these two phases of glycogen deposition and metabolism are probably extensively regulated at the transcription level, posttranscriptional effects may also prove highly significant: the two genes encoding biosynthetic glycogen branching enzymes both appear to be cotranscribed with genes involved in the degradation of glycogen (Schneider et al., unpublished), and it would be surprising if there were no mechanism to avoid the futile breakdown of glycogen as soon as it is made.

It has been suggested that the extensive mobilization of nutrient reservoirs such as glycogen could contribute to the development of turgor pressure to drive aerial growth (Chater, 1989). However, the phenotype of a mutant blocked in the first phase of glycogen synthesis differed little, if at all, from that of the parental strain, either negating the hypothesis or indicating that there are other sources of turgor (Martín et al., 1997). It is interesting that the first nonregulatory gene to be discovered with a promoter directly dependent on σ^{WhiG} encodes a ProX-like protein (Tan et al., 1998). In *E. coli*, ProX is a periplasmic binding protein for the osmotic balancing agent glycine betaine, and it is part of a more complex osmotically regulated ABC transporter system for glycine betaine transport. It is not clear how regulating the binding protein separately from other components of the system during development would be important, but the fact that a presumptive component of osmotic homeostasis is developmentally regulated does emphasize the likely importance of osmotic control during aerial growth. Alternatively, it is possible that σ^{WhiG} also functions in "nondevelopmental" processes that are only observable in special circumstances.

CONCLUDING OVERVIEW

Our picture of the sporulation of *Streptomyces* spp., as revealed by the study of *bld* (see chapter 1) and *whi* mutants, differs strikingly from that of endospore-forming bacilli as studied with *spo* mutants (see chapters 6 to 9). The extent of these apparent differences may be exaggerated in our current perception, since most of the *Streptomyces* genetic studies have been done with genes whose action leads up to the production of prespore compartments and which are broadly equivalent in that sense to the *spo0* genes of *B. subtilis*. Only one regulatory gene believed to act after sporulation septation in *Streptomyces* has been identified, and this is *sigF*. Since the sigma factor encoded by *sigF* falls into a subfamily that includes the two prespore compartment-specific sigma factors of *B. subtilis*, it could be that the later stages of sporulation

in the two organisms will prove to show more homology than the early stages.

If we try to equate the *bld* and early *whi* genes with *spo0* genes, then we can see that in both organisms these genes mostly encode proteins that function in extracellular and intracellular signaling. The *bld* genes, whose activity is important in substrate hyphae that inhabit a common hydrated milieu, seem to be responsible (mostly indirectly) for an elaborate cascade of extracellular signals, which presumably themselves reflect changes in intracellular physiology such as might arise from nutrient limitation (see chapter 1). An equivalent situation in *B. subtilis* may be represented by the role of extracellular oligopeptide signals generated by the activity of the *rap* operons (though I am unaware of any evidence of a hierarchy of exchange of these signals [see chapter 7] [Perego, 1998]).

Once aerial hyphae have formed, there is little opportunity for chemical signaling among hyphae (though gaseous-phase signal exchange has not been ruled out: the earthy odor of damp soil is caused by a volatile product, geosmin, associated with *Streptomyces* spores but of quite unknown function [Gerber, 1979]). Available evidence suggests that individual aerial hyphae can be considered as autonomously developing entities, except insofar as they are nutritionally dependent on the substrate hyphae to which they are attached. The very significant growth associated with this phase of development marks a clear distinction from *B. subtilis* sporulation, and the need to monitor growth and morphological change during this stage accounts for the evolution of the regulatory mechanisms encoded by the early *whi* genes.

Could the spore-bearing stalk of an aerial hypha fulfill any of the roles of the mother cell during endospore formation? It is difficult to imagine extensive parallels, not least because the separation of the stalk from prespore compartments at different positions in the developing spore chain varies from a single sporulation septum to many tens of septa. Thus, any proposed interaction would have to account for very remote effects, presumably either by efficient signal relays or by extracytoplasmic contact through the hydrated hyphal wall. More likely, the physiological role of the stalk is completed when sporulation septation takes place. At this time, the stalk becomes more transparent in phase-contrast microscopy and less fluorescence is seen in stalks after DAPI (4',6-diamidino-2-phenylindole) staining for DNA. Thus, our present state of knowledge does not suggest that the conversion of prespore compartments into spores should involve very complex genetic regulation or developmental checkpoints.

ACKNOWLEDGMENTS

I am very grateful to Wendy Champness, David Hopwood, Tobias Kieser, Klas Flärdh, and Jose Aínsa for comments on the manuscript and for communicating unpublished data and providing figures. Much of the recent work from my laboratory reported here was supported by grant CAD 04380 from the BBSRC.

REFERENCES

Aínsa, J. A., H. D. Parry, and K. F. Chater. A response regulator-like protein that functions at an intermediate stage of sporulation in *Streptomyces coelicolor* A3(2). *Mol. Microbiol.*, in press.

Blanco, G., A. Pereda, P. Brian, C. Méndez, K. F. Chater, and J. A. Salas. 1993. A hydroxylase-like gene product contributes to synthesis of a polyketide spore pigment in *Streptomyces halstedii*. *J. Bacteriol.* **175:** 8043–8048.

Braña, A. F., C. Méndez, L. A. Díaz, M. B. Manzanal, and C. Hardisson. 1986. Glycogen and trehalose accumulation during colony development in *Streptomyces antibioticus*. *J. Gen. Microbiol.* **132:** 1319–1326.

Brown, D. W., and J. J. Salvo. 1994. Isolation and characterisation of sexual spore pigments from *Aspergillus nidulans*. *Appl. Environ. Microbiol.* **60:** 979–983.

Brown, D. W., F. M. Hauser, and J. J. Salvo. 1993. Structural elucidation of a putative conidial pigment intermediate in *Aspergillus parasiticus*. *Tetrahedron Lett.* **34:** 419–422.

Bruton, C. J., K. A. Plaskitt, and K. F. Chater. 1995. Tissue-specific glycogen branching isoenzymes in a multicellular prokaryote, *Streptomyces coelicolor* A3(2). *Mol. Microbiol.* **18:** 89–99.

Chater, K. F. 1972. A morphological and genetic mapping study of white colony mutants of *Streptomyces coelicolor*. *J. Gen. Microbiol.* **72:** 9–28.

Chater, K. F. 1989. Multilevel regulation of *Streptomyces* differentiation. *Trends Genet.* **5:** 372–376.

Chater, K. F. 1991. Saps, hydrophobins, and aerial growth. *Curr. Biol.* **1:** 318–320.

Chater, K. F. 1998. Taking a genetic scalpel to the *Streptomyces* colony. *Microbiology* **144:** 1465–1478.

Chater, K. F., and R. Losick. 1997. The mycelial life-style of *Streptomyces coelicolor* A3(2) and its relatives, p. 149–182. *In* J. H. Shapiro and M. Dworkin (ed.), *Bacteria as Multicellular Organisms.* Oxford University Press, New York, N.Y.

Chater, K. F., C. J. Bruton, K. A. Plaskitt, M. J. Buttner, C. Méndez, and J. Helmann. 1989. The developmental fate of *S. coelicolor* hyphae depends crucially on a gene product homologous with the motility sigma factor of *B. subtilis*. *Cell* **59:** 133–143.

Cole, S. T., R. Brosch, J. Parkhill, T. Garnier, C. Churcher, D. Harris, S. V. Gordon, K. Eiglmeier, S. Gas, C. E. Barry, F. Tekaia, K. Badcock, D. Basham, D. Brown, T. Chillingworth, R. Conner, R. Davies, K. Devlin, T. Feltwell, S. Gentles, N. Hamlin, S. Holroyd, T. Hornsby, K. Jaels, A. Krogh, J. McLean, S. Moule, L. Murphy, K. Oliver, J. Osborne, M. A. Quail, A. Rajandream, J. Rogers, S. Rutter, K. Segger, J. Skelton, R. Squares, S. Squares, J. E. Sulston, K. Taylor, S. Whitehead, and B. G. Barrell. 1998. Deciphering the biology of *Mycobacterium tuberculosis* from the complete genome sequence. *Nature* **393:** 537–544.

Davis, N. K., and K. F. Chater. 1992. The *Streptomyces coelicolor whiB* gene encodes a small transcription factor-like protein dispensable for growth but essential for sporulation. *Mol. Gen. Genet.* **232:** 351–358.

DiRusso, C. C., and T. Nyström. 1998. The fats of *Escherichia coli* during infancy and old age: regulation by global regulators, alarmones and lipid intermediates. *Mol. Microbiol.* **27:** 1–8.

Donachie, W. D. 1993. The cell cycle of *Escherichia coli*. *Annu. Rev. Microbiol.* **47:** 199–230.

Dukan, S., and T. Nyström. 1998. Bacterial senescence: stasis results in increased and differential oxidation of cytoplasmic proteins leading to developmental induction of the heat shock regulon. *Genes Dev.* **12:** 3431–3441.

Flärdh, K. Personal communication.

Flärdh, K., P. Palacios, and M. Vicente. 1998. Cell division genes *ftsQAS* in *Escherichia coli* require distant *cis*-acting signals upstream of *ddlB* for full expression. *Mol. Microbiol.* **30:** 305–315.

Flärdh, K., K. C. Findlay, and K. F. Chater. Association of early sporulation genes with suggested developmental decision points in *Streptomyces coelicolor* A3(2). *Microbiology*, in press.

Gerber, N. N. 1979. Volatile substances from actinomycetes: their role in the odor pollution of water. *Crit. Rev. Microbiol.* **7**:191–214.

Hao, J., and K. E. Kendrick. 1998. Visualization of penicillin-binding proteins during sporulation of *Streptomyces griseus. J. Bacteriol.* **180**:2125–2132.

Hardisson, C., and M. B. Manzanal. 1976. Ultrastructural studies of sporulation in *Streptomyces. J. Bacteriol.* **127**:1443–1454.

Homerová, D., O. Benada, O. Kofronova, B. Rezuchová, and J. Kormanec. 1996. Disruption of a glycogen branching enzyme gene, *glgB*, specifically affects the sporulation-associated phase of glycogen accumulation in *Streptomyces aureofaciens. Microbiology* **142**:1201–1208.

Hopwood, D. A., H. Wildermuth, and H. M. Palmer. 1970. Mutants of *Streptomyces coelicolor* defective in sporulation. *J. Gen. Microbiol.* **61**:397–408.

Kelemen, G. H., and M. J. Buttner. 1998. Initiation of aerial mycelium formation in *Streptomyces. Curr. Opin. Microbiol.* **1**:656–662.

Kelemen, G. H., M. S. B. Paget, and M. J. Buttner. Personal communication.

Kelemen, G. H., G. L. Brown, J. Kormanec, L. Potúčková, K. F. Chater, and M. J. Buttner. 1996. The positions of the sigma factor genes, *whiG* and *sigF*, in the hierarchy controlling the development of spore chains in the aerial hyphae of *Streptomyces coelicolor* A3(2). *Mol. Microbiol.* **21**:593–603.

Kelemen, G. H., P. Brian, K. Flärdh, L. Chamberlin, K. F. Chater, and M. J. Buttner. 1998. Developmental regulation of transcription of whiE, a locus specifying the polyketide spore pigment in *Streptomyces coelicolor* A3(2). *J. Bacteriol.* **180**:2515–2521.

Kendrick, K. E. Personal communication.

Kim, H.-J., M. J. Calcutt, F. J. Schmidt, and K. F. Chater. Chromosome partitioning during sporulation of *Streptomyces coelicolor* A3(2) involves an *oriC*-linked *parAB* locus. Manuscript in preparation.

Kormanec, J., L. Potúčková, and B. Rezuchová. 1994. The *Streptomyces aureofaciens* homologue of the *whiG* gene encoding a putative sigma factor essential for sporulation. *Gene* **143**:101–103.

Kormanec, J., D. Homerová, L. Potúčková, R. Nováková, and B. Rezuchová. 1996. Differential expression of two sporulation specific σ factors of *Streptomyces aureofaciens* correlates with the developmental stage. *Gene* **181**:19–27.

Kormanec, J., B. Ševčíková, O. Sprusansky, O. Benada, O. Kofronova, R. Nováková, B. Rezuchová, L. Potúčková, and D. Homerová. 1998. The *Streptomyces aureofaciens* homologue of the *whiB* gene is essential for sporulation; its expression correlates with the developmental stage. *Folia Microbiol.* **43**:605–612.

Kormanec, J., R. Nováková, D. Homerová, and B. Ševčíková. 1999. The *Streptomyces aureofaciens* homologue of the sporulation gene *whiH* is dependent on *rpoZ*-encoded σ factor. *Biochim. Biophys. Acta* **1444**:80–84.

Losick, R., and L. Shapiro. 1993. Checkpoints that couple gene expression to morphogenesis. *Science* **262**:1227–1228.

Martín, M. C., D. Schneider, C. J. Bruton, K. F. Chater, and C. Hardisson. 1997. A *glgC* gene essential only for the first of two spatially distinct phases of glycogen synthesis in *Streptomyces coelicolor* A3(2). *J. Bacteriol.* **179**:7784–7789.

McCormick, J., E. P. Su, A. Driks, and R. Losick. 1994. Growth and viability of *Streptomyces coelicolor* mutant for the cell division gene *ftsZ. Mol. Microbiol.* **14**:243–254.

McCormick, J. R., and R. Losick. 1996. Cell division gene *ftsQ* is required for efficient sporulation but not growth and viability in *Streptomyces coelicolor* A3(2). *J. Bacteriol.* **178**:5295–5301.

McVittie, A. M. 1974. Ultrastructural studies on sporulation in wild-type and white colony mutants of *Streptomyces coelicolor. J. Gen. Microbiol.* **81**:291–302.

Miguélez, E. M., C. Martín, M. B. Manzanal, and C. Hardisson. 1992. Growth and morphogenesis in *Streptomyces. FEMS Microbiol. Lett.* **100**:351–360.

Miguélez, E. M., M. García, C. Hardisson, and M. B. Manzanal. 1994. Autoradiographic study of hyphal growth during aerial mycelium development in *Streptomyces antibioticus. J. Bacteriol.* **176**:2105–2107.

Miwa, Y., and Y. Fujita. 1988. Purification and characterization of a repressor for the *Bacillus subtilis gnt* operon. *J. Biol. Chem.* **263**:13252–13257.

Nodwell, J., and R. Losick. Personal communication.

Nodwell, J. R., K. McGovern, and R. Losick. 1996. An oligopeptide permease responsible for the import of an extracellular signal governing aerial mycelium formation in *Streptomyces coelicolor. Mol. Microbiol.* **22**:881–893.

Palacios, P., M. Vicente, and M. Sánchez. 1996. Dependency of *Escherichia coli* cell-division size, and independency of nucleoid segregation on the mode and level of *ftsZ* expression. *Mol. Microbiol.* **20**:1093–1098.

Perego, M. 1998. Kinase-phosphatase competition regulates *Bacillus subtilis* development. *Trends Microbiol.* **6**:366–370.

Plaskitt, K. A., and K. F. Chater. 1995. Influences of developmental genes on localised glycogen deposition in colonies of a mycelial prokaryote, *Strepto-*

myces coelicolor A(3)2: a possible interface between metabolism and morphogenesis. *Phil. Trans. R. Soc. B.* **347**:105–121.

Potúčková, L., G. H. Kelemen, K. C. Findlay, M. A. Lonetto, M. J. Buttner, and J. Kormanec. 1995. A new RNA polymerase sigma factor, σ^F, is required for the late stages of morphological differentiation in *Streptomyces* spp. *Mol. Microbiol.* **17**:37–48.

Rezuchová, B., I. Barak, and J. Kormanec. 1997. Disruption of a sigma factor gene, *sigF*, affects an intermediate stage of spore pigment production in *Streptomyces aureofaciens*. *FEMS Microbiol. Lett.* **153**: 371–377.

Rothfield, L. I., and S. S. Justice. 1997. Bacterial cell division: the cycle of the ring. *Cell* **88**:581–584.

Ryding, N. J., J. Aínsa, and K. F. Chater. Unpublished data.

Ryding, N. J., G. H. Kelemen, C. A. Whatling, K. Flärdh, M. J. Buttner, and K. F. Chater. 1998. A developmentally regulated gene encoding a repressor-like protein is essential for sporulation in *Streptomyces coelicolor* A3(2). *Microbiology* **29**: 343–357.

Ryding, N. J., M. J. Bibb, V. Molle, K. C. Findlay, K. F. Chater, and M. J. Buttner. 1999. New sporulation loci in *Streptomyces coelicolor* A3(2). *J. Bacteriol.* **181**, in press.

Ryding, N. J., J. A. Aínsa, N. Hartley, C. J. Bruton, and K. F. Chater. Unpublished data.

Schneider, D., C. J. Bruton, and K. F. Chater. Unpublished data.

Schwedock, J., J. R. McCormick, E. R. Angert, J. R. Nodwell, and R. Losick. 1997. Assembly of the cell division protein FtsZ into ladder-like structures in the aerial hyphae of *Streptomyces coelicolor*. *Mol. Microbiol.* **25**:847–858.

Shahab, N., F. Flett, S. G. Oliver, and P. R. Butler. 1996. Growth rate control of protein and nucleic acid content in *Streptomyces coelicolor* A3(2) and *Escherichia coli* B/r. *Microbiology* **142**:1927–1935.

Soliveri, J., K. L. Brown, M. J. Buttner, and K. F. Chater Two promoters for the *whiB* sporulation gene of *Streptomyces coelicolor* A3(2), and their activities in relation to development. *J. Bacteriol.* **174**: 6215–6220.

Soliveri, J., C. Granozzi, K. A. Plaskitt, and K. F. Chater. 1993. Functional and evolutionary implications of a survey of various actinomycetes for homologues of two *Streptomyces coelicolor* sporulation genes. *J. Gen. Microbiol.* **139**:2569–2578.

Soliveri, J. A., J. Gomez, W. R. Bishai, and K. F. Chater. Unpublished data.

Storz, G., and J. A. Imlay. 1999. Oxidative stress. *Curr. Opin. Microbiol.* **2**:188–194.

Tan, H., H. Yang, Y. Tian, W. Wu, C. A. Whatling, L. C. Chamberlin, M. J. Buttner, J. Nodwell, and K. F. Chater. 1998. The *Streptomyces coelicolor* sporulation-specific σ^{WhiG} form of RNA polymerase transcribes a gene encoding a ProX-like protein that is dispensable for sporulation. *Gene* **212**:137–146.

Tillotson, R. D., H. A. B. Wösten, M. Richter, and J. M. Willey. 1998. A surface active protein involved in aerial mycelium formation in the filamentous fungus *Schizophillum commune* restores the capacity of a bald mutant of the filamentous bacterium *Streptomyces coelicolor* to erect aerial structures. *Mol. Microbiol.* **30**:595–602.

Ward, J. E., Jr., and J. Lutkenhaus. 1985. Overproduction of FtsZ induces minicell formation in *E. coli*. *Cell* **42**:941–949.

Wheeler, R. T., and L. Shapiro. 1997. Bacterial chromosome segregation: is there a mitotic apparatus? *Cell* **88**:577–579.

Wildermuth, H., and D. A. Hopwood. 1970. Septation during sporulation in *Streptomyces coelicolor*. *J. Gen. Microbiol.* **60**:51–59.

Willey, J., R. Santamaría, J. Guijarro, M. Geistlich, and R. Losick. 1991. Extracellular complementation of a developmental mutation implicates a small sporulation protein in aerial mycelium formation by *Streptomyces coelicolor*. *Cell* **65**:641–650.

Willey, J., J. Schwedock, and R. Losick. 1993. Multiple extracellular signals govern the production of a morphogenetic protein involved in aerial mycelium formation by *Streptomyces coelicolor*. *Genes Dev.* **7**:895–903.

Yu, T.-W., Y. Shen, R. McDaniel, H. G. Floss, C. Khosla, D. A. Hopwood, and B. S. Moore. 1998. Engineered biosynthesis of novel polyketides from *Streptomyces* spore pigment polyketide synthases. *J. Am. Chem. Soc.* **120**:7749–7759.

CYANOBACTERIA

CYANOBACTERIAL PHYLOGENY AND DEVELOPMENT: QUESTIONS AND CHALLENGES

David G. Adams

3

Cyanobacteria are a large and morphologically diverse group of phototrophic prokaryotes with such wide ecological tolerance that they occur in almost every habitat on earth. This versatility may explain the remarkable lack of morphological (and presumably physiological) change to be seen when comparing 3.5 billion-year-old fossilized cyanobacteria and their modern-day counterparts (Schopf, 1994, 1996) (see chapter 5). Cyanobacteria may have evolved from the prebiotic soup in as little as 10 million years (Lazcano and Miller, 1994), and they were the first organisms to evolve oxygenic photosynthesis of the type found in higher plants (Castenholz and Waterbury, 1989; Tandeau de Marsac and Houmard, 1993). Indeed, the plastids of higher plants and other photosynthetic eukaryotes are thought possibly to have arisen from a single common ancestor, itself the result of an endosymbiosis between a phagotrophic host and a cyanobacterium (Douglas, 1992, 1994; Löffelhardt and Bohnert, 1994a, 1994b; Helmchen et al., 1995; Bhattacharya and Medlin, 1995; Delwiche and Palmer, 1997). The success of these early cyanobacteria had a profound effect on the earth's atmosphere, changing it from anaerobic to aerobic, and as a consequence the cyanobacteria replaced anoxygenic photoautotrophs throughout much of the earth's photic zone.

All cyanobacteria are photoautotrophs, deriving the ATP and NADPH required for CO_2 fixation from electrons released from water by photolysis. Many can also grow photoheterotrophically, obtaining energy from light and getting carbon and reducing power from organic compounds. Some strains can also grow chemoheterotrophically in the dark at the expense of organic compounds, although such growth is usually much slower than in the light (Castenholz and Waterbury, 1989; Tandeau de Marsac and Houmard, 1993). A small number of cyanobacteria can switch from oxygenic to anoxygenic photosynthesis, in which the reducing power is supplied by H_2S rather than H_2O (Padan and Cohen, 1982; Schmidt, 1988).

Many cyanobacteria are truly multicellular organisms that respond to environmental change by developing a variety of cell and filament types (Fig. 1 and 2). The best studied of these is the heterocyst, which is highly specialized for nitrogen fixation, but other forms include akinetes, hormogonia, baeocytes, and hairs. The purpose of this chapter is to provide a selective overview of present knowledge of

D. G. Adams, Division of Microbiology, School of Biochemistry and Molecular Biology, University of Leeds, Leeds LS2 9JT, United Kingdom.

Prokaryotic Development, edited by Y. V. Brun and L. J. Shimkets,
© 2000 American Society for Microbiology, Washington, DC 20005-4171

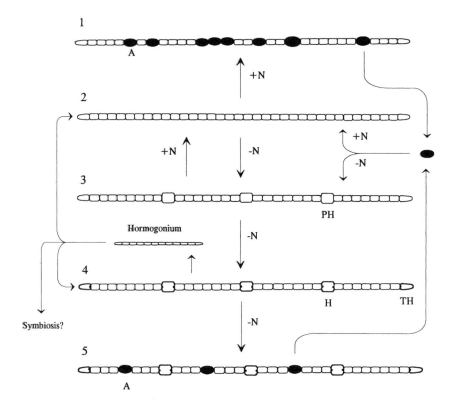

FIGURE 1 Schematic diagram illustrating some of the possibilities for morphological development in filamentous cyanobacteria. When grown in the presence of a source of combined nitrogen, the filament consists entirely of undifferentiated vegetative cells (2). At the end of the exponential growth phase, when light (energy) becomes limiting, some vegetative cells can differentiate into the spore-like cells called akinetes (A), which, in the absence of heterocysts, are randomly placed within the filament (1). In the absence of combined nitrogen, the vegetative filament differentiates the highly specialized N_2-fixing cells, heterocysts, at spaced intervals within the filament (H) and in terminal positions (TH). Heterocysts are characterized by their thickened cell walls, relatively agranular cytoplasm, and polar bodies at the point of attachment to vegetative cells (two in heterocysts within the filament but only one in terminal heterocysts) (4). During their development heterocysts pass through an intermediate stage, the proheterocyst (PH), which, unlike the mature cell, does not have the thickened cell wall and polar bodies and is able to dedifferentiate in the presence of combined nitrogen (3). When akinetes develop in N_2-fixing cultures, they do so at locations with a spatial relationship to the heterocysts (see the text for details), such as midway between them (5). Akinetes can germinate and give rise to filaments with or without heterocysts, depending on the availability of combined nitrogen. Hormogonia are short, motile filaments lacking heterocysts, which develop as a result of a variety of stimuli (see the text for details). Their formation usually involves the rapid, synchronized division of vegetative cells without concomitant growth, followed by fragmentation of the filament to release heterocysts and motile hormogonia. The latter can give rise to heterocystous or nonheterocystous filaments, depending on the availability of combined nitrogen. Hormogonia can also serve as the infective agents in the establishment of symbiotic associations with plants. (Adapted from Adams [1992].)

these morphological forms and of their development. More detailed information on these topics can be found in the reviews indicated in the text, and heterocyst formation and cyanobacterial evolution are dealt with in chapters 4 and 5. More general information on cyanobacteria can be found in several books: Whitton and Potts, in press; Bryant, 1994; van Baalen and Fay, 1987; and Carr and Whitton, 1982.

CLASSIFICATION

Because of the paucity of physiological variation among cyanobacteria, their classification has relied heavily on their morphological complexity. This has certainly resulted in the grouping of morphologically similar but genetically unrelated organisms, and as a consequence, cyanobacterial classification is in a state of transition (Holt et al., 1994). Descriptions of past and current systems of classification can be found elsewhere (Rippka et al., 1979, 1981; Castenholz and Waterbury, 1989; Castenholz, 1992; Wilmotte, 1994; Holt et al., 1994) and will not be considered in detail here. However, the following brief explanation of a commonly used taxonomic system will provide a framework for the later discussion of the developmental capabilities of cyanobacteria.

The simplest current system of classification is derived from that of Rippka and coworkers and is based on the relatively small number of cyanobacteria in culture (Rippka, 1988; Rippka et al., 1979, 1981). A description of this system is provided in *Bergey's Manual of Determinative Bacteriology* (Holt et al., 1994). The oxygenic phototrophic bacteria are divided into two groups, the cyanobacteria and the prochlorophytes (order *Prochlorales*). The latter is a small group of recently discovered oxygenic photosynthetic bacteria that share many features with the cyanobacteria but lack phycobilin pigments and contain chlorophyll *b* as well as chlorophyll *a* (Mur and Burger-Wiersma, 1992; Post and Bullerjahn, 1994; Matthijs et al., 1994). The cyanobacteria are divided into five subgroups, each with ordinal rank. Subgroups 1 and 2 (orders *Chroococcales* and *Pleurocapsales*) contain unicellular cyanobacteria, some of which may exist as nonfilamentous aggregates of cells. In the order *Chroococcales* the cells are spherical, cylindrical, or oval, and they reproduce by symmetrical or asymmetrical binary fission in one, two, or three planes or by budding. The members of the order *Pleurocapsales* reproduce by internal multiple fission or by multiple and binary fission (see Fig. 6 to 9). The products of multiple fission are as many as 1,000 small, spherical reproductive cells known as baeocytes, some of which display gliding motility (see Fig. 6 to 9 and Pleurocapsalean Cyanobacteria below).

The remaining three subgroups all consist of filamentous cyanobacteria, in which vegetative trichomes are often surrounded by a sheath and both true and false branching can occur. The former results from a shift in the plane of cell division from right angles to the filament axis to a plane parallel to the axis. The resulting cells elongate at right angles to the original filament, producing a branch. False branching results from fragmentation of a trichome within a sheath, followed by elongation of the newly created trichome ends, resulting in the protrusion of one of them through the sheath. Cyanobacteria of subgroup 3 (order *Oscillatoriales*) divide by binary fission in one plane only (at right angles to the filament axis) and are incapable of forming heterocysts. Cell division in members of subgroup 4 (order *Nostocales*) occurs in one plane only, and filaments are capable of producing heterocysts in the absence of combined nitrogen (Fig. 1 and 2). The most morphologically and developmentally complex cyanobacteria are found in subgroup 5 (order *Stigonematales*); heterocysts are produced in the absence of combined nitrogen, and cell division often occurs in more than one plane, giving rise to multiseriate trichomes (filaments with more than one row of cells) or to trichomes with true branches (false branches can also occur), or to both.

CELL STRUCTURE

The cyanobacterial cell wall is of the gram-negative type, consisting of cytoplasmic and outer membranes separated by an electron-

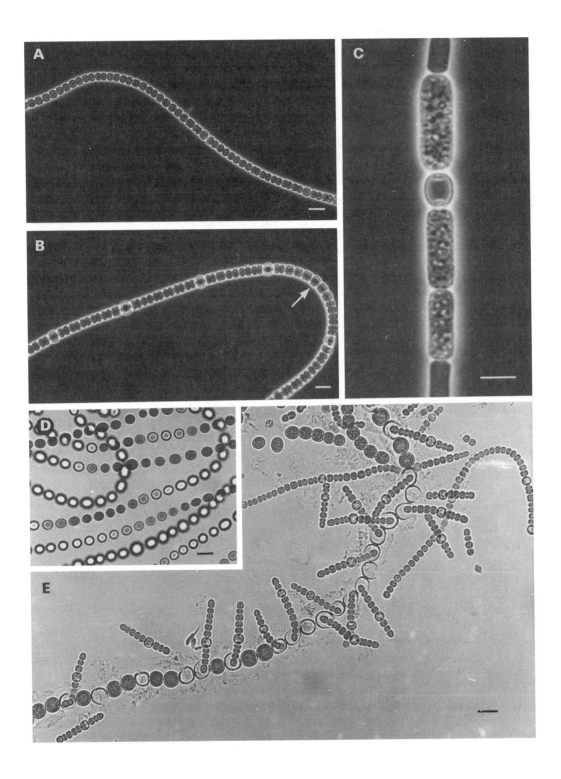

FIGURE 3 Schematic diagram of a thin section of a cyanobacterial cell. C, carboxysome; CPG, cyanophycin granule; T, thylakoid; P, polyphosphate granule; N, nucleoplasmic region; G, glycogen granules; PB, phycobilisome; GV, gas vesicle. Inset A is an enlarged view of a thylakoid, showing paired unit membranes. Inset B is an enlarged view of the cell envelope, showing the outer membrane, the peptidoglycan layer, and the cytoplasmic membrane. (Adapted from Stanier and Cohen-Bazire [1977].)

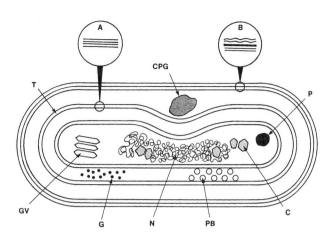

opaque peptidoglycan layer (Fig. 3) (Stanier, 1988; Weckesser and Jurgens, 1988; Castenholz and Waterbury, 1989). However, the thickness of the peptidoglycan, together with its degree of cross-linking and the presence of covalently linked polysaccharide, are more characteristic of gram-positive cell walls (Weckesser and Jurgens, 1988). Although 1 to 10 nm thick in most cyanobacteria, the peptidoglycan layer can be 200 nm thick in *Oscillatoria princeps*, which has cells 100 μm in diameter and filaments millimeters in length. The extremely wide peptidoglycan layer in cyanobacteria such as *O. princeps* is usually traversed by large (70-nm-diameter) pores that bring the cytoplasmic membrane close to the outer membrane (Fig. 4). Small-diameter (5- to 13-nm) pores are found in the walls of all cyano-

bacteria, but their distribution varies greatly. In the longitudinal wall of filamentous strains, they can be distributed over the whole surface or as so-called junctional pores forming a ring around the cross walls (Fig. 4). The cell septum may contain many pores or a single tiny central pore. Many cyanobacteria possess an additional layer outside the outer membrane, consisting predominantly of polysaccharide, although greater than 20% by weight can be polypeptides (Castenholz and Waterbury, 1989). This outer envelope is referred to by a variety of names, including glycocalyx, sheath, capsule, mucilage, and slime, that reflect the variation in the consistency of the layer.

Recent work has revealed the cyanobacterial cell wall to be more complex than once thought. An increasing number of cyanobacte-

FIGURE 2 Photomicrographs illustrating some of the differentiated cell types of cyanobacteria. (A) *Anabaena* sp. strain CA grown in the presence of nitrate, which completely suppresses heterocyst development. (B) *Anabaena* sp. strain CA grown in the absence of combined nitrogen, showing the regular spacing of heterocysts, which are the sites of N_2 fixation, and a developing proheterocyst (arrow). (C) *A. cylindrica*, showing large, granular akinetes developing immediately adjacent to a heterocyst. The two akinetes below the heterocyst show a characteristic gradient of maturity, with the largest and oldest closer to the heterocyst. (D) An old culture of nitrate-grown *Anabaena* sp. strain CA in which all vegetative cells have transformed into spherical akinetes. Although the akinetes have become separated, as a result of pressure created by the coverslip being placed onto the sample, the line of the original filaments can still be seen. Dilution of such a culture leads to germination of the akinetes. If the medium used for dilution does not contain a source of fixed nitrogen, the short filaments which emerge from the akinete coats each contain a heterocyst (E). Bars, 10 μm. Panels C and D were made with phase-contrast optics. Panels A and B reproduced from Adams and Carr [1981] with permission of the publisher (copyright CRC Press Inc.); panel C reproduced from Nichols and Adams [1982] with permission of the publisher; panels D and E reproduced from Adams [1992] with permission of the publisher.

A

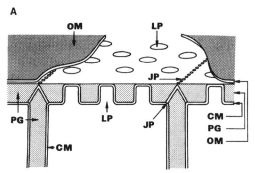

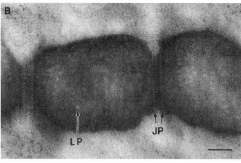

FIGURE 4 (A) Diagram of the cell wall of *Oscilla-toria* sp. The peptidoglycan layer (PG) is perforated by large pore pits (LP) up to 70 nm in diameter, which bring the cytoplasmic membrane (CM) close to the outer membrane (OM). The large pore pits are seen in cross-section in the lower half of the diagram and appear as circles when viewed from above in the top half of the diagram. Junctional pores (JP) traverse the peptidoglycan between cells and meet at the surface, appearing as a single row of small circles when viewed from above in the top half of the diagram. (Adapted from Halfen [1979].) (B) Electron micrograph of a thin longitudinal section of an unidentified member of the order *Oscillatoriales*. The large pore pits (LP) can be seen as bright circles scattered throughout the pepti-doglycan. The plane of the section is within the pepti-doglycan layer, with the result that the junctional pores (JP) appear as parallel rows of small, closely positioned circles at the cell cross walls rather than the single rows that would be seen at the surface of the peptidoglycan. The sample was fixed in osmium tetroxide and stained with uranyl acetate. Bars, 0.2 μm. (Electron micros-copy by Denise Ashworth.)

ria have been shown to possess an S-layer out-side the outer membrane (Rachel et al., 1997). S-layers are paracrystalline monomolecular ar-rays of protein and glycoprotein subunits, the function of which is mostly unknown (Sleytr,

1997). Outside the S-layer in motile strains of the *Oscillatoriaceae* is an additional layer, consist-ing of a fibrillar array of a protein called oscillin (Hoiczyk and Baumeister, 1995, 1997). Al-though required for gliding motility, this layer is not thought to provide the motive force, which is proposed to come from the extrusion of polysaccharide from the junctional pores that encircle cell septa (Hoiczyk and Baumeis-ter, 1998). Some, perhaps all, motile cyanobac-teria also possess a complex fibrillar array be-tween the peptidoglycan and the outer membrane (Fig. 5) (Adams et al., 1999). This fibrillar array may be involved in motility, but direct evidence of this is lacking.

Cyanobacterial photosynthetic membranes (the thylakoids) are conspicuous structures in the cell and consist of two unit membranes (de-rived from the folding of a single unit mem-brane) separated by approximately 3 to 5 nm and usually 60 to 70 nm from adjacent thyla-koids (Fig. 3) (Stanier, 1988; Castenholz and Waterbury, 1989). In many cyanobacteria they occur in rows of three to six, parallel both to each other and to the cell wall (Fig. 3), al-though during heterocyst development they become more convoluted. The chlorophyll *a*-protein complexes, photosynthetic reaction centers, carotenoids, and electron transport sys-tem are contained within the thylakoids, whereas the major light-harvesting pigments of cyanobacteria, the phycobiliproteins, are con-tained within rows of 10- to 12-nm-wide hemidiscoidal structures, known as phycobili-somes, attached to the surfaces of the mem-branes (Fig. 3) (Bryant, 1991). The phycobili-proteins have a large influence on the colors of cyanobacteria, which range from blue-green, red, and brown to almost black.

HETEROCYSTS

The enzyme responsible for nitrogen fixation, nitrogenase, is highly sensitive to oxygen (Gal-lon, 1992). As a consequence, the evolutionary success of cyanobacteria, and the resulting change in the earth's atmosphere from anaero-bic to aerobic, resulted in the confinement of many nitrogen-fixing bacteria to suitable an-

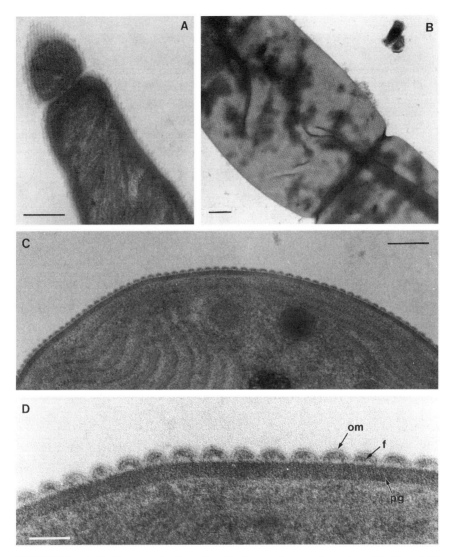

FIGURE 5 A fibrillar array in the cell wall of a motile cyanobacterium. (A, C, and D) Transmission electron micrographs of thin sections of *Oscillatoria* sp. strain FT2 filaments. (A) Longitudinal section. At the top of the figure, beyond a cell septum, the section has grazed the surface of the filament, revealing an array of parallel fibrils running at an angle of approximately 25 to 30° to the filament long axis. Bar, 400 nm. (C) Part of a transverse section showing an end view of the fibrillar array in the cell wall. Bar, 200 nm. (D) Enlarged view of part of panel C, showing the double line of the outer membrane (om) covering the fibrillar array and dipping between each pair of fibril(s) (f) to contact the electron-dense peptidoglycan layer (pg). Bar, 50 nm. (B) Transmission electron micrograph of part of a filament of *Oscillatoria* sp. strain FT3. An actively motile sample was crushed between glass slides and negatively stained. The micrograph shows several cells from which the contents have been extruded and whose walls have been flattened, bringing the fibrils at the front and back of the filament into close contact, allowing them to be viewed simultaneously. The fibrils run helically around the entire surface of the filament, producing the observed criss-cross effect because those in the wall in the foreground run in a different direction than those in the wall in the background. Bar, 500 nm. (Reproduced from Adams et al. [1999] with permission of the publisher.)

aerobic environmental niches. However, because of their oxygenic photosynthesis, the cyanobacteria had to evolve alternative means to protect nitrogenase. In some cases this involved a temporal separation of photosynthesis and nitrogen fixation, with the former occurring during the day and the latter at night (for reviews, see Gallon, 1992; Fay, 1992; and Bergman et al., 1997). However, an alternative strategy evolved in some of the early filamentous cyanobacteria and involved the specialization of a small proportion of cells to provide a suitable anaerobic environment for nitrogen fixation. These cells are heterocysts (Adams and Carr, 1981; Wolk, 1982; Adams, 1992, 1997; Wolk et al., 1994), and they may have evolved from spore-like cells known as akinetes (see below). Heterocysts develop in cyanobacteria of subgroups 4 (order *Nostocales*) and 5 (order *Stigonematales*), and in the *Nostocales* in particular they develop singly at spaced intervals, forming a one-dimensional pattern (Fig. 1 and 2). The development of heterocysts is suppressed in the presence of combined nitrogen.

The generation of an anaerobic interior in the heterocyst requires a series of major modifications to the original vegetative cell (Wolk et al., 1994). These include the loss of photosystem II activity, which eliminates the photosynthetic generation of oxygen, and the laying down of additional cell wall layers (see below) to reduce the diffusion of gases, including oxygen. The small amount of oxygen that diffuses into the cell is removed by several mechanisms, the most important of which is probably respiration (Wolk et al., 1994; Böhme, 1998). Although nitrogen diffusion is also reduced, enough enters the heterocyst to supply the combined-nitrogen needs of the filament. Nitrogen is fixed by nitrogenase to NH_3, which is assimilated via the glutamine synthetase-glutamate synthase pathway (Flores and Herrero, 1994); glutamine is the form in which nitrogen is exported to the vegetative cells (Gallon, 1992; Fay, 1992; Wolk et al., 1994). Heterocysts must receive carbon from vegetative cells because they lack ribulose bisphosphate carboxylase and so do not fix CO_2. This interchange of nutrients between heterocysts and vegetative cells probably influences the frequency and spacing of heterocysts (see below and chapter 4).

Heterocyst Structure

Because of their highly differentiated nature, heterocysts are readily distinguished from vegetative cells, usually being larger, with thickened cell walls and less-granular cytoplasm (Fig. 2). At the points of attachment to neighboring vegetative cells, they possess refractile polar bodies: two in intercalary heterocysts but only one in the terminal heterocysts that develop from the end cell of the trichome (Fig. 1 and 2). The refractility of these polar regions results from both a thickening of the cell wall (see below) and often the deposition of a plug of cyanophycin, which is a nitrogen reserve material unique to cyanobacteria, consisting of a polymer of arginine and aspartic acid (Simon, 1987; Allen, 1988), occasionally with glutamic acid (Merritt et al., 1994). The septum between vegetative cells is traversed by plasma bridges (microplasmodesmata), and as heterocyst development progresses, the number of these decreases three- to fivefold and the septum decreases considerably in size until it becomes a narrow pore channel (Fay, 1992). The thickening of the cell wall during heterocyst development results from the deposition of three extra layers outside the normal cell envelope (Wolk, 1982; Wolk et al., 1994). The innermost of these is the laminated layer, which consists of glycolipid. The next is the homogeneous layer, which consists of polysaccharide, and the outermost is the fibrous layer, which is probably uncompacted strands of the same polysaccharide. Deposition of the fibrous layer is one of the first changes visible in electron micrographs and is followed by deposition of the homogeneous layer and finally the laminated layer. These extra wall layers, particularly the glycolipid, play a major role in maintaining the cell's anaerobic interior by reducing gas diffusion (Walsby, 1985). Indeed, many mutants that can fix nitrogen under anaerobic, but not aerobic, conditions (Fox⁻ mutants) frequently have mutations in the heterocyst envelope that diminish its effectiveness in excluding oxygen (Ernst et al., 1992).

Heterocyst Development and Spacing

Even a superficial microscopic examination of many filamentous, heterocystous cyanobacteria reveals that the heterocysts are not randomly positioned (Fig. 1 and 2) (see chapter 4). Indeed the placement of heterocysts is characteristic for each cyanobacterium (Rippka et al., 1979, 1981; Castenholz and Waterbury, 1989). For example, in the genus *Nostoc* heterocysts occur mostly at intercalary locations within the trichome, whereas in the genus *Anabaena* they can occur at both intercalary and terminal positions, and in *Cylindrospermum* they develop from only the terminal cells. The spacing of intercalary heterocysts ensures an even distribution of fixed nitrogen to the vegetative cells and an adequate supply of fixed carbon for the heterocysts. The spacing and frequency of heterocysts is maintained by the differentiation of a new cell at or near the midpoint of each interheterocyst interval once the number of vegetative cells has approximately doubled. In nitrogen-fixing filaments, therefore, the existing pattern of heterocysts regulates the position of newly developing cells. However, cultures in which heterocyst development has been suppressed by combined nitrogen can form a regular heterocyst pattern de novo following transfer to medium lacking combined nitrogen (Fig. 1 and 2). This requires a flexible process of cell selection, and this is partly provided by an intermediate stage, known as the proheterocyst, which is capable of returning to vegetative growth under the appropriate conditions (Adams, 1992, 1997). Although clearly differentiated from the vegetative cell, the proheterocyst lacks the thickened walls and polar bodies of the fully mature heterocyst. The factors and genes involved in heterocyst development and spacing are discussed in chapter 4.

Heterocyst differentiation is characterized by the temporal synthesis and degradation of a large number of proteins (Fleming and Haselkorn, 1973, 1974) and the breakdown of cellular nitrogen reserves. These reserves consist of phycocyanin, one of the major light-harvesting pigments (Glazer, 1987; Tandeau de Marsac and Houmard, 1993), and cyanophycin. These reserve materials are degraded by specific enzymes (Adams, 1992; Tandeau de Marsac and Houmard, 1993). The products of the nonspecific proteolysis that occurs during the early stages of heterocyst development are mostly excreted from the cell (Thiel, 1990), implying that the function of this proteolysis is primarily the removal of vegetative proteins. The specific degradation of phycocyanin that occurs later in development may supply the amino acids required for new proteins.

When differentiation is induced in aerobically grown cultures by removing combined nitrogen from the medium, the appearance of nitrogenase activity coincides with the development of mature heterocysts, because nitrogen fixation is confined to these cells. When this is done under anaerobic conditions, by using an argon atmosphere and adding 3-(3,4-dichlorophenyl)-1,1-dimethylurea to inhibit the photosynthetic production of oxygen, the characteristic thickened walls of the heterocyst do not develop (Rippka and Stanier, 1978). Even under these anaerobic conditions the Mo-Fe protein of nitrogenase is restricted to the incompletely formed heterocysts and is not present in vegetative cells (Murry et al., 1984). However, the nitrogenase activity is highly sensitive to oxygen, because the heterocysts lack the thickened cell walls required to exclude the gas. These observations imply that the regulation of *nif* gene expression in heterocystous cyanobacteria has both environmental and developmental components, requiring nitrogen starvation, anaerobiosis, and completion (or at least partial completion) of heterocyst development.

This developmental control of *nif* gene expression was cleverly demonstrated by Elhai and Wolk (1990) with the *luxAB* genes, which encode bacterial luciferase, as transcriptional reporters. Bacterial luciferase catalyzes the oxidation of an aldehyde (such as *n*-decanal) and $FMNH_2$, with the production of light. Expression of promoterless *luxAB* genes can be driven by a strong cyanobacterial promoter placed upstream (Elhai and Wolk, 1990). In this way, expression of the cyanobacterial gene can be followed, even in individual cells, by the emission of light, permitting the temporal and spatial expression of specific genes to be followed during heterocyst development.

Using *luxAB* fusions to appropriate cyanobacterial promoters, Elhai and Wolk (1990) were able to confirm existing biochemical data, showing that *glnA* (encoding glutamine synthetase) is expressed in both heterocysts and vegetative cells, *rbcLS* (encoding ribulose bisphosphate carboxylase) is expressed in vegetative cells only, and *nifHDK* (encoding nitrogenase) is expressed in heterocysts only. When heterocyst development in *Anabaena* sp. strain PCC 7120 is induced under anaerobic conditions, expression of *nifHDK* is observed only in the partially formed heterocysts, not in vegetative cells. Similarly, in the mutant strain *Anabaena* sp. strain PCC 7118, which fails to produce heterocysts but does fix nitrogen under anaerobic conditions, expression of *nifHDK* occurs only in well-spaced cells thought to represent partially developed heterocysts (Elhai and Wolk, 1990). These results confirmed that expression of *nifHDK* is under developmental, and not purely environmental, control.

AKINETES

Akinetes (Greek *akinetos*, motionless) are spore-like resting cells that were first described (Carter, 1856; de Bary, 1863; Rabenhorst, 1865) long before the discovery of the bacterial endospore (Cohn, 1877; Koch, 1877). They are produced by many strains of subgroups 4 (order *Nostocales*) and 5 (order *Stigonematales*), usually as cultures approach stationary phase (Nichols and Adams, 1982; Herdman, 1987, 1988; Whitton, 1987a; Adams, 1992).

Many of the properties of akinetes are suited to their role in surviving environmental conditions adverse for vegetative growth. Metabolic activities, such as CO_2 and N_2 fixation, are generally very low or undetectable (Herdman,

1987, 1988), and although not resistant to heat, akinetes have a far greater capacity than vegetative cells for surviving periods of cold, darkness, and desiccation. For example, akinetes of *Anabaena cylindrica* remain viable for at least 5 years in the dark in the dry state, compared with 2 weeks for vegetative cells (Yamamoto, 1975), and akinetes of *Cyanospira* spp. show 90% germination after 7 years of desiccated storage (Sili et al., 1994). Similarly, akinetes of *Nostoc* sp. strain PCC 7524 survive in the dark at 4°C for 15 months, but vegetative cells survive for only 7 days (Sutherland et al., 1979). Desiccation tolerance may be more common in terrestrial strains; for example, akinetes of the planktonic *Anabaena circinalis* are no more resistant to desiccation than vegetative cells (Fay, 1988). Akinetes therefore provide a means of overwintering and, at least in terrestrial strains, surviving dry periods, and there is certainly evidence of their long-term survival in the environment. For example, viable akinetes have been isolated from sediments as old as 64 years (Livingstone and Jaworski, 1980), and the planktonic cyanobacterium *Gloeotrichia echinulata* overwinters as aggregates of akinetes embedded in mucilage in the sediment (Barbiero, 1993). However, there is disagreement about the degree to which germinating akinetes contribute to the spring regrowth of cyanobacteria in natural bodies of water (discussed by Herdman, 1987) because many cyanobacteria can survive in the vegetative state. For example, many species of *Nostocales* overwinter as vegetative filaments (Barbiero and Welch, 1992).

Structure and Composition

Akinetes are usually larger than vegetative cells, with thickened cell walls (Fig. 1 and 2) and multilayered extracellular envelopes that in *Anabaena* spp. contain polysaccharides equivalent to those of the heterocyst envelope (Cardemil and Wolk, 1976, 1979, 1981; Nichols and Adams, 1982; Herdman, 1987, 1988). Akinetes accumulate the nitrogen reserve material cyanophycin as granules in the cytoplasm (Fig. 2C). Although vegetative cells also accumulate cyanophycin after the end of exponential growth (Herdman, 1987), the mean cellular content of cyanophycin in *Nostoc* sp. strain PCC 7524 akinetes is eightfold higher than that of vegetative cells (Sutherland et al., 1979). However, there are exceptions, such as the akinetes of *Cyanospira* spp., that do not contain cyanophycin (Sili et al., 1994). The pigment content of akinetes can vary within and between species, and at least some of this variation may result from changes that occur after the cells have differentiated. For example, akinetes degrade phycocyanin and possibly chlorophyll as they age (Sutherland et al., 1985a). In vegetative cells and akinetes of *Nostoc* sp. strain PCC 7524 the mean cellular contents of RNA, DNA, and protein are similar (Sutherland et al., 1979). By contrast, the akinetes of *A. cylindrica* contain the same amount of RNA but more than twice as much DNA and 10 times as much protein as vegetative cells (Simon, 1977), probably because *A. cylindrica* akinetes have up to 10 times the volume of the vegetative cells (Fay, 1969).

Spatial Relationship to Heterocysts

Akinetes are only produced by heterocystous cyanobacteria, and although they develop in some strains when heterocyst development has been repressed by combined nitrogen, the presence of heterocysts influences the placing of akinetes in a strain-dependent manner. For example, in *Anabaena* sp. strain CA akinetes develop randomly in the absence of heterocysts but midway between heterocysts when they are present. In this particular cyanobacterium, transformation of every vegetative cell into an akinete can occur in the presence of sodium nitrate (Fig. 2D). Akinetes develop adjacent to heterocysts in *A. cylindrica* (Fig. 2C) and develop several cells away in *A. circinalis* and some other planktonic species (Fay et al., 1984; Li et al., 1997). Akinetes usually develop in strings, showing a clear gradient of decreasing maturity away from the first to form (Fig. 2C).

The very different spatial relationships between heterocysts and akinetes in different cy-

anobacteria are difficult to reconcile with a single mechanistic model. The development of akinetes next to heterocysts in *A. cylindrica* might be explained by the need for a supply of fixed nitrogen for the formation of cyanophycin granules. However, this does not explain the presence of akinetes midway between heterocysts in *Anabaena* sp. strain CA and *Nostoc* sp. strain PCC 7524. Indeed, the production of cyanophycin granules is not essential for akinete development because strains of *Nostoc ellipsosporum* carrying a mutation in the arginine biosynthetic gene *argL* can produce akinetes lacking cyanophycin (Leganés et al., 1998) and akinetes formed in *A. cylindrica* in the presence of the arginine analogue canavanine often lack cyanophycin granules (Nichols et al., 1980).

The spatial relationship between akinetes and heterocysts can be altered in the presence of two amino acid analogues, 7-azatryptophan and canavanine. Incubation of *Nostoc* sp. strain PCC 7524 with the tryptophan analogue 7-azatryptophan in the presence of nitrate causes both heterocysts and akinetes to develop further from the centers of the existing heterocyst intervals than in control cultures (Sutherland et al., 1979). This led the authors to conclude that a common control mechanism existed for the placement of both heterocysts and akinetes in *Nostoc* sp. strain PCC 7524. Incubation of nitrogen-fixing cultures of *A. cylindrica* with the arginine analogue canavanine results in the development of akinetes in random positions, in addition to their usual position adjacent to heterocysts (Nichols et al., 1980; Adams, 1992). This random placement of akinetes implies that canavanine acts on vegetative cells rather than heterocysts and that actively growing vegetative cells are inhibited from becoming akinetes by some aspect of their physiology, the heterocyst's function being to remove or negate this inhibition in adjacent cells without directly stimulating their development (Nichols et al., 1980). A similar model had been proposed by Wolk (1965), who used fragmentation of trichomes of *A. cylindrica* to demonstrate that the sporulation of cells adjacent to heterocysts was dependent on attachment to the heterocysts (Wolk, 1966).

Factors That Influence Akinete Development

Akinetes develop at the end of the exponential growth phase, when excess inorganic nutrients are still present in the medium (Fay et al., 1984; Wyman and Fay, 1986), implying that the primary trigger for akinete development is light (energy) limitation (or a metabolic consequence of this), resulting from self-shading as culture density increases (Nichols and Adams, 1982; Herdman, 1987, 1988). Indeed, akinete formation in *A. cylindrica* occurs at higher culture densities when the organism is grown at higher light intensities (Nichols et al., 1980). Similarly, akinete formation in a facultative photoheterotroph such as *Nostoc* sp. strain PCC 7524 can be delayed by prolonging the exponential phase by providing a utilizable carbon source, such as sucrose (Sutherland et al., 1979). Although light energy limitation is the primary trigger for akinete differentiation, limitation of a variety of nutrients, notably phosphate, has also been implicated (Nichols and Adams, 1982; Herdman, 1987, 1988). For example, in *A. circinalis* akinete differentiation is triggered by phosphate limitation but not by limitation of nitrogen, inorganic carbon, iron, trace elements, or light (van Dok and Hart, 1996). Other triggers for akinete formation may include a critical carbon/nitrogen ratio in *Anabaena torulosa* (Sarma and Khattar, 1993) and temperature in a range of planktonic *Anabaena* spp. (Li et al., 1997). It seems likely, therefore, that a variety of triggers operate in different cyanobacteria.

Perhaps surprisingly, there is only one example of the stimulation of akinete formation by extracellular chemical signals. A compound (formula, C_7H_5OSN) found in cell-free supernatants of akinete-containing cultures of *Cylindrospermum licheniforme* stimulates akinete formation in young cultures of the same cyanobacterium (Fisher and Wolk, 1976; Hirosawa and Wolk, 1979a, 1979b). The major function of this may be as an intrafilamentous

signal, but its accumulation in a natural body of water during desiccation may serve to trigger the development of akinetes to survive an impending drought (Fisher and Wolk, 1976).

Akinete Germination

The major stimulus for akinete germination appears to be an increase in light intensity, usually achieved in the laboratory by diluting an akinete-containing culture with fresh or used medium (Herdman, 1987, 1988). However, good germination of *Nodularia spumigena* akinetes occurs after 24 h of exposure to relatively low (<9 μE m^{-2} s^{-1}) light, with good germination obtained with red light alone (Huber, 1985). Although there are exceptions (Neely-Fisher et al., 1989), germination does not usually occur in the dark, even when facultative chemoheterotrophs are supplied with the appropriate sugar (Chauvat et al., 1982). The nutrient requirements for germination are variable, and this may reflect species-specific characteristics. Akinetes of at least *A. circinalis* (van Dok and Hart, 1997) and *N. spumigena* (Huber, 1985) do not require nitrogen for germination; indeed, ammonia (but not nitrate) suppresses germination. In the absence of phosphate *N. spumigena* akinetes fail to germinate (Huber, 1985) and *A. circinalis* germlings are short and stunted compared with controls (van Dok and Hart, 1997).

Sometimes prior to cell division, but more usually after dividing several times, the germinating akinete emerges through a pore that forms at one end of an otherwise intact envelope (Sili et al., 1994; Nichols and Adams, 1982; Herdman, 1987). Empty akinete envelopes are frequently seen following germination (Fig. 2E). On rare occasions the entire envelope dissolves to release the germling. In the absence of combined nitrogen the germling usually develops a single heterocyst at a time and a position characteristic of the cyanobacterium. For example, in *Anabaena* sp. strain CA a single heterocyst forms near the center of the germling when it reaches six to seven cells in length (Fig. 2E) (Nichols and Adams, 1982), whereas in *Nostoc* sp. strain PCC 7524 the first heterocyst develops in a terminal position when the germling is three cells long (Sutherland et al., 1985b). In both of these cyanobacteria further heterocysts develop in intercalary locations as cell division increases the length of the filament.

The biochemical changes associated with germination have been studied in *Nostoc* sp. strain PCC 7524 akinetes induced to germinate synchronously by dilution in fresh medium lacking combined nitrogen (Sutherland et al., 1985a, 1985b). Following the first cell division at 12 h, one of the daughter cells divides again at 16 h while the other develops into a mature heterocyst by 19 h, when N_2 fixation is first detected. By 24 h all germlings have reached this stage and consist of three cells and a terminal heterocyst. Protein synthesis is continuous for the first 11 h of germination and, in the absence of N_2 fixation, the required nitrogen is derived from reserves. Interestingly, however, it is not derived from the two obvious sources, phycocyanin and cyanophycin, but more likely from the thick peptidoglycan layer which is degraded during germination to release a lipopolysaccharide-like laminated layer which accumulates inside the akinete envelope (Herdman, 1987). RNA synthesis continues throughout germination, whereas DNA synthesis does not begin until 80 min after initiation and is continuous thereafter. When DNA synthesis is inhibited by phenethyl alcohol, germination continues if nitrate is present to compensate for the inhibition of heterocyst development (Herdman, 1987). Under these conditions short filaments of 10 vegetative cells are produced because the akinetes contain 10 genome equivalents of DNA (Sutherland et al., 1979), thus allowing cell division to continue in the absence of DNA replication until each newly formed vegetative cell contains one genome copy. A more recent study of the biochemical changes associated with akinete germination in *Cyanospira* spp. made observations broadly similar to those of Sutherland et al. (1985a, 1985b) and confirmed that the likely source of amino acids for protein synthesis was the thick peptidoglycan layer, not cyanophy-

cin, which was not detected in *Cyanospira* akinetes (Sili et al., 1994).

HORMOGONIA

Hormogonia are short, undifferentiated filaments produced by fragmentation of the parent trichome and which often possess gliding motility (Rippka et al., 1979, 1981; Herdman and Rippka, 1988; Adams, 1992, in press; Tandeau de Marsac and Houmard, 1993; Tandeau de Marsac, 1994). Hormogonia formed by members of the nonheterocystous order *Oscillatoriales* are simply short, motile fragments of the parent trichome, which is itself motile (Rippka et al., 1979, 1981). Trichome fragmentation results from the death and lysis of cells known as necridia, which may be sacrificed specifically for this purpose. By contrast, hormogonia produced by the heterocyst-forming cyanobacteria of subgroups 4 (the *Nostocales*) and 5 (the *Stigonematales*) differ greatly from the immotile parent trichomes, being shorter in length with much smaller cells (Fig. 1), gliding motility, and in many cases gas vesicles, and there is no doubt therefore that their formation requires differential gene expression. However, it is not clear if this is true of the *Oscillatoriales*, in which the hormogonia resemble the parent trichome in all but length, and their formation may simply be an indirect result of cell lysis and the consequent fragmentation of the parent trichome.

Function

Cyanobacteria incapable of forming hormogonia are all motile, whereas, apart from the *Oscillatoriales*, the parent trichomes of all hormogonium-forming cyanobacteria are immotile. Hormogonia therefore provide immotile strains with a means of dispersal. Thus, in the order *Nostocales*, members of the genera *Anabaena*, *Nodularia*, *Cylindrospermum*, and *Aphanizomenon* are all permanently motile and do not form hormogonia, whereas members of the genera *Nostoc*, *Scytonema*, and *Calothrix* have immotile trichomes but do form hormogonia. The dispersal of some hormogonia is enhanced by the production of gas vesicles to provide buoyancy (Tandeau de Marsac and Houmard, 1993; Tandeau de Marsac, 1994). In addition, a change in the cell envelope, which is hydrophobic in the vegetative filament but hydrophilic in newly formed hormogonia, may aid dispersal by reducing the tendency of hormogonia to adhere to surfaces (Tandeau de Marsac, 1994). Hormogonia formed by members of the *Nostocales* and *Stigonematales* are only a transient part of the life cycle, eventually losing motility and giving rise to heterocystous trichomes once more. Hormogonia also play a crucial role in the establishment of many cyanobacterial symbioses, particularly those involving plants. They serve as the infective agents in these symbioses because the long, immotile parent filaments are unable to gain entry to the plant structures that house the symbiotic colonies (Campbell and Meeks, 1989; Meeks, 1990, 1998; Johansson and Bergman, 1994; Bergman et al., 1996; Adams, in press).

Environmental Triggering of Hormogonium Formation

The factors that trigger hormogonium formation are diverse but seem to correspond mostly to improved conditions for growth. For example, the addition of phosphate to phosphate-starved *Calothrix* spp. growing on agar can trigger hormogonium formation, as can the addition of iron to iron-deficient cultures (Mahasneh et al., 1990; Whitton, 1989, 1992). Similar responses can be seen in the environment; for example, *Rivularia atra* in intertidal pools in Scotland forms abundant hormogonia when phosphate concentration is high in June (Yelloly and Whitton, 1996). Changes in light quantity and quality can also trigger hormogonium formation. For example, red light (640 to 650 nm) stimulates hormogonium differentiation in *Nostoc muscorum* A (Lazaroff, 1973) and in *Nostoc commune* 584 (Robinson and Miller, 1970), whereas green light reverses the effect. A similar response is seen in *Calothrix* sp. strain PCC 7601, in which hormogonium differentiation can be triggered by resuspension of filaments, to the same cell density, in fresh medium (Damerval et al., 1991; Campbell et

al., 1993; Tandeau de Marsac and Houmard, 1993; Tandeau de Marsac, 1994). When this is done in white light, up to 30% of filaments become hormogonia (assessed as the percentage of cells containing gas vesicles). However, 100% transformation occurs in red light, and up to 75% transformation occurs if cultures are transferred from white to red light without resuspension in fresh medium. Placing cultures in green light, with or without transfer to fresh medium, completely inhibits hormogonium formation. These opposing effects of red and green light arise from differential excitation of photosystem I and photosystem II, respectively (Campbell et al., 1993). Red-light excitation of photosystem I oxidizes the plastoquinone pool, stimulating hormogonium formation (and inhibiting heterocyst development), whereas green-light excitation of photosystem II reduces the plastoquinone pool, inhibiting hormogonium formation (and stimulating heterocyst development).

Hormogonium formation can be triggered by the transfer of old cultures of some *Nostoc* spp. to fresh medium, implying that an inhibitory factor is being diluted by the fresh medium (Rippka et al., 1979; Herdman and Rippka, 1988). Indeed, late-exponential- or stationary-phase cultures of *Nostoc* (formerly *Anabaena*) sp. strain PCC 7119 excrete a compound that inhibits hormogonium formation even when diluted up to 20-fold (Herdman and Rippka, 1988). The identity of the inhibitor is unknown, although it is destroyed by autoclaving and is dialyzable. By contrast, old cultures of *Mastigocladus laminosus* produce an inhibitor of hormogonium formation that is resistant to autoclaving (Hernández-Muñiz and Stevens, 1994).

The dilution of such inhibitors may provide an explanation for the triggering of hormogonium formation in many cyanobacteria following transfer from solid to liquid medium or dilution in liquid medium and may represent a mechanism for triggering hormogonia only at the periphery of a colony or mat, where they have the best chance of successful dispersal and survival. An example of the importance of this for colonization can be seen in *M. laminosus*, in which hormogonium formation is triggered by the transfer of filaments from liquid medium to the surface of medium solidified with agar (Hernández-Muñiz and Stevens, 1987, 1994). This may be a reflection of the behavior of the cyanobacterium in its native habitat, where it often grows in streams where new areas of bare rock can be created by the scouring effect of fast-flowing water (Hernández-Muñiz, personal communication). The production of hormogonia from the edge of the mat allows *Mastigocladus* to rapidly colonize new sites for growth. A similar effect can often be seen in fish tanks in which the glass is colonized by permanently motile cyanobacteria, such as *Oscillatoria* spp. If a small area of this is cleared, it will be recolonized within hours by the movement of filaments from surrounding areas. The only way in which nonmotile cyanobacteria can compete in such circumstances is to rapidly produce motile hormogonia in the manner of *Mastigocladus*.

As well as being inhibited, hormogonium formation can be stimulated by extracellular chemical factors. For example, hormogonium-containing cultures of *N. muscorum* A and *N. commune* 584 excrete a heat-labile substance that induces hormogonium formation when added to dark-grown cultures (Robinson and Miller, 1970). Such extracellular factors may facilitate the coordination of hormogonium formation within colonies of cyanobacteria. Plants also excrete chemical signals that trigger hormogonium development, and this increases the chances of establishing symbiotic colonies (Meeks, 1998; Adams, in press). For example, hormogonium formation in *Nostoc* strains is stimulated by a low-molecular-mass, heat-labile product released by the hornwort *Anthoceros punctatus* when nitrogen starved (Campbell and Meeks, 1989). A transposon mutant of the symbiotic *Nostoc* sp. strain 29133, with increased sensitivity to this hormogonium-inducing factor, has a 50-fold-higher initial frequency of infection of the hornwort than the wild type (Cohen et al., 1994). *A. punctatus* also produces a water-soluble hormogonium-

repressing factor that presumably prevents existing symbiotic colonies from transforming into hormogonia (Cohen and Meeks, 1997). This factor induces expression of two *Nostoc punctiforme* genes, *hrmA* and *hrmU*, the gene products of which block hormogonium formation, perhaps by the production of an inhibitor or by the catabolism of an activator (Cohen and Meeks, 1997). Mucilage from the stem gland of the angiosperm *Gunnera* also contains a hormogonium-inducing signal, in the form of a heat-labile protein of less than 12 kDa (Rasmussen et al., 1994). Hydroponically grown wheat roots also release a hormogonium-inducing factor, produced at a higher level in nitrogen-free medium (Gantar et al., 1993).

For efficient infection of most plants, the hormogonia must be guided into the plant by chemoattraction (Gorelova et al., 1997; Adams, in press). When starved of combined nitrogen, the liverwort *Blasia* releases extracellular signals that not only trigger hormogonium formation (Babic, 1996) but also serve as very effective chemoattractants (Knight and Adams, 1996). However, the hormogonium-triggering mucilage of *Gunnera* stem glands appears not to contain a chemoattractant (Rasmussen et al., 1994).

Hormogonium Development

Hormogonium development in the orders *Nostocales* and *Stigonematales* begins with a period of rapid, synchronized cell division in the absence of cell growth, resulting in a characteristic decrease in cell size (Tandeau de Marsac et al., 1988; Campbell and Meeks, 1989; Adams, 1992, 1997; Tandeau de Marsac, 1994). These cell divisions occur with little or no DNA replication (Herdman and Rippka, 1988; Damerval et al., 1991); this is possible because filamentous cyanobacteria commonly have multiple genome copies (Waterbury and Stanier, 1978; Sutherland et al., 1979; Herdman and Rippka, 1988; Haselkorn, 1991), and this permits several rounds of cell division without DNA replication, as can be seen in the germination of akinetes. This reductive cell division in the absence of DNA synthesis has an interesting parallel in the formation of baeocytes by multiple fission in the pleurocapsalean cyanobacteria (see below).

In *Calothrix* sp. strain PCC 7601 the gas vesicle genes are expressed specifically during hormogonium formation and therefore provide a useful marker for the process. Gas vesicles are rigid hollow cylinders with conical ends, and they consist primarily of the protein GvpA, with small amounts of a second protein, GvpC (Walsby, 1994). The genes for these proteins, *gvpA* and *gvpC*, have been cloned and sequenced (Tandeau de Marsac et al., 1985; Damerval et al., 1987; Hayes et al., 1988; Walsby, 1994) and used as probes to follow the transcription of the corresponding genes during hormogonium formation (Damerval et al., 1991; Tandeau de Marsac and Houmard, 1993; Tandeau de Marsac, 1994). The *gvp* gene transcripts are first detected 1.5 h after the induction of hormogonia under red light and remain abundant until 9 h, after which they begin to disappear and become undetectable by 24 h. However, *gvp* gene expression is repressed and hormogonium formation is arrested by exposure to green light after 3 h of induction under red light (Damerval et al., 1991; Tandeau de Marsac and Houmard, 1993; Tandeau de Marsac, 1994).

As hormogonia lose motility they develop heterocysts at positions characteristic of the genus (Rippka et al., 1979, 1981). In the genus *Nostoc* a heterocyst develops from each of the terminal cells, whereas in the genera *Scytonema* and *Calothrix* a single heterocyst develops from one of the terminal cells. The resumption of growth and cell division is accompanied by the development of many intercalary heterocysts in *Nostoc* and *Scytonema*, but the single terminal heterocyst remains in *Calothrix*. The return to vegetative growth usually results in the loss of any gas vesicles formed during hormogonium development (Tandeau de Marsac and Houmard, 1993; Tandeau de Marsac, 1994).

PLEUROCAPSALEAN CYANOBACTERIA

Members of the pleurocapsalean cyanobacteria group are distinguished from all other cyanobacteria by their ability to undergo multiple fission (Waterbury and Stanier, 1978; Waterbury, 1989). The products of this process are small spherical cells known as baeocytes (Greek for "small cell" but sometimes referred to as endospores or nannocytes) (Waterbury, 1989) that serve as a means of dispersal. The small size of baeocytes is a consequence of a fundamental difference between multiple and binary division. In the latter, a period of growth always follows division, ensuring that cell size is maintained even following repeated binary fission. However, multiple fission involves rapidly repeated rounds of cell division without cell growth, resulting in the production of at least four daughter cells, and often many hundreds, that are considerably smaller than the parent cell (Waterbury and Stanier, 1978; Waterbury, 1989). This form of division has a parallel in hormogonium formation, in which rapidly repeated binary fission occurs without cell growth, producing small-celled motile filaments as a means of dispersal. Because pleurocapsalean vegetative cells have many genome copies, multiple fission, in common with the synchronous cell divisions involved in akinete germination and hormogonium formation, probably does not require de novo DNA synthesis (Waterbury and Stanier, 1978).

Cell division in the genera *Dermocarpa* and *Xenococcus* always involves multiple fission (Fig. 6 to 9; see also Table 1), whereas in *Dermocarpella* multiple fission is preceded by one to three binary fissions (Fig. 6 and 8B). In other genera binary fission in three planes gives rise to complex aggregates of cells that can be regular cubes, as in *Myxosarcina* (Fig. 6 and 7), or irregular masses with filamentous extensions, as with the *Pleurocapsa* group (Fig. 6 and 7). In both cases, some or all of the vegetative cells in the aggregates undergo multiple fission to release baeocytes. By contrast with filamentous cyanobacteria, in which cells are separated only by their respective cytoplasmic membranes and a layer of peptidoglycan, the cells of the multicellular masses produced by pleurocapsalean cyanobacteria are separated from one another by the fibrous outer-wall layers that also en-

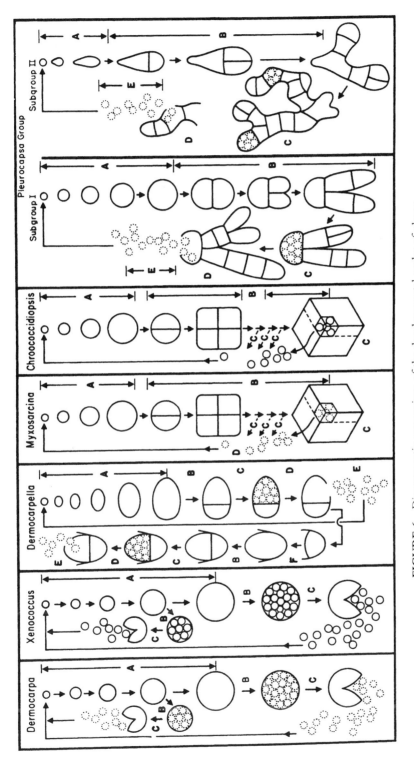

FIGURE 6 Diagrammatic comparison of the developmental cycles of pleuro-capsalean cyanobacteria. Stages A to F are defined in Table 1. The small dotted circles represent baeocytes that are not surrounded by a fibrous outer wall layer at the time of release and are consequently motile. The small solid circles represent baeocytes that are surrounded by a fibrous layer and are immotile. (Reproduced from Waterbury and Stanier [1978] with permission of the publisher.)

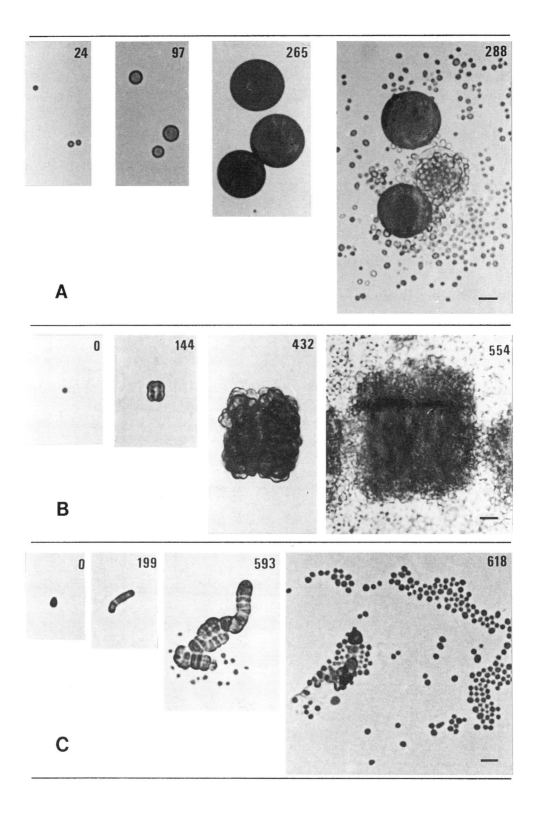

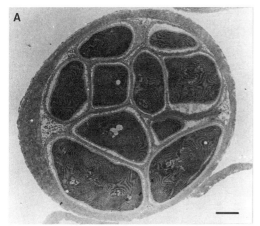

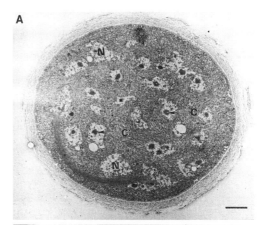

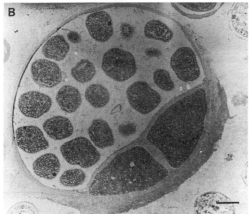

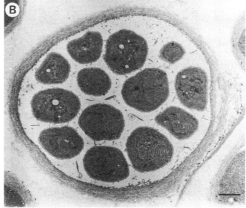

FIGURE 8 (A) Electron micrograph of a thin section of a cell of *Xenococcus* sp. strain PCC 7307 which has divided into numerous baeocytes, each of which is surrounded by a distinct fibrous layer. (B) Electron micrograph of a thin section of a cell of *Dermocarpella* sp. strain PCC 7326. The two basal cells, which are surrounded by a clear fibrous layer, are the products of the previous binary fission, while the apical cell has undergone multiple fission to produce many baeocytes, which do not have a fibrous layer. Bars, 1 μm. (Reproduced from Waterbury and Stanier [1978] with permission of the publisher.)

FIGURE 9 Electron micrographs of thin sections of *Dermocarpa* sp. strain PCC 7304 grown in liquid medium. (A) Cell immediately prior to multiple fission, containing many separate nucleoids (N) frequently associated with carboxysomes (C). (B) Cell which has completed multiple fission and is filled with baeocytes, each of which is surrounded by a peptidoglycan layer and outer membrane but not by an F layer. Bars, 1 μm. (Reproduced from Waterbury and Stanier [1978] with permission of the publisher.)

FIGURE 7 Development of individual baeocytes of *Dermocarpa* sp. strain PCC 7302 (A), *Myxosarcina* sp. strain PCC 7312 (B), and *Pleurocapsa* sp. strain PCC 7516 (C) on agar medium. The number in each photomicrograph indicates the elapsed time, in hours, after the initial observation. The baeocytes follow the developmental cycles shown diagrammatically in Fig. 6, culminating in multiple fission and the production of more baeocytes. The magnification is the same for all photomicrographs. Bars, 10 μm. (Reproduced from Waterbury and Stanier [1978] with permission of the publisher.)

TABLE 1 Comparison of the developmental cycles of pleurocapsalean cyanobacteria[a]

Genus	Characteristics of developmental stage:					
	A	B	C	D	E	F
Dermocarpa and *Xenococcus*	Symmetric baeocyte enlargement	Multiple fission	Release of baeocytes[b]			
Dermocarpella	Asymmetric baeocyte enlargement	Binary fission	Multiple fission of apical cell	Baeocyte release	Baeocyte motility	Basal cell enlargement
Myxosarcina and *Chroococcidiopsis*	Symmetric baeocyte enlargement	Repeated binary fission in 3 planes	Multiple fission of nearly all cells and baeocyte release	Baeocyte motility[c]		
Pleurocapsa groups I and II	Baeocyte enlargement[d]	Binary fission in many irregular planes	Multiple fission of some cells	Baeocyte release	Baeocyte motility	

[a] See Fig. 6.
[b] Baeocytes are motile in *Dermocarpa* and nonmotile in *Xenococcus*.
[c] Baeocytes are motile in *Myxosarcina* and nonmotile in *Chroococcidiopsis*.
[d] Symmetric in subgroup I; asymmetric in subgroup II.

compass the whole cell mass (Fig. 8A). In most pleurocapsalean cyanobacteria the synthesis of this fibrous layer is suppressed during multiple fission, producing baeocytes surrounded only by the gram-negative cell wall (Fig. 9B) (Waterbury and Stanier, 1978; Waterbury, 1989). In a few cases the fibrous layer is synthesized immediately following each successive cell division, so that each baeocyte is surrounded by a complete cell wall together with the fibrous layer and resembles a vegetative cell in all but size (Fig. 8A). The increasing volume of the growing baeocytes eventually causes the fibrous outer-wall layer of the parental cell to rupture. Baeocytes (and vegetative cells) surrounded by the fibrous layer are immotile, whereas those produced without this layer exhibit phototactic gliding motility for a short period following their release. Motility is lost once the baeocytes begin to enlarge into vegetative cells and synthesize the fibrous layer.

HAIRS

Hair formation is characteristic of members of the family *Rivulariaceae* (section C of subgroup 4 [Holt et al., 1994]). These cyanobacteria pro-

duce trichomes that taper from a broad base to a narrow apex, often ending in a long, multicellular hair that can account for two-thirds of a total trichome length of 800 μm or more (Whitton, 1989). Hairs develop by the narrowing and considerable elongation of existing cells, resulting in a great increase in surface area. As the hair develops, the cells lose their pigmentation, granular inclusions, and much of the cytoplasm and form large intrathylakoidal vacuoles (Whitton, 1989). Because the degree of trichome tapering in the *Rivulariaceae* is very variable, a better diagnostic feature of this family is the presence of a single terminal heterocyst (often referred to as a basal heterocyst) that develops at the wider end of the trichome in medium deficient in combined nitrogen. Trichome growth is typically meristematic, being restricted to a specific region of the trichome, usually either close to the basal heterocyst or to the hair, depending on the genus or species (Whitton, 1987b). The availability of combined nitrogen inhibits heterocyst development and often reduces the degree of tapering of trichomes, sometimes resulting in a reduction in the frequency and length of hairs (Sinclair and Whitton, 1977a, 1977b). Indeed, there is often a considerable difference in the morphologies of individual strains of the *Rivulariaceae* in nature and in laboratory culture. Some members of the *Rivulariaceae* have never been seen to produce hairs either in the laboratory or in the field, although it is not clear if this reflects a true genetic difference or an inability to identify the correct environmental trigger (Whitton, 1989).

Although the function of hairs is unclear, a number of observations point to a role in maximizing phosphate uptake (Whitton, 1989; Yelloly and Whitton, 1996). For example, hairs possess the ability to hydrolyze organic phosphate, and their production is stimulated by phosphate deficiency (and less commonly by a deficiency of iron, or rarely magnesium [Whitton, 1987b]). As phosphate concentration decreases in batch cultures of *Calothrix parietina* D184 (Livingstone et al., 1983) and *Calothrix viguieri* D253 (Mahasneh et al., 1990), inducible phosphatase activity increases in parallel with hair formation (Whitton, 1987b; Whitton et al., 1991). Phosphomonoesterase activity is present on the walls of hair cells, the chlorophyll-containing cells adjacent to them, and the part of the sheath furthest from the heterocyst.

QUESTIONS AND CHALLENGES

Heterocysts, Akinetes, and Evolution

Although the ability to form heterocysts probably evolved over 3 billion years ago, this was relatively late in the evolutionary history of filamentous cyanobacteria (Giovannoni et al., 1988). Dating the evolution of heterocystous cyanobacteria has proved difficult, with misidentification of putative heterocyst and akinete fossils a common problem (Golubic et al., 1995). Members of the orders *Nostocales* and *Stigonematales* are virtually unknown in rocks of the Proterozoic (2.5 billion to 550 million years old), yet atmospheric oxygen concentrations high enough to select for the evolution of heterocysts existed at least 2.1 billion years ago (Holland and Beukes, 1990). More convincing fossil evidence for akinetes has been found in mesoproterozoic rocks, setting a minimum age for akinete evolution (Golubic et al., 1995). However, such evidence is unlikely to resolve the interesting possibility that heterocysts evolved from akinetes or that the two cell types share a common ancestor (Wolk et al., 1994; Adams, 1997). There is certainly increasing evidence of a close relationship between the heterocyst and the akinete. For example in *N. ellipsosporum*, an intact *hetR* gene, which is essential for heterocyst development, is also required for akinete differentiation, and *hetR* is actively expressed in akinetes (Leganés et al., 1994). The arginine biosynthetic gene *argL* (which is probably involved in cyanophycin production) is also required for both heterocyst and akinete development and function (Leganés et al., 1998). In *Anabaena variabilis* ATCC 29413 the gene *hepA* is required for the synthesis of both the akinete and the heterocyst envelope polysaccharide (Leganés, 1994). In *Cyanospira rippkae* the same glycolipids are found in both heterocysts and akinetes (Soriente et al., 1993), and in *Anabaena* spp. the envelopes of both cell types contain equivalent polysaccharides (Cardemil and Wolk, 1976, 1979, 1981). Introduction of a plasmid containing wild-type *devR* into cells of *Nostoc* sp. strain 29133 results in stimulation of akinete formation in both ammonium-supplemented and nitrogen-fixing cultures, implying that akinete and heterocyst differentiation may be influenced by similar phosphorelay systems and that cross talk may occur between the two processes (Campbell et al., 1996).

Although heterocyst development is the most studied aspect of cyanobacterial differentiation, there are many unanswered questions. Of particular interest is the relationship between the cell cycle and heterocyst development. The spaced heterocyst pattern that forms following removal of combined nitrogen is achieved by the selection of cells from the vegetative filament by mechanisms about which we can only speculate (see chapter 4) (Wolk et al., 1994; Adams, 1992, 1997). However, one important question regarding heterocyst development has been at least partly answered recently; the long-predicted diffusible inhibitor of heterocyst development has been identified as the product of the gene *patS*, a small 17-amino-acid peptide (Yoon and Golden, 1998) (see chapter 4). A synthetic peptide consisting of the five carboxy-terminal amino acids of PatS inhibits heterocyst development when added at micromolar concentrations to culture medium. This important discovery promises more rapid progress in our understanding of the molecular mechanisms involved in heterocyst differentiation and spacing.

In comparison with heterocysts, akinetes have been poorly studied in the laboratory, probably because of the difficulty in predicting and controlling their formation and the relatively slow progress of the differentiation process itself. Heterocysts clearly influence the position of akinetes in the filament, yet it has been impossible to provide a unifying mechanistic model to explain the very different patterns in, for example, *Nostoc* sp. strain PCC 7524 and *A. cylindrica*, in which akinetes form midway between and adjacent to heterocysts, respectively. Indeed, as with the varied triggers for akinete development, it may be that akinete positioning is controlled by different mechanisms in different cyanobacteria. Most of the little we know of the molecular genetics of akinetes and their formation is derived from work on heterocyst development. Our increasing understanding of the latter may clarify the evolutionary and physiological relationships between these two cell types.

Functional Homologues of Heterocysts and Akinetes?

Although the most clearly differentiated cells are heterocysts and akinetes, there is increasing evidence for more subtle forms of differentiation in at least two genera, *Trichodesmium* and *Chroococcidiopsis*.

TRICHODESMIUM

Many nonheterocystous cyanobacteria are capable of N_2 fixation, mostly only under microaerobic or anaerobic conditions (Bergman et al., 1997). A few strains can fix N_2 aerobically, and in most cases there is a temporal separation

of oxygenic photosynthesis and N_2 fixation, with the former occurring during the day and the latter at night. However, a number of nonheterocystous cyanobacteria can fix N_2 aerobically in the light, and of these the planktonic marine strains of *Trichodesmium* are of particular interest (Bergman et al., 1997). Immunolocalization studies have shown that the nitrogenase protein is present in only approximately 15% of the cells of this filamentous cyanobacterium (Lin et al., 1998). These cells are most commonly found at the center of the trichome and show reduced or no CO_2 fixation. Despite these superficial similarities to heterocysts, no extreme morphological changes are apparent in the N_2-fixing cells. A similar localization of nitrogenase has recently been shown for the nonheterocystous filamentous cyanobacterium *Simploca* sp. strain PCC 8002, in which N_2 fixation occurs in the light part of a light-dark cycle and nitrogenase production is localized to approximately 9% of cells, which unlike *Trichodesmium*, contain ribulose bisphosphate carboxylase and presumably fix CO_2 (Fredriksson et al., 1998). Interestingly, both of these strains possess homologues of the gene *hetR*, which is essential for heterocyst development in other cyanobacteria. It remains to be seen how these cyanobacteria compartmentalize nitrogenase and protect it from oxygen and what the function of *hetR* is in the absence of heterocysts.

CHROOCOCCIDIOPSIS

The genus *Chroococcidiopsis* (subgroup 2, order *Pleurocapsales* [Holt et al., 1994]) consists of unicellular cyanobacteria that frequently possess great resistance to desiccation and are often found as the sole photosynthetic organisms in extremely arid deserts (Friedmann and Ocampo-Friedmann, 1984, 1985). Spore-like cells have been reported in laboratory desiccated cultures (Grilli Caiola et al., 1993) and in desiccated samples on hot desert rocks (Grilli Caiola et al., 1996a). These cells possess thick multilayered envelopes and are also found in liquid cultures grown under nitrogen-limited or nitrogen-starved (but not nitrogen-replete) conditions (Billi and Grilli Caiola, 1996). They may represent a functional equivalent of the akinete, providing a dormant state for survival of prolonged nitrogen starvation, desiccation, or temperature extremes. Survival may be aided by the presence of iron superoxide dismutase in the multilayered envelope, as a defense against free O_2 radicals (Grilli Caiola et al., 1996b).

Gliding Motility

Many bacteria move by gliding, which requires contact with a solid or semisolid substrate, yet no conclusive evidence of motor structures has been obtained despite extensive ultrastructural studies of members of the wide range of gliding bacteria, including cyanobacteria (Häder and Hoiczyk, 1992). Hormogonia, baeocytes, and permanently motile cyanobacteria provide excellent model systems for the study of gliding. The transient motility of hormogonia and baeocytes would require either the wasteful presence of a motor structure during the whole of the mostly immotile life cycle or, more likely, the production of the motor only when needed. It follows (although evidence is lacking) that the mechanism of motility in the permanently motile cyanobacteria and their hormogonia might differ from that of the transient motility of the nostocalean and stigonematalean hormogonia. In the case of the latter type of hormogonia, the motile phase is generally correlated with the presence of pilus-like structures (Tandeau de Marsac, 1994), but the significance of these for the mechanism of motility is unknown.

The use of electron microscopy to search for ultrastructural components of the cell wall that might provide the motor for gliding has provided ambiguous results because of the difficulty of preserving delicate structures without the production of artifacts. However, an interesting picture is begining to emerge. During electron microscopy studies Halfen and Castenholz (1971) observed parallel fibrils beneath the outer membrane of motile cyanobacteria and proposed a model for gliding, based on the theories of Jarosch (1964). They speculated that rhythmical contraction or distortion of the fibrils, which might consist of protein, could provide the motor for gliding in the *Oscillatoriaceae*, and the helical arrangement of the fibrils might explain the characteristic rotation of the filaments as they glide. However, in a recent ultrastructural examination of the cell walls of four gliding filamentous cyanobacteria, Hoiczyk and Baumeister (1997) were unable to identify any fibrillar structures beneath the outer membrane. However, outside the outer membrane and S layer they observed an array of helically arranged fibrils consisting of a single calcium-binding protein they called oscillin. They concluded that the helical surface fibrils served as a screw thread guiding the rotation of the trichome, the power being derived from extrusion of slime from the ring of junctional pores around each cell septum (Hoiczyk and Baumeister, 1995, 1998). However, a complex of parallel fibrils has been found beneath the outer membrane of several *Oscillatoria* spp. (Fig. 5) (Adams et al., 1999). The individual fibrils are wider (25 to 30 nm) than those detected by Halfen and Castenholz (8 to 12 nm [Halfen, 1979]) but do form a helical array. The significance of this complex array of fibrils remains to be established, although its location and helical arrangement imply a possible involvement in gliding.

It may well be that the mechanism of motility is different in different gliding bacteria. Nevertheless, the cyanobacteria have great potential to provide an understanding of at least one mechanism of force generation in prokaryotic gliding. The possible existence of force-generating proteins in cyanobacteria has implications beyond gliding motility, because, as discussed below, these proteins could be involved in other aspects of cell function, such as chromosome segregation, and as potential components of a bacterial "cytoskeleton."

Cell Division and the Bacterial Cytoskeleton

Prokaryotic equivalents of eukaryotic cytoskeletal structures, such as microtubules and microfilament bundles, have not been visualized in *Escherichia coli* (Nanninga, 1998), although there have been some convincing reports of cytoplasmic tubules in, for example, *Azotobacter vinelandii* and other prokaryotes, including cyanobacteria (Bermudes et al., 1994). It should also be remembered that prominent structures in the cell walls of cyanobacteria have been missed until recently (Fig. 5) (Hoiczyk and Baumeister, 1995, 1997, 1998; Adams et al., 1999). The existence of a cytoskeleton-like structure in prokaryotes, therefore, remains a possibility.

Possible candidates for a prokaryotic homologue of a cytoskeletal protein are elongation factor Tu (Nanninga, 1998) and MukB (Schmid and von Freiesleben, 1996), although the likeliest is the essential cell division protein FtsZ, which has features in common with eukaryotic tubulins (Erickson, 1995). In *E. coli* approximately 10,000 to 20,000 FtsZ molecules are evenly distributed in the cytoplasm of each nondividing cell, but prior to cell division, a proportion of these assemble into a ring-like structure at the site of septum formation (Bi and Lutkenhaus, 1991; Lutkenhaus and Mukherjee, 1996). This ring is thought to contract and provide the force for septal invagination. The ability of FtsZ to interact with a range of other proteins, and with itself to produce both filaments and two-dimensional sheets, adds weight to its potential as a prokaryotic cytoskeletal protein (Mukherjee and Lutkenhaus, 1994; Erickson et al., 1996; Vicente and Errington, 1996).

FtsZ is widely conserved among prokaryotes (Erickson, 1995; Lutkenhaus, 1993) and is found in cyanobacteria, including heterocystous strains (Zhang et al., 1995; Doherty and Adams, 1995). In addition to *ftsZ*, homologues of other cell division genes are present in the genome of *Synechocystis* sp. strain PCC 6803 (the complete genome sequence of this cyanobacterium is available at http://www.kasuza.or.jp/cyano/cyano.html). However, in comparison with most other bacteria, cyanobacteria show a remarkable variety in their control of cell division. They can alter the plane of septation to produce branched or multiseriate filaments and clumps, cubes, or two-dimensional

layers of cells, and they can undergo multiple fission and synchronized binary fission. In many cases they can switch among these different forms of division.

This remarkable ability to vary the timing and location of septum formation makes the cyanobacteria of considerable interest for cell division studies, and there are some intriguing questions to be answered. For example, what regulates septum position in cyanobacteria? In *E. coli* this is controlled by the products of the *minC*, *minD*, and *minE* genes, which ensure that the septum is placed centrally (Lutkenhaus and Mukherjee, 1996). A similar system appears to operate in *Bacillus subtilis*, which has homologues of *minC* and *minD* (Rothfield and Justice, 1997) and possesses a protein, DivIVA, with MinE-like characteristics (Cha and Stewart, 1997; Edwards and Errington, 1997). This system is thought to also regulate the switch to polar septum formation during sporulation (Barák et al., 1998). The genome of the cyanobacterium *Synechocystis* sp. strain PCC 6803 contains *minCDE* homologues, and there is preliminary evidence that heterocystous cyanobacteria also have these genes (Kirwan and Adams, unpublished observations). It will be interesting to see if the regulation of septum formation is similar in cyanobacteria, in which the symmetry of cell division is thought to play an important part in cell selection during heterocyst development (Adams, 1992, 1997).

Perhaps the most intriguing question of all is what mechanism induces the rapid, synchronized cell divisions in hormogonium and baeocyte formation? This rapid division of all cells results in a considerable decrease in cell volume, and this is presumably important for the function of hormogonia and baeocytes, although it is not clear why. It may be that the increased surface area-to-volume ratio is important in scavenging nutrients or that the reduced cell size aids dispersal. Cell division in bacteria normally occurs when chromosome replication is complete, to avoid the production of DNA-less cells (Lutkenhaus and Mukherjee, 1996). However, following the triggering of hormogonium formation, all cells in a filament commence division within approximately 10% of the generation time (Babic, 1996). This is far too short a period for any but a very small number of cells to complete DNA replication, yet DNA-less cells are not produced because these filamentous cyanobacteria have multiple genome copies (Herdman and Rippka, 1988; Haselkorn, 1991). The triggering of hormogonium formation therefore overrides the normal control of cell division and its relationship to genome replication; an understanding of this process will shed considerable light on the control of the cell cycle and cell division in bacteria.

Baeocyte formation in the *Pleurocapsales* is another fascinating system about which little is known. Indeed, these cyanobacteria are an understudied group that pose many intriguing questions. How, for example, is multiple fission regulated, and how is chromosome partition achieved when a single cell divides in three dimensions to produce hundreds of baeocytes? Accelerated and delayed cell divisions, possibly analogous to those in the *Pleurocapsales*, have been observed in other cyanobacteria. For example, in field samples of *Merismopedia*-like strains (subgroup 1 [Holt et al., 1994]), which grow as rectangular, flat aggregates of cells, some individual cells may become enlarged ("megacytes") due to delayed cell division (Palinska and Krumbein, 1998). Observations such as this serve to illustrate how little we understand of the true complexity of cyanobacterial cell cycles, because much of what we know is based on the very small number of experimentally amenable strains in culture.

Cell Signaling

Another poorly understood aspect of cyanobacteria is that of intercellular signaling, examples of which have been outlined above. Both akinete and hormogonium development are responsive to cyanobacterial chemical signals. The recent discovery of the diffusible peptide PatS provides the first clear evidence of the intrafilamentous signal that has long been thought necessary for the establishment of heterocyst pattern. Hormogonium development

and chemotaxis in the *Nostocales* are triggered by chemical signals from plants. However, in none of these cases is anything known of the signal transduction mechanisms presumably involved in sensing and responding to the signals.

Concluding Comments

Cyanobacteria pose many unanswered questions, some of which have been discussed here. They are multicellular bacteria that can help us to understand a wide range of important biological problems, yet their capabilities as experimental systems in these areas have been barely tested. Increasingly effective systems for the genetic manipulation of cyanobacteria have provided the necessary tools, and with the will to use them we should see rapid advances in our understanding of these complex prokaryotes.

ACKNOWLEDGMENTS

I am very grateful to Birgitta Bergman, Jeff Elhai, Peter Lindblad, Jack Meeks, and Peter Wolk for communicating data prior to publication.

REFERENCES

Adams, D. G. 1992. Multicellularity in cyanobacteria, p. 341–384. *In* S. Mohan, C. Dow, and J. A. Cole (ed.), *Prokaryotic Structure and Function: a New Perspective. Society for General Microbiology Symposium*, vol. 47. Cambridge University Press, Cambridge, United Kingdom.

Adams, D. G. 1997. Cyanobacteria, p. 109–148. *In* J. Shapiro and M. Dworkin (ed.), *Bacteria as Multicellular Organisms*. Oxford University Press, New York, N.Y.

Adams, D. G. Symbiotic interactions. *In* B. Whitton and M. Potts (ed.), *Ecology of Cyanobacteria: Their Diversity in Time and Space*, in press. Kluwer Academic Publishers, Dordrecht, The Netherlands.

Adams, D. G., and N. G. Carr. 1981. The developmental biology of heterocyst and akinete formation in cyanobacteria. *Crit. Rev. Microbiol.* **9:**45–100.

Adams, D. G., D. Ashworth, and B. Nelmes. 1999. Fibrillar array in the cell wall of a gliding filamentous cyanobacterium. *J. Bacteriol.* **181:**884–892.

Allen, M. M. 1988. Inclusions: cyanophycin. *Methods Enzymol.* **167:**207–213.

Babic, S. 1996. Hormogonia formation and the establishment of symbiotic associations between cyanobacteria and the bryophytes *Blasia* and *Phaeoceros*. Ph.D. thesis, University of Leeds, Leeds, United Kingdom.

Barák, I., P. Prepiak, and F. Schmeisser. 1998. MinCD proteins control the septation process during sporulation of *Bacillus subtilis*. *J. Bacteriol.* **180:** 5327–5333.

Barbiero, R. P. 1993. A contribution to the life-history of the planktonic cyanophyte, *Gloeotrichia echinulata*. *Arch. Hydrobiol.* **127:**87–100.

Barbiero, R. P., and E. B. Welch. 1992. Contribution of benthic blue-green algal recruitment to lake populations and phosphorus translocation. *Freshwater Biol.* **27:**249–260.

Bergman B., A. Matveyev, and U. Rasmussen. 1996. Chemical signalling in cyanobacterial-plant symbioses. *Trends Plant Sci.* **1:**191–197.

Bergman, B., J. R. Gallon, A. N. Rai, and L. J. Stal. 1997. N_2 fixation by non-heterocystous cyanobacteria. *FEMS Microbiol. Rev.* **19:**139–185.

Bermudes, D., G. Hinkle, and L. Margulis. 1994. Do prokaryotes contain microtubules? *Microbiol. Rev.* **58:**387–400.

Bhattacharya, D., and L. Medlin. 1995. The phylogeny of plastids: a review based on comparisons of small-subunit ribosomal RNA coding regions. *J. Phycol.* **31:**489–498.

Bi, E., and J. Lutkenhaus. 1991. FtsZ ring structure associated with division in *Escherichia coli*. *Nature* **354:**161–164.

Billi, D., and M. Grilli Caiola. 1996. Effects of nitrogen limitation and starvation on *Chroococcidiopsis* sp. (Chroococcales). *New Phytol.* **133:**563–571.

Böhme, H. 1998. Regulation of nitrogen fixation in heterocyst-forming cyanobacteria. *Trends Plant Sci.* **3:**346–351.

Bryant, D. A. 1991. Cyanobacterial phycobilisomes: progress towards complete structural and functional analysis via molecular genetics, p. 257–300. *In* L. Bogorad and I. K. Vasil (ed.), *Cell Culture and Somatic Cell Genetics of Plants*. Academic Press Inc., New York, N.Y.

Bryant, D. A. (ed.). 1994. *The Molecular Biology of Cyanobacteria*. Kluwer Academic Publishers, Dordrecht, The Netherlands.

Campbell, D., J. Houmard, and N. Tandeau de Marsac. 1993. Electron transport regulates cellular differentiation in the filamentous cyanobacterium *Calothrix*. *Plant Cell* **5:**451–463.

Campbell, E. L., and J. C. Meeks. 1989. Characteristics of hormogonia formation by symbiotic *Nostoc* spp. in response to the presence of *Anthoceros punctatus* or its extracellular products. *Appl. Environ. Microbiol.* **55:**125–131.

Campbell, E. L., K. D. Hagen, M. F. Cohen, M. L. Summers, and J. C. Meeks. 1996. The *devR* gene product is characteristic of receivers of two-component regulatory systems and is essential for heterocyst development in the filamentous cyano-

bacterium *Nostoc* sp. strain ATCC 29133. *J. Bacteriol.* **178:**2037–2043.

Cardemil, L., and C. P. Wolk. 1976. The polysaccharides from heterocyst and spore envelopes of a blue-green alga. Methylation analysis and structure of the backbones. *J. Biol. Chem.* **251:**2967–2975.

Cardemil, L., and C. P. Wolk. 1979. The polysaccharides from heterocyst and spore envelopes of a blue-green alga. Structure of the basic repeating unit. *J. Biol. Chem.* **254:**736–741.

Cardemil, L., and C. P. Wolk. 1981. Polysaccharides from the envelopes of heterocysts and spores of the blue-green algae *Anabaena variabilis* and *Cylindrospermum licheniforme*. *J. Phycol.* **17:**234–240.

Carr, N. G., and B. A. Whitten (ed.). 1982. *The Biology of Cyanobacteria*. Blackwell Scientific Publications, Oxford, United Kingdom.

Carter, H. J. 1856. Notes on the freshwater infusoria of the island of Bombay. No. 1. Organisation. *Ann. Mag. Nat. Hist.* (2nd series) **18:**115–132, 221–249.

Castenholz, R. W. 1989a. Order *Nostocales*, p. 1780–1793. *In* J. T. Staley, M. P. Bryant, N. Pfennig, and J. G. Holt (ed.), *Bergey's Manual of Systematic Bacteriology*, vol. 3. Williams and Wilkins, Baltimore, Md.

Castenholz, R. W. 1989b. Order *Stigonematales*, p. 1794–1799. *In* J. T. Staley, M. P. Bryant, N. Pfennig, and J. G. Holt (ed.), *Bergey's Manual of Systematic Bacteriology*, vol. 3. Williams and Wilkins, Baltimore, Md.

Castenholz, R. W. 1992. Species usage, concept, and evolution in the cyanobacteria (blue-green algae). *J. Phycol.* **28:**737–745.

Castenholz, R. W., and J. B. Waterbury. 1989. Preface, p. 1710–1727. *In* J. T. Staley, M. P. Bryant, N. Pfennig, and J. G. Holt (ed.), *Bergey's Manual of Systematic Bacteriology*, vol. 3. Williams and Wilkins, Baltimore, Md.

Cha, J. H., and G. C. Stewart. 1997. The *divIVA* locus of *Bacillus subtilis*. *J. Bacteriol.* **179:**1671–1683.

Chauvat, F., B. Corre, M. Herdman, and F. Joset-Espardellier. 1982. Energetic and metabolic requirements for the germination of akinetes of the cyanobacterium *Nostoc* PCC 7524. *Arch. Microbiol.* **133:**44–49.

Cohen, M. F., and J. C. Meeks. 1997. A hormogonium regulating locus, *hrmUA*, of the cyanobacterium *Nostoc punctiforme* strain ATCC 29133 and its response to an extract of a symbiotic plant partner *Anthoceros punctatus*. *Mol. Plant-Microbe Interact.* **10:** 280–289.

Cohen, M. F., J. G. Wallis, E. L. Campbell, and J. C. Meeks. 1994. Transposon mutagenesis of *Nostoc* sp. strain ATCC 29133, a filamentous cyanobacterium with multiple cellular differentiation alternatives. *Microbiology* **140:**3233–3240.

Cohn, F. 1877. Untersuchungen über Bakterien. IV.

Beitrage zur Biologie der Bacillen. *Beitr. Biol. Pflanz.* **2:**249–276.

Damerval, T., J. Houmard, G. Guglielmi, K. Csiszar, and N. Tandeau de Marsac. 1987. A developmentally regulated *gvpABC* operon is involved in the formation of gas vesicles in the cyanobacterium *Calothrix* 7601. *Gene* **54:**83–92.

Damerval, T., G. Guglielmi, J. Houmard, and N. Tandeau de Marsac. 1991. Hormogonium differentiation in the cyanobacterium *Calothrix*: a photoregulated developmental process. *Plant Cell* **3:**191–201.

de Bary, A. 1863. Beitrag zur Kenntnis der Nostocaceen insbesondere der *Rivularien*. *Flora* (Jena) **35:**553–560.

Delwiche, C. F., and J. D. Palmer. 1997. The origin of plastids and their spread via secondary symbiosis. *Plant Syst. Evol.* **S11:**53–86.

Doherty, H. M., and D. G. Adams. 1995. Cloning and sequence of *ftsZ* and flanking regions from the cyanobacterium *Anabaena* PCC 7120. *Gene* **163:**93–96.

Douglas, S. E. 1992. Eukaryote-eukaryote endosymbioses: insights from studies of a cryptomonad alga. *BioSystems* **28:**57–68.

Douglas, S. E. 1994. Chloroplast origins and evolution, p. 91–118. *In* D. A. Bryant (ed.), *The Molecular Biology of Cyanobacteria*. Kluwer Academic Publishers, Dordrecht, The Netherlands.

Edwards, D. H., and J. Errington. 1997. The *Bacillus subtilis* DivIVA protein targets to the division septum and controls the site specificity of cell division. *Mol. Microbiol.* **24:**905–915.

Elhai, J., and C. P. Wolk. 1990. Developmental regulation and spatial pattern of expression of the structural genes for nitrogenase in the cyanobacterium *Anabaena*. *EMBO J.* **9:**3379–3388.

Erickson, H. P. 1995. FtsZ, a prokaryotic homolog of tubulin? *Cell* **80:**367–370.

Erickson, H. P., D. W. Taylor, K. A. Taylor, and D. Bramhill. 1996. Bacterial cell division protein FtsZ assembles into protofilament sheets and mini-rings, structural homologs of tubulin polymers. *Proc. Natl. Acad. Sci. USA* **93:**519–523.

Ernst, A., T. Black, Y. Cai, J.-M. Panoff, D. N. Tiwari, and C. P. Wolk. 1992. Synthesis of nitrogenase in mutants of the cyanobacterium *Anabaena* sp. strain PCC 7120 affected in heterocyst development or metabolism. *J. Bacteriol.* **174:**6025–6032.

Fay, P. 1969. Cell differentiation and pigment composition in *Anabaena cylindrica*. *Arch. Mikrobiol.* **67:**62–70.

Fay, P. 1988. Viability of akinetes of the planktonic cyanobacterium *Anabaena circinalis*. *Proc. R. Soc. Lond. B* **234:**283–301.

Fay, P. 1992. Oxygen relations of nitrogen fixation in cyanobacteria. *Microbiol. Rev.* **56:**340–373.

Fay, P., J. A. Lynn, and S. C. Majer. 1984. Akinete development in the planktonic blue-green alga *Anabaena circinalis*. *Br. Phycol. J.* **19:**163–173.

Fisher, R. W., and C. P. Wolk. 1976. Substance stimulating the differentiation of spores of the blue-green alga *Cylindrospermum licheniforme*. *Nature* **259:**394–395.

Fleming, H., and R. Haselkorn. 1973. Differentiation in *Nostoc muscorum*: nitrogenase is synthesized in heterocysts. *Proc. Natl. Acad. Sci. USA* **70:**2727–2731.

Fleming, H., and R. Haselkorn. 1974. The program of protein synthesis during heterocyst differentiation in nitrogen-fixing blue-green algae. *Cell* **3:**159–170.

Flores, E., and A. Herrero. 1994. Assimilatory nitrogen metabolism and its regulation, p. 487–517. *In* D. A. Bryant (ed.), *The Molecular Biology of Cyanobacteria*. Kluwer Academic Publishers, Dordrecht, The Netherlands.

Foulds, I. J., and N. G. Carr. 1981. Unequal cell division preceding heterocyst development in *Chlorogloeopsis fritschii*. *FEMS Microbiol. Lett.* **10:**223–226.

Fredriksson, C., G. Malin, P. J. A. Siddiqui, and B. Bergman. 1998. Aerobic nitrogen fixation is confined to a subset of cells in the non-heterocystous cyanobacterium *Symploca* PCC 8002. *New Phytol.* **140:**531–538.

Friedmann, E. I., and R. Ocampo-Friedmann. 1984. Endolithic microorganisms in extreme dry environments: analysis of a lithobiontic microbial habitat, p. 177–185. *In* M. J. Klug and C. A. Reddy (ed.), *Current Perspectives in Microbial Ecology*. American Society for Microbiology, Washington, D.C.

Friedmann, E. I., and R. Ocampo-Friedmann. 1985. Blue-green algae in arid cryptoendolithic habitats. *Arch. Hydrobiol. Suppl.* **71:**349–350.

Gallon, J. R. 1992. Tansley review no. 44. Reconciling the incompatible: N_2 fixation and O_2. *New Phytol.* **122:**571–609.

Gantar, M., N. W. Kerby, and P. Rowell. 1993. Colonization of wheat (*Triticum vulgare* L.) by N_2-fixing cyanobacteria. III. The role of a hormogonia-promoting factor. *New Phytol.* **124:**505–513.

Giovannoni, S. J., R. Turner, G. J. Olsen, S. Barns, D. J. Lane, and N. R. Pace. 1988. Evolutionary relationships among cyanobacteria and green chloroplasts. *J. Bacteriol.* **170:**3584–3592.

Glazer, A. N. 1987. Phycobilisomes: assembly and attachment, p. 69–94. *In* P. Fay, and C. van Baalen (ed.), *The Cyanobacteria*. Elsevier Science Publishers B. V., Amsterdam, The Netherlands.

Golubic, S., V. N. Sergeev, and A. H. Knoll. 1995. Mesoproterozoic *Archaeoellipsoides*: akinetes of heterocystous cyanobacteria. *Lethaia* **28:**285–298.

Gorelova, O. A., O. I. Baulina, T. G. Korzhenevskaya, and M. V. Gusev. 1997. Formation of hormogonia and their taxis during the interaction of cyanobacteria and plants. *Microbiology* **66:** 669–675.

Grilli Caiola, M., R. Ocampo-Friedmann, and E. I. Friedmann. 1993. Cytology of long-term desiccation in the desert cyanobacterium *Chroococcidiopsis* (Chroococcales). *Phycologia* **32:**315–322.

Grilli Caiola, M., D. Billi, and E. I. Friedmann. 1996a. Effect of desiccation on the envelopes of the cyanobacterium *Chroococcidiopsis* sp. (Chroococcales). *Eur. J. Phycol.* **31:**97–105.

Grilli Caiola, M., A. Canini, and E. I. Friedmann. 1996b. Superoxide dismutase (Fe-SOD) localization in *Chroococcidiopsis* sp. (Chroococcales). *Phycologia* **35:**90–94.

Häder, D.-P., and E. Hoiczyk. 1992. Gliding motility, p. 1–38. *In* M. Melkonian (ed.), *Algal Cell Motility.* Chapman and Hall, New York, N.Y.

Halfen, L. N. 1979. Gliding movements, p. 250–267. *In* W. Haupt and M. E. Feinleib (ed.), *Encyclopedia of Plant Physiology, New Series,* vol. 7. Springer-Verlag, Heidelberg, Germany.

Halfen, L. N., and R. W. Castenholz. 1971. Gliding in a blue-green alga, *Oscillatoria princeps. J. Phycol.* **7:**133–145.

Haselkorn, R. 1991. Genetic systems in cyanobacteria. *Methods Enzymol.* **204:**418–430.

Hayes, P. K., C. M. Lazarus, A. Bees, J. E. Walker, and A. E. Walsby. 1988. The protein encoded by *gvpC* is a minor component of gas vesicles isolated from the cyanobacteria *Anabaena flos-aquae* and *Microcystis* sp. *Mol. Microbiol.* **2:**545–552.

Helmchen, T. A., D. Bhattacharya, and M. Melkonian. 1995. Analyses of ribosomal RNA sequences from glaucocystophyte cyanelles provide new insights into the evolutionary relationships of plastids. *J. Mol. Evol.* **41:**203–210.

Herdman, M. 1987. Akinetes: structure and function, p. 227–250. *In* P. Fay and C. van Baalen (ed.), *The Cyanobacteria.* Elsevier Science Publishers B. V., Amsterdam, The Netherlands.

Herdman, M. 1988. Cellular differentiation: akinetes. *Methods Enzymol.* **167:**222–232.

Herdman, M., and R. Rippka. 1988. Cellular differentiation: hormogonia and baeocytes. *Methods Enzymol.* **167:**232–242.

Hernández-Muñiz, W. Personal communication.

Hernández-Muñiz, W., and S. E. Stevens, Jr. 1987. Characterization of the motile hormogonia of *Mastigocladus laminosus. J. Bacteriol.* **169:**218–223.

Hernández-Muñiz, W., and S. E. Stevens, Jr. 1994. Development of motility in cultures of the cyanobacterium *Mastigocladus laminosus. FEMS Microbiol. Ecol.* **15:**259–264.

Hirosawa, T., and C. P. Wolk. 1979a. Factors controlling the formation of akinetes adjacent to heterocysts in the cyanobacterium *Cylindrospermum licheniforme* Kütz. *J. Gen. Microbiol.* **114:**423–432.

Hirosawa, T., and C. P. Wolk. 1979b. Isolation and characterization of a substance which stimulates the formation of akinetes in the cyanobacterium *Cylindrospermum licheniforme* Kütz. *J. Gen. Microbiol.* **114:**433–441.

Hoiczyk, E., and W. Baumeister. 1995. Envelope structure of four gliding filamentous cyanobacteria. *J. Bacteriol.* **177:**2387–2395.

Hoiczyk, E., and W. Baumeister. 1997. Oscillin, an extracellular, Ca^{2+}-binding glycoprotein essential for the gliding motility of cyanobacteria. *Mol. Microbiol.* **26:**699–708.

Hoiczyk, E., and W. Baumeister. 1998. The junctional pore complex, a prokaryotic secretion organelle, is the molecular motor underlying gliding motility in cyanobacteria. *Curr. Biol.* **8:**1161–1168.

Holland, H. J., and N. J. Beukes. 1990. A paleo-weathering profile from Griqualand West, South Africa: evidence for a dramatic rise in atmospheric oxygen between 2.2 and 1.9 bybp. *Am. J. Sci.* **290A:**1–34.

Holt, J. G., N. R. Krieg, P. H. A. Sneath, J. T. Staley, and S. T. Williams (ed.). 1994. *Bergey's Manual of Determinative Bacteriology,* vol. 9. Williams and Wilkins, Baltimore, Md.

Houmard, J. 1994. Gene transcription in filamentous cyanobacteria. *Microbiology* **140:**433–441.

Huber, A. L. 1985. Factors affecting the germination of akinetes of *Nodularia spumigena* (Cyanobacteriaceae). *Appl. Environ. Microbiol.* **49:**73–78.

Jarosch, R. 1964. Gleitbewegung und Torsion von Oscillatorien. *Österreich Bot. Z.* **111:**143–148.

Johansson, C., and B. Bergman. 1994. Reconstruction of the symbiosis of *Gunnera manicata* Linden: cyanobacterial specificity. *New Phytol.* **126:** 643–652.

Kirwan, I. G., and D. G. Adams. Unpublished observations.

Knight, C. D., and D. G. Adams. 1996. A method for studying chemotaxis in nitrogen-fixing cyanobacterium-plant symbioses. *Physiol. Mol. Plant Pathol.* **49:**73–77.

Koch, R. 1877. Untersuchungen über Bakterien. V. Die Aetiologie der Milzbrand-Krankheit, begrundet auf der Entwicklungsgeschichte des *Bacillus anthracis. Beitr. Biol. Pflanz.* **2:**227–310.

Lazaroff, N. 1973. Photomorphogenesis and nostocacean development, p. 279–319. *In* N. G. Carr and B. A. Whitton (ed.), *The Biology of Blue-Green Algae.* Blackwell Scientific Publications, Oxford, United Kingdom.

Lazcano, A., and S. L. Miller. 1994. How long did it take for life to begin and evolve to cyanobacteria? *J. Mol. Evol.* **39:**546–554.

Leganés, F. 1994. Genetic evidence that *hepA* gene is involved in the normal deposition of the envelope of both heterocysts and akinetes in *Anabaena variabilis* ATCC 29413. *FEMS Microbiol. Lett.* **123:** 63–68.

Leganés, F., F. Fernández-Piñas, and C. P. Wolk. 1994. Two mutations that block heterocyst differentiation have no effect on akinete differentiation in *Nostoc ellipsosporum. Mol. Microbiol.* **12:**679–684.

Leganés, F., F. Fernández-Piñas, and C. P. Wolk. 1998. A transposition-induced mutant of *Nostoc ellipsosporum* implicates an arginine-biosynthetic gene in the formation of cyanophycin granules and of functional heterocysts and akinetes. *Microbiology* **144:**1799–1805.

Li, R., M. Watanabe, and M. M. Watanabe. 1997. Akinete formation in planktonic *Anabaena* spp. (cyanobacteria) by treatment with low temperature. *J. Phycol.* **33:**576–584.

Lin, S. J., S. Henze, P. Lundgren, B. Bergman, and E. J. Carpenter. 1998. Whole-cell immunolocalization of nitrogenase in marine diazotrophic cyanobacteria, *Trichodesmium* spp. *Appl. Environ. Microbiol.* **64:**3052–3058.

Livingstone, D., and G. H. M. Jaworski. 1980. The viability of akinetes of blue-green algae recovered from the sediments of Rostherne Mere. *Br. Phycol. J.* **15:**357–364.

Livingstone, D., T. M. Khoja, and B. A. Whitton. 1983. Influence of phosphorus on physiology of a hair-forming blue-green alga (*Calothrix parietina*) from an upland stream. *Phycologia* **22:**345–350.

Löffelhardt, W., and H. J. Bohnert. 1994a. Structure and function of the cyanelle genome. *Int. Rev. Cytol.* **151:**29–65.

Löffelhardt, W., and H. J. Bohnert. 1994b. Molecular biology of cyanelles, p. 65–89. *In* D. A. Bryant (ed.), *The Molecular Biology of Cyanobacteria.* Kluwer Academic Publishers, Dordrecht, The Netherlands.

Lutkenhaus, J. 1993. FtsZ ring in bacterial cytokinesis. *Mol. Microbiol.* **9:**404–409.

Lutkenhaus, J., and A. Mukherjee. 1996. Cell division, p. 1615–1626. *In* F. C. Neidhardt, R. Curtiss III, J. L. Ingraham, E. C. C. Lin, K. B. Low, B. Magasanik, W. S. Reznikoff, M. Riley, M. Schaechter, and H. E. Umbarger (ed.), *Escherichia coli and Salmonella: Cellular and Molecular Biology*, 2nd ed. American Society for Microbiology, Washington, D.C.

Mahasneh, I. A., S. L. J. Grainger, and B. A. Whitton. 1990. Influence of salinity on hair formation and phosphatase activities of the blue-green alga (cyanobacterium) *Calothrix viguieri* D253. *Br. Phycol. J.* **25:**25–32.

Matthijs, H. C. P., G. W. M. van der Staay, and L. R. Mur. 1994. Prochlorophytes: the 'other' cyanobacteria? p. 49–64. *In* D. A. Bryant (ed.), *The*

Molecular Biology of Cyanobacteria. Kluwer Academic Publishers, Dordrecht, The Netherlands.

Meeks, J. C. 1990. Cyanobacterial-bryophyte associations, p. 43–63. *In* A. N. Rai (ed.), *CRC Handbook of Symbiotic Cyanobacteria.* CRC Press Inc., Boca Raton, Fla.

Meeks, J. C. 1998. Symbiosis between nitrogen-fixing cyanobacteria and plants. *BioScience* **48:** 266–276.

Merritt, M. V., S. S. Sid, L. Mesh, and M. M. Allen. 1994. Variations in the amino acid composition of cyanophycin in the cyanobacterium *Synechocystis* sp. PCC 6308 as a function of growth conditions. *Arch. Microbiol.* **162:**158–166.

Mukherjee, A., and J. Lutkenhaus. 1994. Guanine nucleotide-dependent assembly of FtsZ into filaments. *J. Bacteriol.* **176:**2754–2758.

Mur, L. R., and T. Burger-Wiersma. 1992. The order Prochlorales, p. 2105–2110. *In* A. Balows, H. G. Trüper, M. Dworkin, W. Harder, and K.-H. Schleifer (ed.), *The Prokaryotes*, vol. II. Springer-Verlag, Heidelberg, Germany.

Murry, M. A., P. C. Hallenbeck, and J. R. Benemann. 1984. Immunochemical evidence that nitrogenase is restricted to the heterocysts of *Anabaena cylindrica. Arch. Microbiol.* **137:**194–199.

Nanninga, N. 1998. Morphogenesis of *Escherichia coli. Microbiol. Mol. Biol. Rev.* **62:**110–129.

Neely-Fisher, D., W. White, and R. Fisher. 1989. Fructose-induced dark germination of *Anabaena* akinetes. *Curr. Microbiol.* **19:**139–142.

Nichols, J. M., and D. G. Adams. 1982. Akinetes, p. 387–412. *In* N. G. Carr and B. A. Whitton (ed.), *The Biology of Cyanobacteria.* Blackwell Scientific Publications, Oxford, United Kingdom.

Nichols, J. M., D. G. Adams, and N. G. Carr. 1980. Effect of canavanine and other amino acid analogues on akinete formation in the cyanobacterium *Anabaena cylindrica. Arch. Microbiol.* **127:**67–75.

Padan, E., and Y. Cohen. 1982. Anoxygenic photosynthesis, p. 215–235. *In* N. G. Carr and B. A. Whitton (ed.), *The Biology of Cyanobacteria.* Blackwell Scientific Publications, Oxford, United Kingdom.

Palinska, K. A., and W. E. Krumbein. 1998. Patterns of growth in coccoid, aggregate forming cyanobacteria. *Ann. Bot. Fennici* **35:**219–227.

Post, A. F., and G. S. Bullerjahn. 1994. The photosynthetic machinery in Prochlorophytes: structural properties and ecological significance. *FEMS Microbiol. Rev.* **13:**393–414.

Rabenhorst, L. 1865. *Flora Europaea Algarum*, volume 2. Leipzig, Germany.

Rachel, R., D. Pum, J. Šmarda, D. Šmajs, J. Komrska, V. Krzyzánek, G. Rieger, and K. O. Stetter. 1997. Fine structure of S-layers. *FEMS Microbiol. Rev.* **20:**13–23.

Rasmussen, U., C. Johansson, and B. Bergman. 1994. Early communication in the *Gunnera-Nostoc* symbiosis: plant-induced cell differentiation and protein synthesis in the cyanobacterium. *Mol. Plant-Microbe Interact.* **7**:696–702.

Rippka, R. 1988. Recognition and identification of cyanobacteria. *Methods Enzymol.* **167**:28–67.

Rippka, R., and R. Y. Stanier. 1978. The effects of anaerobiosis on nitrogenase synthesis and heterocyst development by Nostocacean cyanobacteria. *J. Gen. Microbiol.* **105**:83–94.

Rippka, R., J. Deruelles, J. B. Waterbury, M. Herdman, and R. Y. Stanier. 1979. Generic assignments, strain histories and properties of pure cultures of cyanobacteria. *J. Gen. Microbiol.* **111**: 1–61.

Rippka, R., J. B. Waterbury, and R. Y. Stanier. 1981. Provisional generic assignments for cyanobacteria in pure culture, p. 247–256. *In* M. P. Starr, H. Stolp, H. G. Truper, A. Balows, and H. G. Schlegel (ed.), *The Prokaryotes*, vol. 1. Springer-Verlag, Heidelberg, Germany.

Robinson, B. L., and J. H. Miller. 1970. Photomorphogenesis in the blue-green alga *Nostoc commune* 584. *Physiologia Plantarum* **23**:461–472.

Rothfield, L. I., and S. S. Justice. 1997. Bacterial cell division; the cycle of the ring. *Cell* **88**:581–584.

Sarma, T. A., and J. I. S. Khattar. 1993. Akinete differentiation in phototrophic, photoheterotrophic and chemoheterotrophic conditions in *Anabaena torulosa. Folia Microbiol.* **38**:335–340.

Schmid, M. B., and U. von Freiesleben. 1996. Nucleoid segregation, p. 1662–1671. *In* F. C. Neidhardt, R. Curtiss III, J. L. Ingraham, E. C. C. Lin, K. B. Low, B. Magasanik, W. S. Reznikoff, M. Riley, M. Schaechter, and H. E. Umbarger (ed.), *Escherichia coli and Salmonella: Cellular and Molecular Biology*, 2nd ed. American Society for Microbiology, Washington, D.C.

Schmidt, A. 1988. Sulfur metabolism in cyanobacteria. *Methods Enzymol.* **167**:572–583.

Schopf, J. W. 1994. Disparate rates, differing fates: tempo and mode of evolution changed from the Precambrian to the Phanerozoic. *Proc. Natl. Acad. Sci. USA* **91**:6735–6742.

Schopf, J. W. 1996. Are the oldest fossils cyanobacteria? p. 23–61. *In* D. M. Roberts, P. Sharp, G. Alderson, and M. Collins (ed.), *Evolution of Microbial Life*. Cambridge University Press, Cambridge, United Kingdom.

Sili, C., A. Ena, R. Materassi, and M. Vincenzini. 1994. Germination of desiccated aged akinetes of alkaliphilic cyanobacteria. *Arch. Microbiol.* **162**: 20–25.

Simon, R. D. 1977. Macromolecular composition of spores from the filamentous cyanobacterium *Anabaena cylindrica. J. Bacteriol.* **129**:1154–1155.

Simon, R. D. 1987. Inclusion bodies in the cyanobacteria: cyanophycin, polyphosphate, polyhedral bodies, p. 199–225. *In* P. Fay and C. van Baalen (ed.), *The Cyanobacteria*. Elsevier Science Publishers B. V., Amsterdam, The Netherlands.

Sinclair, C., and B. A. Whitton. 1977a. Influence of nutrient deficiency on hair formation in the *Rivulariaceae. Br. Phycol. J.* **12**:297–313.

Sinclair, C., and B. A. Whitton. 1977b. Influence of nitrogen source on morphology of *Rivulariaceae* (Cyanophyta). *J. Phycol.* **13**:335–340.

Sleytr, U. B. 1997. Basic and applied S-layer research: an overview. *FEMS Microbiol. Rev.* **20**:5–12.

Soriente, A., A. Gambacorta, A. Trincone, C. Sili, M. Vincenzini, and G. Sodano. 1993. Heterocyst glycolipids of the cyanobacterium *Cyanospira ripphkae. Phytochemistry* **33**:393–396.

Stanier, G. 1988. Fine structure of cyanobacteria. *Methods Enzymol.* **167**:157–172.

Stanier, R. Y., and G. Cohen-Bazire. 1977. Phototrophic prokaryotes: the cyanobacteria. *Annu. Rev. Microbiol.* **31**:225–274.

Sutherland, J. M., M. Herdman, and W. D. P. Stewart. 1979. Akinetes of the cyanobacterium *Nostoc* PCC 7524: macromolecular composition, structure and control of differentiation. *J. Gen. Microbiol.* **115**:273–287.

Sutherland, J. M., J. Reaston, W. D. P. Stewart, and M. Herdman. 1985a. Akinetes of the cyanobacterium *Nostoc* PCC 7524: macromolecular and biochemical changes during synchronous germination. *J. Gen. Microbiol.* **131**:2855–2863.

Sutherland, J. M., W. D. P. Stewart, and M. Herdman. 1985b. Akinetes of the cyanobacterium *Nostoc* PCC 7524: morphological changes during synchronous germination. *Arch. Microbiol.* **142**: 269–274.

Tandeau de Marsac, N. 1994. Differentiation of hormogonia and relationships with other biological processes, p. 825–842. *In* D. A. Bryant (ed.), *The Molecular Biology of Cyanobacteria*. Kluwer Academic Publishers, Dordrecht, The Netherlands.

Tandeau de Marsac, N., and J. Houmard. 1993. Adaptation of cyanobacteria to environmental stimuli: new steps towards molecular mechanisms. *FEMS Microbiol. Rev.* **104**:119–190.

Tandeau de Marsac, N., D. Mazel, D. A. Bryant, and J. Houmard. 1985. Molecular cloning and nucleotide sequence of a developmentally regulated gene from the cyanobacterium *Calothrix* PCC 7601: a gas vesicle protein gene. *Nucleic Acids Res.* **13**:7223–7236.

Tandeau de Marsac, N., D. Mazel, T. Damerval, G. Guglielmi, V. Capuano, and J. Houmard. 1988. Photoregulation of gene expression in the filamentous cyanobacterium *Calothrix* sp. PCC

7601: light-harvesting complexes and cell differentiation. *Photosynth. Res.* **18**:99–132.

Thiel, T. 1990. Protein turnover and heterocyst differentiation in the cyanobacterium *Anabaena variabilis. J. Phycol.* **26**:50–54.

van Baalen, C., and P. Fay (ed.). 1987. *The Cyanobacteria.* Elsevier Science Publishers, New York, N.Y.

van Dok, W., and B. T. Hart. 1996. Akinete differentiation in *Anabaena circinalis* (Cyanophyta). *J. Phycol.* **32**:557–565.

van Dok, W., and B. T. Hart. 1997. Akinete germination in *Anabaena circinalis* (Cyanophyta). *J. Phycol.* **33**:12–17.

Vicente, M., and J. Errington. 1996. Structure, function and controls in microbial division. *Mol. Microbiol.* **20**:1–7.

Walsby, A. E. 1985. The permeability of heterocysts to the gases nitrogen and oxygen. *Proc. R. Soc. Lond. B* **226**:345–366.

Walsby, A. E. 1994. Gas vesicles. *Microbiol. Rev.* **58**:94–144.

Waterbury, J. B. 1989. Order *Pleurocapsales* Geitler 1925, emend. Waterbury and Stanier 1978, p. 1746–1770. *In* J. T. Staley, M. P. Bryant, N. Pfennig, and J. G. Holt, (ed.), *Bergey's Manual of Systematic Bacteriology*, vol. 3. Williams and Wilkins, Baltimore, Md.

Waterbury, J. B., and R. Y. Stanier. 1978. Patterns of growth and development in pleurocapsalean cyanobacteria. *Microbiol. Rev.* **42**:2–44.

Waterbury, J. B., J. M. Willey, D. G. Franks, F. W. Valois, and S. W. Watson. 1985. A cyanobacterium capable of swimming motility. *Science* **230**:74–76.

Weckesser, J., and U. J. Jurgens. 1988. Cell walls and external layers. *Methods Enzymol.* **167**:173–188.

Werner, D. 1992. *Symbiosis of Plants and Microbes.* Chapman and Hall, London, United Kingdom.

Whitton, B. A. 1987a. Survival and dormancy of blue-green algae, p. 109–167. *In* Y. Henis (ed.), *Survival and Dormancy of Microorganisms.* Wiley, New York, N.Y.

Whitton, B. A. 1987b. The biology of *Rivulariaceae*, p. 513–534. *In* P. Fay and C. van Baalen (ed.), *The Cyanobacteria.* Elsevier Science Publishers B. V., Amsterdam, The Netherlands.

Whitton, B. A. 1989. Genus I. *Calothrix* Agardh 1824, p. 1791–1793. *In* J. T. Staley, M. P. Bryant, N. Pfennig, and J. G. Holt (ed.), *Bergey's Manual of*

Systematic Bacteriology, vol. 3. Williams and Wilkins, Baltimore, Md.

Whitton, B. A. 1992. Diversity, ecology, and taxonomy of the cyanobacteria, p. 1–51. *In* N. H. Mann and N. G. Carr (ed.), *Photosynthetic Prokaryotes.* Plenum Press, New York, N.Y.

Whitton, B. A., and M. Potts (ed.). *The Ecology of Cyanobacteria: Their Diversity in Time and Space*, in press. Kluwer Academic Publishers, Dordrecht, The Netherlands.

Wilmotte, A. 1994. Molecular evolution and taxonomy of the cyanobacteria, p. 1–25. *In* D. A. Bryant (ed.), *The Molecular Biology of Cyanobacteria.* Kluwer Academic Publishers, Dordrecht, The Netherlands.

Wolk, C. P. 1965. Control of sporulation in a blue-green alga. *Dev. Biol.* **12**:15–35.

Wolk, C. P. 1966. Evidence of a role of heterocysts in the sporulation of a blue-green alga. *Am. J. Bot.* **53**:260–262.

Wolk, C. P. 1982. Heterocysts, p. 359–386. *In* N. G. Carr and B. A. Whitton (ed.), *The Biology of Cyanobacteria.* Blackwell Scientific Publications, Oxford, United Kingdom.

Wolk, C. P., J. Elhai, T. Kuritz, and D. Holland. 1993. Amplified expression of a transcriptional pattern formed during development of *Anabaena. Mol. Microbiol.* **7**:441–445.

Wolk, C. P., A. Ernst, and J. Elhai. 1994. Heterocyst metabolism and development, p. 769–823. *In* D. A. Bryant (ed.), *The Molecular Biology of Cyanobacteria.* Kluwer Academic Publishers, Dordrecht, The Netherlands.

Wyman, M., and P. Fay. 1986. Interaction between light quality and nutrient availability in the differentiation of akinetes in the planktonic cyanobacterium *Gloeotrichia echinulata. Br. Phycol. J.* **21**:147–153.

Yamamoto, Y. 1975. Effect of desiccation on the germination of akinetes of *Anabaena cylindrica. Plant Cell Physiol.* (Tokyo) **16**:749–752.

Yelloly, J. M., and B. A. Whitton. 1996. Seasonal changes in ambient phosphate and phosphatase activities of the cyanobacterium *Rivularia atra* in intertidal pools at Tyne Sands, Scotland. *Hydrobiologia* **325**:201–212.

Yoon, H.-S., and J. W. Golden. 1998. Heterocyst pattern formation controlled by a diffusible peptide. *Science* **282**:935–938.

Zhang, C.-C., S. Huguenin, and A. Friry. 1995. Analysis of genes encoding the cell division protein FtsZ and a glutathione synthetase homologue in the cyanobacterium *Anabaena* sp. PCC 7120. *Res. Microbiol.* **146**:445–455.

HETEROCYST FORMATION IN
ANABAENA

C. Peter Wolk

4

Heterocysts are differentiated cyanobacterial cells whose principal known function is the fixation of dinitrogen (N_2), an oxygen (O_2)-sensitive process, under aerobic conditions. Fossil evidence suggests that heterocysts represent an ancient solution to the challenge resulting from a major environmental change, the development of an atmosphere with abundant O_2 produced by photosynthesis from proliferating cyanobacteria. Heterocysts, in which N_2 fixation is sequestered from the O_2 production taking place in vegetative cells, may then have remained largely unchanged as eukaryotes originated and spread.

Cyanobacteria are either unicellular or filamentous. Because the role of heterocysts is to provide nitrogen to vegetative cells that are unable to fix N_2 aerobically, and because heterocysts are totally dependent upon vegetative cells for a supply of electrons, all heterocyst-forming cyanobacteria are filamentous (Fig. 1). (In the root nodules of legumes, bacteroids are similarly dependent on their hosts for reductant and provide those hosts with the products of N_2 fixation. Whether, like bacteroids, even detached heterocysts in some N_2-fixing symbioses of cyanobacteria with plants can fix N_2 has

not been established.) Heterocysts normally form following nitrogen stepdown, i.e., in response to nitrogen deprivation (Fogg, 1949; reviewed in Wolk, 1973, 1982). However, there are indications that other environmental imbalances, e.g., of light or temperature, may also stimulate their formation (Adams and Carr, 1981; Campbell et al., 1993; Sato and Wada, 1996; Wilcox, 1970).

In species of *Anabaena* and *Nostoc*, depriving vegetative filaments of fixed nitrogen elicits simultaneous differentiation of heterocysts at semiregular intervals along the filaments, generating a spacing pattern. This pattern is maintained during subsequent growth on N_2 by the formation of a new heterocyst approximately midway between two preexisting heterocysts as the number of cells that separate those two approximately doubles. In contrast, heterocysts in certain other species (see below) are found nearly exclusively at the ends of filaments.

Gene transfer was not reported in a heterocyst-forming cyanobacterium until 1984, and isolation of a gene involved in heterocyst differentiation was first reported in 1988, although nitrogenase genes had been cloned by 1980 (reviewed in Wolk et al., 1994). Recent years have seen the identification of numerous other genes that are involved in the regulation of differentiation and in the synthesis of prod-

C. P. Wolk, MSU-DOE Plant Research Laboratory, Michigan State University, East Lansing, MI 48824.

Prokaryotic Development, edited by Y. V. Brun and L. J. Shimkets,
© 2000 American Society for Microbiology, Washington, DC 20005-4171

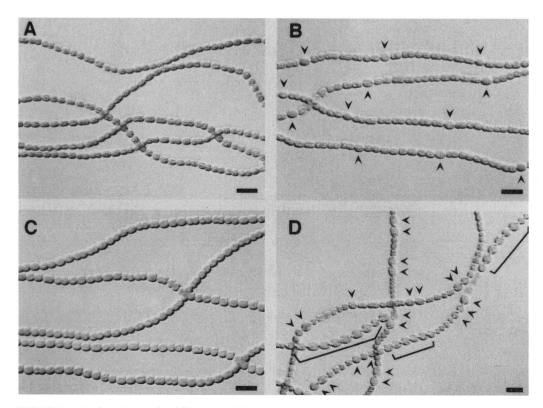

FIGURE 1 Light micrograph of filaments of *Anabaena* sp. strain PCC 7120. (A to C) Wild-type cells; those in panel C contain constitutively promoted extra copies of the *patS* gene on a pDU1-based shuttle vector. (D) *patS* mutant. The cells were grown with (A) and without (B to D) fixed nitrogen. Heterocysts (arrowheads) are absent from the nitrate-grown wild type, are present at semiregular intervals along the filaments of the wild type deprived of fixed nitrogen, are repressed by PatS, and are present in clusters and at abnormally close intervals in the nitrogen-deprived *patS* mutant. (An electron micrograph of a heterocyst of wild-type *Anabaena* sp. strain PCC 7120 may be seen in Fig. 6D.) (Reprinted from Yoon and Golden, 1998.)

ucts that are characteristic of differentiated cells. Two questions concerning the developmental biology of heterocysts appear finally to be yielding to experimentation: how is it decided which cells will differentiate (and which will not); and once it is determined that a particular cell is to differentiate, how is the progression of the differentiation process regulated? Studies that have sought answers to these questions will be discussed first. They will surely be aided by the results of a current effort to sequence the genome of *Anabaena* sp. strain PCC 7120. A complete answer to the first question will probably include a combination of some role of cell lineage (Mitchison and Wilcox, 1972) and cell-

cell interactions (reviewed in Wolk et al., 1994). Experimental evidence indicates that the pattern of spaced heterocysts results, at least in part, from an inhibition by mature and developing heterocysts of the differentiation of nearby cells into heterocysts. A popular model suggests that the inhibition is mediated by the elaboration of some differentiation-inhibiting substance (or substances) that moves outward along a filament (Wolk, 1964, 1967; Mitchison et al., 1976; Wilcox et al., 1973a; Wolk and Quine, 1975). An oligopeptide has been identified that may mediate the critical intercellular inhibition upon nitrogen stepdown (Yoon and Golden, 1998) (see below). There is evidence

(Wilcox et al., 1973b; Wolk and Quine, 1975) that differentiation is dependent on the connection of differentiating cells to vegetative cells. Therefore, substances produced by vegetative cells, perhaps principally providing a source of reductant and carbon (Wolk, 1968), may also be needed if differentiation is to progress.

Three broad classes of genes whose roles are related to the progression of heterocyst differentiation will be discussed (Table 1 and Fig. 2): class 1, those that are activated by nitrogen stepdown but are not involved solely in differentiation (those that are not involved specifically we may call class 0); class 2, those related to differentiation and evidently regulatory, whose activation precedes differentiation detectable by bright-field microscopy; and class 3, those whose products are involved in the metabolic and/or morphological differentiation of the heterocyst. A very crude attempt will be

TABLE 1 Classes of genes activated upon nitrogen stepdown and/or required for aerobic N_2 fixation

Class		Characteristics	Example(s)	Role(s) of examples
0		Activated upon nitrogen deprivation; roles not related to differentiation	*nirA-nrtABCD-narB*; TLN6 locus	Uptake and reduction of nitrate and nitrite; possible role in ammonia uptake
1		Activated upon nitrogen deprivation; roles not restricted to differentiation	*ntcA*	Nitrogen regulation
2		Developmentally regulatory; activation precedes differentiation discernible by bright-field microscopy	*hetR, patS*	Autoregulation of differentiation; heterocyst spacing
3		Required for heterocyst maturation		
	3a	Regulatory	*hetC, hetP*	Regulation of a very early step in heterocyst differentiation
	3b	Involved in metabolic differentiation of the heterocyst protoplast	α71 locus; *xis* and *nif* genes (*argL*)	Respiratory glucose 6-phosphate dehydrogenase; control of a DNA rearrangement; nitrogenase; role in synthesis of cyanophycin, a nitrogen reserve polymer
	3c	Involved in synthesis of the heterocyst envelope		
	$3c_i$	Involved in synthesis of envelope glycolipids	*hetINM* (*hglB*)-*hglCD, hglE*	Structural genes for glycolipid biosynthesis and perhaps some role in pattern formation
	$3c_{ii}$	Involved in export and deposition of envelope glycolipids	*hglK, devABC* (*rfbP*)	Transport and deposition of heterocyst envelope glycolipid
	$3c_{iii}$	Involved in synthesis of envelope polysaccharide	*hepA, -B, -C, -K*	Structural genes for, and regulation of synthesis of, heterocyst envelope polysaccharide

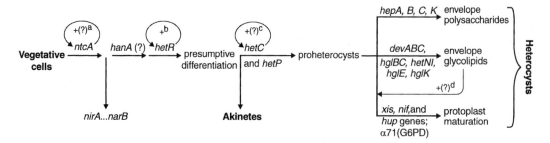

FIGURE 2 Working model of certain genetic dependency relationships expressed in response to nitrogen stepdown. The thick arrows indicate a likelihood that that which follows the arrow depends upon that which precedes the arrow. For example, *ntcA*, but not *hanA*, is required for uptake and reduction of nitrate, which are catalyzed by products of the *nirA-nrtABCD-narB* sequence. Both are required for heterocyst formation. Because the *hanA* gene is preceded by an NtcA-binding site, its transcription may depend upon activity of NtcA. There is suggestive evidence that HanA participates, directly or indirectly, in the transcriptional regulation of *hetR* and that *ntcA* acts also later in heterocyst differentiation. A *hetR* mutation prevents both heterocyst formation and (in *N. ellipsosporum*) akinete formation. The similar phenotypes of *hetC* and *hetP* mutants (the two genes map close together) are suggestive of a blockage in a very early step in the differentiation of cells that have been chosen to become heterocysts but, in *N. ellipsosporum*, no blockage in the formation of akinetes. Beyond the *hetC-hetP* step, formation of the two layers of the heterocyst envelope and development of the protoplast of the heterocyst appear to proceed independently, except that a combination of completion of the glycolipid layer and enhanced respiration may lead to microaerobic conditions in the protoplast, and microaerobiosis is required for such late biochemical changes as the appearance of nitrogenase activity. *hep* genes may also be active during formation of the akinete envelope (Leganés, 1994; Wolk et al., 1994). Where *patS* falls in this scheme is unclear. Thin curved arrows marked a to d refer to regulation that is probable (a), demonstrated (b), and possible (c and d). Temporal relationships, e.g., that in any particular region of the heterocyst envelope, glycolipid deposition follows polysaccharide deposition and that nitrogenase activity follows excisase activity, are not directly illustrated (for references, see the text). An earlier version of this model has appeared (Wolk et al., 1999).

made to enumerate genes involved specifically in differentiation. Finally, the results of initial attempts to discern dependency relationships among genes involved in heterocyst development will be described.

Certain topics about which there is no recent evidence are covered here lightly if at all. The interested reader is invited to examine earlier reviews of *Anabaena* development (e.g., Böhme, 1998; Buikema and Haselkorn, 1993; Golden, 1998; Haselkorn, 1978, 1992, 1995; Haselkorn and Buikema, 1992; Wolk, 1973, 1982, 1991, 1996; Wolk et al., 1994) and of pertinent genetic technology (Thiel, 1994).

THE SPATIAL DISTRIBUTION OF DIFFERENTIATION

Wilcox et al. (1973a) first called attention to evidence of clusters (strings) of cells initiating

differentiation. They concluded that a cluster is then resolved to a single cell that completes differentiation while the other cells in the original cluster revert to being vegetative cells. Compelling though their observations appeared, I have not found more recent direct corroboration by other laboratories that that is the normal pathway of development in *Anabaena* spp. However, Elhai et al. (1998) have suggested that intercellular interactions that affect heterocyst pattern formation may consist solely of the process of resolution of such clusters, i.e., that longer-range interactions may be unnecessary. They allude to evidence that the positions of cells in the cell cycle may be important in determining which cells initiate differentiation. Despite the potential importance of the cell cycle as a possible influence on differentiation, it has been little studied in *Ana-*

baena spp. Two groups have cloned and sequenced homologues of the gene *ftsZ*, whose activity is required for cell division in *Escherichia coli* and other bacteria (Doherty and Adams, 1995; Zhang et al., 1995). Whether, as expected, transcription of *ftsZ* is turned off during heterocyst differentiation has not yet been determined.

Genes denoted *pat* affect the pattern of spacing of heterocysts in *Anabaena* spp. Strains bearing mutations in *patA*, whose product resembles response regulator-type proteins that lack a DNA-binding domain, form heterocysts at the ends of filaments rather than (as is typical of *Anabaena* spp.) at intercalary positions (Liang et al., 1992; see also Wilcox et al., 1975). No protein kinase has yet been identified as acting in concert with PatA.

Although it is not known how a *patA* mutant so drastically alters the pattern of heterocyst formation, the resulting pattern resembles the natural patterns of a variety of other cyanobacteria. That is, in *Anabaenopsis* spp. and certain *Cylindrospermum* spp., the ends of filaments normally form as follows: pairs of vegetative cells differentiate into heterocysts, and as they do so they separate, becoming filament-terminating heterocysts. Other *Cylindrospermum* spp. show cell death far from preexisting heterocysts, with heterocyst formation then ensuing at both of the resultant ends. The branches of *Mastigocoleus testarum* (Geitler, 1932) and the tapering filaments of *Rivularia* spp. and *Gloeotrichia* spp. normally terminate in heterocysts. Even in *Anabaena* spp., heterocysts form at the ends of filaments derived from germination of akinetes (a kind of spore) (Fay et al., 1968; Sutherland et al., 1985) or from experimental fragmentation (e.g., Wolk, 1967). However, too extensive fragmentation, caused by mutations in *fra* genes (Bauer et al., 1995), can result in a Het⁻ phenotype, corresponding to a lack of evident heterocyst formation even after a protracted period of nitrogen deprivation.

The designation *het* has been given to genes whose mutation leads to a nonfragmenting Het⁻ phenotype. However, certain mutant alleles of *hetN*, as well as mutations in *patB* and

patS, result in the formation of clusters of heterocysts (Black and Wolk, 1994; Buikema and Haselkorn, 1991; Liang et al., 1993; Yoon and Golden, 1998). If, as suggested by Wilcox et al. (1973a), the default developmental process is one in which clusters of heterocyst initials are resolved to single heterocysts, such mutations may lead to interference with the default pathway. Deviations from a pattern of single, spaced heterocysts also result from the presence of certain antimetabolites, e.g., 7-azatryptophan, transient exposure to light of high intensity, and the presence of 1 mM cyclic AMP. *Mastigocladus laminosus* normally forms clusters of heterocysts (reviewed in Wolk et al., 1994).

patS (Fig. 1C and D), a recently discovered gene, is important because the following data appear to suggest that its product, PatS, plays a role in the intercellular inhibition of heterocyst formation. *patS* was found by identifying subclones of a cosmid which, like the parent cosmid, blocked heterocyst formation and by the localization of sites of mutation in those subclones that then allowed heterocyst formation (Yoon and Golden, 1998). PatS proved to be a polypeptide of 17 amino acids (or perhaps 13 or 11 amino acids: *patS* has several potential in-frame initiation codons). A synthetic pentapeptide representing the five C-terminal amino acids of PatS inhibits heterocyst formation at a concentration of 0.1 μM, a concentration at which its constituent amino acids do not inhibit heterocyst formation. A *patS* mutant shows extensive clustering of heterocysts, and the average distance between groups of one (i.e., solitary within filaments) or more (i.e., clustered) heterocysts, is approximately half of the distance found in the wild-type strain. It was shown by means of a *patS* fusion to green fluorescent protein that *patS* is activated strongly in spaced cells prior to the appearance of morphologically discernible proheterocysts; at 18 h, its activation is localized to differentiating proheterocysts. If driven by a *hepA* promoter (which is active starting 4.5 to 7 h after nitrogen stepdown and shows far more activity in developing heterocysts than in any other cells [Wolk et al., 1993]), *patS* confers an almost wild-type pattern of heterocyst spacing on a *patS* deletion mutant. The mechanism by which a cell in which *patS* is actively expressed is immune to inhibition of differentiation by PatS is not yet clear. The fact that heterocysts are semiregularly spaced in a *patS* deletion mutant grown in the presence of nitrate (Xu and Wolk, unpublished) implies that some *patS*-independent mechanism is also involved in spacing.

The route by which morphogenetically active substances may move from cell to cell also remains unknown. It is unclear whether transcellular strands called "microplasmodesmata" (Giddings and Staehelin, 1981) constitute conduits, whether intercellular movement takes place via the end membranes of adjacent cells and the wall between those membranes, or whether such substances are transported to and from the periplasm and move through the periplasm that is common to adjacent cells (Wolk and Zarka, 1998; Yoon and Golden, 1998).

The same set of possibilities applies to the movement of reductant, likely in the form of sucrose (Schilling and Ehrnsperger, 1985) or sucrose phosphate (Porchia and Salerno, 1996), from vegetative cells to heterocysts and of fixed nitrogen, partially as glutamine (Thomas et al., 1977) and perhaps partially as arginine or aspartic acid (discussed by Wolk et al., 1994), from heterocysts to vegetative cells. Glutamate, from which the glutamine is derived, may be generated in the vegetative cells (Florencio et al., 1998; Thomas et al., 1977) or possibly in the heterocysts (Häger et al., 1983; discussed in Wolk et al., 1994); unfortunately, techniques giving different answers have been applied to different strains.

Cyclic AMP, which plays a role in signal transduction in prokaryotes as well as eukaryotes, is present in *Anabaena* sp., is responsive to nitrogen deprivation (reviewed in Wolk et al., 1994), and can disrupt the normal pattern of heterocyst formation (Smith and Ownby, 1981). Cyclic AMP is the product of adenylate cyclase. The adenylate cyclase-encoding gene from *Anabaena cylindrica* has been cloned and sequenced (Katayama et al., 1995). *Anabaena* sp. strain PCC 7120 contains at least five genes (*cyaA*, *-B1*, *-B*, *-C*, and *-D*) that have domains that are similar to catalytic domains of eukaryotic adenylate cyclases. The transcript of *cyaC* is predominant. Starting from its N terminus, CyaC has four domains: one that resembles a response regulator, a second that is similar to protein histidine kinases, a second response regulator-like domain, and finally a domain that is similar to adenylate cyclases (Katayama and Ohmori, 1997).

REGULATION OF THE PROGRESSION OF DIFFERENTIATION

The attempt to explain the progression of differentiation began by testing whether the *Bacillus* paradigm of a cascade of sigma factors (Stragier and Losick, 1996) could be invoked. Although multiple species of sigma factors have been reported (Brahamsha and Haselkorn, 1991, 1992; Khudyakov and Golden, 1998), the provisional answer is that in *Anabaena* sp., that model is inadequate to explain the regulation of the differentiation process. Other models remain possible, including the following: a sequence of positive autoregulatory processes; responses to ever more stringent nitrogen deprivation; and, more generally, responses to altered intracellular conditions occasioned by antecedent physiological changes, e.g., responses to the anaerobiosis resulting within developing heterocysts.

GENES INVOLVED IN INITIAL RESPONSES TO DEPRIVATION OF FIXED NITROGEN

The DNA-binding protein NtcA (Ntc stands for nitrogen control), a regulator of transcription that binds at the -35 site of promoters, plays a major role in nitrogen regulation in unicellular cyanobacteria (Luque et al., 1994). The fact that a very similar NtcA protein is required for the formation of heterocysts in filamentous cyanobacteria (Frías et al., 1994; Wei et al., 1994) suggests that heterocyst formation is integrated within a suite of responses to nitrogen stepdown that take place in unicellular as well as filamentous cyanobacteria.

In *Anabaena* sp., nitrate assimilation responds rapidly to nitrogen deprivation. The contiguous genes *nirA*, encoding nitrite reductase; *nrtA*, -*B*, -*C*, and -*D*, encoding nitrate transport proteins; and *narB*, encoding nitrate reductase, are rapidly and strongly induced in response to nitrogen stepdown (Cai and Wolk, 1997a, 1997b; Frías et al., 1997). Another such response involves induction of the gene for a protein that may possibly enhance uptake of ammonium (Cai and Wolk, 1997a). *ntcA*, which encodes NtcA, is repressed by ammonium, is activated in response to removal of ammonium from the growth medium, and is required for nitrate assimilation. NtcA binds upstream from *xisA* (encoding the excisase of an element that is cut out from the genome of *Anabaena* sp. during heterocyst differentiation) and is a candidate for a regulator of transcription of that gene. NtcA also binds strongly near the *nif*-like promoter of *glnA* (*nif* refers to nitrogen fixation genes; *glnA* encodes glutamine synthetase) and weakly to the promoter regions of *nifH* (encoding dinitrogenase reductase) and *rbcLS* (encoding ribulose bisphosphate carboxylase/oxygenase) (Chastain et al., 1990; Ramasubramanian et al., 1994; Tumer et al., 1983). Binding by NtcA to DNA fragments that contain NtcA-binding sites can be increased by increasing concentrations of dithiothreitol (Jiang et al., 1997). Mutational analysis of the relevant promoter regions will indicate whether, as anticipated, NtcA binding affects transcription of these genes significantly.

A protein encoded by *hanA* and designated HanA that is similar to the *E. coli* protein HU that binds single-stranded DNA is required for heterocyst formation and may be required for activation of *hetR*. HanA is not required for activation of nitrate-assimilatory genes and therefore mediates a step in the pathway of response to nitrogen stepdown that leads to heterocyst formation. However, since a *hanA* mutation has pleiotropic effects (Khudyakov and Wolk, 1996), HanA is certainly not specific to the differentiation pathway.

hetR AND AUTOREGULATION

hetR mutants are Het⁻, whereas overexpression of *hetR* or expression of supernumerary copies of *hetR* leads to the formation of heterocysts in the presence of fixed nitrogen and to clusters of heterocysts in the absence of fixed nitrogen (Buikema and Haselkorn, 1991; Haselkorn, 1995). Moreover, *hetR* shows positive autoregulation, i.e., HetR is required for activation of *hetR* (Black et al., 1993). As has been noted (Wolk et al., 1994), this is just the kind of behavior that can stabilize developmental decisions. Enhanced expression of *hetR* has been visualized within 1 to 2 h of nitrogen stepdown; within 3.5 h, its enhanced expression is clearly localized to spaced cells, presumably developing heterocysts (Fig. 3). Immunoblotting shows a twofold increase in HetR within 3 h after nitrogen stepdown (Zhou et al., 1998a).

The requirement for *hetR* in differentiation of akinetes in *Nostoc ellipsosporum* (Leganés et al., 1994) and the presence of *hetR*-homologous DNA in at least one filamentous strain from a family that lacks heterocysts and akinetes (Janson et al., 1998; see also Buikema and Haselkorn, 1991) indicate that *hetR* is not involved exclusively in heterocyst formation. (Because akinetes of *N. ellipsosporum* differ only rather inconspicuously from vegetative cells, it would be desirable to have the effects of *hetR* and of other genes on akinete formation tested in other strains with more conspicuously differentiated akinetes.) Zhou et al. (1998b) have presented evidence that HetR is an unusual serine-type protease that can degrade itself and perhaps other proteins.

ntcA of *Anabaena* sp. is transcribed from multiple start sites and is preceded by a binding site for its own product, suggesting that it, too, is autoregulatory (Ramasubramanian et al., 1996). In addition, evidence suggestive of positive autoregulation of *hetC* has been observed (Khudyakov and Wolk, 1997) (see discussion

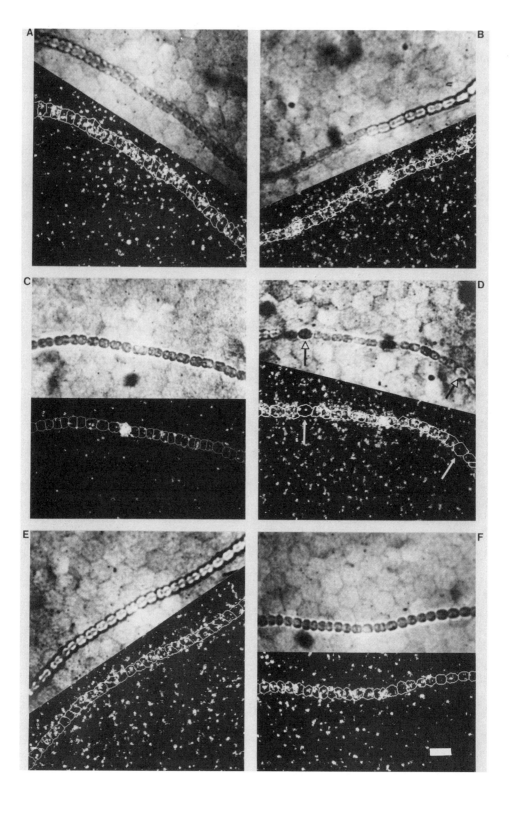

below). Therefore, heterocyst differentiation involves, and may even be regulated by, a series of positive-autoregulation events. Also, Fiedler et al. (1998) have proposed that diminished intracellular partial O_2 pressure (pO_2), dependent upon *devA*, *devB*, and *devC* (*dev* refers to development), is what "triggers" heterocyst maturation, although it has not been reported whether expression of those *dev* genes is limited in a *devA*, *-B*, or *-C* mutant.

GENES THAT ARE INVOLVED IN THE MORPHOLOGICAL AND METABOLIC MATURATION OF HETEROCYSTS

A third class of genes, whose products are involved in the metabolic and/or morphological differentiation of the heterocyst, may be further subdivided into those whose products are regulatory (3a), involved in the differentiation of the heterocyst protoplast (3b), and involved in the synthesis of the envelope of the heterocyst (3c). Specifically, class 3c includes genes whose products are involved in the synthesis ($3c_i$) and the export and deposition ($3c_{ii}$) of the envelope glycolipids into a layer (Winkenbach et al., 1972) that plays a crucial role in the diminution of the pO_2 of the heterocyst protoplast (Murry and Wolk, 1989; Walsby, 1985), and those whose products are involved in the synthesis of the polysaccharide layer of the heterocyst envelope ($3c_{iii}$) (Cardemil and Wolk, 1979, 1981a). The role of the polysaccharide layer appears to be to provide physical protection for the glycolipid layer.

hetC and *hetP*

Genes in ever-increasing numbers are known to have products that are involved in the differentiation process. Those that act earliest are *hetC* (Khudyakov and Wolk, 1997) and *hetP* (Fernández-Piñas et al., 1994), which occur about 1 kb apart in the genome of *Anabaena* sp. strain PCC 7120. When viewed by bright-field microscopy, mutants defective in these genes show no, or delayed, differentiation of heterocysts upon nitrogen stepdown. However, when viewed by fluorescence microscopy with particular excitation and emission wavelengths, semiregularly spaced cells are observed whose fluorescence is diminished relative to the fluorescence of the other cells. No such pattern is seen with *hanA*, *hetR*, or PatS-repressed strains. If, as I consider likely, these spaced cells are cells which in the wild-type organism would have become heterocysts, their reduced fluorescence suggests that an early stage of metabolic differentiation has taken place. Because certain *hetP* mutants can eventually form heterocysts, and rare pseudo-revertants of *hetC* mutants have been found (Khudyakov and Wolk, unpublished), it appears that neither HetC nor HetP is absolutely required for heterocyst formation. Whereas HetP shows no similarity to other proteins in the database, HetC resembles a wide spectrum of ATP-binding cassette (ABC) proteins, with especial similarity to exporters of toxic, often pore-forming, proteins. Such a role could account for the phenotype of a *hetC* mutant. That is, if a pore-forming protein must be secreted from the protoplast of a presumptive heterocyst to allow an inhibitor of differentiation to escape from that same cell or to allow nutrients required for heterocyst differentiation to move from vegetative cells to the developing cell, differentiation might be blocked if that protein were not secreted (Wolk and Zarka, 1998).

FIGURE 3 Localization of expression of *hetR*, visualized as *hetR*∷*luxAB* transcriptional fusions in the presence (A to D) and absence (E and F) of intact *hetR* after 0 (A and E), 3.5 (B), 6, (C), and 24 (D and F) h of deprivation of fixed nitrogen in *Anabaena* sp. strain PCC 7120. In panel D, the arrows indicate mature heterocysts that appear nonluminescent due, at least in part, to a decrease in the concentration of O_2, a substrate of luciferase, within the heterocyst (Elhai and Wolk, 1991; Fernández-Piñas and Wolk, 1994). In each panel, the upper image is a bright-field micrograph with an instrumentally generated honeycomb background and the lower image represents luminescence integrated for 20 min. Bar, 10 μm. (Reproduced from Black et al., 1993, with permission.)

Heterocyst Envelope Glycolipids

In experiments by Bradley and Carr (1977), nitrogen stepdown was followed, at a range of time intervals, with reprovision of fixed nitrogen. Under certain such conditions, some cells appeared to attain an early stage of heterocyst differentiation, but their differentiation then progressed no further. A *devA* mutant appeared to have a similar phenotype (Maldener et al., 1994), in which further differentiation of both the protoplast and the envelope might be considered blocked (Wolk et al., 1999). It was subsequently found, however, that mutants affected in any of the genes of the *devB-devC-devA* operon form envelope glycolipids but do not use them to synthesize an envelope glycolipid layer. Because the proteins encoded by that operon have many of the characteristics of an ABC exporter, Fiedler et al. (1998) proposed that the principal role of those proteins is to export heterocyst envelope glycolipids.

Such an interpretation leads one to reopen the question of the role of a gene referred to as *hglK* (Black et al., 1995) (*hgl* refers to the heterocyst envelope glycolipids), for which a similar role as exporter had been envisioned but which was also noted to affect the form of vegetative cells. The carboxy terminus of HglK consists of many repetitions (with variations) of a pentapeptide and so might alternatively be viewed as a scaffold for the extracellular deposition of heterocyst envelope glycolipid. The Fox⁻ phenotype is defined as an inability to fix nitrogen in the presence of oxygen. Transposon mutant M8 (Ernst et al., 1992) also proved to be mutated in *hglK*, and its Fox⁻ phenotype was shown by reconstruction of the mutation to be closely linked to the transposon insertion (Wolk and Kong, unpublished). Examination of M8 and of the original *hglK* mutant showed more extensive defects than originally reported, including a common but not ubiquitous appearance of the heterocyst envelope polysaccharide layer as an open-ended barrel. HglK may, therefore, play an extensive role in the structuring of the heterocyst envelope. Bateman et al. (1998), who identified such repeats in predicted proteins from *Synecho-cystis* and several heterotrophic bacteria, proposed that the polypeptide repeats of HglK form a superhelical structure in which each pentapeptide forms one β-strand and three of the pentapeptides make a single turn of the β-helix.

Sodano and coworkers (Gambacorta et al., 1996, 1998; Soriente et al., 1992, 1993, 1995; see also Davey and Lambein, 1992a, 1992b) have in recent years provided a thorough structural definition of the class of molecules that make up the glycolipid layers of the heterocysts of numerous species of cyanobacteria. All are long-chain (usually C_{26} and C_{28} but sometimes C_{30} or C_{32}) alcohols glycosidically linked to sugars, usually α-linked glucose but less often α-linked galactose or mannose or β-linked glucose. The lipid moieties have a hydroxyl at the ω-1 position and an oxygen-containing group, either a hydroxyl or a ketone, at the C-3 position. The C_{30} and C_{32} alcohols, and some of the C_{28} alcohols, have a third oxygen-containing group, a hydroxyl or a ketone, at the ω-3 position. At most, one of the two, C-3 or ω-3, bears a ketone. The heterocyst envelope glycolipids of *Anabaena* sp. strain PCC 7120 are shown in Fig. 4.

The fatty acyl portion of such lipids might be expected to be generated by a fatty acid or polyketide synthase. Indeed, it was found that mutations in closely positioned genes denoted *hetN* (Black and Wolk, 1994; Ernst et al., 1992) and *hglB* (*hetM*), *hglC*, and *hglD* (Bauer et al., 1997; Maldener, personal communication) that encode proteins that show a potential func-

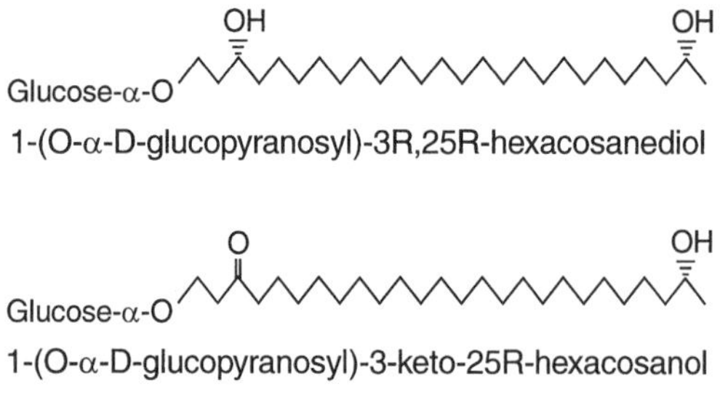

FIGURE 4 Heterocyst envelope glycolipids of *Anabaena* sp. strain PCC 7120.

tional relationship to such synthases resulted in a lack of heterocyst envelope glycolipids. However, some of those same mutations also resulted in alterations in pattern, either to clustered heterocysts or to an absence of heterocysts upon nitrogen stepdown (Bauer et al., 1997; Black and Wolk, 1994), suggesting the possible interpretation that heterocyst envelope glycolipids may also form parts of molecules that play a role in pattern formation (Bauer et al., 1997; Haselkorn, 1995). HetI, encoded 3′ from *hetN*, is a member of a recently discovered superfamily of phosphopantetheinyl transferases (Lambalot et al., 1996). Campbell et al. (1997) identified a *Nostoc punctiforme* ATCC 29133 gene, *hglE*, whose product appears to be involved in biosynthesis of heterocyst envelope glycolipids and is also predicted to resemble a polyketide-biosynthetic protein. *hglE* hybridizes with DNA from *Anabaena* sp. strain PCC 7120 and other heterocyst-forming strains but not with DNA from a unicellular strain, *Synechococcus* sp.

Heterocyst Envelope Polysaccharide

The repeating unit of the structure of the heterocyst envelope polysaccharide has been defined in detail for *A. cylindrica* and provisionally for two other cyanobacteria (Cardemil and Wolk, 1979, 1981a) (Fig. 5). One can predict a need for the following functions for the syn-

thesis and polymerization of that repeating unit. Four or five enzymes are needed to activate the individual sugars and an equal number are needed to transfer the activated sugars to polyprenol monophosphate glycolipids. Two to four enzymes are needed to assemble the mannose-(glucose)$_3$ backbone of the repeating unit, presumably linked to a polyprenol diphosphate. Three to six more are needed to decorate that backbone with side branches, and another is required to transfer the repeating subunit to an elongating polysaccharide chain. Finally, a bacitracin-sensitive phosphatase is probably required for resynthesis of the oligosaccharide repeating unit (Cardemil and Wolk, 1981b). Thus, one can predict that on the order of 20 genes are needed, exclusive of regulatory genes.

Certain corresponding genes have been identified. *hepA* (Holland and Wolk, 1990; Wolk et al., 1988, 1993; renamed by Ernst et al. [1992] to refer to heterocyst envelope polysaccharide) shows extensive similarity to a wide range of ABC transporters. Its function has not been further defined, except that high-resolution electron microscopy (Fig. 6D) has shown that in contrast to the earlier interpretation based on light microscopy (Wolk et al., 1988) and lower-resolution electron microscopy (Murry and Wolk, 1989), the envelope of *hepA* mutant DR1069 consists solely of heterocyst

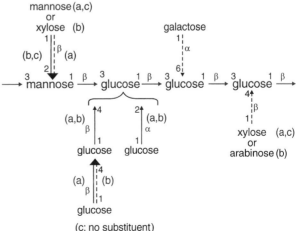

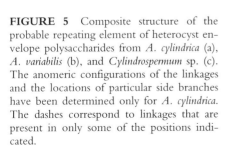

FIGURE 5 Composite structure of the probable repeating element of heterocyst envelope polysaccharides from *A. cylindrica* (a), *A. variabilis* (b), and *Cylindrospermum* sp. (c). The anomeric configurations of the linkages and the locations of particular side branches have been determined only for *A. cylindrica*. The dashes correspond to linkages that are present in only some of the positions indicated.

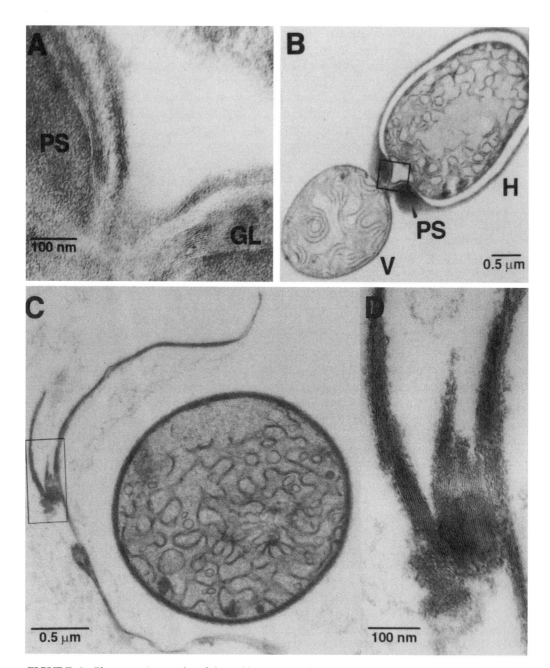

FIGURE 6 Electron micrographs of the wild type (A and B) and *hepA* mutant DR1069 (C and D) of *Anabaena* sp. strain PCC 7120. (A and B) In a heterocyst (H) of wild-type *Anabaena* sp. strain PCC 7120, the laminated layer of glycolipids (GL) is enveloped by a layer of polysaccharide (PS). In contrast, the only envelope layer seen in heterocysts of the *hepA* mutant (C) is the laminated layer of glycolipids. (D) Enlargement of box in panel C. V, vegetative cell. (Panels A and B are reproduced from Zhu et al., 1998, with permission.)

envelope glycolipids. The predicted product of *hepB* (Wolk et al., 1988, 1999) (GenBank accession no. ASU68035) shows similarity to a variety of glycosyl transferases that are involved in the synthesis of extracellular polysaccharides, but its sugar specificity has not yet been clarified. HepC (Zhu et al., 1998), finally, shows that most protracted similarity, among functionally defined proteins, to undecaprenoyl phosphate galactose phosphotransferase (galactosyl-PP-undecaprenol synthetase). If the envelope polysaccharide from heterocysts of *Anabaena* sp. strain PCC 7120, like other such polysaccharides that have been investigated, has galactose present only as a side branch on the backbone of the repeating oligosaccharide, the *hepC* product may be required for the charging of its donor molecule.

Because transcription of *hepA* has been localized to differentiating heterocysts (Wolk et al., 1993), we have chosen to try to define its regulation as an approach to identifying the control of one highly significant stage in heterocyst differentiation. A transcriptional start site of *hepA* was localized 104 bp 5′ from its translational initiation codon. *hepC*, the next gene upstream from *hepA*, was required for synthesis of heterocyst envelope polysaccharide and for normal *repression* of *hepA* during growth on fixed nitrogen, but it was not required for synthesis of the glycolipid layer of the heterocyst envelope. HepK, predicted to resemble a sensory protein histidine kinase of a two-component regulatory system, is required for synthesis of the heterocyst envelope polysaccharide and for *induction* of *hepA*. Interestingly, HepK from *Anabaena* sp. strain PCC 7120 shows the most protracted similarity to ArcB, an *E. coli* protein that regulates metabolic responses to diminished pO_2 (Zhu et al., 1998). However, its homologue from *N. punctiforme* ATCC 29133 does not show such a similarity (Meeks, personal communication).

Transcription of *hepA* is also blocked by a mutation in a gene whose product shows strong similarity (BLAST Expect value, 2×10^{-28}; 41% identical and 57% positive over a sequence of 176 amino acid residues [Altschul et al., 1997]) to a major autolysin, YqiI, *N*-

acetylmuramoyl-ʟ-alanine amidase, of *Bacillus subtilis* (Zhu et al., unpublished). Perhaps degradation of vegetative cell peptidoglycan in developing heterocysts is a prerequisite for activation of envelope polysaccharide biosynthetic genes whose oligosaccharide products must traverse the peptidoglycan layer of the wall. However, the possibility has not been excluded that the phenotype results from a polar effect of the mutation. A different envelope layer of the vegetative cell also appears to affect heterocyst differentiation. That is, mutation of either of two genes of lipopolysaccharide synthesis, *rfbZ* and *rfbP*, renders the host resistant to cyanophages A-1(L) and A-4(L) (perhaps the phages cannot adsorb). The resulting strains proved to be Fox⁻ but Fix⁺ (capable of anaerobic fixation of N_2). The lesions in the lipopolysaccharide biosynthetic genes may lead to defects in the deposition of heterocyst envelope glycolipids (Xu et al., 1997).

Analysis of the region between *hepC* and *hepA* showed the presence of two DNA sequences required for the induction of *hepA* upon nitrogen stepdown. Evidence was obtained that at least one protein binds specifically to one of these sequences (Zhu et al., 1998). The major binding site has now been delimited to 20 bp, and efforts are under way to identify the protein that binds to this sequence (Koksharova and Wolk, unpublished).

In a *devR* mutant of *N. punctiforme*, the heterocyst envelope appears to be reduced to spurlike projections of presumptive envelope material at the heterocyst poles. DevR, encoded by *devR*, is highly similar to receiver domains of response regulator proteins of two-component regulatory systems (Campbell et al., 1996). Whether DevR is part of the same regulatory system as HepK remains to be determined.

Metabolic Differentiation of the Heterocyst Protoplast

RESPIRATORY APPARATUS

No analysis has been made of the heterocyst "honeycomb" membranes whose cellular location correlates with the locus of oxidation of diaminobenzidine in the dark (Wolk et al.,

1994). A great increase in the specific activity of glucose 6-phosphate dehydrogenase (G6PD) in heterocysts relative to that in vegetative cells has long since been documented (Winkenbach and Wolk, 1973). The interpretation that G6PD plays a major role in nitrogen fixation was provided with definitive support by analysis of an insertional *zwf* mutant (thus lacking G6PD) of *Nostoc* sp. strain ATCC 29133 (Summers et al., 1995). Mutants whose heterocysts are deficient in respiration, like those in which the physical barrier to the entry of oxygen into heterocysts is defective, would be expected to exhibit O_2-sensitive nitrogen fixation. Mutant α71 (Ernst et al., 1992), whose rate of uptake of O_2 (measured in the dark at 130 μM O_2 after 2 days of deprivation of fixed nitrogen) was less than 25% of that of other mutants studied or of the wild-type strain, appears to be a mutant of the former type. Reconstruction of this transposon mutant (Ernst and Wolk, unpublished) showed that the phenotype is closely linked to the transposon insertion. DNA sequencing outwards from the transposon has shown that the intercepted gene, sequenced by Bauer (GenBank accession no. ASU14553), appears to encode a G6PD that is not related in amino acid sequence to the G6PD found by Summers et al. (1995)! Thus, it may be that G6PD is on the main route of electrons to both nitrogenase and cytochrome oxidase but that different G6PDs are involved in the two pathways.

Isocitrate dehydrogenase, encoded by the gene *idh*, is an alternative, potentially major source of both NADPH and (via α-ketoglutarate) glutamate in heterocysts and shows a five-fold increase in transcription under N_2-fixing conditions (Böhme, 1998). Because attempts to insertionally inactivate *idh* from *Anabaena* sp. strain PCC 7120 failed to achieve full segregation of a mutant allele (Muro-Pastor and Florencio, 1994), it remains unclear what roles isocitrate dehydrogenase may play in heterocysts.

Data of Peschek et al. (1995a, 1995b) suggest that heme A is synthesized from heme O by an obligately O_2-dependent process, giving rise to cytochrome a_3 as opposed to cytochrome o_3, both of which probably combine with the same cytochrome *c* oxidase apoprotein (Peschek et al., 1995a). Whether cytochrome o_3 is present in heterocysts in vivo remains to be determined.

NITROGENASE

As a result of the synthesis of the polysaccharide and (especially) glycolipid layers of the heterocyst envelope, the rate of entry of oxygen into the protoplast of the heterocyst is greatly diminished. That diminution, together with increased respiratory activity, leads to an extensive decrease in the concentration of oxygen in the cytoplasm of the heterocyst (Elhai and Wolk, 1991). As a consequence, nitrogenase synthesized in the heterocyst is not subject to rapid inactivation. Synthesis of nitrogenase is transcriptionally regulated (Elhai and Wolk, 1990; Haselkorn et al., 1983), but little information is available about how it is regulated.

For reasons that are unclear (Meeks et al., 1994), the *nif* operons in the vegetative cells of *Anabaena* sp. strain PCC 7120 are interrupted by two excisable elements, 11 and 55 kb in length, that are removed from the chromosome of *Anabaena* sp. late during heterocyst differentiation (reviewed in Golden, 1998). As the result of those genetic rearrangements, transcription of the intact *nifD* (which encodes the α-subunit of dinitrogenase) and of *fdxN* (which encodes a bacterial-type ferredoxin whose function is not known) becomes possible (Golden et al., 1985, 1988). When the *fdxN* element and the *nifD* element are present but their excisases are inactivated by mutation, development of nitrogenase activity is blocked (Golden, 1998; Kuritz et al., 1993). *hupL* encodes the large subunit of a membrane-bound NiFe uptake hydrogenase that can reassimilate hydrogen produced when nitrogenase reduces protons. Within the SalD fragment of the chromosome of *Anabaena* sp. strain PCC 7120 (Matveyev et al., 1994) but not in *Anabaena variabilis* ATCC 29413 (Böhme, 1998) or *Nostoc* sp. strain PCC 73102 (strain ATCC 29133) (Oxelfelt et al., 1998), *hupL* is interrupted by a third excisable element 10.5 kb in length. The element within *hupL* contains a presumptive site-specific recombinase gene denoted *xisC* that shows sequence similarity to *xisA*, the recombinase of the *nifD* element. The *hupL* element is flanked by 16-bp direct repeats that are not similar to the recombination sites for the *nifD* or *fdxN* elements (Carrasco et al., 1995).

Most heterocyst-forming cyanobacteria tested possess the 11-kb element, whereas possession of the 55-kb element is less common (Carrasco and Golden, 1995; Golden, 1998; Meeks et al., 1994). Investigation of the regulation of the recombinases that are involved in the genetic rearrangements has been initiated (Ramasubramanian et al., 1994; Ramaswamy et al., 1997), but regulation at the level of the promoters of the genes (or operons) that are interrupted by the excisable elements has yet to be elucidated.

In *A. variabilis* ATCC 29413 incubated under anaerobic conditions, nitrogenase mRNA is detectable within 1.5 to 2 h of nitrogen deprivation in filaments that lack any morphological signs of differentiation (Helber et al., 1988). It was subsequently shown that in addition to a nitrogenase that is expressed specifically in heterocysts (Fay et al., 1968; Peterson and Wolk, 1978), as is the single nitrogenase of *Anabaena* sp. strain PCC 7120 (Elhai and Wolk, 1990), *A. variabilis* has a set of *nif* genes that is expressed also in vegetative cells under anaerobic conditions (Thiel et al., 1995; see also Schrautemeier et al., 1995, and Thiel et al., 1997). The demonstration by Thiel et al. (1995) was based on elegant use of fluorescent products of action of β-galactosidase. It seems highly likely that, like nitrate reductase, these genes are environmentally, rather than developmentally, regulated. *A. variabilis* also has a vanadium-dependent nitrogenase (Thiel, 1993; see also Thiel, 1996), but in which cells the corresponding genes are active has yet to be determined.

Genes that encode eukaryotic-type protein kinases and a eukaryotic-type protein phosphatase have been cloned from *Anabaena* sp. strain PCC 7120 (Zhang, 1993, 1996; Zhang and Libs, 1998; Zhang et al., 1998). Strains with these genes mutated show 5- to 10-fold-diminished nitrogenase activities compared to the wild-type strain, i.e., these genes affect nitrogenase activity quantitatively, and in certain cases (Zhang et al., 1998), they affect heterocyst differentiation qualitatively.

The product of *argL* resembles certain *N*-acetyl glutamate dehydrogenases (in the pathway of arginine biosynthesis). A mutation in *argL*, in the probable presence of another gene whose sequence predicts a functionally equivalent product, leads to the formation of heterocysts that are unable to fix N_2 aerobically. The phenotype may be mediated by an effect on the very arginine-rich reserve polymer, cyanophycin, which also fails to form in the mutated strain in the absence of L-arginine (Leganés et al., 1998).

Uncertain Numerology

Envelope formation, enhanced respiration, and *nif* gene function are conspicuous aspects of differentiation and involve a substantial investment of material and energy. One can argue that the following groups of genes may be required for structural and metabolic differentiation: about 20 genes (see above) dedicated to the synthesis and deposition of the envelope polysaccharide layer; consistent with known series of genes required for syntheses of fatty acids and polyketides (Hopwood et al., 1993), probably not more than about 20 genes (including *hetINM*, *hglCD*, and *devBCA*) for glycolipid synthesis and deposition; perhaps a similar number for synthesis of respiration-related proteins, including cytochrome oxidase, G6PD, and proteins required for generation of honeycomb membranes; and perhaps another 20 or so nitrogenase-related genes, including, in addition to the nitrogenase structural genes *nifHDK*, *nifBENSTUVWX* (not in this order in the genome!), plus other genes in the *nif* region of the genome that encode electron donors to nitrogenase or in other ways enhance the nitrogen fixation process; and perhaps an additional ca. 20 genes, including *hetR*, *hetCP*, *patS*, and *patA*. In short, it is not difficult to visualize the exclusive dedication of 100 genes to the process of heterocyst differentiation, a number similar to those estimated for other bacteria that differentiate (Wolk et al., 1994).

A different argument suggests that it is unlikely that the process of heterocyst differentiation requires the dedicated input of more than about 200 genes. The 7.13-Mbp genome of *Anabaena* sp. strain PCC 7120 (Wolk, 1996) bears ca. 7,000 genes. Whereas mutations in many constitutive genes would be lethal, mutations in differentiation-specific genes should not be lethal under conditions of growth on fixed nitrogen. Because Fox$^-$ mutations constitute only 1 to 2% of transposon mutations under those conditions (Wolk, 1991), a maximum of ca. 140 development-specific genes (2% of 7,000) should be required for a Fox$^+$ phenotype.

However, Lynn et al. (1986) have presented evidence that ca. 15 to 25% of the genome of *A. variabilis* becomes active specifically within heterocysts. If *A. variabilis* has the same density of genes as the unicellular cyanobacterium *Synechocystis* sp. strain PCC 6803, 0.9 genes/kb (Kotani and Tabata, 1998), and the same genetic complexity as *Anabaena* sp. strain PCC 7120, that range of percentages would correspond to 960 to 1,600 genes. How is one to account for that many genes being activated upon heterocyst formation? One possible answer is that since heterocysts are virtually anaerobic, they are loci of anaerobic activation of a plethora of genes, functional under anaerobic conditions, that may have been lost from obligately O$_2$-producing unicellular cyanobacteria. One wonders whether the capacity of heterocysts to generate a low intracellular pO$_2$ in an aerobic environment might be put to biotechnological use.

Dependency Relationships (Fig. 2)

As noted above, transcription of both *nirA* (and thus assimilation of nitrate and nitrite) and *hetR* (and thus heterocyst formation and aerobic fixation of N$_2$) are dependent on *ntcA* (Frías et al., 1994; Wei et al., 1994). In contrast, a mutation in *hanA* is permissive of nitrate assimilation but blocks the differentiation of heterocysts and aerobic fixation of N$_2$ and may block expression of *hetR*. Therefore, *ntcA* and *hanA* bracket a branchpoint in physiological responses to N deprivation. *hetR*, required for differentiation

of heterocysts and (in *N. ellipsosporum*) akinetes (Leganés et al., 1994), is expressed in spaced cells within several hours, whereas much later there is nitrogenase activity in heterocysts but not in akinetes (Fay, 1969). Therefore, temporally downstream of the activation of *hetR* lies a branchpoint between the metabolic differentiation of heterocysts and akinetes. There is controversy as to whether activation of *hetC* and *hetP* is dependent upon activation of *hetR*: the luminescence resulting from a *hetC::luxAB* fusion near the 3′ end of *hetC* shows 100-fold dependence upon *hetR* (Khudyakov and Wolk, 1997) (*luxAB* encodes luciferase), whereas Northern blotting shows an absence of significant dependence close to the 5′ end of the gene (Muro-Pastor et al., 1998). The difference between the results may be due to the presence of an *ntcA*-binding site within the gene (Xu, personal communication) or may reflect the fact that when the plasmid bearing the *lux* reporter gene recombines into the genome, certain sequences are duplicated.

As expected for the product of a gene whose mutation prevents heterocyst formation, HetR evidently has a variety of effects on transcriptional activity: activation of *devA*, *hetM* (Cai and Wolk, 1997b), and *hepA* (Black et al., 1993) are all blocked by a *hetR* mutation. In *Anabaena* sp. strain PCC 7120, whose nitrogenase structural genes appear to be developmentally regulated (Elhai and Wolk, 1990; see also Thiel et al., 1995), nitrogenase activity and immunologically detectable NifH are also lacking in a *hetR* mutant (Ernst et al., 1992). Transcription of *hepA* is also known to be under the control of *hepK* (as noted above, encoding a member of the family of protein-histidine kinases of two-component regulatory systems) and *hepC*. Because HepC is extensively similar to galactosyl transferases, the latter dependency is probably indirect. In contrast to the above-mentioned cases of extensive dependence on *hetR*, the locus of transposon insertion in strain TLN6, distant from *nirA*, was activated hours after activation of *hetR*, and with at most a quantitative dependence on *hetR*, in response to nitrogen stepdown (Cai and Wolk, 1997b).

One might have anticipated that synthesis

of nitrogenase would depend upon the intracellular pO_2 first reaching a level sufficiently low that the enzyme would not at once be extensively inactivated. However, Ernst et al. (1992) observed that an internal pO_2 sufficiently high to inactivate nitrogenase failed to prevent at least some synthesis of NifH. One might also expect that expression of the genes involved in the deposition of the polysaccharide-containing outermost envelope layer would stop once the glycolipid layer was synthesized, because the polysaccharide would be unable to pass the barrier of glycolipid. Fiedler et al. (1998) suggested that the apparent abortion of the differentiation process in a *devA* (or *devB* or *devC*) mutant is an effect of the absence of the glycolipid layer and of the resulting higher pO_2 in the developing heterocyst. Further experimentation should elucidate possible regulatory roles of the low intracellular pO_2 that results from the development of a normal glycolipid layer and normal heterocyst respiration.

Conclusions

The recent discovery of *patS*, whose product appears likely to mediate a critical intercellular interaction, and of other *pat* genes moves the study of pattern formation in *Anabaena* sp. from the realm of cell biology to the molecular level. By means of a diversity of genetic approaches, including transposon mutagenesis, complementation of mutations induced by UV light and chemicals, response to added cosmids, and site-specific mutagenesis, numerous other genes that participate in differentiation of *Anabaena* sp. have been identified. Dependency relationships between them have also been identified. However, with the possible exception of interactions with *ntcA*, for which presumptive regulatory sites have been identified, mechanisms underlying such dependencies are only now starting to be identified.

ntcA appears to regulate a broad range of responses to nitrogen supply. *hetR* plays a central role early in differentiation. *hetC* and *hetP* appear to play roles in the regulation of all but the earliest stages of differentiation. Finally, there may be independent regulation of the

formation of the glycolipid and polysaccharide portions of the heterocyst envelope and of the (as yet little studied) maturation of the interior of the heterocyst.

ACKNOWLEDGMENTS

I am grateful to J. W. Golden, J. C. Meeks, and T. Thiel, as well as to my colleagues at Michigan State University, for their very helpful criticisms of the manuscript.

Support by the U.S. Department of Energy under grant DOE-FG02-91ER20021 and by NSF grant MCB-9723193 is gratefully acknowledged.

REFERENCES

Adams, D. G., and N. G. Carr. 1981. Heterocyst differentiation and cell division in the cyanobacterium *Anabaena cylindrica*: effect of high light intensity. *J. Cell Sci.* **49:**341–352.

Altschul, S. F., T. L. Madden, A. A. Schaffer, J. Zhang, Z. Zhang, W. Miller, and D. J. Lipman. 1997. Gapped BLAST and PSI-BLAST: a new generation of protein database search programs. *Nucleic Acids Res.* **25:**3389–3402.

Bateman, A., A. G. Murzin, and S. A. Teichmann. 1998. Structure and distribution of pentapeptide repeats in bacteria. *Protein Sci.* **7:**1477–1480.

Bauer, C. C., W. J. Buikema, K. Black, and R. Haselkorn. 1995. A short-filament mutant of *Anabaena* sp. strain PCC 7120 that fragments in nitrogen-deficient medium. *J. Bacteriol.* **177:**1520–1526.

Bauer, C. C., K. S. Ramaswamy, S. Endley, L. A. Scappino, J. W. Golden, and R. Haselkorn. 1997. Suppression of heterocyst differentiation in *Anabaena* PCC 7120 by a cosmid carrying wild-type genes encoding enzymes for fatty acid synthesis. *FEMS Microbiol. Lett.* **151:**23–30.

Black, K., W. Buikema, and R. Haselkorn. 1995. The *hglK* gene is required for localization of heterocyst-specific glycolipids in the cyanobacterium *Anabaena* sp. strain PCC 7120. *J. Bacteriol.* **177:**6440–6448.

Black, T. A., and C. P. Wolk. 1994. Analysis of a Het⁻ mutation in *Anabaena* sp. strain PCC 7120 implicates a secondary metabolite in the regulation of heterocyst spacing. *J. Bacteriol.* **176:**2282–2292.

Black, T. A., Y. Cai, and C. P. Wolk. 1993. Spatial expression and autoregulation of *hetR*, a gene involved in the control of heterocyst development in *Anabaena. Mol. Microbiol.* **9:**77–84.

Böhme, H. 1998. Regulation of nitrogen fixation in heterocyst-forming cyanobacteria. *Trends Plant Sci.* **3:**346–351.

Bradley, S., and N. G. Carr. 1977. Heterocyst development in *Anabaena cylindrica*: the necessity for light as an initial trigger and sequential stages of commitment. *J. Gen. Microbiol.* **101**:291–297.

Brahamsha, B., and R. Haselkorn. 1991. Isolation and characterization of the gene encoding the principal sigma factor of the vegetative cell RNA polymerase from the cyanobacterium *Anabaena* sp. strain PCC 7120. *J. Bacteriol.* **173**:2442–2450.

Brahamsha, B., and R. Haselkorn. 1992. Identification of multiple RNA polymerase sigma factor homologs in the cyanobacterium *Anabaena* sp. strain PCC 7120: cloning, expression, and inactivation of the *sigB* and *sigC* genes. *J. Bacteriol.* **174**:7273–7282.

Buikema, W. J., and R. Haselkorn. 1991. Characterization of a gene controlling heterocyst differentiation in the cyanobacterium *Anabaena* 7120. *Genes Dev.* **5**:321–330.

Buikema, W. J., and R. Haselkorn. 1993. Molecular genetics of cyanobacterial development. *Annu. Rev. Plant Physiol. Plant Mol. Biol.* **44**:33–52.

Cai, Y., and C. P. Wolk. 1997a. Nitrogen deprivation of *Anabaena* sp. strain PCC 7120 elicits rapid activation of a gene cluster that is essential for uptake and utilization of nitrate. *J. Bacteriol.* **179**:258–266.

Cai, Y., and C. P. Wolk. 1997b. *Anabaena* sp. strain PCC 7120 responds to nitrogen deprivation with a cascade-like sequence of transcriptional activations. *J. Bacteriol.* **179**:267–271.

Campbell, D., J. Houmard, and N. Tandeau de Marsac. 1993. Electron transport regulates cellular differentiation in the filamentous cyanobacterium *Calothrix.* *Plant Cell* **5**:451–463.

Campbell, E. L., K. D. Hagen, M. F. Cohen, M. L. Summers, and J. C. Meeks. 1996. The *devR* gene product is characteristic of receivers of two-component regulatory systems and is essential for heterocyst development in the filamentous cyanobacterium *Nostoc* sp. strain ATCC 29133. *J. Bacteriol.* **178**:2037–2043.

Campbell, E. L., M. F. Cohen, and J. C. Meeks. 1997. A polyketide-synthase-like gene is involved in the synthesis of heterocyst glycolipids in *Nostoc punctiforme* strain ATCC 29133. *Arch. Microbiol.* **167**:251–258.

Cardemil, L., and C. P. Wolk. 1979. The polysaccharides from heterocyst and spore envelopes of a blue-green alga. Structure of the basic repeating unit. *J. Biol. Chem.* **254**:736–741.

Cardemil, L., and C. P. Wolk. 1981a. The polysaccharides from the envelopes of heterocysts and spores of the blue-green algae *Anabaena variabilis* and *Cylindrospermum licheniforme*. *J. Phycol.* **17**:234–240.

Cardemil, L., and C. P. Wolk. 1981b. Isolated heterocysts of *Anabaena variabilis* synthesize envelope polysaccharide. *Biochim. Biophys. Acta* **674**:265–276.

Carrasco, C. D., and J. W. Golden. 1995. Two heterocyst-specific DNA rearrangements of *nif* operons in *Anabaena cylindrica* and *Nostoc* sp. strain Mac. *Microbiology* **141**:2479–2487.

Carrasco, C. D., J. A. Buettner, and J. W. Golden. 1995. Programed DNA rearrangement of a cyanobacterial *hupL* gene in heterocysts. *Proc. Natl. Acad. Sci. USA* **92**:791–795.

Chastain, C. J., J. S. Brusca, T. S. Ramasubramanian, T.-F. Wei, and J. W. Golden. 1990. A sequence-specific DNA-binding factor (VF1) from *Anabaena* sp. strain PCC 7120 vegetative cells binds to three adjacent sites in the *xisA* upstream region. *J. Bacteriol.* **172**:5044–5051.

Davey, M. W., and F. Lambein. 1992a. Semipreparative isolation of individual cyanobacterial heterocyst-type glycolipids by reverse-phase high-performance liquid chromatography. *Anal. Biochem.* **206**:226–230.

Davey, M. W., and F. Lambein. 1992b. Quantitative derivatization and high-performance liquid chromatographic analysis of cyanobacterial heterocyst-type glycolipids. *Anal. Biochem.* **206**:323–327.

Doherty, H. M., and D. G. Adams. 1995. Cloning and sequence of *ftsZ* and flanking regions from the cyanobacterium *Anabaena* PCC 7120. *Gene* **163**:93–96.

Elhai, J., and C. P. Wolk. 1990. Developmental regulation and spatial pattern of expression of the structural genes for nitrogenase in the cyanobacterium *Anabaena*. *EMBO J.* **9**:3379–3388.

Elhai, J., and C. P. Wolk. 1991. Hierarchical control by oxygen in heterocysts of *Anabaena*, abstr. 114B. *In Abstracts of the 7th International Symposium on Photosynthetic Prokaryotes*.

Elhai, J., A. V. Matveyev, and C. S. Nielsen. 1998. Two-stage model of patterned heterocyst differentiation in *Anabaena*, abstr. B7. *In Abstracts of the 6th Cyanobacterial Workshop*.

Ernst, A., and C. P. Wolk. Unpublished observations.

Ernst, A., T. Black, Y. Cai, J.-M. Panoff, D. N. Tiwari, and C. P. Wolk. 1992. Synthesis of nitrogenase in mutants of the cyanobacterium *Anabaena* sp. strain PCC 7120 affected in heterocyst development or metabolism. *J. Bacteriol.* **174**:6025–6032.

Fay, P. 1969. Metabolic activities of isolated spores of *Anabaena cylindrica*. *J. Exp. Bot.* **20**:100–109.

Fay, P., W. D. P. Stewart, A. E. Walsby, and G. E. Fogg. 1968. Is the heterocyst the site of nitrogen fixation in blue-green algae? *Nature* (London) **220**:810–812.

Fernández-Piñas, F., and C. P. Wolk. 1994. Expression of *luxCD-E* in *Anabaena* sp. can replace the use of exogenous aldehyde for in vivo localization of transcription by *luxAB*. *Gene* **150**:169–174.

Fernández-Piñas, F., F. Leganés, and C. P. Wolk.

1994. A third genetic locus required for the formation of heterocysts in *Anabaena* sp. strain PCC 7120. *J. Bacteriol.* **176**:5277–5283.

Fiedler, G., M. Arnold, S. Hannus, and I. Maldener. 1998. The DevBCA exporter is essential for envelope formation in heterocysts of the cyanobacterium *Anabaena* sp. strain PCC 7120. *Mol. Microbiol.* **27**:1193–1202.

Florencio, F. J., M. Garcia-Dominguez, E. Martin-Figueroa, and J. L. Crespo. 1998. The GS-GOGAT pathway in cyanobacteria. Regulation and complexity of a central metabolic route, abstr. A16. *In Abstracts of the 6th Cyanobacterial Workshop.*

Fogg, G. E. 1949. Growth and heterocyst production in *Anabaena cylindrica* Lemm. II. In relation to carbon and nitrogen metabolism. *Ann. Bot. N. Ser.* **13**:241–259.

Frías, J. E., E. Flores, and A. Herrero. 1994. Requirement of the regulatory protein NtcA for the expression of nitrogen assimilation and heterocyst development genes in the cyanobacterium *Anabaena* sp. PCC 7120. *Mol. Microbiol.* **14**:823–832.

Frías, J. E., E. Flores, and A. Herrero. 1997. Nitrate assimilation gene cluster from the heterocyst-forming cyanobacterium *Anabaena* sp. strain PCC 7120. *J. Bacteriol.* **179**:477–486.

Gambacorta, A., I. Romano, A. Trincone, A. Soriente, M. Giordano, and G. Sodano. 1996. Heterocyst glycolipids from five nitrogen-fixing cyanobacteria. *Gazz. Chim. Ital.* **126**:653–656.

Gambacorta, A., E. Pagnotta, I. Romano, G. Sodano, and A. Trincone. 1998. Heterocyst glycolipids from nitrogen-fixing cyanobacteria other than Nostocaceae. *Phytochemistry* **48**:801–805.

Geitler, L. 1932. *Cyanophyceae*, Aufl. 2. Akademische Verlagsgesellschaft, Leipzig, Germany.

Giddings, T. H., and L. A. Staehelin. 1981. Observation of microplasmodesmata in both heterocyst-forming and non-heterocyst forming filamentous cyanobacteria by freeze-fracture electron microscopy. *Arch. Microbiol.* **129**:295–298.

Golden, J. 1998. Programmed DNA rearrangements in cyanobacteria, p. 162–173. *In* F. J. de Bruijn, J. R. Lupski, and G. M. Weinstock (ed.), *Bacterial Genomes Physical Structure and Analysis.* Chapman and Hall, New York, N.Y.

Golden, J. W., S. J. Robinson, and R. Haselkorn. 1985. Rearrangement of nitrogen fixation genes during heterocyst differentiation in the cyanobacterium *Anabaena*. *Nature* **314**:419–423.

Golden, J. W., C. D. Carrasco, M. E. Mulligan, G. J. Schneider, and R. Haselkorn. 1988. Deletion of a 55-kilobase-pair DNA element from the chromosome during heterocyst differentiation of *Anabaena* sp. strain PCC 7120. *J. Bacteriol.* **170**:5034–5041.

Häger, K.-P., G. Danneberg, and H. Bothe. 1983.

The glutamate synthase in heterocysts of *Nostoc muscorum*. *FEMS Microbiol. Lett.* **17**:179–183.

Haselkorn, R. 1978. Heterocysts. *Annu. Rev. Plant Physiol.* **29**:319–344.

Haselkorn, R. 1992. Developmentally regulated gene rearrangements in prokaryotes. *Annu. Rev. Genet.* **26**:111–128.

Haselkorn, R. 1995. Molecular genetics of nitrogen fixation in photosynthetic prokaryotes, p. 29–36. *In* I. A. Tikhonovich, N. A. Provogov, V. I. Romanov, and W. E. Newton (ed.), *Nitrogen Fixation: Fundamentals and Applications.* Kluwer Academic Publishers, Dordrecht, The Netherlands.

Haselkorn, R., and W. J. Buikema. 1992. Nitrogen fixation in cyanobacteria, p. 166–190. *In* G. Stacey, R. H. Burris, and H. J. Evans (ed.), *Biological Nitrogen Fixation.* Chapman and Hall, New York, N.Y.

Haselkorn, R., D. Rice, S. E. Curtis, and S. J. Robinson. 1983. Organization and transcription of genes important in *Anabaena* heterocyst differentiation. *Ann. Microbiol.* **134B**:181–193.

Helber, J. T., T. R. Johnson, L. R. Yarbrough, and R. Hirschberg. 1988. Regulation of nitrogenase gene expression in anaerobic cultures of *Anabaena variabilis*. *J. Bacteriol.* **170**:552–557.

Holland, D., and C. P. Wolk. 1990. Identification and characterization of *hetA*, a gene that acts early in the process of morphological differentiation of heterocysts. *J. Bacteriol.* **172**:3131–3137.

Hopwood, D. A., C. Khosla, D. H. Sherman, M. J. Bibb, S. Ebert-Khosla, E.-S. Kim, R. McDaniel, W. P. Revill, R. Torres, and T.-W. Yu. 1993. Toward an understanding of the programming of aromatic polyketide synthases: a genetics-driven approach. *In* R. H. Baltz, G. D. Hegeman, and P. L. Skatrud (ed.), *Industrial Microorganisms: Basic and Applied Molecular Genetics.* American Society for Microbiology, Washington, D.C.

Janson, S., A. Matvayev, and B. Bergman. 1998. The presence and expression of *hetR* in the non-heterocystous cyanobacterium *Symploca* PCC 8002. *FEMS Microbiol. Lett.* **168**:173–179.

Jiang, F., B. Mannervik, and B. Bergman. 1997. Evidence for redox regulation of the transcription factor NtcA, acting both as an activator and a repressor, in the cyanobacterium *Anabaena* PCC 7120. *Biochem. J.* **327**:513–517.

Katayama, M., and M. Ohmori. 1997. Isolation and characterization of multiple adenylate cyclase genes from the cyanobacterium *Anabaena* sp. strain PCC 7120. *J. Bacteriol.* **179**:3588–3593.

Katayama, M., Y. Wada, and M. Ohmori. 1995. Molecular cloning of the cyanobacterial adenylate cyclase gene from the filamentous cyanobacterium *Anabaena cylindrica*. *J. Bacteriol.* **177**:3873–3878.

Khudyakov, I., and C. P. Wolk. Unpublished observations.

Khudyakov, I., and C. P. Wolk. 1996. Evidence that the *hanA* gene coding for HU protein is essential for heterocyst differentiation in, and cyanophage A-4(L) sensitivity of, *Anabaena* sp. strain PCC 7120. *J. Bacteriol.* **178:**3572–3577.

Khudyakov, I., and C. P. Wolk. 1997. *hetC*, a gene coding for a protein similar to bacterial ABC protein exporters, is involved in early regulation of heterocyst differentiation in *Anabaena* sp. strain PCC 7120. *J. Bacteriol.* **179:**6971–6978.

Khudyakov, I. Y., and J. W. Golden. 1998. Involvement of group 2 alternative sigma factors in the regulation of diazotrophic growth of *Anabaena* sp. PCC 7120, abstr. B20. *In Abstracts of the 6th Cyanobacterial Workshop.*

Koksharova, O., and C. P. Wolk. Unpublished observations.

Kotani, H., and S. Tabata. 1998. Lessons from sequencing of the genome of a unicellular cyanobacterium, *Synechocystis* sp. PCC6803. *Annu. Rev. Plant Physiol. Plant Mol. Biol.* **49:**151–171.

Kuritz, T., A. Ernst, T. A. Black, and C. P. Wolk. 1993. High-resolution mapping of genetic loci of *Anabaena* PCC 7120 required for photosynthesis and nitrogen fixation. *Mol. Microbiol.* **8:**101–110.

Lambelot, R. H., A. M. Gehring, R. S. Flugel, P. Zuber, M. LaCelle, M. A. Marahiel, R. Reid, C. Khosla, and C. T. Walsh. 1996. A new enzyme superfamily—the phosphopantetheinyl transferases. *Chem. Biol.* **3:**923–936.

Leganés, F. 1994. Genetic evidence that *hepA* gene is involved in the normal deposition of the envelope of both heterocysts and akinetes in *Anabaena variabilis* ATCC 29413. *FEMS Microbiol. Lett.* **123:**63–67.

Leganés, F., F. Fernández-Piñas, and C. P. Wolk. 1994. Two mutations that block heterocyst differentiation have different effects on akinete differentiation in *Nostoc ellipsosporum*. *Mol. Microbiol.* **12:**679–684.

Leganés, F., F. Fernández-Piñas, and C. P. Wolk. 1998. A transposition-induced mutant of *Nostoc ellipsosporum* implicates an arginine-biosynthetic gene in the formation of cyanophycin granules and of functional heterocysts and akinetes. *Microbiology* **144:**1799–1805.

Liang, J., L. Scappino, and R. Haselkorn. 1992. The *patA* gene product, which contains a region similar to CheY of *Escherichia coli*, controls heterocyst pattern formation in the cyanobacterium *Anabaena* 7120. *Proc. Natl. Acad. Sci. USA* **89:**5655–5659.

Liang, J., L. Scappino, and R. Haselkorn. 1993. The *patB* gene product, required for growth of the cyanobacterium *Anabaena* sp. strain PCC 7120 under nitrogen-limiting conditions, contains ferre-doxin and helix-turn-helix domains. *J. Bacteriol.* **175:**1697–1704.

Luque, I., E. Flores, and A. Herrero. 1994. Molecular mechanism for the operation of nitrogen control in cyanobacteria. *EMBO J.* **13:**2862–2869.

Lynn, M. E., J. A. Bantle, and J. D. Ownby. 1986. Estimation of gene expression in heterocysts of *Anabaena variabilis* by using DNA-RNA hybridization. *J. Bacteriol.* **167:**940–946.

Maldener, I. Personal communication.

Maldener, I., G. Fiedler, A. Ernst, F. Fernández-Piñas, and C. P. Wolk. 1994. Characterization of *devA*, a gene required for the maturation of proheterocysts in the cyanobacterium *Anabaena* sp. strain PCC 7120. *J. Bacteriol.* **176:**7543–7549.

Matveyev, A. V., E. Rutgers, E. Söderbäck, and B. Bergman. 1994. A novel genome rearrangement involved in heterocyst differentiation of the cyanobacterium *Anabaena* sp. PCC 7120. *FEMS Microbiol. Lett.* **116:**201–208.

Meeks, J. C. Personal communication.

Meeks, J. C., E. L. Campbell, and P. S. Bisen. 1994. Elements interrupting nitrogen fixation genes in cyanobacteria: presence and absence of a *nifD* element in clones of *Nostoc* sp. strain Mac. *Microbiology* **140:**3225–3232.

Mitchison, G. J., and M. Wilcox. 1972. Rule governing cell division in *Anabaena*. *Nature* (London) **239:**110–111.

Mitchison, G. J., M. Wilcox, and R. J. Smith. 1976. Measurement of an inhibitory zone. *Science* **191:**866–868.

Muro-Pastor, M. I., and F. J. Florencio. 1994. NADP$^+$-isocitrate dehydrogenase from the cyanobacterium *Anabaena* sp. strain PCC 7120: purification and characterization of the enzyme and cloning, sequencing, and disruption of the *icd* gene. *J. Bacteriol.* **176:**2718–2726.

Muro-Pastor, A., A. Valladares, M. F. Vazquez, A. Herrero, and E. Flores. 1998. The cyanobacterial NtcA regulon: role of NtcA in heterocyst development, abstr. S22. *In Abstracts of the 6th Cyanobacterial Workshop.*

Murry, M. A., and C. P. Wolk. 1989. Evidence that the barrier to the penetration of oxygen into heterocysts depends upon two layers of the cell envelope. *Arch. Microbiol.* **151:**469–474.

Oxelfelt, F., P. Tamagnini, and P. Lindblad. 1998. Hydrogen uptake in *Nostoc* sp. strain PCC 73102. Cloning and characterization of a *hupSL* homologue. *Arch. Microbiol.* **169:**267–274.

Peschek, G. A., M. Wastyn, S. Fromwald, and B. Mayer. 1995a. Occurrence of heme O in photoheterotrophically growing, semi-anaerobic cyanobacterium *Synechocystis* sp. PCC6803. *FEBS Lett.* **371:**89–93.

Peschek, G. A., D. Alge, S. Fromwald, and B.

Mayer. 1995b. Transient accumulation of heme O (cytochrome *o*) in the cytoplasmic membrane of semi-anaerobic *Anacystis nidulans*. Evidence for oxygenase-catalyzed heme O/A transformation. *J. Biol. Chem.* **270:**27937–27941.

Peterson, R. B., and C. P. Wolk. 1978. High recovery of nitrogenase activity and of [55]Fe-labeled nitrogenase in heterocysts isolated from *Anabaena variabilis. Proc. Natl. Acad. Sci. USA* **75:**6271–6275.

Porchia, A. C., and G. L. Salerno. 1996. Sucrose biosynthesis in a prokaryotic organism: presence of two sucrose-phosphate synthases in *Anabaena* with remarkable differences compared with the plant enzymes. *Proc. Natl. Acad. Sci. USA* **93:**13600–13604.

Ramasubramanian, T. S., T.-F. Wei, and J. W. Golden. 1994. Two *Anabaena* sp. strain PCC 7120 DNA-binding factors interact with vegetative cell- and heterocyst-specific genes. *J. Bacteriol.* **176:**1214–1223.

Ramasubramanian, T. S., T.-F. Wei, A. K. Oldham, and J. W. Golden. 1996. Transcription of the *Anabaena* sp. strain PCC 7120 *ntcA* gene: multiple transcripts and NtcA binding. *J. Bacteriol.* **178:**922–926.

Ramaswamy, K. S., C. D. Carrasco, T. Fatma, and J. W. Golden. 1997. Cell-type specificity of the *Anabaena fdxN*-element rearrangement requires *xisH* and *xisI. Mol. Microbiol.* **23:**1241–1249.

Sato, N., and A. Wada. 1996. Disruption analysis of the gene for a cold-regulated RNA-binding protein, *rbpA1*, in *Anabaena*: cold-induced initiation of the heterocyst differentiation pathway. *Plant Cell Physiol.* **37:**1150–1160.

Schilling, N., and K. Ehrnsperger. 1985. Cellular differentiation of sucrose metabolism in *Anabaena variabilis. Z. Naturforsch.* **40c:**776–779.

Schrautemeier, B., U. Neveling, and S. Schmitz. 1995. Distinct and differently regulated Mo-dependent nitrogen-fixing systems evolved for heterocysts and vegetative cells of *Anabaena variabilis* ATCC 29413: characterization of the *fdxH1/2* gene regions as part of the *nif1/2* gene clusters. *Mol. Microbiol.* **18:**357–369.

Smith, G., and J. D. Ownby. 1981. Cyclic AMP interferes with pattern formation in the cyanobacterium *Anabaena variabilis. FEMS Microbiol. Lett.* **11:**175–180.

Soriente, A., G. Sodano, A. Gambacorta, and A. Trincone. 1992. Structure of the "heterocyst glycolipids" of the marine cyanobacterium *Nodularia harveyana. Tetrahedron* **48:**5375–5384.

Soriente, A., A. Gambacorta, A. Trincone, C. Sili, M. Vincenzini, and G. Sodano. 1993. Heterocyst glycolipids of the cyanobacterium *Cyanospira rippkae. Phytochemistry* **33:**393–396.

Soriente, A., T. Bisogno, A. Gambacorta, I. Romano, C. Sili, A. Trincone, and G. Sodano. 1995. Reinvestigation of heterocyst glycolipids from the cyanobacterium, *Anabaena cylindrica. Phytochemistry* **38:**641–645.

Stragier, P., and R. Losick. 1996. Molecular genetics of sporulation in *Bacillus subtilis. Annu. Rev. Genet.* **30:**297–341.

Summers, M. L., J. G. Wallis, E. L. Campbell, and J. C. Meeks. 1995. Genetic evidence of a major role for glucose-6-phosphate dehydrogenase in nitrogen fixation and dark growth of the cyanobacterium *Nostoc* sp. strain ATCC 29133. *J. Bacteriol.* **177:**6184–6194.

Sutherland, J. M., W. D. P. Stewart, and M. Herdman. 1985. Akinetes of the cyanobacterium *Nostoc* PCC 7524: morphological changes during synchronous germination. *Arch. Microbiol.* **142:**269–274.

Thiel, T. 1993. Characterization of genes for an alternative nitrogenase in the cyanobacterium *Anabaena variabilis. J. Bacteriol.* **175:**6276–6286.

Thiel, T. 1994. Genetic analysis of cyanobacteria, p. 581–611. *In* D. A. Bryant (ed.), *The Molecular Biology of Cyanobacteria.* Kluwer Academic Publishers, Dordrecht, The Netherlands.

Thiel, T. 1996. Isolation and characterization of the *vnfEN* genes of the cyanobacterium *Anabaena variabilis. J. Bacteriol.* **178:**4493–4499.

Thiel, T., E. M. Lyons, J. C. Erker, and A. Ernst. 1995. A second nitrogenase in vegetative cells of a heterocyst-forming cyanobacterium. *Proc. Natl. Acad. Sci. USA* **92:**9358–9362.

Thiel, T., E. M. Lyons, and J. C. Erker. 1997. Characterization of genes for a second Mo-dependent nitrogenase in the cyanobacterium *Anabaena variabilis. J. Bacteriol.* **179:**5222–5225.

Thomas, J., J. C. Meeks, C. P. Wolk, P. W. Shaffer, S. M. Austin, and W.-S. Chien. 1977. Formation of glutamine from [^{13}N]ammonia, [^{13}N]dinitrogen, and [^{14}C]glutamate by heterocysts isolated from *Anabaena cylindrica. J. Bacteriol.* **129:**1545–1555.

Tumer, N. E., S. J. Robinson, and R. Haselkorn. 1983. Different promoters for the *Anabaena* glutamine synthetase gene during growth using molecular or fixed nitrogen. *Nature* (London) **306:**337–342.

Walsby, A. E. 1985. The permeability of heterocysts to the gases nitrogen and oxygen. *Proc. R. Soc. Lond. B* **226:**345–366.

Wei, T.-F., T. S. Ramasubramanian, and J. W. Golden. 1994. *Anabaena* sp. strain PCC 7120 *ntcA* gene required for growth on nitrate and heterocyst development. *J. Bacteriol.* **176:**4473–4482.

Wilcox, M. 1970. One-dimensional pattern found in blue-green algae. *Nature* (London) **228:**686–687.

Wilcox, M., G. J. Mitchison, and R. J. Smith. 1973a. Pattern formation in the blue-green alga,

Anabaena. I. Basic mechanisms. *J. Cell Sci.* **12**: 707–725.

Wilcox, M., G. J. Mitchison, and R. J. Smith. 1973b. Pattern formation in the blue-green alga, *Anabaena*. II. Controlled proheterocyst regression. *J. Cell Sci.* **13**:637–649.

Wilcox, M., G. J. Mitchison, and R. J. Smith. 1975. Mutants of *Anabaena cylindrica* altered in heterocyst spacing. *Arch. Microbiol.* **103**:219–223.

Winkenbach, F., and C. P. Wolk. 1973. Activities of enzymes of the oxidative and the reductive pentose phosphate pathways in heterocysts of a blue-green alga. *Plant Physiol.* **52**:480–482.

Winkenbach, F., C. P. Wolk, and M. Jost. 1972. Lipids of membranes and of the cell envelope in heterocysts of a blue-green alga. *Planta* **107**:69–80.

Wolk, C. P. 1964. Experimental studies on the development of a blue-green alga. Ph.D. thesis. Rockefeller Institute, New York, N.Y.

Wolk, C. P. 1967. Physiological basis of the pattern of vegetative growth of a blue-green alga. *Proc. Natl. Acad. Sci. USA* **57**:1246–1251.

Wolk, C. P. 1968. Movement of carbon from vegetative cells to heterocysts in *Anabaena cylindrica*. *J. Bacteriol.* **96**:2138–2143.

Wolk, C. P. 1973. Physiology and cytological chemistry of blue-green algae. *Bacteriol. Rev.* **37**:32–101.

Wolk, C. P. 1982. Heterocysts, p. 359–386. *In* N. G. Carr and B. A. Whitton (ed.), *The Biology of Cyanobacteria*. Blackwell, Oxford, United Kingdom.

Wolk, C. P. 1991. Genetic analysis of cyanobacterial development. *Curr. Opin. Genet. Dev.* **1**:336–341.

Wolk, C. P. 1996. Heterocyst formation. *Annu. Rev. Genet.* **30**:59–78.

Wolk, C. P., and R. Kong. Unpublished observations.

Wolk, C. P., and M. P. Quine. 1975. Formation of one-dimensional patterns by stochastic processes and by filamentous blue-green algae. *Dev. Biol.* **46**: 370–382.

Wolk, C. P., and K. Zarka. 1998. Genetic dissection of heterocyst differentiation, p. 191–196. *In* G. Subramanian, B. D. Kaushik, and G. S. Venkataraman (ed.), *Cyanobacterial Biotechnology. Proceedings of the International Symposium, Sept. 18–21, 1996*. Oxford and IBH Publishing Co., New Delhi, India.

Wolk, C. P., Y. Cai, L. Cardemil, E. Flores, B. Hohn, M. Murry, G. Schmetterer, B. Schrautemeier, and R. Wilson. 1988. Isolation and complementation of mutants of *Anabaena* sp. strain PCC 7120 unable to grow aerobically on dinitrogen. *J. Bacteriol.* **170**:1239–1244.

Wolk, C. P., J. Elhai, T. Kuritz, and D. Holland. 1993. Amplified expression of a transcriptional pattern formed during development of *Anabaena*. *Mol. Microbiol.* **7**:441–445.

Wolk, C. P., A. Ernst, and J. Elhai. 1994. Heterocyst metabolism and development, p. 769–823. *In* D. Bryant (ed.), *Molecular Genetics of Cyanobacteria*. Kluwer Academic Publishers, Dordrecht, The Netherlands.

Wolk, C. P., J. Zhu, and R. Kong. 1999. Genetic analysis of heterocyst formation, p. 509–515. *In* G. A. Peschek, W. Löffelhardt, and G. Peschek (ed.), *The Prototrophic Prokaryotes*. Kluwer Academic Publishers, New York, N.Y.

Xu, X. Personal communication.

Xu, X., and C. P. Wolk. Unpublished observations.

Xu, X., I. Khudyakov, and C. P. Wolk. 1997. Lipopolysaccharide dependence of cyanophage sensitivity and aerobic nitrogen fixation in *Anabaena* sp. strain PCC 7120. *J. Bacteriol.* **179**:2884–2891.

Yoon, H.-S., and J. W. Golden. 1998. Heterocyst pattern formation controlled by a diffusible peptide. *Science* **282**:935–938.

Zhang, C.-C. 1993. A gene encoding a protein related to eukaryotic protein kinases from the filamentous heterocystous cyanobacterium *Anabaena* PCC 7120. *Proc. Natl. Acad. Sci. USA* **90**: 11840–11844.

Zhang, C.-C. 1996. Bacterial signalling involving eukaryotic-type protein kinases. *Mol. Microbiol.* **20**: 9–15.

Zhang, C.-C., and L. Libs. 1998. Cloning and characterisation of the *pknD* gene encoding an eukaryotic-type protein kinase in the cyanobacterium *Anabaena* sp. PCC7120. *Mol. Gen. Genet.* **258**:26–33.

Zhang, C.-C., S. Huguenin, and A. Friry. 1995. Analysis of genes encoding the cell division protein FtsZ and a glutathione synthetase homologue in the cyanobacterium *Anabaena* sp. PCC 7120. *Res. Microbiol.* **146**:445–455.

Zhang, C.-C., A. Friry, and L. Peng. 1998. Molecular and genetic analysis of two closely linked genes that encode, respectively, a protein phosphatase 1/2A/2B homolog and a protein kinase homolog in the cyanobacterium *Anabaena* sp. strain PCC 7120. *J. Bacteriol.* **180**:2616–2622.

Zhou, R., Z. Cao, and J. Zhao. 1998a. Characterization of HetR protein turnover in *Anabaena* sp. PCC 7120. *Arch. Microbiol.* **169**:417–423.

Zhou, R., X. Wei, N. Jiang, H. Li, Y. Dong, K.-L. Hsi, and J. Zhao. 1998b. Evidence that HetR protein is an unusual serine-type protease. *Proc. Natl. Acad. Sci. USA* **95**:4959–4963.

Zhu, J., K. Zarka, and C. P. Wolk. Unpublished observations.

Zhu, J., R. Kong, and C. P. Wolk. 1998. Regulation of *hepA* of *Anabaena* sp. strain PCC 7120 by elements 5′ from the gene and by *hepK*. *J. Bacteriol.* **180**:4233–4242.

THE PALEOBIOLOGIC RECORD OF CYANOBACTERIAL EVOLUTION

J. William Schopf

5

During recent decades, major strides have been made toward deciphering the fossil record of prokaryotic life. Hundreds of fossiliferous units have been discovered, containing a large number of formally described microbial taxa—dominantly, but not exclusively, cyanobacterial—and the documented antiquity of life has been extended to an age more than three-quarters that of the Earth. Taken together, the paleontological, geological, and isotopic geochemical evidence indicates that mat-building (stromatolitic) microbial ecosystems, probably including cyanobacteria, photosynthetic bacteria, and other members of the bacterial domain, were extant ~3.5 billion years ago, methanogenic archaeans existed by about 2.8 billion years ago, and gram-negative sulfate-reducing bacteria existed at least as early as 2.7 billion years ago (Fig. 1). On what are these inferences based? How firm are these dates? After addressing those interrelated issues, this chapter considers one of the broadest questions of microbial evolutionary development, namely, what does the paleobiological record suggest about the tempo and mode of prokaryotic, especially cyanobacterial, evolution?

J. W. Schopf, IGPP Center for the Study of Evolution and the Origin of Life, Department of Earth and Space Sciences, and Molecular Biology Institute, University of California, Los Angeles, CA 90095-1567

TRACING THE ROOTS OF PROKARYOTIC LINEAGES

Historical Perspective

By the mid-1800s, when Darwin unveiled his grand theory, major features of the Phanerozoic history of life, spanning the most recent several hundred million years, had already been deciphered. The familiar progressions from seaweeds to land plants, from marine invertebrates to higher mammals, provided Darwin a fossil-based foundation for his unifying thesis. But what were the forerunners of the early algae and primitive invertebrates at the base of Darwin's evolutionary tree? How did evolution proceed during the Precambrian Eon, the seven-eighths of Earth history that came before the oldest fossils then known? To these questions, Darwin had no answers. Precambrian biological history was a total mystery. But Darwin did know that if answers were not forthcoming, his theory was in jeopardy:

> If the theory [of evolution] be true it is indisputable that before the lowest Cambrian stratum was deposited . . . the world swarmed with living creatures. [Yet] to the question why we do not find rich fossiliferous deposits belonging to these earliest periods . . . I can give no satisfactory answer. The case at present must remain inexplicable; and may be truly urged as a valid argument against the views here entertained. (Darwin, 1859, chapter X)

Prokaryotic Development, edited by Y. V. Brun and L. J. Shimkets,
© 2000 American Society for Microbiology, Washington, DC 20005-4171

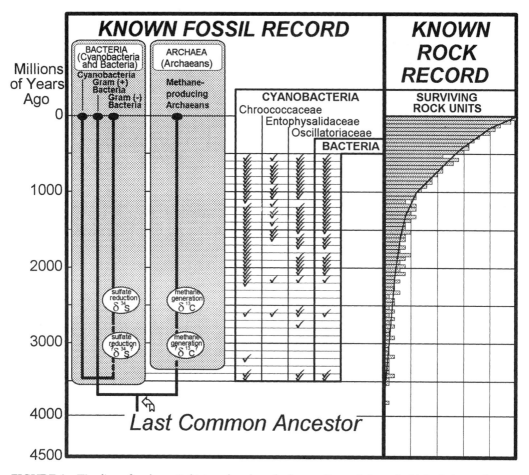

FIGURE 1 Timeline of prokaryotic history, based on the known Precambrian paleobiological record, compared with an estimate of the temporal distribution of geologic units that have survived to the present; check marks denote the presence of fossils inferred to probably belong to the groups listed in each of the 50-million-year-long segments indicated. (Data from Hofmann and Schopf, 1983; Schopf and Walter, 1983; Schopf et al., 1983; Mendelson and Schopf, 1992; and Schopf, 1992b, 1992c.)

Though Darwin posed the problem, it was not until a century later—in the mid-1960s with the birth of a new field of science, Precambrian paleobiology—that this earliest missing fossil record began to be uncovered (Cloud, 1983; Schopf, 1992a, 1999). Progress since the 1960s has been impressive. The documented record of life has been extended steadily and now reaches to nearly 3.5 billion years ago, a date in the geologic past approaching the age (~4.5 billion years) of the Earth itself. As the early fossil record came increasingly into focus, a parallel series of advances was being made in molecular biology. By the late 1980s, these had coalesced to produce the first rRNA phylogenic trees, powerful new means to decipher evolutionary relations among organisms of the modern world (Woese, 1987). Now, as the 1990s draw to a close, the time is ripe to couple the paleobiological and molecular data and bring the two into consonance. Toward that end, assessment is made here of what the available fossil evidence seems to show about the timing and nature of early evolutionary development.

Limitations of the Molecular and Fossil Records

As grouped in rRNA phylogenic trees, all organisms living today are members of only three major clusters (formally, domains [Woese et al., 1990])—the *Eucarya* (defined generally as organisms having cells in which chromosomes are encapsulated in a membrane-bounded nucleus); the *Archaea* (nonnucleated microorganisms, including methanogenic prokaryotes and many extremeophiles, microbes that thrive in exceptionally acidic high-temperature settings); and the *Bacteria*, the domain that includes the cyanobacteria and bacteria of traditional classifications. Despite uncertainties of interpretation of the relevant molecular data (introduced, for example, by lateral gene transfer), rRNA trees can be regarded as providing a plausible, and for the most part reliable, index of relatedness among living organisms and, thus, of the order of branching of lineages that have survived to the present. But because of numerous complexities (including variation in the apparent rate of evolutionary change of the various lineages [Woese, 1987]), there is no clear-cut way to use rRNA trees as chronometers to indicate precisely when in the geologic past particular branches emerged. And, of course, such trees can provide no direct data about either the existence or rate of change of lineages now extinct, long-lost branches of the evolutionary tree that (among eukaryotes, at least) are the rule rather than the exception.

The fossil record is also not without its problems. Accurate dating of the times of origin of the major biological groups has been a long-standing goal in Precambrian paleobiology (Schopf, 1970, 1992b; Schopf et al., 1983). This is a young field, however, and the early fossil record is too incompletely known to provide precise answers. Moreover, even under the best circumstances, fossil evidence can record only the first detected occurrence of a lineage, not its first actual occurrence. Together, fossils and associated geological and geochemical indicators of biological activity (that is, paleobiological evidence) can establish a minimum age for a lineage, but they cannot reveal how much earlier the lineage actually existed.

How firmly established is the early fossil evidence? How far back in time can the major lineages be traced?

PRECAMBRIAN CYANOBACTERIA

The best-documented early branch of the tree of life is the cyanobacterial lineage, represented in the Precambrian fossil record by naked or ensheathed solitary and colonial unicells (referred chiefly to the living cyanobacterial family *Chroococcaceae* [Fig. 2]) and cellular microscopic filaments (for the most part regarded as belonging to the *Oscillatoriaceae*, taxonomically the most diverse extant cyanobacterial family [Fig. 3]). Well-preserved examples are essentially indistinguishable in morphology from modern cyanobacteria, a near identity first recognized some three decades ago on the basis of detailed studies of the stromatolitic microbial community of the ~850-million-year-old Bitter Springs Formation of central Australia (Schopf, 1968; Schopf and Blacic, 1971). Since that time, such similarities have been observed so often that it is now common practice among Precambrian paleobiologists to name fossil taxa after their modern morphological counterparts by adding appropriate prefixes (palaeo- or eo-) or suffixes (-opsis or -ites) to the names of present-day cyanobacterial genera (Mendelson and Schopf, 1992; Schopf, 1994a). More than 40 such namesakes (for example, *Palaeoanacystis*, *Eophormidium*, *Aphanocapsaopsis*, and *Oscillatorites*) are in use worldwide.

The marked similarity of Precambrian microfossils to modern cyanobacteria is characteristic of a wide range of morphological types, not only of taxa referred to the *Chroococcaceae* (Fig. 4F) and *Oscillatoriaceae* (Fig. 4B and D) but to other families as well. For example, the 2.15-billion-year-old colonial fossil

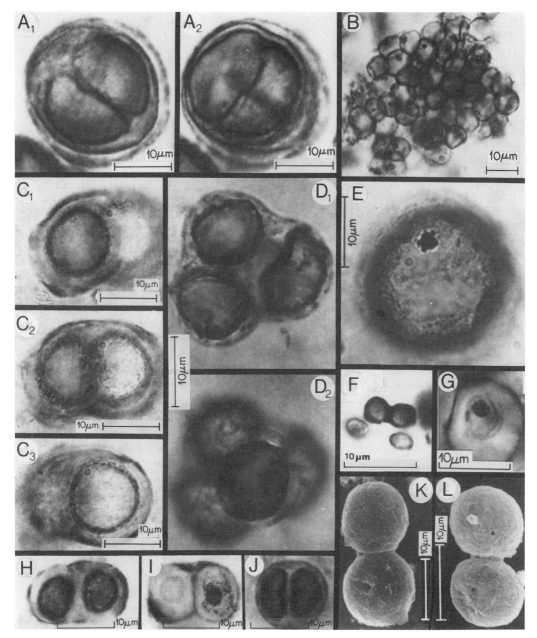

FIGURE 2 Permineralized Precambrian chroococcaceans in petrographic thin sections (A and C to J) and acid-resistant residues (B, K, and L) of flat-laminated stromatolitic chert from the ~850-million-years-old Bitter Springs Formation of Northern Territory, Australia, shown in optical photomicrographs (A to J) and scanning electron micrographs (K and L); (A_1 and A_2, C_1 to C_3, and D_1 and D_2) single specimens at differing focal depths. (A) *Bigeminococcus lamellosus*; (B) *Glenobotrydion aenigmatis*; (C) *Eozygion grande*; (D) *Eotetrahedrion princeps*; (E) *Globophycus rugosum*; (F) *Sphaerophycus parvum*; (G) *Caryosphaeroides pristina*; (H and J) *Eozygion minutum*; (I) *Caryospheroides tetras*; (K and L) unnamed paired cells. (After Schopf, 1968, 1972; Schopf and Blacic, 1971.)

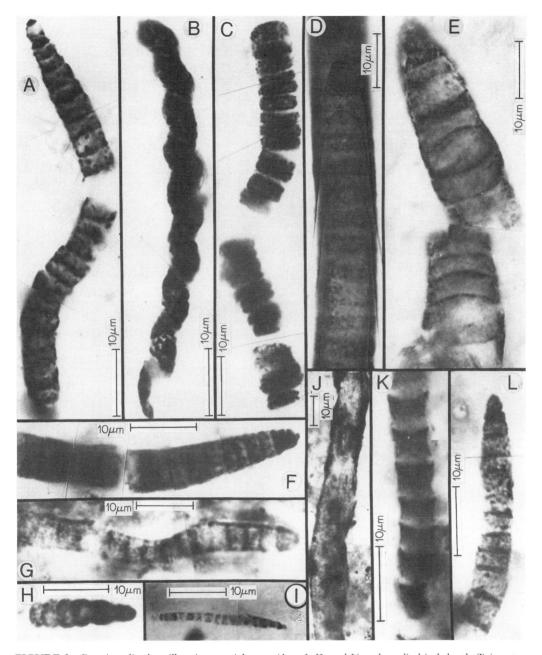

FIGURE 3 Permineralized oscillatoriacean trichomes (A to I, K, and L) and a cylindrical sheath (J) in petrographic thin sections of flat-laminated stromatolitic chert from the ~850-million-year-old Bitter Springs Formation of Northern Territory, Australia. Because of the sinuous, three-dimensional preservation of these petrified carbonaceous microfossils, all except the specimen in panel J are shown in composite photomicrographs. (A, F, and L) *Cephalophytarion laticellulosum*; (B) *Heliconema funiculum*; (C) *Oscillatoriopsis breviconvexa*; (D) unnamed *Oscillatoria*-like trichome; (E) *Obconicophycus amadeus*; (G) *Oscillatoriopsis obtusa*; (H) *Filiconstrictosus diminutus*; (I) *Cephalophytarion minutum*; (J) *Siphonophycus kestron*; (K) *Halythrix nodosa*. (After Schopf, 1968, 1974; Schopf and Blacic, 1971.)

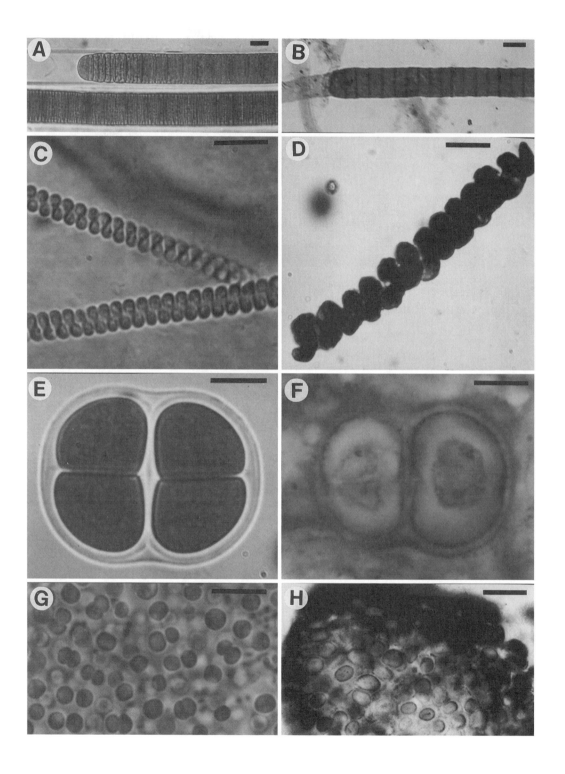

Eoentophysalis belcherensis (Fig. 4H) of North-west Territories, Canada (Hofmann, 1976), is so similar to modern *Entophysalis* (Fig. 4G) that it has been placed in the extant cyanobacterial family *Entophysalidaceae* (Golubic and Hofmann, 1976). The similarities are indeed striking: not only are the fossil and modern species morphologically indistinguishable (in cell shape and in form and arrangement of originally mucilaginous cell-encompassing envelopes), but they exhibit similar frequency distributions of dividing cells and essentially identical patterns of cellular development (resulting from cell division in three perpendicular planes), form microtexturally similar stromatolitic structures in comparable intertidal to shallow marine environmental settings, undergo essentially identical postmortem degradation sequences, and occur in comparable microbial communities, similar both in species composition and in overall diversity (Golubic and Hofmann, 1976).

Fossil and living pleurocapaceans have also been compared in detail. *Polybessurus bipartitus*, first reported from ~770-million-year-old stromatolites of South Australia (Fairchild, 1975; Schopf, 1977), is a morphologically distinctive, gregarious, cylindrical fossil pleurocapsacean composed of nested cup-shaped envelopes often extended into long tubes oriented perpendicular to the substrate (Fig. 5). Specimens of this taxon occurring in rocks of about the same age in East Greenland have been described as being "a close morphological, reproductive, and behavioral counterpart" to populations of a species of the pleurocapsacean *Cyanostylon* present "in Bahamian environments similar to those in which the Proterozoic fossils occur" (Green et al., 1987, p. 928). A second fossil pleurocapsacean described from the ~770-million-year-old South Australian deposit (*Palaeopleurocapsa wopfnerii*) has been compared with its living morphological and ecological analogue (*Pleurocapsa fuliginosa*) and interpreted as "futher evidence of the evolutionary conservatism of [cyanobacteria]" (Knoll et al., 1975, p. 2492). And two other species of morphologically distinct fossil pleurocapsaceans (the endolithic taxa *Eohyella dichotoma* and *Eohyella rectroclada*), regarded as "compelling examples of the close resemblance between Proterozoic prokaryotes and their modern counterparts" (Knoll et al., 1986, p. 857), have been described from the East Greenland geologic sequence as being "morphologically, developmentally, and behaviorally indistinguishable" from living *Hyella* of the Bahama Banks (Green et al., 1988, p. 837–838).

The marked similarity between fossil and living look-alikes is shown also by quantitative morphological comparisons of large assemblages of Precambrian cyanobacterium-like fossils and modern cyanobacterial species and varieties, studies that, because they are based solely on morphometrics, are relatively immune from intepretive biases such as those reflected, for example, by the use of cyanobacterium-derived generic epithets for fossil taxa.

◄——————

FIGURE 4 Modern cyanobacteria (A, C, E, and G) from mat-building stromatolitic communities of Baja, Mexico, compared with their Precambrian morphological counterparts (B, D, F, and H) (bars, 10 μm). (A) *Lyngbya aestuarii* (*Oscillatoriaceae*), encompassed by a cylindrical mucilagenous sheath; (B) *Palaeolyngbya helva*, a similarly ensheathed oscillatoriacean, shown in an acid-resistant residue of carbonaceous siltstone from the ~950-million-year-old Lakhanda Formation of the Khabarovsk region of Siberia, Russia. (C) *Spirulina subsalsa* (*Oscillatoriaceae*); (D) *Heliconema turukhania*, a *Spirulina*-like oscillatoriacean shown in an acid-resistant residue of carbonaceous siltstone from the ~850-million-year-old Miroedikha Formation of the Turukhansk region of Siberia, Russia. (E) *Gloeocapsa* cf. *repestris* (*Chroococcaceae*), a four-celled colony having a thick distinct encompassing sheath; (F) *Gloeodiniopsis uralicus*, a similarly sheath-enclosed four-celled colonial chroococcacean shown in petrographic thin section of bedded chert from the ~1.5-billion-year-old Satka Formation of southern Bashkiria, Russia. (G) *Entophysalis* cf. *granulosa* (*Entophysalidaceae*); (H) *Eoentophysalis belcherensis*, an *Entophysalis*-like colonial entophysalidacean from stromatolitic chert of the ~2.15-billion-year-old Belcher Group of Northwest Territories, Canada. (After Schopf, 1994a.)

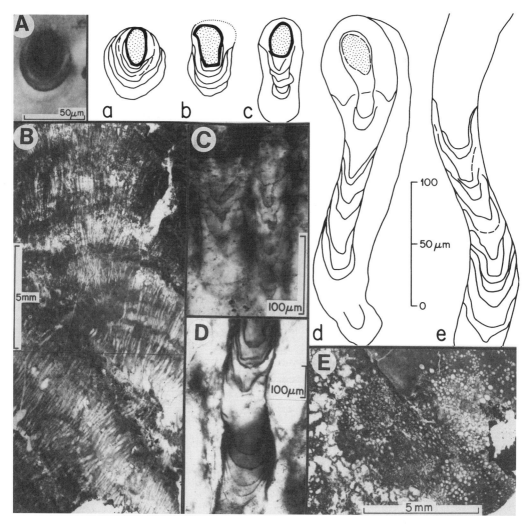

FIGURE 5 Permineralized carbonaceous specimens of the colonial pleurocapsacean *Polybessurus bipartitus* in petrographic thin sections of domical stromatolitic chert from the ~770-million-year-old Skillogalee Dolomite of South Australia, showing stages in its life cycle (a through e). (A) Single ellipsoidal cell; vertical (B) and horizontal (E) sections though a pincushion-like colony; (C and D) petrified, originally mucilagenous stalks like those reconstructed in panels d and e. (After Schopf, 1999.)

Morphometric data for 10 characters present in well-preserved fossils (such as cell size, shape, and range of variability; colony form; and sheath thickness and structure) were compiled for more than 600 species and varieties of living cyanobacteria (Schopf, 1992c) as well as for a worldwide sample of Precambrian cyanobacterium-like fossils (Mendelson and Schopf, 1992), both spheroidal (1,400 occurrences of

named taxa in 260 geologic formations) and filamentous (650 occurrences in 160 formations). To avoid redundancy introduced by variations in taxonomic practice, fossils having the same or similar morphologies (regardless of their formally assigned Latin binomials) were clustered together as informal species-level groupings designed to have ranges of morphological variability comparable to those of living cyanobacterial species (Schopf, 1992d). Virtually all of the 263 fossil morphotypes thus recognized are referable to living genera of cyanobacteria, and more than 60% are essentially identical to modern species—25% to modern coccoid taxa, chiefly chroococcaceans, and 37% to living oscillatoriaceans (Schopf, 1992d).

The fossils included in this extensive morphometric study are preserved either in a petrified (permineralized) state in cherty portions of carbonate stromatolites or as compressed organic films in clastic shales or siltstones, strata that, where studied in appropriate detail, have been interpreted to represent relatively shallow-water, coastal, and for the most part marine environments (ranging from sabkhas and lagoons to mud flats and intertidal carbonate platforms). Microbial mat-building communities in such settings today are almost always dominated by members of the *Oscillatoriaceae* and *Chroococcaceae* (Golubic, 1976a, 1976b; Pierson et al., 1992), the same two families inferred to be present in the same environments during the Precambrian. Evidently, "the biological constitution of [ancient] microbial mat communities was probably quite similar to that of modern communities in comparable environments" (Knoll, 1985, p. 411).

Studies of microbial morphometrics have also provided means to differentiate between fossil cyanobacteria and most noncyanobacterial prokaryotes. For example, virtually all of the 43 species of archaeans now recognized (Stetter, 1996) are morphologically distinctive (occurring as lobed cocci, rods in clusters, discs, discs with fibers, and so forth) and readily distinguishable from cyanobacteria. Archaeans also tend to be decidedly smaller than most cyanobacteria, as do most other noncyanobacterial prokaryotes. The median diameter of extant coccoidal noncyanobacterial members of the bacterial domain is <1 μm, about one-fourth that of cyanobacterial analogues; that of cellular filamentous taxa is ~1.6 μm, less than one-third that of the cellular trichomes of filamentous cyanobacteria; and the median diameter of the tubular sheaths of noncyanobacterial prokaryotic filaments (and of elongate bacterial threadlike cells that, preserved as fossils, might be confused with such sheaths) is ~1.1 μm, about one-sixth that of the tubular sheaths of extant oscillatoriaceans (Schopf, 1992c, 1996a). Because some small cyanobacteria closely resemble "large" noncyanobacterial microbes (with the largest falling well within size ranges typical of cyanobacteria), referral of Precambrian fossils to the cyanobacteria can be quite uncertain. This caveat, however, applies to only a few percent of fossils now known; microbial mimicry poses a problem, but a minor one (Schopf, 1999).

However, and despite the cascade of fossil discoveries over recent years, the known record of Precambrian microbes remains spotty and incomplete, especially from earliest Earth history (Fig. 1). Even Precambrian cyanobacteria, the best-documented branch of early life, have a scanty fossil record, for spheroidal species (mostly chroococcaceans) amounting to fewer than 50 taxonomic occurrences per 50-million-year-long interval and for filamentous forms (oscillatoriaceans) amounting to fewer than 25 occurrences (Mendelson and Schopf, 1992; Schopf, 1992d). Also, several extant cyanobacterial groups are all but unknown among early fossils. Nitrogen-fixing heterocystous varieties, for example, are especially poorly represented, known only indirectly from elongate spore-like fossils such as *Archaeoellipsoides* (Horodyski and Donaldson, 1980) that date back to about 1.5 billion years ago and closely resemble akinetes of living heterocystous taxa (Golubic et al., 1995). The existence by the mid-Precambrian of other distinctive developmental cell types can similarly be indirectly inferred. For example, fossil baeocytes of this age have not been identified per se, but endolithic *Hyella*-like pleurocapsaceans, presumably baeocyte producers, date back to at least ~1.7 billion years ago (Zhang and Golubic, 1987). As noted above, such oldest detected occurrences are certain to be younger than the first actual occurrences, but the known fossil record is too sparse to reveal how much earlier such structures and the cyanobacterial lineages they reflect existed. Nevertheless, and though much more evidence would be needed to sort out rapid evolution like that of the Phanerozoic, even the slim fossil record now known is fully sufficient to show *lack* of change, maintenance of a status quo, over geologically long periods.

HOW ANCIENT ARE THE CYANOBACTERIA?

As shown in Fig. 1, the documented microbial fossil record is abundant, diverse, and continuous back to about 2.2 billion years ago. However, only a few fossiliferous deposits have been discovered in older geologic terrains, and their microbial assemblages tend to be meager and poorly preserved (see, for example, Altermann and Schopf, 1995). Though much of the older rock record is paleontologically unexplored, and lack of study therefore partly explains the lack of data, these deficiencies are largely a result of normal geologic processes. Because of erosion and geological recycling, few rock units from these earlier times have survived to the present (Fig. 1), and most of these are so severely metamorphosed that whatever fossil remnants they may have once contained have been obliterated. Significantly, however, the oldest microbial community now known—nearly 3.5 billion years in age—is also the most diverse and among the best preserved of these very ancient assemblages. This fossil find, from the Apex chert of northwestern Western Australia, holds the key to understanding early stages in the history of life and, perhaps, of the cyanobacterial lineage.

Microbial Fossils of the ~3.5-Billion-Year-Old Apex Chert

Though quite a number of reports of exceedingly ancient "microfossils" are open to question (Schopf and Walter, 1983), such uncertainty does not apply to the fossil microbes of the conglomeratic Apex chert (Schopf, 1992e, 1993).

- The geographic and stratigraphic source of the fossiliferous horizon is known with certainty (fossil-bearing samples having been collected from outcrops at the locality by numerous workers on multiple occasions).

- The ~3.465-billion-year age of the bedded chert (specifically, >3.458 billion ± 1.9 million and <3.471 billion ± 5 million years) is tightly constrained, based on U-Pb zircon dating of immediately overlying and stratigraphically underlying units of the rock sequence (Blake and McNaughton, 1984; Thorpe et al., 1992).

- The petrified cellular fossils are demonstrably indigenous to the chert (as shown by their presence in petrographic thin sections [Fig. 6]), preserved in transported and redeposited rounded pebbles that are unquestionably syngenetic with deposition of the silicified sedimentary unit (a mode of occurrence indicating that the fossils themselves are even older than the conglomeratic bed in which the pebbles occur).

- The morphological complexity and carbonaceous composition of the fossils establish their biogenicity: 11 filamentous species have been identified, ranging from 0.5 to 19.5 μm in diameter and exhibiting flat, rounded, conical, or muffin-shaped terminal cells; quadrate, disc-shaped, or barrel-shaped medial cells; taxon-specific degrees of filament attenuation; and evidence of cell division evidently identical to that of living filamentous prokaryotes.

Because of its great age, cellular preservation, and evident biological diversity, the Apex assemblage is especially noteworthy—indeed, the next-oldest cellularly preserved comparably diverse microbiotas (Barghoorn and Tyler, 1965; Hofmann, 1976; Altermann and Schopf, 1995) are a billion or more years younger. Thus, the Apex fossils provide an unparalleled glimpse of the evolutionary status of the very early biosphere. Of the 11 species identified in the assemblage, 7 (comprising ~63% of measured specimens; Schopf, 1993) seem particu-

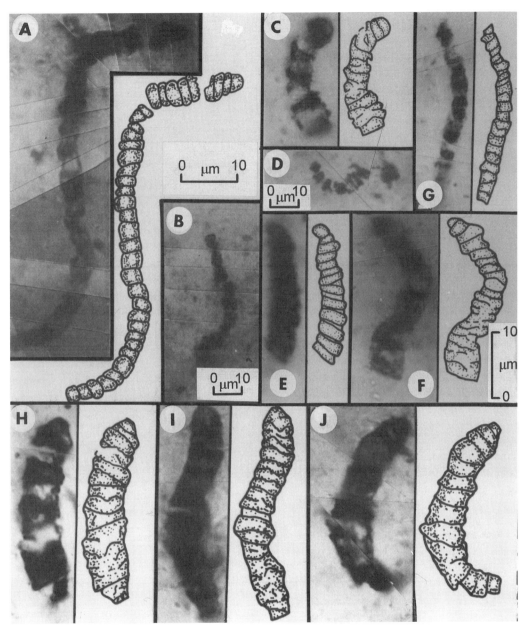

FIGURE 6 Optical photomicrographs and interpretive drawings showing cellular, carbonaceous, cyanobacter-ium-like filaments in petrographic thin sections of the ~3.465-billion-year-old Apex chert of northwestern Western Australia. Because the petrified microbes are three dimensional and sinuous, composite photomicrographs have been used to show the specimens in panels A, B, D, and F through J; the scale in panel F shows the magnification of all fossils, except as otherwise indicated in panels A, B, and D. (A and B) *Primaevifilum amoenum*; (C through F) *Archaeoscillatoriopsis disciformis*; (G) *Primaevifilum delicatulum*; (H, I, and J) *Primaevifilum conicoterminatum*. (After Schopf, 1993.)

larly similar in cellular organization to oscillatoriacean cyanobacteria, having size ranges, median dimensions, and patterns of size distribution that are much more like oscillatoriaceans than other filamentous prokaryotes. Several of these are essentially indistinguishable from common especially well-known taxa of the family, both fossil (*Oscillatoriopsis* spp.) and modern (*Oscillatoria* spp.).

The possible presence of oscillatoriaceans in the Apex assemblage is also consistent with four other lines of evidence. First, models of the early global ecosystem and trace element (cerium and europium) concentrations in Apex-age sediments indicate that aerobic respiration and O_2-producing photosynthesis—processes characteristic of cyanobacteria, with the latter not exhibited by any other prokaryotes—had probably evolved by this early stage in Earth history (Towe, 1990, 1991). Second, as discussed later in this chapter, the isotopic compositions of organic and carbonate carbon in the Apex chert and associated sedimentary units evidence the occurrence of photosynthetic CO_2 fixation like that occurring in extant cyanobacterial populations grown in CO_2-rich environments (Schopf, 1993, 1994b). Third, the occurrence within numerous Apex filaments of bifurcated cells and cell pairs almost certainly reflects the original presence of partial septations and, thus, cell division like that of extant oscillatoriaceans (Schopf, 1992e, 1993). Fourth, rRNA trees of the cyanobacterial lineage place the *Oscillatoriaceae* among the earliest evolved of the extant families (Giavannoni et al., 1988; Wilmotte, 1994).

Taken together, these various lines of evidence seem persuasive. There is little doubt that if the Apex filaments had been discovered in younger Precambrian sediments, where fossil oscillatoriaceans are well known and widespread, or if they had been detected in a modern microbial community and morphology were the only criterion by which to infer biological relations, most would be interpreted as oscillatoriacean cyanobacteria. Nevertheless, at the present state of knowledge they are marooned in time, separated from the next-youngest fossils interpreted as oscillatoriacean by hundreds of millions of years (Fig. 1). For this reason, the Apex taxa were formally classified as "prokaryotes incertae sedis" (Schopf, 1993); unless (or until) new discoveries fill in the gap between them and the younger fossil record and show them to be linked firmly to the cyanobacterial evolutionary continuum that by current data begins about 2.2 billion years ago, even the Apex taxa that most closely resemble oscillatoriaceans are probably best regarded as only cyanobacterium-like.

PALEOBIOLOGY: FOSSILS, GEOLOGY, AND ISOTOPIC GEOCHEMISTRY

Traditionally, the science of paleontology has dealt solely with fossils of the Phanerozoic (the most recent 545 million years of Earth history) and focused almost entirely on aspects of organismal biology—morphology, anatomy, community structure, paleoecology, evolutionary relations, and the like. Other paleobiological issues have drawn less attention, among them biotic interactions with long-term environmental change and the physiology and biochemistry of ancient life. The Precambrian Eon, however, is nearly an order of magnitude longer than the Phanerozoic, and events that are minor or even imperceptible over Phanerozoic time can have cumulative impact if they span appreciable portions of earlier Earth history. Similarly, unlike the Phanerozoic history of life, Precambrian evolution centered largely on metabolism and the development of intracellular biochemical processes (Schopf, 1996b). But because prokaryotes are highly diverse physiologically, and those having similar morphologies can differ markedly in metabolism, the sorting out of such capabilities in Precambrian microbes requires more than traditional fossil-focused paleontology. This need has been met in Precambrian paleobiology by the addition of data and insights drawn from geology (mineralogy) and organic and stable-isotopic geochemistry.

Geologic Evidence of Cyanobacterial Photosynthesis

The capability to carry out oxygen-producing photosynthesis is a universal characteristic of cyanobacteria that distinguishes them from all other prokaryotes. If the Apex assemblage actually includes cyanobacteria, geologic sequences of Apex age can be expected to contain evidence both of the reactants required for oxygenic photosynthesis (H_2O and CO_2) and the products produced (reduced organic carbon and O_2).

The reactants are well evidenced in the early rock record. Liquid water was abundant: the Apex chert, for example, is a water-laid conglomeratic deposit, part of an interbedded sequence of volcanic and sedimentary rock (Groves et al., 1981) laid down in a broad shallow sea fringed with scattered volcanic islands, mud flats, and evaporitic lagoons (Barley et al., 1979). There is also ample evidence of the other reactant of photosynthesis, carbon dioxide, in the form of $CaCO_3$-rich limestones deposited by reaction of dissolved CO_2 (present as bicarbonate, HCO_3^-) and Ca^{2+}, derived from weathering of the land surface.

The two products of cyanobacterial photosynthesis, organic matter and O_2, are also evidenced in the early rock record. Particulate degraded organic carbon (kerogen) is ubiquitous and abundant in sediments of the Apex sequence (ranging from ~0.5 to ~0.8% by weight [Strauss and Moore, 1992]), and at least small amounts of free oxygen were present in the environment, reflected by the occurrence of iron-oxide-rich sedimentary units known as banded iron formations (BIFs), the world's major source of iron ore. Units of this type are widespread in geologic terrains older, but not younger, than about 2.0 billion years; together with uranium-rich pyritic conglomerates, they provide telling evidence of the early history of the atmosphere.

The banding in BIFs is produced by an alternation of iron-rich and iron-poor layers, and because the ferruginous beds are composed of fine rustlike particles of hematite (Fe_2O_3) and, in some deposits, magnetite (Fe_3O_4), BIFs have a characteristic dull- to bright-red color. The iron evidently owes its origin to the hydrothermal circulation of seawater through oceanic crust, primarily at deep submarine ridge systems. In a dissolved (ferrous) state, it then circulated upward into shallower portions of the water column—in some basins evidently seasonally, giving rise to the distinctive millimetric banding—where it was oxidized to ferric iron by reaction with dissolved molecular oxygen and precipitated as a fine rusty rain of minute insoluble iron oxide particles.

The Earth's atmosphere is chiefly an accumulated product of volcanic outgassing of the planetary interior over geologic time, but unlike other principal components of the atmosphere (N_2, H_2O, and CO_2), molecular oxygen is not released from rocks when they are heated. Of various possible sources of O_2 (such as thermal dissociation or UV-induced photodissociation of water vapor), oxygenic photosynthesis appears to be the only quantitatively plausible source for the enormous amount of oxygen sequestered in Precambrian BIFs (Holland, 1994). The existence of these deposits, however, does not indicate that the environment was oxygen rich. On the contrary, BIF-containing basins are large, typically several hundred kilometers in length and breadth, and dissolved ferrous iron could not have been distributed over such extensive areas unless at least the lower portions of the water column were anoxic. Thus, the continuous presence of abundant widespread BIFs >3.5 to ~2.0 billion years ago has been interpreted to indicate that although O_2 was being pumped into the environment by oxygenic (cyanobacterial) photosynthesis, neither the oceanic water column nor the overlying atmosphere was fully oxygenated because of removal of oxygen from the system by its burial in sedimented iron oxides (Klein and Beukes, 1992; Holland, 1994).

The existence of a low-O_2 environment up to ~2.0 billion years ago is also consistent with other geologic evidence. The resistance of some minerals to weathering is strongly affected by the presence of molecular oxygen. Uraninite (UO_2) and pyrite (FeS_2), both minerals which are oxidized and then dissolved in the presence of O_2, are good examples. Extensive detrital accumulations of these minerals do not occur in sediments today because the minerals are easily destroyed during weathering in the present oxygen-rich atmosphere. Nevertheless, large sedimentary ore deposits that contain detrital pebbles of uraninite and pyrite occur in geologic units older, but not younger, than ~2.3 billion years. The persistence of pebbles of uraninite and pyrite during weathering, riverine transport, and deposition in conglomeratic deposits more than 2.3 billion years ago is consistent with a low-O_2 atmosphere up to that time, as are other geologic data, most notably the mineralogy of paleosols (ancient soil horizons) of this age. These and related lines of evidence support the conclusion that the O_2 content of the atmosphere increased dramatically between 2.2 and 1.9 billion years ago, evidently from ≤ 1 to $\geq 15\%$ of the present atmospheric level (Holland, 1994).

Taken together, the geologic and paleontologic evidence thus seems to point to the presence of oxygen-producing cyanobacteria as early as ~3.5 billion years ago. As discussed below, the available isotopic geochemical data add support to this interpretation and, like the known record of cellularly preserved microbial fossils (Fig. 1), evidence the presence of other early-evolved prokaryotic lineages as well.

Isotopic Inferences of Ancient Biochemistries

Studies of the stable-isotopic geochemistry of several biologically important elements (C, H, O, N, and S) have proven indispensable in paleobiology. Among these, analyses of carbon and sulfur stand out as providing particularly useful insight into the metabolic capabilities of ancient life.

In biological systems, the carbon-isotopic composition of organic constituents is determined both by the composition of the carbon source(s) utilized and by isotope effects associated with carbon assimilation. Primary among such effects in autotrophs are enzyme-mediated reactions that result in biosynthesized organic compounds becoming enriched in the lighter stable isotope, ^{12}C, relative to the ^{13}C-^{12}C composition of the source carbon. Autotrophic prokaryotes typically exhibit ranges of carbon-isotopic compositions that are phylogenetically characteristic but somewhat variable due to the varying influences of pH, temperature, metal ion availability, and, especially, CO_2 concentrations (O'Leary, 1981) (Fig. 7). Hence, relative to the carbon isotope Peedee Belemnite (PDB) standard, oxygenic photosynthetic cyanobacteria tend to be ^{12}C enriched by $16 \pm 8\%_0$, anoxygenic photosynthetic bacteria are enriched by $24 \pm 6\%_0$, and chemoautotrophic methanogenic archaeans are enriched by $28 \pm 10\%_0$ (Fig. 7). In relatively young geologic units, organic components can be sufficiently well preserved that the isotopic signatures of such autotrophs can be identified. A prime example is the Eocene (47-million-year-old) Messel Shale of Germany, in which biomarker indicators of oxygenic photosynthesis (chlorophyll derivatives), anoxygenic photoautotrophy (degradation products of bacteriochlorophyll *d*), and methanogenic chemoautotrophy (derivatives of phytyl and biphytyl ethers) are ^{12}C enriched relative to PDB by 22, 24, and 30‰, respectively (Hayes et al., 1987).

Unfortunately, because biomarkers such as those extracted from the Messel Shale are chemically unstable over geologically long periods of time, isotopic studies of extractable metabolically distinctive molecules cannot be carried out on most Precambrian sediments (Summons and Hayes, 1992). Moreover, although ion microprobe-based carbon-isotopic analyses of the kerogenous cell walls of individual Precambrian microfossils hold promise for determining the metabolic characteristics of ancient microbes, such studies are in their infancy (House et al., 1999). Almost all relevant data are therefore derived from analyses of bulk samples, typically of acid-resistant residues of kilogram-sized carbonaceous rocks (Strauss et al., 1992), analyses of the carbon-isotopic compositions of Precambrian kerogens relative to PDB (and hence expressed as $\delta^{13}C_{PDB}$ values) that provide strong evidence for the presence

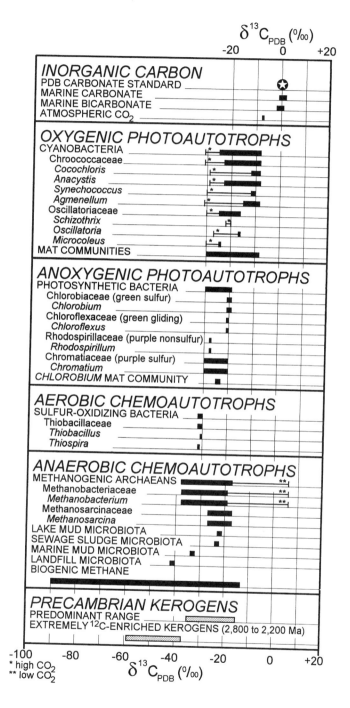

FIGURE 7 Carbon-isotopic compositions of inorganic-carbon reservoirs, prokaryotic autotrophs, and Precambrian kerogens. The fractionation ranges for cultures and anaerobic microbiotas are recalculated for a CO_2 source with $\delta^{13}C_{PDB}$ of $-7‰$. (Data from Abelson and Hoering, 1961; Calder and Parker, 1973; Wong et al., 1975; Games and Hayes, 1976; Pardue et al., 1976; Barghoorn et al., 1977; Fuex, 1977; Sirevåg et al., 1977; Fuchs et al., 1979; Mizutani and Wada, 1982; Belyaev et al., 1983; Hayes, 1983; Hayes et al., 1983; Schidlowski et al., 1983; Holo and Sirevåg, 1986; Ruby et al., 1987; Des Marais et al., 1992; Strauss and Moore, 1992; and Strauss et al., 1992.)

early in Earth history both of photoautotrophic and chemoautotrophic prokaryotes.

PHOTOAUTOTROPHIC PROKARYOTES

The carbon-isotopic compositions of oxygenic (cyanobacterial) and anoxygenic (photosynthetic-bacterial) prokaryotic photoautotrophs are fairly well characterized. Data are available for a large number of modern cyanobacterium-dominated stromatolitic communities (Des Marais et al., 1992) as well as for individual species of chroococcaceans (*Agmenellum*, *Anacystis*, *Coccochloris*, and *Synechococcus*) and oscillatoriaceans (*Microcoleus*, *Oscillatoria*, and *Schizothrix*) cultured under diverse conditions (Fig. 7). Data available for photosynthetic bacteria include analyses in cultures of representatives of the Chlorobiaceae (*Chlorobium*), Chloroflexaceae (*Chloroflexus*), Rhodospirillaceae (*Rhodospirillum*), and Chromatiaceae (*Chromatium*), as well as of photosynthetic-bacterium-dominated natural biocoenoses (Fig. 7).

As a result of isotopic discrimination during CO_2 fixation, photosynthetically produced organic matter sequestered in sedimentary rocks as kerogen is ^{12}C enriched relative to the carbon-isotopic composition of syngenetic carbonate minerals (which reflect the isotopic composition of atmospheric and dissolved CO_2), a difference typically of 25 ± 10‰ (Schidlowski et al., 1983; Strauss et al., 1992). The presence of such ^{12}C-enriched kerogen is thus regarded as strong evidence of the existence of photoautotrophy, an isotopic signature that, based on analyses of more than 1,000 Precambrian kerogen samples, has been traced to at least ~3.5 billion years ago (Schidlowski et al., 1983; Strauss et al., 1992; Hayes, 1994; Schopf, 1994b) and may extend as much as 300 million years earlier (Schidlowski, 1988; Mojzsis et al., 1996). In and of itself, the isotopic signal does not provide a firm basis for distinguishing between kerogens of cyanobacterial and photosynthetic-bacterial origin, but coupled with rRNA-based phylogenies and the geological and paleontological evidence discussed above, available data suggest that

these photoautotrophic lineages probably both date from at least 3.5 billion years ago.

SULFATE-REDUCING BACTERIA

Geochemical evidence also records the presence early in Earth history of gram-negative dissimilatory sulfate-reducing bacteria (such as *Desulfovibrio*, *Desulfotomaculum*, and *Desulfomonas*) that derive energy from hydrogenation of oceanic sulfate to hydrogen sulfide (that is, from reduction of S^{6+} to S^{2-}), a form of anaerobic respiration. As in photosynthesis, this process involves an enzymatic fractionation of two stable isotopes, but of sulfur rather than carbon such that the sulfide generated is enriched in the lighter isotope, ^{32}S, relative to the $^{34}S{:}^{32}S$ composition of the SO_4^{2-} sulfur source (Schidlowski et al., 1983). The bacterially generated H_2S reacts with dissolved ferrous iron in anoxic sediments to form pyrite (FeS_2), which consequently is also enriched in the lighter sulfur isotope. Biogenic pyrite grains in modern sediments typically have ratios of sulfur isotopes that are highly variable, encompassing a ~60‰ range of $\delta^{34}S$ values, and are thus distinguishable from the sulfide minerals in igneous rocks, which generally fall within a narrow range of about 5‰. The isotopic signature of microbial sulfate reduction can be traced back to at least 2.7 billion years ago (Schidlowski et al., 1983) and possibly extends to as early as ~3.4 billion years ago (Ohmoto et al., 1993; Kakegawa et al., 1994).

METHANE-PRODUCING ARCHAEANS

The archaeal lineage also evidently dates from very early in Earth history. Although reported $\delta^{13}C_{PDB}$ values of the cell walls and cytoplasm of anaerobic chemoautotrophic methanogens overlap with those of photoautotrophic prokaryotes (Fig. 7) and cannot, therefore, be used to differentiate between the groups, the methane produced by methanogenic archaeans can leave a telltale signature in the geologic record.

The various carbon sources utilized in archaeal methane production include carbon monoxide, carbon dioxide, formic acid, methanol, and acetate (Wolfe, 1971), with hydrogenation of the first two being energetically most productive:

$$4CO + 2H_2O \rightarrow CH_4 + 3CO_2 + 50\,kcal$$

$$CO_2 + 4H_2 \rightarrow CH_4 + 2H_2O + 32\,kcal$$

$$4HCO_2H \rightarrow CH_4 + 3CO_2 + 2H_2O$$

$$4CH_3OH \rightarrow 3CH_4 + CO_2 + 2H_2O$$

$$\star CH_3CO_2H \rightarrow \star CH_4 + CO_2$$

The methane thus produced exhibits the largest variation of $^{13}C:^{12}C$ known in biogenic materials, a range of $\delta^{13}C_{PDB}$ from -13 to -90‰ (Fuex, 1977). The maximum isotopic fractionation attainable during photosynthesis, both observed (Wong et al., 1975; Pardue et al., 1976; Sirevåg et al., 1977) and calculated (Summons and Hayes, 1992; Hayes, 1993), is about 36‰. Hence, highly ^{12}C-enriched archaeal methane can be distinguished readily from products of photosynthesis and, if incorporated into sedimented organic matter by methane-metabolizing methanotrophs, can leave an unmistakable signature in the preserved kerogen (Kaplan and Nissenbaum, 1966). The presence of kerogens with $\delta^{13}C_{PDB}$ values ranging from -38 to -59‰ in at least 16 geologic units ~2.2 to ~2.8 billion years in age (Strauss and Moore, 1992) has therefore been interpreted as strong evidence for the presence of biogenic methane and, thus, of methanogenic archaeans dating to at least 2.8 billion years ago (Hayes, 1983, 1994).

THE PRECAMBRIAN PROKARYOTIC RECORD

Considered as a whole, available paleontological, geological, and isotopic geochemical data provide convincing evidence of the antiquity of prokaryotic lineages, both bacterial and archaeal. In particular, the data indicate that stromatolitic microbial ecosystems, probably including cyanobacteria, photosynthetic bacteria, and other members of the bacterial domain, were extant ~3.5 billion years ago, methanogenic archaeans existed by ~2.8 billion years ago, and gram-negative sulfate-reducing bacteria existed at least as early as ~2.7 billion years ago.

Among these lineages, the record of cyanobacteria is especially voluminous and particularly well known, a record notable not only for its exceptional longevity but for the marked degree of long-term evolutionary stasis it appears to document, an absence or near absence of evolutionary change over thousands of millions, literally billions, of years. It remains to be shown whether such extreme stasis is typical of free-living prokaryotes generally, of particularly early-evolved lineages, or of Precambrian microbes as a whole, but the evolutionary stasis of cyanobacteria seems so evident, the evidence so compelling, that it begs for explanation. How did cyanobacteria maintain the status quo over geologically enormous spans of time?

CYANOBACTERIAL EVOLUTIONARY STASIS

In his classic volume of 1944, *Tempo and Mode in Evolution*, G. G. Simpson introduced terms for three different rate distributions of evolution shown by comparing the morphologies (size, shape, structural makeup, and so forth) of fossils and their living relatives: (i) tachytelic, for fast-evolving species; (ii) horotelic, the most common rate of change; and (iii) bradytelic, for especially slowly evolving taxa.

Slow, bradytelic evolvers are famous as living fossils. Good examples are horseshoe crabs, coelacanth fish, crocodiles, opossums, and primitive linguloid brachiopods, pegged as bradytelic because they belong to "groups that survive today and show relatively little change since the very remote time when they first appeared in the fossil record" (Simpson, 1944, p. 125). So defined, bradytely is essentially identical to the concept of "arrested evolution" proposed in 1918 by paleontologist Rudolf Ruedemann (1918, 1922a, 1922b). Both ideas are based on comparison of fossil and modern organisms that are nearly indistinguishable yet are separated by a hundred million years or more.

Like the concepts of bradytely and arrested evolution, inference of cyanobacterial evolutionary rates is based on morphologic comparison of fossils and their living relatives. But cyanobacteria stand out as having changed much more slowly even than bradytelic life (a rate distribution hence termed hypobradytelic; Schopf, 1987). Moreover, although bradytelic living fossils are rare, curiosities conspicuous because of their highly atypical lack of change, cyanobacteria constitute one of the most abundant and successful of life's early branches and include diverse families (the *Chroococcaceae*, *Oscillatoriaceae*, *Entophysalidaceae*, and *Pleurocapsaceae*) in which an almost imperceptibly sluggish rate of evolution is evidently the rule, not the exception.

Evolution's Most Successful Ecologic Generalists

Reasons underlying the striking stasis of cyanobacterial evolution are touched on in Simpson's pioneering *Tempo and Mode in Evolution*, for though in 1944 he had no way to guess what the Precambrian fossil record would eventually show, he was much intrigued by living fossils of the Phanerozoic. To explain the unusually slow evolution of these, Simpson proposed two principal factors, large populations and an ability to thrive in varied environments, of which he regarded the more important to be ecological versatility—the capability of an organism to survive in a range of settings that together would make up a "continuously available environment" (Simpson, 1944, p. 140–141). He reasoned that unusually slow evolution involves "not only exceptionally low rates of [change] but also survival for extraordinarily long periods of time" (p. 138) and noted that "more specialized [organisms] tend to become extinct before less specialized" (p. 143). To emphasize these observations, Simpson coined what he called the Rule of the Survival of the Relatively Unspecialized (p. 143).

Simpson's rule was intended for Phanerozoic living fossils, mostly animals, but it fits Precambrian cyanobacteria even better, possibly because their evolutionary stasis can be viewed as having three rather than only two contributing causes.

First, cyanobacteria are strictly asexual, lacking even the parasexual forms of reproduction present in some other prokaryotes. Though this want presumably played a role in their status quo evolution, given the long history of these microbes and even moderate rates of mutation to move evolution along, absence of sex and the genetic variability it generates cannot be the sole explanation for their lack of change.

Second, cyanobacterial populations are of course huge, like those of most microorganisms composed of an astronomical number of individuals. And because of their minute size, cyanobacteria are spread readily, by swirling waters, winds, tornadoes, and hurricanes, so that many species have more or less global distributions. Though Simpson did not consider such gigantic cosmopolitan populations, his observation still applies: taxa having especially large populations evolve especially slowly, a consideration perhaps particularly relevant to cyanobacteria, in which novel adaptive traits cannot be spread by sexual means.

Third, considered as a group, cyanobacteria can live almost anywhere. This versatility, surpassing any Simpson might have imagined, seems the keystone to their success.

Cyanobacterial Versatility

As summarized in Table 1, the versatility of cyanobacteria is shown especially by the *Chroococcaceae* and *Oscillatoriaceae*. Members of these families live, even flourish, in almost total darkness to extreme brightness; in pure, salty, or highly saline waters; in acid hot springs or exceptionally alkaline lakes; in scalding ponds or frigid icefields; in the near absence, presence, or huge overabundance of oxygen or carbon dioxide; in perhaps the most arid locale on Earth, the Chilean Atacama Desert, where rainfall has never been recorded; and even in the lethal radiation of a thermonuclear blast. Many can fix atmospheric nitrogen; provided with light, CO_2, a few trace elements, and a source of hydrogen (H_2O or, for some, even H_2S, H_2, or organic matter), these often are among the first colonizers of newly formed volcanic islands. In short, as a group cyanobac-

TABLE 1 Growth and survival of modern cyanobacteria of the *Chroococcaceae* and *Oscillatoriaceae*[a]

Environment	Conditions	Chroococcaceae	Oscillatoriaceae
Light intensity			
Extremely dim (1–5 mEs^{-1} m^{-1})	Cultures	√	√
Normal light (50–60 mEs^{-1} m^{-1})	Optimum growth	√	√
Exceedingly bright (>2,000 mEs^{-1} m^{-1})	Intertidal zone	√	√
Salinity			
<0.001–0.1%	Freshwater	√	√
3.5%	Marine	√	√
27.5%	Great Salt Lake	√	
100–200%	Salterns	√	√
Acidity/alkalinity			
Acid (pH 4)	Hot springs	√	
Neutral (pH 7–9)	Optimum growth	√	√
Alkaline (pH 11)	Alkaline lakes	√	√
High temperature			
70°C	Hot springs	√	√
74°C	Hot springs	√	
111°C	Dried		√
112°C	Dried		√
Low temperature			
−269°C	Liquid helium		√
−196°C	Liquid H$_2$		√
−55°C	Frozen		√
−2 to 4°C	Antarctic lakes		√
Desiccation			
88 years	Dried		√
82 years	Dried	√	
Absence of rainfall	Atacama Desert	√	√
Oxygen			
<0.01%	Anoxic lakes	√	√
1%	Blooms	√	√
20%	Ambient CO$_2$	√	√
100%	Cultures	√	
Carbon dioxide			
0.001%	Cultures	√	
0.035%	Ambient CO$_2$	√	√
3.5%	Cultures	√	√
40%	Cultures	√	
Radiation			
UV	290–400 nm[b]	√	√
X-rays	200 krad[c]		√
γ Rays	2,560 krad[d]		√
Highly ionizing	Thermonuclear bomb		√

[a] For various of the *Chroococcaceae* entries, species of *Agmenellum, Anacystis, Aphanocapsa, Coccochloris, Microcystis,* and *Synechococcus*; for various of the *Oscillatoriaceae* entries, species of *Lyngbya, Microcoleus, Oscillatoria, Phormidium, Schizothrix,* and *Spirulina.* Data from Desikachary, 1959; Shields and Drouet, 1962; Cameron, 1963; Vallentyne, 1963; Flowers and Evans, 1966; Forest and Weston, 1966; Drouet, 1968; Fuhs, 1968; Castenholz, 1969; Abeliovich and Shilo, 1972; Davis, 1972; Frémy, 1972; Drouet and Daily, 1973; Fogg, 1973; Fogg et al., 1973; Schopf, 1974; Lloyd et al., 1977; Brock, 1978; Langworthy, 1978; Parker et al., 1981; Ciferri, 1983; Grant and Tindall, 1986; Knoll and Bauld, 1989; Davidson, 1991; Garcia-Pichel and Castenholz, 1991; and Vincent et al., 1993.

[b] Absorbed by scytonemin pigment in sheaths.

[c] Twice as resistant as eukaryotic microalgae.

[d] Ten times as resistant as eukaryotic microalgae.

teria are highly successful generalists, able to survive and grow under the most varied conditions. In a teleological sense they thus appear to have had no "need" to evolve, for even if they were outcompeted in a local setting they could find refuge in other locales their competitors could not endure. Interestingly, this "jack-of-all-trades" survival strategy seems not only to explain their status quo evolution but to be itself a product of their evolutionary history.

Life survives by fitting its surroundings. But surroundings change as the global environment evolves, and such changes have been especially great for lineages that date from the distant Precambrian past. Cyanobacteria adapted as the environment evolved but evidently never lost their mastery of settings faced before. Early in Earth history, when the lineage originated, free oxygen was in short supply; cyanobacterially produced O_2 was rapidly scavenged and sedimented in the iron oxide minerals of banded iron formations, so the atmosphere contained only traces of UV-absorbing ozone. In such a setting, the ability to photosynthesize at low light intensities (Table 1) coupled with the presence of gas vesicles to control buoyancy (Jensen, 1993) would have permitted planktonic cyanobacteria to avoid harmful UV by inhabiting the deep oceanic photic zone, a strategy used today by the cosmopolitan and exceptionally abundant marine chroococcacean *Synechococcus*.

Over time, environmental levels of oxygen and ozone began to build, but concentrations remained low and UV was a threat for hundreds of millions of years. To colonize shallow-water settings, cyanobacteria developed biochemical means to repair UV-caused cellular damage; diverse chroococcaceans ensured the protective cover of overlying waters by cementing themselves to the shallow seafloor with gelatinous mucilage, some infused with UV-absorbing scytonemin (Garcia-Pichel and Castenholz, 1991); and benthic phototactic oscillatoriaceans entwined themselves in feltlike stromatolitic mats that blanketed shallow basins.

The remarkable hardiness of cyanobacteria may also reflect their success in competing with other early-evolved microbes for photosynthetic space. Because biosynthesis of bacteriochlorophyll is inhibited by molecular oxygen (Olsen and Pierson, 1987), O_2-producing cyanobacteria would have easily outcompeted and supplanted oxygen-sensitive photosynthetic bacteria throughout much of the global photic zone. Coupled with the ease of their global dispersal, this "microbial gas warfare" seems likely to have enabled them to have spread into a broad range of habitats—perhaps during the early rapid phase of adaptive radiation suggested by the branching pattern of cyanobacterial rRNA trees (Giavannoni et al., 1988)—and to have thereby evolved to become exceptional ecological generalists. Viewed this way, the ecological versatility of cyanobacteria appears to hark back to an early stage in Precambrian history when the group first became established as the dominant primary producers of the global ecosystem.

EVOLUTION EVOLVED!

Simpson's Rule of the Survival of the Relatively Unspecialized fits cyanobacteria to a tee. Moreover, though the evolutionary stasis of this ancient highly successful stock may seem an oddity in comparison with the familiar fast-changing progression of Phanerozoic life, mounting evidence suggests that maintenance of the status quo was actually the norm, rather than the exception, over the vast majority of life's long history (Schopf, 1999). Evidently, evolution itself evolved!

But the rules of early evolution, substantively different from those of later biological advance, are only beginning to come into focus. Acceptance of the pivotal role in early biotic history of "prepackaged," endosymbiotically spurred evolution dates from the 1970s. Yet what then seemed a rather simple story has become increasingly complex, arguably abetted by lateral gene transfer, an evolutionary mechanism until recently widely dismissed. And as it has become recognized that microorganisms initially thought "primitively amitochondriate" are probably not so at all (Roger et al., 1996, 1998), the very foundation of rRNA trees has been opened to question. To this mix, the paleobiological record adds evidence of microbial lineages that are many hundreds of millions of years too old as judged by molecular data (Doolittle et al., 1996) and of fossil microbes that quite unexpectedly seem not to have evolved at all over literally billions of years. The challenge today is to decipher the rules and solve the riddles of life's earliest development. What shape the solutions will ultimately take remains to be defined, but of one thing we can be certain: when the final answers are in, paleobiology and the molecular evidence will be shown to mesh—both, after all, deal with the same evolutionary progression, the same history of life.

ACKNOWLEDGMENTS

This chapter is based on concepts developed at greater length, at a level intended for a general audience and backed by extensive illustrations (some in color), in *Cradle of Life: the Discovery of Earth's Earliest Fossils* (Schopf, 1999).

REFERENCES

Abeliovich, A., and M. Shilo. 1972. Photooxidative death in blue-green algae. *J. Bacteriol.* **111:**682–689.

Abelson, P. H., and T. Hoering. 1961. Carbon isotope fractionation in formation of amino acids by photosynthetic organisms. *Proc. Natl. Acad. Sci. USA* **47:**623–632.

Altermann, W., and J. W. Schopf. 1995. Microfossils from the Neoarchean Campbell Group, Griqualand West Sequence of the Transvaal Supergroup, and their paleoenvironmental and evolutionary implications. *Precambrian Res.* **75:**65–90.

Barghoorn, E. S., and S. A. Tyler. 1965. Microorganisms of the Gunflint chert. *Science* **147:**563–577.

Barghoorn, E. S., A. H. Knoll, H. Dembricki, and W. G. Meinschein. 1977. Variation in stable carbon isotopes in organic matter from the Gunflint Iron Formation. *Geochim. Cosmochim. Acta* **41:**425–430.

Barley, M. E., J. S. R. Dunlop, J. E. Glover, and D. I. Groves. 1979. Sedimentary evidence for an Archean shallow-water volcanic-sedimentary facies, eastern Pilbara Block, Western Australia. *Earth Planet. Sci. Lett.* **43:**74–84.

Belyaev, S. S., R. Wolkin, W. R. Kenealy, M. J. DeNiro, S. Epstein, and J. G. Zeikus. 1983. Methanogenic bacteria from the Bondyuzhskoe Oil Field: general characterization and analysis of stable-carbon isotopic fractionation. *Appl. Environ. Microbiol.* **45:**691–697.

Blake, T. S., and N. J. McNaughton. 1984. A geochronological framework for the Pilbara region, p. 1–22. *In* D. K. Muhling, D. K. Groves, and R. S. Blake (ed.), *Archean & Proterozoic Basins of the Pilbara, Western Australia: Solution and Mineralization Potential.* Publication 9. University of Western Australia Geology Department and University Extension, Perth, Australia.

Brock, T. D. 1978. *Thermophilic Microorganisms and Life at High Temperatures.* Springer, New York, N.Y.

Calder, J. A., and P. L. Parker. 1973. Geochemical implications of induced changes in C^{13} fractionation by blue-green algae. *Geochim. Cosmochim. Acta* **37:**133–140.

Cameron, R. E. 1963. Morphology of representative blue-green algae. *Ann. N. Y. Acad. Sci.* **108:**412–420.

Castenholz, R. W. 1969. Thermophilic blue-green algae and the thermal environment. *Bacteriol. Rev.* **33:**476–504.

Ciferri, O. 1983. *Spirulina*, the edible microorganism. *Microbiol. Rev.* **47:**551–578.

Cloud, P. 1983. Early biogeologic history: the emergence of a paradigm, p. 14–31. *In* J. W. Schopf (ed.), *Earth's Earliest Biosphere, Its Origin and Evolution.* Princeton University Press, Princeton, N.J.

Darwin, C. R. 1859. *The Origin of Species by Means of Natural Selection.* John Murray, London, England.

Davidson, I. R. 1991. Environmental effects on algal photosynthesis: temperature. *J. Phycol.* **27:**2–8.

Davis, J. S. 1972. Survival records of the algae, and the survival role of certain algal pigments, fat, and mucilagenous substances. *Biologist* **54:**52–93.

Desikachary, T. V. 1959. *Cyanophyta.* Indian Council Agricultural Research, New Delhi, India.

Des Marais, D. J., J. Bauld, A. C. Palmisano, R. E. Summons, and D. M. Ward. 1992. The biogeochemistry of carbon in modern microbial mats, p. 299–308. *In* J. W. Schopf and C. Klein (ed.), *The Proterozoic Biosphere, A Multidisciplinary Study.* Cambridge University Press, New York, N.Y.

Doolittle, R. F., D.-F. Feng, S. Tsang, G. Cho, and E. Little. 1996. Determining divergence times of the major kingdoms of living organisms with a protein clock. *Science* **271:**470–477.

Drouet, F. 1968. *Revision of the Classification of the Oscillatoriaceae.* Academy of Natural Sciences, Philadelphia, monograph 15. Fulton, Lancaster, Pa.

Drouet, F., and W. A. Daily. 1973. *Revision of the Coccoid Myxophyceae.* Hafner, New York, N.Y.

Fairchild, T. R. 1975. *The Geologic Setting and Paleobiology of a Late Precambrian Stromatolitic Microflora from*

South Australia. Ph.D. thesis, University of California, Los Angeles.

Flowers, S., and F. R. Evans. 1966. The flora and fauna of the Great Salt Lake Region, Utah, p. 367–393. *In* H. Boyko (ed.), *Salinity and Aridity.* Junk, The Hague, The Netherlands.

Fogg, G. E. 1973. Physiology and ecology of marine blue-green algae, p. 368–378. *In* N. G. Carr and B. A. Whitton (ed.), *The Biology of Blue-Green Algae.* University of California Press, Berkeley, Calif.

Fogg, G. E., W. D. P. Stewart, P. Fay, and A. E. Walsby. 1973. *The Blue-Green Algae.* Academic, New York, N.Y.

Forest, H. S., and C. R. Weston. 1966. Blue-green algae from Atacama Desert of northern Chile. *J. Phycol.* 2:163–164.

Frémy, P. 1972. *Cyanophycées des Côtes D'Europe.* Asher, Amsterdam, The Netherlands.

Fuchs, G., R. Thauer, H. Ziegler, and W. Stichler. 1979. Carbon isotopic fractionation by *Methanobacterium thermoautotrophicum.* Arch. Microbiol. 120:135–139.

Fuex, A. N. 1977. The use of stable carbon isotopes in hydrocarbon exploration. *J. Geochem. Explor.* 7:155–158.

Fuhs, G. W. 1968. Cytology of blue-green algae: light microscopic aspects, p. 213–233. *In* D. F. Jackson (ed.), *Algae, Man, and the Environment.* Syracuse University Press, Syracuse, N.Y.

Games, L. M., and J. M. Hayes. 1976. On the mechanisms of CO_2 and CH_4 production in natural anaerobic environments, p. 51–73. *In* J. O. Nriagu (ed.), *Environmental Biogeochemistry,* vol. 1. *Carbon, Nitrogen, Phosphorus, Sulfur and Selenium Cycles.* Ann Arbor Science, Ann Arbor, Mich.

Garcia-Pichel, F., and R. W. Castenholz. 1991. Characterization and biological implications of a scytonemin, a cyanobacterial sheath pigment. *J. Phycol.* 27:395–409.

Giavannoni, S. J., S. Turner, G. J. Olsen, S. Barns, D. J. Lane, and N. R. Pace. 1988. Evolutionary relationships among cyanobacteria and green chloroplasts. *J. Bacteriol.* 170:3584–3592.

Godward, M. B. E. 1962. Invisible radiations, p. 551–566. *In* R. A. Lewin (ed.), *Physiology and Biochemistry of Algae.* Academic, New York, N.Y.

Golubic, S. 1976a. Organisms that build stromatolites, p. 113–126. *In* M. R. Walter (ed.), *Stromatolites, Developments in Sedimentology 20.* Elsevier, Amsterdam, The Netherlands.

Golubic, S. 1976b. Taxonomy of extant stromatolite-building cyanophytes, p. 127–140. *In* M. R. Walter (ed.), *Stromatolites, Developments in Sedimentology 20.* Elsevier, Amsterdam, The Netherlands.

Golubic, S., and H. J. Hofmann. 1976. Comparison of Holocene and mid-Precambrian Entophysalidaceae (Cyanophyta) in stromatolitic mats: cell division and degradation. *J. Paleontol.* 50:1074–1082.

Golubic, S., V. N. Sergeev, and A. H. Knoll. 1995. Mesoproterozoic *Archaeoellipsoides:* akinetes of heterocystous cyanobacteria. *Lethaia* 28:285–298.

Grant, W. D., and B. J. Tindall. 1986. The alkaline saline environment, p. 25–54. *In* R. A. Herbert and G. A. Codd (ed.), *Microbes in Extreme Environments.* Academic, New York, N.Y.

Green, J. W., A. H. Knoll, S. Golubic, and K. Swett. 1987. Paleobiology of distinctive benthic microfossils from the Upper Proterozoic Limestone-Dolomite "Series," central East Greenland. *Am. J. Bot.* 74:928–940.

Green, J. W., A. H. Knoll, and K. Sweet. 1988. Microfossils from oolites and pisolites of the Upper Proterozoic Eleonore Bay Group, central East Greenland. *J. Paleontol.* 62:835–852.

Groves, D. I., J. S. R. Dunlop, and R. Buick. 1981. An early habitat of life. *Sci. Am.* 245:64–73.

Hayes, J. M. 1983. Geochemical evidence bearing on the origin of aerobiosis: a speculative hypothesis, p. 291–301. *In* J. W. Schopf (ed.), *Earth's Earliest Biosphere, Its Origin and Evolution.* Princeton University Press, Princeton, N.J.

Hayes, J. M. 1993. Factors controlling ^{13}C contents of sedimentary organic compounds: principles and evidence. *Marine Geol.* 113:111–125.

Hayes, J. M. 1994. Global methanotrophy at the Archean-Proterozoic transition, p. 220–236. *In* S. Bengtson (ed.), *Early Life on Earth.* Columbia University Press, New York, N.Y.

Hayes, J. M., I. R. Kaplan, and K. M. Wedeking. 1983. Precambrian organic geochemistry, preservation of the record, p. 93–134. *In* J. W. Schopf (ed.), *Earth's Earliest Biosphere, Its Origin and Evolution.* Princeton University Press, Princeton, N.J.

Hayes, J. M., R. Takigiku, R. Ocampo, H. J. Callot, and P. Albrecht. 1987. Isotopic compositions and probable origins of organic molecules in the Eocene Messel Shale. *Nature* 329:48–51.

Hofmann, H. J. 1976. Precambrian microflora, Belcher Islands, Canada: significance and systematics. *J. Paleontol.* 50:1040–1073.

Hofmann, H. J., and J. W. Schopf. 1983a. Early Proterozoic microfossils, p. 321–360. *In* J. W. Schopf (ed.), *Earth's Earliest Biosphere, Its Origin and Evolution.* Princeton University Press, Princeton, N.J.

Holland, H. D. 1994. Early Proterozoic atmospheric change, p. 237–244. *In* S. Bengtson (ed.), *Early Life on Earth.* Columbia University Press, New York, N.Y.

Holo, H., and R. Sirevåg. 1986. Autotrophic growth and CO_2 fixation of *Chloroflexus auranticus.* Arch. Microbiol. 145:173–180.

Horodyski, R. J., and J. A. Donaldson. 1980. Microfossils from the Middle Proterozoic Dismal Lakes Group, Arctic Canada. *Precambrian Res.* **11:** 125–159.

House, C. H., J. W. Schopf, T. M. Harrison, and K. O. Stetter. 1999. Carbon isotopic analyses of individual microscopic fossils: a novel tool for astrobiology, abstr. P5.8, p. 105. *In Abstracts, 12th International Conference on the Origin of Life and 9th Meeting, International Society for the Study of the Origin of Life.*

Jensen, T. E. 1993. Cyanobacterial ultrastructure, p. 7–51. *In* T. Berner (ed.), *Ultrastructure of Microalgae.* CRC, London, England.

Kakegawa, T., H. Kawai, and H. Ohmoto. 1994. Biological activities and hydrothermal activity recorded in the ~2.5 Ga Mount McRae Shale, Hamersley District, Western Australia. II. Sulfur isotopic composition of pyrite. *Resource Geol.* **44:** 284–285.

Kaplan, I. R., and A. Nissenbaum. 1966. Anomalous carbon isotope ratios in nonvolatile organic material. *Science* **153:**744–745.

Klein, C., and N. J. Buekes. 1992. Time distribution, stratigraphy, sedimentologic setting, and geochemistry of Precambrian iron-formations, p. 139–146. *In* J. W. Schopf and C. Klein (ed.), *The Proterozoic Biosphere, a Multidisciplinary Study.* Cambridge University Press, New York, N.Y.

Knoll, A. H. 1985. A paleobiological perspective on sabkhas, p. 407–425. *In* G. M. Friedman and W. E. Krumbein (ed.), *Ecological Studies.* vol. 53. *Hypersaline Ecosystems.* Springer, New York, N.Y.

Knoll, A. H., and J. Bauld. 1989. The evolution and ecologic tolerance of prokaryotes. *Trans. R. Soc. Edinburgh: Earth Sci.* **80:**209–223.

Knoll, A. H., E. S. Barghoorn, and S. Golubic. 1975. *Palaeopleurocapsa wopfnerii* gen. et sp. nov., a late-Precambrian blue-green alga and its modern counterpart. *Proc. Natl. Acad. Sci. USA* **72:** 2488–2492.

Knoll, A. H., S. Golubic, J. Green, and K. Swett. 1986. Organically preserved microbial endoliths from the late Proterozoic of East Greenland. *Nature* **321:**856–857.

Langworthy, T. A. 1978. Microbial life in extreme pH values, p. 279–315. *In* D. J. Kushner (ed.), *Microbial Life in Extreme Environments.* Academic, New York, N.Y.

Lloyd, N. D. H., D. T. Cavin, and D. A. Culver. 1977. Photosynthesis and photorespiration in algae. *Plant Physiol.* **59:**936–940.

Mendelson, C. V., and J. W. Schopf. 1992. Proterozoic and selected Early Cambrian microfossils and microfossil-like objects, p. 865–951. *In* J. W. Schopf and C. Klein (ed.), *The Proterozoic Biosphere, a Multidisciplinary Study.* Cambridge University Press, New York, N.Y.

Mizutani, H., and E. Wada. 1982. Effect of high atmospheric CO_2 on $\delta^{13}C$ of algae. *Origins Life* **12:** 377–390.

Mojzsis, S. J., G. Arrhenius, K. D. McKeegan, T. M. Harrison, A. P. Nutman, and C. R. Friend. 1996. Evidence of life on Earth before 3,800 million years ago. *Nature* **384:**55–59.

Ohmoto, H., T. Kakegawa, and D. R. Lowe. 1993. 3.4-billion-year-old biogenic pyrites from Barberton, South Africa: sulfur isotope evidence. *Nature* **262:**555–557.

O'Leary, M. H. 1981. Carbon isotopic fractionation in plants. *Phytochemistry* **20:**553–567.

Olson, J. M., and B. K. Pierson. 1987. Evolution of reaction centers of photosynthetic prokaryotes. *Int. Rev. Cytol.* **108:**209–248.

Pardue, J. W., R. S. Scalan, C. Van Baalen, and P. L. Parker. 1976. Maximum carbon isotope fractionation in photosynthesis by blue-green algae and a green alga. *Geochim. Cosmochim. Acta* **40:** 309–312.

Parker, B. C., G. M. Simmons, Jr., G. Love, R. A. Wharton, and K. G. Seaburg. 1981. Modern stromatolites in Antarctic Dry Valley lakes. *BioScience* **31:**656–661.

Pierson, B. K., J. Bauld, R. W. Castenholz, E. D'Amelio, D. J. Des Marais, J. D. Farmer, J. P. Grotzinger, B. B. Jørgensen, D. C. Nelso, A. C. Palmisano, J. W. Schopf, R. E. Summons, M. R. Walter, and D. M. Ward. 1992. Modern mat-building microbial communities: a key to the interpretation of Proterozoic stromatolitic communities, p. 245–342. *In* J. W. Schopf and C. Klein (ed.), *The Proterozoic Biosphere, a Multidisciplinary Study.* Cambridge University Press, New York, N.Y.

Roger, A. J., C. G. Clark, and W. F. Doolittle. 1996. A possible mitochondrial gene in the amitochondriate protist *Trichomonas vaginalis. Proc. Natl. Acad. Sci. USA* **93:**14618–14622.

Roger, A. J., S. G. Sãvard, J. Tovar, C. G. Clark, M. W. Smith, F. D. Gillin, and M. L. Sogin. 1998. A mitochondrial-like chaperonin 60 gene in *Giardia lamblia*: evidence that diplomonads once harbored an endosymbiont related to the progenitor of mitochondria. *Proc. Natl. Acad. Sci. USA* **95:** 229–234.

Rubey, E. G., H. W. Jannasch, and W. G. Deuser. 1987. Fractionation of stable carbon isotopes during chemoautotrophic growth of sulfur-oxidizing bacteria. *Appl. Environ. Microbiol.* **53:**1940–1943.

Ruedemann, R. 1918. The paleontology of arrested evolution. *N. Y. State Mus. Bull.* **196:**107–134.

Ruedemann, R. 1922a. Additional studies of arrested evolution. *Proc. Natl. Acad. Sci. USA* **8:**54–55.

Ruedemann, R. 1922b. Further notes on the paleontology of arrested evolution. *Am. Nat.* **56:**256–272.

Schidlowski, M. 1988. A 3,800-million-year isotopic record of life from carbon in sedimentary rocks. *Nature* **333:**313–318.

Schidlowski, M., J. M. Hayes, and I. R. Kaplan. 1983. Isotopic inferences of ancient biochemistries: carbon, sulfur, hydrogen, and nitrogen, p. 149–186. *In* J. W. Schopf (ed.), *Earth's Earliest Biosphere, Its Origin and Evolution.* Princeton University Press, Princeton, N.J.

Schopf, J. W. 1968. Microflora of the Bitter Springs Formation, Late Precambrian, central Australia. *J. Paleontol.* **42:**651–688.

Schopf, J. W. 1970. Precambrian micro-organisms and evolutionary events prior to the origin of vascular plants. *Biol. Rev. Cambridge Phil. Soc.* **45:**651–688.

Schopf, J. W. 1972. Evolutionary significance of the Bitter Springs (Late Precambrian) microflora, p. 68–77. *In Proceedings of the 24th International Geologic Congress,* Sect. 1. *Precambrian Geology.*

Schopf, J. W. 1974. The development and diversification of Precambrian life. *Origins Life* **5:**119–135.

Schopf, J. W. 1977. Biostratigraphic usefulness of stromatolitic Precambrian microbiotas: a preliminary analysis. *Precambrian Res.* **5:**143–173.

Schopf, J. W. 1987. "Hypobradytely": comparison of rates of Precambrian and Phanerozoic evolution. *J. Vertebr. Paleontol.* **7**(Suppl. 3):25.

Schopf, J. W. 1992a. Historical development of Proterozoic micropaleontology, p. 179–183. *In* J. W. Schopf and C. Klein (ed.), *The Proterozoic Biosphere, a Multidisciplinary Study.* Cambridge University Press, New York, N.Y.

Schopf, J. W. 1992b. Times of origin and earliest evidence of major biologic groups, p. 587–593. *In* J. W. Schopf and C. Klein (ed.), *The Proterozoic Biosphere, a Multidisciplinary Study.* Cambridge University Press, New York, N.Y.

Schopf, J. W. 1992c. Proterozoic prokaryotes: affinities, geologic distribution, and evolutionary trends, p. 195–218. *In* J. W. Schopf and C. Klein (ed.), *The Proterozoic Biosphere, a Multidisciplinary Study.* Cambridge University Press, New York, N.Y.

Schopf, J. W. 1992d. Informal revised classification of Proterozoic microfossils, p. 1119–1168. *In* J. W. Schopf and C. Klein (ed.), *The Proterozoic Biosphere, a Multidisciplinary Study.* Cambridge University Press, New York, N.Y.

Schopf, J. W. 1992e. Paleobiology of the Archean, p. 25–39. *In* J. W. Schopf and C. Klein (ed.), *The Proterozoic Biosphere, a Multidisciplinary Study.* Cambridge University Press, New York, N.Y.

Schopf, J. W. 1993. Microfossils of the Early Archean Apex chert: new evidence of the antiquity of life. *Science* **260:**640–646.

Schopf, J. W. 1994a. Disparate rates, differing fates: tempo and mode of evolution changed from the Precambrian to the Phanerozoic. *Proc. Natl. Acad. Sci. USA* **91:**6735–6742.

Schopf, J. W. 1994b. The oldest known records of life: stromatolites, microfossils, and organic matter from the Early Archean of South Africa and Western Australia, p. 193–206. *In* S. Bengtson (ed.), *Early Life on Earth.* Columbia University Press, New York, N.Y.

Schopf, J. W. 1996a. Cyanobacteria: pioneers of the early Earth. *Nova Hedwigia* **112:**13–32.

Schopf, J. W. 1996b. Metabolic memories of Earth's earliest biosphere, p. 73–107. *In* C. R. Marshall and J. W. Schopf (ed.), *Evolution and the Molecular Revolution.* Jones & Bartlett, Boston, Mass.

Schopf, J. W. 1998. Tracing the roots of the universal tree of life, p. 336–362. *In* A. Brack (ed.), *The Molecular Origins of Life: Assembling the Pieces.* Cambridge University Press, Cambridge, United Kingdom.

Schopf, J. W. 1999. *Cradle of Life, The Discovery of Earth's Earliest Fossils.* Princeton University Press, Princeton, N.J.

Schopf, J. W., and J. M. Blacic. 1971. New microorganisms from the Bitter Springs Formation (Late Precambrian) of the north-central Amadeus Basin, Australia. *J. Paleontol.* **45:**925–961.

Schopf, J. W., and M. R. Walter. 1983. Archean microfossils: new evidence of ancient microbes, p. 214–239. *In* J. W. Schopf (ed.), *Earth's Earliest Biosphere, Its Origin and Evolution.* Princeton University Press, Princeton, N.J.

Schopf, J. W., J. M. Hayes, and M. R. Walter. 1983. Evolution of Earth's earliest ecosystems: recent progress and unsolved problems, p. 361–384. *In* J. W. Schopf (ed.), *Earth's Earliest Biosphere, Its Origin and Evolution.* Princeton University Press, Princeton, N.J.

Shields, L. M., and F. Drouet. 1962. Distribution of terrestrial algae within the Nevada test site. *Am. J. Bot.* **49:**547–554.

Simpson, G. G. 1944. *Tempo and Mode in Evolution.* Columbia University Press, New York, N.Y.

Sirevåg, R., B. B. Buchanan, J. A. Berry, and J. H. Troughton. 1977. Mechanisms of CO_2 fixation in bacterial photosynthesis studied by the carbon isotope fractionation technique. *Arch. Microbiol.* **112:**35–38.

Stetter, K. O. 1996. Hyperthermophilic prokaryotes. *FEMS Microbiol. Rev.* **18:**149–158.

Strauss, H., and T. B. Moore. 1992. Abundances and isotopic compositions of carbon and sulfur species in whole rock and kerogen samples, p. 709–798. *In* J. W. Schopf and C. Klein (ed.), *The Proterozoic Biosphere, a Multidisciplinary Study.* Cambridge University Press, New York, N.Y.

Strauss, H., D. J. Des Marais, J. M. Hayes, and R. E. Summons. 1992. The carbon-isotopic record, p. 117–127. *In* J. W. Schopf and C. Klein (ed.), *The Proterozoic Biosphere, a Multidisciplinary Study.* Cambridge University Press, New York, N.Y.

Summons, R. E., and J. M. Hayes. 1992. Principles of molecular and isotopic biogeochemistry, p. 83–93. *In* J. W. Schopf and C. Klein (ed.), *The Proterozoic Biosphere, a Multidisciplinary Study.* Cambridge University Press, New York, N.Y.

Thorpe, R. I., A. H. Hickman, D. W. Davis, J. K. Mortensen, and A. F. Trendall. 1992. U-Pb zircon geochronology of Archaean felsic units in the Marble Bar region, Pilbara Craton, Western Australia. *Precambrian Res.* **56:**169–189.

Towe, K. M. 1990. Aerobic respiration in the Archaean? *Nature* **348:**54–56.

Towe, K. M. 1991. Aerobic carbon cycling and cerium oxidation: significance for Archean oxygen levels and banded iron-formation deposition. *Palaeogeog. Palaeoclimatol. Palaeoecol.* **97:**113–123.

Vallentyne, J. R. 1963. Environment biophysics and microbial ubiquity. *Ann. N. Y. Acad. Sci.* **108:** 342–352.

Vincent, W. F., R. W. Castenholz, M. T. Downes, and C. Howard-Williams. 1993. Antarctic cyanobacteria: light, nutrients, and photosynthesis in the microbial mat environment. *J. Phycol.* **29:**745–755.

Wilmotte, A. 1994. Molecular evolution and taxonomy of the Cyanobacteria, p. 1–25. *In* D. A. Bryant (ed.), *The Molecular Biology of Cyanobacteria.* Kluwer Academic Publishers, Dordrecht, The Netherlands.

Woese, C. R. 1987. Bacterial evolution. *Microbiol. Rev.* **51:**221–271.

Woese, C. R., O. Kandler, and M. L. Whellis. 1990. Towards a natural system of organisms: proposal for the domains Archaea, Bacteria, and Eucarya. *Proc. Natl. Acad. Sci. USA* **87:**4576–4579.

Wolfe, R. S. 1971. Microbial formation of methane. *Adv. Microb. Physiol.* **6:**107–146.

Wong, W., W. M. Sackett, and C. R. B. Benedict. 1975. Isotope fractionation in photosynthetic bacteria during carbon dioxide assimilation. *Plant Physiol.* **55:**475–479.

Zhang, Y., and S. Golubic. 1987. Endolithic microfossils (Cyanophyta) from Early Proterozoic stromatolites, Hebei, China. *Acta Micropaleontol. Sinica* **4:**1–12.

ENDOSPORE-
FORMING BACTERIA

ENDOSPORE-FORMING BACTERIA: AN OVERVIEW

Abraham L. Sonenshein

6

Endospore formation, an ancient and complex developmental cycle, has been found almost exclusively in gram-positive bacteria. It is one of the defining traits of the genera *Bacillus*, *Clostridium*, *Thermoactinomyces*, *Sporolactobacillus*, and *Sporosarcina*. *Desulfotomaculum* and *Sporomusa* have gram-negative staining characteristics but have a gram-positive-type cell envelope and share 16s rRNA sequences with the *Clostridium* subphylum (Holt et al., 1994; Kuhner et al., 1997; Sass et al., 1998). *Sporohalobacter* is the only well-characterized endospore-forming genus currently thought to be truly gram negative (Holt et al., 1994).

Endospores form by intracellular division, develop within the cytoplasm of a mother cell, and, when mature, are metabolically inactive. Exospores, by contrast, are produced at the tips of cells or mycelia by a process that resembles budding; myxospores are formed by reorganization of a cell without concomitant cell division. Bacterial endospores are distinguished by three characteristics: (i) they are metabolically dormant, principally because their cytoplasm is almost totally dehydrated, (ii) they are birefringent under phase-contrast microscopy (a trait usually referred to as "refractility" or "phase brightness"), and (iii) they are resistant to a number of chemical and physical agents that would kill growing cells of nearly all other bacterial species.

HABITAT

Endospore formers are typically associated with a soil habitat, but many species probably occupy a more complex ecological niche. For example, several spore formers are normal inhabitants of the human or animal intestinal tract. In addition, the intestines of a large number of soil arthropods were recently found to harbor filamentous spore-forming bacteria that are essentially identical to laboratory strains of *Bacillus cereus* (Margulis et al., 1998). These bacteria live symbiotically in the insect gastrointestinal tract and live in the soil only when expelled from the insect gut. It remains to be seen whether other soil-dwelling spore formers also have intestinal life cycle stages. Other *Bacillus* spp. are also found in unexpected niches, such as the rumen of cattle (Priest, 1993) and sewage sludge (Scholz et al., 1987).

CLASSIFICATION AND EVOLUTION

The classical and strict distinction between aerobic (*Bacillus*, *Thermoactinomyces*, *Sporolactobacillus*, and *Sporosarcina*) and anaerobic (*Clostrid-*

A. L. Sonenshein, Department of Molecular Biology and Microbiology, Tufts University School of Medicine, Boston, MA 02111-1800.

Prokaryotic Development, edited by Y. V. Brun and L. J. Shimkets,
© 2000 American Society for Microbiology, Washington, DC 20005-4171

ium) spore formers (Gordon et al., 1973) no longer holds; it is now known that, given the right environment, *Bacillus subtilis* and other *Bacillus* species can grow quite well anaerobically (Priest, 1993; Nakano and Zuber, 1998). Moreover, some *Clostridium* species can survive and even grow slowly in oxygen. Nonetheless, unlike any *Bacillus* species, *Clostridium* spp. grow better anaerobically than aerobically and many are very strict anaerobes. Even DNA base composition does not give an unambiguous classification. Although most *Clostridium* spp. have a very low $G + C$ content, the range in *Bacillus* spp. is from 36 to 60% $G + C$ and that in *Clostridium* species is from 25 to 55% (Priest, 1993; Young and Cole, 1993). Evolutionary trees based on rRNA sequences, however, generally support the distinctiveness of the genera *Bacillus* and *Clostridium* (Young and Cole, 1993).

The diversity of endospore formers is broad. The most commonly studied spore formers are rod-shaped bacteria (bacilli), but *Sporosarcina* is a coccus and *Thermoactinomyces* grows as a mycelium. Unlike true exospore-forming actinomycetes, however, *Thermoactinomyces* forms endospores and has a DNA base composition similar to that of certain *Bacillus* spp. (Priest, 1993).

By far the greatest amount of information is available for the *Bacillus* species, especially *B. subtilis*. The early discovery of genetic transformation in this organism stimulated researchers to undertake detailed genetic analysis, which, over the ensuing years, was coupled first to morphological studies and then to a growing understanding of metabolism and regulation at the biochemical and molecular levels. At the present time, more than 75 *B. subtilis* genes whose functions are required for sporulation are known (Stragier and Losick, 1996). Many other genes are expressed uniquely in sporulating cells; some of these encode major components of spores but are not individually essential because of built-in redundancy of function (Stragier and Losick, 1996).

In guessing how and when spore formation evolved, the simplest assumption is that sporu-lation arose in the pre-oxygen era and became characteristic of aerobic bacteria only when *Bacillus* diverged from *Clostridium* and acquired the functions necessary for aerobic survival and growth. Support for the idea that *Bacillus* and *Clostridium* have a common sporulating ancestor comes from the high conservation of sporulation-regulatory proteins in the two genera (Sauer et al., 1994, 1995; Melville and Sonenshein, 1996; Clouart and Sonenshein, 1999). Thus, it is very unlikely that sporulation evolved independently in the two species. It is not yet known whether nonsporulating bacteria are derived from an ancestor of *Clostridium* that had not yet acquired the ability to sporulate or whether they reflect a common derivative of *Clostridium* that had lost sporulation competence. The near absence of endosporulation in gram-negative bacteria suggests that acquisition of spore-forming ability occurred after the divergence of gram-positive and most gram-negative bacteria or that loss of sporulation competence predated such divergence. Yet, many nonsporulating bacteria have retained some of the important attributes of spore formers, such as regulation of adaptive gene expression by multiple RNA polymerase sigma factors and population density sensing. As more prokaryotic genomes are fully sequenced, it will be possible to assess the extent to which individual sporulation genes have been conserved even when the process itself has not been.

Since the intermediate stages of the complex sporulation process do not seem to provide any advantage and may, in fact, be disadvantageous as end points (see below), we have to wonder if cells could have acquired spore-forming ability in a stepwise fashion. It is not easy to imagine what a primitive spore might have looked like. Perhaps the earliest spore-like cell had evolved the ability to surround its desiccated cytoplasm with a thick wall of peptidoglycan-like cortical material (see below). Later steps in evolution might have led to increased resistance to environmental dangers by assembly of proteinaceous spore coats. This step would have required that spore formation occur within a mother cell, leading to the phenomenon of endosporulation and the necessity of differentiating between mother cell and forespore compartments. The wide dispersal of sporulation genes throughout the chromosome in *Bacillus* and *Clostridium* makes it very unlikely that ability to sporulate is ever transmitted horizontally.

The apparent selective advantage of spore formation is survival under adverse environmental conditions. Mature bacterial endospores are among the most resistant biological entities known with respect to a number of environmental insults, such as UV and gamma radiation, reactive oxygen, high and low temperature, strong acid and alkali, organic solvents, hydrolytic enzymes, etc. (Piggot and Coote, 1976; Setlow, 1995; Xue and Nicholson, 1996). The high resistance of bacterial endospores to heat is the basis for their use as indicators of the efficaciousness of autoclaves. Paradoxically, exposure to deleterious conditions does not induce sporulation. Although growing or stationary-phase cells of *B. subtilis* can be induced to resist reactive oxygen or other stresses by exposure to low levels of the dangerous substance, the resultant cells are not directed to the sporulation pathway. Instead, initiation of sporulation seems to be specifically tied to nutritional limitation and cell population density. It seems, therefore, that endosporulation evolved as a means of protecting the genomes of cells that were prone to finding themselves under conditions of prolonged nutrient depletion, since such cells would inevitably be exposed to agents with the potential to cause severe and cumulative damage.

But just as there is an apparent selective advantage to being able to sporulate, there is also an inherent disadvantage. Once a cell becomes committed to sporulate, it must complete the process of becoming dormant before reinitiating growth. Moreover, the germination of spores and the return to active growth are slow processes that require special environmental conditions, not simply any environment that supports growth. Thus, a cell committed to sporulate gives up the potential to grow for a significant period of time. If the environment changes for the better during this period, competitor bacteria may be able to take over the local niche, crowding out the spores and overwhelming them numerically. This may be one reason why commitment to sporulate depends on extracellular signals that indicate that a certain concentration of like bacteria (a quorum) is present in the local environment.

The potential disadvantage of sacrificing short-term growth for long-term survival may be the reason that most bacteria never acquired the ability to sporulate or why their ancestors sloughed off sporulation genes. One might think that local habitats were a determining factor. But spore formers and non-spore formers coexist in the same ecological niches, whether they be dry soil, hot springs, or intestinal tracts. There is no obvious correlation between habitat and ability to sporulate.

COSTS AND BENEFITS OF ENDOSPORULATORS TO HUMANS AND ANIMALS

Bacillus species include important human, animal, and insect pathogens (*Bacillus anthracis*, *Bacillus thuringiensis*, and *Clostridium botulinum*), as well as species of great importance in the detergent, antibiotic, and food industries. The proteases secreted by *B. subtilis* are a common additive to laundry detergents; the α-amylases and glucose isomerase secreted by *Bacillus amyloliquefaciens*, *Bacillus licheniformis*, and *Bacillus stearothermophilus* are very useful in the conversion of starch to corn syrup and dextrose; the xylanase of *B. stearothermophilus* is used in the paper pulp industry; and *B. subtilis* natto is the agent that produces a fermented soy product commonly eaten for breakfast in Japan. *Clostridium* species include some of the most virulent pathogens known (*C. botulinum* and *Clostridium tetani*) but also species of great utility in industrial solvent production (e.g., *Clostridium acetobutylicum*) and cellulose degradation (e.g., *Clostridium thermocellum*).

Endospore formers are also the producers of important antibiotics, including the peptides bacitracin, fengycin, gramicidin, polymyxin, surfaction, and tyrocidine (Zuber et al., 1993). In several of these cases, antibiotic synthesis is tied to the transition from active growth to stationary phase.

MORPHOLOGICAL CHANGES

Sporulation-associated changes in the cell envelope and in the organization of the nucleoid are remarkably similar in *Bacillus* and *Clostridium* (Chapman, 1959; Ryter, 1965; Young and Fitz-James, 1962; Roper et al., 1976). The major differences relate to the position of the developing spore within the mother cell cytoplasm. In some species, the forespore is seen as a bulging at the tip of the cell; in other species it is more centrally located. The morphological stages defined for the sporulation of *B. subtilis* are illustrated in Fig. 1. Several very unusual features are worth noting. The first morphological change specific to sporulation is the formation of a division septum near one pole of the cell (Fig. 1-3). The process by which this event occurs involves a major reorganization of the cell division (septation) machinery (see chapter 8). During normal growth, cells divide in the middle (Fig. 1-1), because a complex of septum-forming proteins is both attracted to the midpoint of the cell and excluded from polar positions. In cells initiating sporulation, however, this specificity is reversed; only polar localization of the septation machinery occurs (Levin and Losick, 1996). As detailed in chapter 8, the mechanism by which a stationary-phase cell relocates the site of septation from the midcell to the cell pole is a major unanswered question. Two regulatory proteins are known to be needed specifically for the septation event. Spo0A, described below, is needed to express an as-yet-unidentified gene whose product is required for polar assembly of the septation apparatus (Levin and Losick, 1996); the σ^H subunit of RNA polymerase probably stimulates transcription of a gene whose product is necessary for the septation event itself (Levin and Losick, 1996). In addition, isocitrate dehydrogenase, the third enzyme of the Krebs cycle, is apparently needed to maintain an appropriate pH and cation concentration (Jin et al., 1997; Matsuno et al., 1999).

The smaller cytoplasmic compartment is destined to become the spore, while the larger cytoplasmic compartment serves as a mother cell within which the spore develops. In fact, the mother cell membrane soon thereafter engulfs the smaller compartment, creating a cell within a cell (Fig. 1-4, 1-5). The differential division of the cytoplasm of the presporulating cell determines the fate of each compartment with respect to gene expression and eventual function, but the specific mechanism of cell fate determination is unknown. The first regulatory event known to occur after septation is activation of the RNA polymerase factor, σ^F, exclusively in the forespore compartment, by a process that involves antagonism of an anti-σ^F protein (Duncan and Losick, 1993). As detailed in chapter 8, the detailed mechanism of this activation pathway is known, but the basis for its exclusive occurrence in the forespore is not. One theory holds that the size of the forespore determines its fate; that is, if some factor were present at equal numbers of molecules in the forespore and mother cell compartments, its concentration in the forespore would be considerably higher.

Determination of cell fate by unequal division of the cytoplasm seems to be an efficient and therefore advantageous process. Surprisingly, though, unequal division is not characteristic of all endospore-forming bacteria. *Sporosarcina ureae* is a coccus; it divides by medial septation during both growth and sporulation (Zhang et al., 1997). How then does this cell decide which compartment is the forespore and which the mother cell? This may be a stochastic process, but it is in any event very efficient. After septation, one half of the cell engulfs the other to create the cell within a cell characteristic of endospore formers (Zhang et al., 1997).

Other species, e.g., *Metabacterium polyspora*, have an even more surprising property. Their forespore compartments undergo several successive asymmetric divisions, creating multiple small compartments within the mother cell,

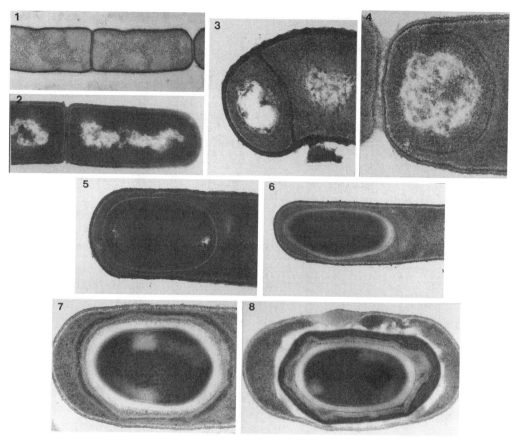

FIGURE 1 Stages of sporulation in *B. subtilis*. Successive stages in spore formation of *B. subtilis* are shown. Vegetative cells (photo 1) divide medially. After the final medial septation at the entry into stationary phase, the two chromosomes of the cell form an axial filament structure (photo 2). Septation occurs near one pole of the cell (photo 3), after which the mother cell cytoplasmic membrane begins to engulf the forespore (photo 4). After completion of engulfment (photo 5), the forespore is surrounded by two membranes derived from the forespore and mother cell and lies fully within the cytoplasm of the mother cell. Cortex, a peptidoglycan-like substance, is synthesized between the two forespore membranes (gray-white material in photo 6). As cortex synthesis nears completion, spore coat proteins begin to assemble around the forespore (photo 7). The inside layers of coat protein are less electron dense than are the outside layers (photo 8). The fully assembled and mature spore is eventually released by lysis of the mother cell. (A version of this figure appeared in the doctoral thesis of S. Jin [1995].)

each of which develops into a spore (Fig. 2) (Angert and Losick, 1998). For this organism, sporulation appears to be the principal mechanism of proliferation, since binary fission occurs slowly and rarely (Angert and Losick, 1998).

The engulfment process also raises some interesting unresolved questions. What is the driving force that causes the mother cell membrane to encircle the forespore compartment? Or, alternatively, what force causes the forespore to push its way into the space occupied by the mother cell, thereby moving toward the center of the cell? At the conclusion of the engulfment process, the two leading ends of the mother cell membrane fuse. Is this a simple, spontaneous reaction, or is it driven by special

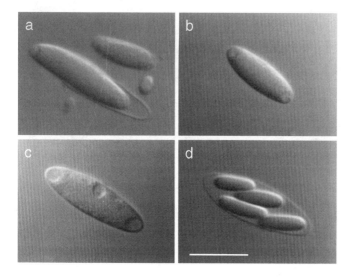

FIGURE 2 Sporulation in *M. polyspora*. Nomarski differential interference contrast micrographs show various stages in the life cycle of *M. polyspora*. (a) A germinated spore emerging from the spore coat; (b) a cell undergoing asymmetric septation at both poles; (c) a cell with three forespore compartments; (d) a mother cell containing four mature spores. (This figure was supplied by E. Angert, Harvard University.)

proteins? To what extent is this form of membrane fusion related to fusion of enveloped viruses with their target cell membranes or fusion of organelle membranes in higher organisms?

After engulfment, the forespore is surrounded by two oppositely oriented membranes. The inner membrane has the normal orientation of a cytoplasmic membrane with respect to the cytoplasm of the forespore, but the outer forespore membrane (derived from the cytoplasmic membrane of the mother cell) has the opposite orientation. That is, the outer face of this membrane, as it was originally assembled, now faces toward the forespore cytoplasm. This arrangement has two potentially advantageous consequences. First, it allows both compartments to secrete proteins and other factors into the intermembrane space. This ability is probably critical for formation of the spore cortex. Second, the completed spore may be protected against deleterious extracellular agents that would normally be transported across a conventionally oriented cytoplasmic membrane.

In fact, synthesis and assembly of cortical material, a macromolecular structure similar but not identical to normal cell wall peptidoglycan, between the two membranes is a critical event in sporulation (Fig. 1-6). The cortex appears to play an essential role in the remarkable heat resistance of endospores, and its assembly contributes to the dehydration of the spore cytoplasm.

The mother cell is not at all a passive partner in the spore formation process. In addition to providing some of the enzymes and constituents for cortex synthesis, the mother cell is also the site of the synthesis and assembly of spore coats. That is, the lamellae of coat protein are laid down from the outside of the spore, as a thin layer of inner coats and as a thicker layer of outer coats (Fig. 1-7, 1-8). Many different coat protein species participate in this assembly. All are synthesized in the mother cell but are sufficiently redundant in function, at least in laboratory tests, that the elimination of almost any one of them by mutation is tolerated well. A few proteins appear to play key roles in this assembly. SpoIVA seems to form a shell around the forespore, upon which a scaffold of coat protein lamellae is laid down (Driks et al., 1994; Price and Losick, 1999). In its absence, sheets of at least partially assembled coat material appear in the mother cell cytoplasm but fail to attach to the surface of the spore (Piggot and Coote, 1976). CotE seems to be the major organizer of the outer coat layer. Spores missing CotE build an inner coat and synthesize at least some outer coat proteins but do not assemble an outer coat layer (Zheng et al., 1988).

When the spore has been fully assembled, the mother cell lyses, releasing the environmentally protected, metabolically dormant spore into the environment. In some species, the released spore is surrounded by a membranous sac known as the exosporium.

Whether lodged in the soil or carried by wind or water or by adherence to a passing animal, the spore will remain dormant until it meets certain very specific conditions that permit germination and return to active growth. For *B. subtilis*, two specific germinants are known: one is L-alanine, and the other is a mixture of L-asparagine, fructose, glucose, and potassium ions (Moir and Smith, 1990; Johnstone, 1994). In either case, specific receptors are thought to be located on the outer surface of the spore membrane, buried under the spore coat. These receptors both recognize the germinants and bring them to transport systems within the membrane. The germination event is rapid and results in loosening of the spore coat structure, rehydration of the spore cytoplasm, activation of certain enzymes, turnover of preexisting protein, and de novo synthesis of RNA and protein (Setlow, 1975; Setlow and Primus, 1975). No growth or breakage of the spore occurs, however, unless additional nutrients are provided. When suspended in a complete medium, germinated spores begin to synthesize DNA within 60 min. The limiting factor seems to be the timing of activation of ribonucleotide reductase to produce deoxyribonucleotides (Setlow, 1973). Subsequent expansion of the cell mass forces the nascent vegetative cell to burst through the remaining spore coats, shedding the cortex as well.

SPORULATION AS A RESPONSE TO NUTRITIONAL LIMITATION AND POPULATION DENSITY

Sporulation in various organisms can be induced by limitation of the carbon source, the nitrogen source, or the phosphorus source, an event provoked in the laboratory either by allowing a culture to exhaust a growth medium or by withdrawing growing cells from a complete medium and resuspending them in a medium with poor nutritional sources (Sterlini and Mandelstam, 1969; Sonenshein, 1989). Also, when a growing culture is maintained for many generations in a poor medium, a substantial fraction of the cells initiates sporulation (Schaeffer et al., 1965; Dawes and Mandelstam, 1970). This result has been interpreted to indicate that sporulation is controlled by one or more threshold phenomena (Schaeffer et al., 1965; Chung et al., 1994). In *B. subtilis*, nitrogen limitation in the presence of glucose directs cells to the genetic competence pathway rather than to the sporulation pathway (Dubnau and Roggiani, 1990).

While sporulation is only initiated under conditions of nutrient limitation, sporulation is not the only or even the first response to such limitation. Instead, nutrient-limited cells (i.e., those that have exhausted their favored nutritional sources) induce a host of adaptive responses (chemotaxis and motility, synthesis of extracellular degradative enzymes, synthesis of antibiotics and toxins, expression of transport systems, induction of catabolic pathways, and activation of the genetic competence cascade) whose function is to help the cell to reach, to liberate (by killing neighboring pro- or eukaryotic cells, if necessary), to take up, and to metabolize potential secondary sources of nutrients (Sonenshein, 1989). The cells only become committed to the sporulation pathway if these adaptive responses fail to provide enough nutrients to support continued growth. How cells delay the commitment to sporulation, eventually recognize that adaptation has been unsuccessful, and then initiate sporulation is still a mystery. Part of the answer seems to reside in a series of inhibitor proteins that transiently interfere with several early steps in sporulation (see below and chapter 7 for more details).

While limitation is critical, the nutrient level cannot be allowed to drop too low; sporulation is an energy-demanding, biosynthetic process in which cells reutilize some of their preexisting macromolecules (e.g., mRNA and some proteins) but still need to generate ATP and some precursors and mRNA and protein synthesis in order to make spore structural proteins. A truly starved cell cannot sporulate. Not surprisingly, mutants lacking certain enzymes of central metabolism, such as Krebs cycle enzymes, sporulate very poorly.

Identifying the nature of the nutritional signal has been the goal of untold person-years of research, whose outcome tells us that there is probably not a unique signal, that there are probably two or more moments during sporulation at which nutritional signals play a role, and that multiple proteins respond to these signals (Sonenshein, 1989). A seductive theory attributed many of the attributes of the nutritional signal to the intracellular concentration of GTP. First, limitation of carbon, nitrogen, or phosphorus would be expected to lead in common to a deficiency in nucleotide synthesis. Second, inhibition of GMP synthesis activates sporulation genes in cells growing in a complete medium (Mitani et al., 1977), as if the consequent decrease in guanine nucleotide pools fools the cells into thinking they are in stationary phase. Third, induction of sporulation has much in common with the stringent response, insofar as amino acid limitation is an effective inducer of sporulation and the stringent response causes GTP to be converted to (p)ppGpp. However, a mutant totally defective in synthesis of (p)ppGpp is able to sporulate at a near normal frequency, albeit more slowly than do wild-type cells (Wendrich and Marahiel, 1997; Sonenshein, 1998). The best candidate to date for the target of the putative GTP effect is Obg, a homolog of small eukaryotic GTPases of the Ras family (Kok et al., 1994). The specific mechanism by which Obg, a protein essential for growth, regulates sporulation in response to GTP availability is unknown, but it is suspected to involve interaction with Spo0B in the Spo0A phosphorelay (Hoch, 1993; Vidwans et al., 1995).

Nutritional limitation is clearly not the only environmental factor that induces sporulation. Efficient sporulation of a population of cells requires that the cell concentration reach a certain minimum value (Mitani et al., 1977; Grossman and Losick, 1988). As detailed in chapter 7, this population density effect can be attributed to the accumulation of certain secreted oligopeptides (Grossman, 1995; Lazazzera and Grossman, 1998; Perego, 1998). The concentrations of these compounds indicate to the cells the local concentration of their siblings. Only when the peptide concentration reaches a critical value do the cells initiate sporulation events. Perhaps, from the cell's point of view, the dual dependence on nutritional limitation and cell population density insures that cells will only sporulate when they are unable to grow and are, at the same time, sufficiently abundant in the local environmental niche that the species will survive even if sporulation is inefficient and even if other bacteria in the local area continue to grow.

Since the apparent rationale for producing spores is to preserve the genome in the face of dangerous environments, it is not surprising that initiation of sporulation is also dependent on the cell having an intact chromosome and the means to segregate it properly to the forespore compartment (Ireton and Grossman, 1994; Ireton et al., 1994).

THE SPO0A PHOSPHORELAY

All of the environmental and intercellular signals that control the onset of sporulation and most of the signals that activate stationary-phase adaptation seem to be perceived by the cell through the multiple components of the Spo0A phosphorelay (Fig. 3) (Hoch, 1993) (chapter 7). Three protein kinases, each active under somewhat different conditions, can phosphorylate the Spo0F protein, whose phosphate is then transferred to Spo0A by Spo0B, a phosphotransferase enzyme (Burbulys et al., 1991; Hoch, 1993; LeDeaux et al., 1995). Phosphorylated Spo0A (Spo0A~P) is the active, DNA-binding form of this critically important regulatory protein (Hoch, 1993). Genes whose products allow the cell to adapt to poor nutritional conditions are in most cases controlled by Spo0A~P indirectly. That is, these genes are repressed by AbrB, one of several proteins that control stationary-phase-induced genes (Strauch and Hoch, 1993). Spo0A~P is a repressor of *abrB* (Strauch and

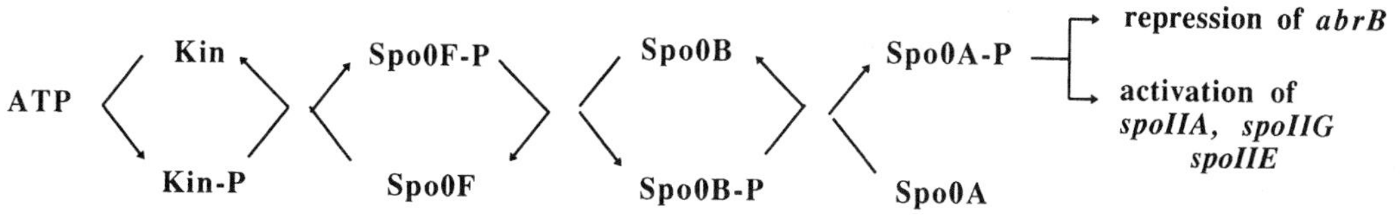

FIGURE 3 The Spo0A phosphorelay. Three different histidine protein kinases autophosphorylate and then transfer their phosphate groups to Spo0F. Through the intermediary of a phosphotransferase, Spo0B, the phosphate is finally transferred to an aspartate residue on Spo0A. Spo0A-phosphate is active as a DNA-binding transcription factor, having both negative and positive effects on gene expression. For additional details, see Burbulys et al., 1991; Hoch, 1993; and LeDeaux et al., 1995.

Hoch, 1993). Thus, activation of the phosphorelay leads to depression of AbrB-repressed genes by reducing synthesis of the direct regulator. The primary signal for this activity of Spo0A~P seems to be accumulation of one or more extracellular signaling peptides (Perego, 1998; Lazazzera and Grossman, 1998). By contrast, the earliest-expressed genes whose products are uniquely required for sporulation are directly controlled by Spo0A~P, acting as a positive transcriptional regulator (Bird et al., 1993; Baldus et al., 1994). For this second activity of the Spo0A phosphorelay, the signal seems to include nutritional limitation.

AbrB is only one of several proteins controlling expression of stationary-phase adaptation genes. For example, CodY represses transcription of many genes involved in secondary nutrient utilization and genetic competence during growth in an enriched medium (Slack et al., 1995; Serror and Sonenshein, 1996). Some of these same genes are also repressed by AbrB. Such dual regulation seems to reflect again the desire of the cell to sense both the concentration of its siblings (quorum sensing), through accumulation of a signaling peptide, and nutrient limitation.

Even though most Spo0A-dependent genes are not required for sporulation, no sporulation-specific morphological changes or gene expression can occur without Spo0A. Even the formation of polar septation complexes depends on this critical transcription factor (Levin and Losick, 1996).

The mechanisms by which specific conditions influence the phosphorelay are not com-

pletely understood, but much evidence has accumulated indicating that the cell synthesizes a series of antagonists of the phosphorelay, each of which could potentially respond to a particular environmental signal and alter the rate of accumulation of Spo0A~P. These antagonists include an inhibitor of kinase A (Wang et al., 1997), phosphatases for Spo0F~P (Perego, 1998) and for Spo0A~P (Ohlson et al., 1994), and a repressor of the *spo0A* gene (Mandic-Mulec et al., 1995). These negative regulatory factors could act either by inhibiting the onset of sporulation and other stationary-phase events or by serving as timing monitors to coordinate specific sporulation events with changes in cell physiology. In one case, an extracellular signaling peptide has been shown to directly inhibit a phosphatase for Spo0F~P, an intermediate in the phosphorelay, but how the accumulation of the signaling peptide is regulated is not completely known (Perego, 1998).

GENE REGULATION BY SEQUENTIAL ACTIVATION OF RNA POLYMERASE SIGMA FACTORS

A second cascade of regulatory proteins is responsible for the temporally and spatially organized expression of genes in the two subcellular compartments (Stragier and Losick, 1996). The primary actors in this cascade are a series of RNA polymerase sigma factors. As mentioned above, the forespore compartment expresses a class of genes whose promoters are recognized by the σ^F-containing form of RNA polymerase and, subsequently, a σ^G-dependent class of genes. By contrast, only σ^E-dependent

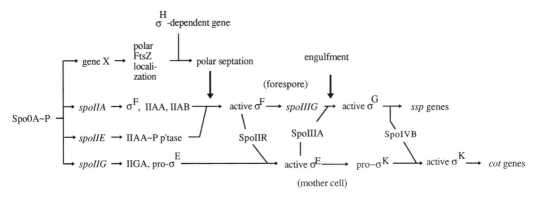

FIGURE 4 The sporulation sigma factor cascade. At the onset of stationary phase, activation of Spo0A by phosphorylation (Fig. 3) leads to expression of the *spoIIA*, *spoIIG*, and *spoIIE* operons. An additional Spo0A~P-dependent gene of unknown identity is required for asymmetric septation. σ^F, a product of the *spoIIA* operon, interacts with core RNA polymerase to direct transcription of the early class of forespore-specific genes. Activation of σ^F requires its release from an inhibitory complex with SpoIIAB, a process that depends on SpoIIAA, after the latter is dephosphorylated by SpoIIE. One of the early forespore-specific genes is *spoIIIG*, which codes for σ^G. When activated, a step that requires the mother cell-expressed *spoIIIA* operon, σ^G directs transcription of late forespore-specific genes, including those that encode internal (Ssp) proteins of the spore. σ^E is encoded in the *spoIIG* operon as an inactive precursor; activation by cleavage depends on an early forespore protein, SpoIIR. Upon activation, σ^E-containing RNA polymerase transcribes genes for early mother cell-specific proteins. Among these proteins is the precursor of σ^K. Activation by cleavage of pro-σ^K depends on a late forespore protein (SpoIVB), as well as on other mother cell proteins. When activated, σ^K recognizes promoters for late mother cell genes, including spore coat protein genes.

and σ^K-dependent genes are expressed in the mother cell compartment (Fig. 4).

A most remarkable aspect of bacterial sporulation is the extent to which morphology and gene expression are coupled (see chapter 8 for more details). At each stage of sporulation, the induction of genes depends on a signal transduced through the membrane. Even more remarkably, the expression of genes in the two compartments is coordinated by signals that pass back and forth by a process dubbed "criss-cross regulation" (Stragier and Losick, 1996). All four sporulation-specific sigma factors are inactive at the moment of synthesis. σ^F and σ^G are held in an inactive state by association with an inhibitor protein; σ^E and σ^K are synthesized as inactive precursor proteins whose proteolytic processing permits activation. Activation of σ^F in the forespore occurs upon formation of the septum and depends on a septum-associated phosphatase, SpoIIE (Arigoni et al., 1996). The septum also serves as a locus of communication between the forespore and the mother

cell. A σ^F-dependent gene transcribed in the forespore encodes a protein that migrates to the septum (forespore side) and stimulates a septum-associated protease that activates σ^E by removal of the N-terminal 27 amino acids of its precursor protein (Hofmeister et al., 1995; Karow et al., 1995; Londoño and Stragier, 1995). This cleavage occurs on both sides of the septum, but σ^E accumulates only in the mother cell compartment, perhaps because it is destroyed in the forespore. This mechanism of activation of σ^E appears, based on sequence DNA analysis, to be conserved in spore formers as disparate as *B. thuringiensis* (Adams et al., 1991) and *C. acetobutylicum* (Sauer et al., 1994).

σ^G is only made in the forespore because its promoter site is recognized by σ^F (and later by σ^G itself). But activation of σ^G depends on a complex of membrane proteins encoded by the *spoIIIA* operon and made in the mother cell under the control of σ^E (Illing and Errington, 1991; Stragier and Losick, 1996). These proteins apparently associate with the mother cell membrane before engulfment and with the outer forespore membrane after engulfment. The detailed mechanism whereby these proteins release σ^G from inhibition is not yet clear.

Expression of σ^K is limited to the mother cell because its gene has a σ^E-dependent promoter. But its activation by cleavage depends on σ^G. A gene expressed in the forespore under σ^G control encodes an inner forespore membrane protein that is thought to interact with three outer forespore membrane proteins (synthesized in the mother cell under the control of σ^E) to promote cleavage of pro-σ^K (Cutting et al., 1991).

The synthesis of σ^K in *B. subtilis* is complicated by the need for a specific DNA rearrangement to create the intact coding sequence (Stragier et al., 1989; Kunkel et al., 1990). That is, the coding sequence for σ^K, as it exists in the genome, is interrupted by an unrelated 48-kb sequence. This element, called *skin* (σ^K intervening element), resembles the genome of a temperate bacteriophage (Kunkel et al., 1990; Takemaru et al., 1995). It is excised at stage III, uniquely from the mother cell copy of the genome, by the action of a site-specific recombination enzyme encoded within the element itself (Popham and Stragier, 1992). The temporal and spatial specificity of excision is determined by its requirement for an accessory protein, SpoIIID (Kunkel et al., 1990). This protein is encoded by a gene transcribed from a σ^E-dependent promoter. Thus, SpoIIID is a mother cell-specific protein that is only made after stage II. The lack of *skin* excision in the forespore compartment guarantees that the next generation of growing cells (derived from the mature spore) will continue to carry *skin* within the σ^K gene.

Surprisingly, the presence of the *skin* element is not required for proper temporal or spatial expression of σ^K. That is, cells of *B. subtilis* engineered to carry a rearranged, uninterrupted σ^K gene sporulate at the normal frequency and express σ^K-dependent genes at the normal time and to the normal extent (Kunkel et al., 1990). Moreover, the σ^K genes of other *Bacillus* species (Adams et al., 1991) and *C. acetobutylicum* (Sauer et al., 1994) have no *skin* element. The most likely sequence of evolutionary events is that an uninterrupted σ^K gene evolved first (presumably in a clostridial progenitor) and was inherited by all derivative *Bacillus* and *Clostridium* species. At a relatively recent moment in evolution, a temperate phage integrated into the σ^K gene of an ancestor of the species we now call *B. subtilis*. The immediate integrant would probably have lost the ability to sporulate, but strong selection for restoration of sporulation would have led to the appearance of a SpoIIID protein that could stimulate excision of *skin*. SpoIIID probably already existed, since it is found in *B. thuringiensis* (Yoshisue et al., 1995) and has a second essential role in *B. subtilis*; it is a transcriptional regulator of some σ^K-dependent promoters (Halberg and Kroos, 1994). That is, an existing DNA-binding protein may have mutated to a form that allowed it to facilitate recombination as well as transcription, both in a temporally regulated, compartment-specific manner. Alternatively, the promoter for the *skin*-removing recombinase gene may have mutated to allow it to be recognized by SpoIIID and the σ^E-containing form of RNA polymerase.

ROLE OF SPORULATION IN PATHOGENESIS

Many spore-forming bacteria are important human and animal pathogens. For instance, *B. anthracis* is the causative agent of anthrax, a devastating disease of cows, sheep, and people and a major concern in the area of biological warfare. *B. cereus* is an important cause of food poisoning. The various forms of *B. thuringiensis* are potent killers of a wide variety of insects. Clostridial pathogens include *C. botulinum* (botulism), *C. perfringens* (gas gangrene and food poisoning), *C. tetani* (tetanus), *Clostridium difficile* (antibiotic-associated and pseudomembranous colitis), and *Clostridium sordellii* (endometritis). In what way is the ability to form a spore a factor in the virulence of these pathogens?

In some cases, spores are passively responsible for the disease. For instance, spores of *C. tetani* are commonly found in the soil. The association of tetanus with stepping on a buried nail is, in fact, the association of spores with the soil in which the nail was resting. The puncture created by the nail allows the spores to penetrate the tissue, germinate, and produce a neurotoxin that travels throughout the body. Similarly, the spores of *C. botulinum* allow the organism to survive in incompletely sterilized food. In fact, heating activates the spores for germination. If the food is kept warm and anaerobic thereafter, as in canned food stored at room temperature, the germinated spores return to vegetative growth and produce botulin toxin. *B. cereus* food poisoning is due to two or more toxins produced by bacteria that survive in improperly cooked food (Granum and Lund, 1997). A classic case is steamed rice, which is often contaminated at a low level with *B. cereus* spores. When the rice is cooked, the spores germinate and, if they are allowed to incubate at a temperature (<48°C) permitting significant growth, the bacteria will reach a dangerous titer. Contamination of dairy products by *B. cereus* has also been recognized as a growing problem (Granum and Lund, 1997).

In the case of *C. difficile* colitis, the spore appears to be the reservoir of disease-causing organisms within the human intestinal tract. That is, many people carry *C. difficile* as part of their normal flora but have no symptoms of disease. Only after completion of treatment with high doses of antibiotics do symptoms appear, coincident with a substantial increase in the titer of *C. difficile*. Yet these bacteria are often sensitive to the antibiotics that were used. Patients can be treated with another antibiotic to stem the disease, but many of these patients relapse after the secondary antibiotic treatment has been stopped. A model to explain this phenomenon is that the normal population of *C. difficile* is kept in check by competition with other bacteria; reduction in the competitor population by antibiotic treatment, coupled with occasional germination of *C. difficile* spores, allows the newly emerging vegetative cells to grow without impediment until they reach a titer at which their level of toxin production (induced during early stationary phase [Dupuy and Sonenshein, 1998]) proves too much for the patient. The relapse phenomenon can also be attributed to the ability to sporulate, since there will always be at least some spores that survive any antibiotic protocol (Walters et al., 1982). This model remains to be proved.

For *C. perfringens*, the relationship of sporulation to food poisoning is more direct. The gene for *C. perfringens* enterotoxin (*cpe*) is under the control of two sporulation promoters that are recognized in the mother cell compartment by the σ^E- and σ^K-containing forms of RNA polymerase (Zhao and Melville, 1998). As a result, ingested spores, which have germinated during passage through the stomach, colonize and grow in the intestinal tract. After they initiate sporulation, they turn on synthesis of enterotoxin and accumulate it in the mother cell. When the mother cell lyses, the released toxin destroys nearby cells of the intestinal epithelium. The result is severe diarrhea. From the point of view of the bacterium, induced diarrhea is advantageous for rapid spread of newly generated spores.

Other aspects of *C. perfringens* pathogenesis have a less direct relationship to spore formation. The various toxins associated with gas gangrene are synthesized during exponential growth phase under the control of a two-component system, VirR–VirS (Rood, 1998). The expression of these genes seems to depend on cell-cell signaling (Rood, 1998).

At least one type of *B. cereus* enterotoxin is produced at the transition from exponential growth to stationary phase, but an emetic toxin is only produced during sporulation, and its gene may be under sporulation control. A hemolytic enterotoxin is synthesized during exponential growth phase (Ryan et al., 1997).

In the case of *B. anthracis*, spores are the usual source of infectious bacteria (Sirard et al., 1996). They are typically ingested or inhaled by pasture animals, penetrate the lining of the intestinal or pulmonary tract, invade local tissues, and then enter the bloodstream. Animals can also be infected through the skin. Humans can be infected by contact, by inhalation of spore-contaminated aerosols, or by ingestion of contaminated meat. The virulence of *B. anthracis* is associated with the presence of two plasmids, which encode two toxins and enzymes for synthesis of a poly-D-glutamic acid capsule. These virulence factors are synthesized during growth. Thus, the dormant spores, while serving as the invading form of infectious organisms, are not pathogenic until they germinate. It has recently been shown that inhaled spores of *B. anthracis* are taken up by phagocytosis into alveolar macrophages and germinate within perinuclear phagolysosomes of the macrophages (Guidi-Rontani et al., 1999). Germination is followed rapidly by the expression of virulence genes. A major unanswered question concerns the nature of the germination signals in macrophages.

Various strains of *B. thuringiensis* produce one or more members of a family of pesticidal proteins that are often so highly expressed that they crystallize in the cytoplasm (Schnepf et al., 1998). In some cases, insecticidal toxins are synthesized during vegetative growth, but most toxin genes are under the control of sporulation-specific sigma factors (σ^E and σ^K) and crystallize in the mother cell cytoplasm of sporulating cells (Adams et al., 1991; Agaisse and Lereclus, 1994; Estruch et al., 1996; Schnepf et al., 1998). Transcription of the *Bacillus popilliae cry18Aa* gene, encoding a crystalline protein toxic for *Coleoptera*, also depends on sporulation-specific sigma factors (Zhang et al., 1998). The sequence of events that leads from ingestion of insecticide to death of the infected insect is not well understood, but certain aspects have been clarified or can be deduced. In some cases, ingestion of the crystals themselves is sufficient to kill the insect; in other cases, spores alone are sufficient or a mixture of spores and crystals is needed (Heimpel and Angus, 1959). Typically, commercial preparations of *B. thuringiensis* insecticide contain a mixture of toxin crystals and spores. A likely scenario for the infection process is that the ingested toxin molecules become solubilized in the intestinal tract and damage the midgut, leading to the break-

down of the barrier between the gut and the hemocoel (Lambert and Peferoen, 1992). At that point, the spores, whose germination would not have been efficient in the gut milieu, could invade the germination-inducing environment of the hemocoel, grow by obtaining nutrients from the dying insect, and produce more toxin and more spores (Lambert and Peferoen, 1992). Interestingly, crystal toxin-spore mixtures have a potential disadvantage. Broadcasting spores can lead to infection of insects or worms whose function is beneficial to humans. For example, spore preparations of *B. thuringiensis* are considered dangerous in Japan because of their potential to destroy the silk industry. This concern has been one of the factors stimulating interest in engineering nonsporulating bacteria or plants to produce *B. thuringiensis* toxin proteins (Lambert and Peferoen, 1992).

In cases where *B. thuringiensis* spores alone are capable of causing a lethal infection, one might guess that a few spores germinate initially; their growth and sporulation would produce a small amount of toxin and other virulence factors that could damage the intestinal tract and allow other spores to invade the hemocoel. Subsequent production of large amounts of toxin would lead to rapid death of the insect.

Strains of *B. thuringiensis* incapable of producing crystal proteins can still cause disease, especially when injected into or fed to lepidopteran larvae. Phospholipase C has been postulated to be one of the factors responsible for this residual virulence. The phospholipase gene, *plcA*, is induced as cells pass from exponential growth phase to stationary phase. This induction has been attributed to the transient induction of a positive regulatory protein, PlcR (Lereclus et al., 1996). Interestingly, PlcR seems to regulate expression of many virulence genes, and the protein and its target site have been found in other pathogenic *Bacillus* species (Agaisse et al., 1999). Specifically, one of the enterotoxin genes of *B. cereus* is regulated by PlcR (Agaisse et al., 1999).

Why are toxins produced preferentially during stationary phase and sporulation? Production early during stationary phase is likely to reflect a desire of the producing bacteria to kill neighboring cells in order to reduce competition for limited nutritional resources and as a means of liberating such resources from no-longer-intact competitor cells. Here the goal is to maintain growth of the toxin-producing cell in the face of limited nutrients. When sporulating cells produce toxin in the mother cell compartment, however, they probably have a different goal. The toxin is released by lysis of the mother cell, an event that corresponds to maturation and release of the spore. These toxins may provoke diarrhea in the infected host, guaranteeing rapid and propulsive dispersal of the spores.

PERSPECTIVES

Despite remarkable advances in our understanding of endospore formation, there remain important and interesting questions to answer. Given that all environmental signals that induce sporulation are transduced through the Spo0A phosphorelay, how are those signals actually perceived by the cell? What chemicals serve as indicators of nutritional limitation, what DNA structures indicate the integrity of the chromosome and the completion of replication rounds, and which components of the phosphorelay are affected most directly by each signal? Similarly, the mechanism by which one of two potential polar sites of septation is chosen in early sporulating cells (as opposed to the medial site or the other polar site) is still mysterious, and the forces that drive the mother cell to engulf the forespore remain to be discovered. The transmembrane signaling that is essential for activation of σ^E, σ^G, and σ^K is at the cusp of full understanding. Most, if not all, of the critical proteins are known; we still need to figure out how they interact. The mechanism of the remarkable assembly of spore coat proteins around the forespore is beginning to be uncovered, but the detailed interactions of assembly and structural proteins will provide fruitful areas of inquiry for many years to come. Finally, the germination of spores will have to be attacked again. After an initial burst of activ-

ity many years ago, focusing on identification of germinants and genes required for germination and outgrowth, the subject has lain essentially dormant for too long. The explosion of new techniques in molecular genetics, bacterial cytology, and biochemistry should stimulate new approaches to an old and fascinating problem.

As spore formers other than *B. subtilis*, especially pathogenic species, are investigated in greater detail, we can anticipate that the vast body of knowledge obtained with the paradigmatic organism will serve well as a model for the less well-studied bacteria. Understanding the specific properties of the pathogenic species will provide important new information about the breadth of endospore formers and the relationship between differentiation and virulence.

The *B. subtilis* system has been a critically useful paradigm not only for the specific knowledge that has emerged but also because it has served as a testing ground for powerful new experimental approaches. By combining genome sequence analysis, molecular studies of gene regulation, and cytological studies (using immunological and fluorescence microscopy techniques), it has become possible to determine the time of synthesis and the spatial localization of any protein during the sporulation process. In this way, it is possible to derive very detailed information about a complex series of events from a whole-cell perspective.

ACKNOWLEDGMENTS

I thank E. Angert for providing Fig. 2, A. Driks for very helpful comments on the manuscript, and H. Agaisse, C. Guidi-Rontani, M. Mock, and P. Hanna for helpful discussions and for making results available before publication.

Unpublished work from my laboratory was supported by a U.S. Public Health Service research grant (GM42219).

REFERENCES

Adams, L. F., K. L. Brown, and H. R. Whiteley. 1991. Molecular cloning and characterization of two genes encoding sigma factors that direct transcription from a *Bacillus thuringiensis* crystal protein gene promoter. *J. Bacteriol.* **173**:3846–3854.

Agaisse, H., and D. Lereclus. 1994. Structural and

functional analysis of the promoter region involved in full expression of the *cryIIIA* toxin gene of *Bacillus thuringiensis*. *Mol. Microbiol.* **13:**97–107.

Agaisse, H., M. Gominet, O. A. Økstad, A.-B. Kolsto, and D. Lereclus. 1999. PlcR is a pleiotropic regulator of extracellular virulence factor gene expression in *Bacillus thuringiensis*. *Mol. Microbiol.* **32:**1043–1053.

Angert, E. R., and R. M. Losick. 1998. Propagation by sporulation in the guinea pig symbiont *Metabacterium polyspora*. *Proc. Natl. Acad. Sci. USA* **95:**10218–10223.

Arigoni, F., L. Duncan, S. Alper, R. Losick, and P. Stragier. 1996. SpoIIE governs the phosphorylation state of a protein regulating transcription factor sigma F during sporulation in *Bacillus subtilis*. *Proc. Natl. Acad. Sci. USA* **93:**3238–3242.

Baldus, J. M., B. D. Green, P. Youngman, and C. P. Moran, Jr. 1994. Phosphorylation of *Bacillus subtilis* transcription factor Spo0A stimulates transcription from the *spoIIG* promoter by enhancing binding to weak 0A boxes. *J. Bacteriol.* **176:**296–306.

Bird, T. H., J. K. Grimsley, J. A. Hoch, and G. B. Spiegelman. 1993. Phosphorylation of Spo0A activates its stimulation of in vitro transcription from the *Bacillus subtilis spoIIG* operon. *Mol. Microbiol.* **9:**741–749.

Burbulys, D., K. A. Trach, and J. A. Hoch. 1991. Initiation of sporulation in *B. subtilis* is controlled by a multicomponent phosphorelay. *Cell* **64:**545–552.

Chapman, G. B. 1959. Electron microscopy of ultrathin sections of bacteria. II. Sporulation of *Bacillus megaterium* and *Bacillus cereus*. *J. Bacteriol.* **71:**348–355.

Chung, J. D., G. Stephanopoulos, K. Ireton, and A. D. Grossman. 1994. Gene expression in single cells of *Bacillus subtilis*: evidence that a threshold mechanism controls the initiation of sporulation. *J. Bacteriol.* **176:**1977–1984.

Clouart, J., and A. L. Sonenshein. 1999. Unpublished results.

Cutting, S., A. Driks, R. Schmidt, B. Kunkel, and R. Losick. 1991. Forespore-specific transcription of a gene in the signal transduction pathway that governs pro-sigma K processing in *Bacillus subtilis*. *Genes Dev.* **5:**456–466.

Dawes, I. W., and J. Mandelstam. 1970. Sporulation of *Bacillus subtilis* in continuous culture. *J. Bacteriol.* **103:**529–535.

Driks, A., S. Roels, B. Beall, C. Moran, and R. Losick. 1994. Subcellular localization of proteins involved in the assembly of the spore coat of *Bacillus subtilis*. *Genes Dev.* **8:**234–244.

Dubnau, D., and M. Roggiani. 1990. Growth medium-independent genetic competence mutants of *Bacillus subtilis*. *J. Bacteriol.* **172:**4048–4055.

Duncan, L., and R. Losick. 1993. SpoIIAB is an anti-σ factor that binds to and inhibits transcription by regulatory protein σ^F from *Bacillus subtilis*. *Proc. Natl. Acad. Sci. USA* **90:**2325–2329.

Dupuy, B., and A. L. Sonenshein. 1998. Regulated transcription of *Clostridium difficile* toxin genes. *Mol. Microbiol.* **27:**107–120.

Estruch, J. J., G. W. Warren, M. A. Mullins, G. J. Nye, and J. A. Craig. 1996. Vip3A, a novel *Bacillus thuringiensis* vegetative insecticidal protein with a wide spectrum of activities against lepidopteran insects. *Proc. Natl. Acad. Sci. USA* **93:**5389–5394.

Gordon, R. E., W. C. Haynes, and C. H.-N. Pang. 1973. *The Genus Bacillus. Agricultural Handbook No. 427.* Agricultural Research Service, U.S. Department of Agriculture, Washington, D.C.

Granum, P. E., and T. Lund. 1997. *Bacillus cereus* and its food poisoning toxins. *FEMS Microbiol. Lett.* **157:**223–228.

Grossman, A. D. 1995. Genetic networks controlling the initiation of sporulation and the development of genetic competence in *Bacillus subtilis*. *Annu. Rev. Genet.* **29:**477–508.

Grossman, A. D., and R. Losick. 1988. Extracellular control of spore formation in *Bacillus subtilis*. *Proc. Natl. Acad. Sci. USA* **85:**4369–4373.

Guidi-Rontani, C., M. Weber-Levy, E. Labruyere, and M. Mock. 1999. Germination of *Bacillus anthracis* spores within alveolar macrophages. *Mol. Microbiol.* **31:**9–17.

Halberg, R., and L. Kroos. 1994. Sporulation regulatory protein SpoIIID from *Bacillus subtilis* activates and represses transcription by both mother-cell-specific forms of RNA polymerase. *J. Mol. Biol.* **243:**425–436.

Heimpel, A. M., and T. A. Angus. 1959. The site of action of crystalliferous bacteria in Lepidoptera larvae. *J. Insect Pathol.* **1:**152–170.

Hoch, J. A. 1993. Regulation of the phosphorelay and the initiation of sporulation in *Bacillus subtilis*. *Annu. Rev. Microbiol.* **47:**441–465.

Hofmeister, A., A. Londoño-Vallejo, E. Harry, P. Stragier, and R. Losick. 1995. Extracellular signal protein triggering the proteolytic activation of a developmental transcription factor in *B. subtilis*. *Cell* **83:**219–226.

Holt, J. G., N. R. Krieg, P. H. A. Sneath, J. T. Staley, and S. T. Williams. 1994. *Bergey's Manual of Determinative Bacteriology*, 9th ed. Williams and Wilkins, Baltimore, Md.

Illing, N., and J. Errington. 1991. The *spoIIIA* operon of *Bacillus subtilis* defines a new temporal class of mother-cell-specific sporulation genes under the control of the σ^E form of RNA polymerase. *Mol. Microbiol.* **5:**1927–1940.

Ireton, K., and A. D. Grossman. 1994. A developmental checkpoint couples the initiation of sporulation to DNA replication in *Bacillus subtilis. EMBO J.* **13:**1566–1573.

Ireton, K., N. W. Gunther IV, and A. D. Grossman. 1994. *spo0J* is required for normal chromosome segregation as well as the initiation of sporulation in *Bacillus subtilis. J. Bacteriol.* **176:**5320–5329.

Jin, S. 1995. Ph.D. thesis. Tufts University, Boston, Mass.

Jin, S., P. A. Levin, K. Matsuno, A. D. Grossman, and A. L. Sonenshein. 1997. Deletion of the *Bacillus subtilis* isocitrate dehydrogenase gene causes a block at stage I of sporulation. *J. Bacteriol.* **179:**4725–4732.

Johnstone, K. 1994. The trigger mechanism of spore germination: current concepts. *J. Appl. Bacteriol. Symp. Suppl.* **76:**17S–24S.

Karow, M. L., P. Glaser, and P. J. Piggot. 1995. Identification of a gene, *spoIIR*, that links the activation of sigma E to the transcriptional activity of sigma F during sporulation in *Bacillus subtilis. Proc. Natl. Acad. Sci. USA* **92:**2012–2016.

Kok, J., K. A. Trach, and J. A. Hoch. 1994. Effects on *Bacillus subtilis* of a conditional lethal mutation in the essential GTP-binding protein Obg. *J. Bacteriol.* **176:**7155–7160.

Kuhner, C. H., C. Frank, A. Griesshammer, M. Schmittroth, G. Acker, A. Gossner, and H. L. Drake. 1997. *Sporomusa sivacetica* sp. nov., an acetogenic bacterium isolated from aggregated forest soil. *Int. J. Syst. Bacteriol.* **47:**352–358.

Kunkel, B., R. Losick, and P. Stragier. 1990. The *Bacillus subtilis* gene for the developmental transcription factor σ^K is generated by excision of a dispensable DNA element containing a sporulation recombinase gene. *Genes Dev.* **4:**525–535.

Lambert, B., and M. Peferoen. 1992. Insecticidal promise of *Bacillus thuringiensis.* Facts and mysteries about a successful biopesticide. *BioScience* **42:**112–122.

Lazazzera, B. A., and A. D. Grossman. 1998. The ins and outs of peptide signalling. *Trends Microbiol.* **6:**288–294.

LeDeaux, J. R., N. Yu, and A. D. Grossman. 1995. Different roles for KinA, KinB, and KinC in the initiation of sporulation in *Bacillus subtilis. J. Bacteriol.* **177:**861–863.

Lereclus, D., H. Agaisse, M. Gominet, S. Salamitou, and V. Sanchis. 1996. Identification of a *Bacillus thuringiensis* gene that positively regulates transcription of the phosphatidylinositol-specific phospholipase C gene at the onset of the stationary phase. *J. Bacteriol.* **178:**2749–2756.

Levin, P. A., and R. Losick. 1996. Transcription factor Spo0A switches the localization of the cell division protein FtsZ from a medial to a bipolar pattern in *Bacillus subtilis. Genes Dev.* **10:**478–488.

Londoño-Vallejo, J. A., and P. Stragier. 1995. Cell-cell signaling pathway activating a developmental transcription factor in *Bacillus subtilis. Genes Dev.* **9:**503–508.

Mandic-Mulec, I., L. Doukhan, and I. Smith. 1995. The *Bacillus subtilis* SinR protein is a repressor of the key sporulation gene *spo0A. J. Bacteriol.* **177:**4619–4627.

Margulis, L., J. Z. Jorgensen, S. Dolan, R. Kolchinsky, F. A. Rainey, and S.-C. Lo. 1998. The *Arthromitus* stage of *Bacillus cereus*: intestinal symbionts of animals. *Proc. Natl. Acad. Sci. USA* **95:**1236–1241.

Matsuno, K., T. Blais, A. W. Serio, T. Conway, T. M. Henkin, and A. L. Sonenshein. 1999. Metabolic imbalance and sporulation in an isocitrate dehydrogenase mutant of *Bacillus subtilis. J. Bacteriol.* **181:**3382–3391.

Melville, S. B., and A. L. Sonenshein. 1996. Unpublished results.

Mitani, T., J. E. Heinze, and E. Freese. 1977. Induction of sporulation in *Bacillus subtilis* by decoyinine or hadacidin. *Biochem. Biophys. Res. Commun.* **77:**1118–1125.

Moir, A., and D. A. Smith. 1990. The genetics of bacterial spore germination. *Annu. Rev. Microbiol.* **44:**531–553.

Nakano, M. M., and P. Zuber. 1998. Anaerobic growth of a "strict aerobe" (*Bacillus subtilis*). *Annu. Rev. Microbiol.* **52:**165–190.

Ohlsen, K. L., J. K. Grimsley, and J. A. Hoch. 1994. Deactivation of the sporulation transcription factor Spo0A by the Spo0E protein phosphatase. *Proc. Natl. Acad. Sci. USA* **91:**1756–1760.

Perego, M. 1998. Kinase-phosphatase competition regulates *Bacillus subtilis* development. *Trends Microbiol.* **6:**366–370.

Piggot, P. J., and J. G. Coote. 1976. Genetic aspects of bacterial endospore formation. *Bacteriol. Rev.* **40:**908–962.

Popham, D. L., and P. Stragier. 1992. Binding of the *Bacillus subtilis spoIVCA* product to the recombination sites of the element interrupting the sigma K-encoding gene. *Proc. Natl. Acad. Sci. USA* **89:**5991–5995.

Price, K. D., and R. Losick. 1999. A four-dimensional view of assembly of a morphogenetic protein during sporulation in *Bacillus subtilis. J. Bacteriol.* **181:**781–790.

Priest, F. G. 1993. Systematics and ecology of *Bacillus*, p. 3–16. *In* A. L. Sonenshein, J. A. Hoch, and R. Losick (ed.), *Bacillus subtilis and Other Gram-Positive Bacteria: Biochemistry, Physiology, and Molecular Genetics.* American Society for Microbiology, Washington, D.C.

Rood, J. I. 1998. Virulence genes of *Clostridium perfringens. Annu. Rev. Microbiol.* **52:**333–360.

Roper, G., J. A. Short, and P. D. Walker. 1976. The ultrastructure of *Clostridium perfringens* spores, p. 279–296. *In* A. M. Barker, J. Wold, D. J. Ellar, G. H. Dring, and G. W. Gould (ed.), *Spore Research.* Academic Press, London, England.

Ryan, P. A., J. D. MacMillan, and B. Zilinskas. 1997. Molecular cloning and characterization of the genes encoding the L_1 and L_2 components of hemolysin BL from *Bacillus cereus. J. Bacteriol.* **179:** 2551–2556.

Ryter, A. 1965. Etude morphologique de la sporulation de *Bacillus subtilis. Ann. Inst. Pasteur* **108:** 40–60.

Sass, H., E. Wieringa, H. Cypionka, H. D. Babenzien, and J. Overmann. 1998. High genetic and physiological diversity of sulfate-reducing bacteria isolated from an oligotrophic lake sediment. *Arch. Microbiol.* **170:**243–251.

Sauer, U., A. Treuner, M. Buchholz, J. D. Santangelo, and P. Durre. 1994. Sporulation and primary sigma factor homologous genes in *Clostridium acetobutylicum. J. Bacteriol.* **176:**6572–6582.

Sauer, U., J. D. Santangelo, A. Treuner, M. Buchholz, and P. Durre. 1995. Sigma factor and sporulation genes in *Clostridium. FEMS Microbiol. Rev.* **17:**331–340.

Schaeffer, P., J. Millet, and J.-P. Aubert. 1965. Catabolite repression of bacterial sporulation. *Proc. Natl. Acad. Sci. USA* **54:**704–711.

Schnepf, E., N. Crickmore, J. van Rie, D. Lereclus, J. Baum, J. Feitelson, D. R. Zeigler, and D. H. Dean. 1998. *Bacillus thuringiensis* and its pesticidal crystal proteins. *Microbiol. Mol. Biol. Rev.* **62:**775–806.

Scholz, T., W. Demharter, R. Hensel, and O. Kandler. 1987. *Bacillus pallidus* sp. nov., a new thermophilic species from sewage. *Syst. Appl. Microbiol.* **9:**91–96.

Serror, P., and A. L. Sonenshein. 1996. CodY is required for nutritional repression of *Bacillus subtilis* genetic competence. *J. Bacteriol.* **178:**5910–5915.

Setlow, P. 1973. Deoxyribonucleic acid synthesis and deoxynucleotide metabolism during bacterial spore germination. *J. Bacteriol.* **114:**1099–1107.

Setlow, P. 1975. Protein metabolism during germination of *Bacillus megaterium* spores. II. Degradation of pre-existing and newly synthesized protein. *J. Biol. Chem.* **250:**631–637.

Setlow, P. 1995. Mechanisms for the prevention of damage to DNA in spores of *Bacillus* species. *Annu. Rev. Microbiol.* **49:**29–54.

Setlow, P., and G. Primus. 1975. Protein metabolism during germination of *Bacillus megaterium* spores. I. Protein synthesis and amino acid metabolism. *J. Biol. Chem.* **250:**623–630.

Sirard, J.-C., M. Malville, A. Fouet, and M.

Mock. 1996. Physiopathologie moléculaire de la maladie du charbon. *Rev. Med. Vet.* **147:**653–670.

Slack, F. J., P. Serror, E. Joyce, and A. L. Sonenshein. 1995. A gene required for nutritional repression of the *Bacillus subtilis* dipeptide permease operon. *Mol. Microbiol.* **15:**689–702.

Sonenshein, A. L. 1989. Metabolic regulation of sporulation and other stationary-phase phenomena, p. 109–130. *In* I. Smith, R. A. Slepecky, and P. Setlow (ed.), *Regulation of Prokaryotic Development.* American Society for Microbiology, Washington, D.C.

Sonenshein, A. L. 1998. Unpublished results.

Sterlini, J. M., and J. Mandelstam. 1969. Commitment to sporulation in *Bacillus subtilis* and its relationship to development of antibiotic resistance. *Biochem. J.* **113:**29–37.

Stragier, P., and R. Losick. 1996. Molecular genetics of sporulation in *Bacillus subtilis. Annu. Rev. Genet.* **30:**297–341.

Stragier, P., B. Kunkel, L. Kroos, and R. Losick. 1989. Chromosomal rearrangement generating a composite gene for a developmental transcription factor. *Science* **243:**507–512.

Strauch, M. A., and J. A. Hoch. 1993. Transition-state regulators: sentinels of *Bacillus subtilis* post-exponential gene expression. *Mol. Microbiol.* **7:** 337–342.

Takemaru, K., M. Mizuno, T. Sato, M. Takeuchi, and Y. Kobayashi. 1995. Complete nucleotide sequence of a *skin* element excised by DNA rearrangement during sporulation in *Bacillus subtilis. Microbiology* **141:**323–327.

Vidwans, S. J., K. Ireton, and A. D. Grossman. 1995. Possible role for the essential GTP-binding protein Obg in regulating the initiation of sporulation in *Bacillus subtilis. J. Bacteriol.* **177:**3308–3311.

Walters, B. A. J., R. Roberts, R. Stafford, and E. Seneviratne. 1982. Relapse of antibiotic-associated colitis: endogenous persistence of *Clostridium difficile* during vancomycin therapy. *Gut* **24:** 206–212.

Wang, L., R. Grau, M. Perego, and J. A. Hoch. 1997. A novel histidine kinase inhibitor regulating development in *Bacillus subtilis. Genes Dev.* **11:** 2569–2579.

Wendrich, T. M., and M. A. Marahiel. 1997. Cloning and characterization of a *relA/spoT* homologue from *Bacillus subtilis. Mol. Microbiol.* **26:** 65–79.

Xue, Y., and W. L. Nicholson. 1996. The two major spore DNA repair pathways, nucleotide excision repair and spore photoproduct lyase, are sufficient for the resistance of *Bacillus subtilis* spores to artificial UV-C and UV-B but not to solar radiation. *Appl. Environ. Microbiol.* **62:**2221–2227.

Yoshisue, H., K. Ihara, T. Nishimoto, H. Sakai,

and T. Komano. 1995. Cloning and characterization of a *Bacillus thuringiensis* homolog of the *spoIIID* gene of *Bacillus subtilis*. *Gene* **154:**23–29.

Young, I. E., and P. C. Fitz-James. 1962. Chemical and morphological studies of bacterial spore formation. IV. The development of spore refractility. *J. Cell Biol.* **12:**115–133.

Young, M., and S. T. Cole. 1993. *Clostridium*, p. 35–52. *In* A. L. Sonenshein, J. A. Hoch, and R. Losick (ed.), *Bacillus subtilis and Other Gram-Positive Bacteria: Biochemistry, Physiology, and Molecular Genetics*. American Society for Microbiology, Washington, D.C.

Zhang, J., H. U. Schairer, W. Schnetter, D. Lereclus, and H. Agaisse. 1998. *Bacillus popilliae cry18Aa* operon is transcribed by sigmaE and sigmaK forms of RNA polymerase from a single initiation site. *Nucleic Acids Res.* **26:**1288–1293.

Zhang, L., M. L. Higgins, and P. J. Piggot. 1997. The division during bacterial sporulation is symmetrically located in *Sporosarcina ureae*. *Mol. Microbiol.* **25:**1091–1098.

Zhao, Y., and S. B. Melville. 1998. Identification and characterization of sporulation-dependent promoters upstream of the enterotoxin gene (*cpe*) of *Clostridium perfringens*. *J. Bacteriol.* **180:**136–142.

Zheng, L. B., W. P. Donovan, P. C. Fitz-James, and R. Losick. 1988. Gene encoding a morphogenic protein required in the assembly of the outer coat of the *Bacillus subtilis* endospore. *Genes Dev.* **2:**1047–1054.

Zuber, P., M. M. Nakano, and M. A. Marahiel. 1993. Peptide antibiotics, p. 897–916. *In* A. L. Sonenshein, J. A. Hoch, and R. Losick (ed.), *Bacillus subtilis and Other Gram-Positive Bacteria: Biochemistry, Physiology, and Molecular Genetics*. American Society for Microbiology, Washington, D.C.

REGULATION OF THE INITIATION OF ENDOSPORE FORMATION IN *BACILLUS SUBTILIS*

William F. Burkholder and Alan D. Grossman

7

Under certain environmental conditions, cells of the bacterium *Bacillus subtilis* can initiate a developmental process leading to the formation of dormant endospores. Much of the work on sporulation in *B. subtilis* has focused on several key questions. What are the environmental conditions that control sporulation? What are the regulatory genes and proteins that are required for sporulation? How are multiple external and internal conditions sensed and signals generated, transduced, and integrated to control the decision to sporulate? Below we describe some of the regulatory circuits that control the initiation of sporulation and the conditions that influence the activity of the regulatory factors.

INTRODUCTION

The responses of the gram-positive soil bacterium *B. subtilis* to changing nutrient conditions have been studied extensively. Cells of *B. subtilis*, like those of many other bacteria, respond to the booms and busts of nutrient availability by growing exponentially when nutrients are plentiful and adjusting metabolism and slowing growth as nutrient levels fall. If nutrients be-

come too scarce—the onset of starvation—the cells cease growing exponentially and enter stationary phase. During the transition to stationary phase, the cells adapt by initiating responses to utilize available nutrients more efficiently, to scavenge for new nutrient sources, and to inhibit the growth of other organisms. Stationary-phase responses of *B. subtilis* can include production of degradative enzymes (proteases, nucleases, and amylases) and antibiotics, the onset of cell motility, the development of genetic competence for uptake of exogenous DNA, and the initiation of spore formation. Overlapping genetic networks determine the choice of response based on environmental and intracellular conditions (reviewed in Dubnau, 1993; Grossman, 1995; Msadek et al., 1995; Smith, 1993; and Strauch and Hoch, 1993).

Under conditions of nutrient depletion at high cell density, the response of last resort for *B. subtilis* is to initiate the process of spore formation, the production of metabolically quiescent cells able to withstand extreme environmental insults and indefinite periods of starvation. In *B. subtilis*, sporulation begins with a significant alteration in the cell division cycle. During exponential growth, cells divide in the middle to produce two indistinguishable daughter cells (the vegetative cell cycle). Sporulating cells, on the other hand, divide asym-

W. F. Burkholder and A. D. Grossman, Department of Biology, Building 68-530, Massachusetts Institute of Technology, Cambridge, MA 02139.

Prokaryotic Development, edited by Y. V. Brun and L. J. Shimkets,
© 2000 American Society for Microbiology, Washington, DC 20005-4171

Vegetative cell cycle

Spore morphogenesis

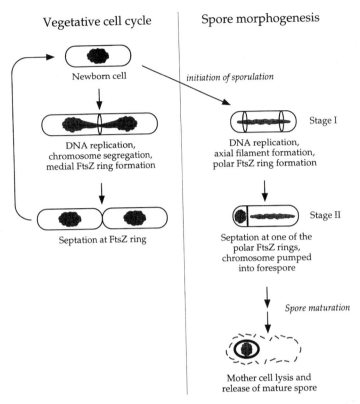

initiation of sporulation

Newborn cell

DNA replication,
chromosome segregation,
medial FtsZ ring formation

Septation at FtsZ ring

Stage I

DNA replication,
axial filament formation,
polar FtsZ ring formation

Stage II

Septation at one of the
polar FtsZ rings,
chromosome pumped
into forespore

Spore maturation

Mother cell lysis and
release of mature spore

FIGURE 1 Cell division cycle during vegetative growth and sporulation. During the vegetative cell cycle, cells grow along their long axes to roughly twice their length at birth and divide at midcell. Sporulating cells switch the site of septation to a polar site, so that septation yields a large and a small cell, the mother cell and the forespore, respectively. The nascent division site is marked in both vegetatively growing and sporulating cells by rings of FtsZ, a tubulin-like protein required for cytokinesis. Rings of FtsZ assemble near both cell poles in sporulating cells, but polar septation occurs at only one of the polar rings. The forespore subsequently develops into a mature spore in a defined series of morphological stages (not shown), aided by the mother cell and culminating in lysis of the mother cell and release of the mature spore. Cells are proficient in making the transition from vegetative growth to the sporulation pathway only during a limited period of the cell cycle following replication initiation; this period probably ends when the cells become committed to another round of the vegetative cell cycle. The nucleoid mass, containing chromosomal DNA, is depicted in light gray, rings of FtsZ are depicted as thin black ovals, the polar septum at stage II of sporulation is depicted as a straight black line, and the mature spore coat is depicted as a thick black oval.

metrically to yield a large cell, the mother cell, and a small cell, the forespore (Fig. 1). The polar division triggers cell-type-specific gene expression, first in the forespore and then in the mother cell (reviewed in chapter 8 and Stragier and Losick, 1996). The mother cell then engulfs the forespore, creating a cell within a cell, with additional genes induced in each type of cell. At the end of development, the mother cell lyses, releasing the mature environmentally resistant spore. When returned to favorable conditions, the spores germinate and resume vegetative growth.

The decision to sporulate is a finely balanced one. If cells fail to sporulate before conditions become too severe, they may perish. On the other hand, if cells sporulate while adaptation and the resumption of growth is still possible, they lose the opportunity to propagate and may be overrun by other organisms in the environment. In addition, sporulation requires synthesis of many new products, completion of ongoing rounds of DNA replication, and cell division. Regulatory mechanisms exist to help ensure that cells do not initiate sporulation unless it is likely that they will be able to complete the process successfully.

INITIATION OF SPORULATION: ACTIVATION OF Spo0A AND INDUCTION OF SIGMA-H

Many genes that are required for sporulation (*spo*) have been identified. The *spo* genes have been named based on the morphological stage at which the gene product is required. Hence, *spo0* genes are required for the initiation of sporulation, and loss-of-function mutations in the *spo0* genes generally prevent the earliest morphological changes associated with sporulation, most notably blocking asymmetric division. *spoII* mutants are able to undergo asymmetric division but do not proceed further in development.

spo0A and *spo0H* encode transcription factors that are required for the initiation of sporulation. *spo0A* encodes a DNA binding protein that functions as a transcriptional activator and repressor, and *spo0H* (also known as *sigH*) encodes a sigma factor, sigma-H, of RNA polymerase. Both *spo0A* and *spo0H* are required for asymmetric division and for transcription of the genes required for establishing cell-type-specific gene expression.

Spo0A

Spo0A is a response regulator protein that is activated by phosphorylation of an aspartate residue in the N-terminal domain (reviewed in Grossman, 1995; Hoch, 1995; and Spiegelman et al., 1995). Spo0A obtains phosphate from a multicomponent phosphotransfer pathway, the phosphorelay (Fig. 2). Histidine kinases autophosphorylate, and phosphate is transferred first to Spo0F, then from Spo0F to Spo0B, and finally from Spo0B to Spo0A. Many of the signals regulating the initiation of sporulation operate by regulating the activity of the kinases, the phosphorelay, and ultimately the amount of phosphorylated Spo0A (Spo0A~P) that accumulates.

Phosphorylation increases the affinity of Spo0A for its DNA binding sites and may enhance its interactions with RNA polymerase holoenzyme (Baldus et al., 1995; Bird et al., 1996; Bramucci et al., 1995; Buckner and Moran, 1998; Buckner et al., 1998; Hatt and Youngman, 1998; Rowe-Magnus and Spiegelman, 1998; Schyns et al., 1997; Spiegelman et al., 1995). Spo0A~P activates the expression of most of its target genes, often by interacting directly with the sigma subunit of RNA polymerase, but inhibits the expression of several

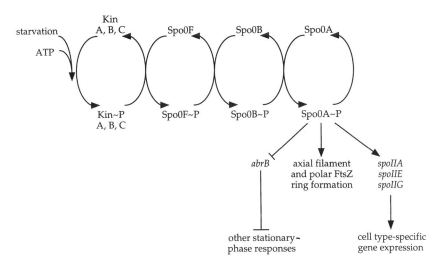

FIGURE 2 The phosphorelay and production of Spo0A~P. The sensor kinases KinA, KinB, and KinC autophosphorylate on a histidine. Phosphate is transferred to Spo0F, then to Spo0B, and finally to Spo0A. Low levels of Spo0A~P are sufficient to repress transcription of *abrB*, derepressing expression of many of the stationary-phase response pathways negatively regulated by AbrB. Higher levels of Spo0A~P stimulate axial-filament formation, polar septation, and transcription of genes (e.g., *spoIIA*, *spoIIE*, and *spoIIG*) required for cell-type-specific gene expression.

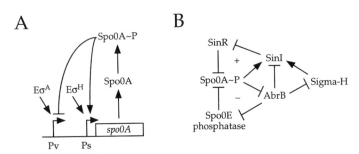

A

B

FIGURE 3 Positive and negative autoregulatory loops involved in production of Spo0A~P. (A) *spo0A* is expressed from two promoters, a vegetative-growth-specific promoter, Pv, transcribed by sigma-A-associated RNA polymerase holoenzyme (Eσ^A), and a sporulation-specific promoter, Ps, transcribed by sigma-H-associated RNA polymerase holoenzyme (Eσ^H). Spo0A~P can negatively autoregulate *spo0A* transcription by binding and inhibiting expression from Pv. Transcription from Ps is activated directly by Spo0A~P and by the induction of sigma-H following starvation. (B) Spo0A~P positively and negatively regulates its expression and activity by controlling the expression or activity of two other proteins, SinR and Spo0E. In a positive feedback loop stimulating *spo0A* transcription, Spo0A~P inhibits the activity of a transcriptional repressor of *spo0A*, SinR, by inducing expression of a SinR inhibitor, SinI. Spo0A~P stimulates transcription of *sinI* by inhibiting expression of AbrB, a transcriptional repressor of both *sinI* and the gene encoding sigma-H, *spo0H*, which drives transcription of *sinI*. Spo0A~P also binds to the promoter region and directly activates transcription of *sinI*. In a negative feedback loop inhibiting Spo0A activity, Spo0A~P induces the expression of the phosphatase Spo0E, which specifically dephosphorylates Spo0A~P. Spo0A~P induces transcription of *spo0E* by inhibiting expression of AbrB, which represses *spo0E* transcription. Arrows indicate activation; barred lines indicate inhibition.

genes. The locations of Spo0A binding sites within a promoter region determine whether Spo0A~P acts as an activator or repressor of transcription.

Many of the genes known to be positively regulated by Spo0A~P are either involved in activating Spo0A (positive autoregulatory loops) (Fig. 3 and 4) or required for the first stages of cell-type-specific gene expression following polar septation (Fig. 2). The genes in the latter category encode the first cell-specific transcription factors, sigma-F (*spoIIAC*) and sigma-E (*spoIIGB*; also designated *sigE*), and factors required for their activation (reviewed in chapter 8 and Stragier and Losick, 1996).

Spo0A~P also controls reorganization of the cell cycle. Early during the initiation of sporulation, the chromosomal DNA forms what is referred to as an axial filament, a structure that extends from one end of the cell to the other (Fig. 1 and 2) (Piggot and Coote, 1976). In addition, the cell division protein FtsZ forms ring structures near the cell poles. During exponential growth, FtsZ forms a ring structure at midcell, marking the site of division and recruiting the cell division machinery. Spo0A~P is needed to reposition the FtsZ rings to polar locations during the initiation of

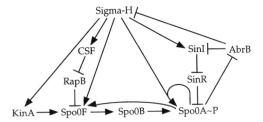

FIGURE 4 Sigma-H is required for the full expression and activation of Spo0A. Sigma-H drives transcription of *kinA*, *spo0F*, and *spo0A*, contributing to the accumulation of high levels of Spo0A~P. Sigma-H also stimulates transcription of *spo0A* by driving expression of *sinI*, activating *spo0A* by inhibiting the activity of a *spo0A* transcriptional repressor, SinR. Sigma-H stimulates the activation of Spo0A by driving expression of the secreted peptide pheromone, CSF, which inhibits the phosphatase, RapB, that dephosphorylates Spo0F~P. Spo0A~P contributes to the full induction of sigma-H by inhibiting AbrB, which inhibits transcription of *spo0H*, encoding sigma-H. Spo0A~P also binds and stimulates transcription from the sigma-H-dependent promoters of *spo0F* and *spo0A*. Consequently, all of the pathways shown here constitute positive feedback loops contributing to the high-level expression and activation of Spo0A. The arrows indicate activation; the barred lines indicate inhibition.

sporulation (Levin and Losick, 1996). The genes controlled by Spo0A that are involved in formation of the axial filament and the polar FtsZ rings are not known.

In addition to its central role as an activator of the sporulation pathway, Spo0A regulates many of the other stationary-phase responses of *B. subtilis*, including cell motility and chemotaxis, competence, and the expression of extracellular proteases, carbohydrate-degrading enzymes, and antibiotics. Spo0A~P regulates these pathways by inhibiting transcription of *abrB* (Fig. 2). *abrB* encodes a transcriptional regulator that represses transcription of many early-stationary-phase genes and both positively and negatively regulates the development of competence (Albano et al., 1987; Hahn et al., 1995; Strauch, 1993; Strauch and Hoch, 1993).

Different amounts of Spo0A~P seem to accumulate based on the specific environmental conditions. Intermediate levels of Spo0A~P cause repression of *abrB* and induction of the stationary-phase responses that may permit cells to adapt and continue growing, such as motility and chemotaxis, the production of extracellular enzymes, and the development of competence (Chung et al., 1994; Siranosian and Grossman, 1994; Trach and Hoch, 1993). Higher levels of Spo0A~P are needed to directly activate gene expression and the initia-

tion of sporulation. This threshold mechanism for controlling stationary-phase gene expression and sporulation is established and maintained by positive and negative autoregulatory loops controlling Spo0A synthesis and activation (Fig. 3 and 4) (Chung et al., 1994; Grossman, 1995; Hoch, 1994).

The Sensor Kinases and the Phosphorelay

At least three histidine protein kinases, KinA, KinB, and KinC, provide phosphate for the activation of Spo0A (Fig. 2 and 5) (Antoniewski et al., 1990; Kobayashi et al., 1995; LeDeaux and Grossman, 1995; Perego et al., 1989; Trach and Hoch, 1993). KinA is a cytosolic protein, whereas KinB and KinC are both predicted to be integral membrane proteins (Fig. 5). Similar to other family members, KinA, KinB, and KinC have well-conserved C-terminal kinase domains and divergent N-terminal domains that are presumed to regulate kinase activity in response to starvation-dependent signals. The nature of these signals is not yet known.

The activities of KinA, KinB, and KinC vary with the growth phase and nutrient conditions, and each kinase probably responds to a different environmental or intracellular signal. KinA and KinB are required for efficient sporulation and have overlapping but not com-

FIGURE 5   Histidine protein kinases that donate phosphate to Spo0F. KinA is a cytosolic protein, and KinB and KinC are both integral membrane proteins. When activated, the kinases autophosphorylate and donate phosphate to Spo0F. Phosphate is then transferred to Spo0B and finally to Spo0A (Fig. 2). KinA activity is inhibited (barred lines) by KipI and activated by KipA, which inhibits KipI. KinB activity requires several additional proteins, including the integral membrane protein KbaA and the extracytoplasmic lipoprotein KapB. Though KinB, KbaA, and KapB are depicted here in a complex, it has not yet been demonstrated that they interact directly. The heavy black line represents the cell membrane.

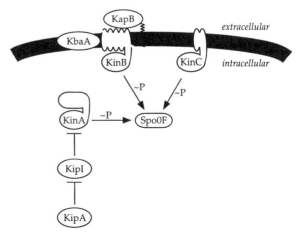

pletely redundant activities under different growth conditions (Dartois et al., 1996; LeDeaux et al., 1995; Trach and Hoch, 1993). KinC, on the other hand, contributes little to sporulation but is required for Spo0A-dependent regulation of *abrB* during exponential growth under some conditions (LeDeaux et al., 1995).

Both KinA and KinC contain PAS domain sequence motifs. These motifs have been identified in a wide variety of proteins that function in prokaryotic and eukaryotic signaling pathways (Ponting and Aravind, 1997; Zhulin and Taylor, 1998; Zhulin et al., 1997). In some cases, PAS domains mediate protein-protein interactions, and in others they bind small ligands, such as heme or flavin adenine dinucleotide cofactors, which may transduce light-, oxygen-, or redox state-dependent signals. It has recently been suggested that bacterial proteins containing PAS domains may be particularly important in redox-sensing pathways (Zhulin and Taylor, 1998), raising the possibility that KinA or KinC is activated in response to starvation-dependent changes in redox levels.

Regulation of Kinase Expression and Activity

It appears that all three kinases are present during exponential growth, and the amounts increase late in exponential growth or early in stationary phase. Interestingly, regulation of transcription of some of the kinase genes is controlled by Spo0A and sigma-H (Dartois et al., 1996; Fujita and Sadaie, 1998; Grossman, 1995).

Additional proteins regulate the activity of the kinases. KinA activity is regulated by two proteins, KipI and KipA (Fig. 5), which are coexpressed from an operon regulated by carbon and nitrogen source availability (Wang et al., 1997). KipI inhibits KinA autophosphorylation, and KipA restores KinA autophosphorylation by inhibiting KipI. KinA activity is also inhibited in vitro by *cis*-unsaturated fatty acids (Strauch et al., 1992). Though this effect has been linked to the inhibition of sporulation

under certain stress conditions, its relevance to the regulation of sporulation under normal growth conditions is unknown.

Several proteins that are required for KinB-dependent activation of sporulation have been identified, including an integral membrane protein, KbaA (Dartois et al., 1996), and the extracytoplasmic-membrane-associated lipoprotein KapB (Fig. 5) that is coexpressed from the *kinB* operon (Dartois et al., 1997a; Trach and Hoch, 1993). In addition, several lines of evidence suggest that methylation may be required for the activity of the KinB signaling pathway, perhaps as a posttranscriptional modification regulating a component of the pathway (Dartois et al., 1997b).

The Phosphorelay Proteins Spo0F and Spo0B

In most two-component signal transduction pathways, a sensor kinase autophosphorylates a histidine residue and phosphate is transferred directly to the aspartate residue of its cognate response regulator. In contrast, phosphate is transferred to Spo0A from its upstream kinases via two phosphotransfer proteins, Spo0F and Spo0B, which form a phosphorelay (Fig. 2) (Burbulys et al., 1991). Spo0F is a single-domain protein homologous to the conserved, phosphorylated domains of response regulators. Spo0F obtains phosphate on an aspartate residue from the upstream histidine kinases. Phosphate is transferred from Spo0F to a histidine residue in Spo0B and finally to an aspartate residue in Spo0A (Burbulys et al., 1991; Hoch, 1993).

Several other examples of two-component signal transduction pathways have been identified in which phosphate is transferred to the response regulator from an upstream kinase via two intermediate aspartate-histidine transfer modules (Appleby et al., 1996). The kinase, the intermediate modules, and the response regulator are either organized as separate polypeptides, as in the Spo0A pathway, or grouped together on one or more polypeptides.

It should soon be possible to understand the structural basis for the specificity of interactions among the sensor kinases, Spo0F, Spo0B, and Spo0A, as well as sensor kinases and response regulators of other signaling pathways. To this end, the structures of Spo0F and Spo0B have recently been determined by X-ray crystallography and nuclear magnetic resonance (Feher et al., 1997; Madhusudan et al., 1996, 1997; Varughese et al., 1998) and detailed kinetic studies of the interactions between wild-type and mutant proteins have been initiated (Feher et al., 1998; Grimshaw et al., 1998; Tzeng and Hoch, 1997; Tzeng et al., 1998a, 1998b)

Sigma-H

The alternative sigma factor sigma-H is required for full expression and activation of Spo0A as well as for formation of the polar septum and the initiation of cell-type-specific gene expression. Sigma-H contributes to the activation of Spo0A by inducing the expression of (i) high levels of Spo0A from the sporulation-specific promoter of *spo0A*, Ps (Fig. 3A); (ii) the sensor kinase KinA and the phosphorelay protein Spo0F (Fig. 4); and (iii) extracellular signaling factors that indirectly promote activation of Spo0A (Fig. 4) (Grossman, 1995; Hoch, 1995). Sigma-H also contributes indirectly to expression of *spo0A* and *kinB* by activating expression of SinI, an antagonist of the repressor protein SinR (Fig. 3 and 4) (Dartois et al., 1996; Grossman, 1995; Hoch, 1995; Lewis et al., 1998; Mandic-Mulec et al., 1995). The requirement that both Spo0A and sigma-H be activated for sporulation to proceed provides another mechanism for integrating signals that regulate the initiation of sporulation.

Sigma-H protein levels increase at the onset of stationary phase due to increased transcription of *spo0H* and increased translation and stability of sigma-H (Healy et al., 1991; Weir et al., 1991). The multilevel control of sigma-H during the onset of sporulation resembles the regulation of the environmentally responsive sigma factors sigma-32 and sigma-S from *Escherichia coli* (reviewed in Bukau, 1993, and Loewen et al., 1998). The mechanisms of sigma-32 and sigma-S regulation are understood in greater detail and may guide the study of sigma-H regulation.

Like Spo0A, sigma-H plays important roles in stationary-phase response pathways in addition to sporulation, including the development of competence. One role of sigma-H common to both sporulation and competence is inducing the expression of extracellular signaling factors that mediate quorum sensing (reviewed in Lazazzera et al., 1999).

CONDITIONS CONTROLLING THE INITIATION OF SPORULATION

A diverse set of extracellular and intracellular signals regulate the initiation of sporulation. These signals are generated by nutrient deprivation, cell density, DNA replication, DNA damage, glucose metabolism, the tricarboxylic acid (TCA) cycle, and chromosome-partitioning proteins. In most cases, these signals seem to regulate the accumulation of Spo0A~P by regulating synthesis of Spo0A and/or activity of the phosphorelay (Fig. 6). In this way, the phosphorelay and accumulation of high levels of Spo0A~P serve to integrate the multiple signals controlling sporulation and help to ensure that sporulation will not begin unless conditions seem proper. In addition, several signals appear to regulate more than one step and control events in addition to accumulation of Spo0A~P.

Nutrient Depletion: Regulation of Sporulation by GDP and GTP Levels?

Starvation for carbon, nitrogen, or phosphate is the key signal for sporulation. Nutrient depletion causes sigma-H levels to increase and activates the kinases leading to activation of Spo0A. How starvation signals are sensed and the mechanism by which they activate Spo0A and induce sigma-H synthesis are unknown.

One physiological change closely linked with the onset of sporulation during starvation is a large and prolonged decrease in GDP and GTP levels (Freese, 1981, 1984; Lopez et al., 1979). In contrast, changes in other nucleotide pools do not correlate with the onset of sporulation under all starvation conditions.

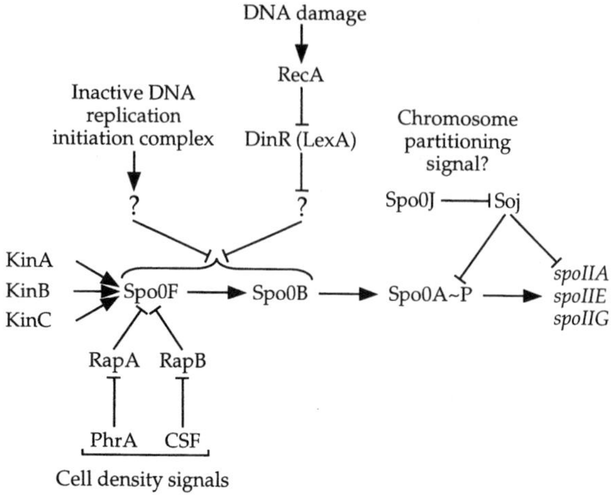

FIGURE 6 Cell density and cell cycle control of the phosphorelay. The physiological function of the phosphorelay is to integrate multiple signals that regulate sporulation. The secreted cell density peptides PhrA and CSF regulate the phosphorelay by inhibiting the activity of the RapA and RapB phosphatases, which negatively regulate Spo0F.

Inhibiting the initiation of DNA replication or inducing the SOS response, which is induced by DNA damage, inhibits the activation of Spo0A. Genetic data indicate that either Spo0F or Spo0B, rather than the histidine kinases or Spo0A itself, is the likely target of this regulation (Ireton and Grossman, 1994). The inhibition of sporulation that results from inhibiting the initiation of DNA replication is mediated by a signaling pathway independent of the SOS response. During the SOS response, RecA is activated and then causes inactivation of the transcriptional repressor DinR (LexA), causing induction of genes in the SOS regulon. At least one gene in the SOS regulon is postulated to be a negative regulator of the phosphorelay. Soj is a negative regulator of sporulation gene expression. Repression by Soj is antagonized by Spo0J, a protein required for accurate chromosome partitioning. The arrows (→) indicate activation; the barred lines (⊣) indicate inhibition.

The decrease in GDP and GTP levels at the onset of sporulation occurs for one of two reasons, depending on the growth conditions used: (i) precursors for purine biosynthesis become limiting or (ii) GMP biosynthesis is specifically inhibited by induction of the stringent response (Lopez et al., 1981; Ochi et al., 1982). The stringent response is induced when amino acid levels are insufficient to maintain the pool of charged tRNAs (Cashel et al., 1996). The observation that GDP and GTP levels are negatively regulated by two distinct mechanisms when cells are starved for different nutrients suggests that a drop in GDP and GTP levels may be a specific starvation-induced signal regulating sporulation. Consistent with this suggestion, depleting GDP-GTP pools by starving an auxotrophic strain for guanine or by specifically inhibiting GDP synthesis with drugs is sufficient to induce sporulation in nutrient-rich conditions (Freese, 1981, 1984).

Several other genes required for sporulation are induced by starvation independently of the phosphorelay and sigma-H. These include *ald*, encoding alanine dehydrogenase, and several of the TCA cycle genes (Grossman, 1995; Sonenshein, 1989).

Cell Density Control of Sporulation

The initiation of sporulation is regulated in part by cell density signals. Cells sporulate poorly when starved at low cell densities (Grossman and Losick, 1988; Ireton et al., 1993; Vasantha and Freese, 1979; Waldburger et al., 1993). Adding conditioned medium from cells grown to high density partially rescues the sporulation defect of cells at low density, indicating that extracellular factors that promote sporulation accumulate as cells grow to high density (Grossman and Losick, 1988). These factors regulate the activation of Spo0A, since the sporulation defect of cells at low density is suppressed by constitutively active *spo0A* mutant alleles (Ireton et al., 1993).

Two exported peptides, PhrA and CSF (competence and sporulation stimulating factor), that stimulate sporulation when present in growth medium have been identified (Perego and Hoch, 1996; Solomon et al., 1996). The precursors of PhrA and CSF are encoded by the *phrA* and *phrC* genes, respectively. Both peptides require the oligopeptide permease (Opp; also called Spo0K) to stimulate sporulation (Perego and Hoch, 1996; Solomon et al., 1996). The oligopeptide permease transports CSF and PhrA into the cell, where both interact with intracellular targets (Lazazzera et al., 1997; Solomon et al., 1996). The target of PhrA is an aspartyl-phosphate phosphatase encoded by a gene upstream of *phrA* in the same operon, *rapA* (Perego and Hoch, 1996). RapA and a second phosphatase, RapB, dephosphorylate Spo0F~P, thereby inhibiting the phosphorelay and accumulation of Spo0A~P (Fig. 6) (Perego et al., 1994). PhrA and CSF stimulate sporulation by antagonizing the activities of the Rap phosphatases. In vitro, PhrA inhibits the activity of RapA and CSF inhibits the activity of RapB (Perego, 1997; Perego and Hoch, 1996).

The expression of extracellular factors regulating sporulation is under the control of two transcription factors, the response regulator ComA and sigma-H, activating expression of *phrA* and *phrC*, respectively (Perego and Hoch, 1996; Solomon et al., 1996). ComA is itself activated by the peptide pheromones ComX and CSF, generally in late exponential or early stationary phase (Solomon et al., 1995).

CSF may be redundant with other signaling factors, since a *phrC* mutation does not confer a sporulation phenotype (Solomon et al., 1996). These other factors may also be under sigma-H control, since a missense mutation in *spo0H* that has pleiotropic effects on extracellular signaling factor production confers a synthetic sporulation phenotype when combined with a *phrC* null mutation (Lazazzera et al., 1997). Five other open reading frames that are predicted to encode extracellular signaling peptides have been identified by sequence analysis (Perego et al., 1996), all of which have putative sigma-H promoters (Lazazzera et al., 1999).

We suspect that the cell density effect on sporulation is an indication of possible competition for scarce nutrients. If cells are starving in a crowded colony, it may be advantageous to sporulate rather than compete for additional nutrients. If cells are dispersed (i.e., at low cell density), then the chances of finding additional nutrients might be higher and sporulation might be less desirable.

The TCA Cycle

As cell metabolism shifts to accommodate growth in poor nutrient conditions, many of the genes of the TCA cycle are induced to increase energy production, utilize poor carbon sources, and synthesize precursors for biosynthesis (Hederstedt, 1993; Sonenshein, 1993). Mutations in most of the TCA cycle genes inhibit sporulation (Freese, 1981; Sonenshein, 1989). The stage of sporulation at which TCA cycle mutants arrest depends on the specific gene mutated and the composition of the growth medium, reflecting the different roles played by each of the TCA cycle genes in metabolism and biosynthesis (Craig et al., 1997; Ireton et al., 1995; Jin et al., 1997). Several TCA cycle mutants arrest early during spore development and are defective either in activation of Spo0A or in a step required for asymmetric division. The early arrest of these mutants suggests that cells may monitor conditions affected by specific enzymes of the TCA cycle to determine if conditions are suitable for sporulation. Alternatively, the mutations may disrupt processes required for Spo0A activation and polar septation rather than generating specific signals that regulate sporulation. Either way, understanding how the mutations inhibit sporulation may reveal new aspects of how the early stages of sporulation are regulated.

Glucose Repression

Glucose inhibits sporulation of starving cells and instead favors the development of competence (Dubnau, 1991; Grossman, 1995; Schaeffer et al., 1965). Glucose appears to inhibit sporulation by reducing the levels of activated Spo0A, since mutations that bypass the phosphorelay or increase levels of activated Spo0A permit sporulation in the presence of glucose (Hoch et al., 1985; Kawamura et al., 1985; Leung et al., 1985; Mueller et al., 1992; Olmedo et al., 1990). How glucose inhibits Spo0A activation is not clear. In some growth media glucose reduces the expression or activities of the TCA cycle genes and sigma-H, which are required for sporulation, but there may be additional targets of catabolite repression important for sporulation.

Several of the genes encoding TCA cycle enzymes required for sporulation are repressed by glucose, including *citZ*, *citB*, and *citC* (Hederstedt, 1993; Jin and Sonenshein, 1994; Sonenshein, 1989). How these genes are repressed by glucose is not yet known. Placing both *citZ* and *citC* under the control of a heterologous promoter that is not repressed by glucose partially relieves the glucose-mediated repression of sporulation (Ireton et al., 1995). Furthermore, conditions that potentiate catabolite repression of the TCA cycle genes—high concentrations of glutamate or glutamine—also potentiate the inhibition of sporulation (Jin and Sonenshein, 1994; Sonenshein, 1989). These results suggest that the catabolite repression of the TCA cycle genes contributes to the catabolite repression of sporulation.

The posttranscriptional induction of sigma-H is another target of glucose-mediated repression of sporulation under some growth conditions. Induction of sigma-H is blocked by low extracellular pH (Cosby and Zuber, 1997). Extracellular pH decreases as cells grow exponentially in the presence of rapidly metabolizable carbon sources such as glucose, due to the secretion of acidic metabolites into the growth medium. Acidification is exacerbated if cells starve in the presence of glucose and glutamine, because of the synergistic repression by glucose and glutamine of the TCA cycle genes and other genes required to metabolize secreted acidic metabolites during early stationary phase. Raising the extracellular pH of late-exponential-phase cultures grown in the presence of excess glucose and glutamine is sufficient to increase sigma-H expression levels and activity (Cosby and Zuber, 1997). Mutations that impair glucose utilization (a *ptsI* null mutation and a mutation closely linked to *ptsG*) also increase sigma-H levels and activity (Frisby and Zuber, 1994). These effects on sigma-H levels appear to be due to increased translation or stability (Cosby and Zuber, 1997; Frisby and Zuber, 1994). It is not yet clear whether inhibiting the induction of sigma-H plays a key role in catabolite repression of sporulation under all growth conditions (Asai et al., 1995).

DNA Replication, DNA Damage, and Cell-Cycle Control

Cells are proficient in sporulating only during a period of the cell cycle closely following replication initiation; otherwise, a new round of replication initiation is required (Dawes and Mandelstam, 1970; Hauser and Errington, 1995; Mandelstam and Higgs, 1974). This window of opportunity may correspond to a period during which the cell is not yet committed to a subsequent round of the vegetative cell cycle. If the replication initiation complex is inactivated when cells are starved, using a conditional mutation in the essential replication initiation factor DnaB, cells fail to induce Spo0A-dependent gene expression (Ireton and Grossman, 1994). This failure to initiate sporulation is bypassed by mutations that allow phosphorylation of Spo0A independently of the phosphorelay, indicating that blocking replication somehow inhibits activity of the phosphorelay (Fig. 6) (Ireton and Grossman, 1994).

DNA damage and inhibition of ongoing DNA replication also appear to inhibit sporulation by inhibiting the phosphorelay (Fig. 6). Both DNA damage and inhibition of ongoing rounds of DNA replication cause induction of the SOS response involved in repair of DNA. Induction of SOS in *B. subtilis* prevents induction of sporulation genes that are activated by Spo0A~P. Again, this block in gene expression is relieved by mutations that allow phosphorylation of Spo0A independently of the phosphorelay, indicating that the SOS response is inhibiting the phosphorelay (Ireton and Grossman, 1992; 1994). The genes expressed during the SOS response that are responsible for inhibiting sporulation have not yet been identified.

Components of the Chromosome Segregation Machinery

In order for spore morphogenesis to proceed correctly, the mother cell and forespore each must receive a complete intact copy of the chromosome. Several proteins have been identified in *B. subtilis* that promote the faithful segregation of daughter chromosomes during vegetative growth. One of these, SpoOJ, acting together with a second protein, Soj, also regulates the initiation of sporulation, suggesting that SpoOJ and Soj may monitor the status of chromosome segregation and regulate expression of sporulation genes accordingly.

Soj and SpoOJ are members of the ParA and ParB families of proteins, respectively (Ireton et al., 1994; Ogasawara and Yoshikawa, 1992). ParA and ParB homologues were first identified as plasmid- and bacteriophage-encoded proteins required for plasmid partitioning (Austin and Nordstrom, 1990; Hiraga, 1992; Nordstrom and Austin, 1989; Williams and Thomas, 1992). ParB proteins bind *cis*-acting sites required for stable plasmid maintenance and are regulated somehow by their cognate ParA proteins. Chromosomally encoded homologues have been identified, including SpoOJ and Soj, that are involved in chromosome partitioning (Glaser et al., 1997; Lewis and Errington, 1997; Lin and Grossman, 1998; Lin et al., 1997; Mohl and Gober, 1997; Ogasawara and Yoshikawa, 1992; Sharpe and Errington, 1996).

In addition to impairing chromosome partitioning, inactivating *spoOJ* inhibits sporulation. Genetic and biochemical data suggest that SpoOJ negatively regulates the activity of Soj and that in the absence of SpoOJ, Soj inhibits sporulation (Fig. 6). Thus, sporulation of a *spoOJ* null mutant is restored to wild-type levels if *soj* is also inactivated (Ireton et al., 1994). Soj inhibits sporulation by binding the *spoOA*, *spoIIA*, *SpoIIE*, and *spoIIG* promoters, repressing transcription (Cervin et al., 1998; Quisel et al., 1999). Consistent with the model that SpoOJ negatively regulates Soj activity, inactivating *spoOJ* increases the binding of Soj to its target promoters in vivo (Quisel et al., 1999). How this activity is controlled during the chromosome-partitioning cycle is not known.

CONCLUSION

A wide variety of extracellular and intracellular signals converge to modulate the activity of SpoOA, thus regulating the decision to sporulate. The conditions generating these signals include starvation, cell density, metabolic states influenced by the metabolism of glucose or the activity of the TCA cycle, cell cycle-dependent events relating to chromosome replication and segregation, and DNA damage. Our understanding of the regulatory circuitry that integrates these diverse signals has grown significantly in the past few years. What has been revealed is a network of regulatory factors, each a potential target of regulation by extracellular and intracellular signals, that are organized into a series of positive and negative feedback loops controlling the level of activated SpoOA. Organized in this way, the regulatory network is able to integrate multiple signals to elicit qualitatively different responses at different threshold levels of activation. However, the nature of the signals that regulate sporulation and the mechanisms by which they are sensed and transduced are still poorly understood. We know relatively little about how cells monitor metabolic and cell cycle-dependent signals.

Many of the genes controlled by specific stationary-phase regulatory proteins have not yet been identified, and very little is known regarding the overlap and interconnections between different signaling pathways monitoring extracellular and intracellular conditions. The recent completion of the *B. subtilis* genome sequence (Kunst et al., 1997) and the promise of new genomics tools, especially DNA array technologies for studying genome-wide expression, should now make it possible to identify entire regulons and to study in detail the interactions between different regulatory pathways. Understanding both the signals that regulate these pathways and the organization of the pathways into networks regulating gene expression will contribute significantly to an integrated view of the interplay among cell physiology, gene expression, adaptation, and development.

ACKNOWLEDGMENTS

We thank Alison Frand, Ilana Goldhaber-Gordon, Petra Levin, and John Quisel for helpful comments on early versions of this chapter.

Work in the Grossman laboratory is supported in part by Public Health Service grants GM50895 and GM41934. W.F.B. was supported in part by a postdoctoral fellowship from the American Cancer Society.

REFERENCES

Albano, M., J. Hahn, and D. Dubnau. 1987. Expression of competence genes in *Bacillus subtilis. J. Bacteriol.* **169:**3110–3117.

Antoniewski, C., B. Savelli, and P. Stragier. 1990. The *spoIIJ* gene, which regulates early developmental steps in *Bacillus subtilis*, belongs to a class of environmentally responsive genes. *J. Bacteriol.* **172:** 86–93.

Appleby, J. L., J. S. Parkinson, and R. B. Bourret. 1996. Signal transduction via the multi-step phosphorelay: not necessarily a road less traveled. *Cell* **86:**845–848.

Asai, K., F. Kawamura, H. Yoshikawa, and H. Takahashi. 1995. Expression of *kinA* and accumulation of sigma-H at the onset of sporulation in *Bacillus subtilis. J. Bacteriol.* **177:**6679–6683.

Austin, S., and K. Nordstrom. 1990. Partition-mediated incompatibility of bacterial plasmids. *Cell* **60:** 351–354.

Baldus, J. M., C. M. Buckner, and C. P. Moran, Jr. 1995. Evidence that the transcriptional activator Spo0A interacts with two sigma factors in *Bacillus subtilis. Mol. Microbiol.* **17:**281–290.

Bird, T. H., J. K. Grimsley, J. A. Hoch, and G. B. Spiegelman. 1996. The *Bacillus subtilis* response regulator Spo0A stimulates transcription of the *spoIIG* operon through modification of RNA polymerase promoter complexes. *J. Mol. Biol.* **256:** 436–448.

Bramucci, M. G., B. D. Green, N. Ambulos, and P. Youngman. 1995. Identification of a *Bacillus subtilis spo0H* allele that is necessary for suppression of the sporulation-defective phenotype of a *spo0A* mutation. *J. Bacteriol.* **177:**1630–1633.

Buckner, C. M., and C. P. Moran, Jr. 1998. A region in *Bacillus subtilis* sigma-H required for Spo0A-dependent promoter activity. *J. Bacteriol.* **180:**4987–4990.

Buckner, C. M., G. Schyns, and C. P. Moran, Jr. 1998. A region in the *Bacillus subtilis* transcription

factor Spo0A that is important for spoIIG promoter activation. *J. Bacteriol.* **180:**3578–3583.

Bukau, B. 1993. Regulation of the *Escherichia coli* heat-shock response. *Mol. Microbiol.* **9:**671–680.

Burbulys, D., K. A. Trach, and J. A. Hoch. 1991. Initiation of sporulation in B. subtilis is controlled by a multicomponent phosphorelay. *Cell* **64:** 545–552.

Cashel, M., D. R. Gentry, V. J. Hernandez, and D. Vinella. 1996. The stringent response, p. 1458–1496. *In* F. C. Neidhardt, R. Curtis III, J. L. Ingraham, E. C. C. Lin, K. B. Low, B. Magasanik, W. S. Reznikoff, M. Riley, M. Schaechter, and H. E. Umbarger (ed.), *Escherichia coli and Salmonella: Cellular and Molecular Biology*, vol. 1. ASM Press, Washington, D.C.

Cervin, M. A., G. B. Spiegelman, B. Raether, K. Ohlsen, M. Perego, and J. A. Hoch. 1998. A negative regulator linking chromosome segregation to developmental transcription in *Bacillus subtilis. Mol. Microbiol.* **29:**85–95.

Chung, J. D., G. Stephanopoulos, K. Ireton, and A. D. Grossman. 1994. Gene expression in single cells of *Bacillus subtilis*: evidence that a threshold mechanism controls the initiation of sporulation. *J. Bacteriol.* **176:**1977–1984.

Cosby, W. M., and P. Zuber. 1997. Regulation of *Bacillus subtilis* sigma-H (Spo0H) and AbrB in response to changes in external pH. *J. Bacteriol.* **179:** 6778–6787.

Craig, J. E., M. J. Ford, D. C. Blaydon, and A. L. Sonenshein. 1997. A null mutation in the *Bacillus subtilis* aconitase gene causes a block in Spo0A-phosphate-dependent gene expression. *J. Bacteriol.* **179:**7351–7359.

Dartois, V., T. Djavakhishvili, and J. A. Hoch. 1996. Identification of a membrane protein involved in activation of the KinB pathway to sporulation in *Bacillus subtilis. J. Bacteriol.* **178:** 1178–1186.

Dartois, V., T. Djavakhishvili, and J. A. Hoch. 1997a. KapB is a lipoprotein required for KinB signal transduction and activation of the phosphorelay to sporulation in *Bacillus subtilis. Mol. Microbiol.* **26:** 1097–1108.

Dartois, V., J. Liu, and J. A. Hoch. 1997b. Alterations in the flow of one-carbon units affect KinB-dependent sporulation in *Bacillus subtilis. Mol. Microbiol.* **25:**39–51.

Dawes, I. W., and J. Mandelstam. 1970. Sporulation of *Bacillus subtilis* in continuous culture. *J. Bacteriol.* **103:**529–535.

Dubnau, D. 1991. Genetic competence in *Bacillus subtilis. Microbiol. Rev.* **55:**395–424.

Dubnau, D. 1993. Genetic exchange and homologous recombination, p. 555–584. *In* A. L. Sonenshein, J. A. Hoch, and R. Losick (ed.), *Bacillus subti-*

lis and Other Gram-Positive Bacteria: Biochemistry, Physiology, and Molecular Genetics. American Society for Microbiology, Washington, D.C.

Feher, V. A., J. W. Zapf, J. A. Hoch, J. M. Whiteley, L. P. McIntosh, M. Rance, N. J. Skelton, F. W. Dahlquist, and J. Cavanagh. 1997. High-resolution NMR structure and backbone dynamics of the *Bacillus subtilis* response regulator, Spo0F: implications for phosphorylation and molecular recognition. *Biochemistry* **36:**10015–10025.

Feher, V. A., Y. L. Tzeng, J. A. Hoch, and J. Cavanagh. 1998. Identification of communication networks in Spo0F: a model for phosphorylation-induced conformational change and implications for activation of multiple domain bacterial response regulators. *FEBS Lett.* **425:**1–6.

Freese, E. 1981. Initiation of bacterial sporulation, p. 1–12. *In* H. S. Levinson, A. L. Sonenshein, and D. J. Tipper (ed.), *Sporulation and Germination.* American Society for Microbiology, Washington, D.C.

Freese, E. 1984. Metabolic and genetic control of bacterial sporulation, p. 101–172. *In* A. Hurst, G. Gould, and J. Dring (ed.), *The Bacterial Spore,* vol. 2. Academic Press, London, England.

Frisby, D., and P. Zuber. 1994. Mutations in *pts* cause catabolite-resistant sporulation and altered regulation of *spo0H* in *Bacillus subtilis. J. Bacteriol.* **176:**2587–2595.

Fujita, M., and Y. Sadaie. 1998. Feedback loops involving Spo0A and AbrB in in vitro transcription of the genes involved in the initiation of sporulation in *Bacillus subtilis. J. Biochem.* (Tokyo) **124:**98–104.

Glaser, P., M. E. Sharpe, B. Raether, M. Perego, K. Ohlsen, and J. Errington. 1997. Dynamic, mitotic-like behavior of a bacterial protein required for accurate chromosome partitioning. *Genes Dev.* **11:**1160–1168.

Grimshaw, C. E., S. Huang, C. G. Hanstein, M. A. Strauch, D. Burbulys, L. Wang, J. A. Hoch, and J. M. Whiteley. 1998. Synergistic kinetic interactions between components of the phosphorelay controlling sporulation in *Bacillus subtilis. Biochemistry* **37:**1365–1375.

Grossman, A. D. 1995. Genetic networks controlling the initiation of sporulation and the development of genetic competence in *Bacillus subtilis. Annu. Rev. Genet.* **29:**477–508.

Grossman, A. D., and R. Losick. 1988. Extracellular control of spore formation in *Bacillus subtilis. Proc. Natl. Acad. Sci. USA* **85:**4369–4373.

Hahn, J., M. Roggiani, and D. Dubnau. 1995. The major role of Spo0A in genetic competence is to downregulate *abrB,* an essential competence gene. *J. Bacteriol.* **177:**3601–3605.

Hatt, J. K., and P. Youngman. 1998. Spo0A mutants of *Bacillus subtilis* with sigma factor-specific

defects in transcription activation. *J. Bacteriol.* **180:**3584–3591.

Hauser, P. M., and J. Errington. 1995. Characterization of cell cycle events during the onset of sporulation in *Bacillus subtilis. J. Bacteriol.* **177:**3923–3931.

Healy, J., J. Weir, I. Smith, and R. Losick. 1991. Post-transcriptional control of a sporulation regulatory gene encoding transcription factor sigma-H in *Bacillus subtilis. Mol. Microbiol.* **5:**477–487.

Hederstedt, L. 1993. The Krebs citric acid cycle, p. 181–197. *In* A. L. Sonenshein, J. A. Hoch, and R. Losick (ed.), *Bacillus subtilis and Other Gram-Positive Bacteria: Biochemistry, Physiology, and Molecular Genetics.* American Society for Microbiology, Washington, D.C.

Hiraga, S. 1992. Chromosome and plasmid partition in *Escherichia coli. Annu. Rev. Biochem.* **61:**283–306.

Hoch, J. A. 1993. Regulation of the phosphorelay and the initiation of sporulation in *Bacillus subtilis. Annu. Rev. Microbiol.* **47:**441–465.

Hoch, J. A. 1994. The phosphorelay signal transduction pathway in the initiation of sporulation, p. 41–60. *In* P. Piggot, C. P. Moran, and P. Youngman (ed.), *Regulation of Bacterial Differentiation.* American Society for Microbiology, Washington, D.C.

Hoch, J. A. 1995. Control of cellular development in sporulating bacteria by the phosphorelay two-component signal transduction system, p. 129–144. *In* J. A. Hoch and T. J. Silhavy (ed.), *Two-Component Signal Transduction.* ASM Press, Washington, D.C.

Hoch, J. A., K. Trach, F. Kawamura, and H. Saito. 1985. Identification of the transcriptional suppressor *sof-1* as an alteration in the spo0A protein. *J. Bacteriol.* **161:**552–555.

Ireton, K., and A. D. Grossman. 1992. Coupling between gene expression and DNA synthesis early during development in *Bacillus subtilis. Proc. Natl. Acad. Sci. USA* **89:**8808–8812.

Ireton, K., and A. D. Grossman. 1994. A developmental checkpoint couples the initiation of sporulation to DNA replication in *Bacillus subtilis. EMBO J.* **13:**1566–1573.

Ireton, K., D. Z. Rudner, K. J. Siranosian, and A. D. Grossman. 1993. Integration of multiple developmental signals in *Bacillus subtilis* through the Spo0A transcription factor. *Genes Dev.* **7:**283–294.

Ireton, K., N. W. Gunther IV, and A. D. Grossman. 1994. *spo0J* is required for normal chromosome segregation as well as the initiation of sporulation in *Bacillus subtilis. J. Bacteriol.* **176:**5320–5329.

Ireton, K., S. Jin, A. D. Grossman, and A. L. Sonenshein. 1995. Krebs cycle function is required for activation of the Spo0A transcription factor in *Bacillus subtilis. Proc. Natl. Acad. Sci. USA* **92:** 2845–2849.

Jin, S., and A. L. Sonenshein. 1994. Transcriptional regulation of *Bacillus subtilis* citrate synthase genes. *J. Bacteriol.* **176:**4680–4690.

Jin, S., P. A. Levin, K. Matsuno, A. D. Grossman, and A. L. Sonenshein. 1997. Deletion of the *Bacillus subtilis* isocitrate dehydrogenase gene causes a block at stage I of sporulation. *J. Bacteriol.* **179:** 4725–4732.

Kawamura, F., L. F. Wang, and R. H. Doi. 1985. Catabolite-resistant sporulation (*crsA*) mutations in the *Bacillus subtilis* RNA polymerase sigma 43 gene (*rpoD*) can suppress and be suppressed by mutations in spo0 genes. *Proc. Natl. Acad. Sci. USA* **82:** 8124–8128.

Kobayashi, K., K. Shoji, T. Shimizu, K. Nakano, T. Sato, and Y. Kobayashi. 1995. Analysis of a suppressor mutation *ssb* (*kinC*) of *sur0B20* (*spo0A*) mutation in *Bacillus subtilis* reveals that *kinC* encodes a histidine protein kinase. *J. Bacteriol.* **177:**176–182.

Kunst, F., N. Ogasawara, I. Moszer, A. M. Albertini, G. Alloni, V. Azevedo, M. G. Bertero, P. Bessieres, A. Bolotin, S. Borchert, R. Borriss, L. Boursier, A. Brans, M. Braun, S. C. Brignell, S. Bron, S. Brouillet, C. V. Bruschi, B. Caldwell, V. Capuano, N. M. Carter, S. K. Choi, J. J. Codani, I. F. Connerton, A. Danchin, et al. 1997. The complete genome sequence of the gram-positive bacterium *Bacillus subtilis*. Nature **390:**249–256.

Lazazzera, B. A., J. M. Solomon, and A. D. Grossman. 1997. An exported peptide functions intracellularly to contribute to cell density signaling in *B. subtilis. Cell* **89:**917–925.

Lazazzera, B. A., T. Palmer, J. Quisel, and A. D. Grossman. 1999. Cell density control of gene expression and development in *Bacillus subtilis*, p. 27–46. *In* G. M. Dunny and S. C. Winans (ed.), *Cell-Cell Signaling in Bacteria*. ASM Press, Washington, D.C.

LeDeaux, J. R., and A. D. Grossman. 1995. Isolation and characterization of *kinC*, a gene that encodes a sensor kinase homologous to the sporulation sensor kinases KinA and KinB in *Bacillus subtilis. J. Bacteriol.* **177:**166–175.

LeDeaux, J. R., N. Yu, and A. D. Grossman. 1995. Different roles for KinA, KinB, and KinC in the initiation of sporulation in *Bacillus subtilis. J. Bacteriol.* **177:**861–863.

Leung, A., S. Rubinstein, C. Yang, J. W. Li, and T. Leighton. 1985. Suppression of defective-sporulation phenotypes by mutations in the major sigma factor gene (*rpoD*) of *Bacillus subtilis. Mol. Gen. Genet.* **201:**96–98.

Levin, P. A., and R. Losick. 1996. Transcription factor Spo0A switches the localization of the cell division protein FtsZ from a medial to a bipolar pattern in *Bacillus subtilis. Genes Dev.* **10:**478–488.

Lewis, P. J., and J. Errington. 1997. Direct evidence for active segregation of *oriC* regions of the *Bacillus subtilis* chromosome and co-localization with the Spo0J partitioning protein. *Mol. Microbiol.* **25:**945–954.

Lewis, R. J., J. A. Brannigan, W. A. Offen, I. Smith, and A. J. Wilkinson. 1998. An evolutionary link between sporulation and prophage induction in the structure of a Repressor: Anti-repressor complex. *J. Mol. Biol.* **283:**907–912.

Lin, D. C., and A. D. Grossman. 1998. Identification and characterization of a bacterial chromosome partitioning site. *Cell* **92:**675–685.

Lin, D. C. H., P. A. Levin, and A. D. Grossman. 1997. Bipolar localization of a chromosome partition protein in *Bacillus subtilis. Proc. Natl. Acad. Sci. USA* **94:**4721–4726.

Loewen, P. C., B. Hu, J. Strutinsky, and R. Sparling. 1998. Regulation in the *rpoS* regulon of *Escherichia coli. Can. J. Microbiol.* **44:**707–717.

Lopez, J. M., C. L. Marks, and E. Freese. 1979. The decrease of guanine nucleotides initiates sporulation of *Bacillus subtilis. Biochim. Biophys. Acta* **587:** 238–252.

Lopez, J. M., A. Dromerick, and E. Freese. 1981. Response of guanosine 5′-triphosphate concentration to nutritional changes and its significance for *Bacillus subtilis* sporulation. *J. Bacteriol.* **146:** 605–613.

Madhusudan, M., J. Zapf, J. M. Whiteley, J. A. Hoch, N. H. Xuong, and K. I. Varughese. 1996. Crystal structure of a phosphatase-resistant mutant of sporulation response regulator Spo0F from *Bacillus subtilis. Structure* **4:**679–690.

Madhusudan, M., J. Zapf, J. A. Hoch, J. M. Whiteley, N. H. Xuong, and K. I. Varughese. 1997. A response regulatory protein with the site of phosphorylation blocked by an arginine interaction: crystal structure of Spo0F from *Bacillus subtilis. Biochemistry* **36:**12739–12745.

Mandelstam, J., and S. A. Higgs. 1974. Induction of sporulation during synchronized chromosome replication in *Bacillus subtilis. J. Bacteriol.* **120:** 38–42.

Mandic-Mulec, I., L. Doukhan, and I. Smith. 1995. The *Bacillus subtilis* SinR protein is a repressor of the key sporulation gene *spo0A. J. Bacteriol.* **177:** 4619–4627.

Mohl, D. A., and J. W. Gober. 1997. Cell cycle-dependent polar localization of chromosome partitioning proteins in Caulobacter crescentus. *Cell* **88:** 675–684.

Msadek, T., F. Kunst, and G. Rapoport. 1995. A signal transduction network in *Bacillus subtilis* includes the DegS/DegU and ComP/ComA two-component systems, p. 447–471. *In* J. A. Hoch and T. J. Silhavy (ed.), *Two-Component Signal Transduction*. ASM Press, Washington, D.C.

Mueller, J. P., G. Bukusoglu, and A. L. Sonenshein. 1992. Transcriptional regulation of *Bacillus subtilis* glucose starvation-inducible genes: control of *gsiA* by the ComP-ComA signal transduction system. *J. Bacteriol.* **174:**4361–4373.

Nordstrom, K., and S. J. Austin. 1989. Mechanisms that contribute to the stable segregation of plasmids. *Annu. Rev. Genet.* **23:**37–69.

Ochi, K., J. Kandala, and E. Freese. 1982. Evidence that *Bacillus subtilis* sporulation induced by the stringent response is caused by the decrease in GTP or GDP. *J. Bacteriol.* **151:**1062–1065.

Ogasawara, N., and H. Yoshikawa. 1992. Genes and their organization in the replication origin region of the bacterial chromosome. *Mol. Microbiol.* **6:**629–634.

Olmedo, G., E. G. Ninfa, J. Stock, and P. Youngman. 1990. Novel mutations that alter the regulation of sporulation in *Bacillus subtilis*. Evidence that phosphorylation of regulatory protein Spo0A controls the initiation of sporulation. *J. Mol. Biol.* **215:**359–372.

Perego, M. 1997. A peptide export-import control circuit modulating bacterial development regulates protein phosphatases of the phosphorelay. *Proc. Natl. Acad. Sci. USA* **94:**8612–8617.

Perego, M., and J. A. Hoch. 1996. Cell-cell communication regulates the effects of protein aspartate phosphatases on the phosphorelay controlling development in *Bacillus subtilis. Proc. Natl. Acad. Sci. USA* **93:**1549–1553.

Perego, M., S. P. Cole, D. Burbulys, K. Trach, and J. A. Hoch. 1989. Characterization of the gene for a protein kinase which phosphorylates the sporulation-regulatory proteins Spo0A and Spo0F of *Bacillus subtilis. J. Bacteriol.* **171:**6187–6196.

Perego, M., C. Hanstein, K. M. Welsh, T. Djavakhishvili, P. Glaser, and J. A. Hoch. 1994. Multiple protein-aspartate phosphatases provide a mechanism for the integration of diverse signals in the control of development in *B. subtilis. Cell* **79:**1047–1055.

Perego, M., P. Glaser, and J. A. Hoch. 1996. Aspartyl-phosphate phosphatases deactivate the response regulator components of the sporulation signal transduction system in *Bacillus subtilis. Mol. Microbiol.* **19:**1151–1157.

Piggot, P. J., and J. G. Coote. 1976. Genetic aspects of bacterial endospore formation. *Bacteriol. Rev.* **40:**938–962.

Ponting, C. P., and L. Aravind. 1997. PAS: a multifunctional domain family comes to light. *Curr. Biol.* **7:**R674–R677.

Quisel, J., D. C.-H. Lin, and A. D. Grossman. 1999. Unpublished data.

Rowe-Magnus, D. A., and G. B. Spiegelman. 1998. Contributions of the domains of the *Bacillus subtilis* response regulator Spo0A to transcription stimulation of the *spoIIG* operon. *J. Biol. Chem.* **273:**25818–25824.

Schaeffer, P., J. Millet, and J. P. Aubert. 1965. Catabolic repression of bacterial sporulation. *Proc. Natl. Acad. Sci. USA* **54:**704–711.

Schyns, G., C. M. Buckner, and C. P. Moran, Jr. 1997. Activation of the *Bacillus subtilis spoIIG* promoter requires interaction of Spo0A and the sigma subunit of RNA polymerase. *J. Bacteriol.* **179:**5605–5608.

Sharpe, M. E., and J. Errington. 1996. The *Bacillus subtilis soi-spo0J* locus is required for a centromere-like function involved in prespore chromosome partitioning. *Mol. Microbiol.* **21:**501–509.

Siranosian, K. J., and A. D. Grossman. 1994. Activation of *spo0A* transcription by sigma-H is necessary for sporulation but not for competence in *Bacillus subtilis. J. Bacteriol.* **176:**3812–3815.

Smith, I. 1993. Regulatory proteins that control late-growth development, p. 785–800. *In* A. L. Sonenshein, J. A. Hoch, and R. Losick (ed.), *Bacillus subtilis and Other Gram-Positive Bacteria: Biochemistry, Physiology, and Molecular Genetics*. American Society for Microbiology, Washington, D.C.

Solomon, J. M., R. Magnuson, A. Srivastava, and A. D. Grossman. 1995. Convergent sensing pathways mediate response to two extracellular competence factors in *Bacillus subtilis. Genes Dev.* **9:**547–558.

Solomon, J. M., B. A. Lazazzera, and A. D. Grossman. 1996. Purification and characterization of an extracellular peptide factor that affects two different developmental pathways in *Bacillus subtilis. Genes Dev.* **10:**2014–2024.

Sonenshein, A. L. 1989. Metabolic regulation of sporulation and other stationary-phase phenomena, p. 109–130. *In* I. Smith, R. A. Slepecky, and P. Setlow (ed.), *Regulation of Procaryotic Development*. American Society for Microbiology, Washington, D.C.

Sonenshein, A. L. 1993. Introduction to metabolic pathways, p. 127–132. *In* A. L. Sonenshein, J. A. Hoch, and R. Losick (ed.), *Bacillus subtilis and Other Gram-Positive Bacteria: Biochemistry, Physiology, and Molecular Genetics*. American Society for Microbiology, Washington, D.C.

Spiegelman, G. B., T. H. Bird, and V. Voon. 1995. Transcription regulation by the *Bacillus subtilis* response regulator Spo0A, p. 159–179. *In* J. A. Hoch and T. J. Silhavy (ed.), *Two-Component Signal Transduction*. ASM Press, Washington, D.C.

Stragier, P., and R. Losick. 1996. Molecular genetics of sporulation in *Bacillus subtilis*. *Annu. Rev. Genet.* **30:**297–241.

Strauch, M. A. 1993. AbrB, a transition state regulator, p. 757–764. *In* A. L. Sonenshein, J. A. Hoch, and R. Losick (ed.), *Bacillus subtilis and Other Gram-Positive Bacteria: Biochemistry, Physiology, and Molecular Genetics*. American Society for Microbiology, Washington, D.C.

Strauch, M. A., and J. A. Hoch. 1993. Transition-state regulators: sentinels of *Bacillus subtilis* post-exponential gene expression. *Mol. Microbiol.* **7:**337–342.

Strauch, M. A., D. de Mendoza, and J. A. Hoch. 1992. *cis*-unsaturated fatty acids specifically inhibit a signal-transducing protein kinase required for initiation of sporulation in *Bacillus subtilis*. *Mol. Microbiol.* **6:**2909–2917.

Trach, K. A., and J. A. Hoch. 1993. Multisensory activation of the phosphorelay initiating sporulation in *Bacillus subtilis*: identification and sequence of the protein kinase of the alternate pathway. *Mol. Microbiol.* **8:**69–79.

Tzeng, Y. L., and J. A. Hoch. 1997. Molecular recognition in signal transduction: the interaction surfaces of the SpoOF response regulator with its cognate phosphorelay proteins revealed by alanine scanning mutagenesis. *J. Mol. Biol.* **272:**200–212.

Tzeng, Y. L., V. A. Feher, J. Cavanagh, M. Perego, and J. A. Hoch. 1998a. Characterization of interactions between a two-component response regulator, SpoOF, and its phosphatase, RapB. *Biochemistry* **37:**16538–16545.

Tzeng, Y. L., X. Z. Zhou, and J. A. Hoch. 1998b. Phosphorylation of the SpoOB response regulator

phosphotransferase of the phosphorelay initiating development in *Bacillus subtilis*. *J. Biol. Chem.* **273:**23849–23855.

Varughese, K. I., Madhusudan, X. Z. Zhou, J. M. Whiteley, and J. A. Hoch. 1998. Formation of a novel four-helix bundle and molecular recognition sites by dimerization of a response regulator phosphotransferase. *Mol. Cell* **2:**485–493.

Vasantha, N., and E. Freese. 1979. The role of manganese in growth and sporulation of *Bacillus subtilis*. *J. Gen. Microbiol.* **112:**329–336.

Waldburger, C., D. Gonzalez, and G. H. Chambliss. 1993. Characterization of a new sporulation factor in *Bacillus subtilis*. *J. Bacteriol.* **175:**6321–6327.

Wang, L., R. Grau, M. Perego, and J. A. Hoch. 1997. A novel histidine kinase inhibitor regulating development in *Bacillus subtilis*. *Genes Dev.* **11:**2569–2579.

Weir, J., M. Predich, E. Dubnau, G. Nair, and I. Smith. 1991. Regulation of *spo0H*, a gene coding for the *Bacillus subtilis* sigma-H factor. *J. Bacteriol.* **173:**521–529.

Williams, D. R., and C. M. Thomas. 1992. Active partitioning of bacterial plasmids. *J. Gen. Microbiol.* **138:**1–16.

Zhulin, I. B., and B. L. Taylor. 1998. Correlation of PAS domains with electron transport-associated proteins in completely sequenced microbial genomes. *Mol. Microbiol.* **29:**1522–1523.

Zhulin, I. B., B. L. Taylor, and R. Dixon. 1997. PAS domain S-boxes in Archaea, Bacteria and sensors for oxygen and redox. *Trends Biochem. Sci.* **22:**331–333.

ASYMMETRIC DIVISION AND CELL FATE DURING SPORULATION IN *BACILLUS SUBTILIS*

Petra Anne Levin and Richard Losick

8

The gram-positive bacterium *Bacillus subtilis* and its family members are unique among the bacteria in their ability to switch the location of their division site during development (Fitz-James and Young, 1969). At the onset of sporulation, *B. subtilis* undergoes a switch from binary fission, the mode of cytokinesis it employs during vegetative growth, to asymmetric division, in which a division septum is laid down close to one of the cell poles (Fig. 1A and C). Polar septation divides the developing cell or sporangium into two unequal-sized cellular compartments that follow dissimilar pathways of differentiation (reviewed in Errington, 1993; Losick et al., 1986; Piggot and Coote, 1976; and Stragier and Losick, 1996). The smaller compartment is called the forespore. It is a germ line cell in that it becomes the mature spore and can, following germination, give rise to subsequent progeny. The larger compartment is called the mother cell. It is a terminally differentiating cell type in that it nurtures the developing forespore, but it is ultimately discarded by lysis when the spore is fully mature. This chapter outlines the genetic pathway governing the switch from binary fission to polar septation in *B. subtilis*, discusses the connection between polar division and the establishment of cell-type-specific gene expression, and finally describes pathways of intercellular communication between the mother cell and forespore chambers of the sporangium.

Sporulation is divided into distinct morphological stages based on electron microscopy (Ryter, 1964; Ryter et al., 1966; Piggot and Coote, 1976) (Fig. 1). Prior to entry into sporulation, cells are said to be in morphological stage 0. The transition from stage 0 to stage I is defined by formation of the axial filament, an elongated nucleoid structure that extends the length of the cell (Fig. 1B). Next, the polar septum is formed, the defining feature of stage II (Fig. 1C). Stages 0 and I are therefore predivisional stages of sporulation, whereas stages II and beyond are postdivisional stages. As sporulation progresses, the polar septum bulges out and migrates around the forespore, thereby engulfing it and eventually pinching it off as free protoplast within the mother cell cytoplasm (Fig. 1D). Engulfment defines stage III of sporulation. Next, the coat and cortex layers of the spore are formed during stages IV and V until finally the fully mature spore (more properly known as an endospore because it is formed within the sporangium) is released by lysis of

P. A. Levin, Department of Biology, Building 68-530, Massachusetts Institute of Technology, Cambridge, MA 02139. *R. Losick*, Department of Molecular and Cellular Biology, The Biological Laboratories, Harvard University, 16 Divinity Avenue, Cambridge, MA 02138.

Prokaryotic Development, edited by Y. V. Brun and L. J. Shimkets,
© 2000 American Society for Microbiology, Washington, DC 20005-4171

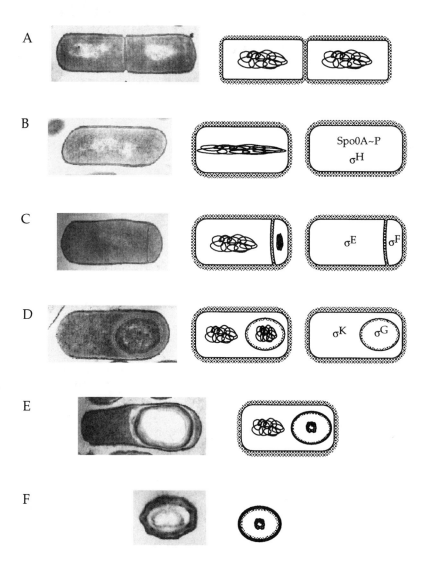

FIGURE 1 Stages of sporulation. On the left in each panel is shown an electron micrograph of the stage of sporulation, and on the right is depicted in cartoon form the disposition of the chromosomes and the time and site of action of the principal regulatory proteins that govern sporulation gene expression. (A) Vegetative cells; (B through E) sporangia at entry into sporulation (stage 0) (B), at polar division (stage II) (C), at engulfment (stage III) (D), and at cortex and coat formation (stage V to VI) (E); (F) a free spore. Notice that at stage V to VI (E) the forespore chromosome is packaged into a doughnut-like shape (Pogliano et al., 1995). (Electron micrographs reprinted from Driks, 1999, with permission of the publisher.)

the mother cell at the end of stage VI (Fig. 1E and F).

Despite the seemingly esoteric nature of the nomenclature for *spo* genes, the system for naming genes required in sporulation is simple and informative (Piggot and Coote, 1976) (see Stragier and Losick, 1996, for a comprehensive list of sporulation genes). The Roman numeral that follows *spo* denotes the stage during which the gene in question is required for further morphogenesis. For example, mutations in *spo0* genes block sporulation at stage 0, that is, prior to the formation of the polar septum, whereas mutations in *spoII* genes prevent engulfment, that is, they permit the formation of the polar septum but block the subsequent engulfment of the forespore by the mother cell. A capital letter follows the Roman numeral to distinguish among the several genes and operons required at a particular morphological stage. Thus, *spoIIA*, *spoIIE*, and *spoIIG* are genes and operons required at morphological stage II. Morphological stage I had fallen into disuse because a mutant blocked at the stage of axial-filament formation had not been identified. However, a recent reexamination of the *spo0H* gene (see below) indicates that it is misnamed and is in fact required at stage I (Levin and Losick, 1996).

CELL DIVISION

The polar septum is a hallmark of the process of sporulation, yet it differs from the medial septum of vegetative growth more in position than in structure or formation. Like the medial septum, the polar septum consists of two septal membranes between which is sandwiched a layer of peptidoglycan, although this cell wall layer is thinner in the sporulation septum than in the medial septum. The sporulation septum is the site of localization of various sporulation gene products, such as SpoIIE, SpoIIGA, and SpoIVA, SpoIVFA, and SpoIVFB, but most of these proteins (with the possible exception of SpoIIE) do not influence the structure or formation of the sporulation septum. On the other hand, all of the genes known to be required for binary fission in *B. subtilis*—*ftsZ*,

ftsA, *divIB*, *divIC*, *pbp2b*, and *ftsL*—are also required for polar septation (Beall and Lutkenhaus, 1989, 1991; Harry and Wake, 1989; Levin and Losick, 1996; Yanouri et al., 1993; Daniel et al., 1998). Among these genes, *ftsZ* is the best characterized.

The *ftsZ* gene product is a highly conserved GTPase that is most probably required for cell division in most bacteria (with the notable exception of *Chlamydia* [Stephens et al., 1998]), archaea, and possibly chloroplasts (Lutkenhaus and Addinall, 1997; Strepp et al., 1998). FtsZ has a tertiary structure similar to that of the eukaryotic cytoskeletal protein tubulin. In response to an unidentified cell cycle signal, FtsZ polymerizes into a ring structure (the Z ring) at the nascent division site (Bi and Lutkenhaus, 1991; Wang and Lutkenhaus, 1993; Levin and Losick, 1996; Margolin, 1998). Although several cell division proteins have been identified, none appear to be responsible for guiding FtsZ to the division site. Z-ring formation is one of the first steps in cytokinesis, and localization of other cell division proteins to the septal site requires FtsZ (Addinall and Lutkenhaus, 1996; Chen et al., 1999; Lutkenhaus and Addinall, 1997; Weiss et al., 1999). Consequently, the Z ring is thought to serve as framework for the assembly of other components of the cell division machinery. Additionally, invagination of the septum may be driven by the force generated by depolymerization of the Z ring following GTP hydrolysis.

The function of the other cell division proteins is not well understood. Some of them, DivIC, DivIB, FtsL, and the *B. subtilis* homologue of the *Escherichia coli* penicillin binding protein 3, Pbp2B, may help alter the direction of peptidoglycan synthesis and contribute to formation of the cross wall (Yanouri et al., 1993; Harry and Wake, 1989; Harry et al., 1993; Levin and Losick, 1994; Daniel et al., 1998). FtsL also appears to be required for the stability of DivIC, suggesting that these two proteins, which each have a single leucine zipper motif, may interact (Daniel et al., 1998). These proteins share similar topological structures in which their long carboxyl termini are facing outward, where they can interact with the cell wall, and their shorter amino termini face into the cytoplasm, anchoring the proteins in the plasma membrane. In contrast, FtsA is a cytoplasmic protein which has been shown to interact directly with FtsZ (Wang et al., 1997). The proper ratios of FtsZ to FtsA are critical for efficient cell division (Dai and Lutkenhaus, 1992; Dewar et al., 1992). FtsA has homology to the DnaK–actin family of ATPases (Bork et al., 1992; Sánchez et al., 1994), and work with *E. coli* suggests that it may serve as a link between the Z ring and peptidoglycan synthesis through the product of the *ftsI* gene, Pbp3 (the *E. coli* Pbp2B homologue) (Tormo et al., 1986).

Development appears to have a more stringent requirement than vegetative growth for the components of the cell division machinery. For example, *B. subtilis* cells carrying a null mutation in *ftsA* are partially viable (in contrast to

E. coli, in which *ftsA* is an essential gene) (Beall and Lutkenhaus, 1992). However, they do not sporulate. In addition, a heat-sensitive allele of *divIC*, *div355*, renders cells viable for growth but not for sporulation at 37°C, and a null mutation in *divIB* blocks sporulation but not vegetative growth in a strain-dependent manner (Harry et al., 1993; Levin and Losick, 1994; Levin, 1996). Thus, polar septation seems to be more sensitive to defects in cell division genes than binary fission. This makes sense in that an occasional failure in cytokinesis would not be expected to cause lethality whereas the formation of the polar septum is obligatory for partitioning the sporangium into forespore and mother cell compartments.

Z-RING FORMATION DURING SPORULATION

The switch from binary fission to polar septation is preceded by a shift in the localization pattern of FtsZ (Fig. 2). During vegetative

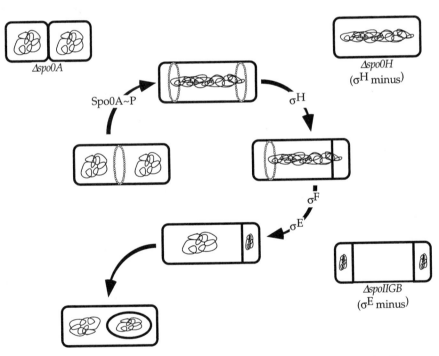

FIGURE 2 Asymmetric division. The arrows depict the course of events involved in asymmetric division and the contribution of sporulation transcription factors. The cartoon on the top left and the two cartoons on the right show the terminal phenotype of mutants lacking the indicated transcription factor.

growth, FtsZ forms a single ring at midcell, thereby establishing the location of the medial division site (Bi and Lutkenhaus, 1991; Wang and Lutkenhaus, 1993). In contrast, at the onset of sporulation, FtsZ polymerizes into two rings, one adjacent to each pole of the cell (Levin and Losick, 1996). This pattern of FtsZ localization suggests that prior to septation each pole is competent for cell division. However, as development progresses, septation occurs at only one of these sites, the choice being essentially random (Hitchins, 1975, 1976; Dunn and Mandelstam, 1977). The Z ring at the distal pole persists until the initiation of cell-type-specific gene expression in the mother cell.

Mutations that prevent cell-type-specific gene expression lead to the formation of so-called "disporic" cells. In these cells, both Z-rings are used and polar septa form sequentially at each end of the sporangium (Lewis et al., 1994; Stragier, 1989). DNA is then packaged into these two forespores, leaving the mother cell anucleate (Fig. 2B). In some disporic mutants, the RNA polymerase sigma factor σ^F becomes activated in both of the polar forespores. (The σ^F factor is the first sporulation-regulatory protein to become activated in a cell-specific fashion during sporulation [see below]. Ordinarily, it is activated exclusively in the forespore compartment formed at one pole of the sporangium.) The unifying feature of mutations causing the disporic phenotype is that in one way or another they prevent the activation of the earliest-acting transcription factor in the mother cell, the RNA polymerase sigma factor σ^E. As we shall see, the activation of σ^E in the mother cell depends on the prior activation of σ^F in the forespore, which in turn depends on the formation of the polar septum. So the formation of a septum at one pole sets in motion a chain of events that blocks the formation of a septum at the other pole. It seems reasonable to postulate that σ^E turns on an unidentified gene that functions to impede the formation of a second polar septum by preventing the use of the second Z ring following asymmetric division. In any event, the existence of disporic mutants reinforces the view

that both polar Z rings are competent to support septum formation.

Why, then, is septation normally restricted to a single pole? Perhaps one or more proteins involved in the conversion of a Z ring into a septum is present in limiting quantities. If so, by a largely stochastic process, septum formation might commence near one pole and recruit this limiting factor to the site of septum formation. Once a septum has formed at one pole, the limiting factor could be recycled for the formation of a septum at the second Z ring (unless prevented from doing so by the product of a gene under the control of σ^E). Limiting quantities of cell division components may also explain the heightened sensitivity of sporulation to defects in cell division genes.

A TALE OF TWO TRANSCRIPTION FACTORS

The switch from binary fission to polar septation is under the control of two transcription factors, Spo0A and σ^H. Alone or in combination, Spo0A and σ^H are responsible for the transcription of all *spoII* genes (Errington, 1993; Stragier and Losick, 1996). Mutations in either *spo0A* or *spo0H*, the gene encoding σ^H, prevent polar septation, and thus each has traditionally been given a *spo0* classification (see below). As discussed below, however, recent evidence from immunofluorescence microscopy indicates that each transcription factor plays a different role in division site selection (Fig. 2).

Spo0A, a member of the response regulator family of transcription factors, is subject to both transcriptional and posttranscriptional forms of regulation. As outlined in the previous chapter, the majority of *spo0* genes define a regulatory pathway known as the phosphorelay that leads to the activation of Spo0A by phosphorylation. Also, the gene for *spo0A* is positively regulated from a σ^H-controlled promoter. Although Spo0A~P can function as a repressor, it generally acts as a transcriptional activator in conjunction with σ^A (the major sigma factor in vegetative cells) or with σ^H (Errington, 1993; Stragier and Losick, 1996). Either the loss of *spo0A* itself or mutations in genes required for its activation result in the complete inhibition of polar septation. The result of this inhibition is that under conditions that normally stimulate sporulation, *spo0A* mutant cells undergo an extra round of binary fission leading to the formation of tiny cells, approximately half the size of wild-type sporangia (Dunn et al., 1976).

Mutations blocking the activation of Spo0A also block the transition from medial to bipolar localization of FtsZ (Levin and Losick, 1996). In contrast to wild-type cells, cells mutant for *spo0A* exhibit medial rings of FtsZ for as long as 3 h after resuspension in sporulation medium. Furthermore, cells engineered to synthesize a constitutively active form of Spo0A (the product of the mutant *spo0A* allele *sad67*) during growth produce polar Z rings and polar septa. These results imply the existence of an unidentified gene (or genes) in the Spo0A regulon that is responsible for the switch to a bipolar pattern of FtsZ localization and asymmetric division.

The σ^H factor is the earliest-acting sporulation-specific sigma factor. Like Spo0A, σ^H is subject to both transcriptional and posttranscriptional control mechanisms. The gene for σ^H is negatively regulated by the transition state regulator AbrB, whose synthesis is in turn repressed by Spo0A (Dubnau et al., 1988; Perego et al., 1988; Strauch, 1993, 1995). Thus, activation of Spo0A stimulates the transcription of *spo0H* by leading to the depletion of AbrB. In addition, the accumulation of σ^H is subject to posttranscriptional control, probably at the level of enhanced protein stability under conditions of nutrient depletion (Healy et al., 1991; Frisby and Zuber, 1994).

In contrast to *spo0A* mutations, mutations in *spo0H* do not block the switch from medial to bipolar localization of FtsZ (Levin and Losick, 1996). Three hours after resuspension in sporulation medium, *spo0H* null mutants exhibit the bipolar pattern of FtsZ localization characteristic of wild-type cells immediately prior to polar septation. However, cytokinesis does not occur at either of the two polar Z rings, a finding that indicates that σ^H is required

for one or more subsequent steps in cytokinesis. The requirement for σ^H in polar septation may simply be to maintain the expression of cell division genes during stationary phase at levels sufficient for cytokinesis. Alternatively, σ^H may direct the transcription of an unidentified gene(s) whose product activates cytokinesis at one of the polar sites. In either case, the phenotype of a *spo0H* mutant suggests that the formation of the Z ring is not the rate-limiting step in polar septation.

The distinct roles of σ^H and Spo0A~P in asymmetric division provide clues as to the nature of the gene(s) ultimately responsible for the switch from binary fission to polar septation. First, the requirement for Spo0A and not σ^H to drive the switch from medial to polar localization of FtsZ suggests that the gene(s) responsible for division site selection is under the control of Spo0A and σ^A, the primary, or "housekeeping," sigma factor of vegetative cells. Moreover, as σ^H stimulates the transcription of *spo0A* (Yamashita et al., 1989; Siranosian and Grossman, 1994), the ability of *spo0H* null mutants to switch the localization pattern of FtsZ suggests that only a low level of Spo0A~P is required to shift the division site from a medial to a polar position.

During sporulation the vegetative pattern of nucleoid segregation is abandoned in favor of axial-filament formation, in which the chromosomes form an elongated structure extending the length of the cell (Ryter, 1964; Setlow et al., 1991). Like the switch from medial to bipolar FtsZ localization, axial-filament formation appears to depend on Spo0A~P and does not require σ^H. A null mutation in *spo0A* blocks axial-filament formation to the same extent that it prevents the switch from medial to polar FtsZ localization (Levin and Losick, 1996). Consequently, during sporulation a *spo0A* null mutant divides in exactly the same manner as does a vegetatively growing cell; that is, it undergoes nucleoid segregation and forms medially positioned septa between newly segregated nucleoids. By comparison, *spo0H* mutations do not block axial-filament formation. Terminally differentiated *spo0H* null mutant cells contain single unsegregated nucleoids flanked by Z rings. These mutant cells are similar in appearance to wild-type cells at morphological stage I, as described by Ryter and co-workers (Ryter, 1964; Ryter et al., 1966). Thus, *spo0H* may more accurately be described as a stage I gene (*spoIH*). Conceivably, the correlation between axial-filament formation and bipolar localization of FtsZ is indicative of a connection between chromosome segregation and division site selection.

Efficient polar septation also requires the actions of the SpoIIE phosphatase. Earlier work demonstrated that SpoIIE is responsible for thinning the cell wall material sandwiched between the mother cell and forespore membranes (Illing and Errington, 1991; Barák and Youngman, 1996). Recent evidence suggests that SpoIIE is also involved in the switch from binary fission to polar septation. Although *spoIIE* is not required for asymmetric division, its disruption causes an ~20-min delay in polar septation (Feucht et al., 1996). Moreover, the formation of polar septa during growth in response to expression of a constitutively active allele of *spo0A* appears to be dependent on *spoIIE* (Khvorova et al., 1998). Consistent with its role in polar septation, immunofluorescence microscopy indicates that SpoIIE colocalizes with FtsZ to the polar sites in an FtsZ-dependent manner (Arigoni et al., 1995; Levin et al., 1997). Thus, SpoIIE may contribute to polar septation by helping to establish and stabilize FtsZ rings at the polar sites.

CHROMOSOME SEGREGATION DURING SPORULATION

In contrast to vegetative growth, in which the nucleoids are segregated from one another prior to cytokinesis, polar septation bisects one end of the axial filament such that only 30% of the chromosome is initially in the forespore compartment. Following septation, the remainder of the chromosome is transferred from the mother cell to the forespore by what appears to be a conjugation-like mechanism (Wu

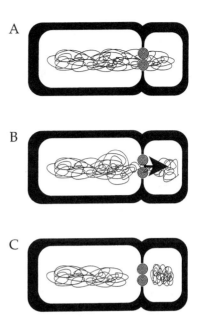

FIGURE 3 Translocation of a chromosome into the forespore. Depicted in the septum is a channel created by the DNA translocase protein SpoIIIE through which the forespore chromosome is translocated.

and Errington, 1994; Wu et al., 1995). This transfer depends on SpoIIIE (Fig. 3).

The *spoIIIE* gene was originally identified as a *spo* gene because it is required at the stage of engulfment. Nevertheless, it is expressed in growing cells as well as sporulating cells and, as we shall see, has a role in chromosome segregation during growth as well as during sporulation. The original *spoIIIE* mutation (the missense mutation *spoIIIE36*) caused a "chromosome position effect" phenotype in which the expression of certain forespore-expressed genes was dependent on the proximity of the gene in question to the chromosomal origin of replication (Sun et al., 1991). In cells carrying the *spoIIIE36* mutation, σ^F-controlled genes in the origin-proximal third of the chromosome were expressed while those in the origin-distal portion of the chromosome were transcriptionally silent. Thus, it appeared that only the origin-proximal third of the chromosome entered the forespore compartment of the *spoIIIE* mutant. This interpretation was confirmed by a variety of cytological methods, demonstrating that it is in fact the origin that is initially packaged into the forespore following polar septation, because the two chromosomes in the sporangium are oriented with their replication origins near opposite poles of the cell (Glaser et al., 1997; Lin et al., 1997; Webb et al., 1997).

SpoIIIE is similar to proteins involved in plasmid transfer in *Streptomyces* and has a high degree of similarity to the carboxyl terminus of the *ftsK* gene product of *E. coli* (Wu and Errington, 1994; Begg et al., 1995). Although not required for cell division, during vegetative growth the *spoIIIE* gene product ensures that accidentally missegregated chromosomes are not "guillotined" during binary fission (Sharpe and Errington, 1995). Immunofluorescence microscopy indicates that SpoIIIE localizes as a discrete dot at the middle of the septum in both sporulating and vegetatively growing cells, a location consistent with its role in DNA translocation (Wu and Errington, 1997).

Evidence indicates that SpoIIIE plays an additional role in sporulation besides its function in chromosome transfer. In contrast to the originally characterized missense mutation *spoIII36*, a *spoIIIE* null mutation abolishes cell-type-specific gene expression. Normally, σ^F- and σ^E-directed gene expression is strictly confined to the forespore and the mother cell, respectively. In the null mutant, however, σ^F-directed gene transcription is observed in the mother cell chamber and σ^E-directed gene expression is observed in the forespore (Wu et al., 1995; Pogliano et al., 1997). This effect cannot easily be attributed to a defect in DNA translocation because the *spoIIIE36* missense mutant is blocked in chromosome transfer but is not defective in the compartment-specific activation of σ^F and σ^E. This suggests that SpoIIIE plays two distinct roles in sporulation, a topic to which we return below.

WHAT DEFINES THE DIVISION SITE?

How a cell specifies its division site remains one of the most poorly understood areas of in-

quiry in prokaryotic biology. During vegetative growth, a new division site must be created de novo at midcell in each generation. As of this writing the mechanisms by which the medial site is established remain a mystery. Although it is formally possible that they are created de novo as well, it appears likely that the polar sites used by *B. subtilis* for asymmetric division are vestiges of the previous round of binary fission.

The idea that the poles can serve as alternative division sites was originally conceived as part of a model for *min* gene function in *E. coli*. Mutations in the *min* genes, *minC* and/or *minD*, lead to the formation of small anucleate cells, the result of polar divisions (Adler et al., 1967). A series of elegant experiments determined that in wild-type cells the products of *minC*, *minD*, and the third gene in the operon, *minE*, function in concert to ensure that cell division occurs only at the middle of the cell (Fig. 4A) (de Boer et al., 1989). The current model for *min* gene function in *E. coli* is based on the idea that there are three potential division sites in the cell, the new site at midcell and the old sites at each pole. MinC and MinD function in tandem to inhibit cell division (Z-ring formation) at all three sites, whereas MinE relieves MinCD division inhibition only at the medial site (de Boer et al., 1989; Bi and Lutkenhaus, 1993). A MinE-GFP fusion localizes to midcell, supporting the idea that its function is restricted to the medial position (Raskin and de Boer, 1997).

B. subtilis also appears to have polar division sites. As in *E. coli*, mutations in the *B. subtilis* homologues of *minC* and *minD* lead to the formation of polar septa and minicells during vegetative growth (Reeve et al., 1973; Coyne and Mendelson, 1974; Levin et al., 1992; Varley and Stewart, 1992; Lee and Price, 1993). The polar septa formed in the *min* mutants are located in essentially the same positions as sporulation septa, and the minicells are approximately the same size as a preengulfment forespore. (In the absence of SpoIIIE, the minicells are no longer anucleate, suggesting that in

A

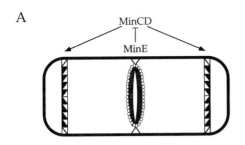

B

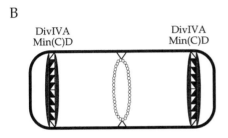

FIGURE 4 Contrasting roles of *E. coli* MinE and *B. subtilis* DivIVA in septum site selection. (A) In *E. coli*, MinE (gray ring) localizes to the nascent septal site in an FtsZ-independent manner. By virtue of its localization pattern, it inhibits MinCD division inhibition at midcell, allowing for FtsZ ring formation (beaded ring) and binary fission. MinCD activity is thus restricted to the cell poles (black and white bands), where it prevents FtsZ ring formation and cell division. (B) In *B. subtilis*, DivIVA (gray) localizes to the cell poles, where it recruits MinD and presumably MinC (black and white bands). By recruiting MinCD to the cell poles, DivIVA leaves the medial site free for FtsZ localization (beaded ring) and cell division.

wild-type cells chromosomal DNA is initially trapped in minicells only to be pumped out through the action of the SpoIIIE DNA translocase [Sharpe and Errington, 1995].) Although no homologue of *minE* has been identified in *B. subtilis*, the *divIVA* gene product appears to serve an analogous function (Cha and Stewart, 1997; Edwards and Errington, 1997). Recent work suggests that DivIVA, by virtue of localizing to the division site late in cytokinesis, sequesters MinD, and most likely MinC as well, to the cell poles (Fig. 4B) (Marston et al., 1998).

Despite their roles in maintaining the medial localization of the vegetative division septa,

neither *minC* nor *minD* appears to play a major role in switching the cell from binary fission to polar septation. First, the loss of *minC* and/or *minD* results in at most a 10-fold decrease in sporulation efficiency (Levin et al., 1992; Lee and Price, 1993; Barák et al., 1998). Moreover, polar divisions occur in only ~10% of cells bearing null mutations in *minC* and *minD*, indicating that simply repressing their functions would not be sufficient for the global switch in septum position required for sporulation (Levin et al., 1998). Finally, and perhaps most importantly, the loss of *minCD* is not sufficient to restore the bipolar pattern of FtsZ localization to a *spo0A* mutant (Levin and Losick, 1996). Nevertheless, given the recently reported polar localization pattern of MinD, some factor present in sporulating cells must serve to inactivate the MinCD division complex at the cell poles prior to polar septation (Marston et al., 1998).

Although they are not essential to development, the drop in sporulation efficiency in the *min* mutants suggests that MinC and MinD participate to some degree in the switch from binary fission to polar septation. Cells with medial septa are frequently observed in electron micrographs of developing *minCD* null mutants. Like wild-type polar septa, these medial septa are targeted by the sporulation protein SpoIIE and, most likely due to the action of SpoIIE at that site, often appear thin (Barák et al., 1998). Thus, it appears that the *min* genes help maintain a polar pattern of division during sporulation and a medial pattern during vegetative growth.

Though they are apparently not directly involved in the switch from binary fission to polar septation, the existence of the *min* genes supports the idea that the polar sites used during sporulation are the remnants of previous rounds of cell division. Thus, the question becomes not how the polar site is selected but rather how medial division is suppressed and polar division activated. More specifically, what is the gene(s) under Spo0A control that inhibits FtsZ nucleation at midcell while promoting ring formation at the polar sites? This remains an open question.

ESTABLISHMENT OF THE FORESPORE AND MOTHER CELL LINES OF GENE EXPRESSION

As we have seen, a hallmark of sporulation in *B. subtilis* is the formation of the polar septum. Part of the reason for interest in its formation is that the progeny of asymmetric division exhibit dissimilar lines of gene transcription. Differential gene transcription between the forespore and the mother cell is established by the transcription factors σ^F and σ^E (Stragier and Losick, 1996). Both proteins are synthesized in the predivisional sporangium, but they do not become active in directing transcription until after polar division (stage II), when the activity of σ^F is confined to the forespore and that of σ^E is confined to the mother cell. As we shall see, the polar septum is intimately involved in the activation of both transcription factors and perhaps can be thought of as an organelle for the establishment of cell fate. Once activated, the two transcription factors set in motion the forespore and mother cell lines of gene expression, with σ^F directing the synthesis of σ^G (see below) in the forespore after the stage of engulfment and σ^E triggering a hierarchical regulatory cascade involving the sigma factor σ^K (see below) and two DNA binding proteins (SpoIIID and GerE) which are not considered further in this review. In the following discussion, we consider the mechanisms governing the activation of σ^F and σ^E and a pathway of intercellular communication linking the two. Then we will consider the appearance of the late-acting sigma factors σ^G and σ^K and two additional pathways of intercellular communication.

THE ACTIVITY OF σ^F IS GOVERNED BY A PROTEIN WITH MULTIPLE FUNCTIONS

Understanding the connection between the activation of σ^F and septum formation and understanding the mechanism by which the action of σ^F is restricted to the forespore are two of the principal unmet challenges in the sporu-

lation field. The activity of σ^F is governed by three regulatory proteins, which, like σ^F, are synthesized in the predivisional sporangium under the control of Spo0A~P. These proteins are SpoIIAA and SpoIIAB, which together with σ^F are encoded by the three-cistron *spoIIA* operon, and SpoIIE, the product of the *spoIIE* gene (Stragier and Losick, 1996, and references therein).

SpoIIAB has three biochemical functions, all of which are important in the regulation of σ^F (Fig. 5A). First, it is an antisigma factor that binds to σ^F, trapping it in an inactive complex. Second, SpoIIAB is a protein kinase that phosphorylates SpoIIAA on a serine residue. SpoIIAB has an adenosine nucleotide-binding pocket, and its capacity to bind to σ^F and to phosphorylate SpoIIAA is dependent on the presence of ATP in the pocket. The third function of SpoIIAB is to form a long-lived complex with SpoIIAA, a reaction that depends on the presence of ADP rather than ATP. When sequestered in the SpoIIAB(ADP)-SpoIIAA complex, SpoIIAB is unable to bind to or inhibit σ^F or to phosphorylate SpoIIAA. Only unphosphorylated SpoIIAA is capable of forming the SpoIIAB(ADP)-SpoIIAA complex. Thus, SpoIIAA and SpoIIAB are mutually antagonistic. SpoIIAB (when it contains ATP in its nucleotide-binding pocket) is capable of inactivating SpoIIAA by phosphorylation, and SpoIIAA is capable of inhibiting SpoIIAB (when SpoIIAB contains ADP) by trapping the antisigma factor-kinase in a long-lived complex. The differential effects of ATP and ADP have given rise to the idea that one input into the mechanisms governing the activation of σ^F could be a decrease in cellular ATP levels (Alper et al., 1994). This idea has not been tested experimentally, but the related transcription factor σ^B is controlled by homologues of SpoIIAA, SpoIIAB and SpoIIE, that respond in part to changes in cellular ATP levels (Alper et al., 1994, 1996).

Recent mechanistic studies indicate that SpoIIAA induces the release of σ^F from SpoIIAB by directly interacting with the SpoIIAB(ATP)-σ^F complex, resulting in the

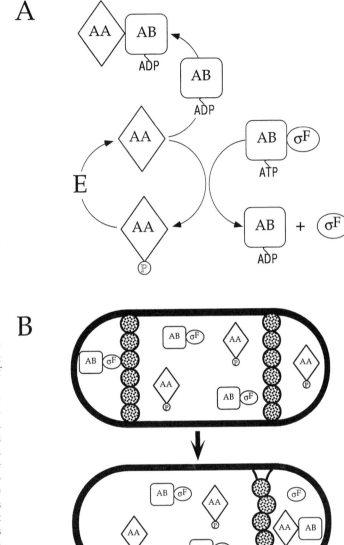

FIGURE 5 Activation of σ^F. (A) Release of σ^F from the antisigma factor-kinase SpoIIAB (AB) by the action of SpoIIAA (AA). SpoIIAA is subject to a cycle of phosphorylation and dephosphorylation governed by the SpoIIE phosphatase (E) and the antisigma factor-kinase (AB). SpoIIAB can form alternative complexes with SpoIIAA when it contains ADP and with σ^F when it contains ATP. (B) Comparison of the state and location of these proteins before and after septation. The SpoIIE phosphatase (stippled circles) is located in rings near each end of the predivisional sporangium and in the septum in the postdivisional sporangium. SpoIIAA (AA) is in the phosphorylated state in the predivisional sporangium and in the mother cell. In the forespore it is unphosphorylated and in a complex with SpoIIAB (AB). Finally, σ^F is held in a complex with SpoIIAB in the predivisional sporangium and in the mother cell and is free of the antisigma factor in the forespore.

discharge of free σ^F and the phosphorylation of SpoIIAA (Duncan et al., 1996; Garsin et al., 1998). Thus, phosphorylation is involved in both the inhibition of σ^F (by inactivating SpoIIAA) and the activation of σ^F (through the induced release reaction), a view that is supported by both biochemical and genetic evidence.

Central to our understanding of the activation of σ^F is the role played by the third regulatory protein, SpoIIE. SpoIIE (not to be confused with the DNA translocase SpoIIIE) is an integral membrane protein with 10 membrane-spanning segments in the N-terminal portion of the protein and a phosphatase domain, which has similarity to the eukaryotic family of PP2C phosphatases, in its C-terminal region (Adler et al., 1997; Arigoni et al., 1999). It is responsible for dephosphorylating SpoIIAA-P. As we have seen, unphosphorylated SpoIIAA can react with the SpoIIAB(ATP)-σ^F complex to effect the release of σ^F or bind to ADP-containing SpoIIAB to sequester the antisigma factor-kinase in an inactive complex. In addition, Yudkin and coworkers (Magnin et al., 1997; Najafi et al., 1997) have found that ADP-containing SpoIIAB generated as a consequence of the phosphorylation of SpoI-IAA is slow to recycle to the ATP-containing form. This slow recycling would contribute to the escape of σ^F from inhibition by SpoIIAB. Evidently, unphosphorylated SpoIIAA can act in several ways to cause the activation of σ^F. The chief point is that by catalyzing the conversion of SpoIIAA-P to SpoIIAA, the SpoIIE phosphatase helps to govern the activation of σ^F.

These findings are the basis for the following model (Fig. 5B). The σ^F factor is held in an inactive complex with SpoIIAB both in the predivisional sporangium and in the mother cell compartment of the postdivisional sporangium because SpoIIAA is in the inactive, phosphorylated state in these two cell types. Somehow, under the control (at least in part) of the SpoIIE phosphatase, unphosphorylated SpoIIAA is selectively generated in the forespore, triggering the release of σ^F. Evidence from immunofluorescence microscopy with antibodies specific to unphosphorylated SpoIIAA seems to support this model (Lewis et al., 1996). However, recent results considered below indicate that unphosphorylated SpoIIAA may not be confined to the forespore and point to the existence of an additional and important regulatory step acting after the dephosphorylation of SpoIIAA-P that governs the activation of σ^F (King et al., 1999).

Although, according to the model shown in Fig. 5B, SpoIIE and SpoIIAA are intimately and indispensably involved in the activation of σ^F, Frandsen et al. (1999) have cleverly devised a strategy that allows a *spoIIE* or a *spoIIAA* mutant to sporulate. These workers have constructed a strain in which the gene for σ^F has

been moved near the origin of replication whereas the gene for SpoIIAB remains at its normal location, far from the origin region. Because the origin-distal region of the chromosome is transported into the forespore after polar septation (see above), the gene for σ^F is present transiently in the forespore of such a strain in the absence of the gene for the antisigma factor. As a consequence, newly synthesized σ^F in the forespore is free of SpoIIAB and therefore gives rise to a burst of σ^F-directed gene transcription. Importantly, such a strain is capable of sporulating, and in a manner that does not depend on either the SpoIIE phosphatase or SpoIIAA! Thus, the normal pathway for the compartmentalization of σ^F activity can be bypassed by causing the gene for σ^F to be present in the forespore in the absence of the gene for its antagonist.

THE SpoIIE PHOSPHATASE LINKS THE ACTIVATION OF σ^F TO THE POLAR SEPTUM

How is the activation of σ^F tied to the formation of the polar septum? Part of the answer emerges from studies of the subcellular localization of SpoIIE (Fig. 5B). Initially, upon its synthesis in the predivisional sporangium, SpoIIE localizes in a bipolar pattern coincident with the bipolar pattern of Z rings in the sporangium (Arigoni et al., 1995; Levin et al., 1997). Next, upon asymmetric division, SpoIIE invades the division septum and disappears from the distal pole of the sporangium. Thus, at the time that σ^F is activated, SpoIIE is sitting at the boundary between the cell (the forespore) in which the transcription factor is activated and the cell (the mother cell) in which it remains inert. (Interestingly, in disporic mutants something different happens. Rather than disappearing from the distal pole, SpoIIE persists near both ends of the sporangium and becomes incorporated into the second polar septum [Arigoni et al., 1995; Pogliano et al., 1997].) Therefore, it is appropriately positioned to trigger the activation of σ^F in both forespore compartments of the disporic sporangium.) Finally, after the activation of σ^F,

SpoIIE disappears from the septum. Thus, it is present at the septum only during a brief interval when the activation of σ^F takes place.

Wu et al. (1998) have recently suggested a different order of events: namely, that SpoIIE initially localizes to the polar septum and then to the distal pole before disappearing from the sporangium entirely. Several lines of evidence speak against this: (i) in sporulating cells and in cells engineered to synthesize SpoIIE during growth, the phosphatase localizes to Z rings prior to septation (Arigoni et al., 1995; Levin et al., 1997); (ii) in a mutant blocked in septation, SpoIIE colocalizes with FtsZ rings during sporulation (Levin et al., 1997); (iii) the use of a vital stain for visualizing membranes in sporulating cells readily reveals predivisional sporangia that have SpoIIE localized near both poles (King et al., 1999); and (iv) time-lapse microscopy shows that sporangia with bipolar SpoIIE undergo conversion into sporangia with SpoIIE at only one pole, rather than the other way around (King et al., 1999). Significantly, despite being localized to FtsZ rings in the division mutant, SpoIIE is not capable of activating σ^F, indicating that septation is critical for its function.

ACTIVATION OF σ^F IS TIED TO SEPTATION AT A STEP AFTER THE DEPHOSPHORYLATION OF SpoIIAA-P

The discovery that SpoIIE localizes to the polar septum links the chain of events from asymmetric division to the activation of σ^F and suggests an explanation for the observation that the activation of σ^F depends on the formation of the polar septum (Levin and Losick, 1994). However, does it in fact matter that SpoIIE localizes to the septum? This question has recently been addressed in experiments in which the phosphorylation state of SpoIIAA was examined in wild-type sporulating cells and in sporulating cells in which polar septation had been blocked by deprivation of the cell division protein FtsZ or use of a heat-sensitive mutant of the cell division protein DivIC. The principal findings are as follows. First, in wild-type sporangia, unphosphorylated SpoIIAA appears earlier, and accumulates to a substantially greater extent, than would be expected if dephosphorylation were confined to the forespore (King et al., 1999). Second, in sporangia that have been deprived of FtsZ and hence are blocked in polar septation, the level of unphosphorylated SpoIIAA is lower than that observed in wild-type sporulating cells. This observation is consistent with the view that full activation of the phosphatase domain of SpoIIE depends on the association of SpoIIE with FtsZ, perhaps at the stage of polar Z-ring formation (King et al., 1999).

Third, and unexpectedly, sporangia blocked at a later stage of polar septation by use of a temperature-sensitive mutant of DivIC accumulate very high levels of unphosphorylated SpoIIAA and yet are strongly impaired in the activation of σ^F (King et al., 1999). A similar observation has been made by J. Errington (personal communication) using a different cell division mutant. These observations suggest that activation of σ^F is governed by a previously unrecognized regulatory step subsequent to the dephosphorylation of SpoIIAA-P. Experiments with modified and mutant forms of SpoIIE implicate SpoIIE itself in this postdephosphorylation regulatory step (King et al., 1999). It is not known how this regulation works, but one possibility is that after dephosphorylating SpoIIAA-P, SpoIIE continues to sequester the resulting molecule of SpoIIAA in a state in which it is incapable of reacting with SpoIIAB-σ^F. Release from this hypothetical SpoIIE-SpoIIAA complex evidently requires completion of polar septum formation. According to this view, asymmetric division is a checkpoint that holds SpoIIAA in check until the formation of the polar septum is complete.

CELL-SPECIFIC ACTIVATION OF σ^F

What has not been addressed in the discussion so far is the mechanism by which σ^F activation is confined to the forespore chamber of the sporangium. Two models have been considered. In one model, SpoIIE is present and active on both faces of the sporulation septum, but because the forespore is so much smaller than the mother cell, the ratio of phosphatase (SpoIIE) to kinase (SpoIIAB) would be substantially higher in the small compartment than in the large compartment (Arigoni et al., 1995). This model has the virtue of simplicity, but a challenge to the model is the existence of an endospore-forming species (*Sporosarcina ureae*) that apparently undergoes division symmetrically rather than asymmetrically during sporulation (Zhang et al., 1997). In such cells, the ratio of phosphatase to kinase would be expected to be the same on both sides of the septum. On the other hand, the vast majority of endospore-forming bacteria form the sporulation septum at an extreme polar position, and hence the mechanisms by which *S. ureae* achieves cell-specific gene activation may not be representative of those employed by *B. subtilis* and its relatives.

The second model holds that the SpoIIE protein or its activity is limited to the forespore face of the septum. Wu et al. (1998) have used lysozyme treatment to tease the forespore and mother cell compartments of the sporangium partially apart and thereby assess whether SpoIIE is restricted to the forespore. They conclude that it is. This raises the question of how the phosphatase could be sequestered to one face of the septum but, if correct, would provide a simple explanation for the basis for the cell-specific activation of σ^F. On the other hand, similar experiments by King et al. (1999) provide evidence for the presence of substantial quantities of SpoIIE in both sporangial compartments. In any event, both models may need to be reconsidered in light of the discovery (see above) that activation of σ^F is governed in part by a SpoIIE-mediated event subsequent to the dephosphorylation of SpoIIAA-P.

Surprisingly, under some circumstances the phosphatase need not be linked to the septum at all. Arigoni et al. (1999) have discovered that cells engineered to synthesize a truncated form of SpoIIE lacking the N-terminal membrane-spanning region but retaining the soluble phosphatase region function to an impressive extent. In these cells, the soluble domain of SpoIIE is evenly distributed throughout the sporangium. The activation of σ^F is delayed for 1 to 2 h in such a strain, but eventually enough σ^F becomes active in a cell-specific fashion to support high levels of spore formation. What are we to make of this enigmatic result? One possibility is that cell-specific activation of σ^F is a composite consequence of multiple convergent pathways, all of which contribute to limiting σ^F-directed gene expression to the forespore. If so, then at least one such pathway operates in a manner that does not depend on the association of SpoIIE with the polar septum.

Finally, we return to the issue of how a null mutation of the *spoIIIE* gene for the DNA translocase is able to break the otherwise-strict compartmentalization of σ^F activity. Normally, during the activation of σ^F, the SpoIIE phosphatase disappears from the forespore-distal pole of the sporangium and perhaps from the mother cell face of the septum as well. Pogliano et al. (1997) and Arigoni et al. (1995) have observed, however, that SpoIIE persists in the mother cell in a *spoIIIE* null mutant. This abnormal persistence could account for the misactivation of σ^F in the mother cell chamber of the mutant sporangium. What is mysterious, however, is the connection between *spoIIIE* and the elimination of SpoIIE from the mother cell. Whatever the basis for the connection, the explanation is unlikely to be a defect in DNA translocation (at least not entirely) because both a *spoIIIE* null mutation and the *spoIIIE* missense mutation *spoIIIE36* block chromosome translocation but only the null mutation breaks the normally strict compartmentalization of σ^F activation (Wu and Errington, 1994; Pogliano et al., 1997).

ACTIVATION OF σ^E IS GOVERNED BY AN INTERCELLULAR SIGNAL TRANSDUCTION PATHWAY OPERATING AT THE LEVEL OF PROPROTEIN PROCESSING

The mother cell transcription factor σ^E is initially synthesized as an inactive proprotein (pro-σ^E) which harbors an N-terminal extension of 27 amino acids (Stragier and Losick,

1996, and references therein). The conversion of pro-σ^E to mature σ^E is mediated by SpoIIGA, which directly interacts with pro-σ^E and is likely to be the proteolytic processing enzyme for the proprotein. Both pro-σ^E and its putative protease SpoIIGA are encoded by the two-cistron *spoIIG* operon, which is induced in the predivisional sporangium under the control of Spo0A~P (Fig. 6).

Evidence indicates that SpoIIGA is inactive in its default state. In order to catalyze the conversion of pro-σ^E to mature σ^E, SpoIIGA must be activated by a signal protein called SpoIIR, which is produced in the forespore under the control of σ^F (Karow et al., 1995; Shazand et al., 1995; Londono-Vallejo and Stragier, 1995). SpoIIR is a secreted protein, and it is thought that it is secreted from the forespore into the space between the two membranes of the polar septum, where it activates SpoIIGA

(Hofmeister et al., 1995). According to this view, SpoIIGA is a receptor-protease that is activated by SpoIIR, although evidence that SpoIIR directly interacts with SpoIIGA is lacking. In keeping with this model, SpoIIGA is known to be localized to the polar septum (Fawcett et al., 1998).

Interestingly, pro-σ^E is a membrane-associated protein (Fig. 6) (Ju et al., 1997; Hofmeister, 1998). Evidence from fluorescence microscopy indicates that it associates with the cytoplasmic membrane in the predivisional sporangium but then migrates to the polar septum at the time of asymmetric division, where it can interact with and undergo cleavage by SpoIIGA. Finally, after proteolytic activation σ^E is released into the cytoplasm, where it binds to and activates transcription by RNA polymerase (Fig. 6). Thus, this mother cell regulatory protein exhibits three subcellular addresses

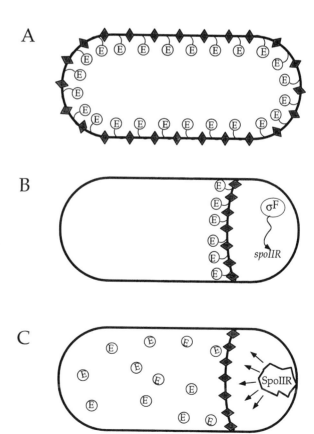

FIGURE 6 An intercellular signal transduction pathway governing the appearance of σ^E. (A) Pro-σ^E and the SpoIIGA-processing enzyme as associated with the cytoplasmic membrane. (B) Both SpoIIGA and pro-σ^E have become located at the polar septum. (B and C) The cartoons show that the signaling protein SpoIIR is produced in the forespore under the control of σ^F and that it then triggers the action of SpoIIGA, causing the release of mature σ^E into the mother cell cytoplasm. The cartoon suggests that pro-σ^E is sequestered on the mother cell face of the septum, thereby explaining the presence of mature σ^E exclusively in the mother cell.

during the course of its conversion from the inactive proprotein to the active transcription factor!

The discovery of an intercellular signal transduction pathway that links the appearance of σ^E in the mother cell to the activation of σ^F in the forespore provides a simple explanation for the timing of pro-σ^E activation. Proteolytic activation of pro-σ^E was known not to commence until stage II and, indeed, to depend on the formation of the polar septum (Beall and Lutkenhaus, 1991). As we have seen, activation of σ^F depends on septum formation through a chain of events that is linked to the polar septum through the SpoIIE phosphatase. Thus, the appearance of σ^E is indirectly tied to septum formation through a signal transduction pathway that couples proprotein processing to σ^F, whose own activation is linked to septum formation.

COMPARTMENTALIZATION OF σ^E-DIRECTED GENE EXPRESSION IS ACHIEVED BY SEQUESTERING THE TRANSCRIPTION FACTOR TO THE MOTHER CELL

What has been ignored in the discussion above is the basis for the compartmentalization of σ^E. Because pro-σ^E is synthesized in the predivisional sporangium, it might have been expected that the proprotein would be distributed to, and present in, both compartments of the postdivisional sporangium. We now know from immunofluorescence microscopy experiments, however, that after polar division σ^E is only present in the mother cell. Interestingly, and in a manner reminiscent of σ^F, this restriction of σ^E to the mother cell depends on the DNA translocase gene *spoIIIE* (Pogliano et al., 1997; Ju et al., 1998). A null mutation breaks compartmentalization and causes the transcription factor to be present and capable of directing gene transcription in the forespore as well as the mother cell. Thus, in a *spoIIIE* null mutant, pro-σ^E is both present in the forespore and converted to the active form of the protein. This shows that the basis for the compartmentalization of σ^E-directed gene expression is exclusion of pro-σ^E from the forespore because,

in a mutant in which pro-σ^E is not excluded from the forespore, it is activated by proteolysis and capable of directing gene transcription in both compartments of the sporangium.

It is both remarkable and mysterious that a gene that has been assigned a central role in chromosome translocation has come to be implicated in the compartmentalization of both σ^F and σ^E. Thus, *spoIIIE* is involved in three decidedly different processes: DNA transport into the forespore, elimination of the SpoIIE phosphatase from the mother cell, and the exclusion or elimination of pro-σ^E from the forespore.

Whatever the nature of the involvement of *spoIIIE*, two models can be put forth for the compartmentalization of σ^E: an elimination model and an exclusion model. In the elimination model, pro-σ^E is distributed to both the forespore and the mother cell upon polar division but is rapidly and selectively destroyed in the forespore by proteolysis (Pogliano et al., 1997). If so, then this compartment-specific proteolysis must depend on *spoIIIE*. In the exclusion model, pro-σ^E molecules are excluded from the forespore during the process by which the septum is formed. The discovery that pro-σ^E molecules migrate from the cytoplasmic membrane to the polar septum during asymmetric division suggests how this could happen: alignment of pro-σ^E molecules selectively on the mother cell face of the enclosing septum would exclude the transcription factor from the forespore (Hofmeister, 1998). In such a model, *spoIIIE* would be involved in restricting pro-σ^E to the mother cell face of the septum, and indeed, SpoIIIE is known to localize to the septum (Wu and Errington, 1997).

Finally, we have seen that for both the case of σ^F and the case of σ^E the polar septum is intimately involved in the activation of the transcription factors and the restriction of their activities to one or another cellular compartment. SpoIIE in the case of σ^F and SpoIIR, SpoIIGA, and pro-σ^E itself in the case of σ^E localize to the polar septum. Also, the SpoIIIE DNA translocase, which is required for strict compartmentalization of the activities of σ^F

and σ^E, is located in the septum. It thus appears that the septum is not simply a partition between two progeny cells but rather an organelle that actively participates in establishing the fate of the forespore and the mother cell compartments of the sporangium.

OTHER INTERCELLULAR CONVERSATIONS

The signal transduction pathway leading from σ^F to the appearance of σ^E is only the first of three conversations that take place between the two compartments of the sporangium (Losick and Stragier, 1992; Stragier and Losick, 1996; and references therein). Following engulfment, σ^F is replaced in the forespore by σ^G. The gene for σ^G is transcribed under the control of σ^F and is also subject to positive autoregulation (Fig. 7B). Additionally, however, the activation of σ^G-directed gene expression in the forespore requires σ^E-directed gene expression in the mother cell, including the transcription of the multicistronic *spoIIIA* operon. The nature of the intercellular pathway leading from σ^E to activation of σ^G is largely mysterious. Like σ^F, σ^G is thought to be subject to negative regulation by SpoIIAB (Kirchman et al., 1993; Kellner et al., 1996). If so, the involvement of SpoIIAB in the regulation of σ^G would pose a paradox. If SpoIIAA enables σ^F to escape from SpoIIAB at the stage of polar septation, then how could σ^G continue to be subject to inhibition by the antisigma factor?

After σ^G is activated in the forespore of the postengulfment sporangium, a third pathway of intercellular signaling involving the mother cell transcription factor σ^K is triggered (Fig. 7B). The gene for σ^K (actually two half genes in certain strains of *B. subtilis* that are stitched together by a DNA rearrangement in the mother cell chromosome) is transcribed in the mother cell under the control of σ^E and the mother cell-specific DNA binding protein called SpoIIID (Losick and Stragier, 1992; Stragier and Losick, 1996; and references therein). The product of the gene for σ^K is not, however, the mature sigma factor but rather an in-

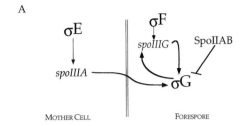

A

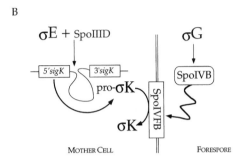

B

FIGURE 7 Intercompartmental pathways governing the activation of σ^G and σ^K. (A) Dependence of the appearance of σ^G on a pathway involving σ^E-dependent gene expression in the mother cell, including expression of *spoIIIA*. The diagram also indicates that the gene for σ^G is transcribed under the direction of both σ^F and its own gene product and that σ^G is subject to inhibition by the antisigma factor SpoIIAB. (B) Pathway governing the processing of pro-σ^K. σ^G directs the synthesis of the signal protein SpoIVB in the forespore, which triggers the action of the membrane-bound putative processing enzyme SpoIVFB. Compare this with the pro-σ^E pathway in Fig. 6 (see the text for further details involving negative regulators of SpoIVFB). The diagram also depicts the creation of the gene for σ^K by a DNA rearrangement in the mother cell chromosome.

active proprotein, pro-σ^K, which harbors an N-terminal extension of 20 amino acids (Lu et al., 1990; Kroos and Losick, 1989). Similar to the signal transduction pathway governing the activation of pro-σ^E, the activation of pro-σ^K is governed by an intercellular pathway involving a signal protein (SpoIVB) produced in the forespore under the control of σ^G (Gomez et al., 1995; Cutting et al., 1991a) and a putative proteolytic processing enzyme, SpoIVFB (Cutting et al., 1991b), which is located in the mother cell membrane that surrounds the forespore (Resnekov et al., 1996).

Unlike the pro-σ^E-processing enzyme SpoIIGA, SpoIVFB is active in its default state. Its activity is negatively regulated by two other mother cell-specific proteins, BofA and SpoIVFA (Cutting et al., 1990; Cutting et al., 1991b; Resnekov and Losick, 1998). The function of the SpoIVB signal protein is to overcome the inhibitory effects of the negative regulators. Not only are the mechanisms by which proteolysis is activated different in pro-σ^E and pro-σ^K, none of the components of the two signal transduction pathways (other than the transcription factors themselves) exhibit significant similarity to each other. Thus, despite the striking analogy between the pathways linking σ^F to pro-σ^E and those linking σ^G to pro-σ^K, the machineries for doing so are decidedly nonhomologous.

It is attractive to believe that contemporary endospore-forming bacteria derive from a more primitive spore former that had only σ^F and σ^E and that over the course of evolution the signal transduction pathway linking σ^F to σ^E was duplicated to give rise to the pathway linking σ^F to σ^K. If so, then after duplication the components of the pro-σ^E pathway evidently came to be replaced by proteins bearing little or no resemblance in mechanism or sequence to the SpoIIR signal protein and the SpoIIGA protease.

POLAR DIVISION, ENGULFMENT, AND THE EVOLUTION OF A NOVEL MODE OF REPRODUCTION

Our principal goal in this chapter has been to consider sporulation in the context of the developmental problem of the establishment of cell fate. Sporulation is also of interest, however, from the point of view of bacterial biodiversity. As we have seen, in *B. subtilis* (and its close relatives) sporulation is a provisional process that is induced under conditions of nutrient limitation and that culminates in the formation of a single endospore. In certain species of gram-positive bacteria, however, sporulation seems to have given rise to a novel and

obligatory means of propagation. In contrast to *B. subtilis*, and as its name implies, the guinea pig symbiont, *Metabacterium polyspora*, produces multiple progeny endospores in a single sporangium. *M. polyspora* achieves this feat in two ways (Angert and Losick, 1998). First, it undergoes septation near both poles of the sporangium and, second, after engulfment, the forespore protoplasts divide in two by a process resembling binary fission. Each progeny forespore from this division process develops into a spore. Sporulation takes place during transit of the bacterium through the gastrointestinal tract of the guinea pig host in an organ-specific fashion. Interestingly, few cells that are undergoing binary fission are observed in the gastrointestinal tract. Instead, germinating spores are frequently observed that have directly entered into sporulation, even before they have shed the husk of the old spore coat. *M. polyspora* is born to sporulate. Instead of undergoing sporulation provisionally as a means of coping with nutritional stress, *M. polyspora* produces multiple endospores, at least in part as a means of reproduction.

Epulopiscium is a symbiont of the surgeonfish of the great barrier reef. It is a huge bacterium (~0.5 mm), but molecular phylogeny indicates that it is most closely related to *Metabacterium* and certain species of *Clostridium* (Angert et al., 1993, 1996). Yet, unlike *Metabacterium* and *Clostridium*, *Epulopiscium* does not produce endospores. Instead, it produces viviparous progeny by an internal process that would appear to resemble engulfment. Evidently, *Epulopiscium* has lost the capacity to produce endospores but has retained the ability to produce progeny cells internally. The simplest interpretation of these findings is that the provisional capacity of *B. subtilis* to produce a single endospore by polar division and engulfment evolved in stepwise fashion into a means of reproduction in *Metabacterium* in which progeny cells are generated in the form of multiple endospores and finally into a mode of reproduction in *Epulopiscium* in which multiple viviparous progeny are produced internally.

ACKNOWLEDGMENTS

We thank A. Driks for kindly providing us with the micrographs reproduced in Fig. 1 and N. King for advice on the manuscript.

P.A.L. is a postdoctoral fellow of the Damon Runyon-Walter Winchell Foundation (DRG1397). Work in the laboratory of R.L. is supported by a grant (GM18568) from the NIH.

REFERENCES

Addinall, S. G., and J. Lutkenhaus. 1996. FtsA is localized to the septum in an FtsZ-dependent manner. *J. Bacteriol.* **178:**7167–7172.

Adler, E., A. Donella-Deana, F. Arigoni, L. A. Pinna, and P. Stragier. 1997. Structural relationship between a bacterial developmental protein and eukaryotic PP2C protein phosphatases. *Mol. Microbiol.* **23:**57–62.

Adler, H. I., W. D. Fisher, A. Cohen, and A. H. Hardigree. 1967. Miniature *Escherichia coli* cells deficient in DNA. *Proc. Natl. Acad. Sci. USA* **57:**321–326.

Alper, S., L. Duncan, and R. Losick. 1994. An adenosine nucleotide switch controlling the activity of a cell type-specific transcription factor in *B. subtilis*. *Cell* **77:**195–205.

Alper, S., A. Dufour, D. A. Garsin, L. Duncan, and R. Losick. 1996. Role of adenosine nucleotides in the regulation of a stress-response transcription factor in *Bacillus subtilis*. *J. Mol. Biol.* **260:**165–177.

Angert, E. R., and R. M. Losick. 1998. Propagation by sporulation in the guinea pig symbiont *Metabacterium polyspora*. *Proc. Natl. Acad. Sci. USA* **95:**10218–10223.

Angert, E. R., K. D. Clements, and N. R. Pace. 1993. The largest bacterium. *Nature* **362:**239–241.

Angert, E. R., A. E. Brooks, and N. R. Pace. 1996. Phylogenetic analysis of *Metabacterium polyspora*: clues to the evolutionary origin of daughter cell production in *Epulopiscium* species, the largest bacteria. *J. Bacteriol.* **178:**1451–1456.

Arigoni, F., K. Pogliano, C. D. Webb, P. Stragier, and R. Losick. 1995. Localization of protein implicated in establishment of cell type to sites of asymmetric division. *Science* **270:**637–640.

Arigoni, F., A.-M. Guérout-Fleury, I. Barák, and P. Stragier. 1999. The SpoIIE phosphatase, the sporulation septum, and the establishment of forespore-specific transcription in *Bacillus subtilis*: a reassessment. *Mol. Microbiol.* **31:**1407–1416.

Barák, I., and P. Youngman. 1996. SpoIIE mutants of *Bacillus subtilis* comprise two distinct phenotypic classes consistent with a dual functional role for the SpoIIE protein. *J. Bacteriol.* **178:**4984–4989.

Barák, I., P. Prepiak, and F. Schmeisser. 1998. MinCD proteins control the septation process during sporulation of *Bacillus subtilis*. *J. Bacteriol.* **180:**5327–5333.

Beall, B., and J. Lutkenhaus. 1989. Nucleotide sequence and insertional inactivation of a *Bacillus subtilis* gene that affects cell division, sporulation, and temperature sensitivity. *J. Bacteriol.* **171:**6821–6834.

Beall, B., and J. Lutkenhaus. 1991. FtsZ in *Bacillus subtilis* is required for vegetative septation and for asymmetric septation during sporulation. *Genes Dev.* **5:**447–455.

Beall, B., and J. Lutkenhaus. 1992. Impaired cell division and sporulation of a *Bacillus subtilis* strain with the *ftsA* gene deleted. *J. Bacteriol.* **174:**2398–2403.

Begg, K. J., S. J. Dewar, and W. D. Donachie. 1995. A new *Escherichia coli* gene, *ftsK. J. Bacteriol.* **177:**6211–6222.

Bi, E., and J. Lutkenhaus. 1991. FtsZ ring structure associated with division in *Escherichia coli. Nature* **354:**161–164.

Bi, E., and J. Lutkenhaus. 1993. Cell division inhibitors SulA and MinCD prevent formation of the FtsZ ring. *J. Bacteriol.* **175:**1118–1125.

Bork, P., C. Sander, and A. Valencia. 1992. An ATPase domain common to prokaryotic cell cycle proteins, sugar kinases, actin, and hsp70 heat shock proteins. *Proc. Natl. Acad. Sci. USA* **89:**7290–7294.

Cha, J.-H., and G. C. Stewart. 1997. The *divIVA* minicell locus of *Bacillus subtilis. J. Bacteriol.* **179:**1671–1683.

Chen, J. C., D. S. Weiss, J. M. Ghigo, and J. Beckwith. 1999. Septal localization of FtsQ, an essential cell division protein in *Escherichia coli. J. Bacteriol.* **181:**521–530.

Coyne, S. I., and N. H. Mendelson. 1974. Clonal analysis of cell division in the *Bacillus subtilis divIV-B1* minicell-producing mutant. *J. Bacteriol.* **118:**15–20.

Cutting, S., V. Oke, A. Driks, R. Losick, S. Lu, and L. Kroos. 1990. A forespore checkpoint for mother cell gene expression during development in *Bacillus subtilis. Cell* **62:**239–250.

Cutting, S., A. Driks, R. Schmidt, B. Kunkel, and R. Losick. 1991a. Forespore-specific transcription of a gene in the signal transduction pathway that governs Pro-sigma K processing in *Bacillus subtilis. Genes Dev.* **5:**456–466.

Cutting, S., S. Roels, and R. Losick. 1991b. Sporulation operon *spoIVF* and the characterization of mutations that uncouple mother-cell from forespore gene expression in *Bacillus subtilis. J. Mol. Biol.* **221:**1237–1256.

Dai, K., and J. Lutkenhaus. 1992. The proper ratio of FtsZ to FtsA is required for cell division to occur in *Escherichia coli. J. Bacteriol.* **174:**6145–6151.

Daniel, R. A., E. J. Harry, V. L. Katis, R. G.

Wake, and J. Errington. 1998. Characterization of the essential cell division gene *ftsL* (*ylld*) of *Bacillus subtilis* and its role in the assembly of the division apparatus. *Mol. Microbiol.* **29**:593–604.

de Boer, P. A. J., R. E. Crossley, and L. I. Rothfield. 1989. A division inhibitor and a topological specificity factor coded for by the minicell locus determine proper placement of the division septum. *Cell* **56**:641–649.

Dewar, D. J., K. J. Begg, and W. D. Donachie. 1992. Inhibition of cell division initiation by an imbalance in the ratio of FtsA to FtsZ. *J. Bacteriol.* **174**:6314–6316.

Driks, A. Spatial and temporal control of gene expression in prokaryotes. *In* V. E. A. Russo, D. J. Cove, L. G. Edgar, R. Jaenisch, and F. Salamini (ed.), *Development: Genetics, Epigenetics and Environmental Regulation*, in press. Springer-Verlag, Berlin, Germany.

Dubnau, E., J. Weir, G. Nair, L. D. Carter, C. Moran, Jr., and I. Smith. 1988. Bacillus sporulation gene *spo0H* codes for sigma 30 (sigma H). *J. Bacteriol.* **170**:1054–1062.

Duncan, L., S. Alper, and R. Losick. 1996. SpoIIAA governs the release of the cell-type specific transcription factor sigma F from its anti-sigma factor SpoIIAB. *J. Mol. Biol.* **260**:147–164.

Dunn, G., and J. Mandelstam. 1977. Cell polarity in *Bacillus subtilis*: effect of growth conditions on spore positions in sister cells. *J. Gen. Microbiol.* **103**:201–205.

Dunn, G., D. M. Torgersen, and J. Mandelstam. 1976. Order of expression of genes affecting septum location during sporulation of *Bacillus subtilis*. *J. Bacteriol.* **125**:776–779.

Edwards, D. H., and J. Errington. 1997. The *Bacillus subtilis* DivIVA protein targets to the division septum and controls the site specificity of cell division. *Mol. Microbiol.* **24**:905–915.

Errington, J. 1993. *Bacillus subtilis* sporulation: regulation of gene expression and control of morphogenesis. *Microbiol. Rev.* **57**:1–33.

Fawcett, P., A. Melnikov, and P. Youngman. 1998. The *Bacillus* SpoIIGA protein is targeted to sites of spore septum formation in a SpoIIE-independent manner. *Mol. Microbiol.* **28**:931–943.

Feucht, A., T. Magnin, M. D. Yudkin, and J. Errington. 1996. Bifunctional protein required for asymmetric cell division and cell-specific transcription in *Bacillus subtilis*. *Genes Dev.* **10**:794–803.

Fitz-James, P., and E. Young. 1969. Morphology of sporulation, p. 39–123. *In* G. W. Gould and A. Hurst (ed.), *The Bacterial Spore*. Academic Press, Inc., New York, N.Y.

Frandsen, N., I. Barák, and P. Stragier. 1999. Transient gene asymmetry during sporulation and establishment of cell specificity in *Bacillus subtilis*. *Genes Dev.* **13**:394–399.

Frisby, D., and P. Zuber. 1994. Mutations in *pts* cause catabolite-resistant sporulation and altered regulation of *spo0H* in *Bacillus subtilis*. *J. Bacteriol.* **176**:2587–2595.

Garsin, D., L. Duncan, D. Paskowitz, and R. Losick. 1998. The kinase activity of the antisigma factor SpoIIAB is required for activation as well as inhibition of transcription factor sigmaF during sporulation in *Bacillus subtilis*. *J. Mol. Biol.* **284**:569–578.

Glaser, P., M. E. Sharpe, B. Raether, M. Perego, K. Ohlsen, and J. Errington. 1997. Dynamic, mitotic-like behavior of a bacterial protein required for accurate chromosome partitioning. *Genes Dev.* **11**:1160–1168.

Gomez, M., S. Cutting, and P. Stragier. 1995. Transcription of *spoIVB* is the only role of sigma G that is essential for pro-sigma K processing during spore formation in *Bacillus subtilis*. *J. Bacteriol.* **177**:4825–4827.

Harry, E. J., and R. G. Wake. 1989. Cloning and expression of a *Bacillus subtilis* division initiation gene for which a homolog has not been identified in another organism. *J. Bacteriol.* **171**:6835–6839.

Harry, E. J., B. J. Stewart, and R. G. Wake. 1993. Characterization of mutations in *divIB* of *Bacillus subtilis* and cellular localization of the DivIB protein. *Mol. Microbiol.* **7**:611–621.

Healy, J., J. Weir, I. Smith, and R. Losick. 1991. Post-transcriptional control of a sporulation regulatory gene encoding transcription factor σ^H in *Bacillus subtilis*. *Mol. Microbiol.* **5**:477–487.

Hitchins, A. D. 1975. Polarized relationship of bacterial spore loci to the "old" and "new" ends of sporangia. *J. Bacteriol.* **121**:518–523.

Hitchins, A. D. 1976. Patterns of spore location in pairs of *Bacillus cereus* sporangia *J. Bacteriol.* **125**:366–368.

Hofmeister, A. 1998. Activation of the proprotein transcription factor pro-σE is associated with its progression through three patterns of subcellular localization during sporulation in *Bacillus subtilis*. *J. Bacteriol.* **180**:2426–2433.

Hofmeister, A. E. M., A. Londoño-Vallejo, E. Harry, P. Stragier, and R. Losick. 1995. Extracellular signal protein triggering the proteolytic activation of a developmental transcription factor in *Bacillus subtilis*. *Cell* **83**:219–226.

Illing, N., and J. Errington. 1991. Roles of σ^E and σ^F in prespore engulfment. *J. Bacteriol.* **173**:3159–3169.

Ju, J., T. Luo, and W. G. Haldenwang. 1997. *Bacillus subtilis* pro-σE fusion protein localizes to the forespore septum and fails to be processed when synthesized in the forespore. *J. Bacteriol.* **179:** 4888–4893.

Ju, J., T. Luo, and W. G. Haldenwang. 1998. Forespore expression and processing of the SigE transcription factor in wild-type and mutant *Bacillus subtilis. J. Bacteriol.* **180:**1673–1681.

Karow, L. M., P. Glaser, and P. J. Piggot. 1995. Identification of a gene, *spoIIR*, which links the activation of σ^E to the transcriptional activity of σ^F during sporulation in *Bacillus subtilis. Proc. Natl. Acad. Sci. USA* **92:**2012–2016.

Kellner, E. M., A. Decatur, and C. P. Moran, Jr. 1996. Two-stage regulation of an anti-sigma factor determines developmental fate during bacterial endospore formation. *Mol. Microbiol.* **21:**913–924.

Khvorova, A., L. Zhang, M. L. Higgins, and P. J. Piggot. 1998. The *spoIIE* locus is involved in the Spo0A-dependent switch in the location of FtsZ rings in *Bacillus subtilis. J. Bacteriol.* **180:** 1256–1260.

King, N., O. Dreesen, P. Stragier, K. Pogliano, and R. Losick. 1999. Septation, dephosphorylation and the activation of σ^F during sporulation in *Bacillus subtilis. Genes Dev.* **13:**1156–1167.

Kirchman, P. A., H. DeGrazia, E. M. Kellner, and C. P. Moran, Jr. 1993. Forespore-specific disappearance of the sigma-factor antagonist SpoIIAB: implications for its role in determination of cell fate in *Bacillus subtilis. Mol. Microbiol.* **8:** 663–671.

Kroos, L., B. Kunkel, and R. Losick. 1989. Switch protein alters specificity of RNA polymerase containing a compartment-specific sigma factor. *Science* **243:**526–529.

Lee, S., and C. W. Price. 1993. The *minCD* locus of *Bacillus subtilis* lacks the *minE* determinant that provides topological specificity to cell division. *Mol. Microbiol.* **7:**601–610.

Levin, P. A. 1996. Asymmetric division during sporulation in *Bacillus subtilis.* Ph.D. thesis. Harvard University, Cambridge, Mass.

Levin, P. A., and R. Losick. 1994. Characterization of a cell division gene from *Bacillus subtilis* that is required for vegetative and sporulation septum formation. *J. Bacteriol.* **176:**1451–1459.

Levin, P. A., and R. Losick. 1996. Transcription factor Spo0A switches the localization of the cell division protein FtsZ from a medial to a bipolar pattern in *Bacillus subtilis. Genes Dev.* **10:**478–488.

Levin, P. A., P. Margolis, P. Setlow, R. Losick, and D. Sun. 1992. Identification of *Bacillus subtilis* genes for septum placement and shape determination. *J. Bacteriol.* **174:**6717–6728.

Levin, P. A., R. Losick, P. Stragier, and F. Arigoni. 1997. Localization of the sporulation protein SpoIIE in *Bacillus subtilis* is dependent upon the cell division protein FtsZ. *Mol. Microbiol.* **25:**839–846.

Levin, P. A., J. J. Shim, and A. D. Grossman. 1998. Effect of *minCD* on FtsZ ring position and polar septation in *Bacillus subtilis. J. Bacteriol.* **180:** 6048–6051.

Lewis, P. J., S. R. Partridge, and J. Errington. 1994. Sigma factors, asymmetry, and the determination of cell fate in *Bacillus subtilis. Proc. Natl. Acad. Sci. USA* **91:**3849–3853.

Lewis, P. J., T. Magnin, and J. Errington. 1996. Compartmentalized distribution of the proteins controlling the prespore-specific transcription factor σ^F of *Bacillus subtilis. Genes Cells* **1:**881–894.

Lin, D. C. H., P. A. Levin, and A. D. Grossman. 1997. Bipolar localization of a chromosome partition protein in *Bacillus subtilis. Proc. Natl. Acad. Sci. USA* **94:**4721–4726.

Londono-Vallejo, J. A., and P. Stragier. 1995. Cell-cell signaling pathway activating a developmental transcription factor in *Bacillus subtilis. Genes Dev.* **9:**503–508.

Losick, R., and P. Stragier. 1992. Crisscross regulation of cell-type-specific gene expression during development in *Bacillus subtilis. Nature* **355:** 601–604.

Losick, R., P. Youngman, and P. J. Piggot. 1986. Genetics of endospore formation in *Bacillus subtilis. Annu. Rev. Genet.* **20:**625–669.

Lu, S., R. Halberg, and L. Kroos. 1990. Processing of the mother-cell sigma factor, sigma K, may depend on events occurring in the forespore during *Bacillus subtilis* development. *Proc. Natl. Acad. Sci. USA* **87:**9722–9726.

Lutkenhaus, J., and S. G. Addinall. 1997. Bacterial cell division and the Z ring. *Annu. Rev. Biochem.* **66:**93–116.

Magnin, T., M. Lord, and M. D. Yudkin. 1997. Contribution of partner switching and SpoIIAA cycling to regulation of sigmaF activity in sporulating *Bacillus subtilis. J. Bacteriol.* **179:**3922–3927.

Margolin, W. 1998. A green light for the bacterial cytoskeleton. *Trends Microbiol.* **6:**233–238.

Marston, A. L., H. B. Thomaides, D. H. Edwards, M. E. Sharpe, and J. Errington. 1998. Polar localization of the MinD protein of *Bacillus subtilis* and its role in selection of the mid-cell division site. *Genes Dev.* **12:**3419–3430.

Najafi, S. M., D. A. Harris, and M. D. Yudkin. 1997. Properties of the phosphorylation reaction catalyzed by SpoIIAB that help to regulate sporulation of *Bacillus subtilis. J. Bacteriol.* **179:**5628–5631.

Perego, M., G. B. Spiegelman, and J. A. Hoch. 1988. Structure of the gene for the transition state regulator *abrB*: regulator synthesis is controlled by the *spo0A* sporulation gene in *Bacillus subtilis. Mol. Microbiol.* **2:**689–699.

Piggot, P. J., and J. G. Coote. 1976. Genetic aspects

of bacterial endospore formation. *Bacteriol. Rev.* **40:** 908–962.

Pogliano, K., E. Harry, and R. Losick. 1995. Visualization of the subcellular location of sporulation proteins in *Bacillus subtilis* using immunofluorescence microscopy. *Mol. Microbiol.* **18:**459–470.

Pogliano, K., A. E. M. Hofmeister, and R. Losick. 1997. Disappearance of the σ^E transcription factor from the forespore and the SpoIIE phosphatase from the mother cell contributes to the establishment of cell-specific gene expression during sporulation in *Bacillus subtilis. J. Bacteriol.* **179:** 3331–3341.

Raskin, D. M., and P. A. J. de Boer. 1997. The MinE ring: an FtsZ-independent cell structure required for selection of the correct division site in *E. coli. Cell* **91:**685–694.

Reeve, J. N., N. H. Mendelson, L. I. Coyne, L. L. Hallock, and R. M. Cole. 1973. Minicells of *Bacillus subtilis. J. Bacteriol.* **114:**860–873.

Resnekov, O., and R. Losick. 1998. Negative regulation of the proteolytic activation of a developmental transcription factor in *Bacillus subtilis. Proc. Natl. Acad. Sci. USA* **95:**3162–3167.

Resnekov, O., S. Alper, and R. Losick. 1996. Subcellular localization of proteins governing the proteolytic activation of a developmental transcription factor in *Bacillus subtilis. Genes Cells* **1:**529–542.

Ryter, A. 1964. Etude morphologique de la sporulation de *Bacillus subtilis. Ann. Inst. Pasteur* **108:** 40–60.

Ryter, A., P. Schaeffer, and H. Ionesco. 1996. Classification cytologique, par leur stade de blocage, des mutants de sporulation de Bacillus subtilis Marburg [Cytologic classification, by their blockage stage, of sporulation mutants of Bacillus subtilis Marburg]. *Ann. Inst. Pasteur* **110:**305–315.

Sánchez, M., A. Valencia, M.-J. Ferrándiz, C. Sander, and M. Vicente. 1994. Correlation between the structure and biochemical activities of FtsA, an essential cell division protein of the actin family. *EMBO J.* **13:**4919–4925.

Setlow, B., N. Magill, P. Febbroriello, L. Nakhimovsky, D. E. Koppel, and P. Setlow. 1991. Condensation of the forespore nucleoid early in sporulation of *Bacillus* species. *J. Bacteriol.* **173:** 6270–6278.

Sharpe, M. E., and J. Errington. 1995. Postseptational chromosome partitioning in bacteria. *Proc. Natl. Acad. Sci. USA* **92:**8630–8634.

Shazand, K., N. Frandsen, and P. Stragier. 1995. Cell-type specificity during development in *Bacillus subtilis*: the molecular and morphological requirements for sigma E activation. *EMBO J.* **14:** 1439–1445.

Siranosian, K. J., and A. D. Grossman. 1994. Activation of *spo0A* transcription by σ^H is necessary

for sporulation but not for competence in *Bacillus subtilis. J. Bacteriol.* **176:**3812–3815.

Stephens, R. S., S. Kalman, C. Lammel, J. Fan, R. Marathe, L. Aravind, W. Mitchell, L. Olinger, R. L. Tatusov, Q. Zhao, E. V. Koonin, and R. W. Davis. 1998. Genome sequence of an obligate intracellular pathogen of humans: *Chlamydia trachomatis. Science* **282:**754–759.

Stragier, P. 1989. Temporal and spatial control of gene expression during sporulation: from facts to speculations, p. 243–254. *In* I. Smith, R. A. Slepecky, and P. Setlow (ed.), *Regulation of Procaryotic Development*. American Society for Microbiology, Washington, D.C.

Stragier, P., and R. Losick. 1996. Molecular genetics of sporulation in *Bacillus subtilis. Annu. Rev. Genet.* **30:**297–341.

Strauch, M. 1995. Delineation of AbrB-binding sites on the *Bacillus subtilis spo0H, kinB, ftsAZ,* and *pbpE* promoters and use of a derived homology to identify a previously unsuspected binding site in the *bsuB1* methylase promoter. *J. Bacteriol.* **177:** 6999–7002.

Strauch, M. A. 1993. AbrB, a transition state regulator, p. 757–764. *In* A. L. Sonenshein, J. A. Hoch, and R. Losick (ed.), *Bacillus subtilis and Other Gram-Positive Bacteria: Biochemistry, Physiology, and Molecular Genetics*. American Society for Microbiology, Washington, D.C.

Strepp, R., S. Scholz, S. Kruse, V. Speth, and R. Reski. 1998. Plant nuclear gene knockout reveals a role in plastid division for the homolog of the bacterial cell division protein FtsZ, an ancestral tubulin. *Proc. Natl. Acad. Sci. USA* **95:**4368–4373.

Sun, D., P. Fajardo-Cavazos, M. D. Sussman, F. Tovar-Rojo, R.-M. Cabrera-Martinez, and P. Setlow. 1991. Effect of chromosome location of *Bacillus subtilis* forespore genes on their *spo* gene dependence and transcription by Eσ^F: identification of features of good Eσ^F-dependent promoters. *J. Bacteriol.* **173:**7867–7874.

Tormo, A., J. A. Ayala, M. A. de Pedro, M. Aldea, and M. Vicente. 1986. Interaction of FtsA and PBP3 proteins in the *Escherichia coli* septum. *J. Bacteriol.* **166:**985–992.

Varley, A. W., and G. C. Stewart. 1992. The *divIVB* region of the *Bacillus subtilis* chromosome encodes homologues of *Escherichia coli* septum placement (MinCD) and cell shape (MreBCD) determinants. *J. Bacteriol.* **174:**6729–6742.

Wang, X., and J. Lutkenhaus. 1993. The FtsZ protein of *Bacillus subtilis* is localized at the division site and has GTPase activity that is dependent upon FtsZ concentration. *Mol. Microbiol.* **9:**435–442.

Wang, X., J. Huang, A. Mukherjee, C. Cao, and J. Lutkenhaus. 1997. Analysis of the interaction of FtsZ with itself, GTP, and FtsA. *J. Bacteriol.* **179:** 5551–5559.

Webb, C. D., A. Teleman, S. Gordon, A. Straight, A. Belmont, D. C. Lin, A. D. Grossman, A. Wright, and R. Losick. 1997. Bipolar localization of the replication origin regions of chromosomes in vegetative and sporulating cells of *B. subtilis. Cell* **88:**667–674.

Weiss, D. S., J. C. Chen, J. M. Ghigo, D. Boyd, and J. Beckwith. 1999. Localization of FtsI (PBP3) to the septal ring requires its membrane anchor, the Z ring, FtsA, FtsQ, and FtsL. *J. Bacteriol.* **181:**508–520.

Wu, L., and J. Errington. 1994. *Bacillus subtilis* SpoIIIE protein required for DNA segregation during asymmetric cell division. *Science* **264:**572–575.

Wu, L. J., and J. Errington. 1997. Septal localization of the SpoIIIE chromosome partitioning protein in *Bacillus subtilis. EMBO J.* **16:**2161–2169.

Wu, L. J., P. J. Lewis, R. Allmansberger, P. M. Hauser, and J. Errington. 1995. A conjugation-like mechanism for prespore chromosome parti-tioning during sporulation in *Bacillus subtilis. Genes Dev.* **9:**1316–1326.

Wu, L. J., A. Feucht, and J. Errington. 1998. Pre-spore-specific gene expression in *Bacillus subtilis* is driven by sequestration of SpoIIE phosphatase to the prespore side of the asymmetric septum. *Genes Dev.* **12:**1371–1380.

Yamashita, S., F. Kawamura, H. Yoshikawa, H. Takahashi, Y. Kobayashi, and H. Saito. 1989. Dissection of the expression signals of the *spo0A* gene of *Bacillus subtilis*: glucose represses sporula-tion-specific expression. *J. Gen. Microbiol.* **135:** 1335–1345.

Yanouri, A., R. A. Daniel, J. Errington, and C. E. Buchanan. 1993. Cloning and sequencing of the cell division gene *pbpB*, which encodes penicil-lin-binding protein 2B in *Bacillus subtilis. J. Bacteriol.* **175:**7604–7616.

Zhang, L., M. L. Higgins, and P. J. Piggot. 1997. The division during bacterial sporulation is sym-metrically located in *Sporosarcina ureae. Mol. Microbiol.* **25:**1091–1098.

MORPHOGENESIS AND PROPERTIES
OF THE BACTERIAL SPORE

Adam Driks and Peter Setlow

9

The end product of sporulation in *Bacillus subtilis* and other members of the genus *Bacillus* is the spore, which is distinguished from growing cells by a number of properties, including extreme dormancy and resistance (see chapter 6). The spore exhibits no detectable metabolism of endogenous compounds even though it contains a number of enzyme-substrate pairs that could rapidly interact and indeed do so in the first minutes of spore germination (Setlow, 1994). The spore also does not metabolize exogenous compounds—or if it does it is an indication that the spore has germinated. One reflection of the spore's extreme dormancy is its lack of ATP and reduced pyridine nucleotides, despite the presence of AMP, ADP, and oxidized pyridine nucleotides (Table 1). However, the spore does have a significant depot of 3-phosphoglyceric acid (3PGA) (Table 1), which is used in the first minutes of spore germination to generate ATP, and there are also other spore molecules whose metabolism supports much of the energy production in the first minutes of spore germination (Setlow, 1983, 1994). The spore's extreme dormancy is undoubtedly important to its long-term survival, and the stability of metabolites in the dormant spore ensures that these metabolites are utilized only when the spore germinates (Setlow, 1983). There may be multiple factors contributing to the dormancy of the spore, but a major one is undoubtedly the significant dehydration of the spore's core, which may preclude enzyme action (Setlow, 1983).

In addition to its dormancy, the spore is also much more resistant than the growing cell to a variety of treatments, including mechanical disruption, wet or dry heat, γ or UV radiation, desiccation, and a variety of chemicals (Gerhardt and Marquis, 1989; Marquis et al., 1994; Setlow, 1992b, 1994, 1995). Like spore dormancy this elevated spore resistance also appears to be due to a number of factors, including the decreased permeability of the spore to a variety of compounds both small and large and the extreme dehydration and mineralization of the spore core (Gerhardt et al., 1972; Gerhardt and Marquis, 1989; Marquis et al., 1994; Setlow, 1994). Spore DNA is also specifically protected from many types of damage by its saturation with a group of small, acid-soluble spore proteins (SASP) of the α/β type (Setlow, 1988, 1994, 1995), and many types of DNA damage acquired during dormancy can be repaired by various DNA repair pathways during

A. Driks and P. Setlow, Department of Microbiology and Immunology, Loyola University Medical Center, 2160 South First Avenue, Maywood, IL 60153. *P. Setlow*, Department of Biochemistry, University of Connecticut Health Center, Farmington, CT 06032.

Prokaryotic Development, edited by Y. V. Brun and L. J. Shimkets,
© 2000 American Society for Microbiology, Washington, DC 20005-4171

TABLE 1 Levels of small molecules in cells and spores of *Bacillus* species[a]

Small molecule	Level of molecules (μmol/g [dry wt])	
	Spores	Cells
NADH	<0.002	1.95
NAD	0.11	0.35
NADPH	<0.001	0.52
NADP	0.018	0.44
ATP	<0.005	3.6
ADP	0.2	1
AMP	1.2–1.3	1
3PGA	5–18[b]	<0.2
DPA	410–470	<0.1
Ca^{2+}	380–916	
Mg^{2+}	86–120	
Mn^{2+}	27–56	
H$^+$	6.3–6.5[c]	7.5–8.2[c]

[a] Unless otherwise noted, values for cells are for *B. megaterium* in mid-log-phase growth (Decker and Lang, 1978; Shioi et al., 1980; Setlow, 1983, 1994; Loshon and Setlow, 1993; Magill et al., 1994, 1996). The values for spores are the range from spores of *B. cereus, B. megaterium,* and *B. subtilis* (Warth et al., 1963; Setlow, 1983, 1994; Loshon and Setlow, 1993; Magill et al., 1994, 1996).
[b] Levels of 2PGA are <1% of 3PGA levels.
[c] Values for H$^+$ are given as pH and are for *B. megaterium* and *B. subtilis* cells and *B. cereus, B. megaterium,* and *B. subtilis* spores.

structures beyond the coat, including a variety of appendages (Hachisuka and Kozuka, 1981; Hachisuka et al., 1984; Kozuka and Tochikubo, 1985; Mizuki et al., 1998) and a separate outer casing, called the exosporium, that surrounds the spore coat (Holt and Leadbetter, 1969; Beaman et al., 1972). If an exosporium is present in *B. subtilis* spores, it is thin and very tightly packed against the coat (Sousa et al., 1976, 1978). The functions of these external layers and appendages are largely unknown,

spore germination and outgrowth (Munakata and Rupert, 1972; Fajardo-Cavazos et al., 1993; Setlow and Setlow, 1995c). However, in contrast to the situation in growing cells, where resistance to some toxic chemicals, such as hydroperoxides, is due to their detoxification by enzymes, spore resistance to toxic chemicals is not affected by the loss of protective enzymes (Casillas-Martinez and Setlow, 1997; Bagyan et al., 1998a; Henriques et al., 1998), consistent with the lack of enzyme activity within the spore core.

The dormant spore differs from the growing cell not only physiologically but also structurally (Fig. 1), and it seems certain that some of the unique structural features of the dormant spore, such as the outer proteinaceous coat and the cortex composed of peptidoglycan (PG), are the causes of some of the unique spore properties. Several species possess external

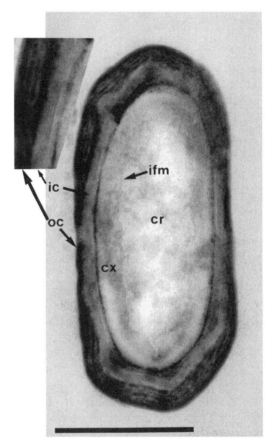

FIGURE 1 Thin-section electron micrograph of a *B. subtilis* spore. Preparation of spores and fixation were performed as described previously (Resnekov et al., 1996). The outer coat (oc), inner coat (ic), cortex (cx), inner forespore membrane (ifm), and core (cr) are indicated. The inset shows a region of the spore coat magnified 1.8 times. The bar represents 500 nm and corresponds to the whole spore.

but it is plausible that they help the spore survive in particular niches.

The most external structure that is clearly common to spores of all known bacilli is the multilayered proteinaceous spore coat; the spore coat structure has no counterpart in growing cells, and the coat proteins are invariably sporulation-specific gene products (see "Spore Coat" below). The spore coat appears to play no role in spore heat or radiation resistance, but it is important in resistance to some chemicals (Bloomfield and Arthur, 1994) and undoubtedly plays a role in the spore's extreme resistance to mechanical disruption. The spore coat, possibly with the associated outer spore membrane, also provides a permeability barrier against large molecules, including enzymes such as lysozyme, thus protecting the spore's PG from attack. Underlying the spore coat is the outer forespore membrane; precise details on the structure of this membrane are not available, nor is it clear if this is a complete membrane in the dormant spore. The cortex underlying the outer forespore membrane is composed of PG with a structure similar but not identical to that of PG in the growing cell; the cortex is thought to be essential for the maintenance and possibly the establishment of the relative dehydration of the central part of the spore, termed the spore core (Gerhardt and Marquis, 1989; Popham et al., 1995b, 1999). Underlying the thick layer of cortical PG there is a thin PG layer with a structure different from that of the cortex; this PG layer is termed the nascent, or germ cell, wall. Between the germ cell wall and the spore core is the inner forespore membrane, and while the details of the structure of this membrane are not known, it appears to be significantly compressed in the dormant spore and may be a major barrier to the entry of small hydrophilic molecules into the spore core (Gerhardt et al., 1972; Stewart et al., 1980). The central spore core is the site of the spore's nucleoid, ribosomes, and many enzymes and is significantly dehydrated, as the spore core water content is ~1.5 g of water/g (dry weight) compared to ~4 g of water/g (dry weight) in a growing cell or in the outer spore

layers (Beaman and Gerhardt, 1986; Gerhardt and Marquis, 1989). This spore core dehydration is thought to be the major factor in spore resistance to wet heat, and there is generally a good inverse correlation across species between spore core water content and spore wet-heat resistance (Gerhardt and Marquis, 1989). Spore core dehydration also seems most likely to be the major cause of spore dormancy, with the low core water content precluding the action of spore enzymes on their substrates, even if the substrates themselves are in the spore core (Setlow, 1994). In addition to a low water content, the spore core also has a number of other unusual or novel features, including (i) new proteins (Setlow, 1988, 1995), some of which alter the structure of the spore's DNA and are important in spore resistance; (ii) an extremely high level of divalent cations, predominantly Ca^{2+}, which also appears to be important in spore wet-heat resistance (Table 1) (Gerhardt and Marquis, 1989; Marquis et al., 1994); (iii) a huge depot (~10% of the spore dry weight) (Table 1) of pyridine-2,6-dicarboxylic acid (dipicolinic acid [DPA]), which chelates much of the spore's divalent cations (DPA allosterically triggers the autoprocessing of a crucial protease which acts during spore germination [Sanchez-Salas and Setlow, 1993; Illades-Aguiar and Setlow, 1994], and spores lacking DPA are extremely unstable and germinate spontaneously; however, the precise role of DPA in maintaining spore dormancy is not clear); and (iv) a pH ~1 unit lower than that in the growing cell (Table 1). While the low pH in the dormant spore does not appear essential for spore resistance, the fall in forespore pH during sporulation appears to be essential for regulation of the activity of phosphoglycerate mutase in the forespore, allowing the accumulation of the spore's depot of 3PGA (Magill et al., 1996). The low forespore pH may also facilitate the autoactivation of the spore protease noted above (Illades-Aguiar and Setlow, 1994).

As noted briefly above, the spore has a number of unique structures or structural components. Since at least some of these unique structures appear to play major roles in a variety of unique spore properties, most importantly spore resistance, three of the spore's unique structures, its nucleoid, cortex, and coat, will be discussed in detail, focusing on work with *B. subtilis* because of the wealth of detailed analyses of this organism. The first section of this discussion will focus on the spore nucleoid, concentrating on the SASP, which saturate the spore chromosome and play an important role in protecting spore DNA from damage. The spore cortex and germ cell wall will be discussed in the second section, which will concentrate on the precise structure of the PG in these two layers, as well as the synthesis and function of these two structures. The spore coat will be discussed in the final section, which will cover the properties and functions of individual coat proteins, as well as the function and assembly of the coat structure itself.

SPORE NUCLEOID

One of the most distinctive morphological changes during sporulation in *B. subtilis* is the asymmetric division at the first to second hour (t_{1-2}) of sporulation that divides the sporulating cell into the larger mother cell and smaller prespore, or forespore, compartments (see chapters 6 and 8). Associated with this unequal cell division is a dramatic change in the nucleoid in the forespore. Although both mother cell and forespore appear to have identical amounts of DNA (i.e., the same number of chromosomes), the forespore nucleoid is greater than twofold more condensed than the mother cell nucleoid (Setlow, 1991; Setlow et al., 1991b). This forespore nucleoid condensation requires the action of a number of *spo0* gene products but not *spoII* gene products (Setlow et al., 1991b), suggesting that asymmetric septum formation is required for forespore nucleoid condensation. Indeed, it appears that asymmetric septum formation largely precedes the appearance of the condensed forespore nucleoid and that much of the forespore chromosome is translocated into the forespore compartment through a pore or annulus in the completed or nearly completed asymmetric septum (Wu and Errington, 1998). The condensation of the

forespore nucleoid early in stage II of sporulation (see chapter 6 for a definition of stages) may then be simply a consequence of fitting a complete nucleoid into the smaller volume of the forespore compartment relative to that of the larger mother cell compartment. In support of this mechanism for forespore nucleoid condensation, no new moderately abundant DNA binding proteins have been found associated with the forespore nucleoid at the time of its condensation (Magill and Setlow, 1992). There are also no definitive data on other possible differences between the mother cell and forespore chromosomes at the time of forespore nucleoid condensation, although there are some older data suggesting that spore DNA may be hypomethylated relative to growing-cell DNA (Bueno et al., 1986). However, this difference has not been definitively established, nor is it known when in sporulation it occurs.

There are currently no data on the biological significance, if any, of the forespore nucleoid condensation early in stage II of sporulation, i.e., whether the condensation is important in sporulation or is simply a consequence of the small size of the forespore. One organism in which this question could be examined is *Sporosarcina ureae*, which is moderately closely related to *B. subtilis* but grows as a coccus. In contrast to *B. subtilis*, in which there is an asymmetric division early in sporulation, in *S. ureae* the sporulation division is symmetric, or nearly so (Zhang et al., 1997). Consequently, if forespore nucleoid condensation takes place early in stage II of sporulation in *S. ureae*, it cannot be due simply to the smaller size of the forespore compartment relative to the size of the mother cell compartment.

Although the importance of forespore nucleoid condensation early in stage II of sporulation is unclear, there is a later dramatic change in forespore nucleoid structure whose function is much clearer. This change takes place in stages III and IV of sporulation, when the forespore chromosome appears to become toroidal, or doughnut shaped (Pogliano et al., 1995). The precise architecture of the DNA in this ring-shaped structure is not known, nor is the significance of this particular structure. However, at the time of this change in chromosome structure, plasmids in the forespore also become ~50% more negatively supercoiled (Nicholson and Setlow, 1990). The formation of the ring-shaped chromosomal structure and the increased negative supercoiling of plasmids in the forespore both require the synthesis of a large amount of novel forespore-specific DNA binding proteins termed SASP of the α/β type. These proteins are named for the two major proteins of this type in *B. subtilis* (SASP-α and -β), and forespores which lack SASP-α and -β do not form the ring-shaped DNA structures and their plasmids do not become more negatively supercoiled (Nicholson and Setlow, 1990; Pogliano et al., 1995).

The α/β-type SASP are synthesized only in the developing forespore during stage III of sporulation (Mason et al., 1988b; Setlow, 1988, 1995). These proteins are small (~60 to 70 residues) and exhibit extremely high amino acid sequence homology both within and across species (Setlow, 1988, 1995) (see below). The genes encoding α/β-type SASP are termed *ssp*, and there are four monocistronic genes encoding α/β-type SASP in *B. subtilis* (*sspA, -B, -C,* and *-D*), although other species (i.e., *Bacillus megaterium*) have more (Setlow, 1988, 1995). Two of the *B. subtilis ssp* genes (*sspA* and *-B*) are expressed at high levels and encode SASP-α and -β, respectively; *sspC* and *-D* are expressed in parallel with *sspA* and *-B* but at much lower levels (Mason et al., 1988b; Setlow, 1988). All *ssp* genes are transcribed by RNA polymerase containing the forespore-specific sigma factor, σ^G, and the major factor regulating *ssp* gene expression appears to be the strength of the σ^G-dependent promoter sequences upstream of the coding regions (Nicholson et al., 1989; Sun et al., 1989, 1991; Fajardo-Cavazos et al., 1991) (see chapter 8). However, there is some additional mechanism keeping the total amount of SASP-α and -β accumulated in the developing spore constant, as overproduction of SASP-α reduces SASP-β production, and vice versa (Mason and Setlow,

1987). While the mechanism of this regulation is not clear (Mason and Setlow, 1987; Mason et al., 1988a), it may be due simply to the fact that the α/β-type SASP are nonspecific DNA binding proteins that saturate the forespore chromosome and can inhibit transcription (Setlow et al., 1992).

The total amount of α/β-type SASP in *B. subtilis* spores is ~4% of total spore protein, and immunoelectron microscopy, immunofluorescence microscopy, and cross-linking studies have localized essentially all of this α/β-type SASP on the forespore/spore nucleoid (Setlow and Setlow, 1979; Francesconi et al., 1988; Pogliano et al., 1995). An α/β-type SASP synthesized in *Escherichia coli* also associates with the cell's nucleoid (Setlow et al., 1991a). Given these in vivo findings, it is not surprising that in vitro α/β-type SASP are nonspecific double-stranded DNA binding proteins, binding ~4 bp per α/β-type SASP (Hayes and Setlow, 1998b; Setlow et al., 1992). Calculations with this in vitro stoichiometry and the known amounts of α/β-type SASP and DNA in *B. subtilis* spores have indicated that there is enough α/β-type SASP in spores to saturate spore DNA, and there is evidence that spore DNA is indeed saturated with α/β-type SASP (Francesconi et al., 1988; Setlow, 1992b, 1995). A variety of in vitro studies have indicated that the binding of α/β-type SASP to DNA is not tremendously strong (K_d for an average DNA is ~5 μM) (Setlow et al., 1992; Hayes and Setlow, 1998b). However, the extremely high concentration of α/β-type SASP in the spore core, as well as the core's dehydration, may promote saturation of the spore chromosome with these proteins. Binding of α/β-type SASP to DNA in vitro is cooperative, with more cooperativity exhibited in binding to AT-rich DNAs than to GC-rich DNAs (Setlow et al., 1992; Griffith et al., 1994). This cooperativity suggests that protein-protein interactions may be important in the binding of α/β-type SASP to DNA, and a number of close contacts between α/β-type SASP bound to DNA have been identified, including contacts between different α/β-type SASP (Hayes and Setlow, 1998a).

As noted above, the α/β-type SASP are small and exhibit a large degree of protein sequence conservation, both within and across species. The sequences of 21 α/β-type SASP from the aerobic line of gram-positive spore formers are known, and 23 residues are identi-

cal in all of these small proteins, with many other residues exhibiting only conservative changes (Fig. 2). Even in the proteins from the evolutionarily more distant clostridia there is striking conservation of primary sequence (Fig. 2). However, the α/β-type SASP exhibit no obvious sequence similarity to any other DNA binding proteins and have none of the struc-

SASP AMINO ACID SEQUENCES

```
                                   ↓
Bam1:          MPSSNRGNQ-QLLVPGAERVLE-F-Y-I-N----S--ADTTA-A---T---I----IRL-QQELNR
Bam2:          MANNNNSSS-QLVAPGAQQAID-M-Y-I-S----Q--ADSTS-A-------I-----QM-EQQLS-FQK
Bce1:      MGKNNSGSRNEVLVRGAEQALDQMKYEIAQEFGVQLGADTTARSNGSVGGEITKRLVAMAEQQLGGRANR
Bce2:          MSRST-KLAVPGARSALD-M-Y-I-Q----Q--ADATA-A-------I-----SL-EQQLG-YQK
BfiA:          MANNNSS-QLVVPGVQQALD-M-Y-I-S----Q--PDATA-A-------I-----QM-EQQMG-YQK
BmeA:          MANT-KLVAPGSAAAID-M-Y-I-S----N--PEATA-A-------I-----QM-EQQLG-K
BmeC:    MANYQNASNRNSS-KLVAPGAQAAID-M-F-I-S----N--PDATA-A-------I-----QL-EQNLG-KY
Bme3:          MANNNSSNN-ELLVYGAEQAID-M-Y-I-S----N--ADTTA-A-------I-----QL-EQQLG-GRF
Bme4:          MANNNKSSNN-ELLVYGAEQAID-M-Y-I-S----N--ADTTA-A-------I-----QL-EQQLG-GRSKTTL
Bme5:          MART-KLLTPGVEQFLD-Y-Y-I-Q----T--SDTAA-S-------I-----QQ-QAHLS-STQK
Bme6:          MANNKSSNN-ELLVYGAEQAID-M-Y-I-S----N--ADTTA-A-------I-----QL-EQQLG-GRF
Bme7:          MANSRNKSS-ELAVHGAQQAID-M-Y-I-S----T--PDTTA-A-------I-----QM-EQQLG-GRSKSLS
Bst1:          MPNQSGSNSS-QLLVPGAAQVID-M-F-I-S----N--AETTS-A-------I-----SF-QQQMG-GVQ
BsuA:          MANNNSGNS-NLLVPGAAQAID-M-L-I-S----N--ADTTS-A-------I-----SF-QQNMG-GQF
BsuB:          MANQNSS-DLLVPGAAQAID-M-L-I-S----N--ADTTS-A-------I-----SF-QQQMG-RVQ
BsuC:      MAQQSRSRSNNN-DLLIPQAASAIE-M-L-I-S----Q--AETTS-A-------I-----RL-QQNMG-QFH
BsuD:          MASR-KLVVPGVEQALD-F-L-V-Q----N--SDTVA-A-------M-----QQ-QSQLN-TTK
Sha1:          MANNNNSS-ELVVPGVQQALD-M-Y-I-Q----Q--ADSTS-A-------I-----QM-EQQFG-QQYGQQQK
Sur1:          MTNNNNSNS-QLLVPGVQQAIN-M-E-I-N----N--PDSTS-A-------I-----RQ-QSQMN-YTK
Sur2:          MPNNNSS-QLLVPGQVQQALN-M-E-I-S----Q--PDASS-A-------I-----RQ-QSQMN-YTK
Tth1:          MAQQGRNRSS-QLLVAGAAQAID-M-F-I-Q----T--ADTTA-A-------I-----SL-QQQLG-GTSF

Cac1:          MSRRN-QTLVPEARGALDKF-M-ASK-V--N( 9 )TS-E------QMV--MIQEYESSLK
Cac2:          MSRRPLVPEAKEGLKKLRE-Y-E-I-AG(11 )IGFIG-P---LM--KMIESVEKKMSDK
Cac3:          MANY-KKLVPEKAERLNRFRM-T-NDI--D( 9 )TSKEA-----KMIDKILQGYEDKIE
Cac4:          MSRNHRVLVPGARGGLQKL-T-ASK-LAEN( 4 )PNDKQYN---QMV-DMIKKVEKNMK
CbiA:          MTTNNN-TKAVPEAKAALK-M-L-I-N-L-IS(10 )TA-Q--Y---YM--K--EM-EQQMS-QQR
CbiB:          MSTKKAVPEAKAALN-M-L-I-N-L-LS(10 )TA-Q--Y---YM--K--EM-ERQMS-K
CbiC:      MANRSS-QLVVPEAKQGLKNL-M-V-N-V-LS(10 )TA-Q--Y---GM--KM-EAYENSLK
Cpe1:          MSKSLVPEAKNGLSKF-N-V-R-L--P( 8 )SS-QC------MV--M-EAYESQIK
Cpe2:          MSQHLVPEAKNGLSKF-N-V-A-M--P( 8 )SSKQC------MV--M-EQYEQGI
Cpe3:          MSQHLVPEAKNGLSKF-N-V-N-M--P( 8 )SS-QC------MV--M-EKYEQSMK
```

FIGURE 2 Amino acid sequences of α/β-type SASP from gram-positive spore formers. Amino acids are given in the one-letter code, and at positions denoted by dashes, the residue present is identical to that in Bce1. The numbers in parentheses in the Cac, Cbi, and Cpe sequences are the number of residues in this region in these clostridial proteins; this number is almost always larger than that in the sequences from aerobic spore formers. The vertical arrow denotes the site of cleavage of α/β-type SASP by the germination-specific protease GPR. Bam, *Bacillus aminovorans*; Bce, *B. cereus*; Bfi, *Bacillus firmus*; Bme, *B. megaterium*; Bst, *Bacillus stearothermophilus*; Bsu, *B. subtilis*; Sha, *Sporosarcina halophila*; Sur, *S. ureae*; Tth, "*Thermoactinomyces thalpophilus*"; Cac, *C. acetobutylicum*; Cbi, *Clostridium bifermentans*; Cpe, *Clostridium perfringens*. (The data are from Cabrera-Martinez et al., 1989; Cabrera-Martinez and Setlow, 1991; Setlow, 1995, 1988; Loshon et al., 1995; GenBank accession no. AF084104; and the unfinished sequence of the *C. acetobutylicum* genome available on the Web at www.cric.com.)

tural motifs, such as a helix–turn–helix, often seen in DNA binding proteins. Consequently, the α/β-type SASP appear to be a new type of DNA binding protein. In support of the importance of the residues conserved throughout evolution in α/β-type SASP function, changes in these conserved residues greatly reduce α/β-type SASP-DNA binding both in vitro and in vivo (Tovar-Rojo and Setlow, 1991).

Studies of α/β-type SASP-DNA binding in vitro have shown that this binding results in dramatic changes in both DNA and protein structures. The α/β-type SASP appear to bind predominantly to the outside of the DNA helix (Griffith et al., 1994), consistent with the lack of sequence specificity in their binding. While the DNA length does not change appreciably on binding of α/β-type SASP, the DNA is tremendously stiffened and straightened, and even sequence-dependent DNA bends are straightened (Griffith et al., 1994). In addition, the DNA adopts an A-like helical conformation when bound to α/β-type SASP, although the precise details of this A-like helical structure are not known (Mohr et al., 1991; Setlow, 1992a) (see below). Plasmids also acquire significant numbers of negative supercoils on binding α/β-type SASP in vitro, and this may reflect a change in DNA structure from a B-helix in naked DNA to an A-like helix when complexed to α/β-type SASP (Nicholson et al., 1990). The A-like helical structure of DNA in the α/β-type SASP-DNA complex is consistent with the preference of α/β-type SASP for binding to different synthetic DNAs, which is (from most to least preferred) poly(dG)·poly(dC); poly(dG-dC)·poly(dG-dC); random-sequence DNA, such as plasmid pUC18; and poly(dA-dT)·poly(dA-dT), with poly(dA)·poly(dT) never bound (Setlow et al., 1992; Hayes and Setlow, 1998b). Poly(dG)·poly(dC) is in or close to an A-helical structure in solution, and the ease of formation of an A-helix for the other DNAs decreases in parallel with their ease of binding to α/β-type SASP, while poly(dA)·poly(dT) has never been seen to adopt an A-helical conformation (Arnott and Selsing, 1974a, 1974b; McCall et al., 1985;

Nishimura et al., 1985; Sarma et al., 1986; Guidibande, 1988; Heineman et al., 1989; Jayasena and Behe, 1991). Although the DNA complexed with α/β-type SASP appears to be in an A-like helix, this is not a classical A-helix, as the length of the DNA helix does not change appreciably (Griffith et al., 1994), and the precise structure of the DNA as well as the protein in the complex is not known. However, the binding of α/β-type SASP to small double-stranded oligonucleotides (Setlow et al., 1992; Hayes and Setlow, 1998b) suggests that it may be possible to use methods such as nuclear magnetic resonance spectroscopy or X-ray crystallography to determine a detailed structure of an α/β-type SASP-DNA complex. In parallel with the changes in the structure of the DNA in the α/β-type SASP-DNA complex, there are also major changes in the structure of the α/β-type SASP. Although the α/β-type SASP are relatively unstructured in solution, they become highly (~60%) α-helical on binding to DNA (Mohr and Setlow, 1990; Hayes and Setlow, 1998b). However, again, the precise details of the structure of the protein when bound to DNA are not known.

Formation of the complex between α/β-type SASP and DNA in vitro changes not only the structures of both members of the complex but also the properties of both the DNA and the protein. When bound to α/β-type SASP (i) the UV photochemistry of DNA changes dramatically; (ii) the DNA becomes extremely resistant to a variety of enzymatic and chemical agents, including nucleases and hydroxyl radicals; and (iii) DNA depurination is greatly slowed (Fairhead and Setlow, 1992; Fairhead et al., 1993; Setlow et al., 1992; Setlow and Setlow, 1993, 1995b; Setlow et al., 1997). The α/β-type SASP in the complex also become resistant to proteolysis, to deamidation at a reactive Asn-Gly sequence, and to oxidation of methionine residues (Setlow and Setlow, 1995a; Hayes and Setlow, 1997; Hayes et al., 1998). Given these dramatic changes in properties seen in vitro, an obvious question is whether the α/β-type SASP-DNA complex studied in vitro is a fair representation of the complex that exists in the spore. The answer to this question appears to be yes, as the properties of both members of the complex observed in vitro, including the DNA's altered UV photochemistry, slow depurination, and lack of reactivity with hydroxyl radicals and the protein's very slow deamidation and methionine oxidation, are also observed in spores (Fairhead et al., 1993, 1994; Setlow and Setlow, 1987, 1993, 1994, 1995b; Setlow, 1995; Hayes and Setlow, 1997; Setlow et al., 1997; Hayes et al., 1998) (see below).

As noted above, the properties of the DNA in an α/β-type SASP-DNA complex in vitro are essentially identical to those of DNA in spores, and this has major implications for mechanisms of spore resistance and long-term survival. The importance of α/β-type SASP binding in spore resistance and survival has become clear through analysis of a *B. subtilis* strain (termed $\alpha^-\beta^-$) with deletions of both the *sspA* and *-B* genes (Mason and Setlow, 1986). This strain sporulates normally, but the spores lack SASP-α and -β; since levels of no other α/β-type SASP increase in $\alpha^-\beta^-$ spores (Mason and Setlow, 1986), their level of total α/β-type SASP is $<25\%$ of that in wild-type spores, well below the amount needed to saturate the spore chromosome. The $\alpha^-\beta^-$ spores are significantly more sensitive than are wild-type spores to being killed by wet heat, desiccation, hydroperoxides, and formaldehyde, but $\alpha^-\beta^-$ spores remain much more resistant than vegetative cells to these treatments (Gerhardt and Marquis, 1989; Setlow and Setlow, 1993; Fairhead et al., 1994; Setlow et al., 1997; Loshon et al., 1999). The $\alpha^-\beta^-$ spores are also much more sensitive than wild-type spores to dry heat, but $\alpha^-\beta^-$ spores and vegetative cells exhibit very similar dry-heat resistance, suggesting that α/β-type SASP are the major factor causing the elevated dry-heat resistance of dormant spores (Setlow and Setlow, 1995b). Treatment of $\alpha^-\beta^-$ spores with dry heat, wet heat, desiccation, hydroperoxides, or formaldehyde is accompanied by significant mutagenesis of the survivors and the generation of a variety of

types of DNA damage, suggesting that these agents kill $\alpha^-\beta^-$ spores in large part by DNA damage (Fairhead et al., 1993; 1994; Setlow and Setlow, 1993, 1994, 1995b; Setlow, 1995; Setlow et al., 1997; Loshon et al., 1999). The sensitivity of $\alpha^-\beta^-$ spores to these treatments is lost if a gene encoding any wild-type α/β-type SASP, even a normally minor protein or one from another species, is expressed at a sufficient level in the developing spore (Mason and Setlow, 1987; Setlow and Setlow, 1993, 1994, 1995b; Fairhead et al., 1994). These data suggest that most if not all α/β-type SASP are largely interchangeable in their functions. However, α/β-type SASP which have changes in residues conserved throughout evolution are ineffective in restoring resistance to $\alpha^-\beta^-$ spores (Tovar-Rojo and Setlow, 1991).

In contrast to the killing of $\alpha^-\beta^-$ spores by dry heat, wet heat, hydroperoxides, and formaldehyde through DNA damage, killing of wild-type spores by these agents (with the exception of dry heat [Setlow and Setlow, 1995b]) is not through DNA damage (Setlow and Setlow, 1993; Setlow, 1995; Setlow et al., 1997; Loshon et al., 1999). This further suggests that binding of α/β-type SASP protects spore DNA against damage, such as depurination and hydroxyl radical attack, and this protection of spore DNA by α/β-type SASP against a variety of types of damage appears to be a significant factor in long-term spore survival (Fairhead et al., 1993; Setlow, 1994). As noted above, the effects of α/β-type SASP on DNA properties in vivo are essentially identical to the effects of α/β-type SASP on DNA properties in vitro. However, factors besides the binding of α/β-type SASP to spore DNA are also important in spore resistance to wet heat, peroxides, and other chemicals. These other factors include the relative dehydration and mineralization of the spore core and the relative impermeability of the spore core to small hydrophilic chemicals (Gerhardt et al., 1972; Gerhardt and Marquis, 1989; Marquis et al., 1994). These latter factors do not appear to provide specific protection to spore DNA against agents such as wet heat and chemicals, as this is provided by α/β-type SASP. However, both types of factors synergistically contribute to the extreme resistance of the spore to these types of treatment.

In contrast to the protection of spore DNA in vivo and in vitro against chemicals such as formaldehyde and peroxides, DNA is not protected against purine alkylation in an α/β-type SASP-DNA complex, suggesting that α/β-type SASP binding does not block access of small alkylating agents such as ethylmethanesulfonate to the purine bases of DNA (Setlow et al., 1998). Gratifyingly, $\alpha^-\beta^-$ spores are no more sensitive to killing or DNA damage by ethylmethanesulfonate than are wild-type spores, again suggesting that the structure of the α/β-type SASP-DNA complex present in spores is quite similar to that formed in vitro (Setlow et al., 1998).

The binding of α/β-type SASP to DNA also plays the major role in the resistance of spores to UV radiation. Spores are 10 to 50 times more resistant to UV radiation at 254 nm than are the corresponding growing cells (Setlow, 1992b, 1995). The reason for the elevated spore resistance is that instead of the cyclobutane-type dimers between adjacent thymines (TT) generated by UV irradiation of growing cells or naked DNA, UV irradiation of spores generates a novel major DNA photoproduct termed 5-thyminyl-5,6-dihydrothymine (initially termed spore photoproduct [SP]) (Donnellan and Setlow, 1965); like TT, SP is also formed between adjacent thymines on the same DNA strand. Studies both in vitro and in vivo have shown that the α/β-type SASP are the major, if not the sole, factor responsible for the dormant spore's unique photochemistry (Setlow and Setlow, 1987; Nicholson et al., 1991; Popham et al., 1995b; Setlow, 1995), and α/β-type SASP binding to DNA also greatly, if not completely, suppresses the formation of other cyclobutane dimers between adjacent pyrimidines as well as the formation of various 6,4-photoproducts (Fairhead and Setlow, 1992). Synthesis of an α/β-type SASP in E. coli also results in significant SP formation following UV irradiation (Setlow et al., 1991a).

Like TT, SP is a potentially mutagenic or lethal lesion, but early in spore germination SP is repaired in a largely error-free process catalyzed by spore photoproduct lyase (Spl) (Munakata and Rupert, 1972; Fajardo-Cavazos et al., 1993). Spl is an iron-sulfur protein synthesized only in the developing spore which monomerizes SP to two thymine residues without incision of the DNA backbone (Pedraza-Reyes et al., 1997; Rebeil et al., 1998).

Although it is clear that α/β-type SASP are extremely important in protecting spore DNA from damage and are the major proteins associated with the dormant spore's nucleoid, there may well be other unique proteins present in much smaller amounts. Dormant spores contain small amounts of a number of other small, basic proteins whose genes are expressed only in sporulation and whose products are found only in the dormant spore. One of these proteins, termed SspF, has some sequence homology to α/β-type SASP, binds to DNA in vitro, and can influence DNA properties in spores (Loshon et al., 1997). However, its association with DNA in spores has not yet been demonstrated. Recently, a number of additional very small, basic proteins with no sequence homology to α/β-type SASP have been discovered uniquely in spores (Bagyan et al., 1998b; Cabrera-Hernandez et al., 1999). However, the functions of these proteins and their possible association with spore DNA have not yet been elucidated.

There is also another DNA binding protein which seems likely to have some function in determining the structure of the forespore and spore nucleoid. This is the B. subtilis HU protein (termed HBsu), which is essential for B. subtilis growth and is associated with DNA in growing cells (Micka et al., 1991; Micka and Mahariel, 1992; Koehler and Mahariel, 1997). Although HBsu has not yet been shown to be associated with DNA in spores, growing cells and spores have similar amounts of HBsu relative to the amount of DNA (Ross et al., 1998). Since binding of HU proteins to DNA in vitro significantly increases DNA bending (Hodges-Garcia et al., 1989) and α/β-type SASP binding greatly straightens and stiffens DNA (Griffith et al., 1994), it is possible that in vivo HBsu binds to spore DNA even in the presence of α/β-type SASP, resulting in enough DNA bending to allow efficient DNA packaging in the spore. Preliminary experiments indicate that in vitro small amounts of HBsu do indeed dramatically increase the DNA bending in an α/β-type SASP-DNA complex (Ross and Setlow, 1998).

SPORE CORTEX AND GERM CELL WALL

The spore cortex surrounds the spore core, between the inner and outer forespore membranes, and is composed predominantly of PG, although there may also be some proteins present. While the identities of such cortical proteins have not been established, good candidates are enzymes involved in cortex lysis during spore germination (Foster and Johnstone, 1989; Johnstone, 1994). Immediately underlying the cortex and adjacent to the inner forespore membrane is a second layer of PG with a structure different from that of the cortex (Buchanan et al., 1994). This second PG layer makes up only a small amount of total spore PG and is the spore's nascent, or germ cell, wall. In contrast to the cortex, which is targeted for degradation early in spore germination, the germ cell wall is not degraded at this time and is thought to provide the template for new cell wall synthesis during spore outgrowth (Buchanan et al., 1994). Unlike the walls of growing cells, in which teichoic acids are a major component (Pooley and Karamata, 1994), teichoic acids are not present in spores (Chen et al., 1968; Johnstone et al., 1982).

There are a wealth of data indicating that the spore cortex plays an extremely important role in determining a number of the unique properties of the dormant spore. Several different types of measurements with spores of a number of species have shown that there is a reasonable correlation between an increased amount of spore cortex and both increased spore wet-heat resistance and a decreased spore core water content (Gerhardt and Marquis,

1989). Thus, the cortex appears extremely important in the maintenance of spore core dehydration, which is in turn an essential element in spore dormancy and spore wet-heat resistance (Gerhardt and Marquis, 1989). During its synthesis the spore cortex may also function in the establishment of spore core dehydration. However, very few details are available on the precise pathway of spore cortex synthesis, despite a fair amount of knowledge about the structure of the cortex in the dormant spore.

The basic structural features of spore cortex and growing cell wall PG in B. subtilis were established almost 30 years ago in work by Warth and Strominger (1969, 1971, 1972). Cell wall PG is composed of repeating disaccharides of N-acetylglucosamine and N-acetylmuramic acid (NAM) in β-1,4 linkage (Fig. 3). Peptide side chains (initially L-alanine-D-glutamate-meso-diaminopimelic acid [Dpm]-D-alanine-D-alanine) are attached to the NAM residues, and in B. subtilis much of the γ-carboxyl group of the Dpm is amidated (Buchanan et al., 1994). As PG synthesis continues, the C-terminal D-alanyl residue is removed either by a DD-carboxypeptidase or in a transpeptidation reaction, forming a cross-link between the penultimate D-alanine of one peptide and the ε-amino group of a Dpm in another peptide (Fig. 3). Approximately 35% of all Dpm in vegetative cell PG participates in a cross-link (Warth and Strominger, 1971; Buchanan et al., 1994) such that the various glycan strands in the PG are cross-linked into one large molecule termed the PG sacculus. While the average structure of growing cell wall PG as given above is well established, the precise three-dimensional structure of the PG sacculus, and in particular the distribution of cross-links and the relative orientation of glycan strands, is not known. The cell wall PG of other Bacillus species can exhibit some differences from that of B. subtilis, one of the more notable being the replacement of Dpm by lysine in some organisms, e.g., Bacillus sphaericus (Imae and Strominger, 1976; Tipper and Linnett, 1976; Buchanan et al., 1994).

In contrast to vegetative cell PG, which can

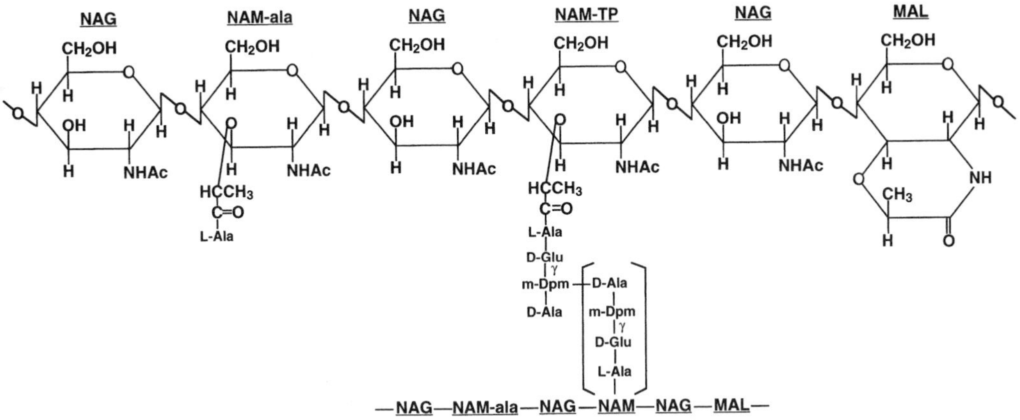

FIGURE 3 Structure of PG in the cortexes of spores of *Bacillus* species. The structure of spore cortex PG was determined by Warth and Strominger (1969, 1972). NAG, *N*-acetylglucosamine. Note that ~13% of the Dpm in NAM-TP is involved in cross-link formation between the ε-amino group of the Dpm and the carboxyl group of a D-Ala in another TP linked to NAM in a glycan strand, as shown with the TP in parentheses.

exhibit significant variation in structure in different *Bacillus* species, the basic features of the structure of cortical PG are similar in all species that have been examined, including the presence of Dpm in the cortex of species in which the vegetative cell wall PG has lysine. However, the structure of cortical PG differs in two major respects from that of vegetative cell PG. First, ~50% of NAM residues in the cortex are present as muramic acid-δ-lactam (MAL), with the majority of the MAL residues spaced at every second muramic acid position in the glycan strands (Fig. 3). Second, ~25% of NAM residues carry only a single L-alanine, and since MAL residues do not carry a peptide side chain, there are only ~1/4 as many Dpm residues available to participate in cross-link formation in spore cortex as in growing-cell PG. As a consequence, cortical PG is much less cross-linked than is vegetative cell wall PG, although there is one report that spore PG is more highly cross-linked than vegetative cell wall PG (Marquis and Bender, 1990). However, recent measurements of cross-linking by high-performance liquid chromatography analyses of cortical PG indicate that only ~12% of total Dpm participates in a cross-link and thus that total cross-linking in cortical PG is ~3%, 10-

to 12-fold below that in vegetative cell wall PG (Table 2) (Atrih et al., 1996; Popham et al., 1996b). Again, the precise arrangement of these cross-links in the cortex, whether uniformly distributed or possibly in some gradient of cross-linking from the inside to the outside of the cortex (Popham et al., 1999), is not clear, nor are the relative orientations of the glycan strands.

Spore PG also contains significant amounts of Dpm-containing tripeptides and tetrapeptides (TP) which are not linked directly to glycan strands (Atrih et al., 1996; Popham et al., 1996b), suggesting that one or more autolytic activities are involved in cortical PG synthesis or remodeling during sporulation. However, the identities of these autolytic enzymes are not known, and spores of strains individually lacking one of a number of different autolysins exhibit heat resistance and cortical PG structures identical to those of wild-type spores (Popham et al., 1996a). The precise pathways of MAL and NAM−L-Ala synthesis during sporulation have also not been established. This is important missing information, as it is possible that there are large changes in the degree and/or the location of cross-linking in cortical PG as sporulation proceeds (Popham et al., 1999).

TABLE 2 Cortex PG cross-linking, core water content, and heat resistance of spores of various *B. subtilis* strains[a]

Strain (genotype)	% PG cross-linking[b]	Core water content (g/ml) relative to wild type[c]	Spore heat resistance relative to wild type[d]
Wild type	3[e]		100
cwlD	8	−0.006	100
dacB	14	Same as wild type to −0.004	20
dacB dacF	19	−0.017	13
dacC	3	Same as wild type	73–100
dacF	3	−0.002	73
dacB spmA spmB	12	−0.041	6
spmA	1.7[f]	−0.022	12
spmB	1.8[f]	−0.021	14

[a] Values are from Atrih et al., 1996; Popham et al., 1995a, 1996a, 1996b, 1999; and Pederson et al., 1998.
[b] Values are calculated as cross-links per NAM plus MAL residues and have been rounded off to whole integers, except for the values from *spmA* and *spmB* spores.
[c] Values are given as the deviation of spore core wet density from that of wild-type spores; values for wild-type spores range from 1.367 to 1.372 g/ml in different spore preparations made in different batches of media. However, all values shown here are relative to the same wild-type value.
[d] Values are expressed relative to the heat resistance of wild-type spores, which is set at 100. The relative values were taken from measurements of the relative amounts of time at 85 or 90°C needed to kill 90% of the spores. The values for all mutant strains were measured in parallel with that for wild-type spores; both types of spores were also prepared and cleaned at the same time.
[e] Different measurements have given values from 2.5 to 3.4%.
[f] These values were obtained in a study where wild-type spores gave a value of 2.5% (Popham et al., 1996b).

Since PG can have important significant properties, in particular that it can expand or shrink depending on its degree of cross-linking and the ionic environment (Ou and Marquis, 1970), it is possible that the cortex may play an active role in spore core dehydration and the attainment of spore dormancy, as has been proposed in several different models (Lewis et al., 1960; Warth, 1985; Gerhardt and Marquis, 1989; Popham et al., 1999). Clearly, knowledge of the pathway of cortical PG synthesis and maturation will provide important information.

Only a small amount of the total spore PG composes the germ cell wall, and the small amount of this PG has made detailed analysis of its structure difficult. However, the germ cell wall PG does not appear to contain MAL and also may have a significant amount of tripeptide attached to NAM residues (Cleveland and Gilvarg, 1975; Atrih et al., 1996; Popham et al., 1996b). This finding, plus the rather high level of spore PG tripeptides which are involved in cross-link formation (Atrih et al., 1996; Popham et al., 1996b), suggest that the germ cell wall PG is much more highly cross-linked than is the cortical PG, and there are some experimental data supporting this suggestion (Cleveland and Gilvarg, 1975). The germ cell wall PG appears to be made predominantly prior to cortex PG synthesis (Vinter, 1965), and enzymes needed specifically for synthesis of the germ cell wall seem to be largely in the developing spore (Tipper and Linnett, 1976).

While a fair amount is known about the average structure of cortical PG, as noted above many of the details of the pathway of the synthesis of this molecule during sporulation are not known. Although germ cell wall PG synthesis appears to be catalyzed largely by enzymes present in the developing forespore, cortical PG appears to be synthesized predominantly (but not exclusively) by enzymes present in the mother cell, with this synthesis taking place largely in stage IV of sporulation (Tipper and Linnett, 1976; Buchanan et al., 1994). It is presumed that the general pathways for glycan strand synthesis, peptide addition, and cross-link formation are similar in growing cells and forespores, but the pathways for MAL and

NAM–L-Ala synthesis are not clear. It seems likely that the latter two components are synthesized from NAM-pentapeptide in glycan strands, but the precise steps and enzymes involved are not known. Many of the reactions in PG synthesis, including synthesis of the glycan strands (through transglycosylase activity), formation of peptide cross-links (through transpeptidase activity), and removal of amino acids from the initial pentapeptide linked to NAM (DD-carboxypeptidase activity), are catalyzed by enzymes which are sensitive to inhibition by penicillin. Consequently, analysis of penicillin-binding proteins (PBPs) in growing and sporulating cells has given some insight into enzymes specifically involved in cortical PG biosynthesis (Todd et al., 1985; Buchanan and Gustafson, 1992; Buchanan and Ling, 1992; Popham and Setlow, 1993, 1996; Buchanan et al., 1994; Popham et al., 1995a, 1995b, 1999; Murray et al., 1997, 1998). However, this analysis has been complicated to a significant degree by the redundancy in function between various PBPs (Popham and Setlow, 1996). Indeed no one of the many PBPs with transglycosylase activity is itself essential for spore cortex biosynthesis. In contrast, the *spoVD* gene, encoding a PBP with putative transpeptidase activity, is essential for spore cortex formation (Daniel et al., 1994). The *spoVD* gene is expressed in the mother cell at t_{2-3} of sporulation under control of the early-acting mother cell-specific sigma factor for RNA polymerase—σ^E (Daniel et al., 1994). *spoVD* mutants can form engulfed prespores, but the subsequent development of spore cortex is severely impaired and the resultant aberrantly developed spores are extremely unstable and have not yet been isolated (Daniel et al., 1994). As a consequence, the precise lesion in cortex formation due to the *spoVD* mutation is not clear. There is also another PBP (4*) which is synthesized largely during sporulation under σ^A control; however, inactivation of the coding gene (*pbpE*) has no effect on sporulation (Popham and Setlow, 1993).

There are two DD-carboxypeptidases specifically involved in cortex synthesis. One is PBP5*, encoded by *dacB* (Todd et al., 1985; Buchanan and Ling, 1992); like *spoVD*, *dacB* is expressed in the mother cell under σ^E control beginning at t_{2-3} of sporulation (Simpson et al., 1974). PBP5* has been shown to be a DD-carboxypeptidase by enzymatic assays in vitro, which is consistent with the effects of a *dacB* mutation on spore PG structure (Todd et al., 1985; Popham et al., 1999). The second DD-carboxypeptidase involved in cortex biosynthesis is the product of *dacF*, which is upstream of the *spoIIA* operon (Wu et al., 1992). In contrast to *dacB*, the *dacF* gene is transcribed only in the forespore at $\sim t_4$ of sporulation under control of σ^F (Wu et al., 1992). There are no data on the activity of DacF in vitro, but analyses of cortex structure in *dacB dacF* spores are consistent with DacF also being a DD-carboxypeptidase acting on the forming spore cortex (Popham et al., 1999) (see below).

A strain with an insertion in *dacB* does form refractile spores (Buchanan and Gustafson, 1992), but analysis of the effects of the *dacB* mutation in this strain is complicated, since *dacB* is the first gene in a three-gene operon and mutations in either of the two downstream genes (*spmA* and -*B*) also have effects on spore properties (Popham et al., 1995a). However, spores produced by a strain with an in-frame deletion in *dacB* have reduced heat resistance and an $\sim$50% increase in peptide side chains on spore cortical PG, with $\sim$35% of the Dpm in the cortex involved in cross-links (Popham et al., 1995a, 1999). This results in a four- to fivefold increase in PG cross-linking in *dacB* spores, consistent with DacB being the major DD-carboxypeptidase acting on cortical PG during sporulation to modulate the degree of PG cross-linking (Table 2). Despite the increase in PG cross-linking in spores of the in-frame *dacB* deletion strain, these spores have a core water content identical to that of wild-type spores (Table 2). However, the *dacB* spores appear to be unable to maintain core dehydration upon heating (Popham et al., 1995a). In contrast to the effects of the *dacB* mutation, loss of the major vegetative cell DD-carboxypeptidase, DacA, or either of two minor sporulation-specific enzymes, DacC and DacF, has no obvious effect on spore properties or cortex structure (Table 2) (Atrih et al., 1996; Popham et al., 1996b, 1999; Pederson et al., 1998). Indeed, DacC appears to play no essential role in spore cortex or germ cell wall synthesis, while DacA may be involved only in germ cell wall synthesis (Atrih et al., 1996; Popham et al., 1996b; Pederson et al., 1998). Although *dacF* mutants make spores with largely wild-type properties, *dacB dacF* spores have defects more extreme than those of *dacB* spores (Popham et al., 1999). Compared to *dacB* spores, the double-mutant spores have lower heat resistance, an increased spore core water content, and more highly cross-linked PG (Table 2) (Popham et al., 1999). These data, plus a significant decrease in MAL content in *dacB dacF* spores (Popham et al., 1999), suggest that DacF does play a role in spore cortex maturation, although its role is less important than that of DacB. A model for how DacB and DacF, acting from the mother cell and forespore sides of the cortex, respectively, might modulate the degree of PG cross-linking in different areas of the cortex has recently been presented (Popham et al., 1999).

In addition to the PBPs mentioned above, there are also a number of other gene products implicated in some aspect of spore cortex formation. Two of these are products of *spmA* and -*B*, mutations in which result in spores with reduced heat resistance and increased spore core water content (Table 2) (Popham et al., 1995a). However, analysis of spore cortex from *spm* spores has revealed no major defects, beyond a small decrease in PG cross-linking (Table 2) (Popham et al., 1996b). Mutations in the *spoVR* locus, a gene transcribed at $\sim t_3$ of sporulation under σ^E control, also result in a large fraction of spores lacking significant cortex; these aberrant spores are phase dark and probably lack DPA (Beall and Moran, 1994). In contrast, *dacB dacF*, *spmA*, and *spmB* spores invariably have normal DPA levels (Popham et al., 1995a, 1999). This result, as well as the observation that up to 30% of spores from *spoVR* cultures have normal cor-

texes (Beall and Moran, 1994), has been interpreted to indicate that *spoVR* may regulate rather than catalyze cortical PG synthesis. A gene implicated in spore coat assembly, *spoIVA*, also plays some role in cortex synthesis (see "Spore Coat" below), as do the products of the *spoVE* and *spoVD* loci (Buchanan et al., 1994). However, at present the mechanism(s) whereby the products of these last three genes are involved in cortex biosynthesis is not clear.

In contrast to the indirect or unclear roles of the previous group of genes in cortical PG synthesis, the *cwlD* gene product (Sekiguchi et al., 1995) has an important role in MAL synthesis, as *cwlD* mutants sporulate normally but the cortexes of *cwlD* spores contain no MAL (Atrih et al., 1996; Popham et al., 1996a). This lack of MAL results in a large increase in the percentage of NAM residues with both L-Ala and TP side chains, and the PG from *cwlD* spores is ~ 2 to 3 times as cross-linked as that in wild-type spores due to the increased levels of NAM-TP (Table 2) (Atrih et al., 1996; Popham et al., 1996a). The presence of elevated NAM-TP and NAM–L-Ala in *cwlD* spores further suggests that like vegetative cell PG, cortical PG is also initially synthesized with a pentapeptide attached to each NAM. Subsequently, other enzymes, including DacB and DacF, act to remove the peptide completely or almost completely from the majority of the NAM residues, with these NAM residues converted to MAL or NAM–L-Ala. However, the precise role of CwlD, as well as the identities of other proteins participating in this process, is not clear. In keeping with the synthesis of the spore cortex predominantly from the mother cell side of the developing spore, *cwlD* transcription begins at $\sim t_3$ of sporulation under control of the mother cell-specific σ^E (Sekiguchi et al., 1995); there is also a later phase of *cwlD* expression in the forespore under control of the forespore-specific σ^G (Sekiguchi et al., 1995). The precise function of *cwlD* expression in the forespore is not clear, since the nascent germ cell wall PG, which is adjacent to the inner forespore membrane, most likely contains no MAL. However, when *cwlD* is expressed in the forespore, it is contrascribed with an upstream gene (*orf1*) which is not transcribed in the mother cell (Sekiguchi et al., 1995). While the function of ORF1 is not known, perhaps it modifies CwlD action in some way, ensuring that CwlD made in the forespore acts only on the developing cortex.

Interestingly, the core water content of *cwlD* spores is normal, as is their heat resistance (Table 2) (Popham et al., 1996a). Consequently, it appears likely that MAL is not crucial for proper spore core dehydration and thus for spore heat resistance. Since the degree of cross-linking in the cortexes of *cwlD* spores is significantly higher than that in wild-type spores (Atrih et al., 1996; Popham et al., 1996a) (Table 2), the precise degree of cortical cross-linking may also not be crucial for spore core dehydration. However, spores with very highly cross-linked cortical PG have increased core water and decreased heat resistance (Table 2); consequently, cortical PG may only function properly within some limits of its normal degree of cross-linking. It is also possible, even perhaps likely, that the degree of cross-linking of cortical PG during its synthesis and maturation, as well as some ordered changes in this cross-linking, may be the most crucial parameter in determining the degree of eventual spore core dehydration. Unfortunately, data relevant to this possibility are not yet available.

Although *cwlD* spores have normal core water content and heat resistance, these spores do not undergo normal spore germination. While *cwlD* spores lose their refractility and release DPA on exposure to germinants, no cortical PG is degraded (Sekiguchi et al., 1995; Popham et al., 1996a). This strongly suggests that a major function of MAL in the spore cortex is to act as a target for enzymes which act early in spore germination to degrade the cortex; since MAL is the target of these lytic enzymes, only the spore cortex and not the spore's germ cell wall is degraded. Consequently, the germ cell wall is retained during spore germination, maintaining the osmotic

stability of the germinated spore as well as a template for new PG synthesis.

SPORE COAT

The outermost structure common to spores of all species is called the coat (Fig. 1). This complex shell consists of a series of one or more morphologically distinct layers (depending on the species) that protect the spore from a variety of noxious molecules and from mechanical damage (Holt and Leadbetter, 1969; Aronson and Fitz-James, 1976; Bloomfield and Arthur, 1994). The striking appearance of the coat in the electron microscope and the resiliency it imparts to the spore drew attention to this structure early on and have inspired a long history of investigation into the mechanisms of coat assembly and function (Strange and Dark, 1956; Halvorson et al., 1966; Murrell, 1967; Gould et al., 1970; Hiragi, 1972; Wood, 1972; Milhaud and Balassa, 1973; Goldman and Tipper, 1978, 1981; Munoz et al., 1978; Pandey and Aronson, 1979; Jenkinson et al., 1980, 1981). These classical studies demonstrated that the coat is largely protein (Kornberg et al., 1968) and that among various *Bacillus* species, the number of coat proteins as well as the architectural complexity of the coat varies considerably. *Bacillus cereus*, with its single major coat protein, has a very elementary coat architecture (Aronson and Horn, 1972), whereas *B. subtilis*, whose coat is composed of upwards of 20 proteins (Goldman and Tipper, 1978; Pandey and Aronson, 1979; Jenkinson et al., 1981), has two major layers that are visible in the electron microscope: a lightly staining lamellar inner coat and a more coarsely layered, darkly staining outer coat (Warth et al., 1963). These experiments also demonstrated that coat proteins are subject to a variety of posttranslational modifications, including cleavage and cross-linking (Aronson and Fitz-James, 1976). In spite of the enormous wealth of data accumulated in this early body of work, it remained to be learned how coat assembly was controlled, how posttranslational modification occurred, and how the individual coat proteins contributed to spore survival.

With the advent of molecular biology, the control of sporulation gene expression in *B. subtilis* received intensive examination (see chapters 6 to 8). This effort identified a large number of *B. subtilis* coat protein genes and elucidated the major controls on their expression (Driks, 1999). As a result, we know the timing of coat protein gene expression and that most, if not all, of these genes are transcribed in the mother cell. This information, in turn, makes it possible to begin to unravel the program of coat assembly and to learn the functions of the individual coat proteins in spore protection.

The proteins that participate in *B. subtilis* coat formation play a variety of distinct roles. Their genes are spread out over the entire chromosome, and most are largely unrelated in sequence (Kunst et al., 1997) (but see Driks, 1999, for a more complete discussion of coat protein sequence similarities). Some appear to be largely structural, in that the deletion of the gene has no readily detectable consequence other than the loss of the gene product from the coat. Some (including several coat structural proteins) are critical for the assembly of several coat components and are therefore designated as morphogenetic proteins. Over 20 coat structural and morphogenetic proteins have been identified, and the assembly of some of them has been studied. As a result, we can propose a preliminary model for coat assembly. We will first review the control of coat protein synthesis and then the control of assembly. We will describe the coat proteins in the order their genes are expressed or become relevant to assembly and not in the order suggested by the nomenclature, which is based solely on the order in which their genes were discovered.

The Coat Protein Genes

All the known coat-related genes are under the control of one or both of the mother cell-specific transcription factors σ^E and σ^K (Driks, 1999) and are, therefore, synthesized exclusively in the mother cell (Fig. 4) (see chapter 8). Expression of the earliest of these genes begins immediately after the formation of the sporula-tion septum, well before the electron-dense coat becomes visible in the electron microscope, as anticipated by previous biochemical and physiological studies (Munoz et al., 1978; Pandey and Aronson, 1979; Jenkinson et al., 1980). Two such early genes are *cotE* (Zheng and Losick, 1990) and the *cotJ* operon (including *cotJA*, *cotJB*, and *cotJC*) (Henriques et al., 1995), transcribed under σ^E control. At least two morphogenetic genes, *spoIVA* and *spoVID*, are also expressed as a result of σ^E (Roels et al., 1992; Stevens et al., 1992; Beall et al., 1993). σ^K, which becomes active after forespore engulfment, directs a large regulon that includes many coat protein genes, such as *cotA*, *cotD*, *cotF*, *cotH*, *cotM*, *cotT*, *cotV*, *cotW*, *cotY*, and *cotZ* (Donovan et al., 1987; Cutting et al., 1991; Zhang et al., 1994; Zheng and Losick, 1990; Henriques et al., 1997). σ^K also directs the expression of *gerE*, encoding a small DNA binding protein that works together with σ^K to activate a third and apparently final phase of mother cell gene expression (Zheng and Losick, 1990; Zheng et al., 1992). This results in the expression of *cotB*, *cotC*, *cotG*, *cotS*, and *cotX*, as well as the continued expression of *cotV*, *cotW*, *cotY*, and *cotZ* (Sacco et al., 1995; Takamatsu et al., 1998).

Control of Assembly

Although the coat protein genes are under strict temporal control, it is very unlikely that the order of coat protein assembly is simply a reflection of the order of gene expression. For example, the probable inner coat proteins CotD (Zheng et al., 1988) and CotS (Takamatsu et al., 1998) are synthesized at intermediate and late times in development (under the control of σ^K and σ^K with GerE, respectively) and the outer coat protein CotE is made early (due to σ^E) (Zheng and Losick, 1990). This is in contrast to the order of assembly predicted if transcription largely drove the order of coat protein deposition onto the forespore. Although the timing of gene expression is certainly important to coat formation, several studies indicate that certain major steps in coat construction are guided primarily by proteins

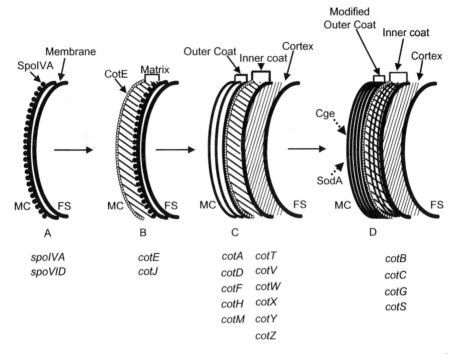

FIGURE 4 A four-step model for the assembly of the spore coat. For each stage, an arc of the forespore and the associated proteins is diagrammed, with the genes for the coat proteins assembled at each step listed below. (A) In the first step, under the control of σ^E, SpoIVA is synthesized and assembles at the mother cell side of the forespore membranes. (B) Next, a layer of CotE forms, separated from SpoIVA by a gap, which is filled with the matrix. (C) In the third stage, under the control of σ^K, inner and outer coat protein synthesis and assembly begins. The cortex is deposited between the two forespore membranes, which are now separated. (D) In the fourth stage, inner and outer coat synthesis and assembly is completed and the coat is further modified by cross-linking and glycosylation. The notion that the products of the GerE-controlled genes are incorporated after the others is especially speculative.

dedicated to assembly and not by the control of gene expression. Our current model for coat assembly is based on the view that assembly requires both the participation of morphogenetic proteins and proper timing of gene expression. This model relies on the extensive molecular genetic analysis of recent years (see chapter 8) as well as classical genetic studies that identified important regulators of morphogenesis (Piggot and Coote, 1976).

In this model, the first step in coat assembly is guided by a protein called SpoIVA (Fig. 4A). Classical studies revealed that a *spoIVA* mutation causes the coat not to assemble around the forespore but rather to accumulate as a swirl

in the mother cell cytoplasm, indicating that SpoIVA plays a role in coat attachment (Piggot and Coote, 1976). Furthermore, as no cortex is detectable in a *spoIVA* mutant, SpoIVA is involved in cortex synthesis as well. It is not known how SpoIVA controls cortex formation, but it does not appear to control the expression of any genes required for cortex synthesis and might, therefore, activate some key step in PG biosynthesis. *spoIVA* expression commences in the mother cell, under the control of σ^E, immediately after the completion of the sporulation septum (Roels et al., 1992; Stevens et al., 1992). Immunoelectron microscopy experiments indicate that SpoIVA forms a layer along the mother cell side of the septum, before any coat assembly can be detected, eventually forming a shell around the engulfed forespore (Driks et al., 1994). Immunofluorescence microscopy has been used to confirm and extend this result (Pogliano et al., 1995; Lewis and Errington, 1996) and to show that SpoIVA assembly continues after engulfment (Price and Losick, 1999). SpoIVA is present in the mature spore, although this is not readily detectable by immunolocalization methods. The presence of SpoIVA at or very near the forespore surface is consistent with a role for SpoIVA in the direct attachment of the coat to the forespore as well as a role in cortex assembly. The finding that SpoIVA plays an early role in both coat and cortex formation raises the possibility that these two structures need to be assembled simultaneously for proper spore function. It appears unlikely, however, that either structure is absolutely required for the assembly of the other. The cortex can be formed in the absence of the coat (as in a *gerE cotE* strain [Driks et al., 1994]), and at least some coat assembly is possible when the cortex is not fully formed (as in the cases of *spoVM* [Levin et al., 1993], *spoVR* [Beall and Moran, 1994], and *spoVS* [Resnekov et al., 1995]). Nonetheless, the existence of mutations that impair both the cortex and coat, and the location of SpoIVA at a site between these structures, may indicate the presence of a mechanism to couple the assembly of coat and cortex. SpoIVA, like the two structures it controls, might be an ancient feature of the spore, as a homologue of SpoIVA is present in *Clostridium acetobutylicum*, a relatively distantly related organism (Price and Losick, 1999).

SpoIVA is not the only protein to participate in coat attachment. Sporulating cells from a *spoVID* strain also produce a swirl of detached coat material in the mother cell cytoplasm (Beall et al., 1993). In contrast to *spoIVA* mutants, *spoVID* cells have an intact cortex and the precoat (see below) in *spoVID* cells is attached to the forespore. Detachment of the coat occurs only after coat assembly is largely

complete. This finding and the fact that the localization of SpoIVA to the forespore surface does not depend on SpoVID suggest that SpoVID acts later than SpoIVA (Driks et al., 1994). The dependency of SpoVID localization on SpoIVA is unknown.

The second step in assembly begins once SpoIVA is positioned at the forespore surface (Fig. 4B). At this time, a protein called CotE forms a layer in the mother cell cytoplasm (and ultimately a shell around the engulfed forespore), alongside SpoIVA but separated from it by a gap of about 75 nm (Driks et al., 1994). The presence of CotE at this location is the hallmark of this second stage of coat formation. The gap between SpoIVA and CotE is presumed to be filled by a material, referred to as the matrix, which is only faintly detectable by electron microscopy and whose composition is currently unknown. The matrix and the layer of CotE form a structure called the precoat. Like *spoIVA*, *cotE* is under the control of σ^E. Presumably, the expression of genes encoding matrix components is also governed by σ^E. The attachment of the precoat to the forespore is dependent on SpoIVA; when SpoIVA is missing, the layer of CotE (and, it is assumed, the matrix) is no longer near the forespore but instead forms a clump in the mother cell cytoplasm, reminiscent of the swirl of coat that forms later on in development. Although the proteins that compose the matrix are unknown, good candidates are the products of the *cotJABC* operon (Henriques et al., 1995). CotJC is present in the coat even in the presence of mutations which eliminate the inner and outer coats but not the matrix. CotJA interacts with CotJC, so it is a plausible matrix protein candidate as well (Seyler et al., 1997).

In the third stage of coat assembly, inner and outer coat assembly begins (and will continue during the fourth stage), as a result of σ^K-directed gene expression (Fig. 4C). The structure of the precoat and the timing of its formation make it a very good candidate for a framework that directs the assembly of the two coat layers. From this perspective, the matrix guides the assembly of the inner coat proteins while CotE serves to nucleate the assembly of the outer coat proteins. As a result, CotE is sandwiched between the inner and outer coat layers, as observed by immunoelectron microscopy (Driks et al., 1994). The proposal that CotE nucleates the outer coat proteins stems from the striking dependency of outer coat assembly on CotE (Zheng et al., 1988). In a *cotE* null mutant, no outer coat is present, as judged by electron microscopy, while the expression of at least two, and most likely all, of the outer coat protein genes remains unaffected. Consistent with the view that the matrix is the site of inner coat assembly, the size of the gap between the rings of SpoIVA and CotE is about the same as the thickness of the inner coat.

The set of likely outer coat proteins assembled during this third step includes CotA, CotD, CotF, CotM, CotV, CotW, CotX, CotY, and CotZ (Donovan et al., 1987; Cutting et al., 1991; Zhang et al., 1993; Henriques et al., 1997). Of these, CotA and CotM have readily detectable effects on coat assembly beyond their own presence in the coat. CotA (previously identified as the product of the *pig* locus) imparts a brown color to mature spores (Sandman et al., 1988). No role for this pigment in spore function or survival is known. CotM is a morphogenetic factor that is required for the correct assembly of the outermost layer of the outer coat (Henriques et al., 1997). *cotM* mutant spores have an altered set of coat proteins and an outer coat with a diffuse structure. Sometimes, portions of the outer coats of these spores appear to have a lamellar structure and to display fine ridges on their outermost surfaces.

The probable inner coat proteins CotD, CotH, and CotT are also likely to be assembled in the third stage (Zheng et al., 1988; Bourne et al., 1991; Naclerio et al., 1996). The deletion of CotD has no readily evident effect on coat function or morphology (Donovan et al., 1987), but *cotH* spores are deficient in several coat proteins and in germination, implying a significant coat defect. Alterations in the level of CotT influence the width of the inner coat

and the response to germinants, indicating that this protein plays a morphogenetic role as well.

The fourth step in coat assembly, under the control of σ^K and GerE, leads to the expression of the likely outer coat protein genes *cotB*, *cotC*, and *cotG* (Zheng and Losick, 1990; Sacco et al., 1995) and the continued expression of *cotV*, *cotW*, *cotX*, *cotY*, and *cotZ* (Zhang et al., 1994) (Fig. 4D). The importance of GerE-controlled expression is underscored by the phenotype of spores of a *gerE* mutant strain. These spores are missing the morphological features of the inner coat, and only a remnant of the outer coat remains (Moir, 1981; Driks et al., 1994).

The outer coat protein CotG has a significant morphogenetic role. When *cotG* is deleted, CotB is not incorporated and the outer coat is reduced to a single darkly staining band (Sacco et al., 1995; Henriques et al., 1998). SodA, encoding a Mn-dependent superoxide dismutase (Casillas-Martinez and Setlow, 1997), plays an important part in CotG assembly. *sodA* spores have essentially normal resistance properties, but the inner and outer coat layers are disrupted and CotG is more readily extracted (Henriques et al., 1998). It is proposed that SodA participates (probably along with other proteins) in cross-linking CotG into the coat, thereby generating much of the darkly layered appearance of the outer coat. CotS is also synthesized during the fourth step in assembly. *cotS* spores do not have a detectable phenotype (Abe et al., 1995), and CotS is a component of the inner coat, as determined by immunoelectron microscopy (Takamatsu et al., 1998). Unexpectedly, for an inner coat protein, CotS assembly depends on CotE. Apparently, CotE controls at least some inner coat formation in addition to its major role in outer coat assembly.

Much of the resiliency of the coat is likely to derive from cross-links and other modifications that appear after the structural proteins are in place. For example, ϵ-(γ-glutamyl)lysine cross-links have been found in the coat and are probably generated by a recently identified spore-associated transglutaminase, Tgl (Kobayashi et al., 1996, 1998a, 1998b). Disulfide cross-links and dityrosine cross-links are likely to be important as well (Aronson and Fitz-James, 1976; Pandey and Aronson, 1979). Intriguingly, preliminary in vitro experiments suggest that CotE may be able to generate dityrosine cross-links (Deits, personal communication). An additional degree of modification may be provided by glycosylation of the coat proteins. *cgeD* (Roels and Losick, 1995), a member of a cluster of genes under the control of σ^K and GerE, has significant homology to glycosyl transferases. Preliminary studies indicate that spores missing *cge* genes are altered in their surface properties. Glycosylation could affect spore permeability and the choice of niche.

The Functions of the Coat

Clearly, the primary role of the coat is to protect the spore. It does this, at the very least, by acting as a sieve that excludes all but the smallest molecules from the spore interior (Scherrer et al., 1971; Nishihara et al., 1989) and by providing mechanical strength. However, protection may also depend on enzymatic activities possessed by individual coat proteins. For example, the coat has been implicated in resistance to iodine and chlorine (Bloomfield and Arthur, 1992; Bloomfield and Megid, 1994). It is possible that some coat proteins help to inactivate toxic halogens. The coat may also have functions unrelated to protection. CotA is intriguing in this regard, as its sequence is similar to that of MnxG of the marine *Bacillus* sp. strain SG-1. MnxG, which is a component of the exosporium of SG-1 (Francis and Tebo, 1999), promotes the formation of a manganese oxide shell around the spore (van Waasbergen et al., 1993, 1996). The formation of metal precipitates by bacteria makes an important contribution to the geochemistry of the earth's soil and waters (Tebo et al., 1997). CotA is unlikely to be involved in the formation of metal deposits, but its similarity to MnxG raises the possibility of a spore-bound oxidase activity in *B. subtilis*.

In addition to its role as a shield, the coat participates in germination. Early studies showed that strains with likely coat protein gene mutations were also somewhat defective in germination (Aronson and Fitz-James, 1975, 1976). More recently, the loss of CotE (Zheng et al., 1988), CotH (Naclerio et al., 1996), CotX, or CotY and CotZ together (Zhang et al., 1993) has been shown to impair germination. The manner in which the coat influences germination is not obvious, as no coat protein gene mutation (or combination of mutations) has been shown to entirely prevent germination, and the machinery for germination is likely to reside in locations other than the coat (Foster and Johnstone, 1989; Moir and Smith, 1990). As the mechanics of germination and coat assembly become clearer, this relationship may come into better focus.

Much of our ignorance of coat function and of the roles of individual coat proteins is due to the significant discrepancies between the manner in which spore survival is tested in the laboratory and the actual situation in the soil. A truly realistic assay for coat function is unlikely to become available soon, as we know relatively little about the specific challenges faced by soil organisms. However, without an assay for spore survival that closely mimics the soil environment it will be very difficult, if not impossible, to determine precisely how the coat protects the spore or to identify any other functions the coat may have. One consequence of the lack of a realistic test is the paradox presented by the large number of apparently dispensable coat proteins. When CotA, CotB, CotC, CotD, CotF, CotJA, CotJB, CotJC, or CotS is deleted, no significant reduction of spore survival or alteration in morphology can be detected in laboratory assays (Donovan et al., 1987; Cutting et al., 1991; Abe et al., 1995; Henriques et al., 1995). In spite of our ignorance, it seems reasonable to presume that these apparently functionally redundant coat proteins do, indeed, provide some selective advantage to the spore. Possibly, the large number of coat proteins in *B. subtilis* corresponds to the diverse microenvironments within the soil, which include the rhizosphere, the surfaces of mineral grains, liquid regions, and the interiors of nematodes, insects, fungi, and other soil-dwelling organisms. The relatively large complement of coat proteins in *B. subtilis* could be part of an accommodation to the diverse niches present in the soil.

ACKNOWLEDGMENTS

We thank Shawn Little for comments on the manuscript and David Guttman for stimulating discussions.

Work in the authors' laboratories has been supported by the Army Research Office (P.S.), the National Institutes of Health (GM-19698 [P.S.] and GM53989 [A.D.]), and The Schweppes Foundation (A.D.).

REFERENCES

Abe, A., H. Koide, T. Kohno, and K. Watabe. 1995. A *Bacillus subtilis* spore coat polypeptide gene, *cotS*. *Microbiology* **141:**1433–1442.

Arnott, S., and E. Selsing. 1974a. Structures for the polynucleotide complexes poly(dA)·poly(dT) and poly(dT)·poly(dA)·poly(dT). *J. Mol. Biol.* **88:** 509–521.

Arnott, S., and E. Selsing. 1974b. The structure of polydeoxyguanylic acid·polydeoxycytidylic acid. *J. Mol. Biol.* **88:**551–552.

Aronson, A. I., and P. Fitz-James. 1976. Structure and morphogenesis of the bacterial spore coat. *Bacteriol. Rev.* **40:**360–402.

Aronson, A. I., and P. C. Fitz-James. 1975. Properties of *Bacillus cereus* spore coat mutants. *J. Bacteriol.* **123:**354–365.

Aronson, A. I., and D. Horn. 1972. Characterization of the spore coat protein of *Bacillus cereus* T, p. 19–27. *In* H. O. Halvorson, R. Hansen, and L. L. Campbell (ed.), *Spores V.* American Society for Microbiology, Washington, D.C.

Atrih, A., P. Zollner, G. Allmaier, and S. F. Foster. 1996. Structural analysis of *Bacillus subtilis* 168 endospore peptidoglycan and its role during differentiation. *J. Bacteriol.* **178:**6173–6183.

Bagyan, I., L. Casillas-Martinez, and P. Setlow. 1998a. The *katX* gene which codes for the catalase in spores of *Bacillus subtilis* is a forespore specific gene controlled by σ^F, and KatX is essential for hydrogen peroxide resistance of the germinating spore. *J. Bacteriol.* **180:**2057–2062.

Bagyan, I., B. Setlow, and P. Setlow. 1998b. New small, acid soluble proteins unique to spores of *Bacillus subtilis:* identification of the coding genes and studies of the regulation and function of two of these genes. *J. Bacteriol.* **180:**6704–6712.

Beall, B., and C. P. Moran, Jr. 1994. Cloning and characterization of *spoVR*, a gene from *Bacillus subtilis* involved in spore cortex formation. *J. Bacteriol.* **176:**2003–2012.

Beall, B., A. Driks, R. Losick, and C. P. Moran, Jr. 1993. Cloning and characterization of a gene required for assembly of the *Bacillus subtilis* spore coat. *J. Bacteriol.* **175:**1705–1716.

Beaman, T. C., and P. Gerhardt. 1986. Heat resistance of bacterial spores correlated with protoplast dehydration, mineralization, and thermal adaptation. *Appl. Environ. Microbiol.* **52:**1242–1246.

Beaman, T. C., H. S. Pankratz, and P. Gerhardt. 1972. Ultrastructure of the exosporium and underlying inclusions in spores of *Bacillus megaterium* strains. *J. Bacteriol.* **109:**1198–1209.

Bloomfield, S. F., and M. Arthur. 1992. Interaction of *Bacillus subtilis* spores with sodium hypochlorite, sodium dichloroisocyanurate and chloramine-T. *J. Appl. Bacteriol.* **72:**166–172.

Bloomfield, S. F., and M. Arthur. 1994. Mechanisms of inactivation and resistance of spores to chemical biocides. *J. Appl. Bacteriol.* **76:**91S–104S.

Bloomfield, S. F., and R. Megid. 1994. Interaction of iodine with *Bacillus subtilis* spores and spore forms. *J. Appl. Bacteriol.* **76:**492–499.

Bourne, N., P. C. Fitz-James, and A. I. Aronson. 1991. Structural and germination defects of *Bacillus subtilis* spores with altered contents of a spore coat protein. *J. Bacteriol.* **173:**6618–6625.

Buchanan, C. E., and A. Gustafson. 1992. Mutagenesis and mapping of the gene for a sporulation-specific penicillin-binding protein in *Bacillus subtilis. J. Bacteriol.* **174:**5430–5435.

Buchanan, C. E., and M.-L. Ling. 1992. Isolation and sequence analysis of *dacB*, which encodes a sporulation-specific penicillin-binding protein in *Bacillus subtilis. J. Bacteriol.* **174:**1717–1725.

Buchanan, C. E., A. O. Henriques, and P. J. Piggot. 1994. Cell wall changes during bacterial endospore formation, p. 167–186. *In* J.-M. Ghuysen and R. Hackenbeck (ed.), *Bacterial Cell Wall.* Elsevier, Amsterdam, The Netherlands.

Bueno, A., J. R. Villanueva, and T. G. Villa. 1986. Methylation of spore DNA in *Bacillus coagulans. J. Gen. Microbiol.* **132:**2899–2905.

Cabrera-Hernandez, A., J.-L. Sanchez-Salas, M. Paidhungat, and P. Setlow. 1999. Regulation of four genes encoding small, acid-soluble spore proteins in *Bacillus subtilis. Gene* **232:**1–10.

Cabrera-Martinez, R., J. M. Mason, B. Setlow, W. M. Waites, and P. Setlow. 1989. Purification and amino acid sequence of two small, acid-soluble proteins from *Clostridium bifermentans* spores. *FEMS Microbiol. Lett.* **61:**139–144.

Cabrera-Martinez, R. M., and P. Setlow. 1991. Cloning and nucleotide sequence of three genes coding for small, acid-soluble proteins of *Clostrid-*

ium perfringens spores. *FEMS Microbiol. Lett.* **77:** 127–132.

Casillas-Martinez, L., and P. Setlow. 1997. Alkyl hydroperoxide reductase, catalase, MrgA, and superoxide dismutase are not involved in resistance of *Bacillus subtilis* spores to heat or oxidizing agents. *J. Bacteriol.* **179:**7420–7425.

Chen, T., J. Younger, and L. Glaser. 1968. Synthesis of teichoic acids. VII. Synthesis of teichoic acids during spore germination. *J. Bacteriol.* **95:** 2044–2050.

Cleveland, E. F., and C. Gilvarg. 1975. Selective degradation of peptidoglycan from *Bacillus megaterium* spores during germination, p. 458–464. *In* P. Gerhardt, R. N. Costilow, and H. L. Sadoff (ed.), *Spores VI.* American Society for Microbiology, Washington, D.C.

Cutting, S., L. Zheng, and R. Losick. 1991. Gene encoding two alkali-soluble components of the spore coat from *Bacillus subtilis. J. Bacteriol.* **173:** 2915–2919.

Daniel, R. A., S. Drake, C. E. Buchanan, R. Scholle, and J. Errington. 1994. The *Bacillus subtilis spoVD* gene encodes a mother-cell-specific penicillin-binding protein required for spore morphogenesis. *J. Mol. Biol.* **235:**209–220.

Decker, S. J., and D. R. Lang. 1978. Membrane bioenergetic parameters in uncoupler-resistant mutants of *Bacillus megaterium. J. Biol. Chem.* **253:** 6738–6743.

Deits, T. Personal communication.

Donnellan, J. E., Jr., and R. B. Setlow. 1965. Thymine photoproducts but not thymine dimers are found in ultraviolet irradiated bacterial spores. *Science* **149:**308–310.

Donovan, W., L. Zheng, K. Sandman, and R. Losick. 1987. Genes encoding spore coat polypeptides from *Bacillus subtilis. J. Mol. Biol.* **196:**1–10.

Driks, A. 1999. The *Bacillus subtilis* spore coat. *Microbiol. Mol. Biol. Rev.* **63:**1–20.

Driks, A., S. Roels, B. Beall, C. P. Moran, Jr., and R. Losick. 1994. Subcellular localization of proteins involved in the assembly of the spore coat of *Bacillus subtilis. Genes Dev.* **8:**234–244.

Fairhead, H., and P. Setlow. 1992. Binding of DNA to α/β-type small, acid-soluble proteins from spores of *Bacillus* or *Clostridium* species prevents formation of cytosine dimers, cytosine-thymine dimers and dipyrimidine photoadducts upon ultraviolet irradiation. *J. Bacteriol.* **174:**2874–2880.

Fairhead, H., B. Setlow, and P. Setlow. 1993. Prevention of DNA damage in spores and in vitro by small, acid-soluble proteins from *Bacillus* species. *J. Bacteriol.* **175:**1367–1374.

Fairhead, H., B. Setlow, W. M. Waites, and P. Setlow. 1994. Small, acid-soluble proteins bound to DNA protect *Bacillus subtilis* spores from killing by freeze-drying. *Appl. Environ. Microbiol.* **60:** 2647–2649.

Fajardo-Cavazos, P., F. Tovar-Rojo, and P. Setlow. 1991. Effect of promoter mutations and upstream deletions on the expression of genes coding for small, acid-soluble spore proteins of *Bacillus subtilis. J. Bacteriol.* **173:**2011–2016.

Fajardo-Cavazos, P., C. Salazar, and W. L. Nicholson. 1993. Molecular cloning and characterization of the *Bacillus subtilis* spore photoproduct lyase (*spl*) gene, which is involved in the repair of UV radiation-induced DNA damage during spore germination. *J. Bacteriol.* **175:**1735–1744.

Foster, S. J., and K. Johnstone. 1989. The trigger mechanism of bacterial spore germination, p. 223–241. *In* I. Smith, R. A. Slepecky, and P. Setlow (ed.), *Regulation of Procaryotic Development.* American Society for Microbiology, Washington, D.C.

Francesconi, S. C., T. J. MacAlister, B. Setlow, and P. Setlow. 1988. Immunoelectron microscopic localization of small, acid-soluble spore proteins in sporulating cells of *Bacillus subtilis. J. Bacteriol.* **170:**5963–5967.

Francis, C., and B. Tebo. 1999. Personal communication.

Gerhardt, P., and R. E. Marquis. 1989. Spore thermoresistance mechanisms, p. 43–63. *In* I. Smith, R. A. Slepecky, and P. Setlow (ed.), *Regulation of Prokaryotic Development.* American Society for Microbiology, Washington, D.C.

Gerhardt, P., R. Scherrer, and S. H. Black. 1972. Molecular sieving by dormant spore structures, p. 68–74. *In* H. O. Halvorson, R. Hanson, and L. L. Campbell (ed.). *Spores V.* American Society for Microbiology, Washington, D.C.

Goldman, R. C., and D. J. Tipper. 1978. *Bacillus subtilis* spore coats: complexity and purification of a unique polypeptide component. *J. Bacteriol.* **135:** 1091–1106.

Goldman, R. C., and D. J. Tipper. 1981. Coat protein synthesis during sporulation of *Bacillus subtilis*: immunological detection of soluble precursors to the 12,200-dalton spore coat protein. *J. Bacteriol.* **147:**1040–1048.

Gould, G. W., J. M. Stubbs, and W. L. King. 1970. Structure and composition of resistant layers in bacterial spore coats. *J. Gen. Microbiol.* **60:** 347–355.

Griffith, J., A. Makhov, L. Santiago-Lara, and P. Setlow. 1994. Electron microscopic studies of the interaction between a *Bacillus* α/β-type small, acid-soluble spore protein with DNA: protein binding is cooperative, stiffens the DNA and induces negative supercoiling. *Proc. Natl. Acad. Sci. USA* **91:** 8224–8228.

Guidibande, S. R., S. D. Jayasena, and M. J.

Behe. 1988. CD studies of double-stranded polydeoxynucleotides composed of repeating units of contiguous homopurine residues. *Biopolymers* **27:** 1905–1915.

Hachisuka, Y., and S. Kozuka. 1981. A new test of differentiation of *Bacillus cereus* and *Bacillus anthracis* based on the existence of spore appendages. *Microbiol. Immunol.* **25:**1201–1207.

Hachisuka, Y., S. Kozuka, and M. Tsujikawa. 1984. Exosporia and appendages of spores of *Bacillus* species. *Microbiol. Immunol.* **28:**619–624.

Halvorson, H. O., J. C. Vary, and W. Steinberg. 1966. Developmental changes during the formation and breaking of the dormant state in bacteria. *Annu. Rev. Microbiol.* **20:**169–188.

Hayes, C. S., and P. Setlow. 1997. Analysis of deamidation of small, acid-soluble spore proteins from *Bacillus subtilis* in vitro and in vivo. *J. Bacteriol.* **179:** 6020–6027.

Hayes, C. S., and P. Setlow. 1998a. Identification of protein-protein contacts between α/β-type small, acid-soluble spore proteins of *Bacillus* species bound to DNA. *J. Biol. Chem.* **273:**17326–17332.

Hayes, C. S., and P. Setlow. 1998b. Unpublished results.

Hayes, C. S., B. Illades-Aguiar, L. Casillas-Martinez, and P. Setlow. 1998. In vitro and in vivo oxidation of methionine residues in small, acid-soluble spore proteins from *Bacillus* species. *J. Bacteriol.* **180:**2694–2700.

Heinemann, U., C. Alings, and H. Lauble. 1989. Structural features of G/C rich DNA going A or B, p. 39–53. *In* R. H. Sarma and M. A. Sarma (ed.), *Structure and Methods: DNA and RNA*, vol. 3. Academic Press, Guilderland, N.Y.

Henriques, A. O., B. W. Beall, K. Roland, and C. P. Moran, Jr. 1995. Characterization of *cotJ*, a sigma E-controlled operon affecting the polypeptide composition of the coat of *Bacillus subtilis* spores. *J. Bacteriol.* **177:**3394–3406.

Henriques, A. O., B. W. Beall, and C. P. Moran, Jr. 1997. CotM of *Bacillus subtilis*, a member of the alpha-crystallin family of stress proteins, is induced during development and participates in spore outer coat formation. *J. Bacteriol.* **179:**1887–1897.

Henriques, A. O., L. R. Melsen, and C. P. Moran, Jr. 1998. Involvement of superoxide dismutase in spore coat assembly in *Bacillus subtilis*. *J. Bacteriol.* **180:**2285–2291.

Hiragi, Y. 1972. Physical, chemical and morphological studies of spore coat of *Bacillus subtilis*. *J. Gen. Microbiol.* **72:**87–99.

Hodges-Garcia, Y., P. J. Hagerman, and D. E. Pettijohn. 1989. DNA ring closure mediated by protein HU. *J. Biol. Chem.* **264:**14621–14623.

Holt, S. C., and E. R. Leadbetter. 1969. Comparative ultrastructure of selected aerobic spore-forming bacteria: a freeze-etching study. *Bacteriol. Rev.* **33:** 346–378.

Illades-Aguiar, B., and P. Setlow. 1994. Autoprocessing of the protease that degrades small, acid-soluble proteins of spores of *Bacillus* species is triggered by low pH, dehydration, and dipicolinic acid. *J. Bacteriol.* **176:**7032–7037.

Imae, Y., and J. L. Strominger. 1976. Relationship between cortex content and properties of *Bacillus sphaericus* spores. *J. Bacteriol.* **126:**907–913.

Jayasena, U. K., and M. J. Behe. 1991. Oligopurine·oligopyrimidine tracts do not have the same conformation as polypurine·polypyrimidines. *Biopolymers* **31:**511–518.

Jenkinson, H. F., D. Kay, and J. Mandelstam. 1980. Temporal dissociation of late events in *Bacillus subtilis* sporulation from expression of genes that determine them. *J. Bacteriol.* **141:**793–805.

Jenkinson, H. F., W. D. Sawyer, and J. Mandelstam. 1981. Synthesis and order of assembly of spore coat proteins in *Bacillus subtilis*. *J. Gen. Microbiol.* **123:**1–16.

Johnstone, K. 1994. The trigger mechanism of spore germination. *J. Appl. Bacteriol.* **76:**17S–24S.

Johnstone, K., F. A. Simion, and D. J. Ellar. 1982. Teichoic acid and lipid metabolism during sporulation of *Bacillus megaterium* KM. *Biochem. J.* **202:** 459–467.

Kobayashi, K., Y. Kumazawa, K. Miwa, and S. Yamanaka. 1996. ε-(γ-Glutamyl)lysine crosslinks of spore coat proteins and transglutaminase activity in *Bacillus subtilis*. *FEMS Microbiol. Lett.* **144:**157–160.

Kobayashi, K., K. Hashiguchi, K. Yokozeki, and S. Yamanaka. 1998a. Molecular cloning of the transglutaminase gene from *Bacillus subtilis* and its expression in *Escherichia coli*. *Biosci. Biotechnol. Biochem.* **62:**1109–1114.

Kobayashi, K., S.-I. Suzuki, Y. Izawa, K. Yokozeki, K. Miwa, and S. Yamanaka. 1998b. Transglutaminase in sporulating cells of *Bacillus subtilis*. *J. Gen. Appl. Microbiol.* **44:**85–91.

Koehler, P., and M. A. Marahiel. 1997. Association of the histone-like protein HBsu with the nucleoid of *Bacillus subtilis*. *J. Bacteriol.* **179:**2060–2064.

Kornberg, A., J. A. Spudich, D. L. Nelson, and M. Deutscher. 1968. Origin of proteins in sporulation. *Annu. Rev. Biochem.* **37:**51–78.

Kozuka, S., and K. Tochikubo. 1985. Properties and origin of filamentous appendages on spores of *Bacillus cereus*. *Microbiol. Immunol.* **29:**21–37.

**Kunst, F., N. Ogasawara, I. Moszer, A. M. Albertini, G. Alloni, V. Azevedo, M. G. Bertero, P. Bessieres, A. Bolotin, S. Borchert, R. Borriss, L. Boursier, A. Brans, M. Braun, S. C. Brignell, S. Bron, S. Brouillet, C. V. Bruschi, B. Caldwell, V. Capuano, N. M. Carter, S. K.

Choi, J. J. Codani, I. F. Connerton, and A. Danchin, et. al. 1997. The complete genome sequence of the gram-positive bacterium *Bacillus subtilis. Nature* **390**:249–256.

Levin, P., A. N. Fan, E. Ricca, A. Driks, R. Losick, and S. Cutting. 1993. An unusually small gene required for sporulation by *Bacillus subtilis. Mol. Microbiol.* **9**:761–771.

Lewis, J. C., N. S. Snell, and H. K. Burr. 1960. Water permeability of bacterial spores and the concept of a contractile cortex. *Science* **132**:544–545.

Lewis, P. J., and J. Errington. 1996. Use of green fluorescent protein for detection of cell-specific gene expression and subcellular protein localization during sporulation in *Bacillus subtilis. Microbiology* **142**:733–740.

Loshon, C. A., and P. Setlow. 1993. Levels of small molecules in dormant spores of *Sporosarcina* species and comparison with levels in spores of *Bacillus* and *Clostridium* species. *Can. J. Microbiol.* **39**:259–262.

Loshon, C. A., E. Hernandez-Alarcon, K. E. Beary, E. Z. Grey, L.-M. Santiago-Lara, and P. Setlow. 1995. Unpublished results.

Loshon, C. A., P. Kraus, B. Setlow, and P. Setlow. 1997. Effects of inactivation or overexpression of the *sspF* gene on properties of *Bacillus subtilis* spores. *J. Bacteriol.* **179**:272–275.

Loshon, C. A., P. C. Genest, B. Setlow, and P. Setlow. 1999. Formaldehyde kills spores of *Bacillus subtilis* by DNA damage, and small, acid-soluble spore proteins of the α/β-type protect spores against this DNA damage. *J. Appl. Microbiol.* **87**:8–14.

Magill, N. G., and P. Setlow. 1992. Properties of purified sporlets produced by *spoII* mutants of *Bacillus subtilis. J. Bacteriol.* **174**:8148–8151.

Magill, N. G., A. E. Cowan, D. E. Koppel, and P. Setlow. 1994. The internal pH of the forespore compartment of *Bacillus megaterium* decreases by about 1 pH unit during sporulation. *J. Bacteriol.* **176**:2252–2258.

Magill, N. G., A. E. Cowan, M. A. Leyva-Vazquez, M. Brown, D. E. Koppel, and P. Setlow. 1996. Analysis of the relationship between the decrease in pH and accumulation of 3-phosphoglyceric acid in developing forespores of *Bacillus* species. *J. Bacteriol.* **178**:2204–2210.

Marquis, R. E., and G. R. Bender. 1990. Compact structure of cortical peptidoglycans from bacterial spores. *Can. J. Microbiol.* **36**:426–429.

Marquis, R. E., J. Sim, and S. Y. Shin. 1994. Molecular mechanisms of resistance to heat and oxidative damage. *J. Appl. Bacteriol.* **76**:40S–48S.

Mason, J. M., and P. Setlow. 1986. Evidence for an essential role for small, acid-soluble, spore proteins in the resistance of *Bacillus subtilis* spores to ultraviolet light. *J. Bacteriol.* **167**:174–178.

Mason, J. M., and P. Setlow. 1987. Different small,

acid-soluble proteins of the α/β-type have interchangeable roles in the heat and ultraviolet radiation resistance of *Bacillus subtilis* spores. *J. Bacteriol.* **169**:3633–3637.

Mason, J. M., P. Fajardo-Cavazos, and P. Setlow. 1988a. Levels of mRNAs which code for small, acid-soluble spore proteins and their *lacZ* gene fusions in sporulating cells of *Bacillus subtilis. Nucleic Acids Res.* **16**:6567–6583.

Mason, J. M., R. H. Hackett, and P. Setlow. 1988b. Studies on the regulation of expression of genes coding for small, acid-soluble proteins of *Bacillus subtilis* spores using *lacZ* gene fusions. *J. Bacteriol.* **170**:239–244.

McCall, M., T. Brown, and O. Kennard. 1985. The crystal structure of d(GGGGCCCC). A model for poly(dG)·poly(dC). *J. Mol. Biol.* **183**:385–396.

Micka, B., and M. A. Marahiel. 1992. The DNA-binding protein HBsu is essential for normal growth and development in *Bacillus subtilis. Biochimie* **74**: 641–650.

Micka, B., N. Groch, U. Heinemann, and M. A. Marahiel. 1991. Molecular cloning, nucleotide sequence, and characterization of the *Bacillus subtilis* gene encoding the DNA binding protein HBsu. *J. Bacteriol.* **173**:3191–3198.

Milhaud, P., and G. Balassa. 1973. Biochemical genetics of bacterial sporulation. IV. Sequential development of resistance to chemical and physical agents during sporulation of *Bacillus subtilis. Mol. Gen. Genet.* **125**:241–250.

Mizuki, E., M. Ohba, T. Ichimatsu, S.-H. Hwang, K. Higuchi, H. Saitoh, and T. Akao. 1998. Unique appendages associated with spores of *Bacillus cereus* isolates. *J. Basic Microbiol.* **38**:33–39.

Mohr, S. C., and P. Setlow. 1990. Unpublished results.

Mohr, S. C., N. V. H. A. Sokolov, C. He, and P. Setlow. 1991. Binding of small acid-soluble spore proteins from *Bacillus subtilis* changes the conformation of DNA from B to A. *Proc. Natl. Acad. Sci. USA* **88**:77–81.

Moir, A. 1981. Germination properties of a spore coat-defective mutant of *Bacillus subtilis. J. Bacteriol.* **146**:1106–1116.

Moir, A., and D. A. Smith. 1990. The genetics of bacterial spore germination. *Annu. Rev. Microbiol.* **44**:531–553.

Munakata, N., and C. S. Rupert. 1972. Genetically controlled removal of "spore photoproduct" from deoxyribonucleic acid of ultraviolet-irradiated *Bacillus subtilis* spores. *Mol. Gen. Genet.* **104**:258–263.

Munoz, L., Y. Sadaie, and R. H. Doi. 1978. Spore coat protein of *Bacillus subtilis. J. Biol. Chem.* **253**: 6694–6701.

Murray, T., D. L. Popham, and P. Setlow. 1997. Identification and characterization of *pbpA* encoding *Bacillus subtilis* penicillin-binding protein 2A. *J. Bacteriol.* **179:**3021–3029.

Murray, T., D. L. Popham, C. B. Pearson, A. R. Hand, and P. Setlow. 1998. Analysis of the outgrowth of *Bacillus subtilis* spores lacking penicillin-binding protein 2a. *J. Bacteriol.* **180:**6493–6502.

Murrell, W. G. 1967. The biochemistry of the bacterial endospore, p. 133–251. *In* A. H. Rose and J. F. Wilkinson (ed.), *Advances in Microbial Physiology*, vol. 1. Academic Press, London, England.

Naclerio, G., L. Baccigalupi, R. Zilhao, M. De Felice, and E. Ricca. 1996. *Bacillus subtilis* spore coat assembly requires *cotH* gene expression. *J. Bacteriol.* **178:**4375–4380.

Nicholson, W. L., and P. Setlow. 1990. Dramatic increase in the negative superhelicity of plasmid DNA in the forespore compartment of sporulating cells of *Bacillus subtilis*. *J. Bacteriol.* **172:**7–14.

Nicholson, W. L., D. Sun, B. Setlow, and P. Setlow. 1989. Promoter specificity of sigma-G-containing RNA polymerase from sporulating cells of *Bacillus subtilis*: identification of a group of forespore-specific promoters. *J. Bacteriol.* **171:** 2708–2718.

Nicholson, W. L., B. Setlow, and P. Setlow. 1990. Binding of DNA *in vitro* by a small, acid-soluble spore protein and its effect on DNA topology. *J. Bacteriol.* **172:**6900–6906.

Nicholson, W. L., B. Setlow, and P. Setlow. 1991. Ultraviolet irradiation of DNA complexed with α/β-type small, acid-soluble proteins from spores of *Bacillus* or *Clostridium* species makes spore photoproduct but not thymine dimers. *Proc. Natl. Acad. Sci. USA* **88:**8288–8292.

Nishihara, T., Y. Takubo, E. Kawamata, T. Koshikawa, J. Ogaki, and M. Kondo. 1989. Role of outer coat in resistance of *Bacillus megaterium* spore. *J. Biochem.* **106:**270–273.

Nishimura, Y., C. Torigoe, and M. Tsuboi. 1985. An A-form poly(dG)·poly(dC) in H_2O solution. *Biopolymers* **24:**1841–1844.

Ou, L.-T., and R. E. Marquis. 1970. Electromechanical interactions in cell walls of gram-positive cocci. *J. Bacteriol.* **101:**92–101.

Pandey, N. K., and A. I. Aronson. 1979. Properties of the *Bacillus subtilis* spore coat. *J. Bacteriol.* **137:** 1208–1218.

Pedersen, L. B., T. Murray, D. L. Popham, and P. Setlow. 1998. Characterization of *dacC*, which encodes a new low-molecular-weight penicillin-binding protein in *Bacillus subtilis*. *J. Bacteriol.* **180:** 4967–4973.

Pedraza-Reyes, M., F. Gutierrez-Corona, and W. L. Nicholson. 1997. Spore photoproduct lyase operon (*splAB*) regulation during *Bacillus subtilis* sporulation: modulation of *splB-lacZ* fusion expression by P1 promoter mutations and by an in-frame deletion of *splA*. *Curr. Microbiol.* **34:**133–137.

Piggot, P. J., and J. G. Coote. 1976. Genetic aspects of bacterial endospore formation. *Bacteriol. Rev.* **40:** 908–962.

Pogliano, K., E. Harry, and R. Losick. 1995. Visualization of the subcellular location of sporulation proteins in *Bacillus subtilis* using immunofluorescence microscopy. *Mol. Microbiol.* **18:**459–470.

Pooley, H. M., and D. Karamata. 1994. Teichoic acid synthesis in *Bacillus subtilis*: genetic organization and biological roles, p. 187–198. *In* J.-M. Ghuysen and R. Hackenbeck (ed.), *Bacterial Cell Wall*. Elsevier, Amsterdam, The Netherlands.

Popham, D. L., and P. Setlow. 1993. Cloning, nucleotide sequence and regulation of the *Bacillus subtilis pbpE* operon, which codes for penicillin-binding protein 4★ and an apparent amino acid racemase. *J. Bacteriol.* **175:**2917–2925.

Popham, D. L., and P. Setlow. 1996. Phenotypes of *Bacillus subtilis* mutants lacking multiple class A high-molecular-weight penicillin-binding proteins. *J. Bacteriol.* **178:**2079–2085.

Popham, D. L., B. Illades-Aguiar, and P. Setlow. 1995a. The *Bacillus subtilis dacB* gene, encoding penicillin-binding protein 5★, is part of a three-gene operon required for proper spore cortex synthesis and spore core dehydration. *J. Bacteriol.* **177:** 4721–4729.

Popham, D. L., S. Sengupta, and P. Setlow. 1995b. Heat, hydrogen peroxide, and UV resistance of *Bacillus subtilis* spores with increased core water content and with or without major DNA binding proteins. *Appl. Environ. Microbiol.* **61:** 3633–3638.

Popham, D. L., J. Helin, C. E. Costello, and P. Setlow. 1996a. Muramic lactam in peptidoglycan of *Bacillus subtilis* spores is required for spore outgrowth but not for spore dehydration or heat resistance. *Proc. Natl. Acad. Sci. USA* **93:**15405–15410.

Popham, D. L., J. Helin, C. E. Costello, and P. Setlow. 1996b. Analysis of the peptidoglycan structure of *Bacillus subtilis* endospores. *J. Bacteriol.* **178:**6451–6458.

Popham, D. L., M. E. Gilmore, and P. Setlow. 1999. Analysis of the roles of low-molecular-weight penicillin-binding proteins in spore peptidoglycan synthesis and spore properties in *Bacillus subtilis*. *J. Bacteriol.* **181:**126–132.

Price, K. D., and R. Losick. 1999. A four-dimensional view of assembly of a morphogenetic protein during sporulation. *J. Bacteriol.* **181:**781–790.

Rebeil, R., Y. Sun, L. Chooback, M. Pedraza-Reyes, C. Kinsland, T. P. Begley, and W. L. Nicholson. 1998. Spore photoproduct lyase from *Bacillus subtilis* spores is a novel iron-sulfur DNA repair enzyme which shares features with proteins such as class III anaerobic ribonucleotide reductases and pyruvate-formate lyases. *J. Bacteriol.* **180:** 4879–4885.

Resnekov, O., A. Driks, and R. Losick. 1995. Identification and characterization of sporulation gene *spoVS* from *Bacillus subtilis. J. Bacteriol.* **177:** 5628–5635.

Roels, S., and R. Losick. 1995. Adjacent and divergently oriented operons under the control of the sporulation regulatory protein GerE in *Bacillus subtilis. J. Bacteriol.* **177:**6263–6275.

Roels, S., A. Driks, and R. Losick. 1992. Characterization of *spoIVA*, a sporulation gene involved in coat morphogenesis in *Bacillus subtilis. J. Bacteriol.* **174:**575–585.

Ross, M. S., and P. Setlow. 1998. Unpublished results.

Ross, M. S., N. M. Magill, and P. Setlow. 1998. Unpublished results.

Sacco, M., E. Ricca, R. Losick, and S. Cutting. 1995. An additional GerE-controlled gene encoding an abundant spore coat protein from *Bacillus subtilis. J. Bacteriol.* **177:**372–377.

Sanchez–Salas, J.-L., and P. Setlow. 1993. Proteolytic processing of the protease which initiates degradation of small, acid-soluble, proteins during germination of *Bacillus subtilis* spores. *J. Bacteriol.* **175:** 2568–2577.

Sandman, K., L. Kroos, S. Cutting, P. Youngman, and R. Losick. 1988. Identification of the promoter for a spore coat protein gene in *Bacillus subtilis* and studies on the regulation of its induction at a late stage of sporulation. *J. Mol. Biol.* **200:**461–473.

Sarma, M. H., G. Gupta, and R. H. Sarma. 1986. 500-MH$_2$ ^{1}H NMR study of poly(dG)·poly(dC) in solution using one-dimensional nuclear Overhauser effect. *Biochemistry* **25:**3659–3665.

Scherrer, R., T. C. Beaman, and P. Gerhardt. 1971. Macromolecular sieving by the dormant spore of *Bacillus cereus. J. Bacteriol.* **108:**868–873.

Sekiguchi, J., K. Akeo, H. Yamamoto, F. K. Khasanou, J. C. Alonso, and A. Kuroda. 1995. Nucleotide sequence and regulation of a new putative cell wall hydrolase gene, *cwlD*, which affects germination in *Bacillus subtilis. J. Bacteriol.* **177:** 5582–5589.

Setlow, B., and P. Setlow. 1979. Localization of low molecular weight basic proteins in *Bacillus megaterium* spores by irradiation with ultraviolet light. *J. Bacteriol.* **139:**486.

Setlow, B., and P. Setlow. 1987. Thymine containing dimers as well as spore photoproducts are found in ultraviolet-irradiated *Bacillus subtilis* spores that lack small acid-soluble proteins. *Proc. Natl. Acad. Sci. USA* **84:**421–423.

Setlow, B., and P. Setlow. 1993. Binding of small, acid-soluble spore proteins to DNA plays a significant role in the resistance of *Bacillus subtilis* spores to hydrogen peroxide. *Appl. Environ. Microbiol.* **59:** 3418–3423.

Setlow, B., and P. Setlow. 1994. Heat inactivation of *Bacillus subtilis* spores lacking small, acid-soluble spore proteins is accomplished by generation of abasic sites in spore DNA. *J. Bacteriol.* **176:** 2111–2113.

Setlow, B., and P. Setlow. 1995a. Binding to DNA protects α/β-type small, acid-soluble spore proteins of *Bacillus* and *Clostridium* species against digestion by their specific protease as well as other proteases. *J. Bacteriol.* **177:**4149–4151.

Setlow, B., and P. Setlow. 1995b. Small, acid-soluble proteins bound to DNA protect *Bacillus subtilis* spores from killing by dry heat. *Appl. Environ. Microbiol.* **61:**2787–2790.

Setlow, B., and P. Setlow. 1995c. Role of DNA repair in *Bacillus subtilis* spore resistance. *J. Bacteriol.* **178:**3486–3495.

Setlow, B., A. R. Hand, and P. Setlow. 1991a. Synthesis of a *Bacillus subtilis* small, acid-soluble spore protein in *Escherichia coli* causes cell DNA to assume some characteristics of spore DNA. *J. Bacteriol.* **173:**1642–1653.

Setlow, B., N. Magill, P. Febbroriello, L. Nakhimovsky, D. E. Koppel, and P. Setlow. 1991b. Condensation of the forespore nucleoid early in sporulation of *Bacillus* species. *J. Bacteriol.* **173:** 6270–6278.

Setlow, B., D. Sun, and P. Setlow. 1992. Studies of the interaction between DNA and α/β-type small, acid-soluble spore proteins: a new class of DNA binding protein. *J. Bacteriol.* **174:**2312–2322.

Setlow, B., C. A. Setlow, and P. Setlow. 1997. Killing bacterial spores by organic hydroperoxides. *J. Ind. Microbiol.* **18:**384–388.

Setlow, B., K. J. Tautvydas, and P. Setlow. 1998. Small, acid-soluble spore proteins of the α/β-type do not protect the DNA in *Bacillus subtilis* spores against base alkylation. *Appl. Environ. Microbiol.* **64:** 1958–1962.

Setlow, P. 1983. Germination and outgrowth, p. 211–254. *In* A. Hurst and G. W. Gould (ed.), *The Bacterial Spore*, vol. 2. Academic Press, London, England.

Setlow, P. 1988. Small acid-soluble, spore proteins of *Bacillus* species: structure, synthesis, genetics, function and degradation. *Annu. Rev. Microbiol.* **42:** 319–338.

Setlow, P. 1991. Changes in forespore chromosome structure during sporulation in *Bacillus* species. *Semin. Dev. Biol.* **2:**55–62.

Setlow, P. 1992a. DNA in dormant spores of *Bacillus*

species is in an A-like conformation. *Mol. Microbiol.* **6:**563–567.

Setlow, P. 1992b. I will survive: protecting and repairing spore DNA. *J. Bacteriol.* **174:**2737–2741.

Setlow, P. 1994. Mechanisms which contribute to the long-term survival of spores of *Bacillus* species. *J. Appl. Bacteriol.* **76:**49S–60S.

Setlow, P. 1995. Mechanisms for the prevention of damage to the DNA in spores of *Bacillus* species. *Annu. Rev. Microbiol.* **49:**29–54.

Seyler, R. W. J., A. O. Henriques, A. J. Ozin, and C. P. Moran, Jr. 1997. Assembly and interactions of *cotJ*-encoded proteins, constituents of the inner layers of the *Bacillus subtilis* spore coat. *Mol. Microbiol.* **25:**955–966.

Shioi, J.-I., S. Matsura, and Y. Imae. 1980. Quantitative measurements of proton motive force and motility in *Bacillus subtilis*. *J. Bacteriol.* **144:**891–897.

Simpson, F. B., T. W. Hancock, and C. E. Buchanan. 1974. Transcriptional control of *dacB*, which encodes a major sporulation-specific penicillin-binding protein. *J. Bacteriol.* **176:**7767–7769.

Sousa, J. C., M. T. Silva, and G. Balassa. 1976. An exosporium-like outer layer in *Bacillus subtilis* spores. *Nature* **263:**53–54.

Sousa, J. C., M. T. Silva, and G. Balassa. 1978. Ultrastructure and development of an exosporium-like outer spore envelope in *Bacillus subtilis*. *Ann. Microbiol.* (Paris) **129:**339–362.

Stevens, C. M., R. Daniel, N. Illing, and J. Errington. 1992. Characterization of a sporulation gene *spoIVA* involved in spore coat morphogenesis in *Bacillus subtilis*. *J. Bacteriol.* **174:**586–594.

Stewart, G. S. A. B., M. W. Eaton, K. Johnstone, M. D. Barrat, and D. J. Ellar. 1980. An investigation of membrane fluidity changes during sporulation and germination of *Bacillus megaterium* K. M. measured by electron spin and nuclear magnetic resonance spectroscopy. *Biochim. Biophys. Acta* **600:**270–290.

Strange, R. E., and F. A. Dark. 1956. The composition of the spore coats of *Bacillus megatherium*, *B. subtilis* and *B. cereus*. *Biochem. J.* **62:**459–465.

Sun, D., P. Stragier, and P. Setlow. 1989. Identification of a new σ-factor involved in compartmentalized gene expression during sporulation of *Bacillus subtilis*. *Genes Dev.* **3:**141–149.

Sun, D., P. Fajardo-Cavazos, M. D. Sussman, F. Tovar-Rojo, R.-M. Cabrera-Martinez, and P. Setlow. 1991. Analysis of the effect of chromosome location of *Bacillus subtilis* forespore specific genes on their *spo* gene dependence and transcription by Eσ^F: identification of features of good Eσ^F dependent promoters. *J. Bacteriol.* **173:**7867–7874.

Takamatsu, H., Y. Chikahiro, T. Kodama, H. Koide, S. Kozuka, K. Tochikubo, and K. Watabe. 1998. A spore coat protein, CotS, of *Bacillus subtilis* is synthesized under the regulation of sigmaK and GerE during development and is located in the inner coat layer of spores. *J. Bacteriol.* **180:**2968–2974.

Tebo, B. M., W. C. Ghiorse, L. G. van Waasbergen, P. L. Siering, and R. Caspi. 1997. Bacterially mediated mineral formation: insights into manganese(II) oxidation from molecular genetic and biochemical studies, p. 225–266. *In* K. Nealson (ed.), *Geomicrobiology.* Mineralogical Society of America, Washington, D.C.

Tipper, D. J., and P. E. Linnett. 1976. Distribution of peptidoglycan synthetase activities between sporangia and forespores in sporulating cells of *Bacillus sphaericus*. *J. Bacteriol.* **126:**213–221.

Todd, J. A., E. J. Bone, and D. J. Ellar. 1985. The sporulation-specific penicillin-binding protein 5a from *Bacillus subtilis* is a DD-carboxypeptidase. *Biochem. J.* **230:**825–828.

Tovar-Rojo, F., and P. Setlow. 1991. Analysis of the effects of mutant small, acid-soluble spore proteins from *Bacillus subtilis* on DNA in vivo and in vitro. *J. Bacteriol.* **173:**4827–4835.

van Waasbergen, L. G., J. A. Hoch, and B. M. Tebo. 1993. Genetic analysis of the marine manganese-oxidizing *Bacillus* sp. strain SG-1: protoplast transformation, Tn*917* mutagenesis, and identification of chromosomal loci involved in manganese oxidation. *J. Bacteriol.* **175:**7594–7603.

van Waasbergen, L. G., M. Hildebrand, and B. M. Tebo. 1996. Identification and characterization of a gene cluster involved in manganese oxidation by spores of the marine *Bacillus* sp. strain SG-1. *J. Bacteriol.* **178:**3517–3530.

Vinter, V. 1965. Spores of microorganisms. XVII. The fate of preexisting diaminopimelic acid-containing structures during germination and postgerminative development of bacterial spores. *Folia Microbiol.* **10:**280–287.

Warth, A. D. 1985. Mechanisms of heat resistance, p. 209–225. *In* G. J. Dring, D. J. Ellar, and G. W. Gould (ed.), *Fundamental and Applied Aspects of Bacterial Spores.* Academic Press Ltd., London, England.

Warth, A. D., and J. L. Strominger. 1969. Structure of the peptidoglycan of bacterial spores: occurrence of the lactam of muramic acid. *Proc. Natl. Acad. Sci. USA* **64:**528–535.

Warth, A. D., and J. L. Strominger. 1971. Structure of the peptidoglycan from vegetative cell walls of *Bacillus subtilis*. *Biochemistry* **10:**4349–4358.

Warth, A. D., and J. L. Strominger. 1972. Structure of the peptidoglycan from spores of *Bacillus subtilis*. *Biochemistry* **11:**1389–1396.

Warth, A. D., D. F. Ohye, and W. G. Murrell.

1963. The composition and structure of bacterial spores. *J. Cell. Biol.* **16**:579–592.

Wood, D. A. 1972. Sporulation in *Bacillus subtilis*. Properties and time of synthesis of alkali-soluble protein of the spore coat. *Biochem. J.* **130**:505–514.

Wu, J.-J., R. Schuch, and P. J. Piggot. 1992. Characterization of a *Bacillus subtilis* sporulation operon that includes genes for an RNA polymerase σ factor and for a putative DD-carboxypeptidase. *J. Bacteriol.* **174**:4885–4892.

Wu, L. J., and J. Errington. 1998. Use of asymmetric cell division and *spoIIIE* mutants to probe chromosome orientation and organization in *Bacillus subtilis*. *Mol. Microbiol.* **27**:777–786.

Zhang, J., P. C. Fitz-James, and A. I. Aronson. 1993. Cloning and characterization of a cluster of genes encoding polypeptides present in the insoluble fraction of the spore coat of *Bacillus subtilis*. *J. Bacteriol.* **175**:3757–3766.

Zhang, J., H. Ichikawa, R. Halberg, L. Kroos, and **A. I. Aronson.** 1994. Regulation of the transcription of a cluster of *Bacillus subtilis* spore coat genes. *J. Mol. Biol.* **240**:405–415.

Zhang, L., M. L. Higgins, and P. J. Piggot. 1997. The division during bacterial sporulation is symmetrically located in *Sporosarcina ureae*. *Mol. Microbiol.* **25**:1091–1098.

Zheng, L., and R. Losick. 1990. Cascade regulation of spore coat gene expression in *Bacillus subtilis*. *J. Mol. Biol.* **212**:645–660.

Zheng, L., W. P. Donovan, P. C. Fitz-James, and R. Losick. 1988. Gene encoding a morphogenic protein required in the assembly of the outer coat of the *Bacillus subtilis* endospore. *Genes Dev.* **2**:1047–1054.

Zheng, L., R. Halberg, S. Roels, H. Ichikawa, L. Kroos, and R. Losick. 1992. Sporulation regulatory protein GerE from *Bacillus subtilis* binds to and can activate or repress transcription from promoters for mother-cell-specific genes. *J. Mol. Biol.* **226**:1037–1050.

MYXOBACTERIA

INTRODUCTION TO THE
MYXOBACTERIA

Martin Dworkin

10

GENERAL ASPECTS OF MYXOBACTERIAL BIOLOGY

Introduction

The myxobacteria are the epitome of prokaryotic multicellular complexity. Their beautiful macroscopic fruiting bodies have attracted the attention of microbiologists ever since Roland Thaxter first described them as bacteria in 1892. They have been referred to as "social prokaryotes" (Reichenbach, 1984), and over the past few decades, interest has indeed centered on the nature of those cell-cell interactions that make such behavior possible. In particular, efforts have focused on understanding the nature, function, and regulation of the signals that play a role in the characteristic social behavior of the myxobacteria. In this context, most experimental attention has been on one species, *Myxococcus xanthus* (see chapters 11, 12, and 13), and to a lesser extent on *Stigmatella aurantiaca* (see chapter 14).

Despite the considerable amount of information that has been obtained over the past few decades, there are many important and intriguing questions that remain unanswered or unexplored. The mechanism of the enigmatic gliding motility of myxobacteria has eluded understanding for many years, many aspects of the physiology and biochemistry of *M. xanthus* remain unexamined, essentially nothing is known about the ecological role of the myxobacteria, the general principles underlying the morphogenesis of their multicellular fruiting bodies are unknown, and the physiology and developmental behavior of the other 10 or so genera remain terra incognita. The goal of this introductory chapter is to set the stage for a better understanding of the four chapters that follow.

Life Cycle

The life cycle of *M. xanthus* is diagrammatically represented in Fig. 1. It consists of two separate but connected stages: feeding, growth, and division on the one hand and development on the other. When the cells are provided with nutrient and appropriate growth conditions, in liquid or solid media, the cells grow and divide with an optimal generation time of about 3.5 h. The cells shift to a developmental mode if three conditions are met. First, they must perceive a nutritional shiftdown. Second, the cells must be on a solid surface to allow their characteristic gliding motility. Finally, the cells must be at a high cell density (Wireman and Dworkin, 1975; Shimkets and Dworkin, 1981).

The cells move into aggregation centers fol-

M. Dworkin, Department of Microbiology, University of Minnesota, Minneapolis, MN 55455-0312.

Prokaryotic Development, edited by Y. V. Brun and L. J. Shimkets,
© 2000 American Society for Microbiology, Washington, DC 20005-4171

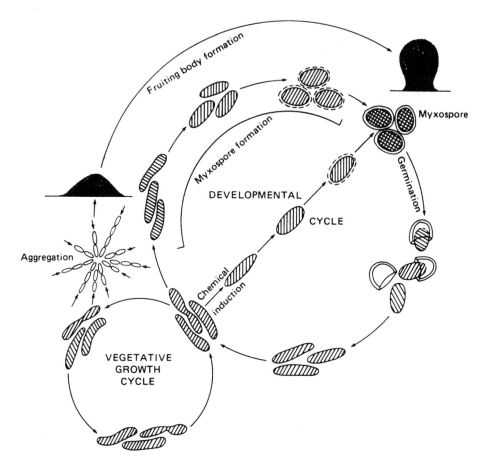

FIGURE 1 Diagram of the life cycle of *Myxococcus xanthus* (Dworkin, 1986). The fruiting body is not drawn to scale but is a few hundredths of a millimeter in diameter. The vegetative cells are about 5 to 7 by 0.7 μm.

lowing clues that are not yet well defined (but see below). The process of aggregation is correlated with a massive lysis of the population (Wireman and Dworkin, 1975, 1977). As many as 65 to 90% of the cells lyse, with the remaining cells converting to myxospores within the fruiting body.

Within the aggregate, the vegetative cells convert to myxospores, which are metabolically quiescent (Dworkin and Niederpruem, 1964), somewhat resistant (Sudo and Dworkin, 1969) resting cells. Under the appropriate conditions, myxospores will germinate to give rise to the motile, metabolically active vegetative cells.

The Fruiting Bodies of the Myxobacteria

The fruiting body is the culmination of myxobacterial development. We presume that it represents the housing for the resting cells, called myxospores. But does it have a function beyond that? Why is it so brightly pigmented? How does one understand the spectrum of morphological complexity, varying from the single, tiny, simple sporangium of *Nannocystis* to the macroscopic complexity of *Chondromyces*?

Figure 2 is a mosaic illustrating the spectrum of complexity of myxobacterial fruiting bodies. The simplest is represented by *Nannocystis exe-*

dens (Fig. 2A), consisting of a single small sporangiole (the sac containing the myxospores). *Sorangium cellulosum*, a cellulose-degrading myxobacterium, produces fruiting bodies that consist of tiny, polyhedral sporangioles, tightly packed together (Fig. 2B). The fruiting bodies of *Cystobacter fuscus* (Fig. 2C) consist of clusters of large, chain-like sporangioles. The genus *Myxococcus* is characterized by rather simple fruiting bodies, consisting of myxospores encased in an undifferentiated matrix of slime. This is particularly the case for species such as *Myxococcus fulvus* (Fig. 2D), but less so for *Myxococcus stiptatus* (Fig. 2E), where the slime-encased ball of myxospores is elevated from the surface by an acellular stalk. Somewhat similar is the genus *Melittangium* (Fig. 2F), characterized by a tiny sporangiole sitting atop a delicate white stalk. The most complex and beautiful of the myxobacterial fruiting bodies are formed by the genera *Stigmatella* (Fig. 2G) and *Chondromyces* (Fig. 2H). In both cases multiple sporangioles are borne on branched or unbranched stalks.

While there are no data or experiments that bear directly on the question of the function of the fruiting body, the density-dependent nature of myxobacterial growth may provide a useful clue. Myxobacteria in general seem to prefer to feed on macromolecules or cells by excreting hydrolytic enzymes (Bender, 1963; Sudo and Dworkin, 1972). The feeding habits of *M. xanthus* have been examined in some detail, and it has been shown that growth on casein requires that the cells be present at a high density. Rosenberg et al. (1977) have shown that this is a reflection of the need to pool the excreted hydrolytic enzymes, a behavior that had been termed the "wolf pack effect" (Dworkin, 1973).

It is possible that the fruiting body provides packaging of the resting myxospores so that upon germination, the ensuing vegetative cell population is poised to begin feeding at an optimal cell concentration and density. The time-lapse films of Reichenbach et al. (1965) show quite clearly that a single germinating sporangiole of *Chondromyces apiculatus* is able to form

an active swarm. The peculiar ecological habitat of each myxobacterium may dictate a particular myxospore-packaging requirement and thus a characteristic fruiting body morphology.

Myxospores

The myxospores are the resting cells of the myxobacteria and are found within the fruiting body.

The myxobacteria are divided into two taxonomic groups—the suborders *Cystobacterineae* and *Sorangineae*. The *Sorangineae* are characterized by short, blunt-ended vegetative cells (Fig. 3) and by myxospores whose morphology is similar to that of the corresponding vegetative cells. The vegetative cells of the *Cystobacterineae* are slender rods with pointed ends (Fig. 4); their myxospores vary from the slightly shortened cells of *Cystobacter* to the refractile ovoids of *Stigmatella* and the round, optically refractile myxospores of *Myxococcus* (Fig. 5).

Only in the case of *M. xanthus* have the myxospores been subjected to a careful description. Thus, we can presume the properties of other myxospores only by analogy. The myxospores of *M. xanthus* are metabolically quiescent (Dworkin and Niederpruem, 1964) and resistant to slightly elevated temperatures, e.g., 55°C (Sudo and Dworkin, 1969), and to sonication and desiccation.

As part of an early attempt to study a portion of the developmental cycle of *M. xanthus* under experimentally well-defined conditions, Dworkin and Gibson (1964) described a process whereby the addition of 0.5 M glycerol resulted in the rapid conversion of vegetative cells of *M. xanthus* to round, optically refractile myxospore-like cells. These cells had some, but not all, of the properties of fruiting body myxospores, e.g., they were mophologically similar, somewhat resistant, and metabolically quiescent, and they could be germinated to form the typical vegetative cells (Sadler and Dworkin, 1966a, 1996b).

Myxobacterial Phylogeny

The myxobacteria have been taxonomically divided into two suborders, the *Cystobacterineae*

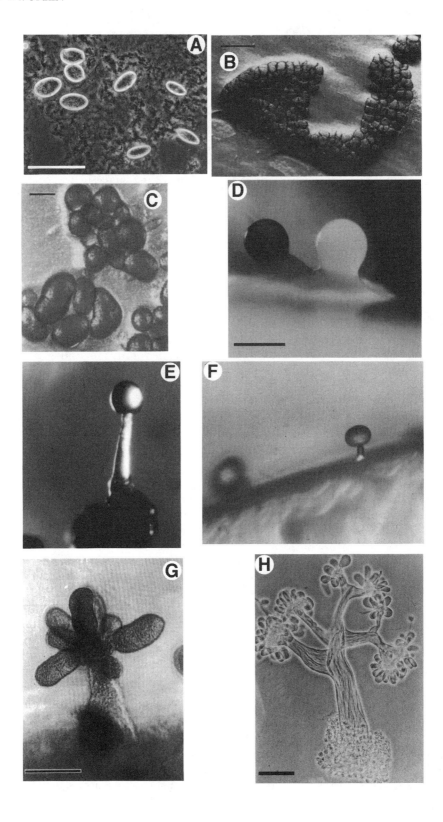

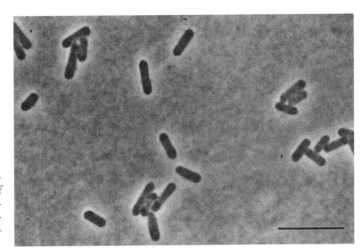

FIGURE 3 Phase-contrast photomicrograph of vegetative cells of *S. cellulosum*, illustrative of the cellular morphology of the suborder *Sorangineae*. Bar, 10 μm. (From Reichenbach, 1993.)

and the *Sorangineae*, based on such phenotypic properties as cell, myxospore, and fruiting body morphologies and fatty acid composition (Reichenbach and Dworkin, 1992). This division has been confirmed and somewhat extended by the molecular phylogenetic analyses of 16S rRNA, which have placed the myxobacteria as a phylogenetically coherent group clustered within the δ subgroup of the proteobacteria (Ludwig et al., 1983; Shimkets and Woese, 1992; Stackebrandt, 1998). The first such analysis (Ludwig et al., 1983) examined five genera of myxobacteria and confirmed that they emerged from one line of descent, that they were closely allied with the proteobacteria, and that the two taxonomic suborders were also distinguished from each other phylogenetically. The second analysis (Shimkets and Woese, 1992) examined 10 genera and 12 species; determined that the genus *Nannocystis* represented a third, separate phylogenetic group; confirmed that the myxobacteria were a phylogenetically coherent group; and showed that the closest relatives among the proteobacteria

were the predatory bacterium *Bdellovibrio* and two anaerobic genera, members of the sulfate-reducing bacteria. The third analysis (Stackebrandt, 1998) examined 67 species of myxobacteria representing nine genera and confirmed that they clustered into three subgroups. Some of these findings are summarized in Fig. 6. Using a different approach, Rice and Lampson (1995) examined 28 species of myxobacteria, using the nature of their retrons and of the genes coding for them as indicators of evolutionary relatedness. Retrons are chromosomal genes coding for unusual multicopy, single-stranded DNA (Temin, 1989). The retron also contains an open reading frame with significant similarity to retroviral reverse transcriptase (Inouye et al., 1989). Retrons have also been found in a number of strains of *Escherichia coli* (Herzer et al., 1990) as well as a few distantly related gram-negative bacteria (Rice et al., 1993). Retrons were found in 27 of the 28 strains of myxobacteria examined. Using the degree of nucleotide sequence similarity, the codon bias of the reverse transcriptase genes,

FIGURE 2 Fruiting bodies of myxobacteria. (A) *N. exedens* (bar, 50 μm); (B) *S. cellulosum* (bar, 100 μm) (Reichenbach, 1993); (C) *C. fuscus* (bar, 100 μm) (Reichenbach, 1993); (D) *M. fulvus* (bar, 100 μm) (Dworkin, 1996); (E) *M. stipitatus* (fruiting body is about 170 μm high); (F) *Mellitangium* sp. strain Hp (fruiting body is about 40 μm high); (G) *S. aurantiaca* (bar, 50 μm); (H) *Chondromyces crocatus* (bar, 100 μm). (All photos taken by Hans Reichenbach.)

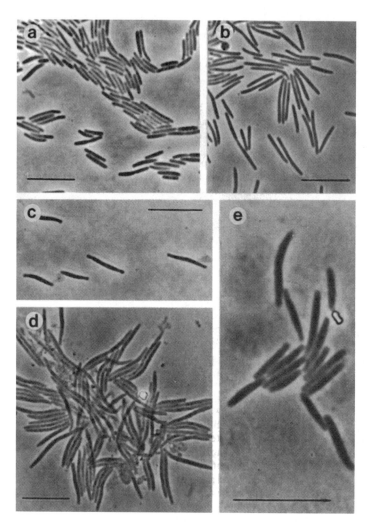

FIGURE 4 Phase-contrast photomicrographs of vegetative cells of the suborder *Cystobacterineae*. (a) *Corallococcus coralloides*; (b) *M. virescens*; (c) *S. aurantiaca*; (d) *C. fuscus*; (e) *Cystobacter velatus*. Bar, 10 μm. (From Reichenbach, 1993.)

and the distribution of the retrons as parameters, the authors concluded that a common ancestor of the myxobacteria acquired the retroelement (Rice and Lampson, 1995).

Myxobacterial Ecology

The study of myxobacterial ecology comprises the following relevant questions. In what habitats do the myxobacteria truly reside? What roles do they play in these habitats? What is the relation between environmental circumstances and their life cycles? How do myxobacteria interact with each other?

There are considerable data that bear on the first question. The last question has been addressed by one study. The remaining questions are unexplored.

Reichenbach and Dworkin (1992) have comprehensively reviewed the literature on the habitats of the myxobacteria. They are generally terrestrial aerobes that grow under mesophilic conditions at essentially neutral pH and normal salt concentrations. However, halophilic myxobacteria have recently been isolated from a marine environment (Iizuka et al., 1998), suggesting that they may be more broadly distributed than hitherto recognized. While they have been isolated from even more

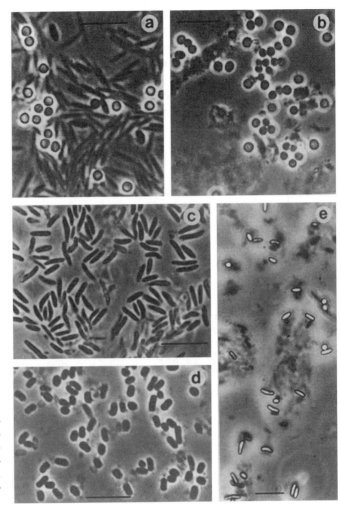

FIGURE 5 Phase-contrast photomicrographs of myxospores of the suborder *Cystobacterineae*. (a) *M. xanthus*; (b) *Angiococcus disciformis*; (c) *Cystobacter ferrugineus*; (d) *Cystobacter velatus*; (e) *S. aurantiaca*. Bar, 10 μm. (From Reichenbach, 1993.)

extreme environments, this is probably a reflection of their ability to form resistant resting cells and to persist under conditions that do not support growth.

While there is little evidence that bears directly on the question of the natural roles of the myxobacteria, what we know about their growth and physiology leads to some reasonable inferences. Every myxobacterium that has been examined has the ability to hydrolyze one or many kinds of macromolecules, including protein (Bender, 1963), polysaccharide (Barreaud et al., 1995; Bourgerie et al., 1994), peptidoglycan (Sudo and Dworkin, 1972), cellu-

lose (Couche, 1969), nucleic acids (Morris and Parish, 1976; Mayer and Reichenbach, 1978), and lipids (Sørhaug, 1974). Their ability to kill and lyse a wide variety of cells has been well documented (Pinoy, 1921; Singh, 1947; Rosenberg and Varon, 1984). Burnham et al. (1981, 1984) have shown that *M. xanthus* and *Myxococcus virescens* are able to form stable cocultures with the cyanobacterium *Phormidium luridum* and, by acting as predators, exert a long-term population control over the cyanobacterium. (For a general review of the hydrolytic and bacteriolytic activities of the myxobacteria, see Rosenberg and Varon, 1984.)

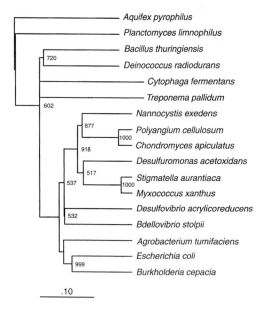

Aquifex pyrophilus
Planctomyces limnophilus
Bacillus thuringiensis
720
Deinococcus radiodurans
Cytophaga fermentans
Treponema pallidum
602
Nannocystis exedens
877
Polyangium cellulosum
1000
Chondromyces apiculatus
918
Desulfuromonas acetoxidans
517
Stigmatella aurantiaca
537
1000
Myxococcus xanthus
Desulfovibrio acrylicoreducens
532
Bdellovibrio stolpii
Agrobacterium tumifaciens
Escherichia coli
999
Burkholderia cepacia

.10

FIGURE 6 Phylogenetic tree showing the position of the myxobacteria within the δ group of the *Proteobacteria* and other major lines of radiation within the domain *Bacteria*. The tree was constructed using the neighbor-joining algorithm from a matrix of pairwise genetic distances as calculated by the Kimura two-parameter method. A total of 1,321 aligned positions, corresponding to *E. coli* positions 125 to 1446, were used in the analysis. *Aquifex pyrophilus*, a member of the domain *Bacteria*, was used as the outgroup. The scale bar represents 0.10 substitutions per base position. The numbers at the nodes of the tree indicate the number of times the group consisting of the species listed to the right of that fork occurred among 1,000 bootstrapped resamplings (values below 500 are not shown). (Courtesy of Mark Wise and Larry Shimkets.)

Thus, the myxobacteria can lyse and grow on living or dead gram-positive and gram-negative bacteria, fungi, yeasts, protozoa, and probably a variety of cells that have not yet been examined.

As a group, the myxobacteria produce a wide spectrum of antibiotics (Reichenbach and Höfle, 1993), and it has been suggested that these antibiotics may play a role in the ability of the myxobacteria to prey on other living cells (Rosenberg and Varon, 1984; Rosenberg and Dworkin, 1996). Given that myxobacteria have been reported at levels of up to 10^5/g of soil (Singh, 1947) it seems reasonable to propose that they may play a significant role in clearing soil and perhaps other habitats of macromolecular debris. Gliding motility may be the most appropriate mechanism for a cell that moves over and among particulate debris rather than feeding on soluble materials in a liquid environment.

How does the ability of the myxobacteria to undergo a complex life cycle relate to their roles in Nature? Or put another way, what is the relationship between environmental circumstances and the life cycle of the myxobacteria?

There are three aspects of the myxobacterial life cycle that can be rationalized in terms of their ecological role. These are their social behavior, the formation of multicellular fruiting bodies, and the resistance of myxospores.

The dominant feature of the myxobacteria is their social behavior—their inclination to remain together as groups of cells. The most explicit example of this is the social motility of *M. xanthus*, namely, the tendency of the cells to move as a swarm despite the fact that they can move as individuals (Hodgkin and Kaiser, 1979). The characteristic swarms of the myxobacteria are illustrated in Fig. 7. It has been proposed that this behavior optimizes feeding that is dependent on the excretion of hydrolytic enzymes. In view of the fact that such feeding is at the mercy of diffusion—of hydrolytic enzymes away from the cell and of products of hydrolysis toward the cell and away from the original substrate—a high cell density will increase the efficiency of this mode of feeding. This has been termed the myxobacterial wolf-pack effect (Dworkin, 1973) and has been experimentally verified by Rosenberg et al. (1977), who showed that the ability of *M. xanthus* to grow on casein was indeed cell density dependent.

One of the functions of the fruiting body may be rationalized in a similar fashion. The fruiting body may be viewed as a device whereby the cells, prior to entering the resting stage, aggregate and are concentrated at a high cell density. The particular organization of the

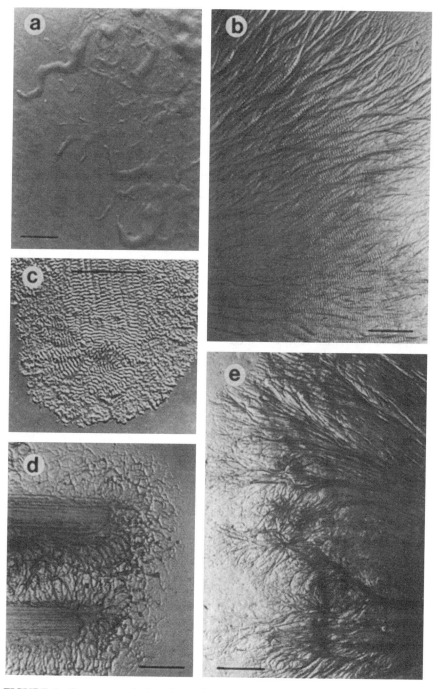

FIGURE 7 Swarming colonies of myxobacteria of the suborder *Cystobacterineae* on agar surfaces. (a) *Corallococcus coralloides*; (b) *Archangium gephyra*; (c) *M. fulvus*; (d) *Corallococcus coralloides*; (e) *Cystobacter violaceus*. Bars, 100 μm (panel a) and 2,000 μm (all others). Note the ripples in panels b and c. (From Reichenbach, 1993.)

sporangioles that is characteristic for each genus may simply reflect the optimal packaging of the myxospores so as to create the most appropriate number and cell density for that particular organism in its characteristic niche. Thus, they are poised, so that upon germination they may immediately be present at a high cell density appropriate for beginning the feeding process.

Finally, the ability of the myxobacteria to form resistant, metabolically quiescent resting cells is of obvious benefit. The myxospores of *M. xanthus* are substantially more resistant to high (55°C) temperature, UV light irradiation, and desiccation than are the corresponding vegetative cells (Sudo and Dworkin, 1969). While the heat resistance of the myxospores is far less dramatic than that of the endospores of *Bacillus*, it is more than sufficient to allow the cells to weather the normal extremes they might encounter. The desiccation resistance is sufficient to have allowed isolation of myxobacteria from soil samples that had been stored at room temperature for as long as 15 years (Reichenbach, 1993).

The relationship between environmental circumstances and the life cycle of the myxobacteria can only be imagined. We know that in the laboratory, a nutritional shiftdown is a necessary trigger for the conversion from growth to development in *M. xanthus*. But what does that mean in the context of a soil particle or a piece of bark? What are the roles of temperature, light-dark cycles, moisture, or any other of the vicissitudes of Nature in the delicate balance between growth and development in the myxobacteria? When and where do the myxobacteria exist as vegetative cells, myxospores, or fruiting bodies? Reichenbach (1993) has written a detailed description of the data available on the distribution of myxobacteria in Nature.

As social organisms, the myxobacteria offer an excellent and intriguing opportunity to ask questions about how they interact with each other under natural conditions. Experiments bearing on this question were stimulated by the common observation that when fruiting body formation is observed on a natural sample, one type of fruiting body invariably dominates the area. Will myxobacteria of different genera form joint aggregates and hybrid fruiting bodies, or are they able to distinguish self from nonself? If they are able to make this distinction, what is the basis of the recognition? How specific is the recognition? Will different but related species mix? What about different strains of the same species? Will they compete with each other? How is territoriality established? Smith and Dworkin (1994) began an investigation of this area by examining the interactions of two closely related species of *Myxococcus*, *M. xanthus* and *M. virescens*. *M. xanthus* was marked with a transposon carrying a kanamycin resistance gene to allow its convenient distinction from *M. virescens*; the two species were mixed at various input ratios and plated under conditions allowing fruiting body formation, and the distribution and composition of the resulting fruiting bodies were determined. The results indicated that the two species did not form hybrid fruiting bodies, that each established separate and delimited fruiting territories, and that they did so by means of an excreted bacteriocin-like material. *M. virescens* competed for territory more effectively than did *M. xanthus*, consistent with its more common isolation from natural materials.

DEVELOPMENTAL ASPECTS OF MYXOBACTERIAL BIOLOGY

Fruiting Body Formation

In the myxobacteria, fruiting body formation and growth are alternative modes and exist in balance with each other. When continued growth is made unlikely by the exhaustion of nutrients, or perhaps by other environmental stresses, the balance shifts towards development.

Aggregation

Reference has been made to the three conditions that must be satisfied for development to occur. These are exhaustion of nutrients, the presence of a solid surface, and a high cell density. The nutrition of *M. xanthus* is essentially amino acid based, and the reduction of any of the required amino acids induces a stringent response, mediated by ReIA and guanosine 3'-di(tri)phosphate 5'-diphosphate [(p)ppGpp] (Singer and Kaiser, 1995). The link between this response and the initiation of aggregation has not yet been characterized.

The presence of a solid surface allows the cells to manifest their characteristic gliding motility. The myxobacteria share the ability to move by gliding across a solid surface with many other prokaryotes (Burchard, 1984). While there has been a considerable amount of work on the genetics of gliding motility in *M. xanthus* (Hartzell and Youderian, 1995), the actual mechanism of gliding remains a mystery. A number of mechanisms have been proposed, varying from the mechanical (Pate and Chang, 1979) to the structural (Freese et al., 1997; Lünsdorf and Reichenbach, 1989) and the biophysical (Keller et al., 1983); none of them has managed to persuade workers in the field.

This movement allows the cells not only to glide into their aggregation centers, following a series of clues that have not yet been defined, but also to align themselves for the efficient exchange of the C-signal, one of the intercellular signals exchanged during development (Kim and Kaiser, 1990a, 1990b).

Unlike their eukaryotic counterparts, the cellular slime molds, with whom they share a number of behavioral properties, aggregation in *M. xanthus* requires that the cells literally be piled atop each other. Aggregation is strictly density-dependent and will not occur unless the cells are at a particular cell density threshold (Wireman and Dworkin, 1975; Shimkets and Dworkin, 1981). There is evidence that the perception and measurement of cell density is accomplished by monitoring of the earliest signal exchanged between the cells, the A-signal (Kaplan and Plamann, 1996), and probably also by contact interactions mediated by extracellular appendages, such as pili (Hodgkin and Kaiser, 1977) and fibrils (Behmlander and Dworkin, 1991).

Evidence from a number of directions (reviewed in Hartzell and Youderian, 1995) has shown convincingly that pili are necessary for the social motility of *M. xanthus*. In addition, end-to-end attachment, presumably mediated by the polar pili, seems to be necessary for the proper exchange of the C-signal (Søgaard-Andersen et al., 1997).

Other extracellular, filamentous appendages called fibrils have been described for *M. xanthus* (Arnold and Shimkets, 1988; Behmlander and Dworkin, 1991). These are described in more detail below.

The nature of the directional clues that lead myxobacteria into their aggregation centers has not been determined. It was tempting, at an earlier stage of myxobacterial studies, to assume that the myxobacteria, by analogy with the cellular slime molds, were responding chemotactically to a self-generated gradient of attractant, similar to the cyclic AMP relaying demonstrated for *Dictystelium discoideum* (Konijn et al., 1967), and a number of early studies claimed to have demonstrated such developmental chemotaxis (e.g., McVittie and Zahler, 1962; Fluegel, 1963a). The interpretations of these studies were equivocal, and while chemotaxis has recently been demonstrated in *M. xanthus* (Kearns and Shimkets, 1998), it remains to be seen if it plays a role in feeding, aggregation, or both. It has been shown that the directional movement of the cells is controlled by the frequency with which the cells reverse their direction (Shi et al., 1995). Furthermore, this has been shown to respond to the exchange of one of the signals, Csg (Sager and Kaiser, 1993a; Søgaard-Andersen et al., 1997). Finally, *M. xanthus* has been shown to respond tactily to the presence of physical objects (Dworkin, 1983). However, how all these factors come together to result in aggregation is not yet understood.

Developmental Autolysis

During an attempt to quantitate cell numbers and types during aggregation and fruiting body formation, Wireman and Dworkin (1977) noticed a dramatic discrepancy between the numbers of vegetative cells at the beginning of the process and the number of myxospores finally present in the fruiting body. A careful accounting, based on both direct cell counts and the release of tritiated thymidine from radiolabeled DNA, indicated that there is a massive autolysis during aggregation, resulting in the loss of 65 to 90% of the vegetative cells. The remaining cells convert to myxospores and end up in the fruiting body. These experiments were repeated with two other species of *Myxococcus*, *M. fulvus* and *M. virescens*, with essentially the same results.

Using a number of other experimental aproaches, a series of experiments have been done, all of which are consistent with the notion that autolysis is necessary for complete development and is not an experimental artifact (reviewed in Dworkin, 1996). It is not clear if the function of developmental autolysis is to provide additional nutrients to allow the completion of development or to externalize some sort of paracrine signal. In any case, while it appears to be an interesting example of myxobacterial apoptosis, until it is possible to specifically block autolysis or to look at the behavior of an autolysis-deficient mutant, a compelling cause and effect relationship cannot be made.

Fruiting Body Morphogenesis

Understanding fruiting body morphogenesis is among the most difficult and challenging aspects of myxobacterial biology. Nevertheless, there have been two efforts in this direction, one from White's laboratory, attempting to understand the role of cell movement in the formation of fruiting bodies of *S. aurantiaca* (Vasquez et al., 1985) and the other in Kaiser's laboratory, using reporter transposons and fluorescent labels to describe fruiting body formation in *M. xanthus* (Sager and Kaiser, 1993a, 1993b).

White has proposed that, in *S. aurantiaca*, aggregates form when cells are locked into a turning mode and become trapped in a circular cul-de-sac. The aggregate heaps up as the cells continue to enter it and spiral over each other. The stalk is formed when cells begin to move under the aggregate, lifting the mass of cells above the surface (White, 1993). The notion that movement in concentric circles is the basis

of the morphogenetic process is based on the microscopic observations that the cells indeed move in concentric circular paths and become trapped in these circles (Vasquez et al., 1985; White, 1993; O'Connor and Zusman, 1989).

Using confocal fluorescence microscopy, Sager and Kaiser showed that the hemispherical fruiting body of *M. xanthus* consists of an outer, densely packed layer, within which vegetative rods are orbiting the fruiting body throughout development, and a less densely packed, concentric inner core containing the nonmotile myxospores. The moving cells in the outer shell have abandoned the back-and-forth movement that is characteristic of randomly gliding cells and instead are circling the fruiting body in a clockwise or counterclockwise direction (Sager and Kaiser, 1993a).

Sager and Kaiser also used transcriptional *lacZ* fusions as reporters for a number of developmental genes to measure their differential expression in time and space during fruiting body formation. Their results led them to propose that C-signaling in the densely packed outer domain induces myxospore formation; the nonmotile myxospores, or their immature precursors, are then passively driven into the less densely packed inner domain by the circling movements of the remaining undifferentiated cells in the outer hemisphere (Sager and Kaiser, 1993b).

The fundamental question of how the one-dimensional nanoscale of genetic information is translated into the macroscopic, three-dimensional structure of the fruiting body reflects a dilemma characteristic of biology as a whole. In the area of morphogenesis, or, literally, shape generation and transmission, there is an absence of the sorts of underlying themes that have yielded explanatory principles in other areas of biology, such as metabolism, regulation, and genetics. The experimental tractability of the myxobacteria would seem to offer a most attractive model for arriving at such a conceptual framework.

Cellular Morphogenesis

An early, important development in modern myxobacterial research was the discovery that it was possible to induce the conversion of vegetative rods of *M. xanthus* to myxospores by the addition of high concentrations of glycerol (Dworkin and Gibson, 1964) or any one of a number of other related compounds (Sadler and Dworkin, 1966a). Subsequently, it was shown that one could, in a similar fashion, induce myxospore formation in *S. aurantiaca* (Reichenbach and Dworkin, 1970; Gerth and Reichenbach, 1978). The availability of such a convenient experimental system for inducing development in a rapid, synchronous, and essentially complete fashion generated a considerable amount of information as to the biochemistry and physiology of the process (reviewed in Zusman, 1984). However, it turned out that myxospores induced with glycerol were not identical with those formed during fruiting body formation. While they shared many properties and, morphologically, underwent the same sequence of events in their formation and germination, it became clear that a number of the regulatory steps leading to myxospore formation had been bypassed and some of the structural features characteristic of fruiting body myxospores were absent from glycerol spores (Zusman, 1984). Nevertheless, they represent a useful system for examining the molecular and biochemical bases of shape change and for looking at the regulation of some of the sporulation genes, whose developmental expression they share with fruiting body myxospores. Most of the studies directed toward the properties of fruiting body myxospores have focused on various cell surface proteins (Dworkin, 1993). The most intensely studied of these is protein S, a 19-kDa major protein of the outer spore coat that contains two high-affinity Ca^{2+}-binding domains and shares sequences both with eukaryotic calmodulin (Inouye et al., 1989) and with bovine crystalline lens protein (Wistow et al., 1985).

Cell-Cell Interactions

Bacteria have traditionally been thought of as exemplifying the concept of the unicellular organism. On the contrary, we are now increasingly aware that since most bacteria exist in populations, they are almost certainly interacting among themselves and with other populations (Shapiro and Dworkin, 1997). Because of their evident multicellularity, the myxobacteria, among the prokaryotes, have become a paradigm for studying the mechanisms through which cells communicate with each other. There are two modes by which this communication may take place: one is by means of the exchange of chemical signals between cells, and the other is by means of physical contact. Five categories of chemical signals, designated A through E, have been identified (Kaiser and Kroos, 1993). The most completely characterized are the A- and C-signals. The five signals are as follows.

A-SIGNAL

This is the earliest of the five signals that play a role in coordinating fruiting body and myxospore development. It is produced when the cells perceive a nutritional shiftdown, and it then functions as a monitor of cell density. The signal itself is a mixture of amino acids and peptides generated by the action of cellular proteases on one or more of the cell's proteins (reviewed in Kaplan and Plamann, 1996). Figure 8 diagrammatically illustrates the A-signaling circuit.

B-SIGNAL

The B-signal acts early in development; however, the actual B-signal itself has not been isolated or characterized. The only gene known to be involved in B-signaling, *bsgA*, has been cloned and sequenced and shown to have substantial homology to the *lon* genes of *E. coli* and *Bacillus subtilis* (Gill et al., 1993). It is a constitutive vegetative gene that encodes a 90.4-kDa protein which copurifies with an intracellular ATP-dependent protease.

C-SIGNAL

The C-signal has been the most extensively characterized of the five signals. It is closely associated with the cell surface (Kim and Kaiser, 1990c; Shimkets and Rafiee, 1990) and is the last acting of the five signals (Kroos and

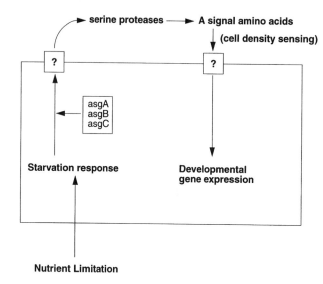

FIGURE 8 Diagram of the A signaling circuit (Dworkin, 1996). The question marks reflect the lack of information as to the nature of the secretory mechanism for the protease and the receptors for the signal amino acids.

Kaiser, 1987; Li et al., 1992). Mutants defective in C-signaling are unable to ripple, to aggregate, to sporulate, or to express many of the developmental genes. The *csgA* gene seems to code for both a 17- and a 25-kDa protein, both of which have been isolated and characterized and shown to function as the C-signal (Kim and Kaiser, 1990c; Lee et al., 1995). The relationship between the two proteins is not clear. Lee et al. (1995) have shown that the 25-kDa gene product of the *csgA* gene is a member of the short-chain alcohol dehydrogenase family, and Shimkets and Crawford (1998) have suggested that it functions somehow to maintain the nongrowing state of the cells during development. Motility is necessary for the exchange of the C-signal (Kroos et al., 1988), suggesting that it also functions to monitor the close spatial interactions among the cells during development (Kim and Kaiser, 1990d).

D-SIGNAL

The D-signal is one of the most enigmatic of the five signals. While single missense mutations in the *dsg* gene only partially disrupt development and have no effect on vegetative growth, Tn*5* insertion mutations are lethal (Cheng and Kaiser, 1989a). The *dsg* gene has been cloned and sequenced (Cheng and Kaiser,

1989b) and has a 50% sequence homology to the translation initiation factor IF3 of *E. coli* (Cheng et al., 1994). Furthermore, it is difficult to see the relationship of these facts to the finding by Rosenbluh and Rosenberg (1989) that mixtures of fatty acids comprising the autocide AM1 (Varon et al., 1986) rescue development in the *dsg* mutant. This may be a bypass phenomenon rather than a bona fide signaling event.

E-SIGNAL

Toal et al. (1995) have shown that the *esg* gene codes for the E1 decarboxylase of the branched-chain keto-acid dehydrogenase, and Downard and Toal (1995) have proposed that long, branched-chain fatty acids are the E-signal itself. They have suggested that early in development a specific phospholipase activity is initiated, which liberates branched-chain fatty acids that serve as short-range signals between the cells. Exchange of the E-signal requires cell-cell contact, and Ramaswamy et al. (1997) have implied that this contact may be mediated by the extracellular fibrils on the cell.

Table 1 summarizes the current state of information about the five signals of *M. xanthus*.

PILI

Pili were first noted in myxobacteria by MacRae et al. (1975, 1977) as fimbriae. In most

TABLE 1 Developmental signals in *M. xanthus*

Signaling group	Complementing fraction	Reference for group	Gene	Gene product	Demonstrated	Inferred from homology	Reference for gene product
A	Mixture of amino acids and peptides generated by a mixture of proteases	Kroos et al., 1990; Kuspa et al., 1992a, 1992b	*asgA*	Transmitter domain of histidine protein kinase	✓	✓	Plamann et al., 1995; Li and Plamann, 1996
				Receiver domain of response regulator		✓	Plamann et al., 1995
			asgB	DNA-binding transcription factor		✓	Plamann et al., 1994
			asgC	Sigma 70		✓	Davis et al., 1995
B	Unknown		*bsgA*	90.4-kDa ATP-dependent protease	✓	✓	Gill et al., 1993; Tojo et al., 1993
C	CsgA protein	Kim and Kaiser, 1990b	*csgA*	Dimer of 17-kDa protein	✓	✓	Kim and Kaiser, 1990c
				Short-chain alcohol dehydrogenase		✓	Lee and Shimkets, 1996
D	Fatty acids (?)	Rosenbluh and Rosenberg, 1989	*dsgA*	Translation initiation	✓	✓	Cheng et al., 1994; Kalman et al., 1994
E	Fatty acids	Downard and Toal, 1995	*esgA* *esgB*	E1 decarboxylase of the branched-chain keto acid dehydrogenase	✓	✓	Toal et al., 1995

organisms where they have been described, pili serve as organelles of attachment—usually to other cells. In *M. xanthus*, however, they serve the dramatic function of mediating social behavior in the form of group or social motility; it was shown that the group motility of *M. xanthus* required that the cells be piliated (Kaiser, 1979); it was further shown that there is a class of mutants that has lost their social motility—the *tgl* mutants, whose social motility is regained if they are placed within pilus range of wild-type cells, whereupon they transiently regain the ability to form their own pili (Hodgkin and Kaiser, 1979). Subsequently, the pili of *M. xanthus* have been photographed (Kaiser, 1979) (Fig. 9) and their biosynthetic genes have been cloned and sequenced and shown to have substantial similarity to the genes of the type IV pilus biosynthetic pathway in *Pseudomonas aeruginosa* (Wu et al., 1997). Wall et al. (1998) have recently suggested that *tgl* plays a role in pilus assembly.

FIBRILS

M. xanthus contains extracellular, filamentous appendages called fibrils arranged peritri-

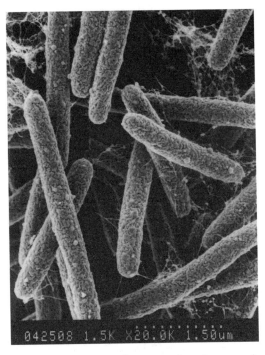

FIGURE 10 Low-voltage scanning electron micrograph of fibrils on vegetative cells of *M. xanthus*, grown on a solid surface in submerged culture. (From Behmlander and Dworkin, 1991.)

chously around the cell (Fig. 10). Fibrils may be as long as 50 μm, and their diameters fall into two approximately equal size distributions, 15 and 30 nm (Behmlander and Dworkin, 1994). These were originally described by Fluegel (1963b) as "myxonemata" and subsequently rediscovered by Arnold and Shimkets (1988), who showed that they play a role in cohesion and social behavior and that the *dsp* mutant, defective in a number of the social functions of *M. xanthus*, lacked fibrils. Chang and Dworkin (1994) showed that a number of the social functions of the *dsp* mutant could be rescued by the addition of the purified fibrils. Behmlander and Dworkin (1994) characterized the fibrils more fully and showed that they are present on living cells and are composed of polysaccharide with a set of tightly adhering proteins referred to as integral fibril proteins (IFP). The fibrils are distinct morphologically, chemically, and functionally from the pili.

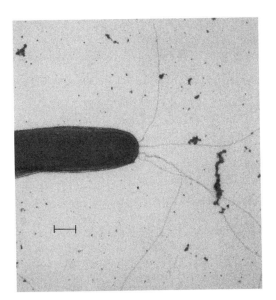

FIGURE 9 Electron micrograph of a negatively stained cell of *M. xanthus* illustrating the polar piliation. Bar, 250 μm. (From Kaiser, 1979.)

Smith and Dworkin (1997) have cloned and sequenced the gene for one of the fibril proteins (IFP-20) and have generated a mutation in the *IFP-20* gene. This mutant is defective in its cohesion properties, cohering in a one-dimensional end-to-end fashion rather than the three-dimensional side-by-side cohesion characteristic of wild-type cells. Accordingly, Smith and Dworkin (1997) have proposed that the fibril proteins play a role in establishing the normal cell-to-cell juxtaposition that is necessary for the optimal exchange of intercellular signals. An appropriate extension of that speculation is that the fibrils may serve as tactile antennae, allowing the cells to sense and monitor the presence of adjacent cells. Hildebrandt et

al. (1997) have shown that isolated, purified fibrils are able to carry out ADP ribosylation, thus serving both as the source of the ADP-ribosyl transferase and as its protein substrate. This unexpected extracellular enzymatic activity could be thought of as the actual sensing mechanism and is illustrated in Fig. 11.

In an attempt to understand the interaction of fibrils with adjacent cells, Chang and Dworkin (1996) isolated a series of secondary mutants of the fibrilless *dsp* mutant, in which cohesion could no longer be rescued by the addition of fibrils. These mutants, termed *fbd* (for fibril binding defective), surprisingly had regained social motility and the ability to aggregate while continuing to lack the ability to form

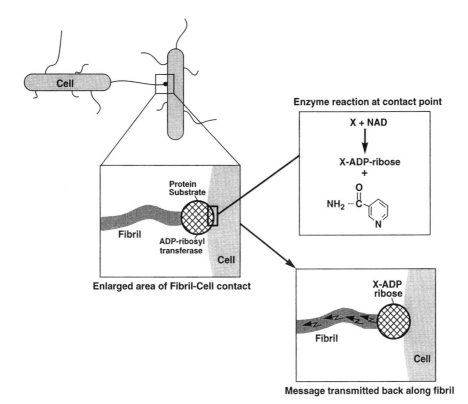

FIGURE 11 Model for the role of ADP ribosylation by fibrils as the sensors of tactile interactions (Dworkin, 1999a). The model proposes that physical proximity of two cells is detected by an interaction between a cellular fibril and a fibril receptor on an apposing cell. The interaction triggers the ADP ribosylation of a fibril protein, mediated by fibril ADP-ribosyl transferase. That is somehow transduced into a signal that is bilaterally transmitted along the fibril and to the apposing cell.

fruiting bodies or myxospores. The authors thus concluded that the physical presence of fibrils is not necessary either for social motility or for the early stages of aggregation. For a recent review of the fibrils of *M. xanthus* see Dworkin (1999).

EVOLUTION OF THE MYXOBACTERIA

Despite the fact that they are well ensconced within the δ group of the proteobacteria (Ludwig et al., 1983), the myxobacteria show a number of eukaryotic properties. Myxospores of *M. xanthus* contain an outer surface protein, protein S, which shares some properties with bovine calmodulin (Inouye et al., 1979) and the β and γ crystallins of the vertebrate eye lens (Wistow et al., 1985). It has, in fact, been suggested that these crystallins may have evolved from myxobacterial proteins (Wistow et al., 1985). Kohl et al. (1983) have found steroids in *Nannocystis*, and Benaïssa et al. (1994) have shown that *S. aurantiaca* has a phosphatidylinositol cycle and a GTPγS-dependent phospholipase C. Inouye et al. (1989) have found reverse transcriptase in *M. xanthus* and in myxobacteria in general.

Inouye's laboratory has also shown the presence of a large number of serine-threonine kinases that play a role in development in *M. xanthus* (Inouye and Inouye, 1993). In addition, the characteristically eukaryotic tyrosine kinase activity has recently been found in *M. xanthus*, and the pattern of tyrosine phosphorylation has been shown to change during fruiting body formation (Frasch and Dworkin, 1996).

Using two different strategies Shimkets (1993) and Kaiser (1986) have approximated the time at which the myxobacteria separated from their nearest bacterial relatives. On the basis of nucleotide sequence divergences of 5S and 16S rRNA, Shimkets calculated that the myxobacteria appeared on the evolutionary scene about 650 to 800 million years ago. Using the 16S rRNA sequence differences between the two major phylogenetic groups of the myxobacteria as a parameter, Kaiser has estimated that they diverged from their common

ancestor at about the time of the appearance of the aerobic atmosphere—about 2 billion years ago (Kaiser 1986). In either case, the myxobacteria may have preceded the emergence of multicellular eukaryotes, determined by fossil evidence to have occurred about 700 million years ago (Glaessner, 1976) or 1.1 billion years ago for soft-bodied triploblastic metazoans (Seilacher et al., 1998). Thus, the myxobacteria may have been among Nature's first experiments in multicellularity.

REFERENCES

Arnold, J. W., and L. J. Shimkets. 1988. Cell surface properties correlated with cohesion in *Myxococcus xanthus*. *J. Bacteriol.* **170:**5771–5777.

Barreaud, J.-P., S. Bourgerie, R. Julien, J. F. Guespin-Michel, and Y. Karamanos. 1995. An endo-*N*-acetyl-β-D-glucosaminidase acting on the di-*N*-acetylchitobiosyl part of N-linked glycans is secreted during sporulation of *Myxococcus xanthus*. *J. Bacteriol.* **177:**916–920.

Behmlander, R. M., and M. Dworkin. 1991. Extracellular fibrils and contact-mediated interactions in *Myxococcus xanthus*. *J. Bacteriol.* **173:**7810–7821.

Behmlander, R. M., and M. Dworkin. 1994. Biochemical and structural analysis of the extracellular matrix fibrils of *Myxococcus xanthus*. *J. Bacteriol.* **176:**6295–6303.

Benaïssa, M., J. Vieyres-Lubochinsky, R. Odéide, and B. Lubochinsky. 1994. Stimulation of inositide degradation in *Stigmatella aurantiaca*. *J. Bacteriol.* **176:**1390–1393.

Bender, H. 1963. Untersuchungen an *Myxococcus xanthus*. I. Bildungsbedingungen, Isolierung und Eigenschaften eines bakteriolytisches Enzymsystems. *Arch. Mikrobiol.* **43:**262–279.

Bourgerie, S., Y. Karamanos, T. Grard, and R. Julien. 1994. Purification and characterization of an endo-*N*-acetyl-β-D-glucosaminidase from the culture medium of *Stigmatella aurantiaca* DW4. *J. Bacteriol.* **176:**6170–6174.

Burchard, R. P. 1984. Gliding motility and taxes, p. 139–161. *In* E. Rosenberg (ed.), *The Myxobacteria*. Springer-Verlag, New York, N.Y.

Burnham, J. C., S. A. Collart, and B. W. Highison. 1981. Entrapment and lysis of the cyanobacterium *Phormidium luridum* by aqueous colonies of *Myxococcus xanthus* PC02. *Arch. Microbiol.* **129:**285–294.

Burnham, J. C., S. A. Collart, and M. J. Daft. 1984. Myxococcal predation of the cyanobacterium *Phormidium luridum* in aqueous environments. *Arch. Microbiol.* **137:**220–225.

Chang, B.-Y., and M. Dworkin. 1984. Isolated fibrils rescue cohesion and development in the *dsp* mutant of *Myxococcus xanthus*. *J. Bacteriol.* **176:** 7190–7196.

Chang, B.-Y., and M. Dworkin. 1996. Mutants of *Myxococcus xanthus dsp* defective in fibril binding. *J. Bacteriol.* **178:**697–700.

Cheng, Y., and D. Kaiser. 1989a. *dsg*, a gene required for cell-cell interaction early in *Myxococcus* development. *J. Bacteriol.* **171:**3719–3726.

Cheng, Y., and D. Kaiser. 1989b. *dsg*, a gene required for *Myxococcus* developmruent, is necessary for cell viability. *J. Bacteriol.* **171:**3727–3731.

Cheng, Y., L. V. Kalman, and D. Kaiser. 1994. The *dsg* gene of *Myxococcus xanthus* encodes a protein similar to translation initiation factor IF3. *J. Bacteriol.* **176:**1427–1433.

Couche, P. 1969. Morphology and morphogenesis of *Sorangium compositum*. *J. Appl. Bacteriol.* **32:**24–29.

Davis, J. M., J. Mayor, and L. Plamann. 1995. A missense mutation in *rpoD* results in an A-signaling defect in *Myxococcus xanthus*. *Mol. Microbiol.* **18:** 943–952.

Downard, J., and D. Toal. 1995. Branched-chain fatty acids—the case for a novel form of cell-cell signaling during *Myxococcus xanthus* development. *Mol. Microbiol.* **16:**171–175.

Dworkin, M. 1973. Cell-cell interactions in the myxobacteria. *Symp. Soc. Gen. Microbiol.* **23:**125–147.

Dworkin, M. 1983. Tactic behavior of *Myxococcus xanthus*. *J. Bacteriol.* **154:**452–459.

Dworkin, M. 1986. *Developmental Biology of the Bacteria*. The Benjamin/Cummings Publishing Co., Inc., Menlo Park, Calif.

Dworkin, M. 1993. Cell surfaces and appendages, p. 63–83. *In* M. Dworkin and D. Kaiser (ed.), *Myxobacteria II*. American Society for Microbiology, Washington, D.C.

Dworkin, M. 1996. Recent advances in the social and developmental biology of the myxobacteria. *Microbiol. Rev.* **60:**70–102.

Dworkin, M. 1999a. Common themes in pathogenesis and development in *Myxococcus xanthus*, p. 5–16. *In* E. Rosenberg (ed.), *Microbial Ecology and Infectious Disease*. American Society for Microbiology, Washington, D.C.

Dworkin, M. 1999. Fibrils as extracellular appendages of bacteria: their role in contact-mediated cell-cell interactions in *Myxococcus xanthus*. *Bioessays* **20:** 590–595.

Dworkin, M., and S. M. Gibson. 1964. A system for studying microbial morphogenesis: rapid formation of microcysts in *Myxococcus xanthus*. *Science* **146:**243–244.

Dworkin, M., and D. J. Niederpruem. 1964. Electron transport system in vegetative cells and microcysts of *Myxococcus xanthus*. *J. Bacteriol.* **87:** 316–322.

Fluegel, W. 1963a. Fruiting chemotaxis in *Myxococcus fulvus* (myxobacteria). *Proc. Minn. Acad. Sci.* **30:** 120–123.

Fluegel, W. 1963b. Simple method for demonstrating myxobacterial slime. *J. Bacteriol.* **85:**1173–1174.

Frasch, S. C., and M. Dworkin. 1996. Tyrosine kinase in *Myxococcus xanthus*, a multicellular prokaryote. *J. Bacteriol.* **178:**4084–4088.

Freese, A., H. Reichenbach, and H. Lünsdorf. 1997. Further characterization and in situ localization of chain-like aggregates of the gliding bacteria *Myxococcus fulvus* and *Myxococcus xanthus*. *J. Bacteriol.* **179:**1246–1252.

Gerth, K., and H. Reichenbach. 1978. Induction of myxospore formation in *Stigmatella aurantiaca* (Myxobacterales). *Arch. Microbiol.* **117:**173–182.

Gill, R. E., M. Karlok, and D. Benton. 1993. *Myxococcus xanthus* encodes an ATP-dependent protease which is required for developmental gene transcription and intercellular signaling. *J. Bacteriol.* **175:** 4538–4544.

Glaessner, M. F. 1976. Early phanerozoic worms and their geological and biological significance. *J. Geol. Soc. Lond.* **132:**259–275.

Hagen, T. J., and L. J. Shimkets. 1990. Nucleotide sequence and transcriptional products of the *csg* locus of *Myxococcus xanthus*. *J. Bacteriol.* **172:**15–23.

Hartzell, P. L., and P. Youderian. 1995. Genetics of gliding motility and development in *Myxococcus xanthus*. *Arch. Microbiol.* **164:**309–323.

Herzer, P. J., S. Inouye, M. Inouye, and T. S. Whittam. 1990. Phylogenetic distribution of branched RNA-linked multicopy, single-stranded DNA among natural isolates of *Escherichia coli*. *J. Bacteriol.* **172:**6175–6181.

Hildebrandt, K., D. Eastman, and M. Dworkin. 1997. ADP-ribosylation by the extracellular fibrils of *Myxococcus xanthus*. *Mol. Microbiol.* **23:**231–235.

Hodgkin, J., and D. Kaiser. 1977. Cell-to-cell stimulation of movement in non-motile mutants of *Myxococcus*. *Proc. Natl. Acad. Sci. USA* **74:** 2938–2942.

Iizuka, T., Y. Jojima, R. Fudou, and S. Yamanaka. 1998. Isolation of myxobacteria from the marine environment. *FEMS Microbiol. Lett.* **169:** 317–322.

Inouye, M., S. Inouye, and D. R. Zusman. 1979. Biosynthesis and self-assembly of protein S, a development specific protein of *Myxococcus xanthus*. *Proc. Natl. Acad. Sci. USA* **76:**209–213.

Inouye, S., and M. Inouye. 1993. Development-specific gene expression: protein serine/threonine kinases and sigma factors, p. 201–212. *In* M. Dworkin and D. Kaiser (ed.), *Myxobacteria II*. American Society for Microbiology, Washington, D.C.

Inouye, S., M.-Y. Hsu, S. Eagle, and M. Inouye. 1989. Reverse transcriptase associated with the biosynthesis of the branched RNA-linked msDNA in *Myxococcus xanthus*. *Cell* **56**:709–717.

Kaiser, D. 1979. Social gliding is correlated with the presence of pili in *Myxococcus xanthus*. *Proc. Natl. Acad. Sci. USA* **76**:5952–5956.

Kaiser, D. 1986. Control of multicellular development: *Dictyostelium* and *Myxococcus*. *Annu. Rev. Genet.* **20**:539–566.

Kaiser, D., and L. Kroos. 1993. Intercellular signaling, p. 257–283. *In* M. Dworkin and D. Kaiser (ed.), *Myxobacteria II*. American Society for Microbiology, Washington, D.C.

Kalman, L. V., Y. L. Cheng, and D. Kaiser. 1994. The *Myxococcus xanthus dsg* gene product performs functions of translation initiation factor IF3 in vivo. *J. Bacteriol.* **176**:1434–1442.

Kaplan, H. B., and L. Plamann. 1996. A *Myxococcus xanthus* cell density-sensing system required for multicellular development. *FEMS Microbiol. Lett.* **139**:89–95.

Kearns, D. B., and L. J. Shimkets. 1998. Chemotaxis in a gliding bacterium. *Proc. Natl. Acad. Sci. USA* **95**:11957–11962.

Keller, K. H., M. Grady, and M. Dworkin. 1983. Surface tension gradients: feasible model for gliding motility of *Myxococcus xanthus*. *J. Bacteriol.* **155**:1358–1366.

Kim, S. K., and D. Kaiser. 1990a. Cell alignment required in differentiation of *Myxococcus xanthus*. *Science* **249**:926–928.

Kim, S. K., and D. Kaiser. 1990b. C-factor: a cell signaling protein required for fruiting body morphogenesis. *Cell* **61**:19–26.

Kim, S. K., and D. Kaiser. 1990c. Purification and properties of *Myxococcus xanthus* C-factor, an intercellular signaling protein. *Proc. Natl. Acad. Sci. USA* **87**:3635–3639.

Kim, S. K., and D. Kaiser. 1990d. Cell motility is required for the transmission of C-factor, an intercellular signal that coordinates fruiting body morphogenesis in *Myxococcus xanthus*. *Genes Dev.* **4**:896–905.

Kohl, W., A. Gloe, and H. Reichenbach. 1983. Steroids from the myxobacterium *Nannocystis exedens*. *J. Gen. Microbiol.* **129**:1629–1635.

Konijn, T. M., J. G. C. Van De Meene, J. T. Bonner, and D. S. Barkley. 1967. The acrasin activity of adenosine 3′,5′-cyclic phosphate. *Proc. Natl. Acad. Sci. USA* **82**:8540–8544.

Kroos, L., and D. Kaiser. 1987. Expression of many developmentally regulated genes in *Myxococcus xanthus* depends on a sequence of cell interactions. *Genes Dev.* **1**:840–854.

Kroos, L., P. Hartzell, K. Stephens, and D. Kaiser. 1988. A link between cell movement and gene expression argues that cell motility is required for

cell-cell signaling during fruiting body development. *Genes Dev.* **2**:1677–1685.

Kroos, L., A. Kuspa, and D. Kaiser. 1990. Defects in fruiting body development caused by Tn5-*lac* insertions in *Myxococcus xanthus*. *J. Bacteriol.* **172**:484–487.

Kuspa, A., L. Plamann, and D. Kaiser. 1992a. Identification of heat-stable A-factor from *Myxococcus xanthus*. *J. Bacteriol.* **174**:3319–3326.

Kuspa, A., L. Plamann, and D. Kaiser. 1992b. A-signaling and the cell density requirement for *Myxococcus xanthus* development. *J. Bacteriol.* **174**:7360–7369.

Lee, B.-U., K. Lee, J. Robles, and L. J. Shimkets. 1995. A tactile sensory system of *Myxococcus xanthus* involves an extracellular NAD(P)$^+$-containing protein. *Genes Dev.* **9**:2964–2973.

Lee, K., and L. J. Shimkets. 1996. Suppression of a signaling defect during *Myxococcus xanthus* development. *J. Bacteriol.* **178**:977–984.

Li, S.-F., B.-U. Lee, and L. J. Shimkets. 1992. *csgA* expression entrains *Myxococcus xanthus* development. *Genes Dev.* **6**:401–410.

Li, Y., and L. Plamann. 1996. Purification and in vitro phosphorylation of *Myxococcus xanthus* AsgA protein. *J. Bacteriol.* **178**:289–292.

Ludwig, W., K. H. Schleifer, H. Reichenbach, and E. Stackebrandt. 1983. A phylogenetic analysis of the myxobacteria *Myxococcus fulvus*, *Stigmatella aurantiaca*, *Cystobacter fuscus*, *Sorangium cellulosum* and *Nannocystis exedens*. *Arch. Microbiol.* **135**:58–62.

Lünsdorf, H., and H. Reichenbach. 1989. Ultrastructural details of the apparatus of gliding motility of *Myxococcus fulvus* (Myxobacterales). *J. Gen. Microbiol.* **135**:1633–1641.

MacRae, T. H., and H. D. McCurdy. 1975. Ultrastructural studies of *Chondromyces crocatus* vegetative cells. *Can. J. Microbiol.* **21**:1815–1826.

MacRae, T. H., W. J. Dobson, and H. D. McCurdy. 1977. Fimbriation in gliding bacteria. *Can. J. Microbiol.* **23**:1096–1108.

Mayer, H., and H. Reichenbach. 1978. Restriction endonucleases: general survey procedure and survey of gliding bacteria. *J. Bacteriol.* **136**:708–713.

McVittie, A., and S. A. Zahler. 1962. Chemotaxis in *Myxococcus*. *Nature* **194**:1299–1300.

Morris, D. W., and J. H. Parish. 1976. Restriction in *Myxococcus xanthus*. *Arch. Mikrobiol.* **108**:227–230.

O'Connor, K. A., and D. R. Zusman. 1989. Patterns of cellular interactions during fruiting body formation in *Myxococcus xanthus*. *J. Bacteriol.* **171**:6013–6024.

Pate, J. L., and L.-Y. E. Chang. 1979. Evidence that gliding motility in prokaryotic cells is driven by rotary assemblies in the cell envelopes. *Curr. Microbiol.* **2:**59–64.

Pinoy, P. E. 1921. Sur les Myxobacteries. *Ann. Inst. Pasteur* **35:**487–495.

Plamann, L., J. M. Davis, B. Cantwell, and J. Mayor. 1994. Evidence that *asgB* encodes a DNA-binding protein essential for growth and development of *Myxococcus xanthus. J. Bacteriol.* **176:** 2013–2020.

Plamann, L., Y. Li, B. Cantwell, and J. Mayor. 1995. The *Myxococcus xanthus asgA* gene encodes a novel signal transduction protein required for multicellular development. *J. Bacteriol.* **177:** 2014–2020.

Ramaswamy, S., M. Dworkin, and J. Downard. 1997. Identification and characterization of *Myxococcus xanthus* mutants deficient in calcofluor white binding. *J. Bacteriol.* **79:**2863–2871.

Reichenbach, H. 1984. Myxobacteria: a most peculiar group of social prokaryotes, p. 1–50. *In* E. Rosenberg (ed.), *Myxobacteria: Development and Cell Interactions.* Springer-Verlag, New York, N.Y.

Reichenbach, H. 1993. Biology of the myxobacteria: ecology and taxonomy, p. 13–62. *In* M. Dworkin and D. Kaiser (ed.), *Myxobacteria II.* American Society for Microbiology, Washington, D.C.

Reichenbach, H., and M. Dworkin. 1970. Induction of myxospore formation in *Stigmatella aurantiaca* (*Myxobacterales*) by monovalent cations. *J. Bacteriol.* **101:**325–326.

Reichenbach, H., and M. Dworkin. 1992. The myxobacteria, p. 3416–3487. *In* A. Balows, H. G. Trüper, M. Dworkin, W. Harder, and K.-H. Schleifer (ed.), *The Prokaryotes,* 2nd ed. Springer-Verlag, New York, N.Y.

Reichenbach, H., and G. Höfle. 1993. Biologically active secondary metabolites from myxobacteria. *Biotechnol. Adv.* **11:**219–277.

Reichenbach, H., H. H. Huenert, and H. Kuczka. 1965. *Schwarmentwicklung und Morphogeneses bei Myxobakterien—Archangium, Myxococcus, Chondromyces.* Film C893. Institut für Wissenschaftlichen Film, Göttingen, Germany.

Rice, S., J. Bleber, J.-Y. Chun, G. Stacey, and B. C. Lampson. 1993. Diversity of retron elements among a population of rhizobia and other gram-negative bacteria. *J. Bacteriol.* **175:**4250–4254.

Rice, S. A., and B. C. Lampson. 1995. Phylogenetic comparison of retron elements among the myxobacteria: evidence for vertical inheritance. *J. Bacteriol.* **177:**37–45.

Rosenberg, E., and M. Dworkin. 1996. Autocides and a paracide, antibiotic TA, produced by *Myxococcus xanthus. J. Ind. Microbiol.* **17:**424–431.

Rosenberg, E., and M. Varon. 1984. Antibiotics and lytic enzymes, p. 109–127. *In* E. Rosenberg (ed.), *Myxobacteria: Development and Cell Interactions.* Springer-Verlag, New York, N.Y.

Rosenberg, E., K. H. Keller, and M. Dworkin. 1977. Cell density-dependent growth of *Myxococcus xanthus* on casein. *J. Bacteriol.* **129:**770–777.

Rosenbluh, A., and E. Rosenberg. 1989. Autocide AMI rescues development in *dsg* mutants of *Myxococcus xanthus. J. Bacteriol.* **171:**1513–1518.

Sadler, W., and M. Dworkin. 1966a. Induction of cellular morphogenesis in *Myxococcus xanthus.* I. General description. *J. Bacteriol.* **91:**1516–1519.

Sadler, W., and M. Dworkin. 1966b. Induction of cellular morphogenesis in *Myxococcus xanthus.* II. Macromolecular synthesis and mechanism of inducer action. *J. Bacteriol.* **91:**1520–1525.

Sager, B., and D. Kaiser. 1993a. Two cell-density domains within the *Myxococcus xanthus* fruiting body. *Proc. Natl. Acad. Sci. USA* **90:**3690–3694.

Sager, B., and D. Kaiser. 1993b. Spatial restriction of cellular differentiation. *Genes Dev.* **7:**1645–1653.

Seilacher, A., P. K. Bose, and F. Pflüger. 1998. Triploblastic animals more than 1 billion years ago: trace fossil evidence from India. *Science* **282:**80–83.

Shapiro, J. A., and M. Dworkin (ed). 1997. *Bacteria as Multicellular Organisms.* Oxford University Press, New York, N.Y.

Shi, W., and D. R. Zusman. 1995. The *frz* signal transduction system controls multicellular behavior in *Myxococcus xanthus,* p. 419–430. *In* J. A. Hoch and T. J. Silhavy (ed.), *Two-Component Signal Transduction.* American Society for Microbiology, Washington, D.C.

Shimkets, L. J. 1993. The myxobacterial genome, p. 85–107. *In* M. Dworkin and D. Kaiser (ed.), *Myxobacteria II.* American Society for Microbiology, Washington, D.C.

Shimkets, L. J., and E. W. Crawford, Jr. 1998. Personal communication.

Shimkets, L. J., and M. Dworkin. 1981. Excreted adenosine is a cell density signal for the initiation of fruiting body formation in *Myxococcus xanthus. Dev. Biol.* **84:**51–60.

Shimkets, L. J., and H. Rafiee. 1990. CsgA, an extracellular protein essential for *Myxococcus xanthus* development. *J. Bacteriol.* **172:**5299–5306.

Shimkets, L., and C. R. Woese. 1992. A phylogenetic analysis of the myxobacteria: basis for their classification. *Proc. Natl. Acad. Sci. USA* **89:** 9459–9463.

Singer, M., and D. Kaiser. 1995. Ectopic production of guanosine penta- and tetraphosphate can initiate early developmental gene expression in *Myxococcus xanthus. Genes Dev.* **9:**1633–1644.

Singh, B. N. 1947. Myxobacteria in soils and composts: their distribution, number and lytic action on bacteria. *J. Gen. Microbiol.* **1:**1–10.

Smith, D. R., and M. Dworkin. 1994. Territorial

interactions between two *Myxococcus* species. *J. Bacteriol.* **176**:1201–1205.

Smith, D. R., and M. Dworkin. 1997. A mutation that affects fibril protein, development, cohesion and gene expression in *Myxococcus xanthus*. *Microbiology* **143**:3683–3692.

Søgaard-Andersen, L., F. J. Slack, H. Kimsey, and D. Kaiser. 1997. Intercellular signaling in *Myxococcus xanthus* involves a branched signal transduction pathway. *Genes Dev.* **10**:740–754.

Sørhaug, T. 1974. Glycerol ester hydrolase, lipase of *Myxococcus xanthus fb*. *Can. J. Microbiol.* **20**:611–615.

Stackebrandt, E. 1998. Personal communication.

Sudo, S., and M. Dworkin. 1972. Bacteriolytic enzymes produced by *Myxococcus xanthus*. *J. Bacteriol.* **110**:236–245.

Sudo, S. Z., and M. Dworkin. 1969. Resistance of vegetative cells and microcysts of *Myxococcus xanthus*. *J. Bacteriol.* **98**:883–887.

Temin, H. M. 1989. Retrons in bacteria. *Nature* **339**:254–255.

Toal, D. R., S. W. Clifton, B. A. Roe, and J. Downard. 1995. The *esg* locus of *Myxococcus xanthus* encodes the E1α and E1β subunits of a branched-chain keto acid dehydrogenase. *Mol. Microbiol.* **16**:177–189.

Tojo, N., S. Inouye, and T. Komano. 1993. The *lonD* gene is homologous to the *lon* gene encoding an ATP-dependent protease and is essential for the development of *Myxococcus xanthus*. *J. Bacteriol.* **175**:4545–4549.

Varon, M., A. Teitz, and E. Rosenberg. 1986. *Myxococcus xanthus* autocide AMI. *J. Bacteriol.* **167**:356–361.

Vasquez, G. M., F. Qualls, and D. White. 1985. Morphogenesis of *Stigmatella aurantiaca* fruiting bodies. *J. Bacteriol.* **163**:515–521.

Wall, D., S. S. Wu, and D. Kaiser. 1998. Contact stimulation of Tgl and type IV pili in *Myxococcus xanthus*. *J. Bacteriol.* **180**:759–761.

White, D. 1993. Myxospore and fruiting body morphogenesis, p. 307–332. *In* M. Dworkin and D. Kaiser (ed.), *Myxobacteria II*. American Society for Microbiology, Washington, D.C.

Wireman, J. W., and M. Dworkin. 1975. Morphogenesis and developmental interactions in myxobacteria. *Science* **189**:516–523.

Wireman, J. W., and M. Dworkin. 1977. Developmentally induced autolysis during fruiting body formation by *Myxococcus xanthus*. *J. Bacteriol.* **129**:796–802.

Wistow, G., L. Summers, and T. Blundell. 1985. *Myxococcus xanthus* spore coat protein S may have a similar structure to vertebrate lens β, γ-crystallins. *Nature* **315**:771–773.

Wu, S. S., J. Wu, and D. Kaiser. 1997. The *Myxococcus xanthus pilT* locus is required for social gliding motility although pili are still produced. *Mol. Microbiol.* **23**:109–121.

Zusman, D. R. 1984. Developmental program of *Myxococcus xanthus*, p. 185–213. *In* E. Rosenberg (ed.), *Myxobacteria: Development and Cell Interactions*. Springer-Verlag, New York, N.Y.

DEVELOPMENTAL AGGREGATION AND FRUITING BODY FORMATION IN THE GLIDING BACTERIUM *MYXOCOCCUS XANTHUS*

Mandy J. Ward and David R. Zusman

11

The myxobacteria have attracted scientific interest for more than a century due to their spectacular morphogenic potential. Since they were first accurately described by Thaxter in 1892, these organisms have provided superb examples of complex multicellular interactions unexpected in prokaryotes. In this chapter we will discuss what is perhaps the most unique facet of myxobacterial morphogenesis, the formation of multicellular aggregates and fruiting bodies by the cooperative action of tens of thousands of individual cells. *Myxococcus xanthus*, unlike many of the other myxobacteria, forms simple fruiting bodies which consist purely of cell mounds within which the myxospores are located. However, in recent years this organism has become the primary focus of research on the myxobacteria, and our understanding of the gliding-motility process, morphogenic potential, and social interactions of these cells has increased dramatically. While we therefore restrict the scope of this chapter to studies performed with *M. xanthus*, the reader is directed to reviews which document the more complex aggregation patterns displayed in the fruiting bodies of other myxobacteria

(Reichenbach and Dworkin, 1981; Shimkets, 1990; Reichenbach, 1984, 1993).

The *M. xanthus* developmental cycle, which results in the formation of spore-filled fruiting bodies, is triggered by starvation, with limitation of carbon, energy, inorganic phosphate, or any single amino acid being sufficient to initiate the process. Many bacteria similarly initiate sporulation during periods of stress, but developmental aggregation is unique to the myxobacteria. What do these bacteria gain from such an energetically expensive and lengthy (2- to 3-day) process? Perhaps both the size and bright coloration of the fruiting bodies attract insects or other animals, thereby aiding dispersal into new environments. Alternatively, the fruiting body might play a defensive role, with cells on the surface providing protection to cells at the center of the aggregate. Certainly, the coloration of the fruiting bodies provides protection from photooxidative damage (Hodgson and Murillo, 1993). However, Reichenbach (1984) stated what is probably the most important role of fruiting body formation when he suggested "the function of the fruiting body seems to be to ensure that a new life cycle is started by a cell community and not by a single cell." Such social behavior undoubtedly aids the feeding strategy of the myxobacteria, since these slow-moving bacteria feed by se-

M. J. Ward and D. R. Zusman, Department of Molecular and Cell Biology, University of California at Berkeley, Berkeley, CA 94720-3204.

Prokaryotic Development, edited by Y. V. Brun and L. J. Shimkets,
© 2000 American Society for Microbiology, Washington, DC 20005-4171

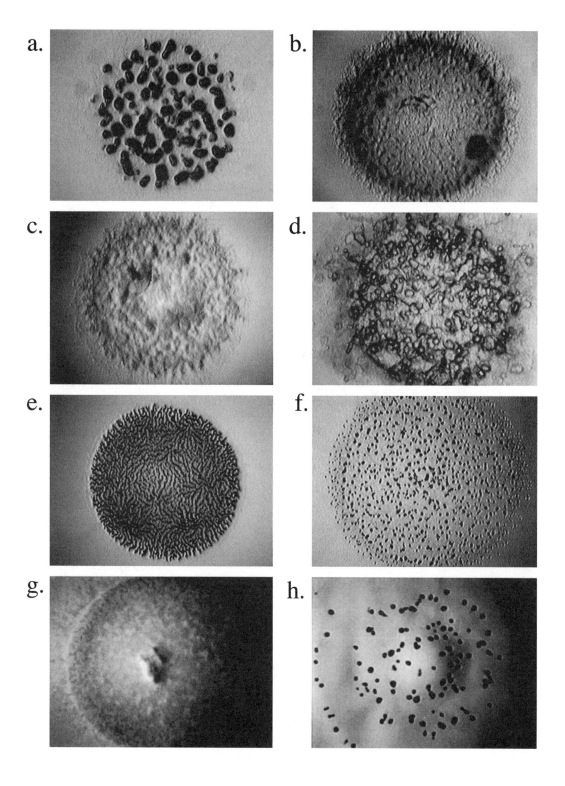

creting enzymes which lyse other organisms or break down biopolymers in their environment. These enzymes should work more efficiently when a large number of cells are involved in their production. This communal feeding strategy is therefore assumed to have resulted in the evolution of complex social behavior, including developmental aggregation, in the myxobacteria.

INDUCTION OF FRUITING BODY FORMATION

One insightful early observation made about *M. xanthus* behavior was that the developmental process would only result in the formation of fruiting bodies (shown in Fig. 1a) when cells were starved at sufficiently high densities that they would be in contact with one another (Wireman and Dworkin, 1975). Thus, the induction of fruiting body formation presumably requires two separate events. First, individual cells must monitor their own nutrient status. Second, each cell must evaluate the nutritional status and size of the entire population. Manoil and Kaiser (1980) first suggested that, since starvation induces a transient rise in the levels of cellular (p)ppGpp (guanosine 3′-di[tri]phosphate-5′-diphosphate), this molecule may be used to monitor the metabolic state of individual cells. More recently, Singer and Kaiser (1995) showed that ectopic elevation of (p)ppGpp levels activates early developmental gene expression. The genes transcriptionally activated by increased (p)ppGpp levels include those required for the production of an early extracellular developmental signal, the A-signal. The A-signal, which begins to be present

in the culture medium approximately 1 h after the onset of starvation, consists of a mixture of amino acids (Kuspa et al., 1992a). Since the level of extracellular A-signal produced is dependent on cell numbers, the A-signal has been proposed to act in a cell density-dependent manner (Kuspa et al., 1992b). This signal could therefore monitor the nutrient status of the entire population of cells. However, the A-signal may not be the sole cell density monitor. Adenosine, for example, has also been implicated as an early, and A-signal-independent, cell density signal (Shimkets and Dworkin, 1981). However, the related events—a transient rise in (p)ppGpp levels within the cells which stimulates production of the cell density-dependent extracellular A-signal—are highly suggestive that these two stages are prerequisites for the initiation of development. Certainly, cells with mutations in the *relA* gene, which cannot accumulate (p)ppGpp in response to starvation, and cells with A-signaling defects are both unable to form fruiting bodies (Harris et al., 1998; Kuspa and Kaiser, 1989). Thus, the accumulation of (p)ppGpp in starved cells may well initiate the developmental process at the individual cell level, while A-signaling, along with other, A-signal-independent factors, may provide the primary population-level developmental signals (see chapter 12).

DIRECTED MOTILITY IS REQUIRED FOR DEVELOPMENTAL AGGREGATION

The first morphological signs of aggregation occur approximately 6 h after the onset of starvation. At this time, observations of developing

FIGURE 1 Wild-type and mutant developmental-aggregation phenotypes. (a) Strain DZ2 (wild type), shown developing at high cell numbers (2×10^{10} CFU/ml) after 48 h of incubation at 34°C; (b) SW505 (*difA*; signal transduction mutant), showing formation of only small mounds after 48 h of incubation; (c) DK2630 (*csgA*; extracellular-signaling mutant), showing formation of only small mounds after 24 h of development; (d) DZF3558 ($\Delta frzA–E$; signal transduction mutant), showing the characteristic frizzy phenotype in the strain FB background after 72 h of incubation; (e) DZF1 (*rpoE1*; ECF sigma factor mutant), showing trails of aggregates rather than discrete mounds after 24 h of development; (f) DZF1 (*espA*; signal transduction mutant), showing many small, closely spaced aggregates (photo courtesy of K. Cho); (g) DZ2 (*tagE*; temperature-dependent aggregation mutant), showing no aggregation at 34°C (provided by K. O'Connor); (h) DZ2 (*tagE*) showing wild-type aggregation at 28°C (provided by K. O'Connor). All cells were plated on CF starvation agar (Hagen et al., 1978).

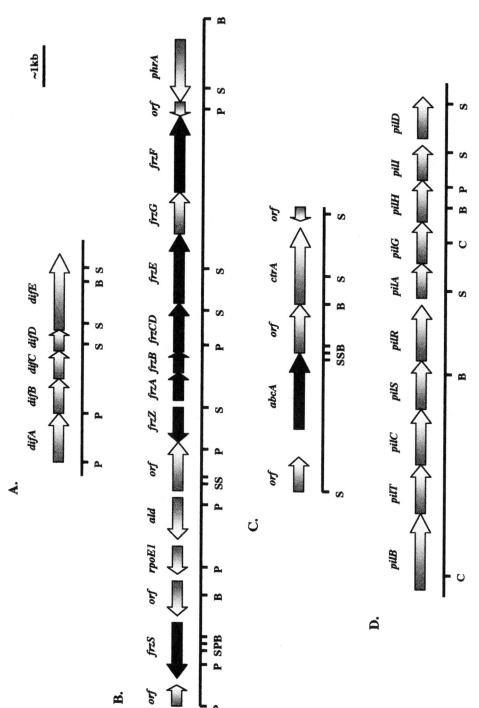

FIGURE 2 Gene organizations within motility- and directed-motility-associated loci. (A) *dif* locus (GenBank accession no. AF076485); (B) *frz* locus (accession no. AF049107, U47814, J04157, M35192, M35200, and U44437); (C) *abcA* locus (accession no. AF047554); (D) *pil* locus (accession no. AF003632, L39904, and L78131). Genes which when mutated result in the frizzy aggregation phenotype during development are shown in dark gray. Restriction enzyme sites: B, *Bam*HI; C, *Cla*I (shown for the *pil* region only); P, *Pst*I; and S, *Sac*I. Arrows designate direction of transcription of genes.

colonies show the cells streaming into aggregation centers. The cells appear to move in spiral patterns within stacked monolayers (O'Connor and Zusman, 1989). These stacked monolayers first form terraces, and then as more and more cells stream into the aggregates, hemispherical mounds are formed. That the formation of these mounds requires chemotaxis towards a self-generated diffusible molecule was first suggested by Lev (1954). McVittie and Zahler (1962) provided additional support for aggregation requiring chemotaxis when they demonstrated that two layers of *M. xanthus* cells would form fruiting bodies directly in alignment with each other when separated by a permeable barrier. While the diffusible molecule, or molecules, involved in this process have never been characterized, we can hypothesize that any such self-generated signaling molecule must be both secreted and detected by developing cells.

The *dif* Locus Plays a Role in Early Aggregation

A signal transduction system that could potentially detect signaling molecules and transduce information to the gliding motility machinery has recently been identified by Yang et al. (1998b). These authors isolated a new class of mutants called *dif* (defect in fruiting) which do

not aggregate past the initial early mound stage (Fig. 1b). Sequence analysis of the gene products from the *dif* locus, which contains five potential open reading frames (difA to -E [Fig. 2a]), suggests that these genes encode a new set of chemotaxis protein homologues (Table 1).

In the enteric bacteria the chemotaxis (Che) proteins form a signal transduction system (Fig. 3) which allows the cells to sense chemical gradients and direct motility towards attractant stimuli and away from repellents (Blair, 1995). The transmembrane receptor proteins in both *Escherichia coli* and *Salmonella typhimurium* are termed methyl-accepting chemotaxis proteins (MCPs). These MCPs bind chemoeffectors (or periplasmic sensory proteins) and transduce information to the cytoplasmic components of the system. CheW is a coupling protein that interacts with both the MCPs and the histidine protein kinase, CheA. The CheA protein can both autophosphorylate and, when stimulated by the receptors, transfer this phosphate to the response regulator, CheY. CheY interacts directly with the "switch" proteins of the flagellar motor, regulating the clockwise-counterclockwise motor bias. Adaptation, which facilitates chemotactic movement up attractant gradients, involves the reversible methylation of the MCPs. CheR, a methyltransferase, and

TABLE 1 Homologues and proposed functions of components of *M. xanthus* Frz and Dif systems

Component of *M. xanthus* Frz system	Homologue in enteric Che system	Component of *M. xanthus* Dif system	Proposed function
FrzA	CheW	DifC	Adapter
FrzB	No homologue		Unknown
FrzCD	Tar (MCP)	DifA	Receptor
FrzE (N terminal)	CheA	DifE	Histidine kinase
(C terminal)	CheY	DifD	Response regulator
FrzF (N terminal)	No homologue		Unknown
(C terminal)	CheR		Methyltransferase
FrzG	CheB		Methylesterase
FrzZ (N terminal)	CheY		Response regulator
(C terminal)	CheY		Response regulator
FrzS	No homologue		S-motility regulator
AbcA	No homologue		Exporter
	No homologue	DifB	Unknown

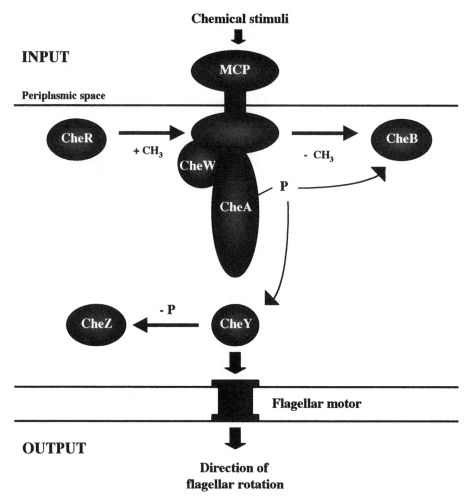

FIGURE 3 Information flow through the enteric chemotaxis system. Chemical stimuli are sensed by the transmembrane MCP receptor proteins. Excitation signals are relayed via CheW to the histidine protein kinase CheA. CheA can both autophosphorylate and, when stimulated by the MCPs, phosphorylate CheY. CheY-P then interacts with the switch proteins of the flagellar motor, regulating the direction of flagellar rotation. Dephosphorylation of CheY-P is stimulated by CheZ, while adaptation to stimuli requires methylation and demethylation of the MCPs by CheR and CheB, respectively.

CheB, a methylesterase, regulate the methylation and demethylation of specific glutamate residues on the cytoplasmic domains of the receptors.

The newly identified *M. xanthus* DifA protein is an MCP homologue, with two potential transmembrane domains which bracket a very short span of amino acids, predicted to be located in the periplasmic space. The cytoplasmic domain of DifA is well conserved with respect to other MCPs and presumably interacts with both the CheW protein homologue, DifC, and the putative histidine protein kinase, DifE. The DifD protein is a CheY homologue, and an alignment with CheY from *S. typhimurium* (Fig. 4a) shows that DifD retains both the phosphorylation site (equivalent to Asp57 in the enteric CheY) and associated amino acids required to form an "acid pocket" for divalent cation binding around this site (Asp12, Asp13, and Lys109). This conservation of important amino acids suggests that the DifD protein may indeed be phosphorylated in a manner analogous to that of the enteric CheY. The DifB protein shows no homology to proteins within the database and currently has no proposed role. However, by analogy to the enteric chemotaxis system, we might presume that DifA acts as the sensor component of the system, relaying information across the inner membrane, via DifC, to the histidine protein kinase, DifE. DifE would then regulate the phosphorylation state of DifD, which would ultimately control output from the system. No Dif-associated homologues of either CheR or CheB have yet been identified.

Yang and coworkers have characterized cells with mutations in the *difA* and *difE* genes. Both mutants show identical developmental aggregation defects, suggesting that the Dif proteins probably do constitute a new signal transduction pathway necessary for aggregation during fruiting body formation, although no biochemical evidence is yet available to show that these proteins participate in such a pathway. Still, a connection between the Dif system and the social (S) motility system (see below) was indicated by both microscopic and vegetative swarm plate analyses. These analyses were performed on both the *difA* and *difE* mutants, and the results suggest that the cells may have defects specifically associated with the regulation of S-motility-associated gliding behavior. Since S-motility requires cell-cell interactions, the *dif* signal transduction system could potentially facilitate responses to cell contact stimuli, as well as to diffusible signaling molecules.

The *frz* Genes Are Required for Aggregation of Cells into Discrete Raised Mounds

M. xanthus encodes a second set of chemotaxis gene homologues which have been designated "frizzy" (*frz*) due to the characteristic aggregation phenotype displayed by cells with mutations in this locus (Fig. 2B) (Zusman, 1982).

Frz mutants form defective aggregates rather than discrete mounds. The characteristic tangled, swirling patterns of these aggregates (Fig. 1d) are presumably caused by a block in the aggregation process at a later stage than that caused by mutations in the *dif* genes, which allow only initial mound formation. Frz mutants also show defects in vegetative swarming, and since cells with mutations in the *frz* genes have been observed to rarely reverse their direction of gliding (Blackhart and Zusman, 1985), both phenotypic effects are presumed to relate to defects in directed-motility behavior (Ward and Zusman, 1997). While the Frz proteins are also homologues of the enteric Che proteins (Table 1) (McBride et al., 1989; McCleary and Zusman, 1990) and make up an analogous signal transduction pathway (Fig. 5), they contain several unique features with respect to the newly identified Dif system of *M. xanthus*, as well as the enteric Che system. The proposed receptor protein (MCP homologue) for the Frz system, FrzCD, does not have membrane-spanning domains and is localized in the cytoplasm. Input to FrzCD must therefore come either through an unidentified membrane-associated receptor-transducer complex or from a cytoplasmic stimulus. The FrzCD protein interacts with FrzB (Astling et al., unpublished), which has no enteric homologue and no known function. FrzA is a CheW homologue and probably serves the same function as the enteric protein, i.e., as an adapter linking the receptor protein FrzCD with the histidine protein kinase domain of FrzE. FrzE is substantially modified with respect to both CheA and DifE histidine protein kinases. While the N-terminal portion of FrzE is homologous to CheA, the C-terminal domain shows homology to the response regulator protein CheY (McCleary and Zusman, 1990). Biochemical analyses of FrzE have shown that the CheA portion of the protein can autophosphorylate and that this phosphorylated intermediate can transfer a phosphate group to the CheY domain (Acuña et al., 1995). An alignment comparison of the CheY domain of FrzE with both DifD and *S. typhimurium* CheY is

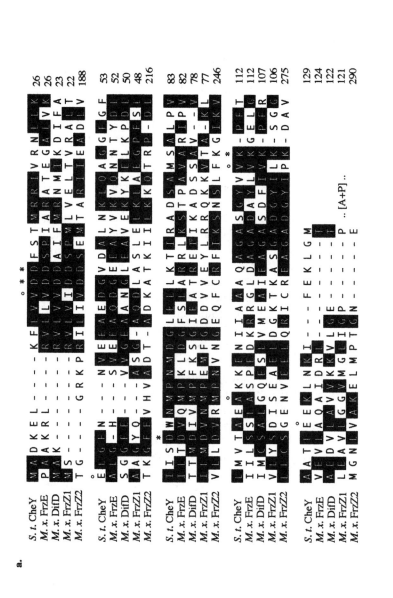

a.

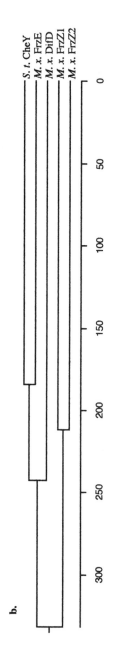

b.

shown in Fig. 4a. Figure 4b shows a dendrogram created from the alignment of these CheY-like proteins, which suggests that the CheY domain of FrzE is most closely related to the enteric CheY protein. This similarity might suggest that FrzE could interact with the directional-switch proteins of the currently unidentified gliding motor, thereby directly regulating the reversal rate of gliding. However, both DifD and FrzE show conservation of specific amino acids known to be required for this interaction with the flagellar motor in the enteric bacteria (Roman et al., 1992), suggesting that perhaps both proteins could interact with the same component of the gliding motor.

Identification of Novel Components of the Frz Signal Transduction System

THE FrzZ PROTEIN IS A NOVEL RESPONSE REGULATOR

Sequencing upstream of the *frz* operon, Trudeau et al. (1996) discovered a new gene, *frzZ*. A *frzZ* null mutant shows abnormal vegetative swarming and produces frizzy aggregates on starvation agar. However, unlike most *frz* mutants, the *frzZ* mutant shows some responses to both attractants and repellents in a spatial chemotaxis assay and is therefore considered a modulator of directed-motility responses rather than a central component of the Frz pathway (Fig. 5). The *frzZ* gene encodes a protein with two domains, both of which show homology to the enteric response regulator CheY and other members of the response regulator family. Figure 4a shows both domains of FrzZ aligned with CheY from *S. typhimurium*, DifD, and the CheY-like domain of FrzE. Again, the putative phosphorylation site and associated amino acids can be seen to be conserved in both domains. Figure 4b, however, suggests that both domains are more distantly related to *S. typhimurium* CheY than either the

C-terminal domain of FrzE or DifD. The two domains are also quite distinct from each other, having only 27.5% identity. Interestingly, neither domain shows conservation of the amino acids known to be required for switch protein interactions in the enteric CheY, suggesting that FrzZ could play an alternative role in the directed-motility process. Several nonenteric bacterial species have been shown to encode multiple copies of the CheY protein, and in *Rhizobium meliloti* these multiple CheY proteins are suggested to play a role in the regulation of chemokinetic behavior (Sourjik and Schmitt, 1996; Armitage and Schmitt, 1997). Since *M. xanthus* is also known to utilize chemokinetic behavior during both stages of its complex life cycle (see below), a similar role for the multiple CheY domains found in the Frz and Dif systems might be considered. In order to determine the role of FrzZ, both CheY-like domains were cloned separately as "bait" into a yeast two-hybrid system and used to screen an *M. xanthus* genomic DNA library for interacting proteins. These studies revealed two possible FrzZ-interacting proteins: an ABC transporter and an extracytoplasmic-function (ECF) sigma factor.

AbcA IS A MEMBER OF THE ABC TRANSPORTER FAMILY

When studied in the yeast two-hybrid system, the first domain of FrzZ appeared to show an interaction with a new protein which, by sequence analysis, is a member of the bacterial exporters of the ABC transporter family (Ward et al., 1998b). Cells with mutations in the *abcA* gene, which is positioned at the start of what appears to be a three-gene operon (Fig. 2c), display the frizzy aggregation phenotype (in the strain FB background), suggesting that this protein might indeed be part of the same signal transduction pathway as FrzZ. The domain

FIGURE 4 Alignment of CheY-like proteins from *M. xanthus* with CheY from *S. typhimurium*. (a) Alignment of the *M. xanthus* CheY homologues in the DifD (Yang et al., 1998b), FrzE (McCleary and Zusman, 1990), and FrzZ (Trudeau et al., 1996) proteins against *S. typhimurium* CheY (Stock et al., 1985). The phosphorylation site (*S. typhimurium* Asp57) and associated amino acids are marked with "*." Amino acids proposed to be involved with switch protein interactions are indicated with "°." (b) Dendrogram created from the CheY protein alignment.

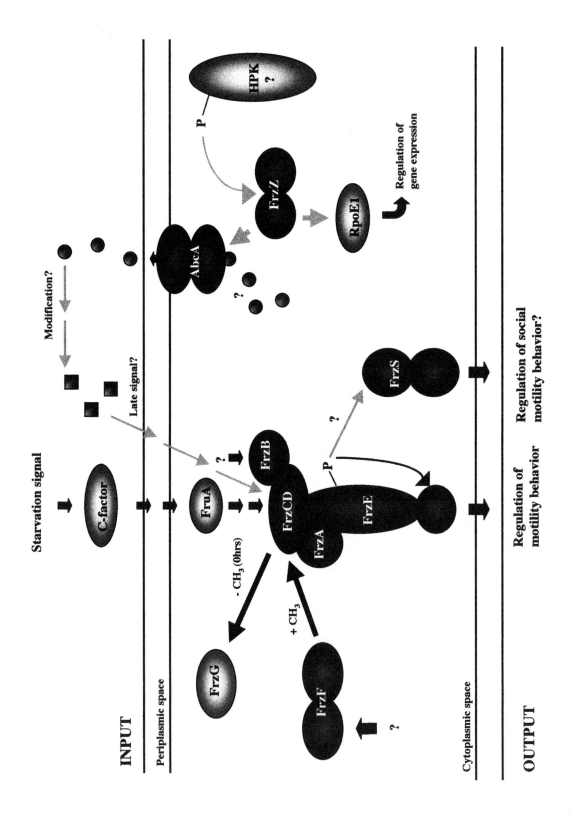

structure of AbcA suggests that this protein may function as an exporter, playing a role in the secretion of a molecule involved in developmental aggregation. This possibility was supported by cell-mixing experiments, in which a nonmotile strain was mixed (1:9) with the AbcA mutant and shown to rescue (complement) the frizzy phenotype of the AbcA mutant. Cells complemented in this way were restored to wild-type fruiting. Since it is not possible to complement the frizzy phenotype of other *frz* mutants, these results suggest that the AbcA protein may function upstream of the Frz signal transduction pathway and play a role in the export of a molecule important in developmental directed-motility responses (Fig. 5). However, it has not proven possible to isolate and characterize this molecule from conditioned medium, suggesting that other interpretations are also feasible. Since the Frz phenotype of the *abcA* mutant was only apparent in the strain FB background (which has a leaky mutation in the social-gliding gene, *pilQ*, formerly *sglA* [Wall et al., 1999]), and not in the fully motile wild-type background, the relevance of the AbcA protein in development is currently unknown. Kashefi and Hartzell (1995), however, have suggested that the PilQ protein itself may be involved with chemotaxis and be able to bypass part of the Frz system during development. A further analysis of the role of PilQ (see below) and its connection to AbcA in directed-motility responses would certainly be of interest.

RpoE1 IS AN ECF SIGMA FACTOR

A second protein suggested to interact with FrzZ was also identified by interaction trap technology with the first domain of FrzZ as bait. This protein shows homology to the family of ECF sigma factors and has been named RpoE1 (Ward et al., 1998a). Analysis of the DNA sequence surrounding the *rpoE1* gene showed it to be located downstream of the *frzZ* gene (Fig. 2b). Mutations in *rpoE1*, however, do not result in cells displaying the frizzy phenotype, although they do result in motility defects during vegetative swarming and an aberrant aggregation pattern at high cell densities (particularly in the strain FB background [Fig. 1e]). While the potential interaction between FrzZ and RpoE1 remains speculative, transcriptional analysis suggests that the *rpoE1* gene is transcribed from a σ^{70} promoter at the start of an operon containing two other genes, the last of which, *frzS*, appears to be another novel component of the Frz network.

FrzS IS REQUIRED FOR S-MOTILITY

The last gene in the *rpoE1* operon is considered a new component of the Frz network, since mutations in this gene result in cells displaying the frizzy phenotype. The gene has been designated *frzS* (Ward et al., unpublished). While *frzS* mutants display the characteristic frizzy phenotype during developmental aggregation, they do not show either the reduced single-cell reversal frequency characteristic of most *frz* mutants or the hyperreversal frequency of *frzD*

FIGURE 5 Model showing the proposed interactions within the Frz signal transduction pathway during developmental aggregation. Components of the pathway that when mutated result in cells displaying the frizzy aggregation phenotype are colored dark gray. The FrzZ and AbcA proteins are suggested to act upstream of the central components, FrzCD, FrzA, and FrzE, and may regulate the export of a developmentally important molecule. This molecule might be modified or might modify a signal associated with late developmental aggregation. The histidine protein kinase (HPK) that phosphorylates FrzZ is unknown. However, FrzZ is proposed to potentially interact with the RpoE1 protein to regulate gene expression. Inputs to the FrzCD receptor may be associated with the C-signaling pathway as well as with the AbcA-associated pathway. Other, unknown inputs may also occur. The role of FrzB is unknown, although the protein is suggested to interact with FrzCD. The FrzF and FrzG proteins methylate and demethylate FrzCD. It is also suggested that FrzF may interact with an unknown regulatory protein. The FrzS protein may be phosphorylated by FrzE and interact directly with the S-motility system. Likely pathways or interactions are shown connected by black arrows. More speculative interactions are shown connected by gray arrows.

mutants (cells with mutations in the 3' end of *frzCD*). Instead, these cells display only a slightly enhanced reversal frequency compared to that of wild-type cells. Interestingly, however, during vegetative swarming *frzS* mutants were shown to have a severe swarming defect on 0.3% agar, suggesting that FrzS might be required for social motility. When the *frzS* mutation is introduced into an A⁻ S⁺ motility mutant, the resulting double-mutant cells were shown to be nonswarming, i.e., A⁻ S⁻. The unique A⁻ *frzS* mutant phenotype suggests that the FrzS protein may be the component of the Frz signal transduction system that interacts directly with the social motility system (see below). The novel structure of FrzS is also consistent with this protein playing a regulatory role in motility. FrzS has an N-terminal receiver domain like those found in the response regulators of two-component signal transduction systems and a C-terminal domain which is predicted to be a coiled-coil structure resembling that of myosin.

C-SIGNALING IS REQUIRED FOR DEVELOPMENTAL AGGREGATION

Directed-motility behavior and extracellular signaling events are both known to be essential for completion of the developmental-aggregation process. While several observations suggest that *M. xanthus* generates and responds to diffusible molecules (see above), recent work has concentrated on the possibility that cell-cell contact signaling could also play a role in aggregation. Cells with mutations in the *csgA* gene (which encodes the cell-associated C-factor protein [Shimkets and Rafiee, 1990]) are arrested in early development, forming only slightly raised mounds instead of true aggregates (Fig. 1c). C-factor has pleiotropic effects on several developmental processes, including rippling, aggregation, sporulation, and gene expression (Shimkets and Kaiser, 1982). Since both rippling and aggregation are also defective in *frz* mutants, the connection between CsgA and FrzCD methylation was explored by Søgaard-Andersen and Kaiser (1996). These authors showed that the addition of purified

C-factor to developing *csgA* mutants resulted in methylation of FrzCD. This methylation response was shown to require *fruA* (Ogawa et al., 1996), as well as the methyltransferase FrzF. The FruA protein is a putative transcription factor which is presumed to be part of the pathway linking C-factor to the Frz system (Fig. 5).

The possibility of a correlation between levels of FrzCD methylation and the reversal frequency of gliding was proposed by Shi et al. (1996). These authors showed that FrzCD is more highly methylated in cells developing at high density than in cells developing at low density, suggesting that higher levels of FrzCD methylation could directly relate to reduced reversal frequencies. To date, no studies have shown that C-signaling causes such changes in reversal frequency. However, a recent and interesting study by Jelsbak and Søgaard-Andersen (1999) has indicated a novel aspect of C-factor involvement in the aggregation process. These authors added purified C-factor to starved wild-type cells and analyzed changes in the motility behavior of isolated, individual cells by using time-lapse video microscopy. Under these conditions, C-factor caused an increase in the speed of gliding rather than the expected change in reversal frequency, suggesting that FrzCD methylation might regulate changes in both the speed and reversal frequency of gliding.

DOES DEVELOPMENTAL AGGREGATION INVOLVE CHEMOTAXIS OR CHEMOKINESIS?

Two recent studies, performed first on swarming vegetative cells (Ward et al., 1998c) and later on developmental cells (Jelsbak and Søgaard-Andersen, 1999), have both indicated the importance of chemokinetic behavior in directed-motility responses in *M. xanthus*. Both studies suggest that cells can respond to environmental conditions by the generation of signaling molecules which result in increased gliding speeds (orthokinetic behavior). During vegetative swarming the ability to move faster would obviously aid colony dispersal. However, an increase in the speed of gliding during developmental aggregation which did not involve sensory adaptation would be unlikely to result in an accumulation of cells and the formation of aggregates. Therefore, we must assume that although cells are stimulated to move faster into aggregates, a second modulating effect on directed-motility behavior must also be required to keep the cells within the enlarging aggregates. This modulating behavior might presumably involve a cascade of different signals or the adaptive behavior of chemotaxis.

Shi et al. (1993) first presented evidence that groups of *M. xanthus* cells move chemotactically towards nutrients and away from repellents in stable gradients, although the design of their experiments did not permit the observation of adaptive behavior within the groups of cells. A recent report by Kearns and Shimkets (1998), however, has documented chemotactic movements of individual cells in response to defined chemical stimuli. These authors observed cells exhibiting directed motility up gradients of phosphatidylethanolamine (PE). Microscopic analysis of individual cells showed that both dilauroyl and dioleoyl PE decreased the reversal frequency of gliding and that sensory adaptation occurred approximately 1 h after the onset of stimulation.

Kearns and Shimkets (1998) also analyzed the role of the Frz signal transduction system in their chemotaxis assay and showed that both FrzCD and FrzE mutants still respond to PE gradients, but only at reduced rates. These mutants were shown to display the excitation responses to PE, but they did not adapt. Since both FrzCD and FrzE are central components of the Frz pathway and are presumably required for Frz-specific excitation responses, this result suggests that other signal transduction pathways must also respond to PE gradients but that these alternative pathways may not have adaptive capacity. An analysis of the response of both Dif mutants and Dif-Frz double mutants to PE gradients would certainly be of interest. Further studies are also required to demonstrate whether *M. xanthus* cells utilize PE compounds during vegetative swarming or during developmental-aggregation responses.

In the enteric bacteria, adaptation to chemoeffectors is modulated by the methylation and demethylation of the MCP receptors. While no analysis of adaptive behavior in *M. xanthus* has been performed during aggregation, the Frz system does contain both a methyltransferase, FrzF, capable of methylating FrzCD and a methylesterase, FrzG, which is presumed to demethylate FrzCD. The presence of both of these enzymes suggests that the Frz system could undergo adaptation in a manner similar to that of the enteric Che system and therefore regulate directed-motility responses chemotactically. However, cells with mutations in the *frzG* locus demethylate FrzCD in response to the repellent isoamyl alcohol (McBride et al., 1992). FrzG mutants are also capable of fruiting, although in the wild type FrzCD is rapidly demethylated at the onset of starvation. These observations suggest that the FrzG protein may be redundant and that a second methylesterase might presumably act to demethylate FrzCD.

MOTILITY AND DEVELOPMENTAL AGGREGATION

Gliding motility in *M. xanthus* involves two distinct forms of movement which together produce wild-type motility behavior (Hodgkin and Kaiser, 1979a, 1979b). Isolated individual cells can only move by using the adventurous (A) motility system, which is generally more efficient on drier surfaces (Shi and Zusman, 1993). A-motility also functions in movement of groups of cells. S-motility, which is more effective under moister conditions, is, however, limited to situations where cells are closely associated, since it requires cell-cell interactions (Kaiser and Crosby, 1983). Cells with mutations in both systems (A⁻ S⁻) are nonswarming. While it might seem logical that cells with mutations in either motility system would be defective in the formation of developmental aggregates and therefore of fruiting bodies, this assumption is not correct. Cells with mutations in the A-motility genes (*agl* and *cgl*) frequently have either minor or no aggregation defects, whereas the majority of cells with mutations in S-motility genes (*pil, sgl,* and *tgl*) have gross fruiting defects (Kroos et al., 1990; Hartzell and Youderian, 1995).

While the mechanism of S-motility remains unknown, it is clear that this form of movement requires the presence of type IV polar pili (Fig. 6) (Kaiser, 1979). Recent studies by Wu and Kaiser (1995, 1996, 1997) and Wu et al. (1997, 1998) have made dramatic advances in our understanding of pilus biosynthesis and secretion, as well as S-motility behavior. The *pilA* gene, which expresses prepilin, the major fimbrial subunit of type IV pili, was localized within a chromosomal region containing several genes (Fig. 2D), either homologous to *pil* genes of *Pseudomonas aeruginosa* or required for S-motility function. In *P. aeruginosa*, type IV pili are suggested to play a role in the surface-dependent translocation mechanism called twitching motility, which is regulated by a set of genes showing homology to the enteric chemotaxis genes (Darzins and Russell, 1997). Social motility in *M. xanthus* and twitching in *P. aeruginosa* appear to be based on a common strategy, and the recently identified links between both the Frz and the Dif systems and the S-motility system (see above) suggest that both S-motility and twitching motility may be regulated on some level by Che-like signal transduction pathways.

Analyses performed on cells with mutations in the known *pil* genes have suggested that they may be classified into several defined groups. Cells with mutations in *pilA* or *pilR* do not produce pilin and are therefore nonpiliated. Conversely, cells with mutations in *pilS* are piliated. Since PilR and PilS are suggested to constitute a two-component signal transduction system, the PilS histidine protein kinase is presumed to be a negative regulator of PilR function (Wu and Kaiser, 1995). Both *pilB* and *pilC* mutants produce pilin but do not have surface pili. The equivalent PilB and PilC proteins of *P. aeruginosa* are themselves homologues of proteins involved in type II secretion, suggesting that the *M. xanthus* PilB and PilC proteins may play this same role. Similarly, the PilG, PilH, and PilI proteins are suggested to act as a transporter system, since they share homology with ABC transporter proteins, although presently they do not have known homologues in other type IV systems (Wu et al., 1998). Cells with mutations in these loci do not produce pili. The *pilT* gene is proposed to encode a protein which functions in pilus retraction, since cells with mutations in PilT have pili but cannot move by S-motility (Wu et al., 1997). The differences in aggregation phenotypes of the *pil* mutants suggest that while pili are required for the formation of normal, round, dark fruiting bodies, S-motility may be required for the normal distribution and size of fruiting bodies.

In a recent manuscript, Wall et al. (1999) have analyzed one further S-motility gene, *sglA*. This gene has been of specific interest because the *sglA1* spontaneous mutation is present in most laboratory strains (FB strains), since it allows dispersed cell growth in liquid while not affecting fruiting body formation. The *sglA* gene was shown to encode a secretin, a member of a group of proteins involved in macromolecular transport across the outer

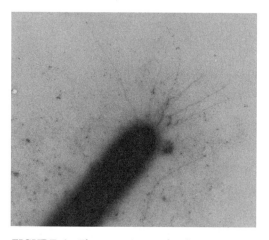

FIGURE 6 Electron micrograph of an *M. xanthus* cell showing the polar associated type IV pili. A strain DZ2 (wild-type) cell is shown negatively stained with uranyl acetate.

membrane, and renamed *pilQ*. *pilQ* was shown to be essential for pilus biogenesis, S-motility, rippling, and fruiting body formation, while the *pilQl* allele was suggested to alter the permeability of the outer membrane, thus resulting in pleiotropic defects.

Several gene products, in addition to those encoded by the *pil, sgl*, and *tgl* (Rodriguez-Soto and Kaiser, 1997) genes, are also required for S-motility behavior. The Dsp locus has been suggested to be part of the S-motility system, since A⁻ Dsp⁻ cells are nearly nonmotile (Shimkets, 1986). In addition to this S-motility defect, cells with mutations in the *dsp* locus show abnormal cell cohesion, aggregation, and sporulation. These pleiotropic effects are suggested to be due to the lack of specific cell surface appendages, termed fibrils. This material, which is composed of protein-associated polysaccharide polymers, can form thick filaments (Behmlander and Dworkin, 1994a, 1994b). These filaments have been observed at both the sides and tips of cells by electron microscopy, appearing to coat the cells and form a cohesive extracellular interconnective matrix (Arnold and Shimkets, 1988). The recently identified *sglK* gene, which encodes a nonessential homologue of the chaperone HSP70 (DnaK), like *dsp*, is involved in both S-motility and cohesion (Yang et al., 1998a; Weimer et al., 1998). Weimer et al. (1998) have shown that *sglK* mutants do not produce extracellular fibrils. However, the role of fibrils in S-motility and development remain unclear, since Chang and Dworkin (1996) have shown that a mutation in the *fbp* locus can restore both S-motility and development to *dsp* mutants without restoring production of fibrils.

The SasA locus has also been shown to be required for S-motility and developmental aggregation (Bowden and Kaplan, 1998). Cells with mutations in any of the three genes within this locus (*wzm, wzt*, and *wbgA*) are defective in lipopolysaccharide O-antigen biosynthesis. Since these mutants retain both type IV pili and fibrils, a role for O-antigen in S-motility also appears likely.

MODIFYING THE COURSE OF FRUITING BODY FORMATION

Temporal aspects of fruiting body formation in *M. xanthus* have received little direct attention, although the conversion of vegetative cells into spores is certainly delayed under normal starvation conditions until the cells are within developmental aggregates. Several genes that may be involved with regulating the timing of fruiting body formation itself have also been identified. One of these genes encodes the cell surface-associated protein MBHA (myxobacterial hemagglutinin). Cells which cannot synthesize this protein show delayed aggregation, suggesting that MBHA may be required to increase the efficiency of cell-cell interactions and fruiting body formation (Romeo and Zusman, 1987). However, perhaps the best studied of the genes which potentially regulate temporal aspects of aggregation are a group of eukaryotic-like serine-threonine kinases. The first prokaryotic Ser-Thr kinase gene, *pkn1*, was identified in *M. xanthus* by Muñoz-Dorado et al. (1991). Construction of a deletion mutant of this gene, Δ*pkn1*, resulted in cells that started to aggregate earlier than the wild type and finally formed small, messy fruiting bodies. This premature aggregation was also shown to be characteristic of a Δ*pkn2* mutant strain (Udo et al., 1996), suggesting that both of these proteins act as negative regulators of fruiting body formation.

At least 13 different Ser-Thr kinases are proposed to be present in *M. xanthus*, and 6 of these appear to be transmembrane proteins suggested to act as receptors for developmental signals (Hanlon et al., 1997). However, not all of these genes work to delay aggregation. The *pkn9* gene is suggested to act as a nutrient sensor which might stimulate aggregation, since a Δ*pkn9* mutation resulted in cells with a delayed-aggregation phenotype. The transcriptionally divergent *pkn5* and *pkn6* genes are interesting, since they appear to play reciprocal roles in *M. xanthus* growth and development (Zhang et al., 1996). While a Δ*pkn5* mutant forms fruiting bodies faster than the wild type,

Δ*pkn6* mutant develops more slowly than the wild type. Pkn5, like Pkn1 and Pkn2, has therefore been suggested to be a negative regulator of development, while Pkn6 may be a transmembrane sensor that detects extracellular developmental signals.

Studies performed with eukaryotic signal transduction systems have indicated that protein phosphatases play roles as important as those of Ser-Thr kinases in regulating specific cellular processes (Bork et al., 1996). Since *M. xanthus* has several Ser-Thr kinases, it is perhaps not surprising that a gene showing homology to the eukaryotic PP2C protein phosphatases was identified in this bacterium (Treuner-Lange et al., unpublished). This gene, *pph1* (protein phosphatase 1), is expressed under both vegetative and developmental conditions from different promoters, with transcription increased at least twofold during development. Most interestingly, however, mutations in the *pph1* gene, like mutations in some of the Pkn mutants, are suggested to cause delayed developmental aggregation.

Recently, Cho and Zusman (in press) identified a new mutant which showed an early-aggregation phenotype, similar to those of the *pkn1, pkn2,* and *pkn6* mutants, along with a second novel phenotype where the mutant produced a large number of very small, closely packed fruiting bodies (Fig. 1f). A microscopic analysis of the *espA* (early sporulation) mutant phenotype during development showed that the entire population of cells, both within and outside of aggregation centers, developed into spores. This phenotype is unusual, since during normal starvation-induced development populations of cells form spores only within aggregates; those cells outside of these aggregation centers remain as a subpopulation called peripheral rods (O'Connor and Zusman, 1991). Thus, it appears that the *espA* gene product, a histidine protein kinase, encodes an inhibitor of sporulation. Cloning and sequencing of the *espA* gene have shown it to be part of an operon. A mutation in the downstream gene, *espB*, delayed sporulation. Expression studies have confirmed that both genes are transcribed

only during the developmental cycle and that they are not expressed in the presence of glycerol sporulation-inducing compounds. This recent study suggests that signal transduction systems involving both Ser-Thr kinases and histidine protein kinases may both be active in the temporal regulation of development in *M. xanthus.*

TEMPERATURE-DEPENDENT AGGREGATION

Studies of *M. xanthus* development suggest that temperature could play an important role in the developmental aggregation process. For example, at 34°C aggregation is accelerated compared to that at 28°C and gene expression shows many differences (O'Connor and Zusman, 1990). A number of mutants were isolated which did not aggregate at 34°C (Fig. 1g) but formed normal aggregates at 28°C (Fig. 1h) (Morrison and Zusman, 1979). One group of these mutants, called Tag (temperature dependent for aggregation) (O'Connor and Zusman, 1990), remained temperature sensitive for aggregation even when the genes were disrupted by transposon mutagenesis or deletions, indicating that the cells have at least two sets of genes for developmental aggregation. The *tag* genes constitute one set: they are required for normal development at 34°C but are not required at 28°C. When the *tag* mutants were observed by scanning electron microscopy, the cells were seen to form small aggregation centers in the monolayers that constitute the early-developing population. These monolayers, however, were unable to initiate the movements perpendicular to the substratum that are necessary for the formation of mounds. However, despite the defects in aggregation, the cells sporulated at the correct time and in the correct numbers. Cloning and sequencing of the *tag* genes showed them to be organized into at least two operons (O'Connor and Zusman, 1990; O'Connor, personal communication). The *tagA* gene is transcribed at a low level during vegetative growth but is induced at the onset of starvation. Expression peaks approximately 5 h after the initiation of development.

Although little is known about the role of *tagA* in the developmental process, the TagA protein is 58% identical to the RpfA protein of *Xanthomonas campestris* (Barber et al., 1997). Both TagA and RpfA also show significant similarities to the iron response element binding proteins (IREBPs) of eukaryotic organisms (Kuhn, 1994). Some IREBPs are known to be responsible for aconitase activity, as is RpfA (Wilson et al., 1998), and a point mutation in the *tagA* gene has been shown to reduce aconitase activity in vegetatively growing *M. xanthus* cells by 50%. This loss of activity does not, however, result in a defect in vegetative growth, suggesting that *M. xanthus* encodes a second aconitase.

The *tagE* gene is predicted to encode a large lipoprotein which is likely to be localized to the outer membrane. Regions of this protein share similarity with PilC of *Neisseria gonorrhoeae* and PilY1 of *P. aeruginosa* (Alm et al., 1996). Both of these proteins are important in twitching motility, again implying a connection between developmental aggregation and S-motility in *M. xanthus*. PilC of *N. gonorrhoeae* is known to be an adhesin which may be localized to the type IV pili (Rudel et al., 1995; Ryll et al., 1997). A similar role for TagE might be proposed, since the Tag mutant phenotype can be extracellularly rescued in mixing experiments with wild-type donor cells. These rescue experiments cannot be performed if the two cell types are separated by a dialysis membrane, suggesting that cell-cell contact (which is required for S-motility) is necessary for complementation. Additional evidence that the TagE protein is required for S-motility was suggested when the *tagE* mutation was introduced into an A$^-$ motility mutant background (A$^-$ S$^+$). The resultant A$^-$ *tagE* mutant cells appear nonmotile on starvation agar at 34°C, although they retain S-motility under other conditions (O'Connor, unpublished).

DEVELOPMENTAL AGGREGATION REQUIRES THE COORDINATED ACTION OF MANY DIFFERENT SYSTEMS

In recent years many advances have been made to our understanding of the developmental aggregation process. Extracellular signaling with both diffusible molecules and cell-cell interactions appears to stimulate aggregation by inducing both chemokinetic and chemotactic directed-motility responses. Currently, two directed-motility-associated signal transduction systems, Dif and Frz, which play dominant roles during different stages of the aggregation process, have been identified. Characterization of the Frz signal transduction system suggests that a highly complex relay of information is required to coordinate the formation of fruiting bodies (Fig. 5). Recent evidence suggests that aggregation may even require the production of a Frz-associated developmental signal, which might be secreted by the AbcA protein. The Frz system is also suggested to interconnect directly with the S-motility system via the FrzS protein, and probably with the A-motility system via FrzE, in order to regulate the directed-motility behavior of cells. Further characterization of these systems, along with the Ser-Thr protein kinase signal transduction systems and other systems which regulate time- and temperature-dependent aspects of aggregation, will undoubtedly provide additional insights into how a simple prokaryote is able to coordinate the highly complex multicellular behavior required for developmental aggregation.

ACKNOWLEDGMENTS

We thank the Shi, Shimkets, and Søgaard-Andersen laboratories for providing us with data that were either unpublished or in press at the time of writing. We also thank members of our laboratory for providing us with recent data and figures.

Research in our laboratory is funded by Public Health Service grant GM20509 from the National Institutes of Health.

REFERENCES

Acuña, G., W. Shi, K. Trudeau, and D. R. Zusman. 1995. The 'CheA' and 'CheY' domains of *Myxococcus xanthus* FrzE function independently in vitro as an autokinase and a phosphate acceptor, respectively. *FEBS Lett.* **358**:31–33.

Alm, R. A., J. P. Hallinan, A. A. Watson, and J. S. Mattick. 1996. Fimbrial biogenesis genes of *Pseudomonas aeruginosa*: *pilW* and *pilX* increase the similarity of type 4 fimbriae to the GSP protein-secretion systems and *pilY1* encodes a gonococcal PilC homologue. *Mol. Microbiol.* **22:**161–173.

Armitage, J. P., and R. Schmitt. 1997. Bacterial chemotaxis: *Rhodobacter sphaeroides* and *Sinorhizobium meliloti*—variations on a theme? *Microbiology* **143:**3671–3682.

Arnold, J. W., and L. J. Shimkets. 1988. Cell surface properties correlated with cohesion in *Myxococcus xanthus. J. Bacteriol.* **170:**5771–5777.

Astling, D. P., M. J. Ward, and D. R. Zusman. Unpublished data.

Barber, C. E., J. L. Tang, J. X. Feng, M. Q. Pan, T. J. G. Wilson, H. Slater, J. M. Dow, P. Williams, and M. J. Daniels. 1997. A novel regulatory system required for pathogenicity of *Xanthomonas campestris* is mediated by a small diffusible signal molecule. *Mol. Microbiol.* **24:**555–566.

Behmlander, R. M., and M. Dworkin. 1994a. Biochemical and structural analyses of the extracellular matrix fibrils of *Myxococcus xanthus. J. Bacteriol.* **176:**6295–6303.

Behmlander, R. M., and M. Dworkin. 1994b. Integral proteins of the extracellular matrix fibrils of *Myxococcus xanthus. J. Bacteriol.* **176:**6304–6311.

Blackhart, B. D., and D. R. Zusman. 1985. "Frizzy" genes of *Myxococcus xanthus* are involved in control of frequency of reversal of gliding motility. *Proc. Natl. Acad. Sci. USA* **82:**8767–8770.

Blair, D. F. 1995. How bacteria sense and swim. *Annu. Rev. Microbiol.* **49:**489–522.

Bork, P., N. P. Brown, H. Hegyi, and J. Schultz. 1996. The protein phosphatase 2C (PP2C) superfamily: detection of bacterial homologues. *Prot. Sci.* **5:**1421–1425.

Bowden, M. G., and H. B. Kaplan. 1998. The *Myxococcus xanthus* lipopolysaccharide O-antigen is required for social motility and multicellular development. *Mol. Microbiol.* **30:**275–284.

Chang, B.-Y., and M. Dworkin. 1996. Mutants of *Myxococcus xanthus dsp* defective in fibril binding. *J. Bacteriol.* **178:**697–700.

Cho, K., and D. R. Zusman. Sporulation timing in *Myxococcus xanthus* is controlled by the *espAB* locus. *Mol. Microbiol.*, in press.

Darzins, A., and M. A. Russell. 1997. Molecular genetic analysis of type-4 pilus biogenesis and twitching motility using *Pseudomonas aeruginosa* as a model system—a review. *Gene* **192:**109–115.

Hagen, D. C., A. P. Bretscher, and D. Kaiser. 1978. Synergism between morphogenic mutants of *Myxococcus xanthus. Dev. Biol.* **64:**284–296.

Hanlon, W. A., M. Inouye, and S. Inouye. 1997. Pkn9, a Ser/Thr protein kinase involved in the development of *Myxococcus xanthus. Mol. Microbiol.* **23:**459–471.

Harris, B. Z., D. Kaiser, and M. Singer. 1998.

The guanosine nucleotide (p)ppGpp initiates development and A-factor production in *Myxococcus xanthus. Genes Dev.* **12:**1022–1035.

Hartzell, P. L., and P. Youderian. 1995. Genetics of gliding motility and development in *Myxococcus xanthus. Arch. Microbiol.* **164:**309–323.

Hodgkin, J., and D. Kaiser. 1979a. Genetics of gliding motility in *Myxococcus xanthus* (Myxobacterales): genes controlling movement of single cells. *Mol. Gen. Genet.* **171:**167–176.

Hodgkin, J., and D. Kaiser. 1979b. Genetics of gliding motility in *Myxococcus xanthus* (Myxobacterales): two gene systems control movement. *Mol. Gen. Genet.* **171:**177–191.

Hodgson, D. A., and F. J. Murillo. 1993. Genetics of regulation and pathway of synthesis of carotenoids, p. 157–181. *In* M. Dworkin and D. Kaiser (ed.), *Myxobacteria II.* American Society for Microbiology, Washington, D.C.

Jelsbak, L., and L. Søgaard-Andersen. 1999. The cell-surface associated intercellular C-signal induces behavioral changes in individual *Myxococcus xanthus* cells during fruiting body morphogenesis. *Proc. Natl. Acad. Sci. USA* **96:**5031–5036.

Kaiser, D. 1979. Social gliding is correlated with the presence of pili in *Myxococcus xanthus. Proc. Natl. Acad. Sci. USA* **76:**5952–5956.

Kaiser, D., and C. Crosby. 1983. Cell movement and its coordination in swarms of *Myxococcus xanthus. Cell Motil.* **3:**227–245.

Kashefi, K., and P. L. Hartzell. 1995. Genetic suppression and phenotypic masking of a *Myxococcus xanthus frzF⁻* defect. *Mol. Microbiol.* **15:**483–494.

Kearns, D. B., and L. J. Shimkets. 1998. Chemotaxis in a gliding bacterium. *Proc. Natl. Acad. Sci. USA* **95:**11957–11962.

Kroos, L., A. Kuspa, and D. Kaiser. 1990. Defects in fruiting body development caused by Tn5lac insertions in *Myxococcus xanthus. J. Bacteriol.* **172:**484–487.

Kuhn, L. C. 1994. Molecular regulation of iron proteins. *Baillieres Clin. Haematol.* **7:**763–785.

Kuspa, A., and D. Kaiser. 1989. Genes required for developmental signalling in *Myxococcus xanthus:* three *asg* loci. *J. Bacteriol.* **171:**2762–2772.

Kuspa, A., L. Plamann, and D. Kaiser. 1992a. Identification of heat-stable A-factor from *Myxococcus xanthus. J. Bacteriol.* **174:**3319–3326.

Kuspa, A., L. Plamann, and D. Kaiser. 1992b. A-signalling and the cell density requirement for *Myxococcus xanthus* development. *J. Bacteriol.* **174:**7360–7369.

Lev, M. 1954. Demonstration of a diffusible fruiting factor in myxobacteria. *Nature* (London) **173:**501.

Manoil, C., and D. Kaiser. 1980. Guanosine pentaphosphate and guanosine tetraphosphate accumulation and induction of *Myxococcus xanthus* fruiting body development. *J. Bacteriol.* **141**:305–315.

McBride, M. J., R. A. Weinberg, and D. R. Zusman. 1989. 'Frizzy' aggregation genes of the gliding bacterium *Myxococcus xanthus* show sequence similarities to the chemotaxis genes of enteric bacteria. *Proc. Natl. Acad. Sci. USA* **86**:424–428.

McBride, M. J., T. Köhler, and D. R. Zusman. 1992. Methylation of FrzCD, a methyl-accepting taxis protein of *Myxococcus xanthus*, is correlated with factors affecting cell behavior. *J. Bacteriol.* **174**:4246–4257.

McCleary, W. R., and D. R. Zusman. 1990. FrzE of *Myxococcus xanthus* is homologous to both CheA and CheY of *Salmonella typhimurium*. *Proc. Natl. Acad. Sci. USA* **87**:5898–5902.

McVittie, A., and S. A. Zahler. 1962. Chemotaxis in *Myxococcus*. *Nature* (London) **194**:1299–1300.

Morrison, C. E., and D. R. Zusman. 1979. *Myxococcus xanthus* mutants with temperature-sensitive, stage-specific defects: evidence for independent pathways in development. *J. Bacteriol.* **140**:1036–1042.

Muñoz-Dorado, J., S. Inouye, and M. Inouye. 1991. A gene encoding a protein serine/threonine kinase is required for normal development of *M. xanthus*, a gram-negative bacterium. *Cell* **67**:995–1006.

O'Connor, K. A. Personal communication.

O'Connor, K. A. Unpublished data.

O'Connor, K. A., and D. R. Zusman. 1989. Patterns of cellular interactions during fruiting-body formation in *Myxococcus xanthus*. *J. Bacteriol.* **171**:6013–6024.

O'Connor, K. A., and D. R. Zusman. 1990. Genetic analysis of *tag* mutants of *Myxococcus xanthus* provides evidence for two developmental aggregation systems. *J. Bacteriol.* **172**:3868–3878.

O'Connor, K. A., and D. R. Zusman. 1991. Development in *Myxococcus xanthus* involves differentiation into two cell types, peripheral rods and spores. *J. Bacteriol.* **173**:3318–3333.

Ogawa, M., S. Fujitani, X. Mao, S. Inouye, and T. Komano. 1996. FruA, a putative transcription factor essential for the development of *Myxococcus xanthus*. *Mol. Microbiol.* **22**:757–767.

Reichenbach, H. 1984. Myxobacteria: a most peculiar group of social prokaryotes, p. 1–50. *In* E. Rosenberg (ed.), *Myxobacteria: Development and Cell Interactions*. Springer-Verlag, Berlin, Germany.

Reichenbach, H. 1993. Biology of the Myxobacteria: ecology and taxonomy, p. 13–62. *In* M. Dworkin and D. Kaiser (ed.), *Myxobacteria II*. American Society for Microbiology, Washington, D.C.

Reichenbach, H., and M. Dworkin. 1981. The order Myxobacterales, p. 328–355. *In* M. P. Starr,

H. Stolp, H. G. Truper, A. Balows, and H. G. Schlegel (ed.), *The Prokaryotes: a Handbook on Habitats, Isolation, and Identification of Bacteria*. Springer-Verlag, Berlin, Germany.

Rodriguez-Soto, J. P., and D. Kaiser. 1997. The *tgl* gene: social motility and stimulation in *Myxococcus xanthus*. *J. Bacteriol.* **179**:4361–4371.

Roman, S. J., M. Meyers, K. Volz, and P. Matsumura. 1992. A chemotactic signaling surface on Che Y defined by suppressors of flagellar switch mutations. *J. Bacteriol.* **174**:6247–6255.

Romeo, J. M., and D. R. Zusman. 1987. Cloning of the gene for myxobacterial hemagglutinin and isolation and analysis of structural gene mutations. *J. Bacteriol.* **169**:3801–3808.

Rudel, T., I. Scheurerpflug, and T. F. Meyer. 1995. *Neisseria* PilC protein identified as type-4 pilus tip-located adhesin. *Nature* (London) **373**:357–359.

Ryll, R. R., T. Rudel, I. Scheurerpflug, R. Barten, and T. F. Meyer. 1997. PilC of *Neisseria meningitidis* is involved in class II pilus formation and restores pilus assembly, natural transformation competence and adherence to epithelial cells in PilC-deficient gonococci. *Mol. Microbiol.* **23**:879–892.

Shi, W., and D. R. Zusman. 1993. The two motility systems of *Myxococcus xanthus* show different selective advantages on various surfaces. *Proc. Natl. Acad. Sci. USA* **90**:3378–3382.

Shi, W., T. Köhler, and D. R. Zusman. 1993. Chemotaxis plays a role in the social behavior of *Myxococcus xanthus*. *Mol. Microbiol.* **9**:601–611.

Shi, W., F. Ngok, and D. R. Zusman. 1996. Cell density regulates cellular reversal frequency in *Myxococcus xanthus*. *Proc. Natl. Acad. Sci. USA* **93**:4142–4146.

Shimkets, L. J. 1986. Role of cell cohesion in *Myxococcus xanthus* fruiting body formation. *J. Bacteriol.* **166**:842–848.

Shimkets, L. J. 1990. Social and developmental biology of the Myxobacteria. *Microbiol. Rev.* **54**:473–501.

Shimkets, L. J., and M. Dworkin. 1981. Excreted adenosine is a cell density signal for the initiation of fruiting body formation in *Myxococcus xanthus*. *Dev. Biol.* **84**:51–60.

Shimkets, L. J., and D. Kaiser. 1982. Induction of coordinated movement in *Myxococcus xanthus* cells. *J. Bacteriol.* **152**:451–461.

Shimkets, L. J., and H. Rafiee. 1990. CsgA, an extracellular protein essential for *Myxococcus xanthus* development. *J. Bacteriol.* **172**:5299–5306.

Singer, M., and D. Kaiser. 1995. Ectopic production of guanosine penta- and tetraphosphate can initiate early developmental gene expression in *Myxococcus xanthus*. *Genes Dev.* **9**:1633–1644.

Søgaard-Andersen, L., and D. Kaiser. 1996. C factor, a cell-surface-associated intercellular signaling protein, stimulates the cytoplasmic Frz signal transduction system in *Myxococcus xanthus*. *Proc. Natl. Acad. Sci. USA* **93**:2675–2679.

Sourjik, V., and R. Schmitt. 1996. Different roles of CheY1 and CheY2 in the chemotaxis of *Rhizobium meliloti*. *Mol. Microbiol.* **22**:427–436.

Stock, A., D. E. Koshland, Jr., and J. Stock. 1985. Homologies between the *Salmonella typhimurium* CheY protein and proteins involved in the regulation of chemotaxis, membrane protein synthesis and sporulation. *Proc. Natl. Acad. Sci. USA* **82**:7989–7993.

Thaxter, R. 1892. On the Myxobacteriaceae, a new order of Schizomycetes. *Bot. Gaz.* **17**:389–406.

Treuner-Lange, A., M. J. Ward, and D. R. Zusman. Unpublished data.

Trudeau, K. G., M. J. Ward, and D. R. Zusman. 1996. Identification and characterization of FrzZ, a novel response regulator necessary for swarming and fruiting-body formation in *Myxococcus xanthus*. *Mol. Microbiol.* **20**:645–655.

Udo, H., M. Inouye, and S. Inouye. 1996. Effects of overexpression of Pkn2, a transmembrane protein serine/threonine kinase, on development of *Myxococcus xanthus*. *J. Bacteriol.* **178**:6647–6649.

Wall, D., P. E. Kolenbrander, and D. Kaiser. 1999. The *Myxococcus xanthus pilQ* (*sglA*) gene encodes a secretin homolog required for the type IV pilus biogenesis, social motility, and development. *J. Bacteriol.* **181**:24–33.

Ward, M. J., and D. R. Zusman. 1997. Regulation of directed motility in *Myxococcus xanthus*. *Mol. Microbiol.* **24**:885–893.

Ward, M. J., H. Lew, and D. R. Zusman. Unpublished data.

Ward, M. J., H. Lew, A. Treuner-Lange, and D. R. Zusman. 1998a. Regulation of motility behavior in *Myxococcus xanthus* may require an extracytoplasmic-function sigma factor. *J. Bacteriol.* **180**:5668–5675.

Ward, M. J., K. C. Mok, D. P. Astling, H. Lew, and D. R. Zusman. 1998b. An ABC transporter plays a developmental aggregation role in *Myxococcus xanthus*. *J. Bacteriol.* **180**:5697–5703.

Ward, M. J., K. C. Mok, and D. R. Zusman. 1998c. *Myxococcus xanthus* displays Frz-dependent chemokinetic behavior during vegetative swarming. *J. Bacteriol.* **180**:440–443.

Weimer, R. M., C. Creighton, A. Stassinopoulos, P. Youderian, and P. L. Hartzell. 1998. A chaperone in the HSP70 family controls production of extracellular fibrils in *Myxococcus xanthus*. *J. Bacteriol.* **180**:5357–5368.

Wilson, T. J. G., N. Bertrand, J.-L. Tang, J.-X. Feng, M.-Q. Pan, C. E. Barber, J. M. Dow, and M. J. Daniels. 1998. The *rpfA* gene of *Xanthomonas campestris* pathovar *campestris*, which is involved in the regulation of pathogenicity factor production, encodes an aconitase. *Mol. Microbiol.* **28**:961–970.

Wireman, J. W., and M. Dworkin. 1975. Morphogenesis and developmental interactions in myxobacteria. *Science* **189**:516–523.

Wu, S. S., and D. Kaiser. 1995. Genetic and functional evidence that type IV pili are required for social gliding motility in *Myxococcus xanthus*. *Mol. Microbiol.* **18**:547–558.

Wu, S. S., and D. Kaiser. 1996. Markerless deletions of *pil* genes in *Myxococcus xanthus* generated by counterselection with the *Bacillus subtilis sacB* gene. *J. Bacteriol.* **178**:5817–5821.

Wu, S. S., and D. Kaiser. 1997. Regulation of expression of the *pilA* gene in *Myxococcus xanthus*. *J. Bacteriol.* **178**:7748–7758.

Wu, S. S., J. Wu, and D. Kaiser. 1997. The *Myxococcus xanthus pilT* locus is required for social gliding motility although pili are still produced. *Mol. Microbiol.* **23**:109–121.

Wu, S. S., J. Wu, Y. L. Cheng, and D. Kaiser. 1998. The *pilH* gene encodes an ABC transporter homologue required for type IV pilus biogenesis and social gliding motility in *Myxococcus xanthus*. *Mol. Microbiol.* **29**:1249–1261.

Yang, Z., Y. Geng, and W. Shi. 1998a. A DnaK homolog in *Myxococcus xanthus* is involved in social motility and fruiting body formation. *J. Bacteriol.* **180**:218–224.

Yang, Z., Y. Geng, D. Xu, H. B. Kaplan, and W. Shi. 1998b. A new set of chemotaxis homologs is essential for *Myxococcus xanthus* social motility. *Mol. Microbiol.* **30**:1123–1130.

Zhang, W., M. Inouye, and S. Inouye. 1996. Reciprocal regulation of the differentiation of *Myxococcus xanthus* by Pkn5 and Pkn6, eukaryotic-like Ser/Thr protein kinases. *Mol. Microbiol.* **20**:435–447.

Zusman, D. R. 1982. 'Frizzy' mutants: a new class of aggregation-defective developmental mutants of *Myxococcus xanthus*. *J. Bacteriol.* **150**:1430–1437.

CELL-INTERACTIVE SENSING OF THE ENVIRONMENT

Dale Kaiser

12

Myxococcus xanthus can respond to nutrient limitation in either of two ways (Fig. 1): (i) it can enter a stationary phase, in which cells grow very slowly, or (ii) it can stop growth and initiate the development of fruiting bodies with the eventual differentiation of spores. Both responses are risky; both are likely to kill many cells. Nevertheless, each has advantages, the strengths of which depend on the conditions.

In stationary phase, the slow growth of some cells is balanced by the death of others. In *Escherichia coli* and *Salmonella typhimurium*, stationary phase triggers differentiation into environmentally resistant rod cells (Hengge-Aronis, 1996), but this resistance is not manifest in *M. xanthus* despite the presence of a stationary phase and a stationary-phase sigma factor, encoded by *sigD* (Ueki and Inouye, 1998). In fact, *M. xanthus* stationary-phase cells are prone to lysis, suggesting that their walls have been weakened (Wireman and Dworkin, 1977). Stationary phase will lead ultimately to the death of the entire population when the nutrients are totally exhausted.

On the other hand, fruiting body development is also biologically costly. Under labora- tory conditions, no more than half of the input cells become morphological spores, and a smaller proportion survive as viable, heat- and sonication-resistant myxospores (Wireman and Dworkin, 1977; O'Connor and Zusman, 1988; Kim and Kaiser, 1990). These propor- tions depend on the level of nutrients available during development to support biosynthesis.

In addition to a biological cost in terms of the number of survivors, the initiation of fruit- ing body development is constrained to antici- pate starvation. At least 30 new proteins are made during fruiting body development, in- cluding spore coat proteins in large quantity (Inouye et al., 1979). Also, a thickened layer of peptidoglycan must be built within the myx- ospore. Consequently, development must be set in motion while some energy and some ca- pacity for protein synthesis remain. As the nu- trient level is progressively decreased, fruiting body development is initiated as soon as the nutrient falls below about 0.3% casitone, or even at higher levels when a less favorable source of amino acids and energy, such as 0.8% nutrient broth, is utilized (McVittie et al., 1962).

Which of the two possibilities is the more prudent choice hinges on a prediction of future levels of nutrients—whether nutrients are on their way to total extinction at the time the

D. Kaiser, Department of Biochemistry and Department of Developmental Biology, Stanford University School of Medi- cine, Beckman Center, B300, Stanford, CA 94305–5329.

Prokaryotic Development, edited by Y. V. Brun and L. J. Shimkets,
© 2000 American Society for Microbiology, Washington, DC 20005-4171

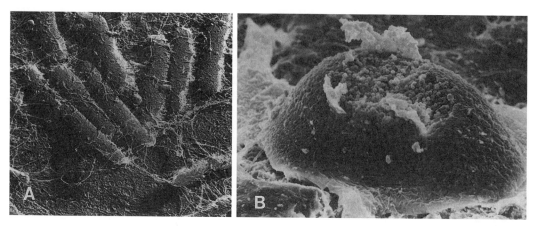

FIGURE 1 Alternatives. Depending on the nutritional regime, the cells may respond to starvation by growing slowly (A) or they may develop a fruiting body (B).

choice must be made, or whether the nutrient shortage is likely to be temporary and the cells have simply "forgotten their lunch." How do the cells make this important choice? A fruiting body contains about 10^5 cells, and the aim of this chapter is to show how cells can make this decision as a group.

To build a fruiting body from a roughly uniform initial distribution of cells, some of the cells first move into small focal aggregates, as shown in Fig. 2, 0 h and 4 h. Initially, those foci are asymmetric (4 h), but as more cells accumulate there, they layer on top of each other, and the aggregate becomes steep sided and hemispherical. Starting at about 20 h within such a mound, rods begin to differentiate into spherical, desiccation-resistant dormant myxospores (Fig. 2, lower right panel). Spore maturation continues over several days. A mature fruiting body is regularly and densely packed with spores. Compared with an isolated spore, the larger surface area of a fruiting body and its elevation above the substratum would favor transport by wind, water, or a passing animal. "Knowing" that there is little or no food where they are, myxobacteria appear to have adopted a survival strategy of sporulating and preparing the spores to be carried to a new place where food might be available (Reichenbach, 1984). Small animals in the soil, like

worms or insects, generally migrate toward food. A fruiting body also ensures that, having found food, a new life cycle is started by a cell community rather than a single cell (Reichenbach, 1984).

Developmentally regulated promoters (Fig. 3) are transcriptionally fused to a promoterless β-galactosidase gene (Kroos and Kaiser, 1984). The progressive morphological changes of aggregation and sporulation are coupled to progressive changes in gene expression recognized by sequential activation of the promoters (Kroos et al., 1986). The promoters at the end of the developmental sequence, for example, drive genes whose products are needed to build the spore.

A-SIGNALING

The time-ordered sequence of morphological and biochemical changes is coordinated by several signals passed between cells. One of these signals, A-factor, plays an important role in evaluating and responding to starvation. A-factor has been purified by using mutants that fail to release it. When A-factor is deficient, as in the *asg* mutants, development is arrested at a 1- to 2-h check point (Fig. 2, 0 h). The terminal morphology of *asgA* and *asgB* mutants is a flat film of cells with no sign of focal aggregation (Kuspa and Kaiser, 1989; Mayo and Kaiser, 1989). A-signal-deficient mutants are capable

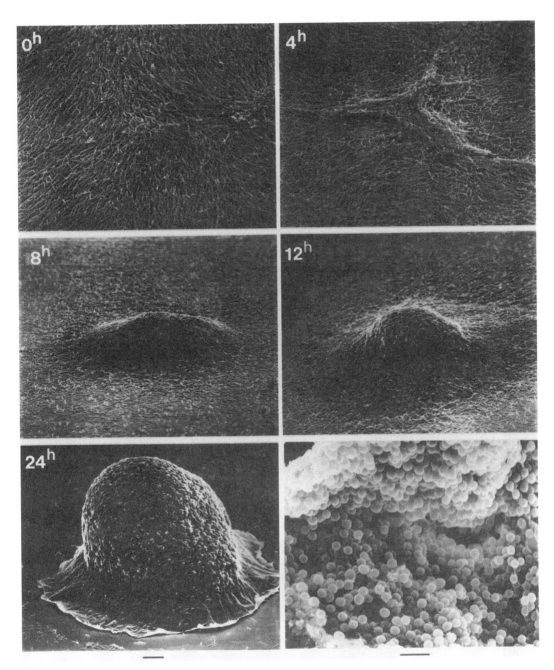

FIGURE 2 Fruiting body development in *M. xanthus*. Development was initiated at 0 h by replacing nutrient medium with a buffer devoid of a usable carbon or nitrogen source. The lower right frame shows a fruiting body which has split open, revealing spores inside. This frame is three times the magnification of the others. (Scanning electron microscopy by J. Kuner. Reproduced from Kaiser et al., 1985.).

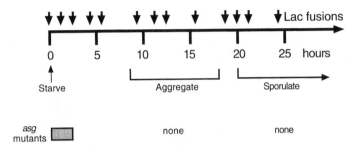

FIGURE 3 A-signal-defective developmental mutants of *M. xanthus*. The vertical arrows point to the times at which each of the *lacZ* fusions to a developmentally regulated promoter begins to be expressed. These fusions are reporters of normal development. The gray horizontal bar indicates the period during which there is normal expression of the reporter fusions in *asg* mutants. To the right of the bar, almost all development is defective (no, or greatly reduced, reporter expression) because these reporters are *asg* dependent. An exception is Ω4469, which is expressed at 4 h but is *asg* independent. The columns to the right indicate the morphological phenotype with respect to aggregation and sporulation of the mutants.

of sensing starvation, however (Singer and Kaiser, 1995). Most promoters later than 2 h are not activated in the *asg* mutants (Fig. 3); cells do not aggregate at all, and spores never differentiate. The fact that the later promoters are not activated in the mutant and that spores do not form illustrates how, by and large, the developmental program is organized as a dependent pathway. There is no evidence for branching of that pathway prior to 6 h.

A-signal production-defective mutations have been found in three genes. One gene, *asgA*, encodes a protein with a two-component receiver domain followed by a histidine protein kinase domain, which has autokinase activity (Davis et al., 1995; Plamann et al., 1994, 1995; Li and Plamann, 1996). The second gene, *asgB*, encodes a helix-turn-helix protein that appears to be a transcription factor for the − 35 region of a promoter; and the third, *asgC*, encodes an RpoD (sigma-70) homologue (Davis et al., 1995). These three proteins are thought to function together with (p)ppGpp in a signal transduction pathway that, in response to starvation, is required for generation of extracellular A-factor (Plamann and Kaplan, 1999) (see below).

A-Factor

Purification of A-factor was based on an A-signal-dependent promoter. To identify A-factor, substances that would allow an A-signal-deficient mutant strain to express the Tn*5 lacZ*

transcriptional fusion Ω4521 were sought from wild-type cells. This bioassay is specific for A-factor because the *asg* mutant cells cannot make or release A-factor, yet they can still respond to it (Kuspa et al., 1986). The *lacZ* fusion Ω4521 is driven by the earliest known A-factor-dependent promoter. Sporulation and fruiting body formation confirmed that A-factor purified by means of this assay has the full native activity.

Purified in this way from conditioned medium, A-factor proved to be a mixture of eight amino acids, peptides containing those amino acids, and trypsin-like extracellular proteases (Kuspa et al., 1992b; Plamann et al., 1992). The amino acids are thought to be the primary signaling factor; peptides and proteases lead to their release. Cell-bound peptidases release A-factor amino acids from the peptides. The proteases release A-factor amino acids and peptides from the mixture of proteins found in 1- to 2-h-conditioned medium. Table 1 lists the specific A-factor activities of these eight amino acids. Multiplying each specific activity by the abundance in conditioned medium shows that Phe, Tyr, Trp, Pro, Leu, Ile, Ala, and Ser account for 93% of the activity due to amino acids.

A-factor is water soluble. Solubility is important for signaling during the early stages of development, when the cells are loosely packed and the message must pass through an aqueous phase which surrounds and separates them.

The process of A-signaling is constrained by

TABLE 1 A-factor activity of amino acids

Amino acid	Sp. act.[a]	Concn[b]	Net activity[c]
Tyr	2.2	15	0.33
Pro	1.5	13	0.20
Phe	1.5		
Trp	1.2	22[d]	0.30
Leu	1.2		
Ile	1.1	55[e]	0.63
Ala	0.5	55	0.28
Ser	0.3	44	0.13
Sum of these eight amino acids			1.87
Sum of all individual amino acids			2.00
Sum of all peptides in conditioned medium			2
Total heat-stable activity in conditioned medium			4

[a] A-factor activity per mole of amino acid.
[b] Concentration in conditioned medium.
[c] Specific activity times concentration.
[d] Phe plus Trp.
[e] Leu plus Ile.

several important features. First, each cell releases a fixed amount of A-factor, and consequently, the concentration is directly proportional to the cell density (Kuspa et al., 1992a). Second, a certain minimum quantity of A-factor is required to produce wild-type levels of expression of A-factor-dependent genes; there is a threshold of 50 μM (Kuspa et al., 1992b). Third, starting from the 50 μM threshold, the signaling concentrations cannot exceed 10 mM. Above that concentration, cells fail to activate their A-factor-dependent genes and grow instead (Kuspa et al., 1992b). The responses of cells to the different constituents of A-factor are roughly additive when measured by Ω4521 expression (Table 1) (Kuspa et al., 1992). Additivity and the A-factor threshold imply a mechanism for perceiving and integrating A-factor constituents, but its molecular nature remains to be found.

There is thus a concentration window for A-signaling. The A-factor threshold at the bottom of the window sets a minimum cell population density that can signal itself. Below critical density, wild-type cells behave like A-

factor-deficient mutants, which can be rescued by added A-factor (Kuspa et al., 1992a). Perhaps the threshold ensures a sufficiently large population of A-factor-producing cells to complete a fruiting body. When a cell produces A-factor, that cell signifies its decision that fruiting body development is justified. A-factor is the way that a cell votes its own assessment of nutritional conditions. When each cell contributes to the total A-factor concentration, the starvation prediction indicated by release of A-factor becomes more reliable. Because fruiting body development involves the death of part of the population, evolutionary selection would have favored an accurate assessment of nutritional status, perhaps more accurate than any individual bacterial cell can achieve.

SENSING STARVATION

On what data do cells base their judgment of starvation? Fruiting body development can be initiated by starvation for carbon, nitrogen, or phosphorus. Despite the variety of initiating conditions, there is specificity: starvation for pyrimidine (Kimsey and Kaiser, 1991), for AMP (Shimkets and Dworkin, 1981), or for GTP (Singer and Kaiser, 1995) does not induce development.

Among the small-molecule regulators that bacteria, including *M. xanthus*, synthesize in response to nutrient deprivation is a pair of highly phosphorylated derivatives of guanosine, ppGpp (guanosine 3'-diphosphate 5'-diphosphate) and pppGpp (guanosine 3'-diphosphate 5'-triphosphate). Figure 4 illustrates how they

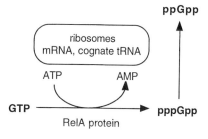

FIGURE 4 Biosynthesis of pppGpp and ppGpp on ribosomes depends on the *relA* protein. Evidence for this scheme is summarized in Cashel et al., 1996.

are made from GTP and ATP. (p)ppGpp is synthesized in vitro when RelA protein (stringent factor) is bound to a ribosome and when mRNA is bound to that ribosome and codon-specified but uncharged tRNA is bound at the acceptor site of the ribosome. Uncharged tRNA binds when the corresponding charged species is unavailable. The synthesis of (p)ppGpp occurs at the GTP-dependent elongation step in protein synthesis. Elongation ordinarily advances one codon, but in this instance it happens without ribosomal movement. Each round of (p)ppGpp synthesis is thought to trigger release of uncharged tRNA, thereby allowing a fresh assessment of starvation to be made even when the rate of peptide elongation is very low (Cashel et al., 1996).

The RelA protein [(p)ppGpp synthetase I] is thus a ribosome-dependent enzyme that catalyzes a synthetic reaction in which ATP is used to donate a pyrophosphate group to the 3′ position of either 5′ GTP or 5′ GDP. The pentaphosphate, pppGpp, may be converted to ppGpp by any of several enzymes, including 5′ phosphohydrolase.

There are several reasons that myxobacterial research has focused on (p)ppGpp synthesis. *M. xanthus* belongs to the bacteriolytic class of myxobacteria, which use the amino acids released by lysis and proteolysis as their major source of carbon, nitrogen, and energy. It is known that starvation for any amino acid induces development. Leu, Ile, and Val are essential amino acids which *M. xanthus* is unable to synthesize. Minimal medium lacking any one of them induces fruiting body development (Manoil and Kaiser, 1980b). Trp or His auxotrophs, isolated as mutants from the wild type, form fruiting bodies in response to a limitation or total withdrawal of these amino acids (Manoil and Kaiser, 1980b). Amino acid analogs, like serine hydroxamate or tyrosinol, that compete with an amino acid for charging its cognate tRNA to form the corresponding aminoacyl-tRNA also induce development. High concentrations of adenine derivatives or of certain amino acids induce fruiting body de-

velopment by creating imbalances among the amino acids that effectively result in starvation for another amino acid; threonine starves cells for lysine, and adenine induces starvation for glycine (Manoil and Kaiser, 1980b). Thus (p)ppGpp is rich in nutritional information: it is made by the ribosome when it stalls for lack of any of the 20 amino acids. Since formation of aminoacyl-tRNA depends on ATP, it can also indicate a limitation of energy or phosphate. Importantly, (p)ppGpp directly assesses the cell's capacity to synthesize any protein for which it has mRNA. Finally, there is as yet no evidence that other starvation-sensing pathways play a role in initiating development, even though the level of cyclic AMP rises twofold after starvation (Yajko and Zusman, 1978).

A stringent response in *E. coli* induces several metabolic changes: it arrests ribosomal and tRNA synthesis (Cashel and Gallant, 1969), it increases proteolysis, and it inhibits DNA and membrane biosynthesis (Cashel et al., 1996). An arrest of net growth as a consequence of (p)ppGpp elevation in *M. xanthus* (Singer and Kaiser, 1995) implies that many of these metabolic changes accompany fruiting body development. The response of *M. xanthus* to amino acid starvation is relatively rapid, the level of (p)ppGpp is elevated within 30 min, as shown in Fig. 5, offering an early as well as general indication of starvation. Though (p)ppGpp is made from GTP, only a slight fall in the GTP pool is detected as (p)ppGpp rises. After rising, it then decays within 2 h to a new steady state, suggesting a recovery of some protein-synthetic capacity. Recovery after an initial loss makes room for synthesis of the development-specific proteins.

Given the attractive features of (p)ppGpp as a continuously and broadly responsive early indicator of starvation, the experimental challenge in 1995 was to test how an elevation of (p)ppGpp might be involved in initiating development. A simple possibility was that this nucleotide is both necessary and sufficient to induce the program of fruiting body development. To test whether (p)ppGpp was sufficient, the *relA*[+] gene from *E. coli*, which has been

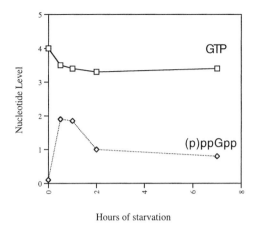

FIGURE 5 Time course of (p)ppGpp and GTP levels in *M. xanthus* after the initiation of starvation by a shift from a nutrient medium rich in amino acids and peptides to a buffered salts medium devoid of any carbon or nitrogen source. (p)ppGpp increases rapidly after starvation.

shown to encode (p)ppGpp synthetase I (Fig. 4), was introduced into *M. xanthus* (Singer and Kaiser, 1995). The *E. coli relA* gene was driven with an *M. xanthus carQRS* promoter (Hodgson, 1993; Martinez-Argudo et al., 1998). Strains into which this gene was introduced and then activated from the *carQRS* promoter proceeded to accumulate protein with the proper electrophoretic mobility and antigenic specificity for the *E. coli* enzyme. Moreover, even when the strains were growing in a medium rich in amino acids, there was (p)ppGpp synthesis; the level of (p)ppGpp nucleotide rose steadily. This accumulation of (p)ppGpp in *M. xanthus* without starvation resembles the effect of overexpression of *relA* in *E. coli* (Schreiber et al., 1991) and may arise from an excess of RelA protein over ribosomes.

Even without starvation but with elevated (p)ppGpp, the early stages of the developmental program could be observed. Normally, during the first 4 h of starvation, nutrient conditions are evaluated and the decision whether to initiate fruiting body development is made. Although there are no morphological changes during this evaluation-initiation phase, expression of several reporter-transcriptional fusions

can be observed. The fusions Ω4408 and Ω4469 are developmentally specific: they are not expressed during growth, they are induced by a starvation regime that is sufficient to induce fruiting body development, and their timing is constant with respect to the later morphological changes. Their expression reflects the normal developmental program. Fusion Ω4408 is expressed at 30 min, and Ω4469 is expressed at 4 h, both prior to any visible aggregation of cells and both independent of A-factor. The finding that ectopic (p)ppGpp accumulation induced expression of both early, development-specific reporters implies that this nucleotide is sufficient to initiate the developmental program (Singer and Kaiser, 1995).

The question remaining at that point was whether (p)ppGpp might also be necessary for development. To address necessity, a mutant which had lost the ability to accumulate (p)ppGpp in response to nutrient deprivation, strain DK527, was studied (Manoil and Kaiser, 1980a, 1980b). This mutant fails to form fruiting bodies or spores in response to starvation (Manoil and Kaiser, 1980a). If (p)ppGpp synthesis is necessary for development and the primary defect in DK527 is in (p)ppGpp synthesis, then the developmental defect of DK527 should be repaired by supplying the proper amount of (p)ppGpp. When the *relA*⁺ gene from *E. coli* was introduced into the DK527 mutant, (p)ppGpp accumulated in response to starvation and the strain recovered its ability to aggregate and to sporulate (Harris et al., 1998).

Moreover, if DK527 fails to recognize starvation by accumulating (p)ppGpp, then its development should be interrupted at a correspondingly early stage. Four well-characterized Tn5 *lac* transcriptional fusion reporters from the early stages of development were examined in the DK527 genetic background. Reporter Ω4408, described above, plus Ω4400, Ω4455, and the early A-factor-dependent reporter Ω4521 begin their β-galactosidase expression at different fusion-specific times in the 0- to 6-h interval. In fact, none of these four reporter fusions were expressed in DK527 while all were expressed in the wild-type parent of DK527 (DK101). Evidently, the mutation in DK527 prevents it from completing the first hour of normal development, when starvation is being recognized and the initial steps in the decision to develop fruiting bodies are being made in wild-type cells.

Finally, the *M. xanthus relA* gene homologue was cloned, and a protein coding sequence that was 49% identical to both the *E. coli* RelA and SpoT proteins (which are very similar to each other [Chakraburtty et al., 1996]) was obtained (Harris et al., 1998). This cloned gene rescued the DK527 mutant fruiting body development and sporulation, suggesting that DK527 is a *relA* mutant of *M. xanthus*. The same cloned gene also rescued the ability of an *E. coli relA* deletion mutant to synthesize (p)ppGpp, demonstrating that the *M. xanthus* gene product has *relA*⁺ [(p)ppGpp synthetase I] function. Disruption of the *M. xanthus relA*⁺ homologue blocked the accumulation of (p)ppGpp in response to amino acid starvation and caused an early developmental arrest, resembling that of DK527. Thus, (p)ppGpp is both necessary and sufficient for the initiation of fruiting body development.

(p)ppGpp AND A-FACTOR

If the first battery of developmentally controlled genes copes with starvation and limitations in the ability to make proteins, the A-signal is necessary to activate the next battery. The second battery includes genes that trigger aggregation; its earliest member is marked by the Ω4521 reporter fusion. *asgB* mutants, which make only 5 to 10% of normal A-factor levels in response to starvation, express Ω4521 at only 10% of normal levels even with (p)ppGpp supplied by an active *E. coli relA* gene in the mutant cells. *asgB*⁺ strains do express Ω4521, but in a *relA*-dependent manner. Thus, Ω4521 expression is controlled by both an A-factor input and a (p)ppGpp input. Other genes of the second battery are, like Ω4521, starvation and A-factor dependent, while those of the first battery are only starvation dependent.

Ω4521 expression is regulated at the level of accumulation of its mRNA (Kaplan et al.,

1991). Its message is specific to development. When the A-signal is absent, Ω4521 mRNA does not accumulate. This message is also absent in an *asg* mutant. However, if wild-type cells, or A-factor partially purified from conditioned medium, is added, the message is restored. Mapping the transcription start site of Ω4521 revealed no sequence resembling a promoter of the sigma-70 family upstream of this site, even though *M. xanthus* has several *rpoD*-like genes. There were, however, sequences at −24 and −12 ahead of the initiation site that resembled established sigma-54 promoters of enteric bacteria.

To substantiate the suggestion that this sequence is the promoter, single-base mutations were constructed to identify the critical bases. The effects of those single-base changes on Ω4521 expression are represented in Fig. 6. Changing bases anywhere within the −24 hexamer decreases Ω4521 expression by 90% or more. The GC dinucleotide found within the −12 pentamer is present in all known sigma-54 promoters, and mutating the G decreased expression more than 90%. Changing bases outside the −24 and −12 regions, including the −7 to −11 region, had little effect on Ω4521 expression. Had the promoter been sigma-70 specific, this region should have been crucial. Changing bases that lay between the −24 and −12 regions also had little effect, but deleting one of those bases, which shrinks the spacing and changes the pitch between the critical −24 and −12 regions, decreases expression by 97% (Fig. 6, T → Δ). Sigma-70 family promoters usually tolerate changes in spacing between their −10 and −35 regions well, but sigma-54 promoters as a group do not tolerate spacing changes between their −24 and −12 regions.

M. xanthus has a typical *rpoN* gene, whose sequence closely resembles that of the *Pseudomonas aeruginosa* gene. Uniquely among bacteria, the *rpoN* gene of *M. xanthus* is vital, even for growth in a medium rich in amino acids and peptides. It is also vital for growth at low, medium, or high temperatures. In addition to the need for sigma-54 during growth, that sigma

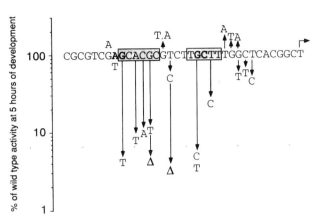

FIGURE 6 The promoter for Ω4521 and the effects on transcription of changing individual bases. Those effects show that the promoter is recognized by sigma-54, encoded by *rpoN*. Boxes enclose the -24 hexamer and -12 pentamer, which are identical in the Ω4521 and *mbhA* promoters. Bases identical to the general sigma-54 consensus are shown in bold type. Arrows indicate the base changed in each mutant. The vertical position of the mutant form of the base indicates the percent wild type β-galactosidase activity in the mutant.

factor is also essential for development. All sigma-54 promoters also require a promoter-specific activator protein and often ATP. The activator protein binds a specific enhancer sequence in the DNA, usually upstream of the promoter, in order to engage that activator with the holoenzyme-promoter complex. Fourteen sigma-54 activator proteins have been identified genetically in *M. xanthus* (Gorski and Kaiser, 1998; Kaufman and Nixon, 1996). At least three of these activators are essential for development because null mutations in the genes that encode them block development (Gorski and Kaiser, 1998). These three activator mutations do not block growth. One of the activators appears to function in the A-

factor response pathway, since the phenotype of the activator null mutant resembles that of an *asg* mutant, except that it cannot be rescued by wild-type cells. Development of the mutant is blocked in an Agg⁻ Spo⁻ state, the A-signaling check point (Gorski and Kaiser, 1998).

The *relA* mutant of *M. xanthus*, like an *asg* mutant, is defective in A-factor production (Table 2). An *asgB* mutant produces 2% of the wild-type level of A-factor, while the *relA* mutant (DK527) strain produces 8% of the wild-type level. Incorporating the *relA*⁺ plasmid into DK527 rescued the A-factor level to 83% of wild type. After another 6 h of development, the level rose to 100% of wild type in the DK527(*relA*⁺ plasmid) strain. These data show

TABLE 2 A-factor production

Donor strain	Production[a]			
	In vivo (cells)		In vitro (conditioned medium)	
DK101 (wild type)	480 ± 86	(100)	36 ± 3.8	(100)
DK480 (*asgB*)	28 ± 12	(6)	0.8 ± 0.3	(2)
DK527 (*relA*)	76 ± 22	(16)	2.2 ± 0.8	(8)
DK527(prelA⁺/*relA*)[b]	375 ± 76	(78)	30 ± 4.6	(83)

[a] Values are shown as units per milliliter of cell suspension, prepared at a cell density of 1,000 Klett units and measured 3 h after the start of development. The values are averages of three independent experiments; additional data can be found in Harris et al., 1998. The percentage of wild-type activity (in parentheses) was calculated by comparing the A-factor activity found in DK480 and DK527 to the amount found in the wild-type strain (100%).
[b] DK527 containing the *relA*⁺ plasmid.

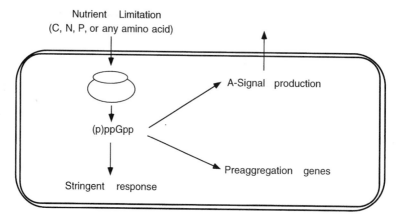

FIGURE 7 Ribosomes, sensing starvation, produce (p)ppGpp. (p)ppGpp initiates a stringent response, the production of A-factor, and expression of early developmentally regulated genes. A stringent response involves the arrest of the synthesis of ribosomes and the major polymers of the cell (Singer and Kaiser, 1995).

that (p)ppGpp is required for A-factor production, as represented in Fig. 7.

CONCLUSIONS

As indicated in Fig. 8, ribosomes, sensing a deficiency in any one of the aminoacyl-tRNAs, which would limit their capacity to synthesize

protein, respond by increasing the intracellular level of (p)ppGpp. That increase, signifying starvation, sets off a stringent response (Fig. 7), which stops the synthesis of new ribosomes and initiates limited breakdown of ribosomal protein, thus releasing amino acids to restore protein synthesis. Ribosome breakdown in *M.*

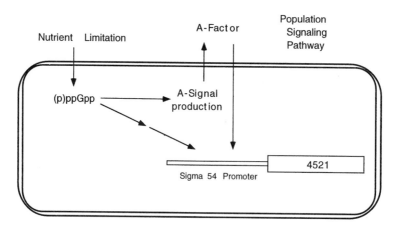

FIGURE 8 A second battery of genes, exemplified by Ω4521, depends on starvation and extracellular A-factor. The first battery consists of starvation-dependent, but A-factor-independent, genes, such as the preaggregation genes shown in Fig. 7. Sigma-54 promoters are receptive to at least two regulatory inputs, one via the sigma factor-polymerase holoenzyme and another via the required sigma-54 activator protein.

xanthus is made evident by comparing the structure of fruiting body spores with glycerol-induced spores, which are made without starvation. Fruiting body spores have fewer ribosomes than glycerol spores (see Fig. 9.3 in Zusman, 1984). That the stringent response also stops membrane synthesis is indicated by the highly infolded membrane in the spores induced in rich medium without starvation, while fruiting body spores have a simple inner membrane (Fig. 9.3 in Zusman, 1984). The (p)ppGpp increase also activates several preaggregation genes, among which are Ω4469 and *sdeK* (the gene inactivated by insertion Ω4408). SdeK is a histidine protein kinase that may be necessary for starvation sensing, since an *sdeK* mutant stops development during the sensing phase (Garza et al., 1998). Finally, (p)ppGpp induces A-factor production, which requires *asgA, asgB,* and *asgC*. A-factor, being soluble, creates a pool outside the cells. Each cell then responds to the total amount of A-factor produced by all cells.

asg-dependent genes, like Ω4521, receive two inputs: an input from (p)ppGpp as well as an integrated input from A-factor (Fig. 8). Both inputs are needed, and sigma-54 promoters, which must be activated by an upstream activator protein, provide a simple way to integrate two inputs. Gulati, Xu, and Kaplan (1995) have shown by deletion experiments that DNA between 120 and 60 bps upstream of the Ω4521 promoter is necessary for the starvation and A-factor dependence of Ω4521 expression. Moreover, a sigma-54 activator gene for Ω4521 has been identified that would require an enhancer element in the Ω4521 upstream DNA (Gorski and Kaiser, 1998). Because the assessment of the A-factor level is made at around 2 h, the two inputs to a sigma-54 promoter, shown in Fig. 8, could provide a comparison between the nutrient level 2 h earlier (represented by the current A-factor level) and the current nutrient level (represented by the concentration of (p)ppGpp). This comparison, performed by the initiation of Ω4521 mRNA, implicitly computes the starvation trend.

For reasons set forth at the beginning of this chapter, the choice of response to depletion of nutrients is critical for cell survival. That choice has to be made before the nutrients are totally depleted so that new developmental proteins can still be synthesized. Assessment of available nutrients is rendered both more accurate and more reliable by the A-signaling process shown in Fig. 8, which includes an extracellular quorum-sensing branch.

REFERENCES

Cashel, M., and M. Gallant. 1969. Two compounds implicated in the function of the RC gene of *Escherichia coli. Nature* **221**:838.

Cashel, M., D. R. Gentry, V. J. Hernandez, and D. Vinella. 1996. The stringent response, p. 1458–1496. *In* F. C. Neidhardt, R. Curtiss III, J. L. Ingraham, E. C. C. Lin, K. B. Low, B. Magasanik, W. S. Reznikoff, M. Riley, M. Schaechter, and H. E. Umbarger (ed.), *Escherichia coli and Salmonella: Cellular and Molecular Biology,* 2nd ed. American Society for Microbiology, Washington, D.C.

Chakraburtty, R., J. White, E. Takano, and M. Bibb. 1996. Cloning, characterization and disruption of a (p)ppGpp synthetase gene (*relA*) of *Streptomyces coelicolor* A3(2). *Mol. Microbiol.* **19**:357–368.

Davis, J. M., J. Mayor, and L. Plamann. 1995. A missense mutation in *rpoD* results in an A-signaling defect in *Myxococcus xanthus. Mol. Microbiol.* **18**:943–952.

Garza, A. G., J. S. Pollack, B. Z. Harris, A. Lee, I. M. Keseler, E. F. Licking, and M. Singer. 1998. *SdeK* is required for early fruiting body development in *Myxococcus xanthus. J. Bacteriol.* **180**:4628–4637.

Gorski, L., and D. Kaiser. 1998. Targetted mutagenesis of sigma-54 activator proteins in *Myxococcus xanthus. J. Bacteriol.* **180**:5896–5905.

Gulati, P., D. Xu, and H. B. Kaplan. 1995. Identification of the minimum regulatory region of a *Myxococcus xanthus* A-signal-dependent developmental gene. *J. Bacteriol.* **177**:4645–4651.

Harris, B. Z., D. Kaiser, and M. Singer. 1998. The guanosine nucleotide (p)ppGpp initiates development and A-factor production in *Myxococcus xanthus. Genes Dev.* **12**:1022–1035.

Hengge-Aronis, R. 1996. Regulation of gene expression during entry into stationary phase, p. 1497–1511. *In* F. C. Neidhardt, R. Curtiss III, J. L. Ingraham, E. C. C. Lin, K. B. Low, B. Magasanik, W. S. Reznikoff, M. Riley, M. Schaechter, and H. E. Umbarger (ed.), *Escherichia coli and Salmonella: Cellular and Molecular Biology,* 2nd ed.

American Society for Microbiology, Washington, D.C.

Hodgson, D. A. 1993. Light-induced carotenogenesis in *Myxococcus xanthus*: genetic analysis of the *carR* region. *Mol. Microbiol.* **7**:471–488.

Inouye, M., S. Inouye, and D. Zusman. 1979. Gene expression during development of *Myxococcus xanthus*: pattern of protein synthesis. *Dev. Biol.* **68**:579–591.

Kaiser, D., L. Kroos, and A. Kuspa. 1985. Cell interactions govern the temporal pattern of *Myxococcus* development. *Cold Spring Harbor Symp. Quant. Biol.* **50**:823–830.

Kaplan, H. B., A. Kuspa, and D. Kaiser. 1991. Suppressors that permit A signal-independent developmental gene expression in *Myxococcus xanthus*. *J. Bacteriol.* **173**:1460–1470.

Kaufman, R. I., and B. T. Nixon. 1996. Use of PCR to isolate genes encoding σ^{54}-dependent activators from diverse bacteria. *J. Bacteriol.* **178**:3967–3970.

Kim, S. K., and D. Kaiser. 1990. Purification and properties of *Myxococcus xanthus* C-factor, an intercellular signaling protein. *Proc. Natl. Acad. Sci.* **87**:3635–3639.

Kimsey, H. H., and D. Kaiser. 1991. Targeted disruption of the *Myxococcus xanthus* orotidine 5′-monophosphate decarboxylase gene: effects on growth and fruiting-body development. *J. Bacteriol.* **173**:6790–6797.

Kroos, L., and D. Kaiser. 1984. Construction of Tn*5 lac*, a transposon that fuses lacZ expression to exogenous promoters, and its introduction into *Myxococcus xanthus*. *Proc. Natl. Acad. Sci. USA* **81**:5816–5820.

Kroos, L., A. Kuspa, and D. Kaiser. 1986. A global analysis of developmentally regulated genes in *Myxococcus xanthus*. *Dev. Biol.* **117**:252–266.

Kuspa, A., and D. Kaiser. 1989. Genes required for developmental signalling in *Myxococcus xanthus*: three asg loci. *J. Bacteriol.* **171**:2762–2772.

Kuspa, A., L. Kroos, and D. Kaiser. 1986. Intercellular signaling is required for developmental gene expression in *Myxococcus xanthus*. *Dev. Biol.* **117**:267–276.

Kuspa, A., L. Plamann, and D. Kaiser. 1992a. A-signalling and the cell density requirement for *Myxococcus xanthus* development. *J. Bacteriol.* **174**:7360–7369.

Kuspa, A., L. Plamann, and D. Kaiser. 1992b. Identification of heat-stable A-factor from *Myxococcus xanthus*. *J. Bacteriol.* **174**:3319–3326.

Li, Y., and L. Plamann. 1996. Purification and phosphorylation of *Myxococcus xanthus* asgA protein. *J. Bacteriol.* **178**:289–292.

Manoil, C., and D. Kaiser. 1980a. Accumulation of guanosine tetraphosphate and guanosine penta-phosphate in *Myxococcus xanthus* during starvation and myxospore formation. *J. Bacteriol.* **141**:297–304.

Manoil, C., and D. Kaiser. 1980b. Guanosine pentaphosphate and guanosine tetraphosphate accumulation and induction of *Myxococcus xanthus* fruiting body development. *J. Bacteriol.* **141**:305–315.

Martinez-Argudo, I., R. Ruiz-Vasquez, and F. J. Murillo. 1998. The structure of an ECF-σ-dependent, light-inducible promoter from the bacterium *Myxococcus xanthus*. *Mol. Microbiol.* **30**:883–894.

Mayo, K., and D. Kaiser. 1989. asgB, a gene required early for developmental signalling, aggregation, and sporulation of *Myxococcus xanthus*. *Mol. Gen. Genet.* **218**:409–418.

McVittie, A., F. Messik, and S. A. Zahler. 1962. Developmental biology of *Myxococcus*. *J. Bacteriol.* **84**:546–551.

O'Connor, K., and D. Zusman. 1988. A reexamination of the role of autolysis in the development of *Myxococcus xanthus*. *J. Bacteriol.* **170**:4103–4112.

Plamann, L., and H. B. Kaplan. 1999. Cell-density sensing during early development in *Myxococcus xanthus*. *In* G. M. Dunny and S. C. Winans (ed.), *Cell-Cell Signaling in Bacteria*. American Society for Microbiology, Washington, D.C.

Plamann, L., A. Kuspa, and D. Kaiser. 1992. Proteins that rescue A-signal-defective mutants of *Myxococcus xanthus*. *J. Bacteriol.* **174**:3311–3318.

Plamann, L., J. M. Davis, B. Cantwell, and J. Mayor. 1994. Evidence that asgB encodes a DNA-binding protein essential for growth and development of *Myxococcus xanthus*. *J. Bacteriol.* **176**:2013–2020.

Plamann, L., Y. Li, B. Cantwell, and J. Mayor. 1995. The *Myxococcus xanthus* asgA gene encodes a novel signal transduction protein required for multicellular development. *J. Bacteriol.* **177**:2014–2020.

Reichenbach, H. 1984. Myxobacteria: a most peculiar group of social prokaryotes, p. 1–50. *In* E. Rosenberg (ed.), *Myxobacteria*. Springer-Verlag, New York, N.Y.

Schreiber, G., S. Metzger, E. Aizenman, S. Roza, M. Cashel, and G. Glaser. 1991. Overexpression of the relA gene in *Escherichia coli*. *J. Biol. Chem.* **266**:3760–3767.

Shimkets, L. J., and M. Dworkin. 1981. Excreted adenosine is a cell density signal for the initiation of fruiting body formation in *Myxococcus xanthus*. *Dev. Biol.* **84**:51–60.

Singer, M., and D. Kaiser. 1995. Ectopic production of guanosine penta- and tetra-phosphate can initiate early developmental gene expression in *Myxococcus xanthus. Genes Dev.* **9:**1633–1644.

Ueki, T., and S. Inouye. 1998. A new sigma factor, SigD, essential for stationary phase is also required for multicellular differentiation in *Myxococcus xanthus. Genes Cells* **3:**371–385.

Wireman, J., and M. Dworkin. 1977. Developmentally induced autolysis during fruiting body for-

mation by *Myxococcus xanthus. J. Bacteriol.* **129:** 796–802.

Yajko, D., and D. Zusman. 1978. Changes in cyclic AMP levels during development in *Myxococcus xanthus. J. Bacteriol.* **133:**1540–1542.

Zusman, D. 1984. Developmental program of *Myxococcus xanthus*, p. 185–213. *In* E. Rosenberg (ed.), *Myxobacteria.* Springer-Verlag, New York, N.Y.

GROWTH, SPORULATION, AND OTHER TOUGH DECISIONS

Lawrence J. Shimkets

13

Many types of bacteria produce dormant spores when faced with nutrient deprivation. The decision to enter the developmental pathway is made when the cell is faced with environmental conditions that are unsuitable for vigorous growth but still possesses the nutritional resources to complete development. If the cell errs on the side of developing while excess nutrients are available, it will become dormant when the competition is still growing, thereby squandering an opportunity to increase in number. This possibility is less costly than the potentially lethal problem of delaying entry into development until insufficient resources remain to complete the process. The choice between growth and development is, therefore, the most serious decision facing the cell.

During development two significant metabolic events occur. First, carbon limitation forces the cell to reduce the catabolism of organic molecules and redirect the flow of carbon to anabolic pathways. Second, the biosynthetic priority shifts from the synthesis of growth-related macromolecular products, such as chromosomes and ribosomes, to structural components of the spore. In short, the synthesis of most cellular macromolecules changes.

METABOLIC CONTROL BY *MYXOCOCCUS XANTHUS*

The myxobacterium *Myxococcus xanthus* is an aerobic chemoorganotroph that derives carbon, nitrogen, and energy through catabolism of amino acids (Dworkin, 1962; Bretscher and Kaiser, 1978). The tricarboxylic acid cycle serves as the major route for reducing NAD and for generating carbon skeletons for biosynthesis. During growth the vast majority of carbon is directed toward energy generation; 90% of the pyruvate in A1 minimal medium is oxidized to CO_2 (Bretscher and Kaiser, 1978). Cells can also incorporate many types of precursor molecules, with the notable exception of monosaccharides. Carbohydrates do not support growth (Bretscher and Kaiser, 1978) and are incorporated very poorly (Hemphill and Zahler, 1968; Watson and Dworkin, 1968).

Macromolecular synthesis has been examined during the cell division cycle. DNA replication occurs during 80% of the cell cycle (Zusman et al., 1978) to replicate the 9.5-Mb chromosome (Chen et al., 1990, 1991). A burst of RNA synthesis precedes septum formation, and then, during the interval between septum formation and physical separation of the daughter cells, there is a decline in DNA, RNA, and protein synthesis (Zusman et al., 1971).

L. J. Shimkets, Department of Microbiology, 527 Biological Sciences Building, University of Georgia, Athens, GA 30602.

Prokaryotic Development, edited by Y. V. Brun and L. J. Shimkets,
© 2000 American Society for Microbiology, Washington, DC 20005-4171

Major changes in macromolecular synthesis occur during sporulation. Although there is little data available for fruiting body spores, macromolecular synthesis has been examined during sporulation induced by adding glycerol to growing cells (Dworkin and Gibson, 1964). Although this type of sporulation bypasses the formation of the multicellular fruiting body, it is highly synchronous, and the morphological transition is complete in about 90 min. During glycerol-induced sporulation, there is no further initiation of chromosome replication (Rosenberg et al., 1967). Net RNA synthesis ceases (Bacon and Rosenberg, 1967) and is coupled to the disappearance of certain vegetative mRNA species and induction of developmental mRNA species (Okano et al., 1970). The synthesis of spore coat polysaccharides is accompanied by elevated levels of uridine 5'-diphosphate-N-acetylgalactosamine (UDPGalNAc) due to an increase in activity of the enzymes required to synthesize UDPGalNAc from fructose 1,6-diphosphate (Filer et al., 1977a, 1977b). Coincident with the appearance of the spore coat is a large increase in the glyoxylate cycle enzymes isocitrate lyase (EC4.1.3.1) and malate synthase (EC4.1.3.2) (Orlowski et al., 1972). The glyoxylate cycle bypasses those tricarboxylic acid cycle steps that result in mineralization of organic carbon to CO_2 and thereby conserves four carbon compounds for anabolism.

From these results it is clear that macromolecular synthesis changes dramatically during development. Unfortunately, relatively little is known about the molecular mechanisms in control of the synthesis of individual types of macromolecules, let alone the global mechanisms for their coordinate regulation.

WHY IS A CHOICE NECESSARY?

Having initiated the developmental pathway, it seems unlikely that the cell will continue both growth-related activities and development-related activities for an extended period. The flexibility afforded by keeping all options open cannot logically last for an entire cell cycle due to the fact that some growth-related functions are incompatible with development. Consider, for example, the peptidoglycan layer. During growth it extends outward to form a long, thin, rod-shaped cell. During sporulation the cell shortens and rounds up to form a spore. These two processes are diametrically opposed. At some point cells must abandon their quest to produce siblings and focus on sporulation. The first step in this process is the cessation of growth-related macromolecular synthesis. Unless this involves a dramatic and irreversible change in the architecture of the cell, it could be reversed later if there was a sudden influx of nutrients. But eventually the decision must become irreversible, especially when the cell wall is being remodeled during sporulation.

MUTANTS IN THE DECISION PROCESS

Since development is an alternative to growth, many types of developmental mutations would not be expected to adversely affect growth. Indeed there are dozens of developmental mutants that exhibit vigorous growth in minimal medium. This mutagenesis strategy, however, might not reveal those interesting regulatory genes that commit a cell to the vegetative pathway; mutations in such genes might induce sporulation even in rich medium, thereby failing to produce viable progeny.

We have identified such a gene, known as *socE*. A transposon insertion in *socE* was originally discovered to suppress a mutation in an essential developmental gene known as *csgA*. The insertion restored some of the phenotypes lost by the *csgA* mutation, such as fruiting body development and sporulation (Rhie and Shimkets, 1989). The *socE* gene sequence analysis revealed no homologous genes in the database. Attempts to disrupt *socE* in *csgA*+ cells by generalized transduction or electroporation proved futile, suggesting that *socE* is essential for growth (Crawford and Shimkets, submitted [b]). This led to an attempt to ectopically express *socE* in wild-type cells by taking advantage of the *carQRS* light-inducible carotenoid biosynthesis promoter described by David Hodgson (1993) and McGowan et al. (1993). The light-inducible promoter was placed in front of a truncated version of *socE* with the last 25% of the gene deleted (Crawford and Shimkets, submitted [a]). The *carQRSp-socE* construct was electroporated into wild-type cells in the presence of light, and homologous recombination between the wild-type *socE* allele and the light-inducible construct produced a light-regulated full-length copy of *socE* in the chromosome (Fig. 1). Light induced ectopic expression of *socE*, and viable progeny were obtained. The cells were grown in suspension in a rich growth medium, and when light was withdrawn, the cells underwent approximately two doublings, diluting the SocE pool, and then ceased growing. Several days later, virtually every cell in the culture converted to a myxospore. The addition of light prior to sporulation allowed rapid resumption of growth. These results suggest that SocE acts as a repressor of development; loss of SocE leads to growth arrest and sporulation even under environmental conditions that do not permit development of wild-type cells.

SocE DEPLETION INDUCES A STRINGENT RESPONSE

Cell elongation and DNA and stable-RNA synthesis cease in SocE-depleted cells. Con-

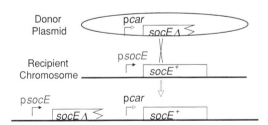

FIGURE 1 Light-inducible expression of *socE*. A *socE* allele lacking essential 3′ coding sequences (*socΔ'*) was placed under control of the *carQRSp* promoter (p*car*) and cloned into pBGS18, a kanamycin-resistant derivative of pUC19 which is unable to replicate in *M. xanthus*, and electroporated into the wild-type strain DK1622. A single homologous integration event produced a merodiploid in which only the functional copy of *socE* is transcribed from *carQRSp* and activated by blue light (Hodgson, 1993; McGowan et al., 1993). (This figure is derived from Crawford and Shimkets, submitted [a]).

versely, certain developmental genes are induced (Crawford and Shimkets, submitted [a]). Preceding these phenomena is a 10-fold increase in the nucleotides guanosine 3′ diphosphate, 5′ triphosphate (pppGpp) and guanosine 3′ diphosphate, 5′ diphosphate (ppGpp), suggesting that the cells are experiencing a stringent response.

In *Escherichia coli*, the stringent response helps regulate the growth rate. When *E. coli* is starved for amino acids, charged tRNA molecules become limiting and ribosomes pause during elongation (reviewed in Cashel, 1996). In response, the ribosome-associated protein RelA uses ATP to phosphorylate the ribose 3′ hydroxyl of GTP to produce pppGpp. A pppGpp 5′ phosphohydrolyase (Gpp) then converts pppGpp to the more abundant derivative ppGpp. These nucleotides, collectively referred to as (p)ppGpp, control macromolecular synthesis and growth rate. The stringent response induces some processes that help alleviate starvation, for example, amino acid biosynthesis and intracellular proteolysis, and inhibit growth-related processes, such as stable-RNA synthesis. *E. coli relA* mutants fail to accumulate (p)ppGpp during aminoacyl-tRNA limitation and uncouple macromolecule synthesis from the cell division cycle.

Introduction of a *relA* null allele (Harris et al., 1998) into SocE-depleted *M. xanthus* cells prevents (p)ppGpp accumulation, growth arrest, and sporulation (Crawford and Shimkets, submitted [a]). These results suggest that production of (p)ppGpp causes growth arrest upon SocE depletion rather than accumulating as a consequence of growth arrest.

GROWTH ARREST REQUIRES THE C-SIGNALING PROTEIN CsgA

C-signaling mutants were originally discovered based on their conditional developmental phenotype. They are unable to develop alone but can form fruiting bodies when placed in contact with wild-type cells or other nondeveloping mutants with a different genotype (Hagen et al., 1978). Five extracellular-complementation groups have been discovered and were designated A, B, C, D, and E. The C group contains a single member, the *csgA* gene, which encodes a 24.6-kDa protein with striking amino acid identity to members of the short-chain alcohol dehydrogenase (SCAD) family (Baker, 1994; Lee et al., 1995; Lee and Shimkets, 1994). These enzymes use NAD(H) or NADP(H) to catalyze the interconversion of secondary alcohols and ketones or to mediate decarboxylation (Neidle et al., 1992; Persson et al., 1991). C-signaling is cell contact dependent and requires cell motility and alignment during presentation of the stimulus in developing cells (Kim and Kaiser, 1990b, 1990c). CsgA purified from wild-type cells (Kim and Kaiser, 1990a, 1990d) or expressed in *E. coli* as a MalE–CsgA fusion (Lee et al., 1995) restores fruiting body formation when added to developing *csgA* cells. Furthermore, anti-CsgA antibodies inhibit wild-type development (Shimkets and Rafiee, 1990). Together these data provide evidence that CsgA stimulates cells from the outside. This presents an interesting paradox, as CsgA, a putative dehydrogenase, would require the intracellular coenzyme NADPH. Thus, it remains unclear whether CsgA performs an essential extracellular function with secreted NADPH or whether its extracellular state is temporary. The relevant substrate has not been identified. Short-chain alcohol dehydrogenases catalyze chemical reactions involving such a wide variety of substrates that it is not possible to deduce the CsgA substrate from amino acid sequence comparisons.

As development progresses, CsgA induces developmental changes which depend on the level of CsgA. At low concentrations, CsgA modulates cell reversal frequency to regulate the directional movement of cells during another starvation-induced behavior known as rippling (Sager and Kaiser, 1994). Higher concentrations of CsgA are required for fruiting body formation, and even higher concentrations are necessary for sporulation (Kim and Kaiser, 1991; Li et al., 1992). One striking role of CsgA is to arrest growth during early development (Crawford and Shimkets, submitted

[a]). It is not clear that CsgA acts in the same way to mediate all these different effects.

The *socE* mutation was originally isolated in a *csgA* mutant background, suggesting that the *csgA* mutation prevents the lethal effect on cell growth (Rhie and Shimkets, 1989). This was verified by introducing a wide variety of *csgA* mutations into the *carQRSp-socE* construct. It appears that the putative dehydrogenase activity of CsgA may be involved. The crystal structure is known for many members of the SCAD family, especially the amino acids lining the NADH or NADPH binding pocket, since the protein crystals also contained crystallized coenzyme. Substitution of a threonine near the N terminus that is anticipated to stabilize coenzyme binding through hydrogen bonding (Ghosh et al., 1994) renders the protein developmentally inactive and unable to bind NAD^+ in vitro (Lee et al., 1995). This same mutation prevents growth arrest and sporulation in SocE-depleted cells (Crawford and Shimkets, submitted [a]). SCAD family members contain a universally conserved SX_{12-14} YXXXK motif near the middle of the protein that mediates chemical catalysis (Persson et al., 1991). Conservative substitution of the CsgA active-site residue, S135T, results in an inactive protein (Lee et al., 1995) and also prevents growth arrest (Crawford and Shimkets, submitted [a]). In fact, all the known *csgA* mutations prevented growth arrest and sporulation upon SocE depletion. In the converse type of experiment, 12 viable revertants were selected from SocE-depleted cells. All of these revertants were complemented by either *relA* or *csgA*. These results argue that the putative CsgA dehydrogenase activity, as well as RelA, are essential for growth arrest and sporulation in SocE-depleted cells.

REGULATION OF *socE* AND *csgA*

The *socE* and *csgA* genes are both regulated by RelA but have opposite transcription patterns during the *M. xanthus* life cycle. Transcription of *socE* is high in vegetative cells and declines during the early stages of development (Crawford and Shimkets, submitted [b]). *socE* transcription is negatively regulated by the stringent response. A *relA* mutation prevents the decline in *socE* expression observed at the onset of development and allows *socE* transcription at levels observed in vegetative cells in spite of the prevailing developmental conditions. In view of the role of SocE as an inhibitor of development (Crawford and Shimkets, submitted [a]), these results can help explain why *relA* mutations inhibit development (Harris et al., 1998).

The *csgA* gene is weakly transcribed in vegetative cells but undergoes a four- to sixfold increase in expression during development that reaches a peak at about the time of sporulation (Hagen and Shimkets, 1990). *csgA* expression appears to be negatively regulated by *socE*, with a *socE* mutation resulting in a 1.5-fold increase in *csgA* expression (Crawford and Shimkets, submitted [b]). However, the decline in *socE* transcription during the early stages of development appears to be insufficient to account for the four- to sixfold developmental induction of *csgA* expression. It seems likely that positive regulation by the stringent response, *bsgA*, and *csgA* helps make up the difference. A *relA* mutation reduces *csgA* expression about 70% (Crawford and Shimkets, submitted [b]), a *bsgA* mutation reduces expression about 50% (Li et al., 1992), and a *csgA* mutation reduces expression about 10 to 40% (Kim and Kaiser, 1991; Li et al., 1992). A combination of *relA* and *bsgA* or *relA* and *csgA* virtually eliminates developmental *csgA* expression (Crawford and Shimkets, submitted [b]). In view of the role of CsgA as an inducer of many developmental behaviors, these results can also help explain why *relA* mutations inhibit development (Harris et al., 1998).

The regulation of these transcriptional units is coupled with the decision process. Induction of the stringent response by amino acid deprivation inhibits *socE* transcription and induces *csgA* transcription, providing coordinate but opposing control over these two transcriptional units. The default choice appears to be growth, which occurs when adequate SocE is available. The decision to develop is normally made at a time when the stringent response leads to depletion of SocE and shifts the balance in favor of CsgA.

SUMMARY: THE DECISION IS A STAGED PROCESS

The decision to develop in *M. xanthus* is made in at least three stages: the stringent pathway, the population-sensing pathway, and the growth arrest pathway (Fig. 2). Each stage involves a simple binary decision that implements new changes in transcription which, combined, form the early developmental pathway. In the previous chapter Dale Kaiser described the first two steps in detail. They are summarized below.

M. xanthus consumes amino acids as the principal carbon, nitrogen, and energy source (Dworkin, 1962). Individual cells assess amino acid availability by using the stringent pathway that samples the availability of charged tRNA molecules (Manoil and Kaiser, 1980a, 1980b). Translational pausing due to lack of a specific charged tRNA causes the ribosome-associated protein RelA to synthesize (p)ppGpp. The stringent response begins to change the levels of the players in the growth arrest pathway by inhibiting transcription of *socE* and inducing transcription of *csgA* (Crawford and Shimkets, submitted [b]). The stringent response also activates the population-sensing pathway (Harris et al., 1998; Singer and Kaiser, 1995) (Fig. 2).

Development requires the coordinated effort of many thousands of cells. Having ascertained that some individuals in the population are starved, the cells need to know whether a sufficient population of starved cells is available to form a fruiting body. Amino acid depletion leads to secretion of proteases by the A-signaling regulatory proteins AsgA, AsgB, and AsgC, which hydrolyze cell surface- and extracellular-matrix-associated proteins (Plamann et al., 1992) (Fig. 2). The concentration of these extracellular proteases is proportional to the density of starved cells such that the free amino acids generated by proteolysis serve as an indicator of starvation and cell density (Kuspa et al., 1992a, 1992b). The cells monitor their density

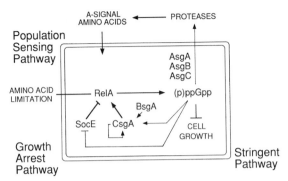

FIGURE 2 Model for signaling events during early development, including the stringent response, A-signaling, and growth control by the C-signaling protein CsgA. Individual cells respond to amino acid starvation by producing the alarmones (p)ppGpp through the stringent pathway (Harris et al., 1998; Singer and Kaiser, 1995). Accumulation of (p)ppGpp activates the population-sensing pathway A-signaling proteins AsgA, AsgB, and AsgC, which leads to protease secretion (Plamann et al., 1992). The (p)ppGpp pool also inactivates the *socE* transcriptional units and activates the *csgA* transcriptional unit (Crawford and Shimkets, submitted [b]). Extracellular-proteases hydrolyze cell surface-associated proteins to generate the A-signal amino acids (Kuspa et al., 1992a, 1992b). The carbon and energy in these imported amino acids is diverted from growth functions to developmental functions by arrest of the vegetative macromolecular synthesis (Crawford and Shimkets, submitted [a]). The CsgA-dependent growth inhibition leads to amino acid-independent (p)ppGpp accumulation through RelA and is antagonized by SocE.

through detection of specific amino acids in this mixture, the A-signal, by a mechanism that is not yet known. Additionally, the entire suite of amino acids serves as a carbon source to fuel development. At this point the cells must set in place a plan for inhibiting growth-related macromolecular synthesis. If this carbon is diverted into vegetative growth, then the cells have squandered a major carbon reserve for development.

The decision to inhibit vegetative macromolecular synthesis and arrest growth, in spite of available amino acids, is determined by the relative level of SocE and CsgA (Crawford and Shimkets, submitted [a]). The most critical feature appears to be the induction of the RelA-mediated synthesis of (p)ppGpp in the presence of excess amino acids (Fig. 2). CsgA clearly acts to stimulate (p)ppGpp synthesis but can only do so if SocE levels are reduced.

CONCLUSIONS

The decision by *M. xanthus* to grow or develop utilizes the universal stringent pathway with a novel twist, SocE and CsgA. The manner in which SocE and CsgA are deployed remains unknown, but the prevailing evidence suggests that they do not stray too far from the traditional stringent reaction. Do they manipulate the supply of amino acids internally to generate an auxotrophy? Do they control the supply of

aminoacyl-tRNA available to translating ribosomes? Or do they regulate the activity of RelA directly? One might imagine that mutations that eliminate any substrates for CsgA would also confer the conditional-developmental phenotype typical of *csgA*, yet *csgA* is the sole known member of this signaling class. One possibility is that the CsgA substrate is essential for translation of vegetative mRNA.

Another question that requires an answer is why CsgA-dependent growth control is coupled to the C-signaling pathway. One possibility is that growth control is just one of several distinct biochemical functions of CsgA and is a function that does not require even a transient extracellular state for CsgA. Indeed, ectopic depletion of SocE was performed with cells in dilute suspension, where cell-cell contact, a necessary condition for C-signaling, is minimal (Crawford and Shimkets, submitted [a]). Though the genetic and environmental conditions of the experiment impose an unnatural state on the cells, the results indicate that C-signaling is not essential for growth arrest and sporulation under these conditions. Another possibility is that cell density sensing is critical in the decision to halt growth and that cells have evolved C-signaling in order to monitor cell proximity by a method that is independent of A-signaling. It is necessary to realize that natural forces can lead to abrupt changes in the environment. A drop of rain or a gust of wind could disperse a population to densities insufficient to complete fruiting body development. If this occurs after A-signaling, then the decision to resume growth would need to be made promptly and in a different manner. If the internal supply of CsgA is due to a cell contact-dependent CsgA import process, then dissociated cells would realize an immediate decline in CsgA and be forced to renegotiate their decision based on the new balance between CsgA and SocE.

ACKNOWLEDGMENTS

This work was supported by National Science Foundation grant 96-01077.

I thank D. B. Kearns and Eugene W. Crawford, Jr., for critical reading of the manuscript.

REFERENCES

Bacon, K., and E. Rosenberg. 1967. Ribonucleic acid synthesis during morphogenesis in *Myxococcus xanthus. J. Bacteriol.* **94:**1883–1889.

Baker, M. 1994. *Myxococcus xanthus* C-factor, a morphogenetic paracrin signal, is similar to *Escherichia coli* 3-oxoacyl-[acyl-carrier-protein] reductase and human 17β-hydroxysteroid dehydrogenase. *Biochem. J.* **301:**311–312.

Bretscher, A. P., and D. Kaiser. 1978. Nutrition of *Myxococcus xanthus*, a fruiting myxobacterium. *J. Bacteriol.* **133:**763–768.

Cashel, M., D. R. Gentry, V. J. Hernandez, and D. Vinella. 1996. The stringent response, p. 1458–1496. *In* F. C. Neidhardt, R. Curtiss III, J. L. Ingraham, E. C. C. Lin, K. B. Low, B. Magasanik, W. S. Reznikoff, M. Riley, M. Schaechter, and H. E. Umbarger (ed.), *Escherichia coli and Salmonella: Cellular and Molecular Biology*, 2nd ed. American Society for Microbiology, Washington, D.C.

Chen, H.-W., I. Keseler, and L. J. Shimkets. 1990. Genome size of *Myxococcus xanthus* determined by pulsed-field gel electrophoresis. *J. Bacteriol.* **172:**4206–4213.

Chen, H.-W., A. Kuspa, I. Keseler, and L. J. Shimkets. 1991. Physical map of the *M. xanthus* chromosome. *J. Bacteriol.* **173:**2109–2115.

Crawford, E. W., Jr., and L. J. Shimkets. The choice between growth and development in *Myxococcus xanthus* is regulated by SocE and the CsgA C-signaling protein. Submitted for publication (a).

Crawford, E. W., Jr., and L. J. Shimkets. The *Myxococcus xanthus socE* and *csgA* genes are regulated by the stringent response. Submitted for publication (b).

Dworkin, M. 1962. Nutritional requirements for vegetative growth of *Myxococcus xanthus. J. Bacteriol.* **84:**250–257.

Dworkin, M., and S. M. Gibson. 1964. A system for studying microbial morphogenesis: rapid formation of microcysts in *Myxococcus xanthus. Science* **146:**243–244.

Filer, D., S. H. Kindler, and E. Rosenberg. 1977a. Myxospore coat synthesis in *Myxococcus xanthus*: enzymes associated with uridine 5′-diphosphate-*N*-acetylgalactosamine formation during myxospore development. *J. Bacteriol.* **131:**745–750.

Filer, D., D. White, S. H. Kindler, and E. Rosenberg. 1977b. Myxospore coat synthesis in *Myxococcus xanthus*: in vivo incorporation of acetate and glycine. *J. Bacteriol.* **131:**751–758.

Ghosh, D., Z. Wawrzak, C. M. Meeks, W. L. Duax, and M. Erman. 1994. The refined three-dimensional structure of 3α,20β-hydroxysteroid dehydrogenase and possible roles of the residues conserved in short-chain dehydrogenases. *Structure* **2:**629–640.

Hagen, T. J., and L. J. Shimkets. 1990. Nucleotide sequence and transcriptional products of the *csg* locus of *Myxococcus xanthus. J. Bacteriol.* **172:**15–23.

Hagen, D. C., A. P. Bretscher, and D. Kaiser. 1978. Synergism between morphogenetic mutants of *Myxococcus xanthus. Dev. Biol.* **64:**284–296.

Harris, B. Z., D. Kaiser, and M. Singer. 1998. The guanosine nucleotide (p)ppGpp initiates development and A-factor production in *Myxococcus xanthus. Genes Dev.* **12:**1022–1035.

Hemphill, H. E., and S. A. Zahler. 1968. Nutrition of *Myxococcus xanthus* FBa and some of its auxotrophic mutants. *J. Bacteriol.* **95:**1011–1017.

Hodgson, D. A. 1993. Light-induced carotenogenesis in *Myxococcus xanthus*: genetic analysis of the *carR* region. *Mol. Microbiol.* **7:**471–488.

Kim, S., and D. Kaiser. 1991. C-factor has distinct aggregation and sporulation thresholds during *Myxococcus* development. *J. Bacteriol.* **173:**1722–1728.

Kim, S. K., and D. Kaiser. 1990a. C-factor: a cell-cell signaling protein required for fruiting body morphogenesis of *M. xanthus. Cell* **61:**19–26.

Kim, S. K., and D. Kaiser. 1990b. Cell alignment required in differentiation of *M. xanthus. Science* **249:**926–928.

Kim, S. K., and D. Kaiser. 1990c. Cell motility is required for the transmission of C-factor, an intercellular signal that coordinates fruiting body morphogenesis of *Myxococcus xanthus. Genes Dev.* **4:**896–905.

Kim, S. K., and D. Kaiser. 1990d. Purification and properties of *Myxococcus xanthus* C-factor, an intercellular signaling protein. *Proc. Natl. Acad. Sci. USA* **87:**3635–3639.

Kuspa, A., L. Plamann, and D. Kaiser. 1992a.

Identification of heat-stable A-factor from *Myxococcus xanthus*. *J. Bacteriol.* **174:**3319–3326.

Kuspa, A., L. Plamann, and D. Kaiser. 1992b. A-signalling and the cell density requirement for *Myxococcus xanthus* development. *J. Bacteriol.* **174:** 7360–7369.

Lee, B.-U., K. Lee, J. Mendez, and L. J. Shimkets. 1995. A tactile sensory system of *Myxococcus xanthus* involves an extracellular NAD(P)$^+$-containing protein. *Genes Dev.* **9:**2964–2973.

Lee, K., and L. J. Shimkets. 1994. Cloning and characterization of the *socA* locus, which restores development to *Myxococcus xanthus* C-signaling mutants. *J. Bacteriol.* **176:**2200–2209.

Li, S., B. Lee, and L. J. Shimkets. 1992. *csgA* expression entrains *Myxococcus xanthus* development. *Genes Dev.* **6:**401–410.

Manoil, C., and D. Kaiser. 1980a. Accumulation of guanosine tetraphosphate and guanosine pentaphosphate in *Myxococcus xanthus* during starvation and myxospore formation. *J. Bacteriol.* **141:** 297–304.

Manoil, C., and D. Kaiser. 1980b. Guanosine pentaphosphate and guanosine tetraphosphate accumulation and induction of *Myxococcus xanthus* fruiting body development. *J. Bacteriol.* **141:**305–315.

McGowan, S. J., H. C. Gorham, and D. A. Hodgson. 1993. Light-induced carotenogenesis in *Myxococcus xanthus*: DNA sequence analysis of the *carR* region. *Mol. Microbiol.* **10:**713–735.

Neidle, E., C. Hartnett, L. N. Ornston, A. Bairoch, M. Rekik, and S. Harayama. 1992. *cis*-Diol dehydrogenase encoded by the TOL pWW0 plasmid *xylL* gene and the *Acinetobacter calcoaceticus* chromosomal *benD* gene are members of the short-chain alcohol dehydrogenase superfamily. *Eur. J. Biochem.* **240:**113–120.

Okano, P., K. Bacon, and E. Rosenberg. 1970. Ribonucleic acid synthesis during microcyst formation in *Myxococcus xanthus*: characterization by de-oxyribonucleic acid-ribonucleic acid hybridization. *J. Bacteriol.* **104:**275–282.

Orlowski, M., P. Martin, D. White, and M. C.-W. Wong. 1972. Changes in activity of glyoxylate cycle enzymes during myxospore development in *Myxococcus xanthus*. *J. Bacteriol.* **111:**784–790.

Persson, B., M. Krook, and H. Jornvall. 1991. Characteristics of short-chain alcohol dehydrogenases and related enzymes. *Eur. J. Biochem.* **200:** 537–543.

Plamann, L., A. Kuspa, and D. Kaiser. 1992. Proteins that rescue A-signal-defective mutants of *Myxococcus xanthus*. *J. Bacteriol.* **174:**3311–3318.

Rhie, H., and L. J. Shimkets. 1989. Developmental bypass suppression of *Myxococcus xanthus csgA* mutations. *J. Bacteriol.* **171:**3268–3276.

Rosenberg, E., M. Katarski, and P. Gottlieb. 1967. Deoxyribonucleic acid synthesis during exponential growth and microcyst formation in *Myxococcus xanthus*. *J. Bacteriol.* **93:**1402–1408.

Sager, B., and D. Kaiser. 1994. Intercellular C-signaling and the traveling waves of *Myxococcus*. *Genes Dev.* **8:**2793–2804.

Shimkets, L. J., and H. Rafiee. 1990. CsgA, an extracellular protein essential for *Myxococcus xanthus* development. *J. Bacteriol.* **172:**5299–5306.

Singer, M., and D. Kaiser. 1995. Ectopic production of guanosine penta- and tetraphosphate can initiate early developmental gene expression in *Myxococcus xanthus*. *Genes Dev.* **9:**1633–1644.

Watson, B. F., and M. Dworkin. 1968. Comparative intermediary metabolism of vegetative cells and microcysts of *Myxococcus xanthus*. *J. Bacteriol.* **96:** 1465–1473.

Zusman, D., P. Gottlieb, and E. Rosenberg. 1971. Division cycle of *Myxococcus xanthus*. III. Kinetics of cell growth and protein synthesis. *J. Bacteriol.* **105:**811–819.

Zusman, D. R., D. M. Krotoski, and M. Cumsky. 1978. Chromosome replication in *Myxococcus xanthus*. *J. Bacteriol.* **133:**122–129.

DEVELOPMENT OF *STIGMATELLA*

David White and Hans Ulrich Schairer

14

Stigmatella is usually found on the bark of rotting wood, where it forms exquisitely shaped bright-orange fruiting bodies consisting of a stalk supporting several sporangioles. Its life cycle, habitat, and isolation are similar to that of other myxobacteria described in chapter 10. Both the stalk and sporangioles are usually covered with a rigid coat. Occasionally, the coat is absent in young fruiting bodies viewed with the scanning electron microscope, revealing a cellular stalk, or at least its outer layer (Fig. 1). In older fruiting bodies the cells in the stalk lyse, as revealed by transmission electron microscopy (Voelz and Reichenbach, 1969; Lünsdorf et al., 1995).

LIFE CYCLE

When cells are subjected to nutrient depletion, they glide into aggregation centers and continue their morphogenetic movements to form fruiting bodies (Fig. 2). Distinct stages in fruiting can be observed. These are the aggregate stage (Fig. 2A), the early stalk stage (Fig. 2B), a later stage referred to as the "mushroom" stage (Fig. 2C), and the final mature-fruit stage (Fig. 2D). The life cycle is shown in Fig. 3. It

is similar to the life cycle of *Myxococcus xanthus* (see chapter 10), except that aggregation is stimulated by light and a pheromone, as described later.

CELL PATTERNING DURING DEVELOPMENT

Vasquez et al. (1985) examined the patterns of cells at various stages of aggregation and fruiting body formation by *Stigmatella aurantiaca*. Scanning electron micrographs of very early stages in aggregation reveal that the cells are arranged in concentric circles or ellipses (Fig. 4). Even at later stages of fruiting body development, the cells are arranged in concentric circles around the developing fruit (Fig. 5). Similar photographs of rings of cells at the bases of aggregates were previously published by Grilione and Pangborn (1975). The photographs suggest that the cells closest to the base of the aggregate leave the circular band of cells and enter the aggregate, moving in a spiral fashion within it (Fig. 6). At a later stage, as the stalk forms, the cells appear to move under the aggregate as it is lifted off the surface (Fig. 7). All of this is, of course, speculation based upon scanning electron micrographs. The movements of live cells have not been sufficiently investigated. More recently, cell movements during fruiting body formation by *M. xanthus*

D. *White*, Department of Biology, Indiana University, Bloomington, IN 47401. *H. U. Schairer*, Zentrum für Molekulare Biologie, Universität Heidelberg, Im Neuenheimer Feld 282, D-69120 Heidelberg, Germany.

Prokaryotic Development, edited by Y. V. Brun and L. J. Shimkets,
© 2000 American Society for Microbiology, Washington, DC 20005-4171

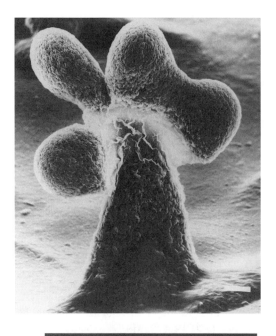

FIGURE 1 Scanning electron micrograph of fruiting body. Note the partial covering of the coat and the size difference between the stalk and sporangiophore cells. Bar = 8 μm. (From Stephens and White, 1980b.)

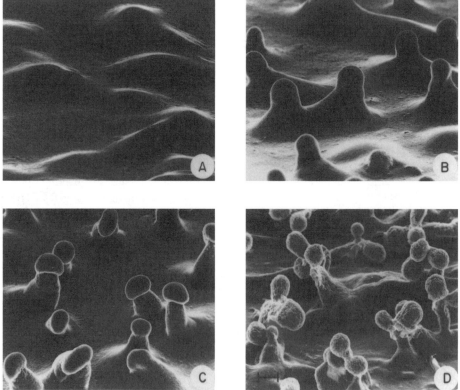

FIGURE 2 Scanning electron micrographs of different stages of fruiting body formation. (A) Early aggregates; (B) early stalk formation; (C) late stalk formation (mushroom stage); (D) mature fruiting bodies. Bar = 20 μm. (From Qualls et al., 1978b.)

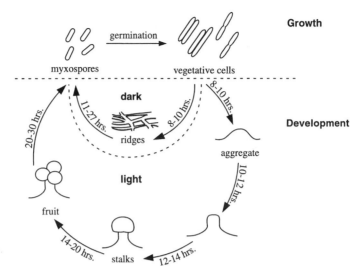

FIGURE 3 Life cycle of *S. aurantiaca*. When cells are placed on non-nutrient agar in the light, aggregates are formed in 8 to 10 h and fully mature fruiting bodies are formed in approximately 20 h. In the dark, the cells tend to aggregate into ridges and fruiting is depressed, although myxospores form. When the cell density is sufficiently low, there is a requirement for light or the addition of exogenous pheromone for aggregation to occur.

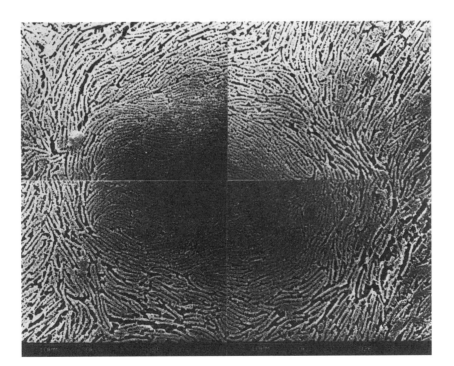

FIGURE 4 Scanning electron micrograph of *S. aurantiaca* aggregation site. This is a very young aggregate. Note that the cells are arranged in concentric rings at the periphery of the aggregate, but not necessarily in the center. (From Vasquez et al., 1985.)

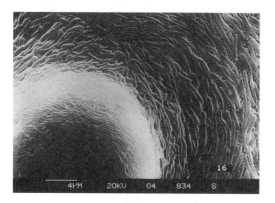

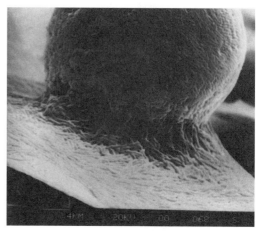

FIGURE 5   An advanced stage of aggregation. Note the circular arrangement of cells at the base of the aggregate and the spiral pattern on the aggregate. (From Vasquez et al., 1985.)

FIGURE 7   Scanning electron micrograph showing cells leaving circular ring around aggregate and entering at bottom as aggregate is lifted off the surface. (From Vasquez et al., 1985.)

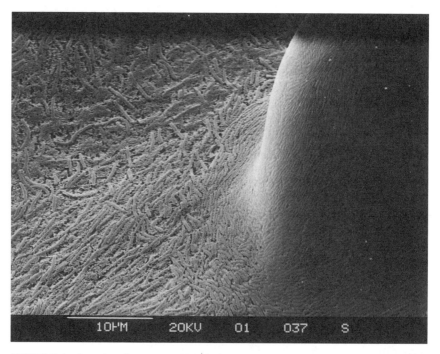

FIGURE 6   Scanning electron micrograph showing cells entering aggregate at base. (From Vasquez et al., 1985.)

have been reported and are described in chapter 11.

Speculation as to How Aggregates Might Form

It is not known how the cells at a distance from the aggregation centers find these centers. It may be that the first cells that enter the aggregate do so randomly but that later arrivals follow slime trails laid down by the cells that have arrived earlier. There is no evidence that the cells follow a chemotactic signal over a large distance towards the aggregate, although as discussed later, they may do so at its base, resulting in nearby cells entering the aggregate.

Just before large aggregates become visible, microscopic aggregates form but then recede. Thus, it appears that something stabilizes the aggregates, allowing them to grow larger. Mechanisms through which this might occur have been discussed previously (White, 1993). One possibility is that the cells within the aggregate become locked into a turning mode (either clockwise or counterclockwise), preventing them from leaving the aggregate. Alternatively, or additionally, a chemical secreted by the cells at the site of aggregation may act as a chemoattractant and prevent the cells from leaving. A pheromone, discussed later, is secreted at the time stable aggregates appear, and this pheromone may play a direct role in aggregate formation. Computer simulations of aggregate formation in myxobacteria have been published (Stevens, 1990, 1991, 1993, 1995; Koch, 1998).

CATION REQUIREMENTS FOR GLIDING MOTILITY, CELL-CELL COHESION, AND DEVELOPMENT

Specific cations have a profound effect on motility, cell-cell cohesion, and development. Of major importance is calcium ion, which has many different effects on *Stigmatella*. For example, calcium ion induces the ability to glide as well as the synthesis of an energy-dependent intercellular cohesion system. In both cases the induction is prevented by inhibitors of protein synthesis (White et al., 1980a; Gilmore and White, 1985; Womack et al., 1989). Calcium ion is required not only for the induction of these systems but also for their operation. (However, cations in the calcium group can substitute for calcium for functioning of the energy-dependent cohesion system.) Possibly related to these effects of calcium on gliding motility and cell-cell cohesion is the fact that calcium ion induces the formation of extracellular fibrils, the appearance in the membrane fractions of a 30-kDa protein as well as other polypeptides, and the translocation to the extracellular matrix of a 55-kDa membrane protein (Chang and White, 1992). The 55-kDa protein is antigenically related to a fibril protein made by the myxobacterium *M. xanthus*.

Cation Requirements for Morphogenesis

Given the fact that cations have major effects on motility and cohesion, as described above, it is not surprising that they also have important influences on the development of fruiting bodies. The presence of calcium, magnesium, and manganese ions is necessary to obtain normal fruiting bodies (White et al., 1980a). When one of these cations is omitted, the cells form abnormal aggregates and incomplete fruiting bodies.

EFFECT OF LIGHT

Fruiting body formation by *Stigmatella* is stimulated by light (Qualls et al., 1978a). (Light had previously been shown by Reichenbach [1975] to stimulate fruiting body formation in another myxobacterium, *Chondromyces*.) In the absence of light, the cells at high density often aggregate into ridges rather than forming individual aggregates, and the ridges rarely develop into fruiting bodies. (Occasionally, high cell densities will enable fruiting bodies to form in the dark. This may be due to the production of sufficient pheromone, as described below.) Under conditions of low cell density, aggregation does not occur in the dark. Interestingly, once aggregates form in the light, fruiting bodies will develop even if the developing cells are placed in the dark, indicating that the light is required only for the formation of the aggregates and not for subsequent stages of fruiting or for myxospore formation.

Effect of Light on the Pattern of Protein Synthesis

Light has a clear effect on the pattern of protein synthesis (Inouye et al., 1980). Several proteins are made in larger amounts when the cells develop in the presence of light.

Action Spectrum

An action spectrum revealed that the most effective light for stimulating aggregation was blue light between 400 and 500 nm (White et al., 1980b). At higher irradiances, other wavelengths, including those in the far-red region, were also effective. The photoreceptor has not been identified.

PHEROMONE

S. aurantiaca secretes a lipoidal pheromone that begins to accumulate in the extracellular medium at the time aggregates appear (Stephens et al., 1982). The addition of purified pheromone to the agar lowers the threshold cell density required for cells to aggregate and form fruiting bodies. Because cells form fruiting bodies at a lower cell density if pheromone is added to the medium, the pheromone appears to play the role of a quorum sensor, and in this way may be analogous to A-signal produced by *M. xanthus* (see chapter 12).

The Pheromone Can Substitute for the Light Requirement

The pheromone decreases the aggregation period required for aggregates to form in the light or in the dark and, very importantly, substitutes for the light requirement. That is to say, cells aggregate and form fruiting bodies in the dark at low cell densities provided extra pheromone is added to the agar.

The Pheromone Probably Acts Early in Development

Since the pheromone can substitute for light, and since light is required only for the aggregation stage, it seems likely that the pheromone acts early in development.

Light Increases the Sensitivity of the Cells to the Pheromone

One might speculate that the role of light is to stimulate the production of pheromone, and this explains why the addition of pheromone can substitute for the light requirement, but it was shown that the amount of pheromone produced in the light and in the dark was the same (Stephens et al., 1982). However, the cells responded to lower amounts of pheromone in the light, suggesting that the role of light is to somehow increase the sensitivity of the cells to the pheromone.

The Pheromone May Affect Gene Transcription

Given the facts that light induces the synthesis of certain proteins, as described above, and the pheromone can substitute for light, it may be that a signaling pathway exists that involves a light-stimulated response to the pheromone, leading to pheromone-dependent gene expression. Any effect that the pheromone would have on gene transcription would no doubt be indirect, since the pheromone is lipoidal and would not be expected to accumulate in the cytoplasm. However, what effect, if any, the pheromone has on gene expression has not yet been investigated.

Light and the Pheromone Have No Effect on *Myxococcus*

Myxococcus aggregation is not stimulated by light, and the *Stigmatella* pheromone has no effect on *Myxococcus* aggregation. Thus, it appears that the signaling pathways that initiate development in these two different myxobacteria are not identical.

An Interesting Parallel with Respect to Light and Intercellular Signaling in the Cellular Slime Mold *Polysphondelium*

Under conditions of nutrient depletion, cellular slime mold amoebae aggregate and form fruiting bodies (Alexander and Rossomando,

1992). Light stimulates aggregation in *Polysphondylium violaceum*. Furthermore, *P. violaceum* produces an aggregation factor (D-factor) that stimulates aggregation in the dark, and light sensitizes the cells to the aggregation factor (Teta and Hanna, 1981.)

Involvement of Guanine Derivatives in the Response to Light and the Pheromone

An intriguing observation is that the addition of guanosine or guanine nucleotides stimulates fruiting body formation at low cell densities in the light or in the dark and can substitute for light or the pheromone in promoting fruiting body formation (Stephens and White, 1980). As with the pheromone, light increases the sensitivity of the cells to the guanine derivatives. This indicates that perhaps guanine nucleotides and the pheromone share a common signaling pathway stimulated by light.

Pheromone Structure

The pheromone has been purified, and its structure has been determined by mass spectrometry, infrared spectroscopy, and ^{1}H and ^{13}C nuclear magnetic resonance (Plaga et al., 1998; Hull et al., 1998). It is a branched aliphatic hydroxy ketone (Fig. 8). The structure of the pheromone has been confirmed by chemical synthesis of a compound that displays all of the activities of the naturally produced pheromone. Nothing is known about the putative pheromone receptor or the pheromone-dependent signaling pathway.

Some Speculation Concerning the Activity of the Pheromone

Given the hydrophobic nature of the pheromone, it is likely that it accumulates in the slime

trails and diffuses very slowly in the agar. (When myxobacteria glide, they leave microscopically visible trails on the agar surface, and these trails are the "highways" on which the myxobacteria move. The trails consist of secreted material called slime, which is thought to be rich in polysaccharide.) Since it is first detectable when aggregates appear and stimulates aggregation, it may accumulate in the slime trails at the sites of aggregation. One possible scenario is that a steep gradient of pheromone exists very close to the aggregate and that cells near the base of the aggregate follow the gradient into the aggregate. Chemotaxis towards a lipoidal substance has a precedent in *M. xanthus*, where it has been demonstrated that the cells move up a gradient of phosphatidylethanolamine in agar (Kearns and Shimkets, 1998). Another possibility is that a high concentration of pheromone within the aggregate keeps the cells turning in either the clockwise or counterclockwise direction, thus trapping newly arrived cells in the aggregate. A third possibility is that the cells adhere to each other more tightly as the amounts of pheromone increase. The effects of the pheromone on cell motility and intercellular adhesion have not been adequately studied.

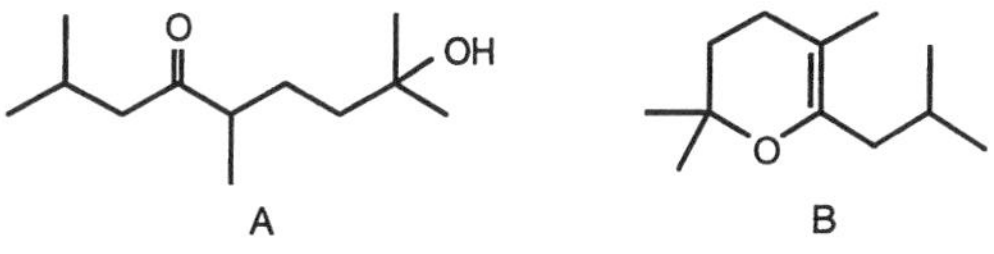

GENETIC METHODS AVAILABLE FOR *S. AURANTIACA*

Genetic methods have been developed to identify genes important for development in *Stigmatella* and to analyze the regulation of their expression (Schairer, 1993). These include the use of transposon mutagenesis with Tn*5* or Tn*5* *lacZ* to generate developmental mutants and to monitor the expression of genes into which Tn*5* *lacZ* has been inserted (Glomp et al., 1988; Pospiech et al., 1993). Given the relatively complex structure of the fruiting body, *Stigmatella* genetics, combined with a biochemical approach, offers the opportunity to analyze the bases of complex morphogenetic movements in a prokaryote.

Genes Homologous to the *Myxococcus csgA* and *mgl* Genes

M. xanthus mutants defective in making C-signal (*csgA* mutants) construct abnormal aggre-

FIGURE 8 Chemical structure of stigmolone (A), which is reversibly converted into a dihydropyrane derivative by dehydration under acidic catalysis (B). (From Plaga et al., 1998, and Hull et al., 1998.)

gates (see chapter 13). CsgA is a membrane-associated protein that is localized at the surface of the cells and appears to be a short-chain alcohol dehydrogenase. With the *M. xanthus csgA* gene as a probe, a homologous gene was isolated from *S. aurantiaca* (Butterfaß, 1992; Schairer, 1993). Inactivation of the gene by insertional mutagenesis caused defective fruiting body formation in *S. aurantiaca*. Similarly, a gene homologous to the *mgl* gene (required for motility in *M. xanthus*) was isolated from *S. aurantiaca* and shown by insertional mutagenesis to be required for motility (Schairer, 1993). These results indicate that the opportunity exists for a methodical survey for homologous developmental and motility genes shared between the two myxobacteria *M. xanthus* and *S. aurantiaca*.

Genes Involved in Fruiting Body Formation

In order to find genes involved in fruiting body formation in *S. aurantiaca*, Tn*5 lacZ* transposon mutagenesis was performed (Glomp et al., 1988). Three classes of mutants have been generated, i.e., class 1, those that do not aggregate; class 2, those that aggregate into clumps; and class 3, those that form anomalous fruiting bodies. Mixtures of class 1 and class 2 mutants allow partial extracellular complementation to form an intermediate-stage fruiting body resembling the mushroom stage shown in Fig. 3C. These results suggest that the mutations result in the inability to make extracellular factor(s). The inactivation of the gene *fbfA* resulted in a mutant phenotype of nonstructured clumps (class 2) (Silakowski et al., 1996). The gene was transcribed after 8 h of development, as determined by measuring the β-galactosidase activity of a *fbfA*-*lacZ* fusion gene and by Northern (RNA) analysis. Sequence analysis of the gene shows homology to the NodC polypeptide of *Rhizobium meliloti*, the hyaluronan synthase of *Streptococcus pyogenes*, and the chitin synthase 3 of *Saccharomyces cerevisiae*. All of these proteins are enzymes involved in extracellular-polysaccharide synthesis, suggesting that *fbfA* may encode an enzyme required for the synthesis of an ex-

tracellular polysaccharide. This may point to a role in morphogenesis for polysaccharides that extend from the cell surface, or are released, into the extracellular matrix. These polysaccharides may play structural roles, may be part of the slime trails, or may be oligomeric signaling factors required for development. A second gene, *fbfB*, encodes a putative polypeptide homologous to galactose oxidase, an enzyme that perhaps modifies galactose residues on polysaccharides (Silakowski et al., 1998). Inactivation of *fbfB* results in defective fruiting body formation. The gene *fbfB* is expressed about 14 h after the beginning of starvation. Thus, two genes, *fbfA* and *fbfB*, that appear to encode enzymes involved in polysaccharide or carbohydrate oligomer synthesis or modification are required for normal fruiting body formation. Whether the products of these genes are somehow involved in developmental intercellular signaling or in the structure of the matrix in the slime trails or between cells remains to be seen.

CONCLUSIONS

What determines that cells enter the aggregates and stalks and what determines the formation of multiple sporangioles from the simple mass of cells at the top of the stalk in the mushroom stage (Fig. 3C) are not understood. Nevertheless, some of the factors that are involved have been discovered. One of these is the hydrophobic pheromone that appears to be required for the formation of stable aggregates. There are three important questions that can presently be asked about the activity of the pheromone. (i) Is the pheromone a chemoattractant? (ii) Does the pheromone influence the turning mode or reversal frequencies of the cells? (iii) Does the pheromone influence the expression of developmental genes? Knowing the answers to these questions will allow the construction of hypotheses regarding the mode of action of the pheromone.

It is interesting that transposon mutagenesis has produced mutants defective in fruiting body formation that have mutations in genes that encode putative enzymes required for exopolysaccharide biosynthesis. This points to an important role for exopolysaccharides (or oligomers) in morphogenesis and emphasizes the need for a systematic attempt to isolate and chemically characterize these molecules. The combination of a biochemical and a genetic approach in this area should produce a greater understanding of what determines aggregation and the shapes of the fruiting bodies formed by *S. aurantiaca*.

REFERENCES

Alexander, S., and E. F. Rossomando. 1992. Regulation of morphogenesis in *Dictyostelium discoideum*, p. 29–61. *In* E. F. Rossomando and S. Alexander (ed.), *Morphogenesis: An Analysis of the Development of Biological Form*. Marcel Dekker, Inc., New York, N.Y.

Butterfaß, H.-J. 1992. Isolierung und charakterisierung des *csgA*-genes aus *Stigmatella aurantiaca*. Ph.D. thesis. University of Heidelberg, Heidelberg, Germany.

Chang, B., and D. White. 1992. Cell surface modifications induced by calcium ion in the myxobacterium *Stigmatella aurantiaca*. *J. Bacteriol.* **174:** 5780–5787.

Gilmore, D. F., and D. White. 1985. Energy-dependent cell cohesion in myxobacteria. *J. Bacteriol.* **161:**113–117.

Glomp, I., P. Saulnier, J. Guespin-Michel, and H. U. Schairer. 1988. Transfer of IncP plasmids into *Stigmatella aurantiaca* leading to insertional mutants affected in spore development. *Mol. Gen. Genet.* **214:**213–217.

Grilione, P. L., and J. Pangborn. 1975. Scanning electron microscopy of fruiting body formation by myxobacteria. *J. Bacteriol.* **124:**1558–1565.

Hull, W. E., A. Berkessel, and W. Plaga. 1998. Structure elucidation and chemical synthesis of stigmalone, a novel type of prokaryotic pheromone. *Proc. Natl. Acad. Sci. USA* **95:**11268–11273.

Inouye, S., D. White, and M. Inouye. 1980. Development of *Stigmatella aurantiaca*: effects of light and gene expression. *J. Bacteriol.* **141:**1360–1365.

Kearns, D. B., and L. J. Shimkets. 1998. Chemotaxis in a gliding bacterium. *Proc. Natl. Acad. Sci. USA* **95:**11957–11962.

Koch, A. L. 1998. The strategy of *Myxococcus xanthus* for group cooperative behavior. *Antonie Leeuwenhoek* **73:**299–313.

Lünsdorf, H., H. U. Schairer, and M. Heidelbach. 1995. Localization of the stress protein SP21 in indole-induced spores, fruiting bodies, and heat-shocked cells of *Stigmatella aurantiaca*. *J. Bacteriol.* **177:**7092–7099.

Plaga, W., I. Stamm, and H. U. Schairer. 1998. Intercellular signaling in *Stigmatella aurantiaca*: purification and characterization of stigmalone, a myxobacterial pheromone. *Proc. Natl. Acad. Sci. USA* **95:** 11263–11267.

Pospiech, A., B. Neumann, B. Silakowski, and H. U. Schairer. 1993. Detection of developmentally regulated genes of the myxobacterium *Stigmatella aurantiaca* with the transposon Tn5 *lacZ*. *Arch. Microbiol.* **159:**201–206.

Qualls, G. T., K. Stephens, and D. White. 1978a. Light-stimulated morphogenesis in the fruiting myxobacterium *Stigmatella aurantiaca*. *Science* **201:** 444–445.

Qualls, G. T., K. Stephens, and D. White. 1978b. Morphogenetic movements and multicellular development in the fruiting myxobacterium *Stigmatella aurantiaca*. *Dev. Biol.* **66:**270–274.

Reichenbach, H. 1975. *The 1975 Second International Symposium on the Biology of Myxobacteria*. Personal communication.

Schairer, H. U. 1993. *Stigmatella aurantiaca*, an organism for studying the genetic determination of morphogenesis, p. 333–346. *In* M. Dworkin and D. Kaiser (ed.), *Myxobacteria II*. American Society for Microbiology, Washington, D.C.

Silakowski, B., A. Pospiech, B. Neumann, and H. U. Schairer. 1996. *Stigmatella aurantiaca* fruiting body formation is dependent on the *fbfA* gene encoding a polypeptide homologous to chitin synthases. *J. Bacteriol.* **178:**6706–6713.

Silakowski, B., H. Ehret, and H. U. Schairer. 1998. *fbfB*, a gene encoding a putative galactose oxidase, is involved in *Stigmatella aurantiaca* fruiting body formation. *J. Bacteriol.* **180:**1241–1247.

Stephens, K., and D. White. 1980a. Morphogenetic effects of light and guanine derivatives on the fruiting myxobacterium *Stigmatella aurantiaca*. *J. Bacteriol.* **144:**322–326.

Stephens, K., and D. White. 1980b. Scanning electron micrographs of fruiting bodies of the myxobacterium *Stigmatella aurantiaca* lacking a coat and revealing a cellular stalk. *FEMS Microbiol. Lett.* **9:** 189–192.

Stephens, K., G. D. Hegeman, and D. White. 1982. Pheromone produced by the myxobacterium *Stigmatella aurantiaca*. *J. Bacteriol.* **149:**739–747.

Stevens, A. 1990. Simulations of the aggregation and gliding behaviour of myxobacteria, p. 548–555. *In* W. Alt and G. Hoffman (ed.), *Biological Motion. Lecture Notes in Biomathematics*, vol. 89. Springer Verlag, Berlin, Germany.

Stevens, A. 1991. A model for gliding and aggregation of myxobacteria, p. 269–276. *In* A. Holden, M. Markus, and H. G. Othmer (ed.), *Nonlinear Wave Processes in Excitable Media. NATO AS1 Series B. Physics*, vol. 244. Plenum Press, New York, N.Y.

Stevens, A. 1993. Aggregation of myxobacteria—a many particle system, p. 519–524. *In* J. Demougeot and V. Capasso (ed.), *First European Conference of Mathematics Applied to Biology and Medicine.* Wuerz Publishing, Winnipeg, Canada.

Stevens, A. 1995. Trail following and aggregation of myxobacteria. *J. Biol. Syst.* **3:**1059–1068.

Teta, L. A., and M. H. Hanna. 1981. Relationship between light and an aggregation stimulating factor in the cellular slime mold *Polysphondelium violaceum,* abstr. I149, p. 111. *In Abstracts of the 81st Annual Meeting of the American Society for Microbiology 1981.* American Society for Microbiology, Washington, D.C.

Vasquez, G. M., F. Qualls, and D. White. 1985. Morphogenesis of *Stigmatella aurantiaca* fruiting bodies. *J. Bacteriol.* **163:**515–521.

Voelz, H. G., and H. Reichenbach. 1969. Fine structure of fruiting bodies of *Stigmatella aurantiaca* (*Myxobacterales*). *J. Bacteriol.* **99:**856–866.

White, D. 1993. Myxospore and fruiting body morphogenesis, p. 307–332. *In* M. Dworkin and D. Kaiser (ed.), *Myxobacteria II.* American Society for Microbiology, Washington, D.C.

White, D., J. A. Johnson, and K. Stephens. 1980a. Effects of specific cations on aggregation and fruiting body morphology in the myxobacterium *Stigmatella aurantiaca. J. Bacteriol.* **144:**400–405.

White, D., W. Shropshire, Jr., and K. Stephens. 1980b. Photocontrol of development by *Stigmatella aurantiaca. J. Bacteriol.* **142:**1023–1024.

Womack, B. J., D. F. Gilmore, and D. White. 1989. Calcium requirement for gliding motility in myxobacteria. *J. Bacteriol.* **171:**6093–6096.

STALKED BACTERIA

THE DIMORPHIC LIFE CYCLE OF *CAULOBACTER* AND STALKED BACTERIA

Yves V. Brun and Raji Janakiraman

15

THE DEVELOPMENTAL STRATEGY OF STALKED BACTERIA

The vast majority of differentiating bacteria initiate their developmental programs in response to environmental cues. In contrast, the development of stalked bacteria like *Caulobacter* is an integral part of cell growth and occurs in each cell division cycle (Poindexter, 1992). Thus, the developmental program of stalked bacteria is tightly coupled to progression through the cell cycle. Most stalked bacteria divide asymmetrically at every cell cycle, producing two dissimilar progeny cells, a motile swarmer cell and a sessile stalked cell (Fig. 1). The development of stalked bacteria poses one of the most fundamental problems in biology: how can the division of a cell yield progeny that carry the same genetic information but differ from one another structurally and functionally? The differences in genetic read-out, structure, and function can result from events that occur at the opposite poles of the cell before division. Such cell differentiation does not occur as a series of isolated events, no matter how complex each of the component parts proves to be. The "clock" that monitors life cycle events is the cumulative result of signal transduction pathways integrated with morphogenetic events. How is this integration accomplished? A second, and equally important question is how a cell organizes itself in three-dimensional space. How does a cell go about changing its structure? What are the mechanisms used by the cell to send proteins to one site rather than another? This chapter will introduce the life cycle of stalked bacteria and then focus on the best-studied example, *Caulobacter crescentus*. Recent progress in understanding the regulation of stalk and holdfast synthesis and of cell division will be reviewed. The following chapters will review the regulation of flagellum biosynthesis (see chapter 16), the DNA replication cycle (see chapter 18), and the role of signal transduction pathways in *Caulobacter* development (see chapter 17).

Stalked bacteria are a diverse group belonging to the α *Proteobacteria* (Sly et al., 1999; Stahl et al., 1992; Weiner et al., in press.). At some stage of their life cycles, all stalked bacteria possess one or more extensions of the cell surface called stalks (Whittenbury and Dow, 1977). Stalks are cylindrical extensions of the cell surface that contain cytoplasm and are bound by the cell membrane and cell wall of the organism. These stalks are also called prostheca, appendages, or hypha, depending on the bacte-

Y. V. Brun, Department of Biology, Jordan Hall 142, Indiana University, 1001 E. 3rd St., Bloomington, IN 47405-3700. R. Janakiraman, Department of Biology, University of Michigan, 830 North University, Ann Arbor, MI 48109-1048.

Prokaryotic Development, edited by Y. V. Brun and L. J. Shimkets,
© 2000 American Society for Microbiology, Washington, DC 20005-4171

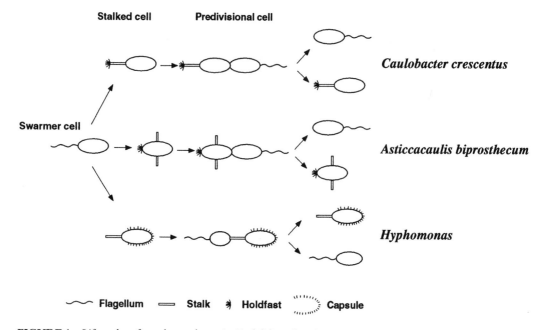

FIGURE 1   Life cycles of prosthecate bacteria. Each life cycle is depicted starting with the swarmer cell. Swarmer cells have a polar flagellum and are chemotactically competent. They are unable to replicate DNA and divide. After remaining in the swarmer stage of the life cycle for a fixed period, the cells differentiate into stalked cells. This differentiation involves release of the flagellum, growth of a stalk and holdfast, and initiation of DNA replication. In *Caulobacter* and *Hyphomonas*, the stalk is synthesized at the pole that shed the flagellum. In *A. biprosthecum*, two stalks are synthesized opposite one another at the midpoint of the cell. In *Caulobacter* and *A. biprosthecum*, the holdfast attachment organelle is synthesized at the previously flagellated pole, resulting in its association with the tip of the stalk in *Caulobacter* but not in *A. biprosthecum*. *Hyphomonas* synthesizes an extracellular polysaccharide capsule involved in surface attachment around the body of the stalked cell. In all cases, cellular growth and synthesis of a new flagellum result in the formation of an asymmetric predivisional cell. In *Hyphomonas*, growth occurs by budding at the tip of the stalk, with a flagellum synthesized at the growing pole of the cell. Growth also appears to be polar in *A. biprosthecum*. Cell division produces a swarmer cell and a stalked cell that can immediately reenter the cell cycle. In *Hyphomonas*, septation occurs at the stalk-daughter cell junction.

rium and the function of the cellular extension. In fact, prostheca is the proper term to refer to these cellular extensions (Poindexter, 1992), but because most of this chapter deals with *Caulobacter* development, we will refer to the prostheca as stalks for simplicity.

The life cycles of different stalked bacteria are very similar (Fig. 1). Cell division produces a stalked cell and a swarmer cell. These two cell types differ not only in their morphology and motility but also in their capacity to enter the next cell cycle. The swarmer cell is unable to initiate DNA replication and is devoted to dispersal for a portion of the cell cycle. The

obligatory time spent as a swarmer cell presumably ensures that progeny cells will colonize a new environmental niche instead of competing with attached stalked cells (Poindexter, 1992). The stalked cell, on the other hand, is sessile and is most often found attached to a surface. The stalked cell can initiate DNA replication and is devoted to producing new swarmer cells. Swarmer cells must undergo an ordered series of events leading to differentiation into a stalked cell to enter a new cell division cycle.

Not all stalked bacteria produce dissimilar progeny (Whittenbury and Dow, 1977). *Prosthecobacter* cells always bear at least one polar stalk and do not synthesize flagella. During growth, a new stalk develops at the younger pole of the cell and results in a symmetric predivisional cell with a stalk at both poles. Division of *Prosthecobacter* thus yields two morphologically similar cells. In *Ancalomicrobium*, cells produce two to eight stalks. Reproduction occurs by budding from the mother cell body, and no flagella are synthesized. Stalks are formed on the bud before division, resulting in the production of two stalked cells.

Dimorphic stalked bacteria can be divided in two major groups based on the function of the stalk. In one group, the stalk is not involved in reproduction (for example, *Caulobacter* and *Asticcacaulis* (see below). In the other group, the budding bacteria, the stalk (often called a hypha) is involved in reproduction. Examples of this group include *Rhodomicrobium vannielii*, *Hyphomonas*, and *Hyphomicrobium*, which grow and reproduce by budding at the tip of the stalk, producing a flagellated swarmer cell (Fig. 1 and 2). In subsequent cell cycles, little if any new cell surface growth occurs in the mother cell and the tip of the stalk is reused repeatedly for future growth. Thus, the mother cell ages, although the consequences of the aging process are as yet unknown. In *Hyphomicrobium neptunium*, DNA replication occurs only in the mother cell (Wall et al., 1980). An interesting consequence of this mode of growth is that the chromosome must traverse the stalk to be inherited by the budding swarmer cell. DNA synthesis begins coincident with the cessation of stalk elongation and the initiation of bud formation. Pseudovesicles formed by an extension of the cytoplasmic membrane in the stalk contain DNA, suggesting that they are involved in the transfer of DNA and perhaps other components to the progeny cell (Zerfas et al., 1997).

THE LIFE CYCLES OF *CAULOBACTER* AND *ASTICCACAULIS BIPROSTHECUM*

The molecular mechanisms that control the developmental cycle of stalked bacteria have been studied most extensively in *C. crescentus*

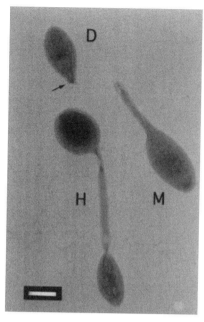

FIGURE 2 Marine *Hyphomonas* species. Transmission electron micrograph of different cell types: D, daughter cell; M, mother cell; H, hypha (stalk) of predivisional cell. The arrow shows the beginning of stalk synthesis in the daughter cell. Bar, 500 nm. (Photo by Ellen Quardokus.)

(Brun et al., 1994; Gober and Marques, 1995). Each division of *Caulobacter* cells gives rise to a swarmer cell and a stalked cell (Poindexter, 1964, 1981; Stove and Stanier, 1962) (Fig. 3). The swarmer cell has a single polar flagellum and is chemotactically competent. After cell division, pili are formed at the flagellated pole of the swarmer cell. During this dispersal stage of the life cycle, swarmer cells do not replicate DNA and do not divide. After approximately one-third of the cell cycle, in response to an unknown internal signal, the swarmer cell sheds its flagellum and the pili are lost. A stalk is synthesized at the pole that previously contained the flagellum, and DNA replication is initiated. The holdfast, the adhesion organelle that allows *Caulobacter* to attach to surfaces, appears at the tip of nascent stalks during swarmer cell differentiation. In natural habitats and in culture, swarmer cells may attach to surfaces prior to cell differentiation. This attachment is

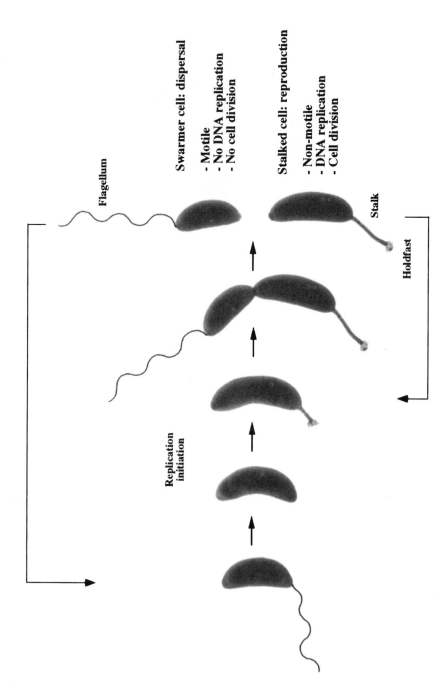

FIGURE 3 Life cycle of *C. crescentus*. The transmission electron micrographs show cells at different stages of the cell cycle. The outline of the flagellum was enhanced for illustration purposes. Pili are not visible but would be present at the flagellated poles of swarmer cells. Polar structures are indicated. The life cycle is described in the text and in the legend to Fig. 1. (Photos by Yves Brun.)

probably mediated by pili and the flagellum, since swarmer cells do not possess a holdfast (as discussed later). However, attachment is not required for the differentiation of swarmer cells. Unlike the swarmer cell, the progeny stalked cell is capable of initiating a new round of DNA replication immediately after cell division. Stalked cells elongate and synthesize a flagellum at the new pole to form an asymmetric predivisional cell.

The life cycle of *Asticcacaulis biprosthecum* is essentially the same as that of *Caulobacter* except for the location of the stalks (Pate et al., 1973) (Fig. 1 and 4). *A. biprosthecum* has two lateral stalks, one on the opposite side of the cell from the other. The stalks of *A. biprosthecum* are identical in structure to those of *Caulobacter* (Pate and Ordal, 1965; Poindexter and Bazire, 1964). However, the stalks of *A. biprosthecum* are not involved in attachment, since the hold-fast is found at the pole opposite the flagellar pole. Comparison of the distance of the stalks from the two poles of the cell in a stalked cell and a predivisional cell suggests that cell growth is polar in *A. biprosthecum* (Pate et al., 1973). This is based on the assumption that stalks do not move during growth, since they are a continuation of the peptidoglycan layer. As seen in Fig. 4, the stalks of an *A. biprosthecum* stalked cell protrude from the midpoint of the cell. In the predivisional cell, the stalks are still at the same distance from the pole of the stalked compartment and are much farther from the flagellar pole. This suggests that the cell has grown mostly on the flagellar side while almost no growth has occurred on the other side of the stalks. This directed growth is analogous to the directed growth of budding bacteria, which occurs from only one pole.

EFFECT OF ENVIRONMENTAL STRESS ON DEVELOPMENT

Nutritional stress can induce specific developmental pathways or affect specific steps within these pathways in stalked bacteria. Exospore formation is induced by nutrient starvation in *R. vannielii* (Whittenbury and Dow, 1977). In late exponential phase, swarmer cell differentiation is inhibited in *Hyphomicrobium*, *R. vannielii* (Morgan and Dow, 1985), and *Caulobacter* (Iba et al., 1975; Janakiraman and Brun, 1997; Morgan and Dow, 1985). Phosphate starvation dramatically induces stalk synthesis in *Caulobacter* (Schmidt and Stanier, 1966), as described later, whereas starvation for nitrogen can block the differentiation of swarmer cells (Chiaverotti et al., 1981). Finally, *Caulobacter* cells undergo drastic morphological changes when cultured in stationary phase for an extended period (Wortinger et al., 1998). During stationary-phase culture, cells gradually acquire an elongated helical shape and display an increased resistance to stress compared to exponentially growing cells. The results of competition experiments indicate that the changes observed in prolonged culture are the result of

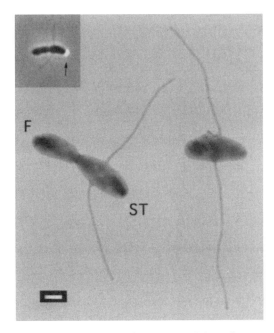

FIGURE 4 *A. biprosthecum*; transmission electron micrograph of a stalked cell and a predivisional cell. F, flagellar pole; ST, stalked pole. The presence of a holdfast at the stalked pole is shown by WGA-FITC binding as a bright fluorescent spot (arrow) in the phase-contrast picture in the inset. Bar, 500 nm. (Photos by Ellen Quardokus.)

entry into a new developmental pathway and are not due to mutation (Wortinger et al., 1998).

THE STALK

Structure of the Stalk

Electron microscopic examination of *Caulobacter* stalks has shown that the cell surface layers of the stalk are continuous with those of the cell body (Poindexter and Bazire, 1964). The core of the stalk contains cytoplasmic material that seems to be continuous with the cytoplasm but is devoid of ribosomes and DNA (Poindexter and Bazire, 1964). The stalk is traversed at intervals by dense rings known as crossbands that are composed, at least in part, of peptidoglycan (Jones and Schmidt, 1973; Pate and Ordal, 1965; Schmidt, 1973; Schmidt and Swafford, 1975). The crossbands are thought to provide rigidity to the stalk by attaching the inner and the outer membranes. The crossbands are probably an indication of cell age, one crossband being synthesized during each cell cycle (Poindexter and Staley, 1996). Recent experiments in which stalk growth was followed in individual cells for more than 150 generations indicate that the stalks elongate during each cell cycle and that stalk length increases linearly with increasing age of the cells (Ackerman and Jenal, personal communication).

The biosynthesis of a stalk is a formidable task for the cell, which has to (i) redirect the biosynthesis of the cell surface to a perpendicular direction at a specific site, (ii) initiate stalk biosynthesis at the proper time in the cell cycle, and (iii) control stalk elongation in response to extracellular phosphate concentration (see below). The biosynthesis of the stalk is localized to its base. This is supported by radiolabeling studies, investigation of the effect of penicillin and mecillinam on growing cells, and studies of the growth of the surface array (Schmidt and Stanier, 1966; Seitz and Brun, 1998; Smit and Agabian, 1982). This implies that the enzymes required for the biogenesis of

stalks, at least for stalk cell wall synthesis, are also localized at the base of the stalk.

Since the formation of the stalk requires peptidoglycan synthesis, penicillin binding proteins (PBPs) which catalyze the cross-linking of peptidoglycan are likely to play an important role in stalk synthesis. Analysis of sheared stalks has revealed that the PBP composition of the *Caulobacter* stalk is different from that of the cell body: PBPs such as PBP1A and PBP3, which are present in the cell envelope, are absent from the stalk, whereas PBPs such as PBPX and PBPY are found predominantly in the stalk (Koyasu et al., 1983). Recently, it has been shown that mecillinam, a β-lactam antibiotic that binds covalently to PBP2 in *Escherichia coli*, inhibits stalk elongation in *Caulobacter* (Seitz and Brun, 1998). The isolation of mecillinam-resistant short-stalked mutants of *Caulobacter* also suggests that PBPs play an important role in stalk synthesis (Koyasu et al., 1983; Seitz and Brun, 1998).

Regulators of Stalk Synthesis

Studies aimed at identifying regulators of flagellum synthesis and the synthesis of polar phage receptors indirectly led to the identification of regulators of stalk synthesis. Mutants of the *rpoN* gene that encodes the σ^{54} subunit of RNA polymerase lack flagella and stalks and have cell division defects (Brun and Shapiro, 1992). Mutants of the *pleC* histidine protein kinase gene are resistant to phage ΦCbK, have inactive flagella, and lack stalks and pili (Ely et al., 1984; Fukuda et al., 1981; Wang et al., 1993). Mutants of the putative response regulator gene *pleD* are motile throughout the cell cycle and fail to elongate stalks properly (Aldridge and Jenal, 1999; Hecht and Newton, 1995). The global response regulator gene *ctrA* is required for stalk synthesis and regulates DNA replication, cell division, and flagellum synthesis (Quon et al., 1996). The functions of these genes are described in more detail in the following chapters. Surprisingly, phosphate starvation induces stalk synthesis in all the known mutants that lack stalks (when phosphate is in excess). This indicates that activation of the phosphate regulon that controls stalk elongation in response to phosphate starvation can bypass the requirement for all the other known regulators of stalk synthesis.

Effect of Phosphate Starvation

Phosphate starvation has a dramatic effect on stalk synthesis (Schmidt, 1968; Schmidt and Stanier, 1966). When *Caulobacter* is starved for phosphate, stalk synthesis is stimulated, resulting in stalks 30 μm or more in length compared to stalk lengths of 1 to 2 μm when phosphate is in excess (compare Fig. 5B and C). Other stalked bacteria also increase the lengths of their stalk when starved for phosphate. Stalked bacteria are typically found in aquatic ecosystems where the most common limiting nutrient is phosphorus (Poindexter, 1984b). Phosphorus is an essential element and is preferentially imported in the form of inorganic phosphate. When starved for phosphate, bacteria increase their ability to take up inorganic phosphate and

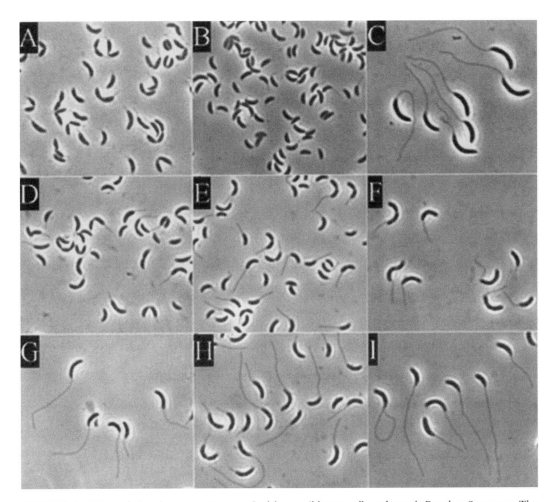

FIGURE 5 Effect of phosphate starvation on *Caulobacter* wild-type cells and on *phoB* and *pstS* mutants. The thin structures at the poles of cells are stalks. Flagella are not visible. (A to C) Wild-type cells; (D to F) *phoB* mutant; (G to H) *pstS* mutants. The left panels show cells grown in a rich peptone-yeast extract medium. The middle panels show cells grown in a minimal salts-glucose medium with a high concentration (10 mM) of phosphate. The right panels show cells grown in a minimal salts-glucose medium with a low concentration (30 μM) of phosphate.

to utilize organic phosphate sources. The increase in surface-to-volume ratio caused by stalk elongation during phosphate starvation is thought to allow *Caulobacter* cells to take up phosphate and other nutrients more efficiently (Poindexter, 1984b). Recent studies have demonstrated that the rate of phosphate uptake (per cell or per unit of dry weight or protein) and the specific activity of alkaline phosphatase increased at a higher carbon-to-phosphorus ratio (Felzenberg et al., 1996). The rate of phosphate uptake was higher in stalked cells than in swarmer cells when calculated per cell or per unit of protein but was similar in the two cell types when it was calculated as activity per cell surface area, including the surface area of the stalk (Felzenberg et al., 1996). Stalk elongation may have additional benefits in the aqueous environment. The buoyancy of stalked cells that results from the increased surface area of the stalk may help keep them at the air-water interface, an obvious advantage for an obligate aerobe (Poindexter, 1981). In addition, stalked cells are often found attached to surfaces in aquatic environments. Stalk elongation under these conditions would allow the cells to extend away from the surface, thus benefiting from more nutrient flow and avoiding competition with other bacteria in a nascent biofilm (Poindexter, 1981).

The Phosphate Regulon and the Control of Stalk Elongation

Insight into the control of stalk elongation by extracellular phosphate concentration came from a mutant screen designed to identify stalk biosynthesis mutants of *Caulobacter*. A subclass of the stalk mutants had a long stalk phenotype irrespective of extracellular phosphate concentration (Fig. 5G to I). The disrupted genes were identified as homologues of the high-affinity phosphate transport genes *pstS*, *pstC*, *pstA*, and *pstB* of *E. coli* (Gonin et al., submitted). In *E. coli*, mutations in *pst* genes result in the constitutive activation of a group of genes called the Pho regulon (Wanner, 1996). The majority of genes of the *E. coli* Pho regulon are involved in the uptake and metabolism of phosphorus

compounds. For example, the Pho regulon includes the *pstSCAB* operon, which encodes the high-affinity phosphate transport system, and *phoA*, the alkaline phosphatase gene required for the utilization of some organic forms of phosphate.

Genes of the *E. coli* Pho regulon contain a *cis*-regulatory sequence, the Pho box, which overlaps with the −35 region of their promoter and is required for activation of these genes under phosphate starvation conditions (Makino et al., 1988). During phosphate starvation in *E. coli*, PhoR, the sensor kinase of the Pho regulon, undergoes autophosphorylation. Phospho-PhoR phosphorylates PhoB, the response regulator of the Pho regulon (Makino et al., 1988, 1989). Phospho-PhoB binds to the Pho box and activates the transcription of the Pho regulon genes by binding to the Pho box and interacting with the σ^{70} subunit of the RNA polymerase holoenzyme (Makino et al., 1993). In addition to their role in phosphate transport, the PstSCAB proteins of *E. coli* form a repression complex with PhoR when phosphate is in excess, thereby inhibiting expression of the Pho regulon (Wanner, 1990). By analogy to *E. coli* *pst* mutants, the *pst* mutants of *Caulobacter* could cause the synthesis of long stalks by activating PhoB. PhoB would then activate, directly or indirectly, as-yet-unknown *Caulobacter* genes involved in stalk elongation. This model is consistent with the fact that *phoB* is required for stalk elongation in response to phosphate starvation (Fig. 5D to F) and that *Caulobacter pst phoB* double mutants have the *phoB* mutant phenotype (Gonin et al., submitted). In a second model, the Pst system would be required to import sufficient phosphate for biosynthesis. Because PhoB is required for the expression of *pst* genes, *phoB* mutant cells might lack the necessary phosphate to synthesize stalks. This is not the case, since *pstS* and *pstB* mutants can make long stalks in both high- and low-phosphate media. Thus, an active Pst system is not required for stalk synthesis. It has also been suggested that the effect of phosphate is to inhibit stalk outgrowth due to its interference with a supportive role of Ca^{2+} ions in stalk synthesis (Poindexter, 1984). However, this is hard to reconcile with the current genetic data.

The precise role of PhoB in stalk elongation remains to be determined. Stalk synthesis involves the synthesis of both cell wall and membrane components and is likely to require the action of many genes. The PhoB-dependent gene(s) involved in stalk synthesis is not known. PhoB could act directly by binding to the promoter(s) of the stalk gene(s) to stimulate its transcription, or it could indirectly regulate the gene by controlling the transcription of other regulatory genes. The identification and characterization of Pho box-containing genes in the *Caulobacter* genomic sequence should help to determine the role of PhoB in stalk elongation and aid in the identification of other genes required for stalk elongation.

We still know very little about stalk synthesis. Some of the regulators of stalk synthesis have been identified, but the genes they control remain unknown. What proteins are involved in the synthesis of the stalk? How does the stalk synthesis machinery become localized to the appropriate subcellular site, the location of which is not conserved between stalked bacteria? What determines the timing of stalk synthesis initiation and the extent of stalk elongation? Clearly, this complex morphogenetic event is subject to many regulatory controls.

THE HOLDFAST

The holdfast is an adhesive organelle, found at the tips of stalks in *Caulobacter*, which mediates attachment to substrates (Fig. 6). However, in certain stalked bacteria, the holdfast is not associated with the stalk. In *A. biprosthecum*, for example, the holdfast is at the pole opposite the flagellar pole while the stalks are at the midpoint of the cell. The precise chemical structure of the holdfast is not known. Staining properties and enzyme sensitivity studies indicate that the holdfast is a complex polysaccharide (Merker and Smit, 1988) with acidic components, such as uronic acids (Umbreit and Pate, 1978). The appearance of the holdfast is temporally regulated during the cell cycle. It

has been proposed that the holdfast first appears at the base of the flagellum in the swarmer pole of the predivisional cell (Kurtz and Smit, 1992; Merker and Smit, 1988; Poindexter, 1981). This is consistent with the observation that swarmer cells collide with and stick more frequently to glass surfaces than nonmotile stalked and dividing cells (Newton, 1972). The presence of the holdfast at the tip of the stalk in *Caulobacter* would then result from the growth of the stalk at the site previously occupied by the flagellum and holdfast.

Recently, the time of appearance of the holdfast during the *Caulobacter* cell cycle was addressed through lectin-binding studies and electron microscopy. The holdfast can be visualized by the specific binding of fluorescein isothiocyanate (FITC)-conjugated wheat germ agglutinin (WGA-FITC) to the holdfast (Merker and Smit, 1988). The use of WGA-FITC staining indicated that while the holdfast is readily detectable in stalked cells and at the stalked pole of predivisional cells, it is not detectable in swarmer cells or at the flagellated pole of predivisional cells (Janakiraman and Brun, 1999). In addition, transmission electron microscopy fails to reveal holdfast material at the base of the flagellum whereas it is clearly visible at the distal tips of short stalks (Janakiraman and Brun, 1999). These results suggest that in *Caulobacter*, synthesis and/or exposure of the holdfast to the outside of the cell occurs during swarmer cell differentiation.

In addition to a single flagellum, the swarmer pole contains pili (Smit, 1987). Pili are involved in mediating attachment to surfaces in many bacteria (Hultgren et al., 1993) and have also been implicated in promoting the primary adhesion event in *Hyphomonas* (Quintero et al., 1998). Thus, perhaps the attachment of the swarmer pole of *Caulobacter* to surfaces is mediated by the pilus and not the holdfast. As predicted by this model, mutants that are unable to synthesize pili but can synthesize holdfasts cannot attach to glass until the swarmer cell has differentiated into a stalked cell (Ackermann et al., personal communication).

Holdfast synthesis genes have not yet been

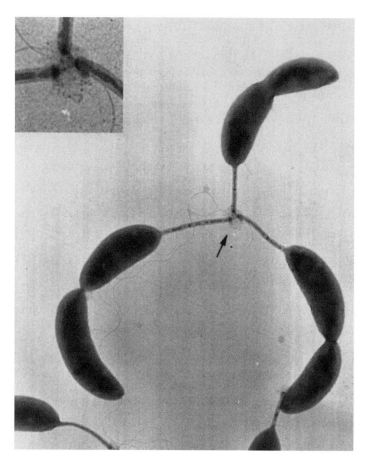

FIGURE 6 Rosette of *Caulobacter* cells. The transmission electron micrograph shows cells attached to one another by their holdfasts (arrow) and forming a rosette. The inset shows an enlargement of the holdfast area. (Photo by Yves Brun.)

studied at the molecular genetic level. However, a cluster of four genes (*hfaABDC*) involved in the attachment of the holdfast to the cell was identified by Tn*5* insertion mutagenesis (Kurtz and Smit, 1992). *hfaA*, *hfaB*, and *hfaD* form an operon (Cole and Brun, 1999). The exact role of each of these genes is unknown. The C-terminal region of HfaA was reported to be similar to the C termini of pilus tip proteins such as the PapG adhesin from *E. coli* and SmfG adhesin from *Serratia marcescens*, and HfaB was reported to be similar to transcriptional activators (Kurtz and Smit, 1992). However, BLAST and PSI-BLAST searches of current sequence databases have failed to reveal any significant similarities to previously identified proteins except for a similarity over 120 amino acids between HfaB and the *E. coli* CsgG

protein involved in *curli* assembly (Loferer et al., 1997; Cole and Brun, unpublished). The sequence of HfaD contains three putative membrane-spanning regions, and it may act as a membrane-associated protein anchor between the holdfast and the cell (Kurtz and Smit, 1994). HfaC is similar to ATP-binding transport related proteins (Kurtz and Smit, 1994). Recent genetic analysis of mutations in each *hfa* gene indicates that *hfaB* mutants have a more severe holdfast-shedding phenotype than *hfaA* and *hfaD* mutants (Cole and Brun, 1999). By contrast, an *hfaC* mutant does not display any defects in holdfast synthesis or attachment.

The transcription of *hfaABD* is temporally regulated during the cell cycle, with maximal levels of transcription occurring in the swarmer compartment of predivisional cells (Janakiraman and Brun, 1999). Because the holdfast does not appear until the differentiation of swarmer cell during the next cell cycle, the reason for the preferential transcription of *hfaA* in the swarmer compartment of the predivisional cell is unclear. This burst of *hfaA* transcription may serve to load *hfaA* mRNA or HfaA itself in the swarmer compartment prior to the beginning of the next cell cycle. This would ensure that the holdfast attachment protein is present in the swarmer cell, ready to be used when the swarmer cell differentiates to become a stalked cell.

REGULATION OF POLAR DEVELOPMENT BY CELL DIVISION CHECKPOINTS

In *Caulobacter*, different stages of development require the completion of previous stages of the replication and division cycles (see chapter 17 for detailed references). The inhibition of DNA replication blocks flagellum synthesis (Huguenel and Newton, 1982; Ohta and Newton, 1996) by preventing the transcription of early flagellar genes that are at the top of the flagellar regulatory hierarchy (Stephens and Shapiro, 1993). Cells that can replicate DNA but that are blocked in cell division are affected in their progression through development. The initiation of cell division plays an essential role in the establishment of differential programs of gene expression that sets up the fates of the progeny cells (Shapiro and Losick, 1997).

The interconnection of cell division and polar development can be illustrated by the fate of one pole of the cell in Fig. 7. In wild-type cells (Fig. 7, top), a flagellum is synthesized at the new pole during the first DNA replication cycle. In the second replication cycle, this same pole loses its flagellum and develops into a

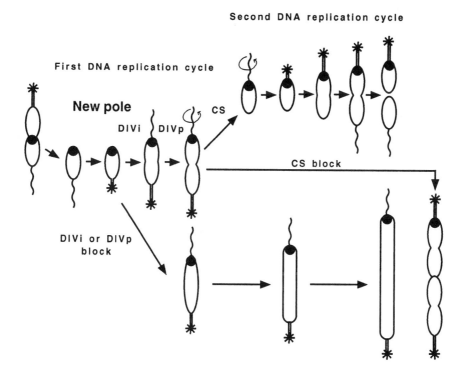

FIGURE 7 Effect of cell division inhibition on polar development. The consequence of inhibition of cell division at DIVi (division initiation) and DIVp (division progression) is compared to a block after these stages but before CS (cell separation). A black dot marks the pole affected by the cell division block.

stalked pole by synthesizing a stalk and a hold-fast. The fate of a cell inhibited at an early stage of cell division is illustrated in the lower branch of Fig. 7. During the first replication cycle, a flagellum is synthesized but its rotation cannot be activated. In the second replication cycle, stalk and holdfast synthesis are blocked. If cell division is blocked after constriction has progressed but before cell separation, long filaments with regularly spaced constrictions are formed. These cells are able to activate the development of the flagellated pole to yield filamentous cells that have a stalk and holdfast at both poles. Thus, polar development is not simply coupled to progression through the DNA replication cycle or to increase in cell mass; it is dependent on the progression of cytokinesis as well (Ohta and Newton, 1996). A similar coupling between the cell cycle and polar development and growth is present in *H. neptunium*. Inhibition of DNA synthesis results in the synthesis of abnormally long stalks and the inhibition of budding (Weiner and Blackman, 1973). Thus, polar growth can still occur but the developmental switch between polar stalk synthesis and budding is inhibited.

How does the initiation of cell division control the development of the flagellated pole of the cell? What is the cell division signal(s) sensed and transduced to the regulators or machinery responsible for the development of the pole? Previous work identified the PleC histidine protein kinase as an important regulator of flagellar rotation and stalk formation. *pleC* mutants are arrested at the same stage of development as cells blocked at early stages of cell division. PleC phosphorylates the single-domain response regulator DivK, and it has been hypothesized that PleC and DivK are members of a signal transduction pathway that couples activation of flagellar rotation and stalk formation to cell division (see chapter 17).

The σ^{54} gene *rpoN* also appears to be involved in the coupling of polar development and cell division (Janakiraman, 1998; Quardokus et al., 1999). In wild-type cells, inhibition of cell division inhibits stalk and holdfast synthesis at the flagellar pole of the cell (Fig. 7).

However, when cell division is inhibited in a *rpoN* mutant, most cells have a holdfast at both poles, as determined by WGA-FITC labeling. In addition, *rpoN* mutants can synthesize stalks at both poles. This was demonstrated by growing an *rpoN* mutant in low-phosphate medium to bypass the stalkless phenotype, allowing stalked-pole development to be observed. In contrast, wild-type cells starved for phosphate were still asymmetric, with stalks at one pole of the cell. Thus, *rpoN* is required to maintain cellular asymmetry when cell division is inhibited.

REGULATION OF CELL DIVISION AND DNA REPLICATION

As described above and in the following chapters, progression through the cell division cycle plays an essential role in the establishment of differential programs of gene expression in the two compartments of the predivisional cell. In addition, the two progeny cells differ in their capacities for cell division and DNA replication. The molecular mechanisms that prevent DNA replication and cell division in swarmer cells but not in stalked cells are now beginning to be understood. Studies of the regulation of cell division in *Caulobacter* have the added advantage that synchronized populations can be easily obtained. DNA replication is initiated only once during each developmental cycle. This allows the unambiguous assignment of specific events to a stage in the DNA replication cycle and constitutes a powerful tool to study the timing and cell cycle dependence of the different steps of cell division.

Cell division is a complex process that requires the action of many proteins. In *E. coli*, at least nine proteins (FtsA, FtsI, FtsK, FtsL, FtsN, FtsQ, FtsW, FtsZ, and ZipA) are required for cell division (Bramhill, 1997; Lutkenhaus and Addinall, 1997; Rothfield and Justice, 1997). In all bacteria examined to date, the abundance and subcellular location of the tubulin-like GTPase FtsZ are critical factors in the initiation of cell division. FtsZ is a highly conserved protein that polymerizes into a ring structure associated with the cytoplasmic membrane at the site of cell division. FtsZ is the first protein to localize to the site of cell division and is required for the localization of all the other known cell division proteins to the site of cell division. In addition, the FtsZ ring may constrict, providing mechanical force for division.

In *Caulobacter*, FtsZ is subject to a tight developmental control. After cell division, only the stalked cell contains FtsZ (Quardokus et al., 1996). FtsZ is absent from swarmer cells and begins accumulating during swarmer cell differentiation, coincident with the initiation of DNA replication. The concentration of FtsZ then increases rapidly and reaches a maximal level around the time when cell constriction first becomes visible. Once cells have started to constrict, the concentration of FtsZ decreases rapidly. Transcriptional and proteolytic controls contribute to the cell cycle and developmental regulation of FtsZ. This two-tiered level of regulation ensures that FtsZ is only present in the cell that will initiate a new cell cycle immediately after division.

Transcriptional Control of *ftsZ*

The *Caulobacter ftsZ* gene is part of a gene cluster which contains genes involved in peptidoglycan biosynthesis and in cell division (Ohta et al., 1997; Quardokus et al., 1996; Sackett et al., 1998). While the gene organization in this region is similar to that of *E. coli*, the transcriptional organization of the genes is different. In *E. coli*, *ftsZ* is transcribed by at least five promoters located in the upstream *ddl*, *ftsQ*, and *ftsA* genes (Donachie, 1993; Vicente and Errington, 1996). In *Caulobacter*, a strong transcriptional terminator separates *ftsA* and *ftsZ* and uncouples their transcription (Fig. 8). *ftsZ* transcription is controlled by a single promoter found downstream of the *ftsA* terminator (Kelly et al., 1998; Sackett et al., 1998).

Because of the complexity of transcriptional organization of cell division genes, definitive answers about their cell cycle regulation have been difficult to obtain for *E. coli*. Many studies of *ftsZ* transcription in *E. coli* have reached conflicting conclusions, but recent experiments in-

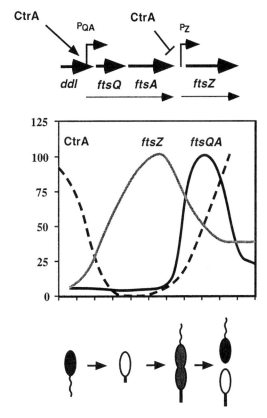

FIGURE 8 Cell cycle regulation of *ftsQA* and *ftsZ* transcription by CtrA. The genetic organization and promoters of the *ddl-ftsQAZ* region are shown at the top. The thick arrows represent genes, the thin arrows represent transcriptional units, and the bent arrows represent promoters. A transcriptional terminator uncouples the transcription of *ftsQA* and *ftsZ*. The arrow and barred line going from CtrA to the promoters are genetic arrows indicating a positive interaction for the *ftsQA* promoter and a negative interaction for the *ftsZ* promoter. The boxed curves represent the rate of transcription for *ftsQA* and *ftsZ* and the protein concentration for CtrA during the cell cycle. The bottom diagram shows the cell cycle. The shaded and dark areas inside the cells represent the concentration of CtrA.

dicate that *ftsZ* is expressed periodically in the cell cycle. One study showed that *ftsZ* transcription is periodically activated during the cell cycle, coincident with the initiation of DNA replication (Garrido et al., 1993). Another study found that *ftsZ* expression is maximal around the middle of the cell cycle and is minimal at the time of cell division (Zhou and Helmstetter, 1994). However, it is not clear what the effect of this variation in *ftsZ* transcription has on the concentration of FtsZ during the cell cycle of *E. coli*.

The cell cycle regulation of *ftsZ* transcription has been clearly established in *Caulobacter*. *ftsZ* is not transcribed in swarmer cells, but its transcription increases at the beginning of the DNA replication period (S phase) (Kelly et al., 1998). Coincident with the end of S phase, at the time when cell division is first apparent, transcription of *ftsZ* begins to decrease. Immediately after the completion of cell division, transcription of *ftsZ* rapidly resumes in stalked cells but remains low in swarmer cells. The global cell cycle regulator CtrA (see chapter 18) is involved in the cell cycle control of FtsZ by repressing its transcription. A sequence with high similarity to the consensus binding site for CtrA overlaps the *ftsZ* transcription start site and the −10 region of the promoter. CtrA binding to this site has been demonstrated by DNase I footprinting (Kelly et al., 1998). The binding of CtrA to a site that overlaps the transcription start site may prevent binding of RNA polymerase to the *ftsZ* promoter. In contrast, promoters activated by CtrA have a CtrA binding site that overlaps the −35 region. The transcription rate of *ftsZ* is inversely proportional to the concentration of CtrA during the cell cycle (Kelly et al., 1998). CtrA is present in swarmer cells where *ftsZ* transcription is repressed, is degraded at the same time as *ftsZ* transcription begins, and reappears when *ftsZ* transcription decreases at the end of the cell cycle (Fig. 8). Removal of the CtrA binding site in the *ftsZ* promoter results in loss of *ftsZ* repression in swarmer cells (Kelly et al., 1998). However, the transcription of the mutant *ftsZ* promoter was still subject to some level of cell cycle control, raising the possibility that other factors are involved in the cell cycle control of *ftsZ* transcription.

CtrA also plays a negative role in the regulation of DNA replication by binding to the origin of replication and repressing the initiation of DNA replication (Domian et al., 1997;

Quon et al., 1996, 1998) (see chapter 18). The degradation of CtrA during swarmer cell differentiation thus coordinates the onset of the replication and division cycles. Late in the cell cycle, when DNA replication is complete and cell division has been initiated, CtrA is synthesized and represses *ftsZ* transcription and initiation of DNA replication. Just before cell separation, CtrA is degraded in the stalked compartment. The absence of CtrA in stalked cells after cell division allows *ftsZ* transcription to resume and DNA replication to be initiated.

Proteolytic Control of FtsZ

Proteolysis plays a major role in the cell cycle regulation of FtsZ concentration and in the cell-type-specific inheritance of FtsZ after cell division. Transcriptional control is clearly important to determine the onset of FtsZ synthesis at the beginning of the cell cycle. However, the decrease in FtsZ concentration at the end of the cell cycle is much too rapid to be explained by simple dilution by mass increase (Quardokus et al., 1996). The importance of proteolysis in the regulation of FtsZ is obvious in strains engineered to transcribe *ftsZ* constitutively during the cell cycle by using heterologous promoters. In these strains, FtsZ is present at a much lower level in swarmer cells than in stalked cells and the cell cycle concentration of FtsZ still varies in a manner similar to that in wild-type strains (Kelly et al., 1998). The cell cycle variation and cell-type-specific localization of FtsZ concentration are thus controlled primarily by a different rate of FtsZ degradation during the cell cycle. At the beginning of the cell cycle and until the cells begin to constrict, FtsZ has a half-life of 80 min, which is equivalent to half of the cell cycle. Once cell division begins, FtsZ becomes highly unstable, with a short half-life of 10 to 20 min (~0.1 cell cycle unit) (Kelly et al., 1998). Thus, FtsZ is stable as it assembles into the cytokinetic ring and is degraded rapidly once the cells have begun to constrict. One of the factors controlling the proteolysis of FtsZ in *Caulobacter* may be its assembly state. It is possible that FtsZ is relatively stable early in the cell cycle because a significant fraction of the protein is polymerized. As the cell constricts during cell division, FtsZ protomers are released into the cytoplasm by the disassembly of the ring. Perhaps the domain of FtsZ recognized by a protease for degradation is inaccessible when FtsZ is assembled but becomes exposed when FtsZ is not assembled. In an alternative but not exclusive model, the synthesis or the activity of the protease responsible for FtsZ degradation is subject to cell cycle control. The fact that FtsZ is localized to the midcell in stalked cells, in which it is relatively stable, is consistent with the assembly model (Kelly et al., 1998).

Ordered Transcription of Cell Division Genes

Transcription of *ftsQ* and *ftsA* is uncoupled from that of *ftsZ* by the transcriptional terminator that separates *ftsA* and *ftsZ*. Most of the transcription of *ftsA* originates from a promoter that also transcribes *ftsQ* (Sackett et al., 1998). The cyclic expression of *ftsQA* is striking (Fig. 8). Although *ftsQA* transcription occurs at a low level throughout the life cycle, there is a strong burst of transcription at the end of the cell cycle. Thus, the bulk of *ftsQ* and *ftsA* transcription begins substantially later than that of *ftsZ* and correlates with the end of the DNA replication period and the beginning of cytokinesis (Sackett et al., 1998). Analysis of a temperature-sensitive allele of *ftsA* (*divE309*) indicates that FtsA is required in the final stages of cytokinesis in *Caulobacter* (Ohta et al., 1997; Osley and Newton, 1977). The order of transcription of *ftsZ* and *ftsA* correlates with their order of action in cell division. The burst of *ftsQA* transcription following the completion of DNA replication may represent one way to coordinate the beginning of cytokinesis with the end of the replication period.

The burst of *ftsQA* transcription coincides with the reappearance of CtrA in predivisional cells, and transcription of the *ftsQA* promoter decreases when CtrA is artificially depleted (Sackett, 1998). Deletion experiments have defined a small region upstream of *ftsQ* containing a CtrA binding site as the promoter for *ftsQA*

(Sackett, 1998; Wortinger and Brun, unpublished). This suggests that CtrA is an activator of *ftsQA* transcription. The *ftsQA* promoter has the same architecture as promoters that are positively regulated by CtrA. One of the CtrA half sites overlaps the −35 region of the promoter; the second half site lies between the −35 and −10 region, the −10 region is less conserved than in typical housekeeping promoters, and the spacing between the −35 and −10 regions is increased. CtrA footprints the CtrA binding site in the *ftsQA* promoter, and mutation of the CtrA binding site abolishes transcription (Wortinger and Brun, unpublished).

In summary, it appears that CtrA is involved in the ordered transcription of *ftsZ* and *ftsQA* after the initiation of DNA replication. CtrA is degraded during swarmer cell differentiation, allowing *ftsZ* transcription but preventing *ftsQA* transcription (Fig. 8). When CtrA begins to accumulate later in the cell cycle, it represses *ftsZ* and activates *ftsQA* transcription. However, regulation by CtrA is not sufficient to explain the complete transcription pattern of the *ftsQA* promoter during the cell cycle. CtrA is present in swarmer cells and yet *ftsQA* transcription is low in swarmer cells. This suggests that additional regulatory elements repress *ftsQA* transcription in swarmer cells.

Initiation of Cell Division and Assembly of the Cell Division Machinery

Recent studies indicate that FtsZ is localized in cells long before the onset of cytokinesis (Addinall et al., 1996; Kelly et al., 1998; Quardokus and Brun, unpublished), suggesting that other events must take place before cytokinesis can be initiated. The ability to easily synchronize *Caulobacter* and the fact that overexpression of FtsZ does not inhibit cell constriction in *Caulobacter*, as it does in *E. coli*, allowed a direct test of this hypothesis. By using an inducible promoter to control the timing of FtsZ synthesis in a synchronized population, FtsZ was synthesized in swarmer cells at a level equivalent to the normal concentration of FtsZ at the initiation of cell division. The results of this experiment indicated that increasing the level of FtsZ at an earlier stage in the swarmer cell cycle does not induce cell division at an earlier time (Din and Brun, 1999). This indicates that the concentration of FtsZ is not the only factor that determines the timing of cell division initiation and that other cell cycle-dependent events have to occur before cell division is initiated. The inability of FtsZ to cause a premature initiation of cell division when it is present at a sufficient concentration early in the cell cycle can be explained by two main models. In the first model, FtsZ is unable to form a Z ring at early stages of the cell cycle. Alternatively, or in addition, other factors required for cell division initiation are only present and/or active at the appropriate time for cell division initiation.

Given the number of components required for cell division, it is likely that the timing of cytokinesis is controlled by assembly of the cell division machinery at the site dictated by the localization of FtsZ. Alternatively, the complete machinery could assemble very early and constriction could be turned on by a cell cycle-dependent signal, perhaps the formation of two separate nucleoids. Understanding the timing of assembly of the different components of the division machinery will be required in order to explain how division is controlled and how it is coordinated with the DNA replication cycle. In this context, an impressive body of work to determine the localization of the different cell division proteins in *E. coli* has accumulated in the last two years. This has led to a model of the dependency of relationships among the different cell division proteins for localization at the site of cell division. A combination of genetic and localization experiments has led to the following model of the order of appearance of cell division proteins at the division site in *E. coli*: FtsZ, FtsA, FtsQ, FtsL, FtsI, and FtsN (Weiss et al., 1999, and references therein). The localization of ZipA and FtsK is dependent on FtsZ, but it is not yet clear where they fit in the model. An important step in understanding the assembly and function of the cell division machinery will be to

determine which cell division proteins interact directly.

Deletion analysis of FtsZ has begun to uncover regions required for its interaction with itself and with FtsA. In most bacteria, FtsZ is composed of three regions based on sequence similarity: a highly conserved N-terminal region of approximately 320 amino acids, a spacer region of variable length and sequence, and a conserved C-terminal region of 8 amino acids. Yeast two-hybrid analysis indicated that the last 24 amino acids of *Caulobacter* FtsZ are required for its interaction with FtsA but not with FtsZ (Din et al., 1998). This suggests that the conserved C-terminal region is responsible for the interaction of FtsZ with FtsA. Furthermore, the interaction of *Bacillus subtilis* FtsZ with FtsA, but not with FtsZ, requires the last 57 amino acids of FtsZ (Wang et al., 1997). Removing the last 24 codons of *Caulobacter ftsZ* causes a dominant-lethal phenotype resulting from an inhibition of the initiation of cell division (Din et al., 1998). These results suggest that the C-terminal region of *Caulobacter* FtsZ is important for FtsZ ring formation or stability, perhaps in part because of its requirement for FtsZ-FtsA interaction. A role for the C-terminal region of FtsZ in the control of FtsZ polymerization was also suggested by experiments with *E. coli* FtsZ (Ma et al., 1996).

A truncated *E. coli* FtsZ-GFP fusion and a GFP fusion of *Rhizobium meliloti* FtsZ1 lacking the linker and the C-terminal conserved region were both able to polymerize, indicating that the C-terminal linker and C-terminal conserved region are not required for FtsZ polymerization (Ma et al., 1996). The large linker region of *Caulobacter* FtsZ is not required for its interaction with FtsZ, but the last 40 amino acids of the N-terminal conserved region are required for this interaction (Din et al., 1998). Similarly, truncation of *B. subtilis* FtsZ to a length of 255 amino acids abolishes its interaction with FtsZ (Wang et al., 1997). Expression of FtsZ that is truncated 40 amino acids from the end of the N-terminal conserved region (FtsZΔC281) in *Caulobacter* prevents FtsZ ring formation (Din et al., 1998). Since interaction data indicate that FtsZΔC281 should not interfere directly with FtsZ polymerization, it may be that FtsZΔC281 competes with FtsZ for a localization factor at the cell division site or with an accessory protein required for FtsZ to interact with the localization factor. Alternatively, FtsZΔC281 could titer a factor required for the stability of the FtsZ ring. For example, mutations in *ftsA*, *ftsQ*, and *ftsI* affect the stability of the FtsZ ring (Addinall et al., 1997; Pogliano et al., 1997). In combination, these data indicate that the last 40 amino acids of the N-terminal conserved region of FtsZ are required for its interaction with itself.

Chromosome Segregation

The analysis of the dynamics of bacterial chromosome segregation has greatly benefited from the study of asymmetric cell division during the differentiation of *B. subtilis* and *Caulobacter* (see chapter 7) (Mohl and Gober, 1997; Sharpe and Errington, 1999). Important insight came from the study of homologues of the ParA and ParB proteins that are required for the partitioning of low-copy-number plasmids. In bacteriophage P1, ParB is a sequence-specific DNA binding protein that binds to repeated sequences called *parS* downstream of the *parB* gene on the P1 plasmid. Binding of ParB to *parS*, possibly in a complex with ParA, is required for efficient partitioning of P1 plasmids. Chromosomally encoded ParA and ParB homologues have been identified in many bacteria and have been shown to play an important role in chromosome segregation.

In *Caulobacter*, ParB binds specifically to DNA sequences downstream of the *parAB* operon, which is located within 80 kb of the origin of DNA replication (Mohl and Gober, 1997). Immunolocalization experiments performed with synchronized populations showed that ParB exhibits a periodic bipolar localization pattern in late predivisional cells. Polar localization of ParB is dependent on DNA replication. ParA was also found to localize to the poles of predivisional cells. Consistent with their hypothesized role in chromosome partitioning, overexpression of ParA and ParB caused aberrant chromosome partitioning. Because ParB binds to sequences near the origin of replication, these results suggest that the replicated origins of replication localize to the two poles of the predivisional cell. Movement of the origins to the poles of the cell has also been observed in *E. coli* and *B. subtilis* (Sharpe and Errington, 1999). *parA* and *parB* homologues have not been found to be essential in other bacteria, and they are not found in *E. coli*. Surprisingly, *parA* and *parB* are required for cell viability in *Caulobacter*, and their overexpression caused defects in cell division (Mohl and Gober, 1997). The essential nature of *parA* and *parB* in *Caulobacter* suggests the exciting possibility that the cellular level and perhaps the proper subcellular location of ParA and ParB are involved in the coordination of cell division and chromosome movement.

PROTEIN LOCALIZATION DURING BACTERIAL GROWTH AND DEVELOPMENT

The study of bacterial development has had a major impact on our understanding of the bacterial cell. In particular, it is now clear that bacterial cells, even those that do not differentiate, are not simply bags of enzymes. Bacterial cells are highly organized at the level of protein localization, and the locations of proteins can change rapidly (Jacobs et al., 1999; Raskin and De Boer, 1999). In nondifferentiating cells, the locations of proteins are mostly controlled by the cell cycle. In differentiating bacteria, the spatial constraints of differentiating cells are combined with temporal constraints that control the ordered progression through the developmental program and the cell cycle. A major challenge will be to determine how temporal and spatial control are integrated during bacterial development. As described in the following chapters, *Caulobacter* development is an ideal experimental system in which to study this problem.

ACKNOWLEDGMENTS

We thank U. Jenal, J. Skerker, and R. Weiner for communicating results prior to publication, E. Quardokus for providing photos, and members of our laboratory for comments on the manuscript.

Our work was supported by National Institutes of Health Predoctoral Fellowship GM07757 (to R.S.J.) and National Institutes of Health grant GM51986 and NSF Career grant MCB-9733958 (to Y.V.B.).

REFERENCES

Ackerman, M., and U. Jenal. Personal communication.

Ackerman, M., U. Jenal, J. Skerker, and L. Shapiro. Personal communication.

Addinall, S. G., E. Bi, and J. Lutkenhaus. 1996. FtsZ ring formation in *fts* mutants. *J. Bacteriol.* **178:** 3877–3884.

Addinall, S. G., C. Cao, and J. Lutkenhaus. 1997. Temperature shift experiments with an *ftsZ84*(Ts) strain reveal rapid dynamics of FtsZ localization and indicate that the Z ring is required throughout septation and cannot reoccupy division sites once constriction has initiated. *J. Bacteriol.* **179:**4277–4284.

Aldridge, P., and U. Jenal. 1999. Cell cycle-dependent degradation of a flagellar motor component requires a novel-type response regulator. *Mol. Microbiol.* **32:**379–392.

Bramhill, D. 1997. Bacterial cell division. *Annu. Rev. Cell Dev. Biol.* **13:**395–424.

Brun, Y., G. Marczynski, and L. Shapiro. 1994. The expression of asymmetry during cell differentiation. *Annu. Rev. Biochem.* **63:**419–450.

Brun, Y. V., and L. Shapiro. 1992. A temporally controlled sigma factor is required for cell-cycle dependent polar morphogenesis in *Caulobacter. Genes Dev.* **6:**2395–2408.

Chiaverotti, T. A., G. Parker, J. Gallant, and N. Agabian. 1981. Conditions that trigger guanosine tetraphosphate accumulation in *Caulobacter crescentus. J. Bacteriol.* **145:**1463–1465.

Cole, J., and Y. V. Brun. Unpublished data.

Cole, J., and Y. V. Brun. 1999. *Abstracts of the Annual Meeting of the American Society for Microbiology*, p. 391.

Din, N., and Y. V. Brun. 1999. *Abstracts of the Annual Meeting of the American Society for Microbiology*, p. 351.

Din, N., E. M. Quardokus, M. J. Sackett, and Y. V. Brun. 1998. Dominant C-terminal deletions of FtsZ that affect its ability to localize in *Caulobacter* and its interaction with FtsA. *Mol. Microbiol.* **27:** 1051–1064.

Domian, I. J., K. C. Quon, and L. Shapiro. 1997. Cell type-specific phosphorylation and proteolysis of a transcriptional regulator controls the G1-to-S transition in a bacterial cell cycle. *Cell* **90:**415–424.

Donachie, W. D. 1993. The cell cycle of *Escherichia coli. Annu. Rev. Microbiol.* **47:**199–230.

Ely, B., R. H. Croft, and C. J. Gerardot. 1984. Genetic mapping of genes required for motility in *Caulobacter crescentus. Genetics* **108:**523–532.

Felzenberg, E. R., G. A. Yang, J. G. Hagenzieker, and J. S. Poindexter. 1996. Physiologic, morphologic and behavioral responses of perpetual cultures of *Caulobacter crescentus* to carbon, nitrogen, and phosphorus limitations. *J. Ind. Microbiol.* **17:**235–252.

Fukuda, A., M. Asada, S. Koyasu, H. Yoshida, K. Yaginuma, and Y. Okada. 1981. Regulation of polar morphogenesis in *Caulobacter crescentus. J. Bacteriol.* **145:**559–572.

Garrido, T., M. Sánchez, P. Palacios, M. Aldea, and M. Vincente. 1993. Transcription of *ftsZ* oscillates during the cell cycle of *Escherichia coli. EMBO J.* **12:**3957–3965.

Gober, J. W., and M. Marques. 1995. Regulation of cellular differentiation in *Caulobacter crescentus. Microbiol. Rev.* **59:**31–47.

Gonin, M., E. M. Quardokus, D. S. O'Donnol, and Y. V. Brun. Submitted for publication.

Hecht, G. B., and A. Newton. 1995. Identification of a novel response regulator required for the swarmer-to-stalked-cell transition in *Caulobacter crescentus. J. Bacteriol.* **177:**6223–6229.

Huguenel, E. D., and A. Newton. 1982. Localization of surface structures during procaryotic differentiation: role of cell division in *Caulobacter crescentus. Differentiation* **21:**71–78.

Hultgren, S. J., S. Abraham, M. Caparon, P. Falk, J. St. Geme III, and S. Normark. 1993. Pilus and nonpilus bacterial adhesins: assembly and function in cell recognition. *Cell* **73:**887–901.

Iba, H., A. Fukuda, and Y. Okada. 1975. Synchronous cell differentiation in *Caulobacter crescentus. Jpn. J. Microbiol.* **19:**441–446.

Jacobs, C., I. J. Domian, J. Maddock, and L. Shapiro. 1999. Cell cycle-dependent polar localization of an essential bacterial histidine kinase controls DNA replication and cell division. *Cell* **97:** 111–120.

Janakiraman, R. 1998. Temporal and spatial control during *Caulobacter* cell differentiation. Ph.D. Thesis. Indiana University, Bloomington, Ind.

Janakiraman, R. S., and Y. V. Brun. 1997. Transcriptional and mutational analyses of the *rpoN* operon in *Caulobacter crescentus. J. Bacteriol.* **179:** 5138–5147.

Janakiraman, R. S., and Y. V. Brun. 1999. Cell cycle control of a holdfast attachment gene in *Caulobacter. J. Bacteriol.* **181:**1118–1125.

Jones, H. C., and J. M. Schmidt. 1973. Ultrastructural study of crossbands occurring in the stalks of *Caulobacter crescentus. J. Bacteriol.* **116:**466–470.

Kelly, A. J., M. J. Sackett, N. Din, E. M. Quardokus, and Y. V. Brun. 1998. Cell cycle-dependent transcriptional and proteolytic regulation of FtsZ in *Caulobacter. Genes Dev.* **12:**880–893.

Koyasu, S., A. Fukuda, Y. Okada, and J. Poin-

dexter. 1983. Penicillin-binding proteins of the stalk of *Caulobacter crescentus. J. Gen. Microbiol.* **129:** 2789–2799.

Kurtz, H. D., Jr., and J. Smit. 1992. Analysis of a *Caulobacter crescentus* gene cluster involved in attachment of the holdfast to the cell. *J. Bacteriol.* **174:** 687–694.

Kurtz, H. D., Jr., and J. Smit. 1994. The *Caulobacter crescentus* holdfast: identification of holdfast attachment complex genes. *FEMS Microbiol. Lett.* **116:** 175–182.

Loferer, H., M. Hammar, and S. Normark. 1997. Availability of the fibre subunit CsgA and the nucleator protein CsgB during assembly of fibronectin-binding *curli* is limited by the intracellular concentration of the novel lipoprotein CsgG. *Mol. Microbiol.* **26:**11–23.

Lutkenhaus, J., and S. G. Addinall. 1997. Bacterial cell division and the Z ring. *Annu. Rev. Biochem.* **66:**93–116.

Ma, X., D. W. Ehrhardt, and W. Margolin. 1996. Colocalization of cell division proteins FtsZ and FtsA to cytoskeletal structures in living *Escherichia coli* cells by using green fluorescent protein. *Proc. Natl. Acad. Sci. USA* **93:**12998–13003.

Makino, K., H. Shinagawa, M. Amemura, S. Kimura, A. Nakata, and A. Ishihama. 1988. Regulation of the phosphate regulon of *Escherichia coli*: activation of *pstS* transcription by PhoB protein *in vitro. J. Mol. Biol.* **203:**85–95.

Makino, K., H. Shinagawa, M. Amemura, T. Kawamoto, M. Yamada, and A. Nataka. 1989. Signal transduction in the phosphate regulon of *Escherichia coli* involves phosphotransfer between PhoR and PhoB proteins. *J. Mol. Biol.* **210:** 551–559.

Makino, K., M. Amemura, S.-K. Kim, A. Nataka, and H. Shinagawa. 1993. Role of the σ^{70} subunit of RNA polymerase in transcriptional activation by activator protein PhoB in *Escherichia coli. Genes Dev.* **7:**149–160.

Merker, R. I., and J. Smit. 1988. Characterization of the adhesive holdfast of marine and freshwater *Caulobacters. Appl. Environ. Microbiol.* **54:**2078–2085.

Mohl, D. A., and J. W. Gober. 1997. Cell cycle-dependent polar localization of chromosome partitioning proteins in *Caulobacter crescentus. Cell* **88:** 675–684.

Morgan, P., and C. S. Dow. 1985. Environmental control of cell-type expression in prosthecate bacteria, p. 131–169. *In* M. Fletcher and G. D. Floodgate (ed.), *Bacteria in Their Natural Environments.* Academic Press, London, England.

Newton, A. 1972. Role of transcription in the temporal control of development in *Caulobacter crescentus. Proc. Natl. Acad. Sci. USA* **69:**447–451.

Ohta, N., and A. Newton. 1996. Signal transduction in the cell cycle regulation of *Caulobacter* differentiation. *Trends Microbiol.* **4:**326–332.

Ohta, N., A. Ninfa, N. Allaire, L. Kulick, and A. Newton. 1997. Identification, characterization, and chromosomal organization of cell division cycle genes in *Caulobacter crescentus. J. Bacteriol.* **179:** 2169–2180.

Osley, M. A., and A. Newton. 1977. Mutational analysis of developmental control in *Caulobacter crescentus. Proc. Natl. Acad. Sci. USA* **74:**124–128.

Pate, J. L., and E. J. Ordal. 1965. The fine structure of two unusual stalked bacteria. *J. Cell. Biol.* **27:** 133–150.

Pate, J. L., J. S. Porter, and T. L. Jordan. 1973. Asticcacaulis biprosthecum sp.nov.: life cycle, morphology, and cultural characteristics. *Antonie Leeuwenhoek* **39:**569–583.

Pogliano, J., K. Pogliano, D. Weiss, R. Losick, and J. Beckwith. 1997. Inactivation of FtsI inhibits constriction of the FtsZ cytokinetic ring and delays the assembly of FtsZ rings at potential division sites. *Proc. Natl. Acad. Sci. USA* **94:**559–564.

Poindexter, J. L. S., and G. C. Bazire. 1964. The fine structure of stalked bacteria belonging to the family Caulobacteraceae. *J. Cell Biol.* **23:**587–607.

Poindexter, J. S. 1964. Biological properties and classification of the *Caulobacter* group. *Bacteriol. Rev.* **28:**231–295.

Poindexter, J. S. 1981. The Caulobacters: ubiquitous unusual bacteria. *Microbiol. Rev.* **45:**123–179.

Poindexter, J. S. 1984a. The role of calcium in stalk development and in phosphate acquisition in *Caulobacter crescentus. Arch. Microbiol.* **138:**140–152.

Poindexter, J. S. 1984b. Role of prostheca development in oligotrophic aquatic bacteria, p. 33–40. *In* M. J. Klug and C. A. Reddy (ed.), *Current Perspectives in Microbial Ecology.* ASM Press, Washington, D.C.

Poindexter, J. S. 1992. Dimorphic prosthecate bacteria: the genera *Caulobacter, Asticcacaulis, Hyphomicrobium, Pedomicrobium, Hyphomonas,* and *Thiodendron,* p. 2176–2196. *In* H. G. T. A. Balows, M. Dworkin, W. Harder, and K.-H. Schleifer (ed.), *The Prokaryotes.* Springer-Verlag. New York, N.Y.

Poindexter, J. S., and J. T. Staley. 1996. *Caulobacter* and *Asticcacaulis* stalk bands as indicators of stalk age. *J. Bacteriol.* **178:**3939–3948.

Quardokus, E. M., and Y. V. Brun. Unpublished data.

Quardokus, E. M., N. Din, and Y. V. Brun. 1996. Cell cycle regulation and cell type-specific localization of the FtsZ division initiation protein in *Caulobacter. Proc. Natl. Acad. Sci. USA* **93:** 6314–6319.

Quardokus, E. M., R. Janakiraman, and Y. V.

Brun. 1999. *Abstracts of the Annual Meeting of the American Society for Microbiology*, p. 390.

Quintero, E. J., K. Busch, and R. M. Weiner. 1998. Spatial and temporal deposition of adhesive extracellular polysaccharide capsule and fimbriae by *Hyphomonas sp.* strain MHS-3. *Appl. Environ. Microbiol.* **64:**1246–1255.

Quon, K. C., G. T. Marczynski, and L. Shapiro. 1996. Cell cycle control by an essential bacterial two-component signal transduction protein. *Cell* **84:**83–93.

Quon, K. C., B. Yang, I. J. Domian, L. Shapiro, and G. T. Marczynski. 1998. Negative control of DNA replication by a cell cycle regulatory protein that binds at the chromosome origin. *Proc. Natl. Acad. Sci. USA* **95:**120–125.

Raskin, D. M., and P. A. J. De Boer. 1999. Rapid pole-to-pole oscillation of a protein required for directing division to the middle of *Escherichia coli. Proc. Natl. Acad. Sci. USA* **96:**4971–4976.

Rothfield, L., and S. Justice. 1997. Bacterial cell division: the cycle of the ring. *Cell* **88:**581–584.

Sackett, M. J. 1998. Regulation of the cell division genes *ftsQ, ftsA*, and *ftsZ* in *Caulobacter crescentus*. Ph.D. thesis. Indiana University, Bloomington, Ind.

Sackett, M. J., A. J. Kelly and Y. V. Brun. 1998. Ordered expression of *ftsQA* and *ftsZ* during the *Caulobacter crescentus* cell cycle. *Mol. Microbiol.* **28:**421–434.

Schmidt, J. M. 1968. Stalk elongation in mutants of *Caulobacter crescentus. J. Gen. Microbiol.* **53:**291–298.

Schmidt, J. M. 1973. Effect of lysozyme on crossbands in stalks of *Caulobacter crescentus. Arch. Mikrobiol.* **89:**33–40.

Schmidt, J. M., and R. Y. Stanier. 1966. The development of cellular stalks in bacteria. *J. Cell Biol.* **28:**423–436.

Schmidt, J. M., and J. R. Swafford. 1975. Ultrastructure of crossbands in prosthecae of *Asticcacaulis* species. *J. Bacteriol.* **124:**1601–1603.

Seitz, L. C., and Y. V. Brun. 1998. Genetic analysis of mecillinam-resistant mutants of *Caulobacter crescentus* deficient in stalk biosynthesis. *J. Bacteriol.* **180:**5235–5239.

Shapiro, L., and R. Losick. 1997. Protein localization and cell fate in bacteria. *Science* **276:**712–718.

Sharpe, M. E., and J. Errington. 1999. Upheaval in the bacterial nucleoid: an active chromosome segregation mechanism. *Trends Genet.* **15:**70–74.

Sly, L. I., T. L. Cox, and T. B. Beckenham. 1999. The phylogenetic relationships of *Caulobacter, Asticcacaulis* and *Brevundimonas* species and their taxonomic implications. *Int. J. Syst. Bacteriol.* **49:**483–488.

Smit, J. 1987. Localizing the subunit pool for the temporally regulated polar pili of *Caulobacter crescentus. J. Cell Biol.* **105:**1821–1828.

Smit, J., and N. Agabian. 1982. Cell surface patterning and morphogenesis: biogenesis of a periodic surface array during *Caulobacter* development. *J. Cell. Biol.* **95:**41–49.

Stahl, D. A., R. Key, B. Flesher, and J. Smit. 1992. The phylogeny of marine and freshwater caulobacters reflects their habitat. *J. Bacteriol.* **174:**2193–2198.

Stephens, C. M., and L. Shapiro. 1993. An unusual promoter controls cell-cycle regulation and dependence on DNA replication of the *Caulobacter fliLM* early flagellar operon. *Mol. Microbiol.* **9:**1169–1179.

Stove, J. L., and R. Y. Stanier. 1962. Cellular differentiation in stalked bacteria. *Nature* **196:**1189–1192.

Umbreit, T. H., and J. L. Pate. 1978. Characterization of the holdfast region of wild-type cells and holdfast mutants of Asticcacaulis biprosthecum. *Arch. Microbiol.* **118:**157–168.

Vicente, M., and J. Errington. 1996. Structure, function and controls in microbial division. *Mol. Microbiol.* **20:**1–7.

Wall, T. M., G. R. Hudson, D. A. Danald, and R. M. Weiner. 1980. Timing of swarmer cell cycle morphogenesis and macromolecular synthesis by *Hyphomicrobium neptunium* in synchronous culture. *J. Bacteriol.* **144:**406–412.

Wang, S. P., P. L. Sharma, P. V. Schoenlein, and B. Ely. 1993. A histidine protein kinase is involved in polar organelle development in *Caulobacter crescentus. Proc. Natl. Acad. Sci. USA* **90:**630–634.

Wang, X., J. Huang, A. Mukherjee, C. Cao, and J. Lutkenhaus. 1997. Analysis of the interaction of FtsZ with itself, GTP, and FtsA. *J. Bacteriol.* **179:**5551–5559.

Wanner, B. L. 1990. Phosphorus assimilation and its control of gene expression in *Escherichia coli*, p. 152–163. *In* G. Hauska and R. Thauer (ed.), *The Molecular Basis of Bacterial Metabolism*. Springer-Verlag, Heidelberg, Germany.

Wanner, B. L. 1996. Phosphorus assimilation and control of the phosphate regulon, p. 1357–1381. *In* F. C. Neidhardt, R. Curtiss III, J. L. Ingraham, E. C. C. Lin, K. B. Low, B. Magasanik, W. S. Reznikoff, M. Riley, M. Schaechter, and H. E. Umbarger (ed.), *Escherichia coli and Salmonella: Cellular and Molecular Biology*, 2nd ed. ASM Press, Washington, D.C.

Weiner, R. M., and M. A. Blackman. 1973. Inhibition of deoxyribonucleic acid synthesis and bud formation by nalidixic acid in *Hyphomicrobium neptunium. J. Bacteriol.* **116:**1398–1404.

Weiner, R. M., M. Melick, K. O'Neill, and E.

Quintero. *Hyphomonas adhaerens* sp. nov., *Hyphomonas johnsonii* sp. nov., and *Hyphomonas rosenbergii* sp. nov., marine budding and prosthecate bacteria. *Int. J. Syst. Bacteriol.,* in press.

Weiss, D. S., J. C. Chen, J.-M. Ghigo, D. Boyd, and J. Beckwith. 1999. Localization of FtsI(PBP3) to the septal ring requires its membrane anchor, the Z-ring, FtsA, FtsQ, and FtsL. *J. Bacteriol.* **181:** 508–520.

Whittenbury, R., and C. S. Dow. 1977. Morphogenesis and differentiation in *Rhodomicrobium vannielii* and other budding and prosthecate bacteria. *Bacteriol. Rev.* **41:**754–808.

Wortinger, M., and Y. V. Brun. Unpublished data.

Wortinger, M. A., E. M. Quardokus, and Y. V. Brun. 1998. Morphological adaptation and inhibition of cell division during stationary phase in *Caulobacter crescentus. Mol. Microbiol.* **29:**963–973.

Zerfas, P. M., M. Kessel, E. J. Quintero, and R. M. Weiner. 1997. Fine-structure evidence for cell membrane partitioning of the nucleoid and cytoplasm during bud formation in *Hyphomonas* species. *J. Bacteriol.* **179:**148–156.

Zhou, P., and C. E. Helmstetter. 1994. Relationship between *ftsZ* gene expression and chromosome replication in *Escherichia coli. J. Bacteriol.* **176:** 6100–6106.

REGULATION OF FLAGELLUM BIOSYNTHESIS AND MOTILITY IN *CAULOBACTER*

James W. Gober and Jennifer C. England

16

Caulobacter crescentus divides to form two distinct daughter cells: a motile swarmer cell, which possesses a single polar flagellum, and a nonmotile stalked cell (Fig. 1). These two cell types differ in their global programs of gene expression and DNA replication (reviewed in Brun et al., 1994; Gober and Marques, 1995; Wu and Newton, 1997). For example, the nascent stalked cell possesses the capacity to reinitiate DNA replication following cell division, whereas DNA replication is repressed for a defined period of time in the progeny swarmer cell. Following this period of silencing, the swarmer cell differentiates into a stalked cell through the programmed loss of the flagellum and the biogenesis of a stalk. Differentiation into a stalked cell is accompanied by the initiation of DNA replication. As replication and cell growth proceed, a cascade of events is initiated which results in the biogenesis of a polar flagellum at the opposite pole from the stalk. Flagellar biogenesis is governed both by cell cycle events and the progression of flagellar assembly. The aggregate consequence of these overlapping regulatory pathways is the eventual generation of a motile swarmer cell. This chapter will describe, in detail, the molecular events that lead to the biogenesis of the *C. crescentus* flagellum and how these events are integrated into the generation of a new daughter cell type.

GENETIC ANALYSIS OF MOTILITY IN *C. CRESCENTUS*

Flagellar biogenesis is the best-known aspect of cellular differentiation in *C. crescentus*. The *C. crescentus* flagellum has been observed by electron microscopy and possesses a structure that is typical of flagella from gram-negative bacteria (Wagenknecht et al., 1981; Trachtenberg and Derosier, 1988; Stallmeyer et al., 1989). As with these organisms, the *C. crescentus* flagellum consists of three major subassemblies (Fig. 2). The basal body, embedded in and traversing the inner and outer membranes, serves as both an anchor and a rotor. A flexible hook, which lies outside the cell, is attached to the rods of the basal body. Finally, a rigid flagellar filament, which functions to propel the organism through the medium, is attached to the hook.

Nonmotile mutants have been isolated by either chemical or transposon mutagenesis (Johnson and Ely, 1977, 1979; Johnson et al.,

J. W. Gober and J. C. England, Department of Chemistry and Biochemistry and Molecular Biology Institute, University of California, Los Angeles, CA 90095-1569.

Prokaryotic Development, edited by Y. V. Brun and L. J. Shimkets,
© 2000 American Society for Microbiology, Washington, DC 20005-4171

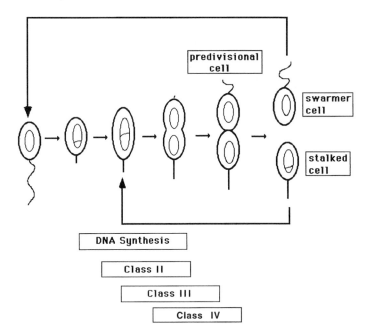

FIGURE 1 *C. crescentus* cell division cycle and temporal expression pattern of flagellar genes. The chromosome is represented by an oval, with the "theta" structure indicating DNA replication. The times during which DNA synthesis occurs and the major classes of flagellar genes expressed are delineated beneath the relevant structural changes taking place during the cell cycle. The expression of the single known class I gene, *ctrA*, is apparently initiated concurrently with DNA replication.

1979, 1983; Ely et al., 1984; Ely and Ely, 1989). Exhaustive analysis has revealed that approximately 50 genes are required for motility, although only 18 encode the structural components of the flagellum itself (Ely and Ely, 1989). The vast majority of these genes have been cloned and sequenced and are highly homologous to the flagellar structural genes found in other organisms. The remaining genes are required for either secretion or assembly of flagellar components or are *trans*-acting regulatory factors. The flagellum is sequentially assembled from the interior towards the exterior of the cell (Fig. 3), with the first component part of the basal body thought to be the MS ring, which is encoded by the *fliF* gene. The *fliF* gene product is an integral membrane protein which is presumed to anchor the entire flagellum to the inner cytoplasmic membrane. The *fliF* gene lies in an operon which contains genes encoding the flagellar switch proteins (*fliG* and *fliN*), which function to switch the rotation of the flagellum in response to chemotactic signals (Ohta et al., 1984; Chen et al., 1986). Additionally, this early operon contains the genes for two *trans*-acting factors, *flbD* and *flbE* (Ra-

makrishnan and Newton, 1990). The third switch protein, encoded by *fliM*, lies in a genetically unlinked operon, *fliLM* (Hahnenberger and Shapiro, 1987; Yu and Shapiro, 1992). The product of the *fliL* gene is required for flagellar function, but not for assembly (Jenal et al., 1994).

Other components of the flagellum that are assembled early are required for the export of proteins that constitute the external structures of the flagellum. These proteins are encoded by *flhA*, *flhB*, *fliI*, *fliJ*, *fliO*, *fliP*, *fliQ*, and *fliR* and are homologous to the class III secretion genes of pathogenic bacteria (Ramakrishnan et al., 1991; Sanders et al., 1992; Gober et al., 1995; Zhuang and Shapiro, 1995; Lee and Ely, 1996; Stephens et al., 1997). In pathogens, the type III secretion system is required for the translocation into the host cell of virulence factors which lack the typical signal sequence that is the most common determinant for export of bacterial proteins (reviewed in Hueck, 1998). This system consists of protein components, approximately 15 to 20 different polypeptides, that are located in both the inner and outer membranes. The flagellar secretion system ap-

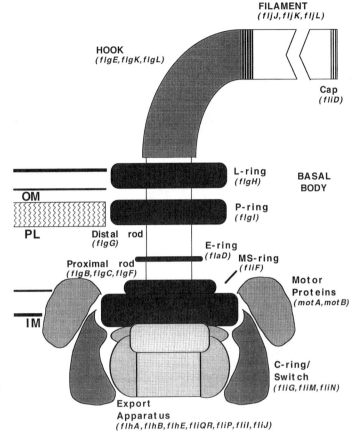

FIGURE 2 Structure of the *C. crescentus* flagellum. In this schematic illustration, the components that make up the flagellum are indicated with respect to position in the membrane, shown in cross section. The names of the genes that encode the relevent structural proteins are shown in parentheses below each component name. The individual components are grouped into three major superstructures: the basal body, the hook, and the filament. The basal body includes the cytoplasmic structures and the structures that span the inner (cytoplasmic) membrane (IM), the peptidoglycan layer (PL) in the periplasmic space, and the outer membrane (OM) of the cell. Outside the cell, the flexible hook connects the basal body to the rigid flagellar filament.

parently consists only of components that lie within the inner membrane of the cell, but analogously to virulence factors secreted by the type III secretion systems of pathogens, many of these exported flagellar proteins lack a discernible, conserved sequence determinant for export. The flagellum secretion system is thought to be associated with the MS ring-switch complex of the basal body.

The assembly of the MS, switch, and secretory apparatus is required for both the expression (see below) and assembly of most of the remaining components of the flagellum. The next components to be assembled are the rods and outer rings of the basal body. Four proteins that require the flagellum-specific secretion system for export compose the rod structure of the basal body and are encoded by the *flgBCFG*

genes (Dingwall et al., 1992a; Boyd and Gober, 1999). The assembly of the rod structure is, in turn, required for the assembly of the P and L rings, encoded by the *flgI* (Khambaty and Ely, 1992) and *flgH* (Dingwall et al., 1990) genes, respectively, which are integral membrane proteins located in the outer membrane (Mohr et al., 1996). The *C. crescentus* flagellum possesses an additional E ring, not found in enteric bacteria, which is encoded by the *flaD* gene and lies between the inner membrane and the peptidoglycan layer (Hahnenberg and Shapiro, 1988).

The assembly of the flexible hook, encoded by *flgE* (Langenaur et al., 1978; Sheffery and Newton, 1979; Johnson et al., 1979; Ohta et al., 1982; Kornacker and Newton, 1994), and hook-associated proteins (*flgKL*) (Mullin and

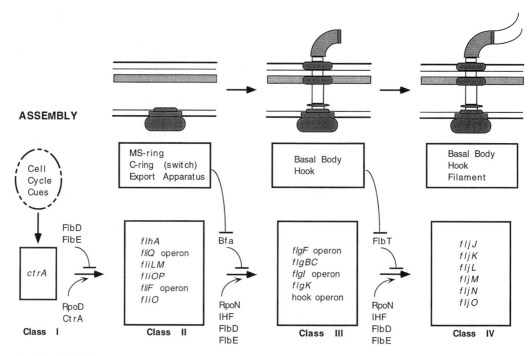

ASSEMBLY

REGULATION

FIGURE 3 Assembly of the *C. crescentus* flagellum. The upper diagram shows the flagellum at the major stages of cell-proximal-to-cell-distal assembly. Below, the regulatory hierarchy is outlined, with the genes expressed at each step in the cascade shown in boxes beneath the corresponding structures. The roles of specific proteins controlling expression of each gene class are illustrated with arrows for positive regulation and lines with bars for negative regulation; the influence of assembled structures on expression is similarly indicated. Assembly begins with the production and then activation, via phosphorylation, of CtrA in response to (as-yet-unknown) cell cycle cues. The expression of class III genes is dependent upon class II gene products and assembly of the early basal-body structures. Mutations at the *bfa* locus bypass this requirement. It is hypothesized that *bfa* regulation of class III genes occurs, because the Bfa protein is exported from the cell solely by the assembled early basal-body structures and can no longer be exported once the class III structures are added. Thus, during this time Bfa concentrations would be low in the cell, allowing class III gene transcription. Before and after this window, high levels of the Bfa protein would serve to repress class III genes. Similarly, synthesis of the class IV flagellins is coupled to successful assembly of class III structures, although in this case the regulatory mechanism is known to be posttranscriptional. Apparently, FlbT destabilizes the flagellin mRNA, either directly or indirectly, thus significantly reducing what is otherwise an extremely long half-life. Completion of the basal body and hook assembly seems to counteract FlbT activity and allows translation of the flagellin mRNA.

Newton, 1993) follows the completion of the basal-body complex. These so-called axial proteins, like the components of the rod, require the type III secretion system for assembly. Once the hook complex has been successfully assembled, the flagellins are then secreted and assembled into the rigid flagellar filament. *C. crescentus* possesses six different flagellin genes located in two different clusters (Langenaur and

Agabian, 1976, 1978; Osley et al., 1977; Fukuda et al., 1976; Gill and Agabian, 1982, 1983; Ely and Leclerc, 1998). The alpha cluster contains *fljL* (27.5 kDa), *fljK* (25 kDa), and *fljJ* (28.5 kDa) (Ely and Gerardot, 1988; Minnich et al., 1988). The genetically unlinked beta cluster contains three additional genes encoding a 25-kDa flagellin, *fljMNO* (Ely and Leclerc, 1998). Biochemical dissection and immunoelectron microscopy of the flagellar filament have revealed that the flagellins are assembled in a specific order, with the the minor, 28.5-kDa flagellin proximal to the hook, then the 27.5-kDa flagellin, and finally, most abundant, the 25-kDa flagellin (Weissborn et al., 1982; Driks et al., 1989). Although the 27.5- and 28.5-kDa flagellins are present in relatively minor amounts, gene replacement experiments have determined that both *fljL* and *fljJ* are required for normal motility (Minnich et al., 1988).

Additional gene products that regulate the synthesis and assembly of the external hook and filament structures have also been identified. For example, the hook operon contains the *fliK* gene (Mullin et al., 1998), which is involved in regulating the number of FlgE monomers incorporated into the hook structure. Several genes are also required for proper flagellin synthesis. The expression of *flmA*, *flmD*, *flmE*, *flmG*, *flmH* (formerly *flaA*, *flaR*, *flaZ*, *flbA*, and *flaG*, respectively), *flmB*, and *flmC* is required for the synthesis of normal flagellin (Leclerc et al., 1998). The *flmABCD* genes encode proteins that are homologous to polysaccharide biosynthetic genes found in other organisms. Similarly, *flmEFGH* encode a putative methyltransferase, enzymes similar to tryptophan monooxygenase, an O-linked acetylglucosamine transferase, and an acetyl transferase, respectively. Mutations in these genes lead to the production of flagellin proteins with aberrant molecular masses, suggesting that they are involved in the posttranslational modification of flagellin. Two additional genes, *flbT* and *flaF*, when mutated, affect the level of flagellin synthesis (Shoenlein and Ely, 1989; Shoenlein et al., 1992) and are presumed to be posttranscriptional regulators of flagellin synthesis (Anderson and Gober, 1998; Mangan et al., 1999) (see below).

TEMPORAL TRANSCRIPTIONAL REGULATION OF FLAGELLAR BIOGENESIS

Growth of the newly divided or differentiated stalked cell is accompanied by the formation of a swarmer cell at the pole opposite the stalk.

Flagellar biogenesis in the stalked and predivisional cell is regulated by both cell cycle and flagellar assembly events. The molecular dissection of the cell cycle events resulting in the timed expression of flagellar genes has yielded important information about how bacteria utilize the cell cycle to regulate gene expression. *C. crescentus* is well suited for the assay of cell cycle events because large synchronized populations of swarmer cells can be isolated by density gradient centrifugation. Epistasis experiments (i.e., assaying the expression of a given flagellar gene in different mutant strains) have demonstrated that flagellar biogenesis in *C. crescentus* is regulated by a *trans*-acting regulatory hierarchy similar to that operating in enteric bacteria (Komeda, 1986; Kutsukake et al., 1990), in which the expression of early flagellar genes (class II) is required for the expression of genes that encode structures that are assembled later (Bryan et al., 1984; Ohta et al., 1985; Champer et al., 1987; Minnich and Newton, 1987; Newton et al., 1989; Xu et al., 1989; Ramakrishnan et al., 1994) (Fig. 3). For example, a mutation in any one of the class II genes (e.g., those encoding the MS ring, switch, or type III secretion system) results in the repression of transcription of class III genes, such as those encoding the basal-body rods (*flgBCFG*), rings (*flaD* and *flgHI*), and hook structure (*flgEK*). Likewise, the successful expression and assembly of these structures is required for the expression of the class IV flagellin genes. Therefore, two checkpoints that couple assembly to gene expression operate during flagellar biogenesis in *C. crescentus* (see below).

The results of epistasis experiments, described above, suggested that the genes encoding early, class II flagellar structures would be expressed in the cell cycle prior to the expression of genes encoding early structures. This prediction has been borne out by assays of the cell cycle expression of these genes, using either transcription fusions to promoterless reporter genes, such as β-galactosidase (*lacZ*) or neomycin phosphotransferase (NPTII; encoded by *neo*), or by measuring the mRNA directly (Mullin et al., 1987; Dingwall et al., 1992a;

Yu and Shapiro, 1992; Stephens and Shapiro, 1993; VanWay et al., 1993; Gober et al., 1995; Mohr et al., 1998). In general, class II promoters are expressed in the early predivisional cell, shortly after the initiation of chromosomal DNA replication. Indeed, the one known cell cycle cue that triggers the expression of class II genes is apparently DNA replication activity. Early experiments had shown that the inhibition of DNA replication resulted in an inhibition of hook and flagellin synthesis (Osley et al., 1977; Sheffery and Newton, 1981). Subsequently, the transcription of the class II, *fliF*, *fliQR*, and *fliLM* promoters has been shown to be sensitive to treating cells with the DNA synthesis inhibitor hydroxyurea (HU) (Dingwall et al., 1992b; Stephens and Shapiro, 1993). Sensitivity to HU treatment varied, depending on the stage of the cell cycle. For example, if cells were treated before the initiation of DNA replication, transcription of *fliLM* was exquisitely sensitive to HU. If the cells had progressed well past the initiation of DNA replication, then transcription was relatively insensitive to HU treatment (Stephens and Shapiro, 1993). This suggests that the initiation of replication or some early critical event that is coupled to S-phase triggers the timed transcription of early class II promoters.

How does the cell couple DNA replication to gene expression? Class II flagellar promoters all have significant sequence homology, suggesting that transcription is regulated by common *trans*-acting factors (Mullin et al., 1987; Dingwall et al., 1992b; Yu and Shapiro, 1992; Stephens and Shapiro, 1993; VanWay et al., 1993; Gober et al., 1995; Mohr et al., 1998). Notably, these promoters are also similar to several nonflagellar promoters: *ccrM*, which encodes an essential DNA methyltransferase (Stephens et al., 1996); *ftsZ*, an essential cell division gene (Kelly et al., 1998); and *hemE*, a heme biosynthetic gene whose promoter activity is linked to the initiation of DNA replication (Marczynski et al., 1995; Quon et al., 1998). The similar class II flagellar promoters are transcribed by σ^{70}-containing RNA polymerase holoenzyme (Wu et al., 1998), the most abundant form of RNA polymerase in bacterial cells. Temporal regulation has been shown to be controlled by the essential transcription factor CtrA (Quon et al., 1996) (see chapter 18 for a detailed description of the role of CtrA in regulating global cell cycle events). The amino terminus of CtrA is homologous to the domains which typify response regulators in the large family of bacterial two-component regulatory systems. Typically, this class of proteins is phosphorylated on a critical aspartate residue by a cognate sensor histidine kinase (reviewed in Parkinson and Koifoid, 1992). The sensor histidine kinase, in response to an environmental cue, autophosphorylates and subsequently transfers the phosphate to its response regulator, which in this case would lead to the activation of transcription. *ctrA* was originally identified as a conditional temperature-sensitive lethal mutation which led to an increase in the expression of the *fliQR* promoter (Quon et al., 1996). Pure CtrA protein has been shown to bind to several class II flagellar promoters (Quon et al, 1996; Mohr et al., 1998). In addition, in vitro transcription experiments have demonstrated that CtrA can activate the transcription of these promoters in the presence of σ^{70}-containing RNA polymerase and a sensor histidine kinase, DivJ (Wu et al., 1998). Interestingly, depletion experiments indicate that CtrA apparently has a dual role in the regulation of class II promoters (Quon et al., 1996). In these experiments, the rate of expression of a class II-*lacZ* reporter fusion was assayed in a strain containing *ctrA* under the control of an inducible promoter. Removal of the inducer from the medium depleted the cell of ctrA, resulting first in an increase and then, in some cases, a sharp decrease in *lacZ* expression. This result implies that CtrA functions as a negative regulator of transcription at high concentrations and an activator of transcription at lower concentrations. The cell cycle transcription of class II flagellar promoters is accomplished in two ways (Domian et al., 1997). First, the availability of CtrA is temporally regulated. CtrA, which is present in swarmer cells, is degraded upon differentiation into stalked cells. In late stalked-early predivisional cells, CtrA is then synthesized. The temporal phosphorylation of CtrA at this time also probably contributes to the cell cycle transcription pattern of class II promoters. Two-component regulatory systems are designed to sense environmental cues, so it is likely that CtrA, via its cognate kinase, senses a cell cycle cue. However, the nature of this cue is currently unknown, but as noted above, it must be linked to the onset of DNA replication. Recently, a sensor histidine kinase, encoded by *divJ*, and a response regulator, encoded by *divK*, have been implicated as part of the CtrA phosphorylation pathway (Wu et al., 1998), although the kinase that is responsible for directly phosphorylating CtrA remains unknown (see chapter 18).

The transcription of class III and IV flagellar promoters follows the expression and assembly of class II gene products (Bryan et al., 1984; Minnich and Newton, 1987; Ohta et al., 1985; Newton et al., 1989; Xu et al., 1989; Gober and Shapiro, 1992; Wingrove et al., 1993; Marques and Gober, 1995). Class III promoters have marked similarity in their sequence elements and are thought to be subject to common regulation (Minnich and Newton, 1987; Mullin et al., 1987; Mullin and Newton, 1989; Ninfa et al., 1989; Dingwall et al., 1990, 1992b; Gober et al., 1991b; Khambaty and Ely, 1992; Gober and Shapiro, 1992; Wingrove et al., 1993; Benson et al., 1994a, 1994b; Marques and Gober, 1995). The class III promoters all require RNA polymerase containing the alternative sigma factor σ^{54} (Brun and Shapiro; 1992; Anderson et al., 1995). σ^{54}-containing RNA polymerase holoenzyme, when bound to promoter DNA, is incapable of initiating transcription unless it interacts with an activator protein (reviewed in Kustu et al., 1989). The binding site for the activator protein in this type of promoter is usually located approximately 100 bp upstream from the transcription initiation site. Most class III flagellar promoters possess a conserved sequence at this location called *ftr*, for flagellar transcriptional regulation (Minnich and Newton, 1987; Mullin and Newton, 1989, 1993). Site-directed mutagen-

esis of this region has demonstrated that as little as a 2-bp change can completely abolish promoter activity (Minnich and Newton, 1987; Mullin and Newton, 1989, 1993; Wingrove et al., 1993; Marques and Gober, 1995). Epistasis experiments showed that Tn5 insertions in the gene encoding the σ^{54} transcriptional activator, *flbD*, resulted in a complete lack of class III gene expression (Newton et al., 1989; Xu et al., 1989; Ramakrishnan and Newton, 1990; Wingrove et al., 1993). Furthermore, DNA footprinting, gel mobility shift, methyl interference, and transcription assays have demonstrated that pure FlbD can bind to and activate the transcription of many of these promoters in vitro (Wingrove et al., 1993; Wingrove and Gober, 1994; Benson et al., 1994a, 1994b; Mullin et al., 1994).

The class III flagellar promoters also contain a binding site for the DNA-bending protein, integration host factor (IHF) (Gober and Shapiro, 1990, 1992; Gober and Marques, 1995). Both gel shift and electron microscopy experiments with bacteriophage λ DNA have shown that IHF binding introduces a large bend, estimated at up to 140° by gel mobility shift assay, in its DNA substrate (reviewed in Landy, 1989). The IHF binding site in these flagellar promoters is located between the promoter sequences and the FlbD binding site. In vitro transcription experiments with enteric bacterial nitrogenase σ^{54} promoters (*nif*) have shown that the IHF-induced bend significantly enhances transcriptional activation (Hoover et al., 1990; Santero et al., 1992). It has been proposed that the bend actuated by IHF binding serves to increase the frequency of interactions between RNA polymerase bound at the promoter and its transcriptional activator bound at the distant enhancer site (Hoover et al., 1990). Mutations in the IHF binding site in the *C. crescentus flgE* operon promoter (formerly *flbG*) decreased the expression of a promoterless *lacZ* reporter gene two- to fivefold, suggesting that the IHF-mediated bend in these promoters is also required for maximal promoter activity (Gober and Shapiro, 1990). Immunoblots performed from extracts of synchronized cultures have shown that the levels of IHF vary during the cell cycle. Maximal amounts of IHF are present in predivisional cells at a time when class III promoters are at peak transcriptional activity (Gober and Shapiro, 1990). However, it is not known whether the cell cycle modulation in IHF levels influences the temporal pattern of class III promoter activity or simply enhances the level of transcription in predivisional cells.

It is likely that both σ^{54}-containing RNA polymerase and IHF can be viewed as general transcription factors and that the major regulator of cell cycle transcription is FlbD (Ramakrishnan and Newton, 1990; Ramakrishnan et al., 1991; Wingrove et al., 1993; Benson et al., 1994a, 1994b; Mullin et al., 1994). Interestingly, unlike IHF, the levels of FlbD do not vary during the cell cycle; the protein is present in equal amounts regardless of the cell type (Wingrove et al., 1993). As noted above, FlbD, like CtrA, possesses a conserved, two-component regulatory system response regulator domain at its amino terminus. In vivo phosphorylation experiments have shown that FlbD is phosphorylated exclusively in predivisional cells at a time when class III promoters are transcribed (Wingrove et al., 1993). Experiments with the *Salmonella typhimurium* response regulator, NtrC, have shown that phosphorylation promotes both oligomerization of NtrC at its enhancer sequences and the stimulation of open complex formation (Popham et al., 1989; Porter et al., 1993). By analogy, it has been proposed that FlbD phosphorylation is the major regulatory event in activating the transcription of class III promoters.

The FlbD kinase has not been firmly identified; however, the product of the *flbE* gene has been implicated as having a role in controlling FlbD phosphorylation (Wingrove and Gober, 1996). Strains with deletions in *flbE*, which is located within the *fliF* operon along with *flbD*, do not express class III flagellar promoters. Furthermore, *flbE* is not homologous to any known flagellar genes in other organisms but does contain several sequences that are conserved in sensor histidine kinases. In vitro experiments have demonstrated that pure FlbE is capable of autophosphorylation and of transferring phosphate to FlbD (Wingrove and Gober, 1996). Based on this genetic and biochemical data, it was proposed that FlbE functioned as the cognate histidine kinase for FlbD. Recent experiments, however, indicate that FlbE, when mutated at the critical histidine residue, is capable of complementing motility and restoring class III gene expression in a strain containing a chromosomal deletion in *flbE* (Muir and Gober, 1998). Therefore, although FlbE is required for maximal FlbD activity, another, as-yet-unidentified, histidine kinase probably regulates FlbD activity.

FlbD also has a role in regulating early class II gene transcription. Strains containing Tn5 insertions in *flbD* exhibit an increase in the expression of several class II promoters, suggesting that FlbD might negatively regulate their expression (Benson et al., 1994; Wingrove and Gober, 1994; Muir and Gober, 1999). Analysis of the *fliF* promoter has revealed *ftr* sequences (e.g., FlbD binding sites) located between and adjacent to the critical −35 and −10 sequences (Fig. 4). DNase footprinting and methylation interference assays have shown that FlbD binds to this region in vitro (Benson et al., 1994; Wingrove and Gober, 1994; Mullin et al., 1994). Furthermore, mutagenesis of the FlbD binding site results in a fourfold increase in promoter activity as assayed with *lacZ* reporter fusions (Wingrove and Gober, 1994). FlbD binding would be predicted to occlude the binding of RNA polymerase and possibly CtrA to the *fliF* promoter, resulting in transcriptional repression. Apparently, FlbD repression has an important role in regulating the timing of *fliF* transcription. An *fliF* mutant promoter that could not be bound by FlbD in vitro exhibited a significant delay in the cessation of transcription in the cell cycle (Wingrove and Gober, 1994). This experimental observation indicates that FlbD mediates a dynamic interplay between the expression of class II and class III promoters.

Recently, it has been found that several flagellar promoters are organized such that an early class II promoter transcript diverges from a later class III transcript (Mohr et al., 1998)

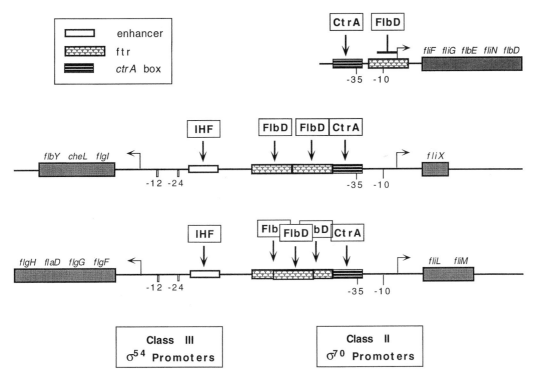

FIGURE 4 Comparison of known *cis*-acting regulatory elements of class II and class III genes. Shown are the intergenic regions between the *fliX* gene and the *flgI* operon and between the *fliL* and *flgF* operons, as well as the upstream regulatory region of the *fliF* operon. The positions of the promoters (σ^{70} for class II and σ^{54} for class III) and *cis*-acting sequences are also diagrammed, with the positive or negative regulation of the corresponding binding factor indicated by an arrow or barred line, respectively. FlbD has been shown to down-regulate the class II *fliF* operon, perhaps by competing with the activator CtrA. FlbD activates the class III operons, presumably brought into proximity with the promoters by an IHF-induced bending of the intervening DNA. It has been speculated that the close, or partially overlapping, organization of *ftr* and CtrA box elements in the intergenic regions provides a means of coordinating the expression of these class II and class III genes.

(Fig. 4). In each case, the class III genes encode products assembled into structures that immediately follow the assembly of the MS ring-switch secretory complex. These include *flgBC* (Gober et al., 1995; Boyd and Gober, 1999) (basal-body rods), *flgFG, flaD,* and *flgH* (Dingwall et al., 1992; Marques and Gober, 1995) (encoding basal-body rods and rings), and *flgI* (Khambaty and Ely, 1992; Mohr et al., 1998) (basal-body ring). Thus, the entire flagellar subassembly that lies between the MS ring and the external hook structure possesses this type of promoter organization. It has been suggested that the organization of *cis*-acting elements in these divergent promoter systems may serve to coordinately regulate class II and class III gene expression (Mohr et al., 1998). Interestingly, in support of this view, the CtrA binding site of the class II promoters either is adjacent to or partially overlaps the FlbD binding site of the divergent class III promoter, raising the possibility that the binding of one transcription factor could influence the binding of the other.

REGULATION OF FLAGELLAR-GENE EXPRESSION BY ASSEMBLY

As described above, strains containing mutations in class II structural genes do not transcribe class III flagellar promoters, suggesting that early flagellar assembly events are coupled to class III gene expression. This coupling of assembly to gene expression is analogous to flagellar biogenesis in *S. typhimurium*, where the assembly of a hook structure is required for the transcription of both flagellin and chemotaxis (*che*) genes (Komeda, 1986; Kutsukake et al., 1990). In this case, the flagellin and *che* genes are transcribed by a minor form of RNA polymerase containing σ^{28}. A mutation that restored σ^{28}-dependent transcription in a strain that did not assemble a hook structure mapped to the *flgM* gene, which was found to encode an anti-sigma factor (Gillen and Hughes, 1991a, 1991b). The *flgM* gene product can bind to σ^{28} and thereby prevent polymerase from transcribing flagellin and *che* promoters (Onishi et al., 1992). Assembly is "sensed" from the ability of the completed basal-body–hook complex to export FlgM out of the cell, resulting in a decrease in its cytoplasmic concentration and a subsequent relief of repression (Hughes et al., 1993).

In contrast to enteric bacteria, the flagellar hierarchy in *C. crescentus* possesses two points at which flagellar assembly is coupled to gene expression: the assembly of the MS ring–switch secretory complex (class II) is required for the expression of genes encoding the remainder of the basal body and the hook (class III), and the expression and assembly of class III gene products is required for the expression of the class IV flagellins. In order to investigate the coupling of class III gene expression to early assembly events, a genetic screen was undertaken which selected for mutants that restore class III gene expression in a class II $\Delta fliQR$ mutant (Mangan et al., 1995). A single, still-uncharacterized genetic locus, called *bfa*, for bypass of flagellar assembly, was identified. The *bfa* mutation could bypass the transcriptional requirement for all of the class II flagellar mutants tested except those strains which had mutations in critical transcription factors, such as *flbD* or *rpoN* (Mangan et al., 1995). What is the function of Bfa in wild-type cells? One possibility is that Bfa serves to negatively regulate class III gene expression after the appropriate structure has been built. For example, in one

scenario, the cellular levels or activity of Bfa may be low when a complete class II structure is present, thereby permitting class III gene expression. Following the assembly of a class III structure they may increase because this structure is unable to export (analogously to FlgM regulation) or inactivate Bfa. This, in effect, would serve to shut off class III gene expression. In enteric bacteria, it has been proposed that FlgM functions analogously, perhaps regulating the number of flagella synthesized by each cell. In support of this view, *bfa* mutations in *C. crescentus* fail to turn off class III gene expression late in the cell cycle (Mangan et al., 1995).

Assembly of the hook structure also regulates flagellin expression. In the absence of a hook, cells produce no detectable flagellin protein, as assayed by immunoblotting. In contrast to *bfa* or *flgM* regulation, the repression of flagellin synthesis in these mutants is mediated through a posttranscriptional mechanism (Mangan et al., 1995; Anderson et al., 1997). Class III flagellar mutants were shown to express flagellin-*lacZ* transcription fusions but not flagellin-LacZ protein fusions (Anderson et al., 1997). In the case of *fljK*, a major 25-kDa flagellin, the sequences required for regulation apparently lie within the 5′ untranslated region of the transcript. Deletion of this region partially alleviated negative regulation (Anderson et al., 1997). It has now been demonstrated that a mutation in *flbT*, a flagellar gene of unknown function, can restore flagellin protein synthesis in class III mutants (Mangan et al., 1999). Mutations in *flbT* were previously shown to have pleiotropic effects (Schoenlein and Ely, 1989; Schoenlein et al., 1992), including the overproduction of flagellin protein and defects in the timing of release of the flagellum at the swarmer-to-stalked-cell transition. Immunoblot analysis revealed that an *flbT* mutation could restore the synthesis of all flagellins in class III mutants, indicating that all six flagellin genes are subject to common regulation by FlbT (Mangan et al., 1999). Class III mutant strains were shown to produce little flagellin mRNA (Anderson et al., 1997; Mangan et al.,

1999). This is apparently attributable to a decrease in mRNA stability. Flagellin mRNA possesses an unusually long half-life for a bacterial transcript, approximately 15 min when assayed in a mixed population. In a class III hook mutant, the half-life of *fljK* mRNA decreases to 3 min (Mangan et al., 1999). The stability of the *fljK* mRNA is restored in a class III hook-*flbT* double mutant, exhibiting a half-life of over 45 min. Therefore, FlbT has a role in regulating the stability of flagellin mRNA in response to the assembly of class III flagellar structures. From these experiments, it is not possible to discern whether FlbT directly destabilizes flagellin mRNA or whether an inhibition of translation results in a shorter half-life, as has been reported for some bacterial mRNAs. Therefore, the mechanism by which FlbT accomplishes this negative regulation is as yet unknown. Given its long half-life in wild-type cells, it is possible that a factor(s) actively stabilizes *fljK* mRNA. FlbT may simply function to antagonize the activity of this positive regulator. Alternatively, FlbT binding to the mRNA may directly promote degradation or inhibit translation. In vitro gel mobility shift assays with the *fljK* transcript have shown that FlbT interacts with the 5′ untranslated region (Anderson and Gober, 1998). Furthermore, antibodies directed against FlbT can supershift the RNA-protein complexes in this type of assay when cell extracts are incubated with radiolabeled FljK transcript. No supershift occurs when the extract is derived from *flbT* mutant cells. Interaction is apparently indirect, as purified FlbT is unable to bind to the *fliK* transcript. The 5′ untranslated region of the *fljK* transcript is predicted to fold into two alternative secondary structures. In one of these predicted structures, the ribosome binding site would be obscured by base pairing with other sequences within the transcript. Mutations in this stem-loop structure relieve FlbT-mediated regulation (Anderson and Gober, 1998). Interestingly, anti-FlbT antibodies cannot supershift RNA-protein complexes in wild-type cell extracts when this mutant transcript is used for the gel mobility shift experiment.

The role of FlbT-mediated regulation in wild-type cells is apparently analogous to the function of Bfa. Cell cycle experiments in an *flbT* mutant strain show that a FljK-LacZ fusion protein is expressed completely throughout the *C. crescentus* cell cycle (Mangan et al., 1999). Even though *fljK* is not transcribed in swarmer and stalked cells, the *flbT* mutant results in a long-lived mRNA that persists well after transcription has abated. Therefore, mRNA produced in the predivisional cell remains in the swarmer cell after division and is still present after these cells have differentiated into stalked cells, where it continues to be translated. The implication is that both regulatory systems, *bfa* and *flbT*, which couple assembly to gene expression, are also utilized to influence the cell cycle timing of flagellar gene expression.

ASYMMETRIC POSITIONING OF FLAGELLAR PROTEINS

A hallmark of the *C. crescentus* developmental program is the asymmetric positioning of proteins in the predivisional cell. For example, flagellar proteins synthesized in the predivisional cell are specifically targeted to its swarmer pole. What are the molecular mechanisms that establish this asymmetry? Although the complete details of this aspect of cellular differentiation are not known, it is evident that *C. crescentus* uses two distinct mechanisms to target flagellar gene products: pole-specific transcription and pole-specific protein targeting.

Pole-Specific Transcription of Flagellar Genes

Early experiments in which the fate of flagellin mRNA was assayed showed that newly formed swarmer cells specifically inherited flagellin mRNA (Milhausen and Agabian, 1983). Flagellin mRNA synthesis does not occur in the swarmer cell but rather before division in the predivisional cell type (Minnich and Newton, 1987). This result indicated that newly formed flagellin mRNA segregated to the swarmer pole of the predivisional cell. Two possible mechanisms could result in the segregation of flagellin mRNA. First, sequences within the mRNA could specifically direct targeting to the swarmer pole of the predivisional cell. Second, the flagellin genes could be specifically transcribed in the swarmer compartment of the predivisional cell and, as a consequence of the unusual stability of flagellin mRNA, these transcripts would persist in the progeny swarmer cells. To distinguish between these two mechanisms, the segregation of the products of flagellar gene-*neo* transcription reporter fusions was tested (Gober et al., 1991a). In this experiment, cells harboring the flagellar-*neo* transcription fusion were synchronized and the culture was permitted to progress to the predivisional stage, where gene expression was maximal. The proteins were then pulse labeled, the label was chased, and following cell division, the newly formed progeny swarmer and stalked cells were separated, and labeled NPTII was immunoprecipitated. The presence of labeled reporter gene product in either progeny cell type would indicate the location of synthesis in the predivisional cell. For example, if more labeled NPTII was present in the swarmer cell, it would indicate that expression of the reporter fusion occurred in the swarmer compartment of the predivisional cell. Utilizing this type of experiment, a *fljK-neo* transcription fusion was found to be exclusively expressed in the swarmer compartment of the predivisional cell (Gober et al., 1991a). Since the *fljK-neo* fusion was an operon fusion, not a protein fusion, this result provided a strong indication that transcription of the *fljK* gene was restricted to the swarmer compartment of the predivisional cell.

As noted above, the flagellin genes, *fljK* and *fljL*, as well as all the class III flagellar genes are regulated by common *trans*-acting factors, including σ^{54}-containing RNA polymerase, IHF, FlbD, and FlbE. If compartmentalization of transcription of *fljK* were occurring, then it seemed likely that these other commonly regulated genes would also exhibit swarmer pole-specific transcription when tested with reporter gene transcription fusions. Indeed, all the class III flagellar genes tested to date exhibit compartmentalized expression in the swarmer pole of the predivisional cell (Gober et al., 1991a). To test whether promoter activity itself was

sufficient to drive compartmentalized gene expression, a fusion containing 118 bp of the hook operon promoter (formerly *flbG*) with no *C. crescentus* mRNA sequences was placed upstream of a promoterless *neo* reporter gene (Gober et al., 1991a). This fusion also exhibited swarmer compartment-specific expression and demonstrated that promoter sequences were solely responsible for polar localization of mRNA. This fusion contained, in addition to the promoter, an IHF binding site and a single *ftr* enhancer (FlbD binding site).

The most likely candidate for the regulator of compartmentalized transcription is the transcriptional activator, FlbD. As noted above, FlbD activates the temporal transcription of class III and class IV flagellar promoters. FlbD activity itself has been shown to be regulated by temporal phosphorylation. Two possible mechanisms could result in compartmentalized transcription. First, FlbD may be specifically targeted to the swarmer pole of the predivisional cell. Alternatively, active, presumably phosphorylated FlbD may be restricted to the swarmer pole. Immunoblot analysis revealed that FlbD protein is present in all cell types, including newly divided progeny swarmer and stalked cells, indicating that FlbD is not specifically targeted to the swarmer pole (Wingrove et al., 1993). To test whether activated FlbD was responsible for swarmer pole-specific transcription, a constitutive mutant was constructed in which a serine residue in the conserved central transcriptional activation domain was changed to a phenylalanine residue (Wingrove et al., 1993). This type of mutation has been shown to render the σ^{54} activator, NtrC, active in the absence of phosphorylation (Popham et al., 1989). When the constitutive FlbD was expressed in *C. crescentus* cells, there was a loss of compartmentalized flagellin expression. In this strain, flagellin protein was expressed in both the swarmer and stalked compartments (Wingrove et al., 1993). This result indicates that, in wild-type cells, phosphorylated, active FlbD is restricted to the swarmer compartment of the predivisional cell.

Since FlbD is active before there are two

compartments formed in the predivisional cell, it is presumed that, initially, phosphorylated FlbD is present in both compartments (Fig. 5A). It is not known how an asymmetric distribution of phosphorylated FlbD is established in the predivisional cell. One possibility is that there exists an active phosphatase activity that is restricted to the stalked compartment. Alternatively, a kinase or factor that activates FlbD is trapped in the swarmer compartment following the formation of the cell division plane and that FlbD in the stalked pole, in the absence of persistent phosphorylation, eventually becomes dephosphorylated and inactive. The positive factor FlbE has been implicated in establishing the asymmetric activation of FlbD (Wingrove and Gober, 1996). As noted above, *flbE* is required for the transcription of *flbD*-dependent genes. Experiments with immunofluorescence microscopy have demonstrated that the subcellular localization of FlbE varies during the cell cycle (Wingrove and Gober, 1996). In swarmer and stalked cells, FlbE exhibits no localization, being uniformly distributed throughout the cytoplasm. As predivisional cells begin to pinch to form a cell division plane, FlbE localizes in discreet foci: one located at the midcell and another at the stalked pole. Eventually, when a cell division plane forms, midcell-localized FlbE is trapped in the swarmer compartment of the predivisional cell. This cell cycle-regulated pattern of localization coincides with the temporal pattern of FlbD phosphorylation and class III flagellar gene transcription (Wingrove and Gober, 1996). The localization domain of FlbE was found to reside in the amino-terminal 52 amino acids. When fused to *lacZ*, this region was sufficient to drive the localization of β-galactosidase to the midcell and the stalked pole (Wingrove and Gober, 1996). Overexpression of the fusion interfered with localization of wild-type FlbE to both these sites and behaved as a dominant-negative mutant, decreasing the level of expression of an *fljK-lacZ* transcription fusion. Furthermore, this dominant-negative form of FlbE resulted in a loss of compartmentalized expression of *fljK*. These results, in sum-

mary, indicate that the localization of FlbE is essential for the activation of FlbD-dependent promoters and is apparently essential for compartmentalized transcription. Since FlbE is localized in both cellular compartments, it raises the question of how it functions to activate FlbD exclusively in the swarmer pole. One attractive possibility is that FlbE localization to the midcell initiates a cascade of events that results in class III flagellar gene transcription. In this way, the initiation of cell division would be coupled to flagellar biogenesis. This regulatory scheme is analogous to that operating through CtrA to activate the transcription of early flagellar genes in response to the initiation of chromosomal DNA replication. In this case, morphogenesis of a cellular structure, the cell division plane, triggers the next stage in flagellar biogenesis. Indeed, the formation of the cell division plane may contribute to the asymmetric activation of FlbD by sequestering active FlbE (i.e., those molecules localized at the midcell) in the swarmer compartment of the predivisional cell. The coupling of cellular morphogenesis to developmentally regulated gene expression is not unique to *C. crescentus*. In the sporulating bacterium *Bacillus subtilis*, a serine phosphatase, SpoIIE, localizes to the sporulation septum, where it directs a program of sporangium-specific transcription (see chapter 7).

FlbD not only directs the compartmentalized transcription of late flagellar genes but also that of early class II genes (Fig. 5A). As noted above, FlbD specifically binds to and represses the transcription of the early *fliF* operon, which encodes the earliest-assembled flagellar structure, the MS ring. Cell cycle expression of a mutant promoter that cannot be bound by FlbD showed that there was an overall increase in transcription throughout the cell cycle, as well as a significant delay in the cessation of transcription later in the cell cycle (Wingrove and Gober, 1994). Utilizing the compartmentalized transcription assay described above, it was demonstrated that the wild-type promoter, in contrast to class III flagellar promoters, was expressed in the stalked compartment but not in the swarmer compartment of the predivi-

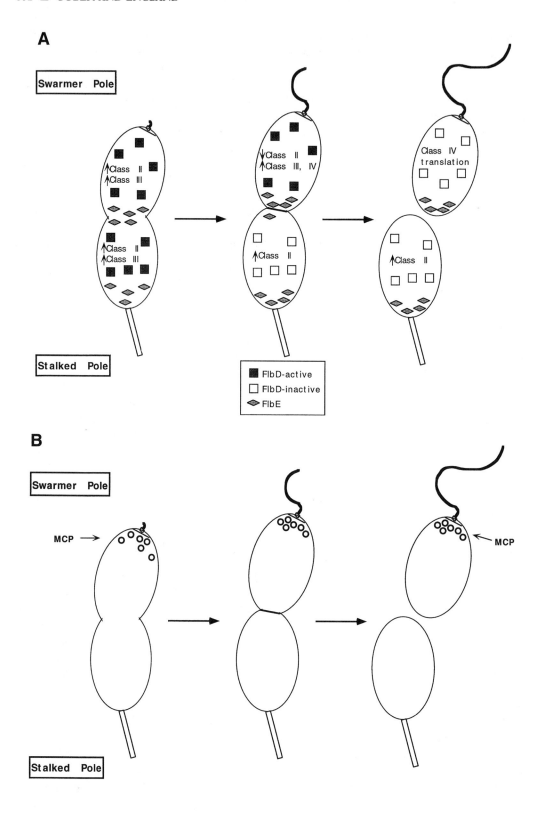

sional cell (Wingrove and Gober, 1994). To determine whether the lack of expression in the swarmer compartment was a consequence of FlbD repression, the mutant promoter which could not be bound by FlbD was also assayed for pole-specific gene expression. Predictably, this promoter was expressed at equal levels in both poles of the predivisional cell, indicating that FlbD functioned to repress transcription in the swarmer pole. Apparently, the asymmetric phosphorylation of FlbD also plays a role in pole-specific repression of the *fliF* promoter. Cells that expressed constitutive FlbD mutants exhibited a marked decrease in *fliF* expression, whereas cells expressing a mutant FlbD that could not be phosphorylated had increased *fliF* expression (Wingrove and Gober, 1994). The simplest interpretation of these results is that the phosphorylation of FlbD regulates binding to the *fliF* promoter. Consistent with this view, it has been shown that the phosphorylation of a related σ^{54} transcriptional activator, NtrC, significantly enhances oligomerization at enhancer sequences (Porter et al., 1993).

Therefore, the asymmetric activation of FlbD accomplishes two distinct programs of gene expression (Fig. 5A). The stalked-cell-specific transcription of the early *fliF* operon results in the initiation of the flagellar cascade in the progeny stalked cell. Activation of late flagellar genes in the swarmer compartment of the predivisional cell ensures that the nascent flagellum is supplied with the precursors for complete assembly. Swarmer pole-specific gene expression has special significance for flagellar filament biogenesis. The 25-kDa flagellin, encoded by *fljK*, is the most abundant polypeptide in the flagellar filament (Driks et al., 1989). Following cell division, filament growth is incomplete. Therefore, newly synthesized flagellin must be incorporated into the filament in the progeny swarmer cell. However, the transcription of *fljK* is rapidly shut off following cell division. This is presumably a consequence of the fact that FlbD requires an early cell division event for activation (see above). *C. crescentus* addresses this dilemma, the need for a supply of flagellin monomers at a time when the flagellin gene is no longer transcribed, through the combined effect of compartmentalized transcription and unusually stable *fljK* mRNA. Compartmentalized transcription of *fljK* essentially supplies the swarmer compartment, and eventually, by virtue of the long *fljK* mRNA half-life, the progeny swarmer cell with a sufficient supply of *fljK* mRNA to be translated into flagellin and assembled into the growing filament.

FIGURE 5 Polarity is established in the *C. crescentus* predivisional cell as a result of compartmentalized gene expression or protein targeting. The schematic diagrams of a portion of the *C. crescentus* cell cycle show the early and late predivisional cells and the resulting progeny cells. (A) Relative expression of flagellar genes in the two poles and progeny cells. Transcription of the late flagellar genes (classes III and IV) exclusively in the swarmer pole of the predivisional cell is due to the swarmer pole-specific activation of the transcriptional activator, FlbD. Following the biogenesis of the cell division plane, FlbD also negatively regulates the class II *fliF* promoter in the swarmer pole, which is therefore only transcribed in the stalked pole, where FlbD is inactive. Because late class IV flagellin genes are only transcribed in the swarmer pole of the predivisional cell after the cell division plane is intact, the long-lived flagellin mRNA is segregated to the swarmer progeny cell, ensuring that the still-growing flagellar filament will have adequate flagellin precursors. FlbE is also necessary for the correct pole-specific program of flagellar gene expression and has been found to localize to the stalked pole and to the cell division plane in the swarmer compartment. The protein, with regions of homology to known sensor histidine kinases, is able to transfer phosphate groups to FlbD in vitro. This evidence, and the correlation between FlbE localization and FlbD activation in the swarmer pole, suggests the attractive possibility that FlbE, triggered by its localization to structures at the cell division plane, maintains FlbD activation by phosphorylation. FlbE localized to the stalked pole is apparently unable to phosphorylate FlbD. (B) Establishment of MCP localization. MCP is synthesized in the predivisional cell and targeted to the swarmer pole. The MCP in the swarmer cell is later degraded at the swarmer-to-stalked-cell transition. Regions of the carboxyl terminus of the protein direct both polar localization and proteolytic degradation.

Protein Targeting

The newly synthesized flagellar components and chemotaxis apparatus are targeted to the swarmer pole of the predivisional cell (Agabian et al., 1979; Gomes and Shapiro, 1984; Nathan et al., 1986; Alley et al., 1992). Although little is known about the mechanisms which govern the positioning of flagellar components, the targeting of the chemotaxis proteins has been examined in some detail. The signal-transducing chemotaxis machinery in *C. crescentus* is similar to that of other bacteria and consists of a methyl-accepting chemotaxis receptor (MCP), an integral membrane protein, and cytoplasmic components which transduce chemotactic signals to the flagellar motor (Alley et al., 1991). The chemotaxis genes (*che*) are expressed under cell cycle control, with peak expression occurring in predivisional cells. Following expression, the components of the chemotaxis machinery are targeted to the swarmer pole (Gomes and Shapiro, 1984; Nathan et al., 1986; Alley et al., 1992). This was initially demonstrated by determining the cell type distribution of methyltransferase activity. In this experiment, the methyltransferase activity, and by analogy the chemotaxis machinery, specifically segregated to progeny swarmer cells (Gomes and Shapiro, 1984; Nathan et al., 1986). This segregation is attributable to protein localization and not compartmentalized gene expression, since the *che* genes are transcribed in the predivisional cell well before a barrier (cell division plane) has formed between the swarmer and stalked compartments. In an experiment similar to that described for assaying compartmentalized gene expression, it was subsequently found that newly synthesized MCP segregated to progeny swarmer cells (Alley et al., 1992). The MCP in the swarmer cell is then specifically degraded upon differentiation into a stalked cell.

The subcellular localization of the MCP has been assayed by both immunofluorescence and immunogold electron microscopy (Alley et al., 1992). The MCP was found to localize at a discrete patch at the swarmer poles of predivisional cells. Progeny swarmer cells were also found to possess a single focus of MCP at the pole. Deletion analysis demonstrated that a highly conserved carboxyl-terminal domain was required for polar localization (Alley et al., 1993). A version of MCP with deletions, which lacked the entire carboxyl terminus, which lies in the cytoplasm, failed to localize to the cell pole. This suggests that interaction with a cytoplasmic protein(s) is required for polar localization. *Escherichia coli* cells were also found to possess polarly localized MCP (Maddock and Shapiro, 1993). In this case localization was dependent on two cytoplasmic chemotaxis components, CheA and CheW, reinforcing the idea that the carboxyl terminus of MCP must interact with cytoplasmic factors in order to be polarly localized. Unlike that of *C. crescentus*, the *E. coli* MCP was found to be localized at both poles of the predivisional cell. Deletion of the 14 carboxyl-terminal amino acids of *C. crescentus* MCP resulted in bipolar localization strikingly similar to that observed in *E. coli* (Alley et al., 1993). The deletion of this region was found to prevent the proteolytic turnover of MCP when swarmer cells differentiated into stalked cells. This truncated protein apparently is also not degraded in the stalked pole of the predivisional cell. These experiments suggest that polar localization of MCP in *C. crescentus* is attributable to two different processes (Alley et al., 1993) (Fig. 5B). First, MCP is synthesized in the predivisional cell, where both poles are competent to localize MCP. Localization to the stalked pole is not detectable in wild-type cells but can be visualized if MCP is overexpressed. In this experiment, the polarly localized MCP at the stalked pole is degraded by a stalked-cell-type-specific protease activity. These experimental observations raise many new questions. How can an integral membrane protein be specifically targeted to a discrete region of cytoplasmic membrane at the cell poles? How is protease activity sequestered to the stalked pole? Careful analysis of MCP localization in *C. crescentus* should provide general models for protein localization and the determination of cell fate in many other organisms.

Little is known about the polar localization of flagellar components. Since the MS ring is the first structure assembled, it is possible that the polar localization signals for the entire flagellum lie within the FliF protein. The cellular distribution of FliF was found to parallel that of MCP. The FliF that is present in swarmer cells is degraded upon differentiation into a stalked cell (Jenal and Shapiro, 1996). The protein is then synthesized in the predivisional cell and eventually segregates to the progeny swarmer cell. Therefore, a temporally regulated stalked-cell protease is responsible for the cell-type-specific distribution of FliF. In contrast to MCP, mutant derivatives of FliF that were resistant to degradation also segregated to progeny swarmer cells (Jenal and Shapiro, 1996). This observation suggests that the mechanism responsible for the asymmetric distribution of FliF differs significantly from that controlling MCP localization.

SUMMARY

C. crescentus utilizes several distinct and overlapping regulatory mechanisms to ensure that the biogenesis of a single polar flagellum is coupled to the formation of the progeny swarmer cell. These include cell cycle-regulated transcription, flagellar assembly-regulated gene expression, cell-type-specific gene expression, and proteolysis. The aggregate effect of these overlapping genetic networks ensures that the timing of expression and assembly of the relatively complex flagellum is coordinated with the *C. crescentus* cell cycle. Thus, the examination of flagellar biogenesis in *C. crescentus* has provided us with a tractable experimental system to understand cellular differentiation in microorganisms. Further experiments that define the mechanisms of these basic developmental processes should not only yield insights into how *C. crescentus* forms a differentiated daughter cell but provide a paradigm for cell differentiation and development in other organisms as well.

ACKNOWLEDGMENTS

Work in our laboratory is supported by grants from the National Institutes of Health (GM48417) and the National Science Foundation (MCB-9513222).

REFERENCES

Agabian, N., M. Evinger, and G. Parker. 1979. Generation of asymmetry during development. Segregation of type-specific proteins in *Caulobacter*. *J. Cell Biol.* **81:**123–136.

Alley, M. R. K., S. L. Gomes, W. Alexander, and L. Shapiro. 1991. Genetic analysis of a temporally transcribed chemotaxis gene cluster in *Caulobacter crescentus*. *Genetics* **129:**333–342.

Alley, M. R. K., J. Maddock, and L. Shapiro. 1992. Polar localization of a bacterial chemoreceptor. *Genes Dev.* **6:**825–836.

Alley, M. R. K., J. Maddock, and L. Shapiro. 1993. Requirement of the carboxyl terminus of a bacterial chemoreceptor for its targeted proteolysis. *Science* **259:**1754–1757.

Anderson, D. K., and A. Newton. 1997. Posttranscriptional regulation of *Caulobacter* flagellin genes by a late flagellum assembly checkpoint. *J. Bacteriol.* **179:**2281–2288.

Anderson, D. K., N. Ohta, J. Wu, and A. Newton. 1995. Regulation of the *Caulobacter crescentus rpoN* gene and function of the purified sigma 54 in flagellar gene transcription. *Mol. Gen. Genet.* **246:**697–706.

Anderson, P., and J. W. Gober. 1998. Abstr. H-112, p. 295. *Abstracts of the 98th Annual Meeting of the American Society for Microbiology 1998*.

Benson, A. K., G. Ramakrishnan, N. Ohta, J. Feng, A. J. Ninfa, and A. Newton. 1994a. The *Caulobacter crescentus* FlbD protein acts at ftr sequence elements both to activate and to repress transcription of cell cycle-regulated flagellar genes. *Proc. Natl. Acad. Sci. USA* **91:**2369–2373.

Benson, A. K., J. Wu, and A. Newton. 1994b. The role of FlbD in regulation of flagellar gene transcription in *Caulobacter crescentus*. *Res. Microbiol.* **145:**420–430.

Boyd, C., and J. W. Gober. 1999. Unpublished data.

Brun, Y. V., and L. Shapiro. 1992. A temporally controlled sigma-factor is required for polar morphogenesis and normal cell division in *Caulobacter*. *Genes Dev.* **6:**2395–2408.

Brun, Y. V., G. Marczynski, and L. Shapiro. 1994. The expression of asymmetry during *Caulobacter* cell differentiation. *Annu. Rev. Biochem.* **63:**419–450.

Bryan, R., M. Purucker, S. L. Gomes, W. Alexander, and L. Shapiro. 1984. Analysis of the pleiotropic regulation of flagellar and chemotaxis gene expression in *Caulobacter crescentus* by using plasmid complementation. *Proc. Natl. Acad. Sci. USA* **81:**1341–1345.

Champer, R., A. Dingwall, and L. Shapiro. 1987. Cascade regulation of *Caulobacter* flagellar and chemotaxis genes. *J. Mol. Biol.* **194:**71–80.

Chen, L. S., D. Mullen, and A. Newton. 1986.

Identification, nucleotide sequence, and control of developmentally regulated promoters in the hook operon region of *Caulobacter crescentus*. *Proc. Natl. Acad. Sci. USA* **83:**2860–2864.

Dingwall, A., J. W. Gober, and L. Shapiro. 1990. Identification of a *Caulobacter* basal body structural gene and a *cis*-acting site required for activation of transcription. *J. Bacteriol.* **172:**6066–6076.

Dingwall, A., J. D. Garman, and L. Shapiro. 1992a. Organization and ordered expression of *Caulobacter* genes encoding flagellar basal body rod and ring proteins. *J. Mol. Biol.* **228:**1147–1162.

Dingwall, A., W. Zhuang, K. Quon, and L. Shapiro. 1992b. Expression of an early gene in the flagellar regulatory hierarchy is sensitive to an interruption in DNA replication. *J. Bacteriol.* **174:**1760–1768.

Domian, I. J., K. C. Quon, and L. Shapiro. 1997. Cell type-specific phosphorylation and proteolysis of a transcriptional regulator controls the G1-to-S transition in a bacterial cell cycle. *Cell* **90:**415–424.

Driks, A., R. Bryan, L. Shapiro, and D. J. DeRosier. 1989. The organization of the *Caulobacter crescentus* flagellar filament. *J. Mol. Biol.* **206:**627–636.

Ely, B., and C. J. Gerardot. 1988. Use of pulsed-field-gradient gel electrophoresis to construct a physical map of the *Caulobacter crescentus* genome. *Gene* **68:**323–333.

Ely, B., and T. W. Ely. 1989. Use of pulsed field gel electrophoresis and transposon mutagenesis to estimate the minimal number of genes required for motility in *Caulobacter crescentus*. *Genetics* **123:**649–654.

Ely, B., and G. Leclerc. 1998. The *Caulobacter crescentus fljMNO* flagellin genes. GenBank Accession no. AF040268.

Ely, B., R. H. Croft, and C. J. Gerardot. 1984. Genetic mapping of genes required for motility in *Caulobacter crescentus*. *Genetics* **108:**523–532.

Fukuda, A., K. Miyakawa, H. Iida, and Y. Okada. 1976. Regulation of polar surface structures in *Caulobacter crescentus*: pleiotropic mutations affect the coordinate morphogenesis of flagella, pili and phage receptors. *Mol. Gen. Genet.* **149:**167–173.

Gill, P. R., and N. Agabian. 1982. A comparative structural analysis of the flagellin monomers of *Caulobacter crescentus* indicates that these proteins are encoded by two genes. *J. Bacteriol.* **150:**925–933.

Gill, P. R., and N. Agabian. 1983. The nucleotide sequence of the M_r = 28,500 flagellin gene of *Caulobacter crescentus*. *J. Biol. Chem.* **258:**7395–7401.

Gillen, K. L., and K. T. Hughes. 1991a. Molecular characterization of *flgM*, a gene encoding a negative regulator of flagellin synthesis in *Salmonella typhimurium*. *J. Bacteriol.* **173:**6453–6459.

Gillen, K. L., and K. T. Hughes. 1991b. Negative regulatory loci coupling flagellin synthesis to flagellar assembly in *Salmonella typhimurium*. *J. Bacteriol.* **173:**2301–2310.

Gober, J. W., and M. Marques. 1995. Regulation of cellular differentiation in *Caulobacter crescentus*. *Microbiol. Rev.* **59:**31–47.

Gober, J. W., and L. Shapiro. 1990. Integration host factor is required for the activation of developmentally regulated genes in *Caulobacter*. *Genes Dev.* **4:**1494–1504.

Gober, J. W., and L. Shapiro. 1992. A developmentally regulated *Caulobacter* flagellar promoter is activated by 3′ enhancer and IHF binding elements. *Mol. Biol. Cell.* **3:**913–926.

Gober, J. W., R. Champer, S. Reuter, and L. Shapiro. 1991a. Expression of positional information during cell differentiation of *Caulobacter*. *Cell* **64:**381–391.

Gober, J. W., H. Xu, A. K. Dingwall, and L. Shapiro. 1991b. Identification of cis and trans-elements involved in the timed control of a *Caulobacter* flagellar gene. *J. Mol. Biol.* **217:**247–257.

Gober, J. W., C. H. Boyd, M. Jarvis, E. K. Mangan, M. F. Rizzo, and J. A. Wingrove. 1995. Temporal and spatial regulation of *fliP*, an early flagellar gene of *Caulobacter crescentus* that is required for motility and normal cell division. *J. Bacteriol.* **177:**3656–3667.

Gomes, S. L., and L. Shapiro. 1984. Differential expression and positioning of chemotaxis methylation proteins in *Caulobacter*. *J. Mol. Biol.* **178:**551–568.

Hahnenberger, K. M., and L. Shapiro. 1987. Identification of a gene cluster involved in flagellar basal body biogenesis in *Caulobacter crescentus*. *J. Mol. Biol.* **194:**91–103.

Hahnenberger, K. M., and L. Shapiro. 1988. Organization and temporal expression of a flagellar basal body gene in *Caulobacter crescentus*. *J. Bacteriol.* **170:**4119–4124.

Hoover, T. R., E. Santero, S. Porter, and S. Kustu. 1990. The integration host factor (IHF) stimulates interaction of RNA polymerase with NifA, the transcriptional activator for nitrogen fixation operons. *Cell* **63:**11–21.

Hueck, C. J. 1998. Type III protein secretion systems in bacterial pathogens of animals and plants. *Microbiol. Mol. Biol. Rev.* **62:**379–433.

Hughes, K. T., K. L. Gillen, M. J. Semon, and J. E. Karlinsey. 1993. Sensing structural intermediates in bacterial flagellar assembly by export of a negative regulator. *Science* **262:**1277–1280.

Jenal, U., and L. Shapiro. 1996. Cell cycle-controlled proteolysis of a flagellar motor protein that is asymmetrically distributed in the *Caulobacter* predivisional cell. *EMBO J.* **15:**2393–2406.

Jenal, U., J. White, and L. Shapiro. 1994. *Caulobacter* flagellar function but not assembly requires FliL: a non-polarly localized membrane protein present in all cell types. *J. Mol. Biol.* **243:** 227–244.

Johnson, R. C., and B. Ely. 1977. Isolation of spontaneously derived mutants of *Caulobacter crescentus.* *Genetics* **86:**25–32.

Johnson, R. C., and B. Ely. 1979. Analysis of nonmotile mutants of the dimorphic bacterium *Caulobacter crescentus. J. Bacteriol.* **137:**627–634.

Johnson, R. C., M. P. Walsh, B. Ely, and L. Shapiro. 1979. Flagellar hook and basal complex of *Caulobacter crescentus. J. Bacteriol.* **138:**984–989.

Johnson, R. C., D. M. Ferber, and B. Ely. 1983. Synthesis and assembly of flagellar components by *Caulobacter crescentus* motility mutants. *J. Bacteriol.* **154:**1137–1144.

Kelly, A. J., M. J. Sackett, N. Din, E. Quardokus, and Y. V. Brun. 1998. Cell cycle-dependent transcriptional and proteolytic regulation of FtsZ in *Caulobacter. Genes Dev.* **12:**880–893.

Khambaty, F. M., and B. Ely. 1992. Molecular genetics of the *flgI* region and its role in flagellum biosynthesis in *Caulobacter crescentus. J. Bacteriol.* **174:**4101–4109.

Komeda, Y. 1986. Transcriptional control of flagellar genes in *Escherichia coli* K-12. *J. Bacteriol.* **168:** 1315–1318.

Kornacker, M. G., and A. Newton. 1994. Information essential for cell-cycle-dependent secretion of the 591-residue *Caulobacter* hook protein is confined to a 21-amino-acid sequence near the N-terminus. *Mol. Microbiol.* **14:**73–85.

Kustu, S., E. Santero, J. Keener, D. Popham, and D. S. Weiss. 1989. Expression of σ^{54} (*ntrA*)-dependent genes is probably united by a common mechanism. *Microbiol. Rev.* **53:**367–376.

Kutsukake, K., Y. Ohya, and T. Iino. 1990. Transcriptional analysis of the flagellar regulon of *Salmonella typhimurium. J. Bacteriol.* **172:**741–747.

Lagenaur, C., and N. Agabian. 1976. Physical characterization of *Caulobacter crescentus* flagella. *J. Bacteriol.* **128:**435–444.

Lagenaur, C., and N. Agabian. 1978. *Caulobacter* flagellar organelle: synthesis, compartmentation, and assembly. *J. Bacteriol.* **135:**1062–1069.

Lagenaur, C., M. De Martini, and N. Agabian. 1978. Isolation and characterization of *Caulobacter crescentus* hooks. *J. Bacteriol.* **136:**795–798.

Landy, A. 1989. Dynamic, structural and regulatory aspects of λ site specific recombination. *Annu. Rev. Biochem.* **38:**913–949.

Leclerc, G., S. P. Wang, and B. Ely. 1998. A new class of *Caulobacter crescentus* flagellar genes. *J. Bacteriol.* **180:**5010–5019.

Lee, F., and B. Ely. 1996. *Caulobacter crescentus* membrane protein homolog (*podW*) gene. GenBank Accession no. U42203.

Maddock, J. R., and L. Shapiro. 1993. Polar localization of the chemoreceptor complex in the *Escherichia coli* cell. *Science* **259:**1717–1723.

Mangan, E., M. Bartamian, and J. W. Gober. 1995. A mutation that uncouples flagellum assembly from transcription alters the temporal pattern of flagellar gene expression in *Caulobacter crescentus. J. Bacteriol.* **177:**3176–3184.

Mangan, E., J. Malakooti, A. Caballero, P. E. Anderson, B. Ely, and J. W. Gober. 1999. Unpublished data.

Marczynski, G. T., K. Lentine, and L. Shapiro. 1995. A developmentally regulated chromosomal origin of replication uses essential transcription elements. *Genes Dev.* **9:**1543–1557.

Marques, M. V., and J. W. Gober. 1995. Activation of a temporally regulated Caulobacter promoter by upstream and downstream sequence elements. *Mol. Microbiol.* **16:**279–289.

Milhausen, M., and N. Agabian. 1983. *Caulobacter* flagellin mRNA segregates asymmetrically at cell division. *Nature* **302:**630–632.

Minnich, S. A., and A. Newton. 1987. Promoter mapping and cell cycle regulation of flagellin gene transcription in *Caulobacter crescentus. Proc. Natl. Acad. Sci. USA* **84:**1142–1146.

Minnich, S. A., N. Ohta, N. Taylor, and A. Newton. 1988. Role of the 25-, 27-, and 29-kilodalton flagellins in *Caulobacter crescentus* cell motility: method for construction of deletion and Tn5 insertion mutants by gene replacement. *J. Bacteriol.* **170:** 3953–3960.

Mohr, C. D., U. Jenal, and L. Shapiro. 1996. Flagellar assembly in *Caulobacter crescentus:* a basal body P-ring null mutation affects stability of the L-ring protein. *J. Bacteriol.* **178:**675–682.

Mohr, C. D., J. K. MacKichan, and L. Shapiro. 1998. A membrane-associated protein, FliX, is required for an early step in *Caulobacter* flagellar assembly. *J. Bacteriol.* **180:**2175–2185.

Muir, R. E., and J. W. Gober. 1999. Unpublished data.

Mullin, D., S. Minnich, L. S. Chen, and A. Newton. 1987. A set of positively regulated flagellar gene promoters in *Caulobacter crescentus* with sequence homology to the *nif* gene promoters of *Klebsiella pneumoniae. J. Mol. Biol.* **195:**939–943.

Mullin, D. A., and A. Newton. 1989. Ntr-like promoters and upstream regulatory sequence ftr are required for transcription of a developmentally regulated *Caulobacter crescentus* flagellar gene. *J. Bacteriol.* **171:**3218–3227.

Mullin, D. A., and A. Newton. 1993. A sigma 54 promoter and downstream sequence elements ftr2 and ftr3 are required for regulated expression of divergent transcription units *flaN* and *flbG* in *Caulobacter crescentus. J. Bacteriol.* **175**:2067–2076.

Mullin, D. A., S. M. Van Way, C. A. Blackenship, and A. H. Mullin. 1994. FlbD has a DNA binding activity near its carboxy terminus that recognizes ftr sequences involved in positive and negative regulation of flagellar gene transcription in *Caulobacter crescentus. J. Bacteriol.* **176**:5971–5981.

Mullin, D. A., A. H. Mullin, and A. Newton. 1998. Organization and expression of Caulobacter crescentus genes needed for assembly and function of the flagellar hook. GenBank Accession no. AF072135.

Nathan, P., S. L. Gomes, K. Hahnenberger, A. Newton, and L. Shapiro. 1986. Differential localization of membrane receptor chemotaxis proteins in the *Caulobacter* predivisional cell. *J. Mol. Biol.* **191**:433–440.

Newton, A., N. Ohta, G. Ramakrishnan, D. Mullin, and G. Raymond. 1989. Genetic switching in the flagellar gene hierarchy of *Caulobacter* requires negative as well as positive regulation of transcription. *Proc. Natl. Acad. Sci. USA* **86**:6651–6655.

Ninfa, A. J., D. A. Mullin, G. Ramakrishnan, and A. Newton. 1989. *Escherichia coli* sigma 54 RNA polymerase recognizes *Caulobacter crescentus flbG* and *flaN* flagellar gene promoters in vitro. *J. Bacteriol.* **171**:383–391.

Ohta, N., L. S. Chen, and A. Newton. 1982. Isolation and expression of cloned hook protein gene from *Caulobacter crescentus. Proc. Natl. Acad. Sci. USA* **79**:4863–4867.

Ohta, N., E. Swanson, B. Ely, and A. Newton. 1984. Physical mapping and complementation analysis of transposon Tn*5* mutations in *Caulobacter crescentus:* organization of transcriptional units in the hook gene cluster. *J. Bacteriol.* **158**:897–904.

Ohta, N., L. S. Chen, E. Swanson, and A. Newton. 1985. Transcriptional regulation of a periodically controlled flagellar gene operon in *Caulobacter crescentus. J. Mol. Biol.* **186**:107–115.

Ohta, N., L. S. Chen, D. A. Mullin, and A. Newton. 1991. Timing of flagellar gene expression in the *Caulobacter* cell cycle is determined by a transcriptional cascade of positive regulatory genes. *J. Bacteriol.* **173**:1514–1522.

Onishi, K., K. Kutsukake, H. Suzuki, and T. Iino. 1992. A novel transcriptional regulatory mechanism in the flagellar regulon of *Salmonella typhimurium:* an anti-sigma factor inhibits the activity of the flagellum-specific sigma factor, σ^F. *Mol. Microbiol.* **6**:3149–3157.

Osley, M. A., M. Sheffery, and A. Newton. 1977. Regulation of flagellin synthesis in the cell cycle of *Caulobacter:* dependence on DNA replication. *Cell* **12**:393–400.

Parkinson, J. S., and E. C. Kofoid. 1992. Commu-nication modules in bacterial signaling proteins. *Annu. Rev. Genet.* **26**:71–112.

Popham, D. L., D. Szeto, J. Keener, and S. Kustu. 1989. Function of a bacterial activation protein that binds to transcriptional enhancers. *Science* **243**:629–635.

Porter, S. C., A. K. North, A. B. Wedel, and S. Kustu. 1993. Oligomerization of NTRC at the *glnA* enhancer is required for transcriptional activation. *Genes Dev.* **7**:2258–2273.

Quon, K. C., G. T. Marczynski, and L. Shapiro. 1996. Cell cycle control by an essential bacterial two-component signal transduction protein. *Cell* **84**:83–93.

Quon, K. C., B. Yang, I. J. Domian, L. Shapiro, and G. T. Marczynski. 1998. Negative control of bacterial DNA replication by a cell cycle regulatory protein that binds at the chromosome origin. *Proc. Natl. Acad. Sci. USA* **95**:120–125.

Ramakrishnan, G., and A. Newton. 1990. FlbD of *Caulobacter crescentus* is a homologue of the NtrC (NRI) protein and activates sigma 54-dependent flagellar gene promoters. *Proc. Natl. Acad. Sci. USA* **87**:2369–2373.

Ramakrishnan, G., J. L. Zhao, and A. Newton. 1991. The cell cycle-regulated flagellar gene *flbF* of *Caulobacter crescentus* is homologous to a virulence locus (*lcrD*) of *Yersinia pestis. J. Bacteriol.* **173**: 7283–7292.

Ramakrishnan, G., J. L. Zhao, and A. Newton. 1994. Multiple structural proteins are required for both transcriptional activation and negative autoregulation of *Caulobacter crescentus* flagellar genes. *J. Bacteriol.* **176**:7587–7600.

Sanders, L. A., S. Van Way, and D. A. Mullin. 1992. Characterization of the *Caulobacter crescentus flbF* promoter and identification of the inferred *flbF* product as a homolog of the LcrD protein from a *Yersinia enterocolitica* virulence plasmid. *J. Bacteriol.* **174**:857–866.

Santero, E., T. R. Hoover, A. K. North, D. K. Berger, S. C. Porter, and S. Kustu. 1992. Role of integration host factor in stimulating transcription from the sigma 54-dependent *nifH* promoter. *J. Mol. Biol.* **227**:602–620.

Schoenlein, P. V., and B. Ely. 1989. Characterization of strains containing mutations in the contiguous *flaF*, *flbT*, or *flbA*-*flaG* transcription unit and identification of a novel Fla phenotype in *Caulobacter crescentus. J. Bacteriol.* **171**:1554–1561.

Schoenlein, P. V., J. Lui, L. Gallman, and B. Ely. 1992. The *Caulobacter crescentus flaFG* region regulates synthesis and assembly of flagellin proteins encoded by two genetically unlinked gene clusters. *J. Bacteriol.* **174**:6046–6053.

Sheffery, M., and A. Newton. 1979. Purification

and characterization of a polyhook protein from *Caulobacter crescentus*. *J. Bacteriol.* **138:**575–583.

Sheffery, M., and A. Newton. 1981. Regulation of periodic protein synthesis in the cell cycle: control of initiation and termination of flagellar gene expression. *Cell* **24:**49–57.

Stallmeyer, M. J., K. M. Hahnenberger, G. E. Sosinsky, L. Shapiro, and D. J. DeRosier. 1989. Image reconstruction of the flagellar basal body of *Caulobacter crescentus*. *J. Mol. Biol.* **205:** 511–518.

Stephens, C., A. Reisenauer, R. Wright, and L. Shapiro. 1996. A cell cycle-regulated bacterial DNA methyltransferase is essential for viability. *Proc. Natl. Acad. Sci. USA* **93:**1210–1214.

Stephens, C., C. Mohr, C. Boyd, J. Maddock, J. Gober, and L. Shapiro. 1997. Identification of the *fliI* and *fliJ* components of the *Caulobacter* flagellar type III protein secretion system. *J. Bacteriol.* **179:**5355–5365.

Stephens, C. M., and L. Shapiro. 1993. An unusual promoter controls cell-cycle regulation and dependence on DNA replication of the *Caulobacter fliLM* early flagellar operon. *Mol. Microbiol.* **9:**1169–1179.

Trachtenberg, S., and D. J. DeRosier. 1988. Three-dimensional reconstruction of the flagellar filament of *Caulobacter crescentus*. A flagellin lacking the outer domain and its amino sequence lacking an internal segment. *J. Mol. Biol.* **202:**787–808.

Van Way, S. M., A. Newton, A. H. Mullin, and D. A. Mullin. 1993. Identification of the promoter and a negative regulatory element, ftr4, that is needed for cell cycle timing of *fliF* operon expression in *Caulobacter crescentus*. *J. Bacteriol.* **175:** 367–376.

Wagenknect, T., D. DeRosier, L. Shapiro, and A. Weissborn. 1981. Three-dimensional reconstruction of the flagellar hook from *Caulobacter crescentus*. *J. Mol. Biol.* **151:**439–465.

Weissborn, A., H. M. Steinman, and L. Shapiro.

1982. Characterization of the proteins of the *Caulobacter crescentus* flagellar filament. Peptide analysis and filament organization. *J. Biol. Chem.* **257:** 2066–2074.

Wingrove, J. A., and J. W. Gober. 1994. A σ^{54} transcriptional activator also functions as a pole-specific repressor in *Caulobacter*. *Genes Dev.* **8:** 1839–1852.

Wingrove, J. A., and J. W. Gober. 1996. Identification of an asymmetrically localized sensor histidine kinase responsible for temporally and spatially regulated transcription. *Science* **274:**597–601.

Wingrove, J. A., E. K. Mangan, and J. W. Gober. 1993. Spatial and temporal phosphorylation of a transcriptional activator regulates pole-specific gene expression in *Caulobacter*. *Genes Dev.* **7:**1979–1992.

Wu, J., and A. Newton. 1997. Regulation of the *Caulobacter* flagellar gene hierarchy; not just for motility. *Mol. Microbiol.* **24:**233–239.

Wu, J., A. K. Benson, and A. Newton. 1995. Global regulation of a sigma 54-dependent flagellar gene family in *Caulobacter crescentus* by the transcriptional activator FlbD. *J. Bacteriol.* **177:**3241–3250.

Wu, J., N. Ohta, and A. Newton. 1998. An essential, multicomponent signal transduction pathway required for cell cycle regulation in *Caulobacter*. *Proc. Natl. Acad. Sci. USA* **95:**1443–1448.

Xu, H., A. Dingwall, and L. Shapiro. 1989. Negative transcriptional regulation in the *Caulobacter* flagellar hierarchy. *Proc. Natl. Acad. Sci. USA* **86:** 6656–6660.

Yu, J., and L. Shapiro. 1992. Early *Caulobacter crescentus* genes *fliL* and *fliM* are required for flagellar gene expression and normal cell division. *J. Bacteriol.* **174:**3327–3338.

Zhuang, W. Y., and L. Shapiro. 1995. *Caulobacter* FliQ and FliR membrane proteins, required for flagellar biogenesis and cell division, belong to a family of virulence factor export proteins. *J. Bacteriol.* **177:** 343–356.

SIGNAL TRANSDUCTION AND CELL CYCLE CHECKPOINTS IN DEVELOPMENTAL REGULATION OF *CAULOBACTER*

Noriko Ohta, Thorsten W. Grebe, and Austin Newton

17

Several prokaryotic organisms have been successfully employed to study the genetic, molecular, and biochemical mechanisms responsible for the temporal and spatial patterns of cell differentiation, as reviewed in this volume. A feature shared by many, if not all, of these organisms is the generation of new cell types by asymmetric cell division (Horvitz and Herskowitz, 1992). In the gram-negative bacterium *Caulobacter crescentus*, differentiation occurs by the repeated division of the nonmotile "mother" stalked cell (A) to produce the A cell plus a new, motile swarmer cell (B) in an A ⇒ A + B pattern of division characteristic of stem cell differentiation (Newton, 1984).

The pattern of DNA segregation to the progeny stalked and swarmer cells indicates that they inherit identical genomes (Osley and Newton, 1974; Marczynski et al., 1990); however, they express different genetic programs. During the stalked-cell cycle, the flagellum, adhesive holdfast, DNA phage receptor sites, and other differentiated structures required to build a new swarmer cell are elaborated in a tightly ordered temporal sequence and targeted to the stalk-distal pole of the dividing cell. The

flagellum begins to rotate immediately before division, the cell gains motility, and the motile swarmer cell is released at cell separation (Fig. 1). The new swarmer cell, by contrast, must undergo an additional sequence of developmental events before it can enter the cell division cycle. These events include pillus formation, loss of motility when the flagellum is ejected into the medium, retraction of the pili, and finally, formation of a membranous stalk at the same cell pole (Fig. 1).

The two progeny cells also inherit different capacities for chromosome replication. The stalked cell initiates DNA synthesis immediately after cell separation. The newly divided swarmer cell, by contrast, enters a presynthetic gap, or G_1 period, during which the cell prepares to initiate chromosome replication and differentiates into a stalked cell (Fig. 1) (Degnen and Newton, 1972; Marczynski et al., 1990).

What sets *C. crescentus* development apart from many prokaryotic developmental systems is its independence from known environmental cues, including cell-cell interactions and nutritional signals. This contrasts with developmental regulation in many other microbial systems, including sporulation in *Bacillus subtilis* (see chapter 7), aggregation and spore formation in *Myxococcus xanthus* (see chapter 12), and

N. Ohta, T. W. Grebe, and A. Newton, Department of Molecular Biology, Princeton University, Princeton, NJ 08544-1014.

Prokaryotic Development, edited by Y. V. Brun and L. J. Shimkets,
© 2000 American Society for Microbiology, Washington, DC 20005-4171

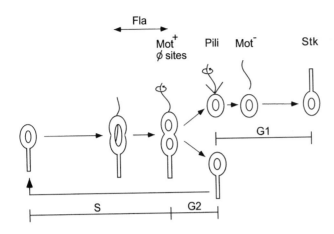

FIGURE 1 *Caulobacter* cell cycle. Developmental events in the wild-type strain CB15 include flagellum formation (Fla), activation of flagellum rotation (Mot$^+$), appearance of polar bacteriophage receptors (ϕ), pilus formation (Pili), flagellum ejection and loss of motility (Mot$^-$), and stalk formation (Stk). The periods corresponding to presynthetic gap (G1), DNA synthesis (S), and postsynthetic gap (G2) are indicated.

heterocyst formation in cyanobacteria (see chapter 4). Morphogenesis in *Caulobacter* appears to be driven instead by internal cues, with stages of the cell division cycle acting as checkpoints for specific developmental events. Genetic studies have identified signal transduction pathways mediated by members of the His-Asp phosphorelay proteins that are essential both for cell cycle control and developmental regulation. As considered in this chapter, these pathways provide a molecular mechanism for the coordination of developmental events with cell cycle progression.

ORGANIZATION OF CELL DIVISION CYCLE GENES

Regulation of the *Caulobacter* cell division cycle has been studied in temperature-sensitive (TS) mutants that grow and divide normally at 30°C but form long filaments at the nonpermissive temperature of 37°C. The strains were classified as either *dna* or *div* mutants, depending on whether DNA replication is blocked at the nonpermissive temperature. *dna* mutants blocked in either DNA chain initiation (DNAi) or DNA chain elongation (DNAe) fail to complete DNA synthesis (DNAc). The *div* mutants replicate DNA at the nonpermissive temperature, but as judged by the degree of pinching at the division site, they are blocked in various stages of cell division: division initiation (DIVi), division progression (DIVp), or

cell separation (CS) (Fig. 2) (Osley and Newton, 1980; Huguenel and Newton, 1982; Newton, 1984).

The organization of *dna* and *div* gene functions has been studied in reciprocal-shift experiments with conditional *dna* or *div* mutants in combination with drugs that reversibly block either DNA synthesis (hydroxyurea) or division (penicillin G) (reviewed in Ohta and Newton, 1996). These studies showed that cell cycle steps are organized into two dependent pathways, a DNA synthetic pathway (DNAi ⇒ DNAe ⇒ DNAc) and a cell division pathway (DIVi ⇒ DIVp [Fig. 2]). The gene function(s) leading to cell separation requires completion of both the DNA synthetic and division pathways.

Several of the *Caulobacter dna* and *div* genes have been identified as homologues of *Escherichia coli* genes (see Fig. 2), and this information provides a biochemical framework for understanding the functional organization of the two dependent pathways. The conditional *dnaC* mutation, which was shown to disrupt DNA chain elongation at the nonpermissive temperature (Osley and Newton, 1977), maps to *holB*. This gene encodes the δ′ subunit of DNA polymerase, and as in *E. coli* (Reynes et al., 1996), the *Caulobacter holB* gene is located immediately downstream of the thymidine kinase (*tmk*) gene (Ohta and Newton, 1999). The *tmk-holB* operon is transcribed early in the G$_1$

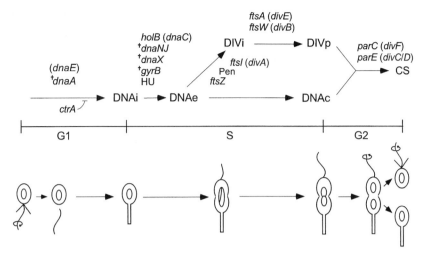

FIGURE 2 Organization of DNA synthetic and cell division pathways. Cell cycle events on the DNA synthetic and cell division pathways are defined in the text. The functional dependency of the gene-mediated steps was determined by reciprocal shift experiments with TS mutations in *dna* or *div* genes (shown in parentheses) in combination with the reversible inhibitor hydroxyurea (HU) or penicillin G (Pen) (Osley and Newton, 1980). †, *Caulobacter* homologues of *E. coli* genes expected to act in initiation of DNA replication and DNA chain elongation (see the text and chapter 18). These genes were not examined in the epistasis experiments. G1, S, G2, see the legend to Fig. 1.

phase (Ohta et al., 1990), when the swarmer cell is preparing for DNA initiation. In addition to Tmk and HolB, other DNA replication proteins identified in *Caulobacter* include DnaX (Winzeler and Shapiro, 1997), DnaN (beta subunit of DNA polymerase [Roberts and Shapiro, 1997]), and GyrB (Rizzo et al., 1993). Although the roles of these genes in replication have not been examined genetically, they display the same early pattern of transcription activation during the late G_1 phase, as observed for the *tmk* operon (see chapter 18).

Two regulatory proteins involved in DNA initiation have been identified. One, the DnaA protein encoded by *dnaA* (Zweiger and Shapiro, 1994), presumably acts as a positive regulator of replication initiation by binding to sites at the *Caulobacter* replication origin (Cori), as does its *E. coli* homologue (Messer and Weigel, 1996). The second protein, CtrA, also binds at Cori, where it acts as a negative regulator of DNA initiation (Quon et al., 1998). The silencing of DNA replication by CtrA in the new

swarmer cell plays a key role in regulating the asymmetric initiation of DNA synthesis in *Caulobacter* (see chapter 18). In addition, the phenotype of the *Caulobacter dnaE* mutation indicates that this gene encodes another positive regulator of DNA initiation (Fig. 2) (Osley and Newton, 1980), although the *dnaE* gene product has not been identified.

In the cell division pathway, the *divA* gene is homologous to *ftsI* (Ohta and Newton, 1999), which encodes a penicillin binding protein (PBP3) required for septation in *E. coli* (Nakamura et al., 1983). Identification of DivA as a PBP3 homologue explains the interdependence of the conditional *divA305* mutation with the penicillin G-sensitive defect (Fig. 2) (Osley and Newton, 1980). Although it might be expected that the FtsI function would be required throughout cell division, the *divA305* allele displays an early execution point and its function is not interdependent with cell division genes required later for division progression (DIVp), namely, *divE* and *divB*. Consistent with this observation, the time at which cells are sensitive to division inhibition by penicillin G is also confined to a short period (the execution point) early in the cell cycle corresponding to DIVi (Fig. 2) (Osley and Newton, 1980).

One of the first genes thought to act in prokaryotic division initiation is *ftsZ*, whose gene product polymerizes into an annular ring at the division site, where it may act as a scaffolding for other division proteins (see chapter 15) (Lutkenhaus and Addinall, 1997). A conditional *ftsZ* mutation has not been reported in *Caulobacter*, but *ftsZ* overexpression leads to hyperconstriction without cell separation (Din et al., 1998). The temperature-sensitive alleles of *divE* and *divB*, which were used to define DIVp, have been mapped to *ftsA* and *ftsW*, respectively (Ohta et al., 1997; Ohta and Newton, 1999). The *E. coli* homologues of these genes are known to act after *ftsZ* in cell division (Donachie, 1993).

Although the *divF* and *divC-divD* mutations were thought to identify late-acting cell division genes required for cell separation (Osley and Newton, 1980), they are now known to map to the *Caulobacter* topoisomerase IV (Topo IV) genes *parC* and *parE*, respectively (Ward and Newton, 1997). Topo IV is required for chromosome segregation in *E. coli*, where it functions to decatenate daughter chromosomes (Zechiedrich and Cozzarelli, 1995). Thus, chromosome decatenation in *Caulobacter* is required for effective cell separation, and the previous genetic analysis (Fig. 2) suggests that Topo IV activity depends on completion of DNA replication and the function of earlier cell division genes, *ftsA* and *ftsW* (Fig. 2). Although *parC* and *parE* mutants of *Caulobacter* form highly pinched filaments at the nonpermissive temperature, they do not divide to form cells lacking DNA (Ward and Newton, 1997), which is typical of *E. coli* and *Salmonella typhimurium* Topo IV mutants (Kato et al., 1988; Schmid, 1990). The abnormal nucleoid segregation pattern of a *parE* mutant can be visualized more clearly in a genetic background with a conditional mutation in the earlier-acting *ftsA* gene. The *parE-ftsA* mutant forms longer fila-

ments than the *parE* mutant, and the cells display the unpinched phenotype characteristic of the *ftsA* mutant (Ward and Newton, 1997). These cell division and growth phenotypes indicate that the *divE309* mutation in *ftsA* is epistatic to the *parE* mutation, which is consistent with the proposed organization of these genes shown in Fig. 2.

CELL CYCLE REGULATION OF POLAR DIFFERENTIATION

Differentiation in *Caulobacter* is tightly coordinated with the cell division cycle. The developmental defects displayed by TS *dna* and *div* mutants blocked at specific stages of the cell cycle suggest that cell cycle progression may provide cues for the sequence of developmental events (reviewed in Ohta and Newton, 1996). Inhibition of either DNA initiation or DNA chain elongation (Fig. 3A) results in filamentous cells that are nonmotile and defective in flagellum assembly (Osley and Newton, 1977). However, when cell division is interrupted in *ftsA* or *ftsW* mutants before completion of division progression DIVp, flagellum biosynthesis proceeds normally but the new flagella assemble at the same stalk-distal pole of the filamenting cells and are paralyzed. These filamentous cells also fail to shed the flagella or to form new stalks (Fig. 3B) (Huguenel and Newton, 1982).

Topo IV mutants blocked at the last stage of division, CS, form highly pinched filamentous cells that are motile, presumably because they complete the DIVp stage of division. They also make stalks and assemble new flagella at the partially completed, internal cell poles (Huguenel and Newton, 1982), but they do not assemble pili (Sommer and Newton, 1988). Thus, as summarized in Fig. 2, successive steps in the cell division cycle appear to act as developmental cues or checkpoints (Huguenel and Newton, 1982; Ohta and Newton, 1996). These checkpoints include (i) DNA synthesis for initiation of flagellum biosynthesis; (ii) completion of DIVp for motility, stalk formation, and formation of new cell poles competent for flagellum assembly; and (iii) comple-

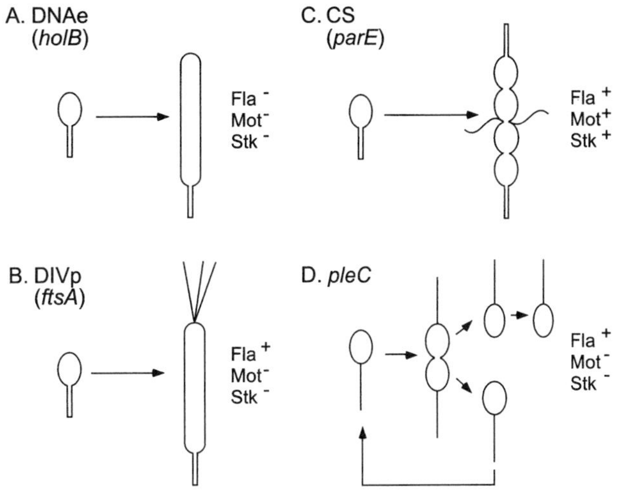

FIGURE 3 Developmental phenotypes of cell division cycle and *pleC* mutant cells. The diagrams depict the cellular morphology of mutations in cell cycle and developmental genes that affect polar morphogenesis. (A) *holB* mutant blocked in DNAe; (B) *ftsA* mutant blocked in DIVp; (C) *parE* mutant blocked in CS; (D) *pleC* mutant blocked in the swarmer-to-stalked-cell transition. The ability or failure of the mutants to execute flagellum biosynthesis (Fla), gain motility (Mot), and make stalks (Stk) is shown by + and −, respectively.

tion of CS for pillus formation. As described below, His-Asp phosphorelay proteins are essential for both developmental and cell cycle regulation.

TWO-COMPONENT SYSTEM PROTEINS IN *CAULOBACTER*

Signal transduction by two-component proteins constitutes a central mechanism in many bacteria, eukaryotic microorganisms, and plants, where they control cellular responses to fluxes in growth conditions, stress, osmolarity, cell density, and other environmental conditions. Mutant screens and inspection of the *Caulobacter* genome have identified a large number of histidine protein kinases (HPKs) and response regulators. In addition to their functions in adaptation to environmental changes, some of these signal transduction proteins also play essential roles in developmental and cell cycle regulation. The sensor kinases contain an input domain and a transmitter domain. The response regulator typically contains a receiver domain and an output domain (Parkinson and Kofoid, 1992). The input domain of the HPK responds to an input signal by autophosphorylation on a conserved histidine residue (H1) in the transmitter or catalytic domain. Transfer of this phosphate to a conserved aspartate (D1) in the receiver domain of the cognate

response regulator changes the activity of the output domain, which may have a variety of functions. These range from enzymatic activity to DNA binding in the case of transcription regulators (Hoch and Silhavy, 1995). Some HPKs also have protein phosphatase activity capable of stimulating dephosphorylation of the phosphorylated response regulator in response to input signals (reviewed in Stock et al., 1995).

A subgroup of response regulators lack output domains and are composed entirely of the receiver domain. Some of these single-domain response regulators, like Spo0F in *B. subtilis*, act as phosphorelay proteins. They transfer phosphate to the conserved histidine (H2) of a histidine phosphotransferase (HPT), which in turn transfers the phosphate to the conserved aspartate (D2) of a second response regulator (see chapter 7) (Burbulys et al., 1991). The H1, D1, H2, and D2 modules of the *B. subtilis* phosphorelay are on separate proteins, but this is not always the case. Hybrid HPKs contain H1 and D1 modules (H1-D1), and some hybrid HPKs also contain an HPT domain, H2 (Appleby et al., 1996). As discussed below, *Caulobacter* appears to contain an unusually large number of H1-D1 hybrid kinases.

Genes encoding two-component signal transduction proteins that regulate *Caulobacter* cell cycle events were identified in a pseudore-version analysis of *pleC* mutations (Sommer and Newton, 1991). Although *pleC* mutants divide normally and assemble flagella, they are similar to *ftsA* and *ftsW* mutants, which are blocked in DIVp, in that they fail to either gain motility, lose flagella, or form stalks (cf. Fig. 3B and D). Thus, both *pleC* and DIVp mutants are effectively blocked in developmental steps required for the swarmer-to-stalked-cell transition (Fig. 1). To explore the possibility that *pleC* and cell division progression regulate motility by a common mechanism, Sommer and Newton (1991) isolated extragenic suppressors of a TS *pleC* mutation that restored motility at 37°C while simultaneously conferring a cell division phenotype at 24°C. These cold-sensitive suppressors were mapped to three new genes, *divK*, *divJ*, and *divL*. An additional suppressor mutation was isolated in backcrosses from one of the pseudorevertant strains and mapped to *pleD*. The *pleD* mutant does not have a cell division defect but is blocked early in the sequence of developmental events required for the swarmer-to-stalked-cell transition. Because these motile cells do not eject

flagella, lose motility, or make stalks, they display a "supermotile" phenotype (Sommer and Newton, 1989).

The four genes identified in this analysis, along with *pleC*, encode homologues of the two-component system proteins (Fig. 4). DivJ (Ohta et al., 1992), DivL (Wu et al., submitted), and PleC (Wang et al., 1993) are HPK homologues with N-terminal hydrophobic regions presumably involved in membrane anchoring. DivL is a novel member of this kinase family in that its function requires phosphorylation of a tyrosine instead of a histidine residue (Wu et al., submitted). DivK is an essential, single-domain response regulator (Hecht et al., 1995). PleD is an unusual response regulator containing two N-terminal receiver domains (D and D′) and a highly conserved C-terminal output domain that contains the so-called GGEEF motif (Fig. 4) (Hecht and Newton, 1995).

CtrA is also an essential response regulator in *Caulobacter* and is similar in the transmitter domain sequence to the *E. coli* OmpR protein (Fig. 4). It was identified by its role in flagellar gene regulation and shown to play multiple

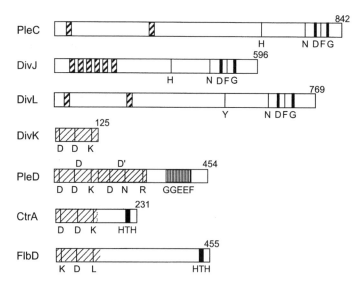

FIGURE 4 Predicted domain organization of *Caulobacter* HPKs and response regulators. The cross-hatched boxes in kinases PleC, DivJ, and DivL indicate predicted transmembrane regions. The conserved sequence motifs in the H, N, D, F, and G boxes of the catalytic domains are shown in Table 1. The receiver domains of response regulators DivK, PleD, CtrA, and FlbD are shown by cross-hatched boxes. Residues corresponding to Asp-57 (D), the presumptive site of phosphorylation, and the conserved Asp-13 (D) and Lys-109 (K) in CheY are indicated for each of the proteins. The helix-turn-helix (HTH) motifs of CtrA and FlbD represent the predicted DNA binding domains. D and D′ (cross-hatched boxes) are the two putative receiver domains, and GGEEF is the carboxy-terminal conserved domain of PleD.

roles in cell cycle regulation (Quon et al., 1996).

SIGNAL TRANSDUCTION PATHWAYS IN CELL CYCLE REGULATION

Flagellum Biosynthesis

Flagellum biosynthesis, like many stage-specific developmental events in *Caulobacter*, is transcriptionally regulated. The 40-odd flagellar genes are transcribed periodically during the cell cycle in an order corresponding to the sequence of gene product assembly (see review by Wu and Newton, 1997, and chapter 16). This complex temporal pattern of gene expression is achieved by the organization of flagellar genes into a hierarchy with four classes of genes (Newton et al., 1989)(I to IV) in which genes at each level must be expressed before genes below them can be expressed (Newton et al., 1989; Xu et al., 1989). Class II genes are near the top of the hierarchy and are transcribed early in the cell cycle. They encode the innermost components of the flagellum, including the MS ring, which is assembled from the *fliF* gene product; the switch complex, or C ring; and proteins making up the type III apparatus for flagellar protein export. The class III genes encode proteins assembled in the outer part of the basal-body–hook complex, and the class IV genes encode the flagellin subunits of the external flagellar filament. Two response regulators play pivotal roles in regulating this transcriptional cascade. CtrA is required to initiate flagellum biosynthesis, and FlbD orchestrates the switch from class II to class III gene transcription.

Class II gene promoters have an unusual -35, -10 architecture that was thought to be recognized by a specialized sigma factor (reviewed in Brun et al., 1994). The long-standing question of how flagellar gene transcription is initiated was answered in large part by the isolation of a mutation in the class I *ctrA* gene (Quon et al., 1996). This TS allele was shown to affect the regulation of the class II *fliQ* promoter at the permissive temperature and to confer a cell division defect at the nonpermissive temperature. CtrA is essential for viability and behaves in vivo as both a negative and positive regulator of class II promoters (Quon et al., 1996). In vitro transcription experiments with RNA polymerase holoenzyme reconstituted from purified *Caulobacter* proteins (Wu et al., 1997) demonstrated that the class II *fliF* promoter is recognized by the principal signal factor σ^{73} (Wu et al., 1998) and that transcription from this promoter is dependent on phosphorylated CtrA. Transcription from the class II *fliL* promoter is also stimulated by the phosphorylated CtrA, but interestingly, *fliL* transcription is inhibited to some extent by unphosphorylated CtrA (Wu et al., 1998). Regulation of CtrA activity results from an intricate pattern of phosphorylation and cell compartment-specific proteolysis during the cell cycle (Domian et al., 1997). HPKs involved in phosphorylation of CtrA are considered in a later section.

Alignment of promoters from other CtrA-regulated genes, including *hemE* (Marczynski et al., 1995), *ccrM* (Stephens et al., 1995), and *divK* (Benson et al., 1999), raises the possibility that they may also be recognized by σ^{73} (Wu et al., 1998). These cell cycle-regulated promoters display some conservation of the σ^{73} consensus at -35, but the spacing between the -10 and -35 elements is 17 to 19 bp, compared to the 10- to 14-bp spacing of the *Caulobacter* σ^{73}-dependent housekeeping promoters (Malakooti et al., 1995). The 5′ TTAAC of the conserved CtrA box (TTAAC N6 TTAAC) overlaps the putative -35 element of the class II flagellar gene promoters, and the 3′ TTAAC element appears to lie within the spacer region (Wu et al., 1998). This unusual promoter architecture, including the novel -10, -35 spacing, may account for the requirement for CtrA in the regulation of class II promoters.

Flagellum biosynthesis is also regulated at the level of DNA replication. Inhibition of DNA synthesis prevents synthesis of the late class III hook protein and the class IV flagellin proteins (Sheffery and Newton, 1981). This regulation is independent of the *recA*-dependent SOS response system (Ohta et al., 1985), and it is executed at the level of class II flagellar gene transcription (Dingwall et al., 1992; Stephens and Shapiro, 1993). Coupling DNA synthesis and transcription is not unique to *Caulobacter*. Inhibition of DNA replication prevents sporulation in *B. subtilis* either by a *recA*-dependent SOS response system (Ireton and Grossman, 1992) or by a *recA*-independent pathway (Ireton and Grossman, 1994). In the bacteriophage T4, Geiduschek and coworkers have elaborated an exquisite mechanism for the regulation of late transcription by DNA replication proteins (Tinker-Kulberg et al., 1996).

FlbD, which is the second response regulator with a crucial role in regulating the flagellar gene hierarchy (Ramakrishnan and Newton, 1990), acts as a transcriptional switch in the transition from class II to class III and IV gene expression. Class III and IV genes are transcribed from σ54-dependent promoters (Mullin and Newton, 1989; Ninfa et al., 1989). Their expression in vivo and in vitro depends on RNA polymerase containing the specialized factor σ54 (Brun and Shapiro, 1992; Anderson et al., 1995) and FlbD (Wu et al., 1995). This response regulator contains a conserved functional domain similar to those in NtrC and other transcriptional activators of σ54 promoters (Kustu et al., 1989). FlbD is also required for the negative autoregulation of the class II *fliF* operon (VanWay et al., 1993; Benson et al., 1994a; Wingrove and Gober, 1994), whose expression is initiated earlier in the cell cycle. Both the positive regulation of class III and IV genes and the negative regulation of *fliF* transcription depend on the binding of FlbD to pairs of *cis*-acting flagellar transcription regulator (*ftr*) elements. Pairs of *ftr* elements responsible for transcription activation are located upstream of class III and IV promoters (Wu et al., 1995), while a single *ftr4* element needed for repression of *fliF* transcription overlaps the *fliF* promoter (Benson et al., 1994a; Wingrove and Gober, 1994).

FlbD contains the presumptive site of phosphorylation (Asp-53), but it lacks other highly conserved residues that are signatures of the receiver domains of response regulators (Ramakrishnan and Newton, 1990). Presumably for this reason, the receiver domain of FlbD could not be assigned to any of the existing subfamilies of response regulators (see below). Phosphorylation of purified FlbD stimulates its activity as a transcription activator in vitro (Benson et al., 1994b), and FlbD is differentially phosphorylated during the cell cycle (Wingrove et al., 1993). FlbE is another class II gene product that is required for the pole-specific activation of class III genes. Although FlbE is apparently not the FlbD kinase, as previously reported (Wingrove and Gober, 1996), it is reported to be required for maximum phosphorylation of FlbD (see chapter 16).

Regulation of Motility and Stalk Formation

Except for continued filament elongation in the swarmer cell, flagellum assembly is essentially complete when *Caulobacter* predivisional cells gain motility (Mot+) shortly before cell separation (Fig. 1). Flagellum biosynthesis and gain of motility are thus separated in time, and different signal transduction components control the two events. As discussed above, the PleC kinase is required for the gain of motility and subsequent stalk formation but not for flagellum biosynthesis (Sommer and Newton, 1989; Wang et al., 1993).

The isolation of mutations in *divK* that suppress the *pleC* motility defect (Sommer and Newton, 1991) and behave as dominant gain-of-function alleles provides genetic evidence that DivK is the cognate response regulator of the PleC kinase. Consistent with this conclusion is the demonstration that the catalytic domain of PleC efficiently carries out the trans-phosphorylation of purified DivK in the presence of ATP and also acts as a phospho-DivK phosphatase (Hecht et al., 1995). These results support a model in which the PleC-DivK signal transduction pathway regulates motility and stalk formation late in the cell cycle, possibly in response to checkpoint control by completion of DIVp (Fig. 5). The isolation of mutations in *pleC* that bypass the motil-

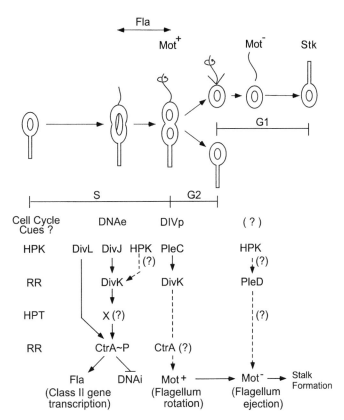

FIGURE 5 Model for the organization of signal transduction pathways regulating differentiation and cell division. This model indicates possible cell cycle cues or checkpoints that may regulate signal transduction and development (see the text). The DivL-CtrA and DivJ-DivK pathways presumably function during mid-S phase. CtrA accumulates at this time in early predivisional cells, when it is required for repression of premature initiation of DNA replication, activation of class II flagellar genes, and regulation of other cell cycle genes (reviewed in chapter 18). Dashed arrows indicate proposed signal transduction pathways.

ity defect of *ftsA* cell division mutants (Fig. 3B) would provide direct evidence for this model.

DivK could act like Spo0F as a phosphorelay protein (Appleby et al., 1996) or like CheY by direct interaction with another protein (Schuster et al., 1998). The effect of changing conserved amino acids in the DivK protein provides insight into the function of this response regulator. The *divK* (D53N) mutation alters DivK at the presumptive site of phosphorylation and, like the *divK341* allele (Hecht et al., 1995), it is a dominant bypass suppressor of all *pleC* point and null mutations examined (Hofmeister and Newton, unpublished). This result suggests that the mutant form of DivK permits activation of flagellar rotation. The flagellar motor components regulated by the pleC-DivK pathway have not been identified. However, the finding that a TS allele of *ctrA* (see below) (Wu et al., 1998) acts as a bypass sup-

pressor of the *pleC* motility phenotype (Fig. 5) (Ohta and Newton, 1999) suggests that the PleC-DivK pathway may function through CtrA.

A second signal transduction component involved in the regulation of *Caulobacter* motility is PleD. This response regulator is required for loss of motility, ejection of the flagellum, and stalk formation. Both *pleD* and *pleC* mutants (Fig. 3D) (Sommer and Newton, 1989) form bipolar flagellated predivisional cells that are blocked in the developmental transition from swarmer to stalked cell (Fig. 1). Unlike the paralyzed *pleC* mutants, however, *pleD* mutants are motile throughout the cell cycle, which accounts for their supermotile phenotype (Sommer and Newton, 1989).

The tandem N-terminal D and D′ domains of PleD contain conserved sequence motifs characteristic of receiver domains of response regulators. The D domain contains the four typically invariant residues of this protein subfamily, including Asp-53 at the predicted site of phosphorylation (Volz, 1993). The D′ domain of PleD is less well conserved, however (Hecht et al., 1995). The 169-residue C terminus of PleD is not similar to the output domains of previously investigated response regulators, but it does contain four regions (I to IV) with high sequence identity to proteins from a variety of other bacteria. The most highly conserved is the GGEEF motif present in region III. Public databases now contain more than 90 open reading frames (ORFs) encoding GGEEF motifs present in organisms ranging from gram-negative and gram-positive bacteria to cyanobacteria and archaebacteria. Other than PleD, the only GGEEF protein with a described function is the *Vibrio anguillarum* VirC protein (Milton et al., 1995). VirC is essential for virulence, but there is no report of the mechanism of action.

When *Caulobacter* cells lose motility, the filament-hook-rod assembly of the flagellum is ejected into the medium (Sheffery and Newton, 1977; Hahnenberger and Shapiro, 1987). However, the FliF protein, which assembles in the plasma membrane to form the basal-body MS ring, is degraded during the swarmer-to-stalked-cell transition (Jenal and Shapiro, 1996). *pleD* mutants, which do not shed the flagella or lose motility, are also defective in FliF turnover. Aldridge and Jenal (1999) have shown that the GGEEF motif of the C-terminal output domain of PleD is required for FliF turnover and that proteolysis does not depend on the presence of any of the basal-body structural genes examined. These results raise the possibility that the PleD-dependent proteolysis of FliF may be responsible for flagellum ejection rather than a consequence of this developmental event.

The original *pleD301* mutation identified is semidominant to the wild-type *pleD+* allele for suppression of the *pleC* motility defect (Sommer and Newton, 1989), which suggests that PleC is required for PleD regulation and could function as the PleD kinase. A *pleD* null muta-

tion also confers a supermotile phenotype and suppresses the Mot⁻ phenotype of *pleC* mutations, but unlike *pleD301*, it is recessive to the wild-type *pleD+* gene (Hecht and Newton, 1995). The null mutant displays no defect in growth or division, indicating that *pleD* is not required for viability or normal cell division. The D53G change (Hecht and Newton, 1995) in the *pleD301* mutant protein suggests that PleD phosphorylation is important in regulating activity of the GGEEF output domain in FliF proteolysis.

In one model, activation of the signal transduction pathway controlling PleD during the swarmer-to-stalked-cell transition leads to flagellum ejection and loss of motility (Fig. 5), and as described above, the PleC-DivK signal transduction pathway functions in late predivisional cells to initiate motility. This model offers a plausible explanation of how *pleD* mutations suppress the nonmotile phenotype of *pleC* strains. If the PleC-DivK pathway functions in some way to counteract the PleD-dependent turnover of FliF, *pleD* mutants would effectively escape the requirement for the PleC-dependent motility pathway. Alternatively, PleD could function twice, first in late predivisional cells when PleC is required for onset of motility, and second for loss of motility during the swarmer-to-stalked-cell transition (see discussion in Hecht and Newton, 1995).

Cell Cycle Regulation

The filamentous phenotype of the cold-sensitive *divK341* mutation under nonpermissive conditions suggests that DivK is also required for an early cell division step. Consistent with a dual function for DivK are the results of genetic complementation showing that while *divK341* is a gain-of-function allele for suppression of the *pleC* Mot phenotype, it is a loss-of-function allele for the cell division phenotype (Hecht et al., 1995).

Conditional mutations in *divJ*, like those in *divK*, display an unpinched, filamentous phenotype. This result and the observation that DivJ functions in vitro as a DivK kinase and a phospho-DivK phosphatase have led to the suggestion that DivJ and DivK are members of a signal transduction pathway regulating cell division (Hecht et al., 1995). A downstream target of this pathway was identified by selecting for pseudorevertants of the cold-sensitive *divK356* allele that suppress the cell division defect of this mutation (Wu et al., 1998). The TS *sokA* (suppressor of divK) allele *sokA301* was isolated as an extragenic suppressor, and the mutation was mapped to the C terminus of the CtrA protein. The *sokA* allele, which displays an early block in cell division at the nonpermissive temperature, also suppresses the lethal phenotype of a *divK* gene disruption and the cold-sensitive cell division phenotype of *divJ* mutations at the permissive temperature (Wu et al., 1998).

The purified catalytic domain of DivJ phosphorylates CtrA and activates specific transcription of class II flagellar gene promoters, but CtrA is unlikely to be the immediate target of DivJ. DivJ preferentially phosphorylates DivK in the presence of CtrA and, given the H1 ⇒ D1 ⇒ H2 ⇒ D2 architecture of multicomponent signal transduction pathways (Appleby et al., 1996), phospho-DivK (D1) would not be expected to phosphorylate CtrA (D2) directly. An additional, unidentified HPT (X) containing the H2 domain may be responsible for phosphoryl group transfer to CtrA in a DivJ ⇒ DivK ⇒ X ⇒ CtrA phosphorelay. This phosphorelay presumably functions in mid-S phase, when CtrA protein has accumulated in predivisional cells (Domian et al., 1997) and phospho-CtrA is required for transcriptional activation of class II flagellar genes (see the legend to Fig. 5) (Quon et al., 1996; Wu et al., 1998). Transcription of *divK*, which contains a CtrA-dependent promoter (Benson et al., 1999), also coincides with this cell cycle period (Hecht et al., 1995).

Although all *divJ* mutations confer cell division defects, DivJ is not essential for viability (DeRose et al., 1999), suggesting that an additional kinase or kinases are required for CtrA regulation. One of these is DivL. Conditional *divL* mutants display a severe cell division defect (Sommer and Newton, 1991), and cells with the disrupted *divL* gene are not viable (Wu et al., submitted). The DivL protein contains signature sequences conserved in HPKs (Fig. 4). However, the so-called H-box motif contains a tyrosine residue at position 550 instead of the normally invariant histidine, which is the site of phosphorylation in all reported HPKs. Several lines of evidence indicate that DivL functions as a tyrosine kinase. The Tyr-550 residue of DivL is essential for the functions of this protein in vivo, and it is autophosphorylated by ATP in vitro (Wu et al., submitted). Interestingly, DivL homologues containing a Tyr residue instead of a His residue within the H-box motif are found in *C. crescentus* CB2, CB4, CB13, and CV115, as well as in the related species *Asticcacaulis excentricus* and *Asticcacaulis biprosthecum*, which display a different pattern of polar differentiation. These results suggest that the novel DivL tyrosine kinase may represent a widespread subfamily of prokaryotic protein kinases.

The ability of the *sokA301* allele of *ctrA* to suppress the cell division phenotype of cold-sensitive mutations in *divL* (Wu et al., submitted) indicates that the DivL kinase regulates CtrA activity. Consistent with this genetic result, purified DivL catalyzes the phosphorylation and activation of CtrA in the presence of ATP, as measured in an in vitro transcription assay (Wu et al., submitted). However, DivL does not catalyze the phosphorylation of DivK, indicating that another, as yet unidentified kinase must be involved in the regulation of DivK. These results suggest that at least two signal transduction pathways function to coordinate the *Caulobacter* cell cycle. One of these is the DivJ ⇒ DivK ⇒ X ⇒ CtrA phosphorelay initiated by DivJ and probably one or more additional sensor kinases, and the other is the DivL ⇒ CtrA pathway regulated directly or indirectly by the tyrosine kinase (Fig. 5).

It is not clear why multiple kinases are required for the regulation of a common response regulator. By analogy to the role of the response regulator Spo0A in *B. subtilis* sporulation (see chapter 7), different cell cycle signals controlling development and cell division cycle progression itself could be integrated by the CtrA protein. Given the multiple cell cycle events regulated by CtrA, it can be anticipated that additional components, including kinases (see chapter 18) and phosphoprotein phosphatases, will be found to regulate its activity.

SURVEY OF *CAULOBACTER* HPKs AND RESPONSE REGULATORS

The identities and functional assignments of HPKs and response regulators in *Caulobacter* cannot be fully assessed until the genome sequence is completed in 1999 and then annotated. However, inspection of the partial sequence being compiled by the Institute for Genomic Research (November 1998 release; three genome equivalent sequences on 3,760 contigs) provides a preliminary and useful catalogue of the number and types of two-component proteins that can be expected. For this

analysis the translated *Caulobacter* DNA database was searched for histidine kinases by using DivJ (Ohta et al., 1992), PleC (Wang et al., 1993), and CheA (Alley, 1998) in TFASTA analysis (Genetics Computer Group program). DivK (Hecht et al., 1995), CtrA (Quon et al., 1996), *B. subtilis* Spo0F protein (Trach et al., 1988), and the *S. typhimurium* CheY protein (Stock et al., 1985) were used in a similar analysis for response regulators.

Protein Kinases

Grebe and Stock (1999) have classified 348 HPKs available from public databases and assigned them to 11 subfamilies (Table 1), based on amino acid sequence alignment in the conserved H, N, D, F, and G boxes (Parkinson and Kofoid, 1992; Stock et al., 1995). We have applied a similar analysis to the translated *Caulobacter* HPKs. Over 60 ORFs were similar

TABLE 1 Subfamilies of histidine protein kinases[a]

Group	H-box	N-box	D and F box	G-box
HPK 1a	+..Fhs.hSHEh+TPL..h	D...h..hh.NLh.NAh+ys	D.G.Gh......hF..F.R..	G.GLGLshh..hh..HGG
HPK 1b	+..FLA.MSHEhRTPL..h	D..+h.Qhh.NLh.NAhKFT	DsG.Gh......hF..F.Q..	G.GLGLshh..hh..MGG
HPK 2a	...F..BhsH-LRTPL..h	D..hh..hh.NLh.NAh+ys	D.G.Gh......hF..F.R..	G.GLGLshh..hh..HGG
HPK 2b	...hh..hsH-LRTPL.Rh	....h..hh.NLh.NAhRyG	D.G.Gh.s.....hF.PF.R..	G.GLGLshh..hh..HGG
HPK 4	hG.hhs.hAHEh..Ph..h	D...h.QhhhNLh.NAhzAh	D.G.Gh......hF.PF.TT+	G.GhGL.hh..hh..HGG
HPK 5	..z.LR...HEy.N.h..h	....hh.hhGNLh-NAh.s.	D.G.Gh......hF..G.STK	..GhGL.hh...h...GG
HPK 6	h.hh..hhRHDhhN.h..h	AB..h..hh-NLh.NAh.HG	D.G.GhP.zh...hF..GF...	G.GLGLyhh+.hh..YGG
HPK 7	-+..hA+-hHD.h...h.h	....h..hh.EAh.NAh+Hs	D.G.Gh *no F-box*	..:.GL.Gh.-+h..hGG
HPK 8	+.LQsQh.PHFhyN.LN.h	.h.hP.h.hQ.hhENAh.ys	D.G.Gh *no F-box*	G.G.GL.Nh..Rh...FG
HPK 9	.h..hFR.hHThKG....h	hh-.h.-PhhHhhRNAhDHG	DDG.Gh.<29>.hhF.PGFST.	GRGVGhDVV+..h..h.G
HPK 10	...-h+.F+HDY.Nhh..h	.hh.h.R.h..hh.NAhE.s	*no D-box* .-hF...FSTK	.RGLGL..h.-hh.....
HPK 11	+-hhh.EhHHRV+NNLQhh	ThhPh.hhh.ELhsNAh+ys	DBG.Gh *F poorly conserved*	..shGh.hh..hh..h.G

[a] Based on alignments compiled by Grebe and Stock (1999). Abbreviations: ., not conserved; h, ACFIMLV; y, FHY; s, AGST; +, KR; −, DE; B, DN; z, DENQ. HPK 3 has been omitted for simplicity, as this group is composed of a diverse spectrum of orthodox kinases. Almost all known hybrid kinases fall into group HPK 1b. Residues in boldface are characteristic for a particular box ("H" for H-box, "N . . . NA" for N-box, "D.G.G." for D-box, "F" for F-box, "GLGL" for G-box), and underlined residues are characteristic for a given kinase subfamily.

TABLE 2 Histidine protein kinases of five bacterial species

Subfamily	No. of kinases[a]				
	C. crescentus[b]	*E. coli*[c]	*B. subtilis*[d]	*M. ther*[e]	*A. fulg*[f]
1	29 (16)	8 (5)	6	0	0
2	3	7	0	0	0
3	2	3	4	1	1
4	5 (2)	3	5	0	0
5	0	2	3	0	0
6	0	0	0	0	10
7	0	3	9	0	0
8	0	2	3	0	0
9	2	1	1	0	1
10	0	0	0	0	0
11	10 (2)	0	0	15	0

[a] Number of hybrid kinases within a subfamily is given in parentheses.
[b] Based on the November 1998 release of *Caulobacter* genome sequences by the Institute for Genomic Research. Only 51 of the 60 HPK transmitter domains identified were analyzed (see the text). However, it is possible that the remaining nine sequences will be found to represent HPKs when the genome sequencing is completed.
[c] Blattner et al., 1997.
[d] Kunst et al., 1997.
[e] *M. thermoautotrophicum* (Smith et al., 1997).
[f] *A. fulgidus* (Klenk et al., 1997).

to at least a part of the HPK transmitter domain. With the BLAST search (National Center for Biotechnology Information), 51 of these sequences were found to contain most of the homology boxes of HPK and not to be redundant (these sequences included 23 complete ORFs). At least 20 of the 51 HPKs were hybrid kinases containing both a kinase and a response regulator domain (H1-D1 proteins). This compares with five hybrid kinases among 29 reported HPKs in *E. coli* (Blattner et al., 1997) and none among 31 *B. subtilis* HPKs (Kunst et al., 1997) (Table 2).

Two-component proteins have also been found in archaebacterial genomes, including those of *Methanobacterium thermoautotrophicum* (Smith et al., 1997) and *Archaeoglobus fulgidus* (Klenk et al., 1997). No hybrid kinases were found among the 16 HPKs of *M. thermoautotrophicum* or the 12 HPKs of *A. fulgidus* (Table 2). Although four of the five hybrid kinases in *E. coli* are H1-D1-H2 domain proteins (Appleby et al., 1996), all of the *Caulobacter* hybrid kinases appear to be H1-D1 proteins (see below).

A multisequence alignment (by the PILEUP program of the Wisconsin Genetics Computer Group package) of the presumptive *Caulobacter* HPKs revealed several distinct groups (Table 2) belonging to subfamilies compiled by Grebe and Stock (1999). The consensus sequences of the subfamilies are shown in Table 1. Two observations are striking. First, 29, or two-thirds, of the kinases have characteristic motifs that place them in subfamily 1 (Table 2); at least 16 of these are hybrid HPKs (subfamily 1b [Table 1]). All eukaryotic and hybrid kinases also belong to subfamily 1b. The *Caulobacter* HPKs PleC (Wang et al., 1993), DivJ (Ohta et al., 1992), and DivL (Wu et al., submitted) discussed above belong to this subfamily, as well as the PhoR kinases of *E. coli*, *B. subtilis*, and other bacterial species. Second, 10 *Caulobacter* HPKs belong to subfamily 11 (Table 2). Interestingly, with the exception of two HPKs from *Sinorhizobium meliloti* (ExsG) and *Mycobacterium tuberculosis* (G2072665 HPK), the 15 HPKs in subfamily 11 as originally compiled (Grebe and Stock, 1999) were represented by *M. thermoau-*

totrophicum proteins. The putative H box of this unusual kinase subfamily displays low similarity to those of other HPK subfamilies; notably, the conserved histidine is followed by an arginine (Table 1).

Two *Caulobacter* CheA homologues (Alley, 1998) have been identified and assigned to subfamily 9, which contains exclusively CheA homologues. The remaining nine *Caulobacter* histidine kinases belong to HPK subfamilies 2, 3, and 4, as indicated in Table 2. Two of the five HPKs of subfamily 4 are hybrid kinases.

Response Regulators

Grebe and Stock (1999) analyzed 298 response regulators that have been associated with one or more specific kinases. Based on their receiver domains, these response regulators were classified into eight distinct subfamilies, A to H. Fifty-eight presumptive receiver (R) domains of response regulators were identified in the *Caulobacter* genome, including the 20 that are part of hybrid kinases (Table 2). Multisequence alignment of the 58 R domains with the 298 response regulators indicates that they belong to subfamilies A, B, C, and F (Table 3). A fifth group with five members could not be classified within the previously described subfamilies, and they were assigned to a new subgroup I; all are homologues of a previously unclassified response regulator (ExsF of *S. meliloti* [Grebe and Stock, in press]). ExsG, which is the cognate kinase of ExsF, is classified as a subfamily 11 HPK. In addition, two of the *Caulobacter* R domains of this new subgroup I are part of hybrid kinases belonging to subfamily 11 (Table 2). Thus, it is tempting to speculate that the other three response regulators in subfamily I also represent cognate response regulators of *Caulobacter* subfamily 11 HPKs.

The R domains of all 16 *Caulobacter* subfamily 1b hybrid kinases belong to the receiver subfamily B. The R domains of two other hybrid kinases (subfamily 4 HPKs) are homologous to the receiver domain of the *Rhizobium leguminosarum* hybrid kinase FixL, which is a subfamily 4 kinase. FlbD, an NtrC-like response regulator with unusual residues in the

TABLE 3 Classification of *Caulobacter* response regulators based on receiver and output domains

Domain	Group	No. of members
Receiver	A	19 (5)[a]
	B	16 (1)
	C	3
	D	0
	E	0
	F	3 (1)
	G	0
	H	0
	I	5 (2)
	Unclassified[b]	2 (1)
Output	None	32[c]
	ActR	2
	CheB	2
	FixJ	2
	NtrC	4
	OmpR	10
	Unclassified[d]	4
	Unclassified[e]	2

[a] The number of single-domain response regulators is indicated in parentheses.
[b] Includes FlbD.
[c] Includes receiver domains of 20 hybrid kinases and 2 additional putative hybrid kinases and 10 single-domain response regulators.
[d] See the text.
[e] Includes PleD.

conserved R domains (Fig. 4) (Ramakrishnan and Newton, 1990), and another presumptive *Caulobacter* response regulator do not fall into any of the subfamilies.

The response regulators were also classified according to whether they have an output domain, e.g., DNA binding domains or enzymatic function. The 32 receiver domains that do not have an output domain include 20 that are part of hybrid kinases and 10 single-domain response regulators (Table 3). The 20 hybrid kinases (H1-D1) (Table 2) may function in phosphorelay pathways, or alternatively, the receiver domains could regulate the activity of the kinase domains as described for VirA (Heath et al., 1995). Some of the 10 single-domain response regulators may be components of phosphorelay pathways (Appleby et al., 1996). They include the essential response regulator DivK (Hecht et al., 1995) and two subfamily I (see above) response regulators. Five single-domain response regulators are the previously described CheYI and CheYII of *Caulobacter* (Alley, 1998) and three other CheY homologues (subfamily A). Four other response regulator homologues, including CheYIII (Alley, 1998), have C-terminal sequences of unknown function.

Twenty of the response regulators contain previously defined output domains, as listed in Table 3. Ten, including CtrA (Quon et al., 1996), have OmpR-type output domains. FlbD (Ramakrishnan and Newton, 1990), TacA (Marques et al., 1997) and two other ORFs have NtrC-type output domains, suggesting that at least four different transcriptional activators may be involved in the regulation of transcription from genes with σ^{54}-dependent promoters.

CONCLUSIONS

The family of His-Asp phosphorelay proteins plays central roles in receptor-mediated signal transduction in bacteria, eukaryotic microorganisms, and plants where they mediate cellular responses to environmental changes. *Caulobacter* contains an unusually large and diverse collection of protein kinases and response regulators, many of which doubtless play similar regulatory roles in this bacterium. In addition, members of this protein family are responsible for orchestrating steps in the cell division and DNA synthetic pathways of *Caulobacter* and coordinate an intricate sequence of developmental events with cell cycle progression. As discussed here and by Hung et al. in chapter 18, CtrA is a pivotal component in this signal transduction network and is subject to regulation by phosphorylation and protein turnover. Members of the histidine kinase and response regulator families implicated in the regulation of CtrA activity include the HPK DivJ, the essential HPK DivL, and the essential response regulator DivK (Fig. 5), but it seems likely that additional signal transduction components are also involved.

Other signal transduction proteins that remain to be identified are those regulating activation of flagellar rotation in response to the PleC kinase and the protein kinase or kinases that presumably regulate the PleD-mediated loss of motility and stalk formation (Fig. 5). The PleC- and PleD-dependent pathways assume particular importance because they are directly responsible for generating the two characteristic *Caulobacter* cell types, the motile swarmer cell and the nonmotile stalked cell. A major goal of future research will be to identify the cell cycle, and possibly environmental, cues to which these and other signal transduction pathways respond.

ACKNOWLEDGMENTS

We thank The Institute for Genome Research for making *Caulobacter* sequences available before publication. We are grateful to Jianguo Wu for critically reading the manuscript.

Work in our laboratory was supported by Public Health Service Grants GM22299 and GM58794 from the National Institutes of Health.

REFERENCES

Aldridge, P., and U. Jenal. 1999. Cell cycle-dependent degradation of a flagellar motor component requires a novel-type response regulator. *Mol. Microbiol.* **32:**379–391.

Alley, M. R. K. 1998. GenBank accession no. AJ006687.

Anderson, D. K., N. Ohta, J. Wu, and A. Newton. 1995. Regulation of the *Caulobacter crescentus rpoN* gene and function of the purified σ^{54} in flagellar gene transcription. *Mol. Gen. Genet.* **246:** 697–706.

Appleby, J. L., J. S. Parkinson, and R. B. Bourret. 1996. Signal transduction via the multi-step phosphorelay: not necessarily a road less traveled. *Cell* **86:**845–848.

Benson, A. K., G. Ramakrishnan, N. Ohta, J. Feng, A. Ninfa, and A. Newton. 1994a. The *Caulobacter crescentus* FlbD protein acts at *ftr* sequence elements both to activate and repress transcription of cell cycle regulated flagellar genes. *Proc. Natl. Acad. Sci. USA* **91:**4989–4993.

Benson, A. K., J. Wu, and A. Newton. 1994b. The role of FlbD in regulation of flagellar gene transcription in *Caulobacter crescentus. Res. Microbiol.* **12:**420–430.

Benson, A. K., N. Ohta, J. Wu, and A. Newton. 1999. Unpublished results.

Blattner, F. R., G. Plunkett III, C. A. Bloch, N. T. Perna, V. Burland, M. Riley, J. Collado-Vides, J. D. Glasner, C. K. Rode, G. F.

Mayhew, J. Gregor, N. W. Davis, H. A. Kirk-patrick, M. A. Goeden, D. J. Rose, B. Mau, and Y. Shao. 1997. The complete genome sequence of *Escherichia coli* K-12. *Science* **277:** 1453–1474.

Brun, Y. V., and L. Shapiro. 1992. A temporally controlled σ-factor is required for polar morphogenesis and normal cell division in *Caulobacter. Genes Dev.* **6:**2395–2408.

Brun, Y. V., G. Marczynski, and L. Shapiro. 1994. The expression of asymmetry during *Caulobacter* cell differentiation. *Annu. Rev. Biochem.* **63:**419–450.

Burbulys, D., K. A. Trach, and J. A. Hoch. 1991. Initiation of sporulation in *B. subtilis* is controlled by a multicomponent phosphorelay. *Cell* **64:**545–552.

Degnen, S. T., and A. Newton. 1972. Chromosome replication during development in *Caulobacter crescentus. J. Mol. Biol.* **64:**671–680.

DeRose, R., N. Ohta, and A. Newton. 1999. Unpublished results.

Din, N., E. M. Quardokus, M. J. Sackett, and Y. V. Brun. 1998. Dominant C-terminal deletions of FtsZ that affect its ability to localize in *Caulobacter* and its interaction with FtsA. *Mol. Microbiol.* **27:** 1051–1063.

Dingwall, A., W. Y. Zhuang, K. Quon, and L. Shapiro. 1992. Expression of an early gene in the flagellar regulatory hierarchy is sensitive to an interruption in DNA replication. *J. Bacteriol.* **174:** 1760–1768.

Domian, I. J., K. C. Quon, and L. Shapiro. 1997. Cell type-specific phosphorylation and proteolysis of a transcriptional regulator controls the G1-to-S transition in a bacterial cell cycle. *Cell* **90:**415–424.

Donachie, W. D. 1993. The cell cycle of *Escherichia coli. Annu. Rev. Microbiol.* **47:**199–230.

Grebe, T., and J. Stock. 1999. The histidine protein kinase superfamily. *Adv. Microb. Physiol.,* **41.**

Hahnenberger, K. M., and L. Shapiro. 1987. Identification of a gene cluster involved in flagellar basal body biogenesis in *Caulobacter crescentus. J. Mol. Biol.* **194:**91–103.

Heath, J. D., T. C. Charles, and E. W. Nester. 1995. Ti plasmid and chromosomally encoded two-component systems important in plant cell transformation by *Agrobacterium* species, p. 367–385. *In* J. A. Hoch and T. J. Silhavy (ed.), *Two-Component Signal Transduction.* ASM Press, Washington, D.C.

Hecht, G. B., and A. Newton. 1995. Identification of a novel response regulator required for the swarmer-to-stalked-cell transition in *Caulobacter crescentus. J. Bacteriol.* **177:**6223–6229.

Hecht, G. B., T. Lane, N. Ohta, J. N. Sommer, and A. Newton. 1995. An essential single domain response regulator required for normal cell division and differentiation in *Caulobacter crescentus. EMBO J.* **14:**3915–3924.

Hoch, J. A., and T. J. Silhavy (ed.). 1995. *Two-Component Signal Transduction.* ASM Press, Washington, D.C.

Hofmeister, T.S., and A. Newton. Unpublished data.

Horvitz, H. R., and I. Herskowitz. 1992. Mechanisms of asymmetric cell division: two B's or not two B's, that is the question. *Cell* **68:**237–255.

Huguenel, E. D., and A. Newton. 1982. Localization of surface structures during procaryotic differentiation: role of cell division in *Caulobacter crescentus. Differentiation* **21:**71–78.

Ireton, K., and A. D. Grossman. 1992. Coupling between gene expression and DNA synthesis early during development in *Bacillus subtilis. Proc. Natl. Acad. Sci. USA* **89:**8808–8812.

Ireton, K., and A. D. Grossman. 1994. A developmental checkpoint couples the initiation of sporulation to DNA replication in *Bacillus subtilis. EMBO J.* **13:**1566–1573.

Jenal, U., and L. Shapiro. 1996. Cell cycle-controlled proteolysis of a flagellar motor protein that is asymmetrically distributed in the *Caulobacter* predivisional cell. *EMBO J.* **15:**2393–2406.

Kato, J., Y. Nishimura, M. Yamada, H. Suzuki, and Y. Hirota. 1988. Gene organization in the region containing a new gene involved in chromosome partition in *Escherichia coli. J. Bacteriol.* **170:** 3967–3977.

Klenk, H. P., R. A. Clayton, J. F. Tomb, O. White, K. E. Nelson, K. A. Ketchum, R. J. Dodson, M. Gwinn, E. K. Hickey, J. D. Peterson, D. L. Richardson, A. R. Kerlavage, D. E. Graham, N. C. Kyrpides, R. D. Fleischmann, J. Quackenbush, N. H. Lee, G. G. Sutton, S. Gill, E. F. Kirkness, B. A. Dougherty, K. McKenney, M. D. Adams, B. Loftus, and J. C. Venter. 1997. The complete genome sequence of the hyperthermophilic, sulphate-reducing archaeon Archaeoglobus fulgidus. *Nature* **390:** 364–370.

Kunst, F., N. Ogasawara, I. Moszer, A. M. Albertini, G. Alloni, V. Azevedo, M. G. Bertero, P. Bessieres, A. Bolotin, S. Borchert, R. Borriss, L. Boursier, A. Brans, M. Braun, S. C. Brignell, S. Bron, S. Brouillet, C. V. Bruschi, B. Caldwell, V. Capuano, N. M. Carter, S. K. Choi, J. J. Codani, I. F. Connerton, and A. Danchin. 1997. The complete genome sequence of the gram-positive bacterium *Bacillus subtilis. Nature* **390:**249–256.

Kustu, S., E. Santero, J. Keener, D. Popham, and D. Weiss. 1989. Expression of σ54 (*ntrA*)-dependent genes is probably united by a common mechanism. *Microbiol. Rev.* **53:**367–376.

Lutkenhaus, J., and S. G. Addinall. 1997. Bacterial cell division and the Z ring. *Annu. Rev. Biochem.* **66:**93–116.

Malakooti, J., S. P. Wang, and B. Ely. 1995. A consensus promoter sequence for *Caulobacter crescentus* genes involved in biosynthesis and housekeeping function. *J. Bacteriol.* **177:**4372–4376.

Marczynski, G. T., A. Dingwall, and L. Shapiro. 1990. Plasmid and chromosomal DNA replication and partitioning during the *Caulobacter crescentus* cell cycle. *J. Mol. Biol.* **212:**709–722.

Marczynski, G. T., K. Lentine, and L. Shapiro. 1995. A developmentally regulated chromosomal origin of replication uses essential transcription elements. *Genes Dev.* **9:**1543–1557.

Marques, M. D. V., S. L. Gomes, and J. W. Gober. 1997. A gene coding for a putative sigma 54 activator is developmentally regulated in *Caulobacter crescentus. J. Bacteriol.* **179:**5502–5510.

Messer, W., and C. Weigel. 1996. Initiation of chromosome replication, p. 1579–1601. *In* F. C. Neidhardt, R. Curtiss III, J. L. Ingraham, E. C. C. Lin, K. B. Low, B. Magasanik, W. S. Reznikoff, M. Riley, M. Schaechter, and H. E. Umbarger (ed.), *Escherichia coli and Salmonella: Molecular and Cellular Biology,* 2nd ed. ASM Press, Washington, D.C.

Milton, D. L., A. Norqvist, and H. Wolf-Watz. 1995. Sequence of a novel virulence-mediating gene, virC, from *Vibrio anguillarum. Gene* **164:**95–100.

Mullin, D. A., and A. Newton. 1989. Ntr-like promoters and upstream regulatory sequence *ftr* are required for transcription of a developmentally regulated *Caulobacter crescentus* flagellar gene. *J. Bacteriol.* **171:**3218–3227.

Nakamura, M., I. N. Maruyama, M. Soma, J. I. Kato, H. Suzuki, and Y. Hirota. 1983. On the process of cellular division in Escherichia coli: nucleotide sequence of the gene for penicillin-binding protein 3. *Mol. Gen. Genet.* **191:**1–9.

Newton, A. 1984. Temporal and spatial control of the *Caulobacter* cell cycle, p. 51–75. *In* P. Nurse and E. Streiblova (ed.), *Microbial Cell Cycle.* CRC Press, Boca Raton, Fla.

Newton, A., N. Ohta, G. Ramakrishnan, D. Mullin, and G. Raymond. 1989. Genetic switching in the flagellar gene hierarchy of *Caulobacter* requires negative as well as positive regulation of transcription. *Proc. Natl. Acad. Sci. USA* **86:**6651–6655.

Ninfa, A. J., D. A. Mullin, G. Ramakrishnan, and A. Newton. 1989. *Escherichia coli* σ-54 RNA polymerase recognizes *Caulobacter crescentus flaK* and *flaN* flagellar gene promoters in vitro. *J. Bacteriol.* **171:**383–391.

Ohta, N., and A. Newton. 1999. Unpublished results.

Ohta, N., and A. Newton. 1996. Signal transduction in the cell cycle regulation of *Caulobacter* differentiation. *Trends Microbiol.* **4:**326–332.

Ohta, N., L. S. Chen, E. Swanson, and A. Newton. 1985. Transcriptional regulation of a periodically controlled flagellar gene operon in *Caulobacter crescentus. J. Mol. Biol.* **186:**107–115.

Ohta, N., M. Masurekar, and A. Newton. 1990. Cloning and cell cycle-dependent expression of DNA replication gene *dnaC* from *Caulobacter crescentus. J. Bacteriol.* **172:**7027–7034.

Ohta, N., T. Lane, E. G. Ninfa, J. M. Sommer, and A. Newton. 1992. A histidine protein kinase homologue required for regulation of bacterial cell division and differentiation. *Proc. Natl. Acad. Sci. USA* **89:**10297–10301.

Ohta, N., A. J. Ninfa, A. D. Allaire, L. Kulick, and A. Newton. 1997. Identification, characterization and chromosomal organization of cell division cycle genes in *Caulobacter crescentus. J. Bacteriol.* **179:**2169–2180.

Osley, M. A., and A. Newton. 1974. Chromosome segregation and development in *Caulobacter crescentus. J. Mol. Biol.* **90:**359–370.

Osley, M. A., and A. Newton. 1977. Mutational analysis of developmental control in *Caulobacter crescentus. Proc. Natl. Acad. Sci. USA* **74:**124–128.

Osley, M. A., and A. Newton. 1980. Temporal control of the cell cycle in *Caulobacter crescentus*: roles of DNA chain elongation and completion. *J. Mol. Biol.* **138:**109–128.

Parkinson, J. S., and E. C. Kofoid. 1992. Communication modules in bacterial signaling proteins. *Annu. Rev. Genet.* **26:**71–112.

Quon, K. C., G. T. Marczynski, and L. Shapiro. 1996. Cell cycle control by an essential bacterial two-component signal transduction protein. *Cell* **84:**83–93.

Quon, K. C., B. Yang, I. J. Domian, L. Shapiro, and G. T. Marczynski. 1998. Negative control of bacterial DNA replication by a cell cycle regulatory protein that binds at the chromosome origin. *Proc. Natl. Acad. Sci. USA* **95:**120–125.

Ramakrishnan, G., and A. Newton. 1990. FlbD of *Caulobacter crescentus* is a homologue of NtrC (NRɪ) and activates sigma-54 dependent flagellar gene promoters. *Proc. Natl. Acad. Sci. USA* **87:**2369–2373.

Reynes, J. P., M. Tiraby, M. Baron, D. Drocourt, and G. Tiraby. 1996. *Escherichia coli* thymidylate kinase: molecular cloning, nucleotide sequence, and genetic organization of the corresponding *tmk* locus. *J. Bacteriol.* **178:**2804–2812.

Rizzo, M. F., L. Shapiro, and J. Gober. 1993. Asymmetric expression of the gyrase B gene from the replication-competent chromosome in the *Caulobacter crescentus* predivisional cell. *J. Bacteriol.* **175:**6970–6981.

Roberts, R. C., and L. Shapiro. 1997. Transcription of genes encoding DNA replication proteins is coincident with cell cycle control of DNA replication in *Caulobacter crescentus*. *J. Bacteriol.* **179:** 2319–2330.

Schmid, M. B. 1990. A locus affecting nucleoid segregation in *Salmonella typhimurium*. *J. Bacteriol.* **172:** 5416–5424.

Schuster, M., W. N. Abouhamad, R. E. Silversmith, and R. B. Bourret. 1998. Chemotactic response regulator mutant Che Y95IV exhibits enhanced binding to the flagellar switch and phosphorylation-dependent constitutive signaling. *Mol. Microbiol.* **27:**1065–1075.

Sheffery, M., and A. Newton. 1977. Reconstitution and purification of flagellar filaments from *Caulobacter crescentus*. *J. Bacteriol.* **132:**1027–1030.

Sheffery, M., and A. Newton. 1981. Regulation of periodic protein synthesis in the cell cycle: control of initiation and termination of flagellar gene expression. *Cell* **24:**49–57.

Smith, D. R., L. A. Doucette-Stamm, C. Deloughery, H. Lee, J. Dubois, T. Aldredge, R. Bashirzadeh, D. Blakely, R. Cook, K. Gilbert, D. Harrison, L. Hoang, P. Keagle, W. Lumm, B. Pothier, D. Qiu, R. Spadafora, R. Vicaire, Y. Wang, J. Wierzbowski, R. Gibson, N. Jiwani, A. Caruso, D. Bush, and J. N. Reeve. 1997. Complete genome sequence of *Methanobacterium thermoautotrophicum* deltaH: functional analysis and comparative genomics. *J. Bacteriol.* **179:** 7135–7155.

Sommer, J. M., and A. Newton. 1988. Sequential regulation of developmental events during polar morphogenesis in *Caulobacter crescentus*: assembly of pili on swarmer cells requires cell separation. *J. Bacteriol.* **170:**409–415.

Sommer, J. M., and A. Newton. 1989. Turning off flagellum rotation requires the pleiotropic gene *pleD*: *pleA*, *pleC*, and *pleD* define two morphogenic pathways in *Caulobacter crescentus*. *J. Bacteriol.* **171:** 392–401.

Sommer, J. M., and A. Newton. 1991. Pseudoreversion analysis indicates a direct role of cell division genes in polar morphogenesis and differentiation in *Caulobacter crescentus*. *Genetics* **129:**623–630.

Stephens, C. M., and L. Shapiro. 1993. An unusual promoter controls cell-cycle regulation and dependence on DNA replication of the *Caulobacter fliLM* early flagellar operon. *Mol. Microbiol.* **9:**1169–1179.

Stephens, C. M., G. Zweiger, and L. Shapiro. 1995. Coordinate cell cycle control of a *Caulobacter* DNA methyltransferase and the flagellar genetic hierarchy. *J. Bacteriol.* **177:**1662–1669.

Stock, A., D. E. Koshland, Jr., and J. Stock. 1985.

Homologies between the *Salmonella typhimurium* CheY protein and proteins involved in the regulation of chemotaxis, membrane protein synthesis, and sporulation. *Proc. Natl. Acad. Sci. USA* **82:** 7989–7993.

Stock, J. B., M. G. Surette, M. Levit, and P. Park. 1995. Two-component signal transduction systems: structure-function relationships and mechanisms of catalysis, p. 25–51. *In* J. A. Hoch and T. J. Silhavy (ed.), *Two-Component Signal Transduction*. American Society for Microbiology, Washington, D.C.

Tinker-Kulberg, R. L., T.-J. Fu, E. P. Geiduschek, and G. E. Kassavetis. 1996. A direct interaction between a DNA-tracking protein and a promoter recognition protein: Implications for searching DNA sequence. *EMBO J.* **15:** 5032–5039.

Trach, K. A., J. W. Chapman, P. Piggot, D. leCoq, and J. A. Hoch. 1988. Complete sequence and transcriptional analysis of the *spo0F* region of the *Bacillus subtilis* chromosome. *J. Bacteriol.* **170:**4194–4208.

VanWay, S. M., A. Newton, A. H. Mullin, and D. A. Mullin. 1993. Identification of the promoter and a negative regulatory element, *ftr4*, that is needed for cell cycle timing of *fliF* operon expression in *Caulobacter crescentus*. *J. Bacteriol.* **175:** 367–376.

Volz, K. 1993. Structural conservation in the CheY superfamily. *Biochemistry* **32:**11741–11753.

Wang, S. P., P. L. Sharma, P. V. Schoenlein, and B. Ely. 1993. A histidine protein kinase is involved in polar organelle development in *Caulobacter crescentus*. *Proc. Natl. Acad. Sci. USA* **90:**630–634.

Ward, D., and A. Newton. 1997. Requirement of topoisomerase IV *parC* and *parE* genes for cell cycle progression and developmental regulation in *Caulobacter crescentus*. *Mol. Microbiol.* **26:**897–910.

Wingrove, J. A., and J. W. Gober. 1994. A σ^{54} transcriptional activator also functions as a pole-specific repressor in *Caulobacter*. *Genes Dev.* **8:** 1839–1852.

Wingrove, J. A., and J. W. Gober. 1996. Identification of an asymmetrically localized sensor histidine kinase responsible for temporally and spatially regulated transcription. *Science* **274:**597–601.

Wingrove, J. A., E. K. Mangan, and J. W. Gober. 1993. Spatial and temporal phosphorylation of a transcriptional activator regulates pole-specific gene expression in *Caulobacter*. *Genes Dev.* **7:**1979–1992.

Winzeler, E., and L. Shapiro. 1997. Translation of the leaderless *Caulobacter dnaX* mRNA. *J. Bacteriol.* **179:**3981–3988.

Wu, J., A. Benson, and A. Newton. 1995. FibD,

a global regulator, is required for activation of transcription from σ^{54}-dependent *fla* gene promoters. *J. Bacteriol.* **177:**3241–3250.

Wu, J., and A. Newton. 1997. Regulation of the *Caulobacter* flagellar gene hierarchy; not just for motility. *Mol. Microbiol.* **24:**233–240.

Wu, J., N. Ohta, A. K. Benson, A. J. Ninfa, and A. Newton. 1997. Purification, characterization, and reconstitution of DNA-dependent RNA polymerases from *Caulobacter crescentus. J. Biol. Chem.* **272:**21558–21564.

Wu, J., N. Ohta, J.-L. Zhao, and A. Newton. A novel tyrosine protein kinase involved in signal transduction and cell cycle regulation in *Caulobacter.* Submitted for publication.

Wu, J., N. Ohta, and A. Newton. 1998. An essential, multicomponent signal transduction pathway required for cell cycle regulation in *Caulobacter. Proc. Natl. Acad. Sci. USA* **95:**1443–1448.

Xu, H., A. Dingwall, and L. Shapiro. 1989. Negative transcriptional regulation in the *Caulobacter* flagellar hierarchy. *Proc. Natl. Acad. Sci. USA* **86:** 6656–6660.

Zechiedrich, E. L., and N. R. Cozzarelli. 1995. Roles of topoisomerase IV and DNA gyrase in DNA unlinking during replication in *Escherichia coli. Genes Dev.* **9:**2859–2869.

Zweiger, G., and L. Shapiro. 1994. Expression of *Caulobacter dnaA* as a function of the cell cycle. *J. Bacteriol.* **176:**401–408.

REGULATION OF THE *CAULOBACTER* CELL CYCLE

Dean Hung, Harley McAdams, and Lucy Shapiro

18

The study of development centers on the question of how one cell gives rise to genetically identical but functionally distinct daughter cells. Such asymmetric cell division forms the basis for development in all multicellular organisms and is a vital field of study in biology. *Drosophila* and *Caenorhabditis elegans* localize cell fate determinants asymmetrically during critical cell division events, as does *Saccharomyces cerevisiae* during bud formation (Horvitz and Herskowitz, 1992; Guo and Kemphues, 1996; Lu et al., 1998). In a remarkable example of conservation of regulatory mechanisms among disparate organisms, bacteria also carry out asymmetric cell divisions in the course of undergoing diverse changes in physiology and morphology (Shapiro and Losick, 1997; Jacobs and Shapiro, 1998). For example, under starvation conditions the *Bacillus subtilis* vegetative cell divides asymmetrically into a transient mother cell and a hardy spore that can wait out the famine before differentiating back into the vegetative state (Grossman, 1995) (see chapter 8). In another example, the blue-green alga (cyanobacterium) *Anabaena*, which grows in chains of cells, periodically divides asymmetrically, yielding a differentiated heterocyte capable of fixing nitrogen and another vegetative cell (Golden and Yoon, 1998).

The focus of this chapter is the aquatic bacterium *Caulobacter crescentus*, which divides asymmetrically during each cell division cycle (Fig. 1), yielding progeny cells that differ both structurally and functionally. The initially motile swarmer cell progeny sheds its flagellum and differentiates into a nonmotile stalked cell. The progeny stalked cell elongates into a predivisional cell, forming a new swarmer pole opposite the existing stalked pole. During this polar differentiation, the proteins of the flagellar machinery localize to the incipient swarmer pole (see chapter 16), as do the methyl-accepting chemotaxis receptors (Alley et al., 1992, 1993). Polar pili are assembled at the flagellated pole of the new progeny swarmer cell (Sommer and Newton, 1988). Because its morphological characteristics facilitate synchronization of cell populations (Evinger and Agabian, 1977), *Caulobacter* is an excellent bacterial system in which to study the genetic regulatory network that controls cell cycle progression and asymmetric cell division.

In addition to morphological differences, the stalked- and swarmer cell progeny inherit different competencies for chromosome repli-

D. Hung, Department of Genetics, Beckman Center B300, Stanford University School of Medicine, Stanford, CA 94305. *H. McAdams and L. Shapiro*, Department of Developmental Biology, Beckman Center B300, Stanford University School of Medicine, Stanford, CA 94305.

Prokaryotic Development, edited by Y. V. Brun and L. J. Shimkets,
© 2000 American Society for Microbiology, Washington, DC 20005-4171

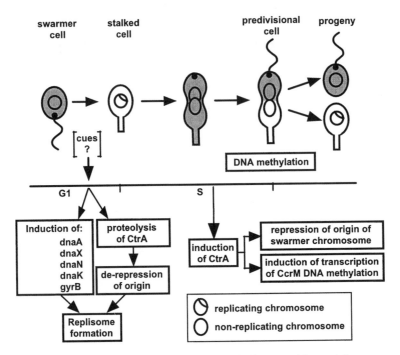

FIGURE 1 Control of the *Caulobacter* G$_1$-to-S-phase transition and the asymmetric regulation of replication initiation in the predivisional cell. The shaded areas indicate the presence of CtrA. Dark circles indicate localization of McpA.

cation (Degnen and Newton, 1972; Marczynski et al., 1990; Marczynski and Shapiro, 1992). The stalked cell immediately initiates the replication of its genome, whereas genome replication is blocked in the swarmer cell until it has differentiated into a stalked cell. The mechanisms responsible for this disparity in replication capabilities between swarmer and stalked cells in now known to depend, at least in part, on temporal and spatial changes in two-component signal transduction proteins (Quon et al., 1996, 1998). With the *Caulobacter* genome sequence to be completed in 1999, genome-wide studies will provide an even more detailed picture of global gene expression patterns and control networks throughout the cell cycle, allowing the identification of the entire complement of genes and mechanisms underlying *Caulobacter*'s cellular asymmetry.

SPATIAL AND TEMPORAL CONTROL OF REPLICATION INITIATION

A central component of any cell cycle is the initiation of chromosome replication coupled with strict controls to prevent repeated rounds of DNA replication without intervening cell divisions. Identification of the factors that regulate entry into S phase and the factors that prevent premature reinitiations during S phase is fundamental to understanding cell cycle control. In eukaryotic cells, it has become abundantly clear that control of DNA replication is mediated by phosphorylation by cyclin-dependent kinases (Hartwell, 1995). Cell cycle-specific proteolysis of the cyclins functions to integrate the control of replication initiation with other cell cycle events. Although much is known about eukaryotic cell cycle control, the biochemical pathways that regulate replication initiation in the bacterial cell and how they are coordinated with other cell cycle events are only now being identified (Messer and Noyer-Weidner, 1988; Lu et al., 1994; Crooke, 1995; Slater et al., 1995; Quon et al., 1996, 1998; Stephens et al., 1996; Domian et al., 1997). In *Escherichia coli*, factors such as the SeqA protein, the DAM DNA methyltransferase, and DnaA orchestrate the cell cycle timing of replication initiation, but little is known about the biochemical mechanisms that link the timing of replication initiation with other cell cycle events. In *Caulobacter*, however, more details are known, and it has become evident that prokaryotes, like the eukaryotes, use phosphorylation and proteolysis of regulatory factors to control the cell cycle. In addition to a DNA methyltransferase (CcrM) and DnaA, components of the two-component histidine kinases and response regulators not only control the initiation of replication but also serve to coordinate multiple cell cycle events. The CtrA response regulator protein, which controls the initiation of DNA replication and is essential for viability and cell cycle progression, is itself temporally controlled by phosphorylation and proteolysis (Domian et al., 1997).

As shown diagrammatically in Fig. 1, the initiation of DNA replication in *Caulobacter* is restricted to the stalked cell and happens once and only once per cell division (Dingwall and Shapiro, 1989; Marczynski et al., 1990; Marczynski and Shapiro, 1992). In contrast, *E. coli* and other rapidly dividing bacteria can initiate several overlapping rounds of replication per cell division (Helmstetter and Leonard, 1987). The *Caulobacter* swarmer cell does not initiate DNA synthesis until it differentiates into a new stalked cell, whereas the progeny stalked cell immediately initiates chromosome replication (Degnen and Newton, 1972; Marczynski et al., 1990; Marczynski and Shapiro, 1992). Thus, the progeny stalked cell functions somewhat like a stem cell, continually elongating and producing new swarmer cells. Independently of generation time, after approximately one-third of its cell cycle, the swarmer cell sheds its flagellum, replaces it with a stalk, and initiates chromosome replication, followed by chromosome segregation and cell division. Important questions regarding these processes are the following. What are the signals that cue the swarmer-to-stalked-cell transition? What are the mechanisms used by the cell to tightly coordinate differentiation events, chromosome replication, chromosome segregation, and cell division? At division, why do the two new chromosomes

have different competencies for replicating DNA?

THE ORIGIN OF REPLICATION

The observation that the two newly replicated chromosomes in the predivisional cell have different replicative abilities raises a fundamental question: are the chromosomes themselves different, or do the swarmer and stalked-cell compartments contain different regulatory factors? The two newly replicated chromosomes in the late predivisional cell segregate randomly to the progeny cells (Osley and Newton, 1974; Marczynski et al., 1990), implying that either of the two chromosomes is capable of responding to changes in the immediate intracellular environment. The emerging picture of this process suggests that compartment-specific intracellular factors act at the origin of replication to repress initiation of replication in the swarmer cell.

The *Caulobacter* origin of replication was identified and cloned by taking advantage of the observation that replication is always initiated in the stalked cell. Synchronized cells pulse labeled with DNA precursors at short time intervals reveal the order of appearance of newly replicated portions of DNA (Dingwall and Shapiro, 1989). In these experiments, the earliest radiolabeled fragment identifies the DNA region containing the origin of replication. By analyzing the order of appearance of newly labeled restriction fragments and using the known physical map (Ely and Gerardot, 1988) of the chromosome and correlating it with the genetic map (Barrett et al., 1982a, 1982b; Lott et al., 1987), it was shown that replication proceeds bidirectionally. In the same experiment, the timing of the appearance of labeled restriction fragments established the rate of chromosome replication during the cell cycle (Dingwall and Shapiro, 1989). The replication origin was cloned from a cosmid bearing the earliest replicating fragment of the chromosome (Marczynski and Shapiro, 1992). A subclone of this cosmid was shown to support replication of plasmids that do not normally replicate in *Caulobacter*. Most importantly, the cell cycle timing of replication of the plasmid-borne ori-

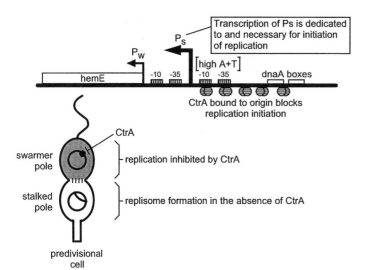

FIGURE 2 *Caulobacter* chromosome origin of replication. The −10 and −35 regions of the weak promoter (P_W) and the strong promoter (P_S) are indicated by hatched bars. The shaded regions in the predivisional cells indicate the presence of CtrA.

gin was the same as the timing of chromosome replication, demonstrating that critical *cis*-acting regulatory elements were present on the cloned origin fragment. The *Caulobacter* origin of replication, Cori (Fig. 2), shows both similarities to and differences from the *E. coli* origin. Both origins contain similar AT-rich regions and binding sites for DnaA, but Cori also has features not found in other known bacterial origins. The most notable are five copies of the GTTAA-N7-TTAA motif that serve as binding sites for the CtrA response regulator (see below). In addition to the binding sites for CtrA, two promoter elements that read away from the origin sequence are essential for the initiation of replication (Marczynski et al., 1995).

An open reading frame adjacent to the replication origin is homologous to the *E. coli hemE* gene, which encodes an enzyme involved in heme biosynthesis (Marczynski et al., 1995). Two functional promoters for the open reading frame were identified, P_W and P_S (Fig. 2). While P_W is active throughout the cell cycle and drives expression of the *hemE* gene, P_S activity correlates with the initiation of DNA replication in the nascent stalked cell and in the stalked portion of the predivisional cell. The RNA transcribed from P_S appears not to be translated into protein (Marczynski et al.,

1995), and it probably plays a role in replisome formation. P_S transcription from the stalked-cell chromosome remains constant as early predivisional cells progress to cell division and is repressed on the chromosome in the swarmer portion of the predivisional cell. Thus, differential transcription of a promoter in the origin of replication appears to contribute to the expression of asymmetry in the two halves of the predivisional cell where the two newly replicated chromosomes manifest differing replication abilities.

The CtrA two-component system response regulator binds to the 9-mers in the origin of replication in the swarmer compartment, where it represses the initiation of DNA replication, but is absent from the stalked portion of the predivisional cell (Quon et al., 1996, 1998; Domian et al., 1997). Specific mutations in the CtrA binding site that overlap P_S (Marczynski et al., 1995) or that disrupt a CtrA binding site upstream of this promoter (Quon et al., 1998) cause unregulated replication initiation. The binding of CtrA to five sites in the origin (Quon et al., 1998) inhibits replisome formation by inhibiting the synthesis of the short transcript from the P_S promoter and by occupying a site that is adjacent to an essential DnaA box (Fig. 2). The presence of five CtrA binding sites within the Cori region suggests that CtrA and the origin of replication form a multimeric nucleoprotein complex that blocks replisome formation (Quon et al., 1998).

TEMPORAL AND SPATIAL CONTROL OF CtrA DURING THE CELL CYCLE

Microbial cells are able to monitor changes in their environment, detect changes in cell density, and communicate with each other and with other organisms through signals. The cells respond to these signals by altering behavior (e.g., chemotaxis) or by altering gene expression patterns leading to specific differentiation pathways (e.g., sporulation). The major microbial signal transduction proteins belong to the superfamily of two-component sensor kinases and response regulators.

Two-component signal transduction systems use a histidine kinase to sense an internal or external change. The sensor histidine kinase then autophosphorylates on a conserved histidine and presents the phosphate group as a substrate for the phosphorylation of an aspartate on the response regulator, which in turn effects a response (Parkinson, 1993; Hoch and Silhavy, 1995). Two-component systems frequently have more than two components, sometimes consisting of several components even in the same protein (Appleby et al., 1996). Sensor kinase-response regulator phosphorelay cascades (Appleby et al., 1996) regulate a variety of processes, such as sporulation initiation in *B. subtilis* (Burbulys et al., 1991), nitrogen-activated protein (MAP) kinase pathways in *Saccharomyces cerevisiae* (Posas et al., 1996; Posas and Saito, 1998), and virulence in *Bordetella* (Uhl and Miller, 1996a, 1996b). *Caulobacter* uses these two-component signal transduction proteins to control the progression of the cell cycle (Hecht et al., 1995; Hecht and Newton, 1995; Ohta and Newton, 1996; Quon et al., 1996, 1998; Domian et al., 1997; Wu et al., 1998) (see also chapter 17). The phosphorelay appears to culminate in the CtrA response regulator, a transcription factor that falls into the *E. coli* OmpR response regulator superfamily (Quon et al., 1996).

CtrA is a central regulatory component in the *Caulobacter* cell cycle. It was first identified because of its role in the temporal control of flagellar biogenesis. Construction of the flagellum involves over 40 genes controlled by a transcriptional cascade (see chapter 16). Although formation of a flagellum is not essential for viability, and there are hundreds of nonmotile mutants, no mutations had been identified in a gene or genes that controlled the initiation of the flagellar gene transcription cascade. Therefore, Quon and his coworkers hypothesized that a regulatory factor must coordinate the start of flagellar construction with other cell cycle controls (Quon et al., 1996). Because of its postulated role in cell cycle progression, this putative protein was also expected to be essential. A genetic screen designed to find genes involved in the initiation of flagellar biogenesis that are also essential for viability identified a temperature-sensitive allele of the *ctrA* gene (Quon et al., 1996). In addition to its role in flagellar biogenesis, CtrA controls several critical cell cycle events, including DNA replication, DNA methylation, and cell division (Quon et al., 1996; Kelly et al., 1998; Reisenauer et al., 1999). Highly conserved CtrA sequence recognition motifs are found in the class II flagellar gene promoters, in the origin of replication, and in the *ccrM* DNA methyltransferase promoter (Stephens et al., 1995; Quon et al., 1996, 1998; Reisenauer et al., 1999). CtrA footprints each of these binding motifs, and mutations in them prevent CtrA binding and affect promoter activity (Quon et al., 1998; Reisenauer et al., 1999). CtrA also footprints the promoter of the *ftsZ* gene and contributes to the control of FtsZ during the cell cycle (Kelly et al., 1998). CtrA homologs that potentially regulate multiple cell cycle-coordinated functions have also been identified in *Rhizobium* meliloti (Hung et al., unpublished) and *Brucella abortis* (Devos and Letesson, unpublished).

Spatial and Temporal Expression of CtrA

As a critical regulator of cell cycle progression, it is not surprising that the CtrA protein is itself under strict temporal and spatial control. Visualization of CtrA by immunofluorescence microscopy shows that CtrA's spatial and temporal expression is consistent with its activities throughout the cell (Domian et al., 1997). CtrA binds to the origin of replication and prevents it from forming a replisome in the swarmer cell. During the swarmer-to-stalked-cell transition (G_1-S), CtrA is cleared from the cell by proteolysis, allowing replisome formation and the initiation of DNA replication in the new stalked cell (Fig. 1). Soon thereafter, proteolysis of CtrA ceases and *ctrA* transcription is induced, allowing threshold levels of CtrA to accumulate (see Fig. 6), thereby preventing reinitiation of DNA replication. In the early predivisional-cell stage (S phase), CtrA is distributed throughout the cell, first activating the flagellar-gene transcriptional cascade and then turning on the transcription of the CcrM DNA methyltransferase. In the late predivisional cell, at division, CtrA is cleared from the stalked compartment but not from the swarmer compartment, where it once again represses initiation of DNA replication (Fig. 1). Thus, asymmetrically regulated proteolysis is a critical factor controlling differential development of the stalked and swarmer progeny cells.

What is the proteolytic machinery that degrades CtrA, and how does it know when and where to operate? ClpXP appears to be the protease that degrades CtrA (Jenal and Fuchs, 1998), but the factors providing temporal and spatial control of CtrA proteolysis are not yet known. In the *E. coli* ClpXP protease system, the ClpP protein provides the proteolytic activity. The substrate protein may be unfolded by the ClpX ATPase that has been shown to interact with C-terminal amino acids of its targets (Laachouch et al., 1996; Levchenko et al., 1997). The CtrA C terminus is required for cell cycle-controlled proteolysis (Domian et al., 1997), and it also has sequence homology with the C termini of ClpX substrates. In *Caulobacter*, both ClpP and ClpX are essential for viability (Jenal and Fuchs, 1998). Shutdown of expression of either protein results in a cell division defect in which cells become filamentous and accumulate mass but do not divide. When either ClpP or ClpX expression is artificially turned off during the stalked-cell stage, CtrA is not cleared but remains present in the cell (Jenal and Fuchs, 1998). Flow cytometric analysis of chromosome content in these Clp-deficient stalked cells shows a predominant single-chromosome peak, suggesting that cells without ClpP or ClpX are at least partially blocked in initiation of DNA replication. DNA replication is not completely blocked, due perhaps to residual ClpXP activity. Because ClpP and ClpX are present throughout the cell cycle (Jenal and Fuchs, 1998) and CtrA is degraded at specific times, some as-yet-unidentified factor must target CtrA for proteolysis. A chaperone may be required to present CtrA to the ClpXP machinery, or alternatively, CtrA may be transiently modified, marking it as a target for proteolysis.

In the budding yeast *S. cerevisiae*, DNA replication is controlled by the cyclin-dependent kinase inhibitor Sic1, an inhibitor of DNA replication (Schwob et al., 1994). Unlike CtrA, which acts as a direct negative regulator of DNA replication, Sic1 acts indirectly to inhibit cyclin-dependent kinase (CDK) triggering of entry into S phase. In a manner parallel to CtrA, the yeast Sic1 protein is also regulated by cell-type-specific proteolysis. Ubiquitin-mediated proteolysis of Sic1 eliminates Sic1 activity at the G_1-S transition (Mendenhall, 1993; Schwob et al., 1994), relieving inhibition of CDK and allowing DNA replication to begin (see Fig. 4). Thus, in both a eukaryotic and a prokaryotic cell cycle, proteolysis rather than modification of an inhibitory factor effects entry into S phase (Amon, 1998).

Regulation of CtrA by Phosphorylation

Although the availability of CtrA during the cell cycle is modulated by cell-type-specific proteolysis, phosphorylation of CtrA is another factor which dictates its activity (Domian et al., 1997). CtrA is not normally present in stalked cells. Surprisingly, in mutant cells in which CtrA was not cleared from the stalked cell by proteolysis, the cell cycle was not perturbed (Domian et al., 1997). This was explained by the observation that the CtrA then present in stalked cells was not phosphorylated and phosphorylation is required to activate CtrA (see Fig. 5). It appears that the signals controlling the flow of phosphate to CtrA are cell cycle regulated. If an active CtrA derivative is forced to be present in stalked cells, then the G_1-S transition is blocked, demonstrating the critical nature of this phosphate flow (Domian et al., 1997).

The signals that cue CtrA-mediated cell cycle events are not yet known. A controlling phosphorelay involving two-component histidine kinases that culminates in CtrA phosphorylation has been suggested by Wu and coworkers (1998). Several nonessential histidine kinases have been identified that have cell division phenotypes, but the internal or external signals that initiate their respective phosphorylation relays are unknown (see chapter 17). A histidine kinase, CckA, that is responsible, directly or indirectly, for phosphorylation of CtrA has recently been identified (Jacobs et al., 1999). CckA is essential for viability, and a *cckA* temperature-sensitive allele has phenotypes identical to those of a *Caulobacter* strain carrying a *ctrA* temperature-sensitive mutation. The CckA kinase, unlike CtrA, is present throughout the cell cycle. However, an unusual feature of CckA is that it changes its location in the cell as a function of the cell cycle. Following the initiation of replication, the membrane-bound CckA kinase is polarly localized in the early predivisional cell, and upon cell division it is dispersed throughout the cellular membrane of both the swarmer and stalked cells.

There are parallels between cell cycle regulation by the *Caulobacter* CtrA response regulator and by the *B. subtilis* SpoOA response regulator that governs differentiation from vegetative cells to spores. In both cases a two-component phosphorelay culminates in a response regulator that controls the expression of a wide variety of genes and cell functions (Shapiro and Losick, 1997). When starved, *B. subtilis* switches from binary fission to an asymmetric division that results in the formation of a dormant endospore within the mother cell. As the spore matures, the mother cell lyses and releases the environmentally resistant spore, which remains dormant until conditions improve (Grossman, 1995). Sporulation relies on the two-component-based phosphorelay to integrate internal and environmental signals and ultimately to render a binary decision to sporulate or not to sporulate (Hoch, 1993). The sporulation program mediated by SpoOA includes forespore-specific transcription, mother cell-specific transcription, and regulation of the site of FtsZ assembly (see chapter 8). SpoOA also acts in an autoregulatory feedback loop where SpoOA~P activates its own transcription. The *ctrA* gene has CtrA binding sites in its promoter region and activates its own transcription in the early predivisional cell stage (Domian et al., 1999). While CtrA and SpoOA regulation and function resemble each other in many ways, CtrA differs from SpoOA in that it is required for the progression of the cell cycle as opposed to adaptive developmental programs.

OTHER FACTORS INVOLVED IN REPLICATION CONTROL

The *Caulobacter* DnaA protein is a likely candidate for a positive regulator of the initiation of DNA replication. The predicted amino acid sequence of *Caulobacter* DnaA shows 37% identity and 80% similarity to its *E. coli* homologue (Zweiger and Shapiro, 1994) and is thought to function in the same manner by binding to DnaA recognition sites in the origin and melting the DNA to create access to the replication machinery (Bramhill and Kornberg, 1988). The *Caulobacter* origin of replication contains several canonical DnaA binding sites, at least one of which is essential for replication (Marczynski et al., 1995). One CtrA binding site is adjacent to this essential DnaA binding site, possibly preventing DnaA from binding and initiating replication. As in *E. coli* (Skarstad and Boye, 1994), overexpression of *Caulobacter* DnaA leads to increased DNA replication initiations.

The *Caulobacter dnaA* gene is transcribed throughout the cell cycle, but the rate of transcription increases approximately fourfold at the stalked-cell stage, coincident with the initiation of DNA replication (Zweiger and Shapiro, 1994). As the swarmer cell differentiates into a stalked cell, the competition between CtrA and DnaA for access to their origin binding sites is mitigated by increased expression of *dnaA* and by proteolysis of CtrA. In *E. coli*, a similar antagonism between positive and negative factors regulates initiation of replication. Once *E. coli* replication is initiated, the negative factor SeqA interacts with hemimethylated *oriC* and prevents methylation and thus reinitiation at the origin for approximately one-third of the cell cycle (Lu et al., 1994; Slater et al., 1995). Loss of SeqA results in multiple rounds of replication initiation in *E. coli*; similarly, a CtrA loss-of-function mutation causes multiple rounds of DNA replication initiation (Quon et al., 1998). Thus, although these proteins have different roles in the control of replication initiation, both act as negative regulators.

In *Caulobacter*, several genes that encode homologues of the *E. coli* proteins involved in DNA replication have been identified, including genes for DnaA (Zweiger and Shapiro, 1994), the Tau clamp loader (DnaX) (Winzeler et al., 1997), the GyrB gyrase (Rizzo et al., 1993), and the β subunit of DNA polymerase (DnaN) (Roberts and Shapiro, 1997). All of the genes encoding these proteins exhibit maximal transcription at the swarmer-to-stalked-cell (G_1-S) transition, at the time in the cell cycle when replication enzymes are needed. Furthermore, *dnaA*, *dnaX*, *gyrB*, *dnaN*, and, to some extent, *dnaKJ* (Gomes et al., 1990) have a conserved 13-mer motif in their promoter regions (Winzeler and Shapiro, 1996; Roberts and Shapiro, 1997). None of these promoters contains a CtrA binding site, nor is the transcription of these genes under CtrA control. These observations suggest that a common factor, independent of the CtrA regulon but probably coordinately regulated, controls expression of these genes in a cell cycle-dependent manner. Site-directed mutagenesis of the 13-mer motif in the *dnaX* promoter results in a fourfold increase in promoter activity, indicating that the 13-mer may function as a repressor binding site (Winzeler and Shapiro, 1996). The factors that regulate these replication genes may be part of another signal transduction network that coordinates progression through the cell cycle.

CtrA REGULATES MULTIPLE CELL CYCLE EVENTS

Regulation of the CcrM DNA Methyltransferase

In addition to negatively regulating initiation of DNA replication, CtrA regulates the transcription of other cell cycle genes. One of CtrA's critical transcriptional targets is the gene encoding the CcrM DNA methyltransferase (Quon et al., 1996; Reisenauer et al., 1999) (Fig. 3), an essential gene required for methylation of the genome (Zweiger et al., 1994). In *Caulobacter*, DNA replication initiates in the stalked cell on a fully methylated chromosome. As replication proceeds, the newly replicated DNA strand remains unmethylated (Stephens et al., 1996) so that at the completion of replication the two chromosomes are hemimethylated. Near the end of DNA replication in the late predivisional cell, CtrA activates transcription of *ccrM* and approximately 3,000 molecules of CcrM per cell are synthesized (Berdis et al., 1998). During a short segment of the cell cycle, CcrM methylates approximately 16,000 GANTC sites, bringing both chromosomes to full methylation. Thus, the timing of DNA remethylation is established by the timing of CcrM expression, which is also under tight cell cycle control (Stephens et al., 1995, 1996; Wright et al., 1997).

The maintenance of the hemimethylated state of the genome during most of the cell cycle is due to restriction of CcrM methylation activity to the time just before cell division, thus contributing to the prevention of reinitiation events. After CcrM completes methylation of the newly replicated chromosomes, CcrM synthesis ceases and it is rapidly lost from

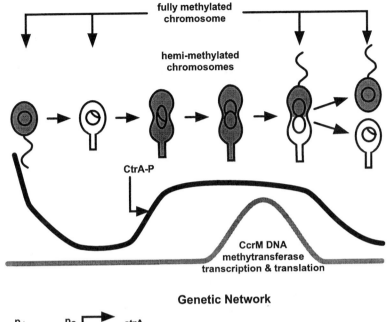

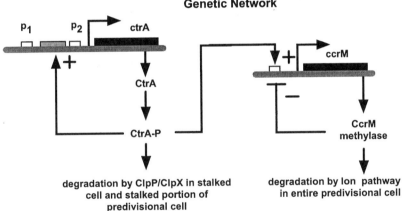

FIGURE 3 Temporal and spatial regulation of the CcrM DNA methyltransferase by the CtrA~P response regulator. The shaded areas indicate the presence of CtrA.

the cell. CcrM proteolysis depends on the presence of wild-type Lon (Wright et al., 1997). Once bound to DNA, CcrM processively methylates the N-6 position of adenine at GANTC sites on the chromosome (Berdis et al., 1998) and is highly labile unless it is bound to DNA, perhaps accounting for the short half-life of CcrM in vivo (Wright et al., 1997). Thus, as shown in Fig. 3, through both transcriptional control by CtrA and Lon-mediated proteolysis, *Caulobacter* restricts CcrM activity to a narrow time window in the cell cycle.

There are several possible mechanisms whereby differential DNA methylation may contribute to cell cycle control. In *E. coli*, methylation of GATC sequences by DAM plays a role (i) in the coordination of mismatch repair with DNA replication (Modrich, 1989), (ii) in the timing of initiation of DNA replication (Bakker and Smith, 1989; Boye and Lobner-Olesen, 1990), and (iii) in the transcriptional control of *pap* pili underlying phase variation (Braaten et al., 1994). The *E. coli pap* operon encodes proteins involved in synthesis and assembly of pili on the bacterial cell surface (Hultgren and Normark, 1991). The promoter of the *pap* operon contains two *dam* methylation sites that are differentially methylated. Methylation of one site inhibits binding of regulatory factors needed for transcription (Braaten et al., 1994). On the other hand, methylation of the other site is thought to prevent repression of *pap* transcription by inhibiting the binding of a negatively acting factor (Braaten et al., 1994). A parallel type of methylation-controlled transcription in *Caulobacter* might function to regulate the expression of some genes during the cell cycle.

An example of the methylation state influencing transcription levels in *Caulobacter* is provided by the *ccrM* gene itself. The GANTC sites recognized by CcrM occur on average once every 256 bp in the *Caulobacter* genome, but there are four sites in the *ccrM* promoter region, two nearby in an inverted repeat upstream of the −35 region and two more in an inverted repeat in the leader region. When either one or both of the GANTC methylation sites in the leader region are mutated, *ccrM* promoter activity increases and continues into the swarmer cell. These observations suggest that, under normal conditions, once the *ccrM* gene is expressed, the CcrM protein product methylates the DNA in the leader region, repressing further transcription of the gene (Stephens et al., 1995).

Because the chromosome replicates only once per cell cycle, and replication of the fully methylated chromosomes begins at a single origin and proceeds bidirectionally towards the terminus 180° away (Dingwall and Shapiro, 1989; Marczynski et al., 1990), hemimethylated regions of the chromosome are generated in an ordered and sequential fashion (Stephens et al., 1996). Thus, the change from fully methylated to hemimethylated regions of the replicating chromosome could serve as a timing mechanism to cue specific cell cycle events, such as temporally regulated transcription initiation or chromosome segregation. For example, once the replication fork passes through a given region of the chromosome, resulting in a change from a fully methylated to a hemimethylated state, that region could become a binding site for a segregation protein or, if a promoter is involved, for the activation or repression of a gene. DNA methylation is thought to have a crucial role in the *Caulobacter* cell cycle because (i) in the absence of DNA methylation the cells are not viable and (ii) in the presence of constitutively methylated chromosomes the cells are morphologically aberrant and the cell cycle is altered. Therefore, the chromosomal position, relative to the origin of replication, of critical regulatory regions with multiple methylation sequence motifs may be important elements for study in *Caulobacter* gene regulation as research progresses to genome-wide analysis.

Regulation of the FtsZ Cell Division Protein

CtrA also regulates the *ftsZ* gene that encodes an essential tubulin-like GTPase that polymerizes and forms a cytokinetic ring around the site of cell division (see chapter 15). The FtsZ ring, in a process somewhat analogous to the drawing tight of a purse string, constricts the cell at the division plane. Following division, FtsZ is lost from the progeny swarmer cell (see Fig. 6) but is present in the progeny stalked cell (Kelly et al., 1998). In the swarmer cell, CtrA represses transcription of the *ftsZ* gene by binding to a CtrA binding site in the *ftsZ* promoter. This CtrA binding site overlaps the *ftsZ* transcriptional start site and may prevent the binding of factors needed for *ftsZ* transcription, similar to the process by which CtrA negatively regulates initiation of DNA replication. When nascent swarmer cells differentiate into stalked cells, ClpXP eliminates CtrA from the cell (Jenal and Fuchs, 1998), thereby relieving transcriptional repression at the *ftsZ* promoter (Kelly et al., 1998). Transcription of *ftsZ* continues as the cell progresses to the late predivisional stage, where FtsZ once again initiates cell division.

Regulation of Flagellar Biogenesis

In addition to regulating *ccrM* and *ftsZ* expression, CtrA also regulates the initiation of the transcriptional cascade that culminates in the assembly of a polar flagellum in the predivisional cell. Flagellar genes are organized in a four-level regulatory hierarchy (see chapter 16). The class II flagellar genes are expressed earlier in the cell cycle and are required for the expression of class III and class IV flagellar genes later in the cell cycle. In addition to regulatory proteins, the class II genes encode multiple components of the type III transport system used to transport the rod, hook, and filament proteins to their site of assembly in the growing flagellum. Also in the class II category are genes encoding the membrane ring that forms the flagellar motor. As described above, *ctrA* was isolated in a screen utilizing promoter sequence motifs found in class II genes and shown to be required for the initiation of flagellum biogenesis (Quon et al., 1996).

There is much to be learned about other roles of the CtrA protein in *Caulobacter* cell regulation. What are other targets of CtrA? Do other CtrA-regulated genes contribute to the regulation of DNA replication, chromosome separation, and cell division?

ASYMMETRIC PROTEIN LOCALIZATION DURING THE CELL CYCLE

A hallmark of the *Caulobacter* cell cycle is the asymmetric deposition of proteins and subcellular structures that culminates in the production of progeny that differ from one another (Shapiro and Losick, 1997). At different times during the cell cycle, the chemotaxis machinery, the flagellum, and several pili are assembled at one pole of the cell (Fig. 4). The localization of any one of these assemblies is independent of the others. Thus, it is likely that, although they all appear at the same pole, they are recognizing different receptors or polar protein complexes.

The McpA chemoreceptor is synthesized in the early predivisional cell. It is then either directly integrated into the membrane at the pole opposite the stalk or is first integrated uniformly around the cell and then moves by diffusion, only to be trapped at the cell pole when it binds to a localized complex (Alley et al., 1993). Clues about chemoreceptor localization have come from studies of chemotaxis machinery localization in *E. coli* (Maddock and Shapiro, 1993). In *E. coli*, the chemoreceptors cluster at both poles of the cell (Maddock and Shapiro, 1993). The cytoplasmic sensor histidine kinase CheA is held in a complex with the transmembrane chemoreceptor by the adapter protein CheW (Gegner et al., 1992). Without CheA and CheW, the membrane-bound chemoreceptor is not localized but is dispersed around the cell (Maddock and Shapiro, 1993). In the absence of the chemoreceptor, CheA and CheW are randomly distributed in the cytoplasm. Thus, it is likely that complex formation is required to trap the chemotaxis machinery at the *E. coli* cell poles. When the *Caulobacter* McpA chemoreceptor is expressed in *E. coli*, it is also found at both poles of the cell. What are the factors that result in the asymmetric deposition of the chemotaxis proteins at only one pole of the *Caulobacter* cell but at both poles in *E. coli*? The unique event during the *Caulobacter* cell cycle that contributes to this asymmetry is the proteolysis of the McpA chemoreceptors at the flagellated pole of the swarmer cell during the swarmer-to-stalked-cell transition (Fig. 4). If the chemoreceptors were to be left in place at that pole, then they would persist at the site of the new stalk. Then, the newly synthesized chemoreceptors localized at the pole opposite the stalk in the predivisional cell would result in chemoreceptors at both poles, as is the case in *E. coli*. In fact, a *Caulobacter* strain bearing a mutant chemoreceptor lacking the C-terminal region required for proteolysis exhibits bipolar chemoreceptors (Alley et al., 1993). In a parallel manner, the polar flagellum is released from the cell pole during the swarmer-to-stalked-cell transition. In this case, the FliF protein, which as a multimer anchors the flagellum in the membrane and forms the M ring of the flagellar motor, is proteolytically destroyed (Jenal and

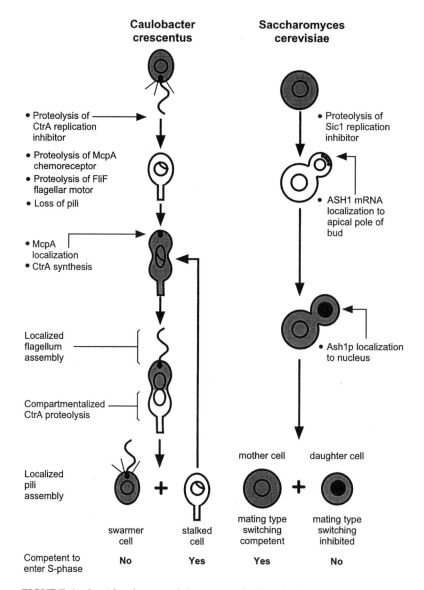

Caulobacter crescentus

- Proteolysis of CtrA replication inhibitor

- Proteolysis of McpA chemoreceptor
- Proteolysis of FliF flagellar motor
- Loss of pili

- McpA localization
- CtrA synthesis

Localized flagellum assembly

Compartmentalized CtrA proteolysis

Localized pili assembly

swarmer cell   stalked cell

Saccharomyces cerevisiae

- Proteolysis of Sic1 replication inhibitor

- ASH1 mRNA localization to apical pole of bud

- Ash1p localization to nucleus

mother cell   daughter cell

mating type switching competent   mating type switching inhibited

Competent to enter S-phase   No   Yes   Yes   No

FIGURE 4 Spatial and temporal deposition of cell cycle determinants during the *Caulobacter* and budding yeast cell cycles. The shaded areas in the *Caulobacter* cell cycle indicate the presence of CtrA, and the dot at the cell pole represents the McpA chemoreceptor. The shaded areas in the yeast cell cycle indicate the presence of the Sic1 replication inhibitor. Neither the *Caulobacter* swarmer cell progeny nor the yeast daughter cell (which is two-thirds the size of the mother cell) is competent for replication initiation (entry into S phase) until later in the cell cycle.

Shapiro, 1996). Consequently, the flagellum is released from the pole and a new stalk grows at the site previously occupied by the flagellum. A new flagellum is then synthesized and assembled at the pole opposite the stalk in the predivisional cell.

Although the protease(s) responsible for McpA and FliF turnover has not yet been identified, proteolysis clearly plays a vital role in *Caulobacter*'s asymmetry. However, how the chemotaxis machinery or the flagellum is localized to the cell pole remains an unsolved problem. Furthermore, several pili are assembled at the flagellated pole of the swarmer cell (Sommer and Newton, 1988), and the mechanism of their localization is also unknown. These pili are lost from the swarmer pole, possibly by retraction, only to be resynthesized later in the cell cycle (Schmidt, 1966; Agabian-Keshishian and Shapiro, 1971; Shapiro et al., 1971; Smit and Agabian, 1982). Although assembly of the flagellum, chemoreceptors, and polar pili in *Caulobacter* coincide spatially, these polar organelles may not utilize common mechanisms to establish their polar deposition. Strains bearing mutations in any one of these nonessential organelles leave intact the polar localization of the remaining organelles.

The generation of dissimilar progeny cells in both prokaryotes and eukaryotes frequently depends on asymmetric localization of regulatory factors prior to division. The asymmetric deposition of regulatory proteins and their cell cycle control in *Caulobacter* has striking parallels with budding yeast (Jacobs and Shapiro, 1998). At division in both *Caulobacter* and budding yeast, there are cell fate determinants that are asymmetrically distributed to the progeny cells (Fig. 4). At division, CtrA is present only in the *Caulobacter* swarmer progeny and the Ash1 protein is present only in the yeast bud. The Ash1 protein represses the expression of the HO endonuclease (Bobola et al., 1996; Sil and Herskowitz, 1996), required for switching between the **a** and α mating types. Only a mother cell can switch mating type; a daughter cell must undergo a division before it can switch. Although the segregation pattern of Ash1 pro-

tein resembles the segregation of CtrA, it occurs by a different mechanism. The localization of Ash1 involves the cytoskeleton-dependent movement of *ASH1* mRNA to the distal tip of the emerging bud (Long et al., 1997; Takizawa et al., 1997). The Ash1 protein is then localized to the nucleus. The protein remains in the new daughter cell but decays during progression through the cell cycle, as the daughter cell becomes a mother cell competent for mating-type switching. Despite the difference in localization mechanisms in eukaryotic and prokaryotic systems, in each case the asymmetric distribution of regulatory factors during the cell cycle controls cell fate (Amon, 1998).

CONCLUSIONS

The periodicity of DNA replication, cell division, and, in *Caulobacter*, cell cycle-dependent morphological and behavioral differences contrasts with the continuous nature of most metabolic reactions that produce cellular growth. The cell cycle regulatory machinery initiates each of these discontinuous events in sequence at the appropriate time in the cycle and somehow coordinates the process with cell growth. Since *Caulobacter* differentiates by asymmetric cell division, it must coordinate cell division not only with DNA replication and cell growth but also with developmental events. Figure 5 shows the known network of interacting genetic- and protein-level reactions that control the sequencing of events in the *Caulobacter* cell cycle. Additional proteins in the network are being actively sought, with particular emphasis on the proteins that control the various replication proteins and the CtrA and CcrM degradation machinery. A key mechanism is control of CtrA activity through phosphorylation. A phosphorylation cascade, possibly mediated by the CckA histidine kinase (Jacobs et al., 1999), is thought to be central to that function (see chapter 17).

As the regulatory and functional components of the *Caulobacter* cell cycle emerge, numerous functional parallels with eukaryotic cell cycle regulation are becoming apparent. The functional parallels include:

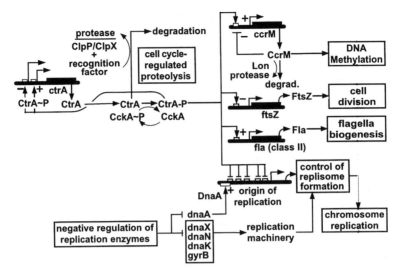

FIGURE 5 A portion of the regulatory circuit that controls the *Caulobacter* cell cycle.

- Phosphorylation and proteolysis as interdependent partners that collaborate to order key biochemical events.
- Use of the chemical irreversibility of proteolysis to provide directionality at critical steps of the cell cycle.
- The central role of proteolysis in regulating the timing of cell cycle transitions.
- Active localization of key regulatory proteins to effect asymmetry in daughter cell progeny.
- Ordered assembly of many proteins at the origins of DNA replication to form a competent prereplicative chromosomal state.
- Use of cell cycle-regulated protein kinase pathways to repress initiation of DNA replication at all times except immediately after entry into the stalked-cell state.

A prominent difference between prokaryotic and eukaryotic cell cycle control is the lack of intrinsic growth limitations in bacterial cells. Bacteria are designed to divide ceaselessly so long as there are sufficient nutrients. In contrast, most mammalian cell types are programmed to eventually cease dividing. Although the evidence is not yet definitive, it is looking more and more as though one can argue for conservation of mechanisms and thus for common ancient origins of cell cycle control in all organisms. The study of these bacterial cell cycle control mechanisms is not only of intrinsic interest, it will also help unravel the complexities of cell cycle regulation in more complex biological systems.

Figure 6 shows the timing of events in the *Caulobacter* cell cycle regulatory network versus the respective phases of the swarmer and stalked-cell cycles. The two cell cycles are shown schematically at the top of the figure. The swarmer progeny's cell cycle is one-third again as long as the stalked progeny's cell cycle. The swarmer progeny cell is flagellated, motile, and incompetent for initiating DNA replication for approximately one-third of its cell cycle. At the functional equivalent of the G_1-S transition, the cell turns on the proteolysis of CtrA, allowing the initiation of DNA replication in the new stalked cell. Once replication has initiated, a critical event is the shutting off of CtrA proteolysis and the induction of *ctrA* transcription, which allows the accumulation of new CtrA. Once this new CtrA is phosphorylated, it acts to prevent the reinitiation of DNA replication.

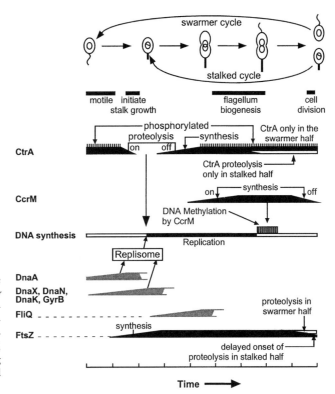

FIGURE 6 Relative timing of multiple cell cycle events during the *Caulobacter* swarmer and stalked-cell cycles. The synthesis of the CtrA, CcrM, and FtsZ proteins is indicated by solid bars, and the induction of transcription of *dnaA*, *dnaX*, *dnaN*, *dnaK*, *gyrB*, and the gene encoding the FliQ class II flagellar gene is indicated by shaded bars.

Phosphorylated CtrA also represses initiation of cell division except in the time window early in the stalked-cell life cycle when proteolysis clears CtrA from the cell. The removal of CtrA repression of *ftsZ* allows the FtsZ protein that is critical to cell division to begin accumulating. As CtrA is newly synthesized in the predivisional cell, flagellum biogenesis is initiated with production of the class II flagellar genes, represented in Fig. 6 by FliQ. Recently, a protein was identified which represses transcription of DnaA, DnaX, and other genes essential for formation of the replisome until the swarmer-to-stalked-cell transition (Fig. 1) (Keiler and Shapiro, unpublished). Repression is relieved in time to initiate DNA replication when origin repression by CtrA is eliminated in the stalked cell. It is not yet known how the cell cycle regulatory network "closes" upon itself, that is, the complete set of coupled cyclical reactions that drive growth and division continuously forward.

A complete understanding of cell cycle regulation must include a description of not only the key events and the signals which trigger them but also the various coupling mechanisms that coordinate replication initiation, nucleoid separation, and septation initiation and that permit the cell to operate with high reliability.

REFERENCES

Agabian-Keshishian, N., and L. Shapiro. 1971. Bacterial differentiation and phage infection. *Virology* **44**:46–53.

Alley, M. R., J. R. Maddock, and L. Shapiro. 1992. Polar localization of a bacterial chemoreceptor. *Genes Dev.* **6**:825–836.

Alley, M. R., J. R. Maddock, and L. Shapiro. 1993. Requirement of the carboxyl terminus of a bacterial chemoreceptor for its targeted proteolysis [see comments]. *Science* **259**:1754–1757.

Amon, A. 1998. Controlling cell cycle and cell fate: common strategies in prokaryotes and eukaryotes. *Proc. Natl. Acad. Sci. USA* **95**:85–86.

Appleby, J. L., J. S. Parkinson, and R. B. Bourret. 1996. Signal transduction via the multi-step phosphorelay: not necessarily a road less traveled. *Cell* **86:**845–848.

Bakker, A., and D. W. Smith. 1989. Methylation of GATC sites is required for precise timing between rounds of DNA replication in *Escherichia coli. J. Bacteriol.* **171:**5738–5742.

Barrett, J. T., R. H. Croft, D. M. Ferber, C. J. Gerardot, P. V. Schoenlein, and B. Ely. 1982a. Genetic mapping with Tn5–derived auxotrophs of *Caulobacter crescentus. J. Bacteriol.* **151:**888–898.

Barrett, J. T., C. S. Rhodes, D. M. Ferber, B. Jenkins, S. A. Kuhl, and B. Ely. 1982b. Construction of a genetic map for *Caulobacter crescentus. J. Bacteriol.* **149:**889–896.

Berdis, A. J., I. Lee, J. K. Coward, C. Stephens, R. Wright, L. Shapiro, and S. J. Benkovic. 1998. A cell cycle-regulated adenine DNA methyltransferase from *Caulobacter crescentus* processively methylates GANTC sites on hemimethylated DNA. *Proc. Natl. Acad. Sci. USA* **95:**2874–2879.

Bobola, N., R. P. Jansen, T. H. Shin, and K. Nasmyth. 1996. Asymmetric accumulation of Ash1p in postanaphase nuclei depends on a myosin and restricts yeast mating-type switching to mother cells. *Cell* **84:**699–709.

Boye, E., and A. Lobner-Olesen. 1990. The role of dam methyltransferase in the control of DNA replication in *E. coli. Cell* **62:**981–989.

Braaten, B. A., X. Nou, L. S. Kaltenbach, and D. A. Low. 1994. Methylation patterns in pap regulatory DNA control pyelonephritis-associated pili phase variation in *E. coli. Cell* **76:**577–588.

Bramhill, D., and A. Kornberg. 1988. Duplex opening by dnaA protein at novel sequences in initiation of replication at the origin of the *E. coli* chromosome. *Cell* **52:**743–755.

Burbulys, D., K. A. Trach, and J. A. Hoch. 1991. Initiation of sporulation in *B. subtilis* is controlled by a multicomponent phosphorelay. *Cell* **64:**545–552.

Crooke, E. 1995. Regulation of chromosomal replication in *E. coli:* sequestration and beyond. *Cell* **82:**877–880.

Degnen, S. T., and A. Newton. 1972. Dependence of cell division on the completion of chromosome replication in *Caulobacter. J. Bacteriol.* **110:**852–856.

Devos, D., and J. J. Letesson. Unpublished data.

Dingwall, A., and L. Shapiro. 1989. Rate, origin, and bidirectionality of *Caulobacter* chromosome replication as determined by pulsed-field gel electrophoresis. *Proc. Natl. Acad. Sci. USA* **86:**119–123.

Domian, I., A. Reisenauer, and L. Shapiro. 1999. Feedback control of a master bacterial cell-cycle regulator. *Proc. Natl. Acad. Sci. USA* **96:**6648–6653.

Domian, I. J., K. C. Quon, and L. Shapiro. 1997. Cell type-specific phosphorylation and proteolysis of a transcriptional regulator controls the G1-to-S transition in a bacterial cell cycle. *Cell* **90:**415–424.

Ely, B., and C. J. Gerardot. 1988. Use of pulsed-field-gradient gel electrophoresis to construct a physical map of the *Caulobacter crescentus* genome. *Gene* **68:**323–333.

Evinger, M., and N. Agabian. 1977. Envelope-associated nucleoid from *Caulobacter crescentus* stalked and swarmer cells. *J. Bacteriol.* **132:**294–301.

Gegner, J. A., D. R. Graham, A. F. Roth, and F. W. Dahlquist. 1992. Assembly of an MCP receptor, CheW, and kinase CheA complex in the bacterial chemotaxis signal transduction pathway. *Cell* **70:**975–982.

Golden, J. W., and H.–S. Yoon. 1998. Heterocyst formation in *Anabaena. Curr. Opin. Microbiol.* **1:**623–629.

Gomes, S. L., J. W. Gober, and L. Shapiro. 1990. Expression of the *Caulobacter* heat shock gene *dnaK* is developmentally controlled during growth at normal temperatures. *J. Bacteriol.* **172:**3051–3059.

Grossman, A. D. 1995. Genetic networks controlling the initiation of sporulation and the development of genetic competence in *Bacillus subtilis. Annu. Rev. Genet.* **29:**477–508.

Guo, S., and K. J. Kemphues. 1996. Molecular genetics of asymmetric cleavage in the early *Caenorhabditis elegans* embryo. *Curr. Opin. Genet. Dev.* **6:**408–415.

Hartwell, L. 1995. Introduction to cell cycle controls, p. 1–15. In C. Hutchison (ed.), *Cell Cycle Control.* Oxford University Press, New York, N.Y.

Hecht, G. B., T. Lane, N. Ohta, J. M. Sommer, and A. Newton. 1995. An essential single domain response regulator required for normal cell division and differentiation in *Caulobacter crescentus. EMBO J.* **14:**3915–3924.

Hecht, G. B., and A. Newton. 1995. Identification of a novel response regulator required for the swarmer-to-stalked-cell transition in *Caulobacter crescentus. J. Bacteriol.* **177:**6223–6229.

Helmstetter, C. E., and A. C. Leonard. 1987. Coordinate initiation of chromosome and minichromosome replication in *E. coli. J. Bacteriol.* **169:**3489–3494.

Hoch, J. A. 1993. Regulation of the phosphorelay and the initiation of sporulation in *Bacillus subtilis. Annu. Rev. Microbiol.* **47:**441–465.

Hoch, J. A., and T. J. Silhavy (ed.). 1995. *Two-Component Signal Transduction.* American Society for Microbiology, Washington, D.C.

Horvitz, H. R., and I. Herskowitz. 1992. Mechanisms of asymmetric cell division: two Bs or not two Bs, that is the question. *Cell* **68:**237–255.

Hultgren, S. J., and S. Normark. 1991. Biogenesis of the bacterial pilus. *Curr. Opin. Genet. Dev.* **1:**313–318.

Hung, D., M. Barnett, R. M. Long, and L. Shapiro. Unpublished data.

Jacobs, C., I. Domian, J. Maddock, and L. Shapiro. 1999. Cell cycle-dependent polar localization of an essential bacterial histidine kinase that controls DNA replication and cell division. *Cell* 97: 111–120.

Jacobs, C., and L. Shapiro. 1998. Microbial asymmetric cell division: localization of cell fate determinants. *Curr. Opin. Genet. Dev.* 8:386–391.

Jenal, U., and T. Fuchs. 1998. An essential protease involved in bacterial cell-cycle control. *EMBO J.* 17:5658–5669.

Jenal, U., and L. Shapiro. 1996. Cell cycle-controlled proteolysis of a flagellar motor protein that is asymmetrically distributed in the *Caulobacter* predivisional cell. *EMBO J.* 15:2393–2406.

Keiler, K., and L. Shapiro. Unpublished data.

Kelly, A. J., M. J. Sackett, N. Din, E. Quardokus, and Y. V. Brun. 1998. Cell cycle-dependent transcriptional and proteolytic regulation of FtsZ in *Caulobacter*. *Genes Dev.* 12:880–893.

Laachouch, J. E., L. Desmet, V. Geuskens, R. Grimaud, and A. Toussaint. 1996. Bacteriophage *Mu* repressor as a target for the *Escherichia coli* ATP-dependent Clp protease. *EMBO J.* 15: 437–444.

Levchenko, I., C. K. Smith, N. P. Walsh, R. T. Sauer, and T. A. Baker. 1997. PDZ-like domains mediate binding specificity in the Clp/Hsp100 family of chaperones and protease regulatory subunits. *Cell* 91:939–947.

Long, R. M., R. H. Singer, X. Meng, I. Gonzales, K. Nasmyth, and R. Jensen. 1997. Mating-type switching in yeast controlled by asymmetric localization of ASH1 mRNA. *Science* 277:383–387.

Lu, B., L. Y. Jan, and Y. N. Jan. 1998. Asymmetric cell division: lessons from flies and worms [in process citation]. *Curr. Opin. Genet. Dev.* 8:392–399.

Lu, M., J. L. Campbell, E. Boye, and N. Kleckner. 1994. SeqA: a negative modulator of replication initiation in *E. coli*. *Cell* 77:413–426.

Maddock, J. R., and L. Shapiro. 1993. Polar location of the chemoreceptor complex in the *Escherichia coli* cell. *Science* 259:1717–1723.

Marczynski, G. T., A. Dingwall, and L. Shapiro. 1990. Plasmid and chromosomal DNA replication and partitioning during the *Caulobacter crescentus* cell cycle. *J. Mol. Biol.* 212:709–722.

Marczynski, G. T., K. Lentine, and L. Shapiro. 1995. A developmentally regulated chromosomal origin of replication uses essential transcription elements. *Genes Dev.* 9:1543–1557.

Marczynski, G. T., and L. Shapiro. 1992. Cell-cycle control of a cloned chromosomal origin of replication from *Caulobacter crescentus*. *J. Mol. Biol.* 226:959–977.

Mendenhall, M. D. 1993. An inhibitor of p34CDC28 protein kinase activity from *Saccharomyces cerevisiae*. *Science* 259:216–219.

Messer, W., and M. Noyer-Weidner. 1988. Timing and targeting: the biological functions of Dam methylation in *E. coli*. *Cell* 54:735–737.

Modrich, P. 1989. Methyl-directed DNA mismatch correction. *J. Biol. Chem.* 264:6597–6600.

Ohta, N., and A. Newton. 1996. Signal transduction in the cell cycle regulation of Caulobacter differentiation. *Trends Microbiol.* 4:326–332.

Osley, M. A., and A. Newton. 1974. Chromosomes segregation and development in *Caulobacter crescentus*. *J. Mol. Biol.* 90:359–370.

Parkinson, J. S. 1993. Signal transduction schemes of bacteria. *Cell* 73:857–871.

Posas, F., and H. Saito. 1998. Activation of the yeast SSK2 MAP kinase by the SSK1 two-component response regulator. *EMBO J.* 17:1385–1394.

Posas, F., S. M. Wurgler-Murphy, T. Maeda, E. A. Witten, T. C. Thai, and H. Saito. 1996. Yeast HOG1 MAP kinase cascade is regulated by a multistep phosphorelay mechanism in the SLN1-YPD1-SSK1 "two-component" osmosensor. *Cell* 86:865–875.

Quon, K. C., G. T. Marczynski, and L. Shapiro. 1996. Cell cycle control by an essential bacterial two-component signal transduction protein. *Cell* 84:83–93.

Quon, K. C., B. Yang, I. J. Domian, L. Shapiro, and G. T. Marczynski. 1998. Negative control of bacterial DNA replication by a cell cycle regulatory protein that binds at the chromosome origin. *Proc. Natl. Acad. Sci. USA* 95:120–125.

Reisenauer, A., K. Quon, and L. Shapiro. 1999. The CtrA response regulator mediates temporal control gene expression during the *Caulobacter* cell cycle. *J. Bacteriol.* 181:2430–2439.

Rizzo, M. F., L. Shapiro, and J. Gober. 1993. Asymmetric expression of the gyrase B gene from the replication-competent chromosome in the *Caulobacter crescentus* predivisional cell. *J. Bacteriol.* 175:6970–6981.

Roberts, R. C., and L. Shapiro. 1997. Transcription of genes encoding DNA replication proteins is coincident with cell cycle control of DNA replication in *Caulobacter crescentus*. *J. Bacteriol.* 179: 2319–2330.

Schmidt, J. M. 1966. Observations on the adsorption of *Caulobacter* bacteriophages containing ribonucleic acid. *J. Gen. Microbiol.* 45:347–353.

Schwob, E., T. Bohm, M. D. Mendenhall, and K. Nasmyth. 1994. The B-type cyclin kinase inhibitor p40SIC1 controls the G1 to S transition in *S. cerevisiae*. *Cell* 79:233–244. (Erratum, 84:175, 1996.)

Shapiro, L., N. Agabian-Keshishian, and I.

Bendis. 1971. Bacterial differentiation. *Science* **173:** 884–892.

Shapiro, L., and R. Losick. 1997. Protein localization and cell fate in bacteria. *Science* **276:**712–718.

Sil, A., and I. Herskowitz. 1996. Identification of asymmetrically localized determinant, Ash1p, required for lineage-specific transcription of the yeast HO gene. *Cell* **84:**711–722.

Skarstad, K., and E. Boye. 1994. The initiator protein DnaA: evolution, properties and function. *Biochim. Biophys. Acta* **1217:**111–130.

Slater, S., S. Wold, M. Lu, E. Boye, K. Skarstad, and N. Kleckner. 1995. *E. coli* SeqA protein binds *oriC* in two different methyl-modulated reactions appropriate to its roles in DNA replication initiation and origin sequestration. *Cell* **82:**927–936.

Smit, J., and N. Agabian. 1982. *Caulobacter crescentus* pili: analysis of production during development. *Dev. Biol.* **89:**237–247.

Sommer, J. M., and A. Newton. 1988. Sequential regulation of developmental events during polar morphogenesis in *Caulobacter crescentus:* assembly of pili on swarmer cells requires cell separation. *J. Bacteriol.* **170:**409–415.

Stephens, C., A. Reisenauer, R. Wright, and L. Shapiro. 1996. A cell cycle-regulated bacterial DNA methyltransferase is essential for viability. *Proc. Natl. Acad. Sci. USA* **93:**1210–1214.

Stephens, C. M., G. Zweiger, and L. Shapiro. 1995. Coordinate cell cycle control of a *Caulobacter* DNA methyltransferase and the flagellar genetic hierarchy. *J. Bacteriol.* **177:**1662–1669.

Takizawa, P. A., A. Sil, J. R. Swedlow, I. Herskowitz, and R. D. Vale. 1997. Actin-dependent localization of an RNA encoding a cell-fate determinant in yeast. *Nature* **389:**90–93.

Uhl, M. A., and J. F. Miller. 1996a. Central role of the BvgS receiver as a phosphorylated intermediate in a complex two-component phosphorelay. *J. Biol. Chem.* **271:**33176–33180.

Uhl, M. A., and J. F. Miller. 1996b. Integration of multiple domains in a two-component sensor protein: the *Bordetella pertussis* BvgAS phosphorelay. *EMBO J.* **15:**1028–1036.

Winzeler, E., and L. Shapiro. 1996. A novel promoter motif for *Caulobacter* cell cycle-controlled DNA replication genes. *J. Mol. Biol.* **264:**412–425.

Winzeler, E., R. Wheeler, and L. Shapiro. 1997. Transcriptional analysis of the *Caulobacter* 4.5 S RNA ffs gene and the physiological basis of an ffs mutant with a Ts phenotype. *J. Mol. Biol.* **272:** 665–676.

Wright, R., C. Stephens, and L. Shapiro. 1997. The CcrM DNA methyltransferase is widespread in the alpha subdivision of proteobacteria, and its essential functions are conserved in *Rhizobium meliloti* and *Caulobacter crescentus. J. Bacteriol.* **179:** 5869–5877.

Wu, J., N. Ohta, and A. Newton. 1998. An essential, multicomponent signal transduction pathway required for cell cycle regulation in *Caulobacter. Proc. Natl. Acad. Sci. USA* **95:**1443–1448.

Zweiger, G., G. Marczynski, and L. Shapiro. 1994. A *Caulobacter* DNA methyltransferase that functions only in the predivisional cell. *J. Mol. Biol.* **235:**472–485.

Zweiger, G., and L. Shapiro. 1994. Expression of *Caulobacter dnaA* as a function of the cell cycle. *J. Bacteriol.* **176:**401–408.

PATHOGENESIS AND SYMBIOSIS

SWARMING MIGRATION BY *PROTEUS* AND RELATED BACTERIA

Gillian M. Fraser, Richard B. Furness, and Colin Hughes

19

Swarming motility is a specialized form of bacterial translocation across solid surfaces. Unlike swimming motility through liquid, swarming is a multicellular process. It involves the differentiation of short vegetative cells into elongated hyperflagellated swarm cells that undergo rapid and coordinated population migration. Swarming has been observed in a number of flagellated genera, both gram negative and gram positive, including *Proteus*, *Vibrio*, *Serratia*, *Bacillus*, and *Clostridium* (Henrichsen, 1972). More recently, swarming by *Salmonella* and *Escherichia coli* has been described (Harshey, 1994; Guard-Petter, 1997). Swarming provides an experimentally accessible example of bacterial differentiation and multicellular behavior and, as with other instances of prokaryotic development, its study addresses mechanisms underlying flagellum biogenesis, cell division, signaling, and migration (Fig. 1). In addition, it highlights how differentiation can be coupled to bacterial pathogenesis. This chapter reviews current knowledge of bacterial swarming and the likely underlying mechanisms, focusing principally on studies of *Proteus mirabilis*.

THE MULTICELLULAR SWARMING OF *P. MIRABILIS*

When swarming of *P. mirabilis* was reported by Hauser in 1885, he named the differentiating genus after *Proteus*, "the Greek god of the ocean who took many shapes to escape questioning" (Williams and Schwarzhoff, 1978). Since then, the bacterium has been a continuing mystery and a familiar nuisance in clinical microbiology, as it obscures other bacterial colonies growing on solid media. Swarming *P. mirabilis* produces very large spreading colonies in which concentric zonation, or terracing, results from periodic cycles of mass migration interspersed with population growth without expansion of the colony edge (Williams and Schwarzhoff, 1978; Allison and Hughes, 1991a; Belas, 1992; Rauprich et al., 1996; Shapiro, 1998) (Fig. 2).

Three stages are apparent. (i) Cells inoculated onto a rich solid growth medium initially grow and divide normally into a colony of short vegetative rods (2 to 4 μm long; 0.8 μm wide). After a variable lag phase (dependent on the density of the inoculum, medium composition, and agar concentration), vegetative cells at the colony margin differentiate into long (20 to 80 μm long; 0.7 μm wide), polyploid, aseptate swarm cells which possess up to 50-fold more flagella per unit cell surface area, i.e., 500

G. M. Fraser, R. B. Furness, and C. Hughes, Department of Pathology, University of Cambridge, Tennis Court Road, Cambridge, CB2 1QP, United Kingdom.

Prokaryotic Development, edited by Y. V. Brun and L. J. Shimkets,
© 2000 American Society for Microbiology, Washington, DC 20005-4171

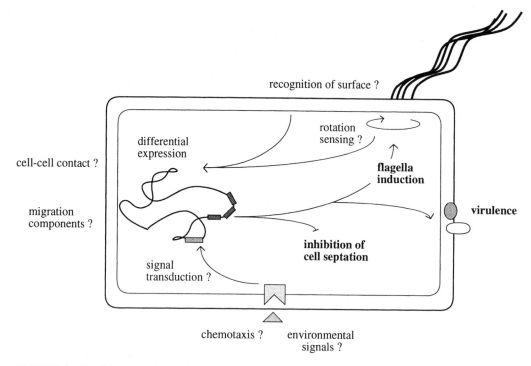

FIGURE 1 Possible mechanisms of swarming. A simple view of a swarm cell indicating putative underlying mechanisms that have been the foci of study.

to 5,000 flagella on elongated swarm cells compared to 1 to 10 flagella on vegetative cells (Klienberger-Nobel, 1947; Hoeniger, 1964, 1965; Jones and Park, 1967a; Bisset, 1973a, 1973b). Differentiation into swarm cells presumably involves recognition of signals that trigger the change from vegetative cells, in particular, the activation of flagellar gene expression and the repression of cell division. (ii) The swarm cells assemble into groups, or rafts, and migrate away from the original colony. There is intense activity at the colony edge as many thousands of flagella move in synchrony, creating oscillating waves as the swarm cells spread rapidly across the surface. There is also constant streaming of differentiated cells behind the advancing front. Swarming is a strictly multicellular phenomenon, as single cells emerging from the moving mass are stranded and cease to migrate (Sturdza, 1973). The velocity and duration of migration are variable among bacterial strains and are dependent upon culture conditions. (iii) After 2 to 4 h, swarm cells eventually stop migrating (consolidate) and revert by division to the short form (Hoeniger, 1964, 1966; Bisset, 1973a, 1973b). During the first consolidation phase, a wave of cell multiplication, involving growth and division of undifferentiated cells as well as dedifferentiation of stationary swarm cells, moves outward from the point of inoculation until it reaches the colony edge. As the second swarming phase begins (between 1 and 3 h after cessation of the first phase), the first wave of consolidation continues to expand outward for a short time and produces the final boundary of the first swarm terrace (Rauprich et al., 1996). During subsequent consolidation phases, a wave of multiplication is not observed but instead there is a uniform thickening of the newly colonized area. Although the processes of migration and consolidation are not absolutely temporally dis-

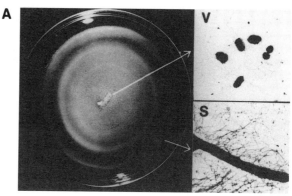

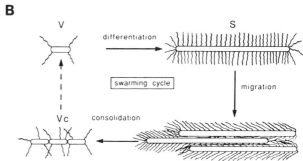

FIGURE 2 Swarming on solid growth medium by *P. mirabilis*. (A) Swarming of *P. mirabilis* inoculated in the center of a Petri plate containing complex growth medium solidified by 2% agar. V, vegetative cells from the center of the colony; S, part of a hyperflagellated swarm cell from the migrating colony edge. (B) Stages in the swarming cycle of *P. mirabilis* growing on solid medium. (From Allison and Hughes, 1991a.)

crete, each migration-plus-consolidation phase can be viewed as a single cycle producing a single swarm terrace (Rauprich et al., 1996).

Common sense suggests that population migration would be a powerful means of rapid colonization of nutrient-rich environments, facilitating the spread of the colony and accelerating the production of biomass. Swarming seems to be especially common in the medically important *Enterobacteriaceae* family. While *P. mirabilis* can swarm over growth medium containing high agar concentrations (up to 2%), *Serratia* species exhibit a comparable, although less vigorous, behavior on media solidified with 0.5 to 0.8% agar (Alberti and Harshey, 1990). Recently, some highly virulent strains of *Salmonella enteritidis* have been reported to differentiate and spread over defined medium containing 1.4 to 2% agar (Guard-Petter, 1997, 1998; Guard-Petter et al. 1997), and low agar concentrations support swarming of laboratory strains of *S. typhimurium* and *E. coli* (Harshey and Matsuyama, 1994). While it

can reasonably be assumed that these closely related genera are undertaking an analogous process, it is also probable that high-percentage agar selects for isolates with cell surface properties that potentiate migration under conditions likely to be encountered outside the laboratory.

The transition of *Proteus* cells through differentiation, migration, and consolidation is marked by a number of physiological and biochemical changes. Swarm cells synthesize less nucleic acid and protein than vegetative cells (Armitage, 1981), and they exhibit reduced oxygen uptake and lower activities of enzymes, such as β-galactosidase, tryptophanase, and alkaline phosphatase (Falkinham and Hoffman, 1984; Allison et al., 1992b). Repression of biosynthetic pathways during swarming may offset the commitment required for the concomitant hyperexpression and function of flagella and, as we will see later, virulence factors.

SWARMING AND VIRULENCE

Proteus species are widely distributed in the environment, e.g., in soil and polluted water, as well as in the intestines of animals, including humans. They can also cause infections, particularly those ascending the urinary tract, i.e., to the kidney (Peerboorns et al., 1985; Mobley and Belas, 1995). Urinary tract infection is especially common in patients with indwelling urinary catheters (Warren et al., 1987) and is frequently persistent and difficult to treat, partly due to bladder and kidney stone formation (Krajden et al., 1984; Rosenstein, 1986). *Proteus* species can also cause opportunistic infections of the respiratory tract, wounds, and burns, as well as gastroenteritis (Penner, 1992; Cooper et al., 1971). There is also speculation that *Proteus* infections may contribute to the autoimmune diseases myasthenia gravis (Stefansson, 1985) and rheumatoid arthritis, particularly as a region of the *P. mirabilis* hemolysin toxin has identity with the HLA-DR4 antigen, expression of which is associated with susceptibility to rheumatoid disease (Ebringer, 1985; Deighton, 1992).

A role for bacterial motility has been demonstrated in a variety of pathogen-host interactions (Otteman and Miller, 1997). In an in vivo model of ascending urinary tract infection, it was shown that motile but swarming-defective mutants of *P. mirabilis* were unable to establish kidney infection (Allison et al., 1994), suggesting that swarming migration additionally facilitates ascending colonization of the urinary tract. This may be coupled to biofilm formation, which can lead to catheter colonization and obstruction (Stickler et al., 1993a, 1993b). While the production of hyperflagellated swarm cells has an important function in establishing infection, flagella are strongly immunogenic bacterial surface antigens which are likely to elicit a host immune response. This survival problem may be overcome by variation of the flagellin antigen in a subset of the swarm cell population (Belas, 1994; Murphy and Belas, 1999). *P. mirabilis* carries three flagellin-encoding genes, closely related to those of *E. coli* and *Salmonella typhimurium*. One is expressed and two are silent (Belas and Flaherty, 1994), and gene conversion of these loci results in the expression of an antigenically distinct flagellin (Belas, 1994; Murphy and Belas, 1999).

The ability of *Proteus* to undertake cell inva-sion (Peerboorns et al., 1984; Rozalski et al., 1986; Allison et al., 1994) may not only serve as a means of avoiding the host immune response, it might also promote dissemination of *Proteus* within host tissues. Invasion studies with cultured human urothelial cells showed that the invasive ability of *P. mirabilis* changed over the differentiation cycle, with swarm cells being 15-fold more invasive than vegetative cells (Allison et al., 1992a, 1992b). Histological analysis of renal tissues from mice infected by wild-type *P. mirabilis* revealed that differentiated cells were the major invasive cell type (Allison et al., 1994). In contrast, in vitro studies with different *P. mirabilis* isolates and with human renal proximal tubular epithelial cells or mouse fibroblasts indicated that short vegetative rods were the invasive cell form (Chippendale et al., 1994; Rozalski et al., 1997).

In addition to enabling rapid population migration and possibly cell invasion, differentiation into swarm cells is coupled to the coordinate hyperexpression of several virulence proteins (Allison et al., 1992b), including a cytolytic hemolysin toxin (HpmA) (Senior and Hughes, 1987; Swihart and Welch, 1990; Uphoff and Welch, 1990; Mobley et al, 1991), a urease that precipitates the formation of bladder and kidney stones (Guo and Lin, 1965; Senior et al., 1980; Rosenstein et al., 1981; Zhao et al., 1998), and an extracellular metalloprotease that hydrolyzes immunoglobulin IgA, immunoglobulin IgG, and type IV collagen (Loomes et al., 1990; Wassif et al., 1995). The virulence factors of *Proteus* have recently been reviewed by Rozalski et al. (1997). During the differentiation cycle, urease and protease activities increase ca. 10-fold, with peak urease expression occurring in swarm cells and maximal protease expression occurring in cells just prior to consolidation (Allison et al., 1992b). Expression of flagella and the HpmA hemolysin are modulated in parallel, with peak hemolytic activity in swarm cells being ca. 20-fold higher than that in vegetative cells (Allison et al., 1992b; Fraser et al., unpublished), and in nonswarming flagellar gene mutants hemolysin is not hyperexpressed (Gygi et al., 1995a, 1997). The close coupling of hemolysin and flagellar gene expression during differentiation suggests that their transcription may be regulated by a common factor. Although there is no documented evidence that *Serratia* virulence is strongly coupled to swarming, it is possible that it is, and in *Serratia liquefaciens*, the *phlAB* operon, encoding phospholipase, is transcribed as part of the flagellar regulon from a putative σ^{28} promoter (Givskov et al., 1995a, 1995b). The view of swarming as a factor in pathogenicity is strengthened by the finding that virulent *Salmonella* commonly exhibit swarming-like behavior (Guard-Petter et al., 1997).

MOLECULAR ANALYSIS OF SWARMING DIFFERENTIATION AND MIGRATION

While molecular and genetic investigations of prokaryotic development and multicellular behavior have focused on nonpathogenic bacteria, such as *Myxococcus xanthus*, *Caulobacter crescentus*, and *Bacillus subtilis*, described elsewhere in this book, genetic loci involved in swarming have been identified in several pathogenic gram-negative bacteria (McCarter et al., 1988; Allison and Hughes, 1991a; Belas et al., 1991a, 1991b; Harshey, 1994; Givskov et al., 1995a, 1995b). Swarming-defective transposon mutants of *P. mirabilis* have been categorized into five broad phenotypic classes, spanning the three phases (differentiation, migration, and consolidation) of swarming behavior (Fig. 3): (i) nonmotile nonswarming mutants that can neither swim through liquid media nor differ-

entiate and migrate across solid agar surfaces; (ii) motile nonswarming mutants that, despite being motile in liquid, cannot undergo surface translocation; (iii) swarmers that produce an aberrant "dendritic" pattern of migration with normally spaced terraces; and (iv and v) frequent and infrequent consolidators, in which either migration velocity or temporal control of the swarm cycle is altered such that a pattern of tightly packed or rare terraces is formed (Allison and Hughes, 1991b). In addition to macroscopic swarming colony defects, many mutants are also profoundly disabled in their ability to differentiate into elongated, hyperflagellated cells that hyperexpress virulence proteins (Gygi et al., 1995a, 1997; Hay et al., 1997), suggesting a close regulatory coupling of cell division, flagellum biogenesis, and virulence gene expression. Swarming-impaired transposon mutants have also been classified in terms of cell elongation, in particular, as elongation negative (Elo$^-$) or constitutively elongated (Elo^C) (Belas et al., 1995).

Mutation frequencies suggest that many genes are involved in directing not only swimming motility but also swarm cell differentiation and migration (Allison and Hughes, 1991a, 1991b; Belas et al., 1991a, 1991b). Analysis of distinct mutant swarm loci by pulse field gel electrophoresis indicated that many are closely linked within a 340-kbp region on the chromosome (Allison and Hughes, 1991b) focused on the flagellar gene clusters. In *Proteus*, flagellar gene regions I and II are contiguous

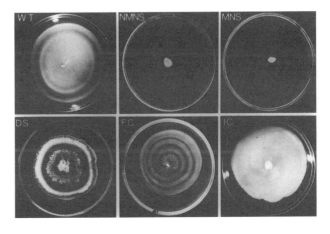

FIGURE 3 *P. mirabilis* swarming-defective mutants. Wild-type *P. mirabilis* (WT) and transposon mutants, nonmotile nonswarming (NMNS), motile nonswarming (MNS), dendritic swarming (DS), frequent consolidation (FC), and infrequent consolidation (IC), are shown. (From Allison and Hughes, 1991a.)

(Gygi et al., 1997) rather than approximately 70 kb apart as in *S. typhimurium* and *E. coli* (Macnab, 1992). At the junction of regions I and II in *P. mirabilis* is a gene of unknown function, *floA* (Gygi et al., 1997), that is not present in *S. typhimurium* or *E. coli*.

PHYSIOLOGICAL, BIOCHEMICAL, AND PHYSICAL STIMULI FOR DIFFERENTIATION

Unlike other examples of bacterial differentiation, such as myxobacterial fruiting body formation, *Bacillus* sporulation, and the dimorphic life cycle of *Caulobacter* species (Roberts et al., 1996), the swarming of *Proteus* is not a starvation response and does not represent an obligatory development stage in the life cycle (Jones and Park, 1967a, 1967b; Williams and Schwarzhoff, 1978; Allison et al., 1993; Rauprich et al., 1996). It is nonetheless a radical and reversible change in bacterial behavior in response to the environment. Early studies by Moltke (1927) and by Lominski and Lendrum (1947) implicated chemotaxis, both negative to toxic metabolites and positive to a nutrient source, in the initiation of *Proteus* differentiation. The chemotaxis response, which has been thoroughly characterized in *S. typhimurium* and *E. coli* (reviewed in Stock and Surette, 1996), alters the bias of the flagellar motor between counterclockwise (swimming) and clockwise (tumbling) rotation. Environmental attractants or repellents are recognized by transmembrane receptors, the methyl-accepting chemotaxis proteins (MCPs), that transduce signals to the cytoplasmic components of the chemotaxis phosphorelay, CheWAY. In the absence of an attractant or in the presence of a repellent, a phosphorylated MCP-CheW-CheA complex is formed which is able to phosphorylate CheY. Phospho-CheY then interacts with the flagellar switch, inducing clockwise rotation and disrupting smooth swimming. Three other components of the chemotaxis system, CheZ, CheR, and CheB, modulate the central Che-WAY pathway and ultimately alter levels of phospho-CheY. *Proteus* swarm cells have been shown to be remarkably insensitive to chemotactic gradients (Williams et al., 1976; Williams and Schwartzhoff, 1978), although it has been suggested that glutamine can act as a chemoattractant for differentiated cells (Allison et al., 1993). A subset of swarm-impaired mutants of *P. mirabilis* are defective in chemotaxis (Belas et al., 1991b), and swarming differentiation of laboratory *E. coli* on low-percentage agar is abolished in mutants defective in *cheA*, *cheW*, *cheR*, and *cheY* (Harshey and Matsuyama, 1994), although nonchemotactic flagellar-switch mutants and MCP mutants were able to swarm (Burkart et al., 1998). These data indicate that while the chemotaxis phosphorelay is required for swarming differentiation and possibly migration of *E. coli*, classical chemotactic behavior is not.

Extracellular oligopeptide signals are involved in several examples of prokaryotic differentiation, including *Bacillus* sporulation (Grossman and Losick, 1988; Rudner et al., 1991) and aerial mycelium development in *Streptomyces coelicolor* (Willey et al., 1993). Peptides and amino acids, especially glutamine, have been shown to stimulate swarming differentiation (Jones and Park, 1967b; Allison et al., 1993). *P. mirabilis* produces a protease with broad substrate specificity (Loomes et al., 1990; Lai, 1994; Wassif et al., 1995), and although its function in differentiation has not been determined, it is possible that an extracellular peptide or amino acid signal, analogous to the *M. xanthus* A-signal (Kaplan and Plamann, 1996) (see chapter 12), could be generated by such a protease, especially as the *P. mirabilis* protease is maximally expressed in consolidating cells (Allison et al., 1992b). The finding (Gaisser and Hughes, 1997) that swarm cell differentiation and migration are impaired by mutations in a locus encoding a nonribosomal peptide or polyketide synthase and a short oligopeptide, reminiscent of the *B. subtilis* ComX pheromone that regulates competence and surfactin production (Magnuson et al., 1994; Cosby et al., 1998), further suggests that several peptide, amino acid, or fatty acid signals may contribute to differentiation.

Clearly the physiological status of *P. mirabilis* cells affects their ability to swarm, as high growth rates of vegetative cells on nutrient-rich solid medium stimulate differentiation (Jones and Park, 1967b). This view is strengthened by studies of *B. subtilis* colony development, a process analogous and possibly related to enterobacterial swarming, in which nutrient concentration profoundly influences the extent and pattern of colony migration (Fujikawa and Matsushita, 1989; Ben-Jacob et al., 1992; Ohgiwari et al., 1992). However, recent studies of *P. mirabilis* have indicated that while metabolic cues are important in the initiation of differentiation, nutrient (glucose) depletion seems not to be involved, as cells at the center of a swarm colony are growing exponentially even as the second cycle of differentiation is initiated (Rauprich et al., 1996). The mechanisms linking the physiological condition of the cell to differentiation have not been defined, although the global transcriptional regulator Lrp (for leucine-responsive regulatory protein) could be a key component in relaying physiological signals. A motile but nonswarming *P. mirabilis lrp* mutant does not elongate, hyperflagellate, or hyperexpress hemolysin toxin (Hay et al., 1997) (see "Integration of Signals during Swarm Cell Development" below), and it is likely that during differentiation Lrp could integrate several signals, particularly as it is known to be responsive to a wide range of metabolites, e.g., amino acids (Calvo and Matthews, 1994). One further signal may be the level of intracellular cations, as an infrequently consolidating mutant deficient in a P-type ATPase cation transporter was found to be impaired in differentiation and to have reduced levels of *lrp* transcript (Lai et al., 1998).

Cell density is an important factor connecting population growth to the differentiation and migration of *P. mirabilis*, as the duration of the lag phase that precedes the first swarming migration phase is influenced by inoculum density (Rauprich et al., 1996; Belas et al., 1998; Gygi and Hay, unpublished). In *S. enteritidis*, the ability to grow to high densities has been correlated with virulence and swarming-like cell elongation and hyperflagellation (Guard-Petter, 1998). It has been demonstrated in a wide range of bacteria that extracellular

N-acyl homoserine lactones (AHLs) mediate cell density-dependent signaling, also known as quorum sensing, and contribute to the regulation of many processes, including bioluminescence and biosynthesis of antibiotics and extracellular virulence proteins (Salmond et al., 1995; Gray, 1997). AHLs are involved in the swarming migration of *S. liquefaciens* (Eberl et al., 1996b) (see below), and AHL production by *P. mirabilis* has been detected (Bainton et al., 1992), but its involvement in swarming differentiation has yet to be demonstrated.

Although initiation of swarm cell differentiation is clearly induced by several environmental and physiological signals, a pivotal stimulus is bacterial contact with a solid surface. Differentiation in *P. mirabilis* and *Vibrio parahaemolyticus* is always induced when wild-type vegetative cells are transferred to a solid or high-viscosity medium or are tethered with polyclonal antiserum raised against bacterial cell surface components (McCarter et al., 1988; Allison et al., 1993; Belas et al., 1995). While surface contact would presumably change the microenvironment in many ways, e.g., increasing nutrient concentration and cell density, there is substantial evidence to suggest that surface-sensing is specifically mediated by the flagellar filament, as flagellar mutants of swarming bacteria display abnormal swarm cell differentiation (McCarter and Silverman, 1990; Harshey and Matsuyama, 1994; Belas and Flaherty, 1994; Gygi et al., 1995a). In swarming *V. parahaemolyticus*, the polar sheathed flagellum acts as a mechanosensor, detecting the viscosity of the environment and regulating the expression of the structurally distinct lateral flagella, which mediate swarming motility (Kawagishi et al., 1996). Mutations that abolish the assembly or function of the polar flagellum result in both constitutive cell elongation and expression of lateral flagella (McCarter et al., 1988). Likewise, chemical inhibition of sodium ion flow through the polar flagellum motor, preventing flagellar rotation, also induces constitutive differentiation (Kawagishi et al., 1996). In contrast, mutations that disrupt assembly of functional flagella in *Proteus* and *Serratia*, in which swimming and swarming motility are mediated by structurally indistinguishable flagella, result in the inability of cells to differentiate (Gygi et al., 1995a, 1997; Eberl et al., 1996a; Furness et al., 1997).

MULTICELLULAR MIGRATION AND CONSOLIDATION

Differentiated *Proteus* swarm cells are capable of multicellular motility across solid surfaces, but what are the factors controlling the initiation, velocity, and duration of migration? As mentioned earlier, individual swarm cells cannot migrate, suggesting that differentiation into swarm cells is not the sole prerequisite for migration. Both microscopic observations of the formation of multicellular swarm cell rafts (Bisset, 1973a, 1973b; Sturdza, 1973) and macroscopic observations of the lag time preceding migration away from the colony edge (Rauprich et al., 1996) indicate that cell density-dependent thresholds are a major factor in the initiation of migration. A mathematical model of *P. mirabilis* swarm colony development has recently been presented (Esipov and Shapiro, 1998) in which swarm cell age and cell-density thresholds for collective motility are used as parameters. The model supposes that there is a minimum age at which swarm cells are capable of migration and that a minimum density of swarm-competent cells must be reached before migration is initiated. Cessation of migration and consolidation are also proposed to be functions of swarm cell age, i.e., there is a maximum age for swarm cell motility, and once this is reached swarm cells cease to migrate and dedifferentiate. This model predicts that initiation of migration and consolidation are governed by population dynamics rather than by responses to nutrient depletion or accumulation of chemotactic repellents, as had first been suggested several decades ago (Moltke, 1927; Lominski and Lendrum, 1947).

Like the population migration of *Myxococcus* during fruiting body development (Dworkin, 1996), the coordinated migration of *P. mirabilis* requires close cell-cell contact, with differentiated swarm cells aligning along their long axes

in multicellular rafts (Dienes, 1946). In a motile but nonswarming mutant, a transposon insertion in the gene *ccmA*, which has no known homologues, caused swarm cells to have a curved shape (Hay et al., submitted). The ability of the *ccmA* mutant to differentiate was not impaired, as cells elongated and hyperflagellated, suggesting that the migration defect resulted from the inability of the deformed swarm cells to align. Two forms of the membrane-associated CcmA protein are hyperexpressed in differentiated wild-type *P. mirabilis*, and they may function to maintain linearity of highly elongated swarm cells. In *M. xanthus*, transmission of the intercellular C-factor signal that prompts multicellular organization requires end-to-end cell contact (Kim and Kaiser, 1990; Sogaard-Andersen and Kaiser, 1996) (see chapter 11), and it may be that alignment of *Proteus* swarm cells is important for the transmission of short-range intercellular signals that initiate migration.

Cell-cell contact is stabilized by the production of cell-surface polysaccharides that form a slime capsule around groups of *P. mirabilis* swarm cells (Stahl et al., 1983; Gygi et al., 1995b). In addition to possibly facilitating intercellular communication, the acidic capsular polysaccharide is thought to act as a lubricant, creating a fluid environment through which *Proteus* can swarm by extracting water from the agar medium beneath the colony (Rauprich et al., 1996). This hypothesis is supported by observations that increased agar concentration or reduced polysaccharide biosynthesis, both of which result in a lowered agar/capsular polysaccharide osmotic activity ratio, reduce migration velocity but do not inhibit differentiation (Gygi et al., 1995b; Rauprich et al., 1996). Surface-active agents are produced by other bacteria that undergo population migration, e.g., myxobacteria produce an extracellular slime during fruiting body development (Dworkin, 1996), groups of hypermotile *B. subtilis* cells are cocooned in exopolymer during surface translocation (Ben-Jacob et al., 1994), and *Serratia* species produce serrawettin, a small cyclic lipopeptide that promotes both swarming migration and flagellum-independent spreading growth (Matsuyama et al., 1989, 1992).

In swarming *S. liquefaciens*, the production of serrawettin is regulated by *N*-butanoyl-L-homoserine lactone (BHL) in a cell density-dependent manner (Lindum et al., 1998). BHL controls the expression of nearly 30 proteins in *S. liquefaciens* (Givskov et al., 1998), and a mutant that does not synthesize BHL undergoes delayed migration on rich medium and cannot swarm on defined minimal medium containing casamino acids (Eberl et al., 1996b). The swarming defect of this mutant, which contains an insertion in the *swrI* autoinducer synthase gene, can be overcome by the addition of surface-active agents to the growth medium (Lindum et al., 1998), suggesting that the primary role of BHL in swarming may be the regulation of surfactant production. Cell-surface lipopolysaccharide (LPS) appears not surprisingly to be required for *P. mirabilis* swarming migration, as several nonswarming mutants were found to have mutated LPS biosynthesis genes (Belas et al., 1995). Disruption of the LPS core structure has many pleiotropic effects and abolishes the motility of *E. coli* K-12 (Parker et al., 1992). LPS defects may reduce the ability of *P. mirabilis* to hyperflagellate. The swarming-like behavior of virulent strains of *S. enteritidis* has been correlated to the production of high-molecular-weight LPS, and data from animal models suggest that swarm-cell differentiation in conjunction with the production of specific LPS structures aids the pervasive spread of this pathogen (Guard-Petter et al., 1997; Rahman et al., 1997). Integrity of the LPS structure is also important in the social motility of *M. xanthus* (Bowden and Kaplan, 1998).

One aspect of *P. mirabilis* swarming migration that is not observed in other swarming bacteria is the Dienes phenomenon (Dienes, 1946, 1947), in which colonies of two "strains" inoculated onto the same agar surface either coalesce (are of the same Dienes type) or form a zone of demarcation where they converge (are of different Dienes types). Several factors have been postulated to effect this antagonistic behavior, including the production of volatile amines (Proom and Woiwod, 1951), a negative chemotactic agent (Naylor, 1964), proticines (Skirrow, 1969; Senior, 1977), and an undefined inhibitor of swarming (Grabow, 1972), and it has been reported that lysogenic conversion of *Proteus* strains can lead to Dienes boundary formation (Coetzee, 1961). Boundary formation is also seen where two out-of-phase swarming colonies meet (Shapiro, 1995), but this is probably mechanistically distinct.

HYPEREXPRESSION OF FLAGELLA IS CENTRAL TO SWARMING DIFFERENTIATION

Hyperflagellation is the most prominent feature common to swarm cells of different species, and mutations that disrupt assembly of functional flagella, i.e., hyperflagellation, cell elongation, and hyperexpression of virulence factors, prevent swarming differentiation (Gygi et al., 1995a, 1997). The peritrichous flagella produced during enterobacterial swarming are structurally indistinguishable from those utilized for swimming motility through liquid medium, for which *E. coli* and *S. typhimurium* are well-characterized models (Harshey, 1994; Macnab, 1996). Of the 40 to 50 genes required for the production, assembly, and operation of flagella, only 18 encode structural proteins, and the large number of export and regulatory proteins reflects the high degree of organization required (Macnab, 1996; Aizawa, 1996). It has been estimated that production and operation of the 2 to 10 flagella needed for swimming motility can require as much as 2% of the cell's energy resources, so the process must be extremely tightly controlled (Macnab, 1992). Clearly, for swarming bacteria such as *Proteus*, able to express 50 times this number of flagella, regulation is even more critical.

The enterobacterial flagellum (Fig. 4) is composed of three parts: a helical filament that, when rotated, acts as a propeller; a curved hook that serves as a flexible coupling between the filament and the third component, the basal body; and the basal body, which is anchored in the cell envelope and consists of a central rod, several rings, and a motor (Fig. 4) (De-

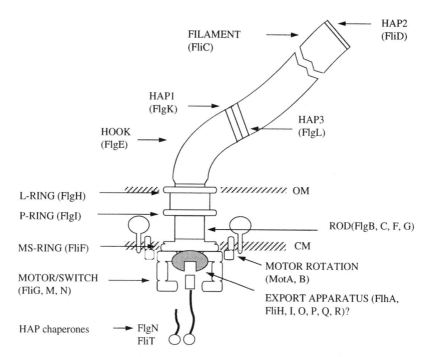

FIGURE 4 The enterobacterial flagellum. A representation of an enterobacterial flagellum is shown, indicating the protein components of each flagellar substructure. The flagellum crosses both the cytoplasmic membrane (CM) and the outer membrane (OM) and is stabilized in the cell envelope by three ring structures: (i) the MS ring, (ii) the P ring, and (iii) the L ring. The axial substructures of the flagellum, i.e., the rod, the hook, the HAPs, and the filament, make up a helical array of polymerized proteins that form a continuous hollow tube. The substructures involved in generating rotation are associated with the CM at the base of the flagellum. The flagellar export apparatus (shaded) is thought to be loosely associated with the base of the flagellum.

Pamphilis and Adler, 1971; Macnab, 1996). The hollow filamentous axial structure, comprising the basal-body rod, the hook, the hook-filament junction, the filament, and the filament cap, traverses both the cytoplasmic and outer membranes and can be up to 15 μm in length (Namba and Vonderviszt, 1997). Two outer rings embedded in the plane of the lipopolysaccharide (L) and peptidoglycan (P) layers form a bushing that protects the cell from the shearing forces of the rotating rod. In contrast to the static L and P rings, the cytoplasmic membrane-supramembrane (MS) ring forms part of the active rotor element that extends into the cytoplasm. The proteins responsible for switching the motor between counterclockwise (swimming) and clockwise (tum-

bling) rotation form a ring at the cytoplasmic face of the basal body (Khan et al., 1991; Francis et al., 1992, 1994). This fourth ring, the C or cytoplasmic ring, interacts with two integral membrane proteins (MotA and MotB) that are necessary for motor rotation but which do not themselves form part of the revolving apparatus. The transmembrane location of these proteins suggests that they have a role in proton conductance, and each one of the 11 MotAB complexes that encircle the basal body contributes independently to the generation of torque (Block and Berg, 1984; Blair and Berg, 1990).

Several components of the flagellum-specific export apparatus have been localized to the patch of cytoplasmic membrane in the middle of the MS ring (Fan et al., 1997) and are possibly associated with the C ring. The export complex transports axial flagellar proteins from the cytosol through the 25- to 30-Å central channel in the flagellum to their point of assembly at the distal end of the growing structure (Namba and Vonderviszt, 1997). It has been suggested that the hook-associated protein (HAP) molecules (FlgK, FlgL, and FliD) are escorted to the export apparatus in a monomeric export-competent conformation by the cytosolic putative chaperones, FlgN and FliT, preventing their premature polymerization (Fraser et al., submitted). Although not essential for swimming motility, these chaperones are required for efficient assembly of flagella, particularly during the transition from undifferentiated vegetative cells to hyperflagellated swarm cells, as *Proteus flgN* mutants are able to swim but cannot swarm (Gygi et al., 1997).

The flagellar master operon, *flhDC*, is the class 1 apex of the three-tiered transcriptional hierarchy governing flagellar biogenesis (Bartlett et al., 1998; Liu and Matsumura, 1994; Macnab, 1996). FlhD and FlhC form a heterotetrameric complex that activates transcription of the seven class 2 operons encoding proteins of the flagellar export apparatus, the structural components of the hook and basal body, and the flagellum-specific sigma factor, σ^{28} or FliA (Liu and Matsumura, 1994). FliA directs transcription of class 3 flagellar genes that encode the flagellar filament protein, the HAPs, the putative HAP chaperones, MotAB, the components of the chemotaxis system, and the antisigma factor, FlgM (Arnosti and Chamberlin, 1989; Kutsukake et al., 1990). The organization of flagellar genes into this transcriptional hierarchy allows expression of flagellar proteins to be temporally regulated; e.g., the filament protein, flagellin, is only expressed once the basal body and the hook have been completed (Kubori et al., 1992). Underlying this regulation is the interaction of FlgM with both FliA and the flagellum structure. Prior to completion of the basal body and the hook, FlgM accumulates in the cytosol (Hughes et al., 1993), where it binds to FliA, holding it in an inactive state (Ohnishi et al., 1992; Kutsukake and Iino,

1994). Once the hook is fully assembled, FlgM is exported via the flagellum-specific export pathway, releasing FliA to direct transcription of class 3 genes (Hughes et al., 1993). Completion of the flagellar basal body and hook therefore acts as a developmental checkpoint in much the same way as septum formation in *B. subtilis* and *C. crescentus* links cell morphology to gene expression (Shapiro and Losick, 1997).

While chromosomal organization of the homologous flagellar genes in *P. mirabilis* differs from that in *Salmonella* and *E. coli* (Gygi et al., 1997), flagellar gene expression is similarly regulated by FlhDC (Furness et al., 1997), and class 2 flagellar mutants display characteristic transcriptional repression of class 3 genes (Gygi et al., 1995a). Although the *P. mirabilis fliA* gene has not yet been isolated, a homologue of *flgM* has been identified (Gygi et al., 1997).

FlhDC, A REGULATORY FULCRUM IN SWARMING DIFFERENTIATION

An apparent regulatory focus during differentiation is the *flhDC* master operon, as *Proteus* swarm cells express 30-fold more *flhDC* mRNA than vegetative cells and artificial overexpression of *flhDC* promotes differentiation in *S. liquefaciens* and *P. mirabilis* cells without the normal requirement for surface contact (Eberl et al., 1996a; Furness et al., 1997). In *Salmonella* and *E. coli*, transcription of *flhDC* is subject to catabolite repression by the cyclic AMP receptor protein, CRP, and is also under the control of the osmoregulator OmpR and the H-NS protein (Silverman and Simon, 1974; Shin and Park, 1995; Bertin et al., 1994). In *P. mirabilis*, increasing glucose concentrations do not, however, affect hyperflagellation or motility, indicating that catabolite repression may be overridden during differentiation (Rauprich et al., 1996; Lai et al., 1997). In addition to the transcriptional regulation of *flhDC* expression, changes in the stability of the labile FlhD and FlhC proteins may also act to modulate levels of FlhDC during the swarm cycle (Claret, unpublished). Analysis of swarming-impaired *P. mirabilis* mutants has revealed an additional network of regulation superimposed on the classical flagellar gene regulatory hierarchy.

A nonmotile nonswarming mutant of *P. mirabilis* deficient in the FlhA flagellar export protein was not only defective in flagellar biogenesis but also exhibited reduced swarm cell elongation and hyperexpression of hemolysin toxin (Gygi et al., 1995a). In addition to a reduction in expression of class 2 and 3 flagellar genes in the *flhA* mutant, the peak level of *flhDC* mRNA is 7- to 12-fold lower than that in wild-type swarm cells (Furness et al., 1997). It is this negative feedback to the transcription of *flhDC* that causes the elongation defect of the mutant, as overexpression of either FlhD alone or FlhDC results in the restoration of elongation, even in the absence of hyperflagellation and swarming (Furness et al., 1997; Furness, unpublished). In *S. liquefaciens*, FlhDC controls expression of extracellular phospholipase (Givskov et al., 1995a). Overexpression of FlhDC in *Proteus* does not, however, restore hyperexpression of hemolysin to the *flhA* mutant (Fraser et al., unpublished), indicating that although expression of flagella and hemolysin are closely coupled, FlhDC is not the sole regulator of hemolysin expression during differentiation. The phenotype of the *flhA* mutant demonstrates that relatively small changes in *flhDC* expression can have extensive effects on swarm cell development.

The negative feedback to *flhDC* transcription does not appear to be mediated by FlgM and FliA, as a class 3 *fliC* mutant of *P. mirabilis*, which presumably exports FlgM at enhanced rates through filamentless hook/basal-body structures, does not elongate (Belas and Flaherty, 1994; Gygi et al., 1997), indicating that feedback is still in operation. Transcription of *flhDC* is also reduced in a motile *P. mirabilis* *flgN* mutant (Furness, unpublished), which is able to construct functional flagella but cannot differentiate or swarm (Gygi et al., 1997). Artificial overexpression of *flhDC* in the *flgN* mutant restores swarming differentiation (Dufour et al., 1998), suggesting that the mutant can support swarming but is normally unable to produce the threshold number of flagella required. The negative feedback has not been reported for other enteric organisms, possibly because it is less obvious in undifferentiated cells, as indeed is the case in *Proteus* vegetative cells.

The primary function of FlhDC is the control of flagella biogenesis, but in *E. coli*, *S. liquefaciens*, and *P. mirabilis* FlhDC also represses cell division (Pruß and Matsumura, 1996; Eberl et al., 1996a; Furness et al., 1997). Analysis of the mechanism of cell division control by FlhD in *E. coli* suggests that induction of the acid response lysine decarboxylase, CadA, may be involved, as a *cadA* transposon mutant has a cell division phenotype similar to that of the *flhD* mutant and transcription of *cadA* is positively regulated by FlhD (Pruß et al., 1997). The mechanism of FlhD and CadA regulation of cell division remains unclear, although it does not appear to involve modulation of *ftsQAZ* cell division gene expression (Pruß et al., 1997). In addition to its roles in flagella biogenesis and cell division, two-dimensional polyacrylamide gel electrophoresis analysis of protein expression in *S. liquefaciens* indicates that FlhDC controls the synthesis of at least 62 proteins (Givskov et al., 1998), suggesting that the master operon is a major regulatory fulcrum or global regulator.

INTEGRATION OF SIGNALS DURING SWARM CELL DEVELOPMENT

It seems likely that *flhDC* is a primary site for the integration of signals inducing swarm cell differentiation, and components have been identified that upregulate the flagellar master operon in swarm cells. As mentioned above, a mutation in the gene encoding the Lrp global transcriptional regulator completely disables swarming differentiation (Hay et al., 1997). Lrp is pivotal to the expression of at least 40 genes in *E. coli*, including many involved in amino acid biosynthesis and degradation, nitrogen assimilation, biosynthesis of one-carbon units, peptide transport, and pilin synthesis, repressing expression of some genes and activating others (Calvo and Matthews, 1994). The effect of Lrp upon hyperflagellation is mediated through *flhDC*, since transcription of the master operon and of the genes below it in the flagellar hierarchy is strongly reduced in the motile *lrp* mutant. Swarming can be restored by artificial induction of *flhDC* expression, indicating that the *lrp* mutant is not physiologically incapable of supporting hyperflagellation but rather fails to induce *flhDC* upregulation. Consistent with a role for Lrp in the induction of swarming, *lrp* transcription is upregulated in differentiated swarm cells compared to the levels seen in liquid culture, with kinetics similar to those of *flhDC*. Lrp-mediated regulation of flagella might become critical during the hyperinduction involved in swarming. It may be that Lrp directly regulates the flagellar and other swarming-associated genes, as putative Lrp binding sites have been identified 5′ of *flhDC* and the hemolysin operon, *hpmBA* (Fraser et al., unpublished). Alternatively, the Lrp effect might be indirect, mediated by an Lrp-dependent factor or factors.

Other proteins involved in the regulation of the *flhDC* master operon during the switch from vegetative to swarm cells have been identified by exploiting the "leaky" flagella assembly phenotype of the *flgN* putative HAP chaperone mutant. The *flgN* mutation does not preclude flagella assembly but reduces its efficiency, and the fine balance of swimming and swarming in the mutant can be tipped by artificially overexpressing *flhDC*, thus increasing the production of flagellar components. Using restoration of the highly visible swarm phenotype as a screen for upregulation of *flhDC*, four distinct randomly cloned *P. mirabilis* genes (*umoA*, *umoB*, *umoC*, and *umoD*) were isolated when the *flgN* mutant was transformed with a *P. mirabilis* chromosomal expression library (Dufour et al., 1998). Each *umo* gene upregulated expression of *flhDC* not only in the *flgN* background but also in the wild type when present in *trans* and in multicopy. Chromosomal mutations in each *umo* gene reduced swarming differentiation of the wild type, dramatically in the case of *umoB*Ω, substantially in *umoD*Ω, and marginally in *umoA*Ω and *umoC*Ω, and

caused corresponding reductions in the level of *flhDC* transcript.

Only UmoB and UmoD show sequence similarity to previously described proteins, *E. coli* YrfF and YcfJ, respectively, but these do not have ascribed functions (Dufour et al., 1998). UmoA and UmoC have no homologues in *E. coli* or, as yet, elsewhere. The predicted locations for all four Umo proteins in the cell membrane or periplasm are possibly indicative of a role for the proteins in sensing environmental stimuli, and it is possible that UmoA and UmoC, without *E. coli* homologues, may be specifically involved with interpreting swarming signals. The transcripts for *umoB* and *umoC* are not detectable, but *umoA* and *umoD* are strongly upregulated during swarming in parallel with *flhDC*, again consistent with their having a swarm-specific role, and they are also subject to feedback from the *flhA* block in flagellar assembly. Furthermore, although the *umo* genes do not appear to act in sequence within a pathway to bring about the upregulation of *flhDC*, both *umoA* and *umoD* are reciprocally regulated by FlhDC, as overexpression of *flhDC* results in substantially higher levels of *umoA* and *umoD* transcript. The mechanism underlying this reciprocal regulation of the master operon is unclear but strengthens the view of the master operon as a major assimilatory checkpoint (Fig. 5).

In genetic analysis of *Proteus* mutants defective in swarm cell elongation it was anticipated that lesions would be found in genes that have been well characterized in *E. coli* as essential for septation, such as *ftsZAQ* and *minAB*, but instead a variety of flagellar, LPS and peptidoglycan, and cell division loci were tentatively identified (Belas et al., 1995), including the cell division gene *gidA*. Mutations in *gidA* cause *E. coli* cells growing on glucose to become filamentous (von Meyenburg et al., 1982; Asai et al., 1990) and may function in *Proteus* to link metabolism and cell septation to differentiation (Belas et al., 1995). This gene has been repeatedly identified in different mutagenesis screens, and its loss results in a leaky motile nonswarming phenotype, i.e., *gidA* mutants only initiate

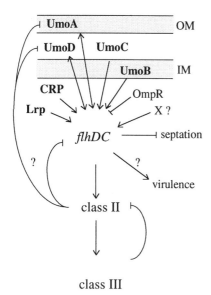

FIGURE 5 FlhDC, a major assimilatory checkpoint. Summary of the observed relationships between known (CRP, UmoA to D, Lrp, and OmpR) and as-yet-uncharacterized (X) regulators of the flagellar gene hierarchy. FlhD and FlhC together serve as a regulatory fulcrum assimilating swarm signals and mediating responses. Negative feedback to *P. mirabilis flhDC, umoA,* and *umoD* arises from defects in flagellar assembly and class 2 gene repression by the accumulation of the anti-sigma factor FlgM, which binds the flagellum-specific σ²⁸. Arrows, positive regulation; barred lines, negative regulation.

swarming after a prolonged incubation period (Hay and Tipper, unpublished). Several mutants which displayed a constitutively elongated phenotype and precocious swarming behavior contained transposon insertions in a gene, *rsbA*, encoding a protein homologous to membrane sensor histidine kinases of the two-component regulatory family (Belas et al., 1998). The *rsbA* gene is adjacent to *rcsB* and *rcsC*, which encode a two-component regulator of capsule synthesis, and it may be that the *rsbA* precocious-swarming phenotype results from polar effects on the expression of *rcsB*, a view supported by the finding that *rcsC* mutants have a similar phenotype (Belas et al., 1998). The RcsB-RcsC proteins have been shown to modulate expression of invasion and

flagellar proteins in *S. typhi* (Arricau et al., 1998), and while the function of this two-component regulatory system in *P. mirabilis* remains unclear, it may influence the expression of virulence and flagellar genes, as well as those involved in capsule synthesis.

A critical stimulus for the initiation of swarming is surface contact, but the mechanism by which a surface is sensed and induces cell differentiation is not known. One possibility is that the viscosity of the medium is interpreted by the flagellum itself, as is the case for the polar flagellum of *V. parahaemolyticus*, which acts as a mechanosensor converting external mechanical stress into the reduction of sodium ion flux (Kawagishi et al., 1996). Reduced rotation or sodium ion flux may be the trigger for differentiation to swarm cells, but lateral flagellar gene expression is induced even when the polar flagellum filament is missing, which would not be expected if differentiation arises through perturbation of the motor. These data support the idea that the cell might be detecting actual swimming speed rather than flagellum rotation, but local ion movement and reverse chemotaxis have also been proposed as alternative mechanisms for assessing viscosity via the flagellum (Kawagishi et al., 1996). As components of the chemotaxis phosphorelay pathway are closely associated with the rotating flagellum, it might be that the response to increased viscosities is transmitted via CheWAY to regulators such as *flhDC*.

CONCLUDING COMMENTS

Recent years have seen substantial progress in the understanding of prokaryotic development and an increased awareness of its value as a source of model systems of broad relevance. Investigation of swarming has confirmed it as an interesting and tractable model of bacterial differentiation and multicellularity within a growing colony, and this is particularly so in *Proteus*, in which swarming is a vigorous and cyclical phenomenon. Knowledge of the processes underlying swarming has advanced through physiological and genetic studies, primarily in *Proteus* and *Serratia*, and the character-

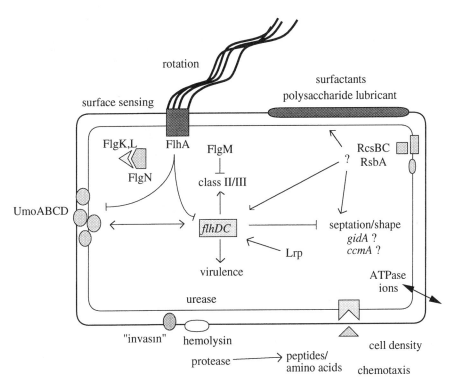

FIGURE 6 Components and factors identified as involved in swarming. Studies of swarming differentiation in gram-negative organisms, including *P. mirabilis*, have substantiated the early view of swarming portrayed in Fig. 1.

ization of swarming behavior in the better-characterized *Salmonella* and *E. coli* opens up new avenues of investigation. It also supports the view that differentiation into swarm cells is based upon widely conserved pathways governing flagellum biogenesis, motility, and septation rather than upon evolution of a distinct developmental program. Nevertheless, these pathways are clearly subject to additional modulation during swarming, e.g., by altered sensitivity to physiological and environmental signals through known regulators or novel cell proteins, such as Lrp and FlhDC, UmoA to D, and GidA, and swarming cells also require additional extracellular components, e.g., polysaccharide and surfactants, to physically facilitate mass migration over difficult terrain (Fig. 6). The hyperexpression of the flagellar gene hierarchy in *Proteus* has highlighted induction and negative regulation barely evident in undifferentiated cells, and the coupling of swarming to virulence, whether through an intrinsic role in colonization or coregulation of motility and virulence genes, adds an additional level of significance. While several signals are believed to induce differentiation, e.g., surface contact, cell density, and amino acids, the pathways of signal integration are not understood, in particular the apparent surface contact sensing by flagellar filaments and the basis of the cell-cell communication assumed to underlie coordinated migration. Continuing analysis of this network of components will further define swarming, a relatively simple example of complex behavior.

ACKNOWLEDGMENTS

This work was supported by a Wellcome Trust Programme grant (C.H.) and Medical Research Council studentships (G.M.F. and R.B.F.).

REFERENCES

Aizawa, S.-I. 1996. Flagellar assembly in *Salmonella typhimurium. Mol. Microbiol.* **19:**1–5.

Alberti, L., and R. M. Harshey. 1990. Differentiation of *Serratia marcescens* 274 into swimmer cells and swarmer cells. *J. Bacteriol.* **172:**4322–4328.

Allison, C., and C. Hughes. 1991a. Bacterial swarming: an example of prokaryotic differentiation and multicellular behaviour. *Sci. Prog.* (Edinburgh) **75:**403–422.

Allison, C., and C. Hughes. 1991b. Closely linked genetic loci required for swarm cell differentiation and multicellular migration by *Proteus mirabilis. Mol. Microbiol.* **5:**1975–1982.

Allison, C., P. Jones, N. Coleman, and C. Hughes. 1992a. Ability of *Proteus mirabilis* to invade human urothelial cells is coupled to motility and swarming differentiation. *Infect. Immun.* **60:**4740–4746.

Allison, C., H.-C. Lai, and C. Hughes. 1992b. Co-ordinate expression of virulence genes during swarm-cell differentiation and population migration of *Proteus mirabilis. Mol. Microbiol.* **6:**1583–1591.

Allison, C., H.-C. Lai, D. Gygi, and C. Hughes. 1993. Cell differentiation of *Proteus mirabilis* is initiated by glutamine, a specific chemoattractant for swarming cells. *Mol. Microbiol.* **8:**53–60.

Allison, C., L. Emody, N. Coleman, and C. Hughes. 1994. The role of swarm cell differentiation and multicellular migration in the uropathogenicity of *Proteus mirabilis. J. Infect. Dis.* **169:**1155–1158.

Armitage, J. P. 1981. Changes in metabolic activity of *Proteus mirabilis* during swarming. *J. Gen. Microbiol.* **125:**445–450.

Arnosti, D. N., and M. J. Chamberlin. 1989. Secondary sigma factor controls transcription of flagellar and chemotaxis genes in *Escherichia coli. Proc. Natl. Acad. Sci. USA* **86:**830–834.

Arricau, N., D. Hermant, H. Waxin, C. Ecobichon, P. S. Duffey, and M. Y. Popoff. 1998. The RcsB-RcsC regulatory system of *Salmonella typhi* differentially modulates the expression of invasion proteins, flagellin and Vi antigen in response to osmolarity. *Mol. Microbiol.* **29:**835–850.

Asai, T., M. Takanmi, and M. Imae. 1990. The AT richness and *gid* transcription determine the left border of the replication origin of the *E. coli* chromosome. *EMBO J.* **9:**4065–4072.

Bainton, N. J., B. W. Bycroft, S. R. Chhabra, P. Stead, L. Gledhill, P. J. Hill, C. E. D. Rees, M. K. Winson, G. P. C. Salmond, G. S. A. B. Stewart, and P. Williams. 1992. A general role for the *lux* autoinducer in bacterial cell signalling: control of antibiotic synthesis in *Erwinia. Gene* **116:**87–91.

Bartlett, D. H., B. B. Frantz, and P. Matsumura. 1988. Flagellar transcriptional activators FlbB and FlaI: gene sequences and 5' consensus sequences of operons under FlbB and FlaI control. *J. Bacteriol.* **170:**1575–1581.

Belas, R. 1992. The swarming phenomenon of *Proteus mirabilis. ASM News* **58:**15–22.

Belas, R. 1994. Expression of multiple flagellin-encoding genes of *Proteus mirabilis. J. Bacteriol.* **176:**7169–7181.

Belas, R., and D. Flaherty. 1994. Sequence and genetic analysis of multiple flagellin-encoding genes from *Proteus mirabilis. Gene* **128:**33–41.

Belas, R., D. Erskine, and D. Flaherty. 1991a. Transposon mutagenesis in *Proteus mirabilis. J. Bacteriol.* **173:**6289–6293.

Belas, R., D. Erskine, and D. Flaherty. 1991b. *Proteus mirabilis* mutants defective in swarmer cell differentiation and multicellular behavior. *J. Bacteriol.* **173:**6279–6288.

Belas, R., M. Goldman, and K. Ashliman. 1995. Genetic analysis of *Proteus mirabilis* mutants defective in swarmer cell elongation. *J. Bacteriol.* **177:**823–828.

Belas, R., R. Schneider, and M. Melch. 1998. Characterization of *Proteus mirabilis* precocious swarming mutants: identification of *rsbA*, encoding a regulator of swarming behavior. *J. Bacteriol.* **180:**6126–6139.

Ben-Jacob, E., H. Shmueli, O. Schochet, and A. Tenenbaum. 1992. Adaptive self-organization during growth of bacterial colonies. *Physica. A* **187:**378–424.

Ben-Jacob, E., A. Tenenbaum, O. Schochet, and O. Avidan. 1994. Holotransformations of bacterial colonies and genomic cybernetics. *Physica. A* **202:**1–47.

Bertin, P., E. Terao, E. H. Lee, P. Lejeune, C. Colson, A. Danchin, and E. Collatz. 1994. The H-NS protein is involved in the biogenesis of flagella in *Escherichia coli. J. Bacteriol.* **176:**5537–5540.

Bisset, K. A. 1973a. The motion of the swarm in *Proteus mirabilis. J. Med. Microbiol.* **6:**33–35.

Bisset, K. A. 1973b. The zonation phenomenon and structure of the swarm colony in *Proteus mirabilis. J. Med. Microbiol.* **6:**429–433.

Blair, D. E., and H. C. Berg. 1990. The MotA protein of *E. coli* is a proton conducting component of the flagellar motor. *Cell* **60:**439–449.

Block, S. M., and H. C. Berg. 1984. Successive incorporation of force-generating units in the bacterial rotary motor. *Nature* **309:**470–472.

Bowden, M. G., and H. B. Kaplan. 1998. The *Myxococcus xanthus* lipopolysaccharide O-antigen is required for social motility and multicellular development. *Mol. Microbiol.* **30:**275–284.

Burkart, M., A. Toguchi, and R. M. Harshey.

1998. The chemotaxis system, but not chemotaxis, is essential for swarming motility in *Escherichia coli*. *Proc. Natl. Acad. Sci. USA* **95:**2568–2573.

Calvo, J. M., and R. G. Matthews. 1994. Leucine-responsive regulatory protein—a global regulator of metabolism in *Escherichia coli*. *Microbiol. Rev.* **58:** 466–490.

Chippendale, G. R., J. W. Warren, A. L. Trifillis, and H. L. T. Mobley. 1994. Internalization of *Proteus mirabilis* by human renal epithelial cells. *Infect. Immun.* **62:**3115–3121.

Claret, L. Unpublished data.

Coetzee, J. N. 1961. Lysogenic conversion in the genus *Proteus*. *Nature* **189:**946–947.

Cooper, K. E., J. Davies, and J. Wieseman. 1971. An investigation of an outbreak of food poisoning associated with organisms of the *Proteus* group. *J. Pathol. Bacteriol.* **52:**91–98.

Cosby, W. M., D. Vollenbroich, O. H. Lee, and P. Zuber. 1998. Altered *srf* expression in *Bacillus subtilis* resulting from changes in culture pH is dependent on the Spo0K oligopeptide permease and the ComQX system of extracellular control. *J. Bacteriol.* **180:**1438–1445.

Deighton, C. M. 1992. P blood group phenotype, *Proteus* antibody titres, and rheumatoid arthritis. *Ann. Rheum. Dis.* **51:**1242–1244.

DePamphilis, M. L., and J. Adler. 1971. Purification of intact flagella from *Escherichia coli* and *Bacillus subtilis*. *J. Bacteriol.* **105:**376–383.

Dienes, L. 1946. Reproductive processes in *Proteus* colonies. *Proc. Soc. Exp. Biol. Med.* **63:**265–270.

Dienes, L. 1947. Further observations on the reproduction of bacilli from large bodies in *Proteus* cultures. *Proc. Soc. Exp. Biol. Med.* **66:**97–98.

Dufour, A., R. B. Furness, and C. Hughes. 1998. Novel genes that upregulate the *Proteus mirabilis* *flhDC* master operon controlling flagellar biogenesis and swarming. *Mol. Microbiol.* **29:**741–751.

Dworkin, M. 1996. Recent advances in the social and developmental behaviour of the Myxobacteria. *Microbiol. Rev.* **60:**70–102.

Eberl, L., G. Christiansen, S. Molin, and M. Givskov. 1996a. Differentiation of *Serratia liquefaciens* into swarm cells is controlled by the expression of the *flhD* master operon. *J. Bacteriol.* **178:**554–559.

Eberl, L., M. K. Winson, C. Sternberg, G. S. A. B. Stewart, G. Christiansen, S. R. Chhabra, B. Bycroft, P. Williams, S. Molin, and M. Givskov. 1996b. Involvement of N-acyl-homoserine lactone autoinducers in controlling the multicellular behaviour of *Serratia liquefaciens*. *Mol. Microbiol.* **20:**127–136.

Ebringer, A. 1985. Antibodies to *Proteus* in rheumatoid arthritis. *Lancet* **i:**305–307.

Esipov, S. E., and J. A. Shapiro. 1998. Kinetic model of *Proteus mirabilis* swarm colony development. *J. Math. Biol.* **36:**249–268.

Falkinham, J. O., and P. S. Hoffman. 1984. Unique developmental characteristics of the swarm and short cells of *Proteus vulgaris* and *Proteus mirabilis*. *J. Bacteriol.* **158:**1037–1040.

Fan, F., K. Ohnishi, N. R. Francis, and R. M. Macnab. 1997. The FliP and FliR proteins of *Salmonella typhimurium*, putative components of the type III flagellar export apparatus, are located in the flagellar basal body. *Mol. Microbiol.* **26:**1035–1046.

Francis, N. R., V. M. Irikura, S. Yamaguchi, D. J. DeRosier, and R. M. Macnab. 1992. Localization of the *Salmonella typhimurium* flagellar switch protein FliG to the cytoplasmic M-ring face of the basal body. *Proc. Natl. Acad. Sci. USA* **89:** 6304–6308.

Francis, N. R., G. E. Sosinsky, D. Thomas, and D. J. DeRosier. 1994. Isolation, characterisation and structure of bacterial flagellar motors containing the switch complex. *J. Mol. Biol.* **235:**1216–1270.

Fraser, G. M. Unpublished data.

Fraser, G. M., J. C. Q. Bennett, and C. Hughes. FlgN and FliT, substrate-specific chaperones that facilitate *Salmonella* flagellum assembly by binding hook-associated proteins. Submitted for publication.

Fraser, G. M., S. Gupta, and R. B. Furness. Unpublished data.

Fujikawa, H., and M. Matsushita. 1989. Fractal growth of *Bacillus subtilis* on agar plates. *J. Phys. Soc. Jpn.* **58:**3875–3878.

Furness, R. B. Unpublished data.

Furness, R. B., G. M. Fraser, N. A. Hay, and C. Hughes. 1997. Negative feedback from a *Proteus* class II flagellum export defect to the *flhDC* master operon controlling cell division and flagellum assembly. *J. Bacteriol.* **179:**5585–5588.

Gaisser, S., and C. Hughes. 1997. A locus coding for putative non-ribosomal peptide/polyketide synthase functions is mutated in a swarming defective *Proteus mirabilis* strain. *Mol. Gen. Genet.* **253:** 415–427.

Givskov, M., L. Eberl, G. Christiansen, M. J. Bendik, and S. Molin. 1995a. Induction of phospholipase and flagellar synthesis in *Serratia liquefaciens* is controlled by expression of the master operon *flhD*. *Mol. Microbiol.* **15:**445–454.

Givskov, M., L. Eberl, and S. Molin. 1995b. Control of exoenzyme production, motility and cell differentiation in *Serratia liquefaciens*. *FEMS Microbiol. Lett.* **148:**115–122.

Givskov, M., J. Ostling, L. Eberl, P. W. Lindum, A. B. Christensen, G. Christensen, S. Molin, and S. Kjelleberg. 1998. Two separate regulatory systems participate in control of swarming motility of *Serratia liquefaciens* MG1. *J. Bacteriol.* **180:** 742–745.

Grabow, W. O. K. 1972. Growth inhibiting metabolites of *Proteus mirabilis*. *J. Med. Microbiol.* **5:** 191–204.

Gray, K. M. 1997. Intercellular communication and group behaviour in bacteria. *Trends Microbiol.* **5:** 184–188.

Grossman, A. D., and R. Losick. 1988. Extracellular control of spore formation in *Bacillus subtilis*. *Proc. Natl. Acad. Sci. USA* **85:**4369–4373.

Guard-Petter, J. 1997. Induction of flagellation and a novel-agar-penetrating flagellar structure in *Salmonella enterica* grown on solid media: possible consequences for serological identification. *FEMS Microbiol. Lett.* **149:**173–180.

Guard-Petter, J. 1998. Variants of smooth *Salmonella enterica* serovar enteritidis that grow to higher cell density than the wild type are more virulent. *Appl. Environ. Microbiol.* **64:**2166–2172.

Guard-Petter, J., D. J. Henzler, M. M. Rahman, and R. W. Carlson. 1997. On-farm monitoring of mouse-invasive *Salmonella enterica* serovar enteritidis and a model for its association with the production of contaminated eggs. *Appl. Environ. Microbiol.* **63:**1588–1593.

Guo, M. M. S., and P. V. Lin. 1965. Serological specificities of ureases of *Proteus* species. *J. Gen. Microbiol.* **38:**417–422.

Gygi, D., and N. A. Hay. Unpublished data.

Gygi, D., M. J. Bailey, C. Allison, and C. Hughes. 1995a. Requirement for FlhA in flagella assembly and swarm cell differentiation by *Proteus mirabilis*. *Mol. Microbiol.* **15:**761–769.

Gygi, D., M. M. Rahman, H.-C. Lai, R. Carlson, J. Guard-Petter, and C. Hughes. 1995b. A cell-surface polysaccharide that facilitates rapid population migration by differentiated swarm cells of *Proteus mirabilis*. *Mol. Microbiol.* **17:**1167–1175.

Gygi, D., G. Fraser, A. Dufour, and C. Hughes. 1997. A motile but non-swarming mutant of *Proteus mirabilis* lacks FlgN, a facilitator of flagella filament assembly. *Mol. Microbiol.* **25:**597–604.

Harshey, R. M. 1994. Bees aren't the only ones: swarming in Gram-negative bacteria. *Mol. Microbiol.* **13:**389–394.

Harshey, R. M., and P. Matsuyama. 1994. Dimorphic transition in *Escherichia coli* and *Salmonella typhimurium*: surface induced differentiation into hyperflagellate swarmer cells. *Proc. Natl. Acad. Sci. USA* **91:**8631–8635.

Hay, N. A., and D. J. Tipper. Unpublished data.

Hay, N. A., D. J. Tipper, D. Gygi, and C. Hughes. A novel membrane protein influencing cell shape and multicellular swarming of *Proteus mirabilis*. Submitted for publication.

Hay, N. A., D. J. Tipper, D. Gygi, and C.

Hughes. 1997. A nonswarming mutant of *Proteus mirabilis* lacks the Lrp global transcriptional regulator. *J. Bacteriol.* **179:**4741–4746.

Henrichsen, J. 1972. Bacterial surface translocation: a survey and a classification. *Bacteriol. Rev.* **36:** 478–503.

Hoeniger, J. F. M. 1964. Cellular changes accompanying the swarming of *Proteus mirabilis*. I. Observations on living cultures. *Can. J. Microbiol.* **10:**1–9.

Hoeniger, J. F. M. 1965. Development of flagella by *Proteus mirabilis*. *J. Gen. Microbiol.* **40:**29–42.

Hoeniger, J. F. M. 1966. Cellular changes accompanying the swarming of *Proteus mirabilis*. I. Observations of stained organisms. *Can. J. Microbiol.* **12:** 113–122.

Hughes, K. T., K. L. Gillen, J. S. Melinda, and J. E. Karlinsey. 1993. Sensing structural intermediates in bacterial flagella assembly by export of a negative regulator. *Science* **262:**1277–1280.

Jones, H. E., and R. W. A. Park. 1967a. The short and long forms of *Proteus*. *J. Gen. Microbiol.* **47:** 359–367.

Jones, H. E., and R. W. A. Park. 1967b. The influence of medium composition on the growth and swarming of *Proteus*. *J. Gen. Microbiol.* **47:**369–378.

Kaplan, H. B., and L. Plamann. 1996. A *Myxococcus xanthus* cell density-sensing system required for multicellular development. *FEMS Microbiol. Lett.* **139:**89–95.

Kawagashi, I., M. Imagawa, Y. Imae, L. McCarter, and M. Homma. 1996. The sodium-driven polar flagellar motor of marine *Vibrio* as the mechanosensor that regulates lateral flagellar expression. *Mol. Microbiol.* **20:**693–699.

Khan, S. I., H. Khan, and T. S. Reese. 1991. New structural features of the flagellar base in *Salmonella typhimurium* revealed by rapid-freeze electron microscopy. *J. Bacteriol.* **173:**2888–2896.

Kim, S. K., and D. Kaiser. 1990. Cell alignment required in differentiation of *Myxococcus xanthus*. *Science* **249:**926–928.

Klienberger-Nobel, E. 1947. Morphological appearances of various stages in *B. proteus* and *E. coli*. *J. Hyg.* **45:**410–412.

Krajden, S., M. Fuksa, W. Lizewski, L. Barton, and A. Lee. 1984. *Proteus penneri* and urinary calculi formation. *J. Clin. Microbiol.* **19:**541–542.

Kubori, T., N. Shimamoto, S. Yamaguchi, K. Namba, and S.-I. Aizawa. 1992. Morphological pathway of flagellar assembly in *Salmonella typhimurium*. *J. Mol. Biol.* **226:**433–446.

Kutsukake, K., and T. Iino. 1994. Role of the FliA-FlgM regulatory system on the transcriptional control of the flagellar regulon and flagellar formation in *Salmonella typhimurium*. *J. Bacteriol.* **176:** 3598–3605.

Kutsukake, K., Y. Ohya, and T. Iino. 1990. Transcriptional analysis of the flagellar regulon of *Salmonella typhimurium*. *J. Bacteriol.* **172:**741–747.

Lai, H.-C. 1994. Ph.D. thesis. University of Cambridge, Cambridge, United Kingdom.

Lai, H.-C., J.-C. Shu, S. Ang, M.-J. Lai, B. Fruta, S. Lin, K.-T. Lu, and S.-W. Ho. 1997. Effect of glucose concentration on swimming motility in Enterobacteria. *Biochem. Biophys. Res. Commun.* **231:**692–695.

Lai, H.-C., D. Gygi, G. M. Fraser, and C. Hughes. 1998. A swarming-defective mutant of *Proteus mirabilis* lacking a putative cation-transporting membrane P-type ATPase. *Microbiology* **144:**1957–1961.

Lindum, P. W., U. Anthoni, C. Christoffersen, L. Eberl, S. Molin, and M. Givskov. 1998. *N*-acyl-L-homoserine lactone autoinducers control production of an extracellular lipopeptide biosurfactant required for swarming motility of *Serratia liquefaciens* MG1. *J. Bacteriol.* **180:**6384–6388.

Liu, X., and P. Matsumura. 1994. The FlhD/FlhC complex, a transcriptional activator of the *Escherichia coli* flagellar class II operons. *J. Bacteriol.* **176:**7345–7351.

Lominski, I., and A. C. Lendrum. 1947. The mechanism of swarming of *Proteus*. *J. Pathol. Bacteriol.* **59:**688–691.

Loomes, L. M., B. W. Senior, and M. A. Kerr. 1990. A proteolytic enzyme secreted by *Proteus mirabilis* degrades immunoglobulins of the immunoglobulin A1 (IgA1), IgA2, and IgG isotypes. *Infect. Immun.* **58:**1979–1985.

Macnab, R. M. 1992. Genetics and biogenesis of bacterial flagella. *Annu. Rev. Genet.* **26:**131–158.

Macnab, R. M. 1996. Flagella and motility, p. 123–145. *In* F. C. Neidhardt, R. Curtiss III, J. L. Ingraham, E. C. C. Lin, K. B. Low, B. Magasanik, W. S. Reznikoff, M. Riley, M. Schaechter, and H. E. Umbarger (ed.), *Escherichia coli and Salmonella: Cellular and Molecular Biology*, 2nd ed., vol. 1. American Society for Microbiology, Washington, D.C.

Magnuson, R., J. Solomon, and A. D. Grossman. 1994. Biochemical and genetic characterisation of a competence pheromone from *B. subtilis*. *Cell* **77:**207–216.

Matsuyama, T., M. Sogawa, and Y. Nakagawa. 1989. Fractal spreading growth of *Serratia marcescens* which produces surface active exolipids. *FEMS Microbiol. Lett.* **61:**243–246.

Matsuyama, T., K. Kaneda, Y. Nakagawa, K. Isa, H. Hara-Hotta, and Y. Isuya. 1992. A novel extracellular cyclic lipopeptide which promotes flagellum-dependent and -independent spreading growth of *Serratia marcescens*. *J. Bacteriol.* **174:**1769–1776.

McCarter, L., and M. Silverman. 1990. Surface induced swarmer cell differentiation of *Vibrio parahaemolyticus*. *Mol. Microbiol.* **4:**1057–1062.

McCarter, L., M. Hilmen, and M. Silverman. 1988. Flagellar dynamometer controls swarmer cell differentiation of *V. parahaemolyticus*. *Cell* **54:**345–351.

Mobley, H. L. T., and R. Belas. 1995. Swarming and pathogenicity of *Proteus mirabilis* in the urinary tract. *Trends Microbiol.* **3:**280–284.

Mobley, H. L. T., G. R. Chippendale, K. G. Swihart, and R. A. Welch. 1991. Cytotoxicity of the HpmA hemolysin and urease of *Proteus mirabilis* and *Proteus vulgaris* against cultured human renal proximal tubular epithelial cells. *Infect. Immun.* **59:**2036–2042.

Moltke, O. 1927. *Contributions to the Characterization and Systematic Classification of Bact. Proteus vulgaris (Hauser)*. Levin and Munksgaard, Copenhagen, Denmark.

Murphy, C. A., and R. Belas. 1999. Genomic rearrangements in the flagellin genes of *Proteus mirabilis*. *Mol. Microbiol.* **31:**679–690.

Namba, K., and F. Vonderviszt. 1997. Molecular architecture of the bacterial flagellum. *Q. Rev. Biophys.* **30:**1–65.

Naylor, P. 1964. The effect of electrolytes or carbohydrates in a sodium chloride deficient medium on the formation of discrete colonies of *Proteus* and the influence of these substances on growth in liquid culture. *J. Appl. Bacteriol.* **27:**422–431.

Ohgiwari, M., M. Mutsushita, and T. Matsuyama. 1992. Morphological changes in growth phenomena of bacterial colony patterns. *J. Phys. Soc. Jpn.* **61:**816–822.

Ohnishi, K., K. Kutsukake, H. Suzuki, and T. Iino. 1992. A novel transcriptional mechanism in the flagellar regulon of *Salmonella typhimurium*: an anti-sigma factor inhibits the activity of the flagellum-specific sigma factor σ^F. *Mol. Microbiol.* **6:**3149–3157.

Otteman, K. M., and J. F. Miller. 1997. Roles for motility in bacterial-host interactions. *Mol. Microbiol.* **24:**1109–1117.

Parker, C. T., A. W. Kloser, C. A. Schnaitman, M. A. Stein, S. Gottesman, and B. W. Gibson. 1992. Role of the *rfaG* and *rafP* genes in determining the lipopolysaccharide core structure and cell surface properties of *Escherichia coli* K-12. *J. Bacteriol.* **174:**2525–2538.

Peerbooms, P. G. H., A. M. J. Verweij, and D. M. MacLaren. 1984. Vero cell invasiveness of *Proteus mirabilis*. *Infect. Immun.* **43:**1068–1071.

Peerbooms, P. G. H., A. M. J. Verweij, and D. M. MacLaren. 1985. Uropathogenic properties of *Proteus mirabilis* and *Proteus vulgaris*. *J. Med. Microbiol.* **19:**55–60.

Penner, J. L. 1992. The genera *Proteus, Providencia*,

and *Morganella*, p. 2849–2853. *In* A. Balows et al., (ed.), *The Prokaryotes*, vol. III. Springer-Verlag KG, Berlin, Germany.

Proom, H., and A. Woiwod. 1951. Amine production in the genus *Proteus*. *J. Gen. Microbiol.* **5:** 930–938.

Pruß, B. M., and P. Matsumura. 1996. A regulator of the flagellar regulon of *Escherichia coli*, *flhD*, also affects cell division. *J. Bacteriol.* **178:**668–674.

Pruß, B. M., D. Maekovic, and P. Matsumura. 1997. The *Escherichia coli* flagellar transcriptional activator *flhD* regulates cell division through induction of the acid response gene *cadA*. *J. Bacteriol.* **179:** 3818–3821.

Rahman, M. M., J. Guard-Petter, and R. W. Carlson. 1997. A virulent isolate of *Salmonella enteritidis* produces a *Salmonella typhi*-like lipopolysaccharide. *J. Bacteriol.* **179:**2126–2131.

Rauprich, O., M. Matsushita, C. J. Weijer, F. Siegert, S. E. Esipov, and J. A. Shapiro. 1996. Periodic phenomena in *Proteus mirabilis* swarm colony development. *J. Bacteriol.* **178:**6525–6538.

Roberts, R. C., C. D. Mohr, and L. Shapiro. 1996. Developmental programs in bacteria. *Curr. Top. Dev. Biol.* **34:**207–257.

Rosenstein, I. J. M. 1986. Urinary calculi: microbiological and crystallographic studies. *Crit. Rev. Clin. Lab. Sci.* **22:**245–277.

Rosenstein, I. J. M., J. M. Hamilton-Miller, and W. Brumfitt. 1981. Role of urease in the formation of infection stones: comparison of ureases from different sources. *Infect. Immun.* **32:**32–37.

Rozalski, A., H. Dlugonska, and K. Kotelko. 1986. Cell invasiveness of *Proteus mirabilis* and *Proteus vulgaris* strains. *Arch. Immunol. Ther. Exp.* **34:** 505–511.

Rozalski, A., Z. Sidorczyk, and K. Kotelko. 1997. Potential virulence factors of *Proteus* bacilli. *Microbiol. Mol. Biol. Rev.* **61:**65–89.

Rudner, D. Z., J. R. LeDeaux, K. Ireton, and A. D. Grossman. 1991. The *spo0K* locus of *Bacillus subtilis* is homologous to the oligopeptide permease locus and is required for sporulation and competence. *J. Bacteriol.* **173:**1388–1398.

Salmond, G. P. C., B. W. Bycroft, G. S. A. B. Stewart, and P. Williams. 1995. The bacterial 'enigma': cracking the code of cell-cell communication. *Mol. Microbiol.* **16:**615–624.

Senior, B. W. 1977. The Dienes phenomenon: identification of the determinants of compatibility. *J. Med. Microbiol.* **102:**235–244.

Senior, B. W., and C. Hughes. 1987. Production and properties of haemolysins from clinical isolates of the *Proteeae*. *J. Med. Microbiol.* **24:**17–25.

Senior, B. W., N. C. Bradford, and D. S. Simpson. 1980. The ureases of *Proteus* strains in relation to virulence for the urinary tract. *J. Med. Microbiol.* **13:**507–512.

Shapiro, J. A. 1995. The significances of bacterial colony patterns. *Bioessays* **17:**597–607.

Shapiro, J. A. 1998. Thinking about bacterial populations as multicellular organisms. *Annu. Rev. Microbiol.* **52:**81–104.

Shapiro, L., and R. Losick. 1997. Protein localization and cell fate in bacteria. *Science* **276:**712–717.

Shin, S., and C. Park. 1995. Modulation of flagellar expression in *Escherichia coli* by acetyl phosphate and the osmoregulator OmpR. *J. Bacteriol.* **177:** 4696–4702.

Silverman, M., and M. Simon. 1974. Characterization of *Escherichia coli* flagellar mutants that are insensitive to catabolite repression. *J. Bacteriol.* **120:** 1196–1203.

Skirrow, M. B. 1969. The Dienes (mutual inhibition) test in the investigation of *Proteus* infections. *J. Med. Microbiol.* **2:**471–477.

Sogaard-Andersen, L., and D. Kaiser. 1996. C factor, a cell-surface-associated intercellular signalling protein, stimulates the cytoplasmic Frz signal transduction system in *Myxococcus xanthus*. *Proc. Natl. Acad. Sci USA* **93:**2675–2679.

Stahl, S. J., K. R. Stewart, and F. D. Williams. 1983. Extracellular slime associated with *Proteus mirabilis* during swarming. *J. Bacteriol.* **154:** 930–937.

Stefansson, K. 1985. Sharing of antigenic determinants between the nicotinic acetylcholine receptor and proteins in *E. coli*, *Proteus vulgaris* and *Klebsiella pneumoniae*: possible role in the pathogenesis of myasthenia gravis. *N. Engl. J. Med.* **312:**221–225.

Stickler, D. J., J. B. King, C. Winters, and S. L. Morris. 1993a. Blockage of uretheral catheters by bacterial biofilms. *J. Infect.* **27:**133–135.

Stickler, D. J., L. Ganderton, J. B. King, J. Nettleton, and C. Winters. 1993b. *Proteus mirabilis* biofilms and the encrustation of urethral catheters. *Urol. Res.* **21:**407–411.

Stock, J. B., and M. Surette. 1996. Chemotaxis, p. 1103–1129. *In* F. C. Neidhardt, R. Curtiss III, J. L. Ingraham, E. C. C. Lin, K. B. Low, B. Magasanik, W. S. Reznikoff, M. Riley, M. Schaechter, and H. E. Umbarger (ed.), *Escherichia coli and Salmonella: Cellular and Molecular Biology*, 2nd ed., vol. 1. American Society for Microbiology, Washington, D.C.

Sturdza, S. A. 1973. La reaction d'immobilisation des filaments des *Proteus* sur les milieux geloses. *Arch. Roum. Pathol. Exp. Microbiol.* **32:**575–580.

Swihart, K. G., and R. A. Welch. 1990. Cytotoxic activity of the *Proteus* hemolysin HpmA. *Infect. Immun.* **58:**1861–1869.

Uphoff, T. S., and R. A. Welch. 1990. Nucleotide sequencing of the *Proteus mirabilis* calcium-independent hemolysin genes (*hpmA* and *hpmB*) reveals sequence similarity with *Serratia marcescens* hemolysin genes (*shlA* and *shlB*). *J. Bacteriol.* **172**:1206–1216.

von Meyenburg, K., B. Jorgensen, J. Neilson, and F. Hansen. 1982. Promoters of the *atp* operon coding for the membrane-bound ATP-synthase of *Escherichia coli* mapped by Tn*10* insertion mutations. *Mol. Gen. Genet.* **188**:240–248.

Warren, J. W., D. Damron, J. H. Tenney, J. M. Hoopes, B. Deforge, and H. J. Muncie. 1987. Fever, bacteremia and death as complications of bacteriuria in women with long-term urethral catheters. *J. Infect. Dis.* **155**:1151–1158.

Wassif, C., D. Cheek, and R. Belas. 1995. Molecular analysis of metalloprotease from *Proteus mirabilis*. *J. Bacteriol.* **177**:5790–5798.

Willey, J., J. Schwedock, and R. Losick. 1993. Multiple extracellular signals govern the production of a morphogenetic protein involved in ariel mycelium formation by *Streptomyces coelicolor*. *Genes Dev.* **7**:895–903.

Williams, F. D., and R. H. Schwarzhoff. 1978. Nature of the swarming phenomenon in *Proteus*. *Annu. Rev. Microbiol.* **32**:101–122.

Williams, F. D., D. M. Anderson, P. S. Hoffman, R. H. Schwarzhoff, and S. Leonard. 1976. Evidence against the involvement of chemotaxis in swarming of *Proteus mirabilis*. *J. Bacteriol.* **127**:237–248.

Zhao, H., R. B. Thompson, V. Lockatell, D. E. Johnson, and H. L. T. Mobley. 1998. Use of green fluorescent protein to assess urease gene expression by uropathogenic *Proteus mirabilis* during experimental ascending urinary tract infection. *Infect. Immun.* **66**:330–335.

THE CHLAMYDIAL DEVELOPMENTAL CYCLE

Daniel D. Rockey and Akira Matsumoto

20

HISTORICAL PERSPECTIVE

The investigation of chlamydial biology has spanned this century, and studies by both early and contemporary scientists provide valuable insight into chlamydial biology. This chapter will begin with an account of what was learned by the earliest and possibly most astute chlamydiologists. We will then describe current molecular and cellular biology approaches that have expanded upon these early observations. Collectively, this discussion will provide a picture of the developmental cycle, a process common to all chlamydial species but unique among prokaryotes.

Initial Descriptions of the Chlamydiae

The investigation of what we now call chlamydiae has its origin in work by Halberstaeder and von Prowazek (cited in excellent historical reviews: Thygeson, 1962; Schachter, 1978; Page, 1966), who identified intracytoplasmic inclusions in individuals with trachoma, a disease that commonly leads to blindness. Over the next 20 years, similar structures were reported associated with a variety of clinical conditions. A focused effort directed at understanding the chlamydial developmental cycle followed the "psittacosis pandemic" of the 1920s and 1930s (Gordon, 1930). At that time serious human respiratory disease (psittacosis) was documented in Europe and the United States and was associated with ownership of imported, primarily psittacine (parrot-like), birds. In many cases the mortality was quite high in both the affected human and bird populations. The initial description of the chlamydial life cycle was provided by Bedson and colleagues working with "psittacosis virus" (Bedson, 1932, 1933) isolated from an infected parrot. Subsequently, several independent research groups, working with agents of trachoma, chlamydial zoonotic disease, and various diseases of animals, assisted in determining the link between these distinct clinical conditions and the infectious agent that caused the diseases.

Chlamydial Disease

These earliest investigations of the chlamydiae were in response to clinical presentations that were of unknown etiology. Many of the diseases investigated by these scientists are still serious problems today. Halberstaeder and von Prowazek were studying patients with trachoma, an ancient disease that is currently the leading cause of preventable blindness world-

D. D. Rockey, Department of Microbiology, Oregon State University, Corvallis, OR 97331-3804. *A. Matsumoto*, Department of Microbiology, Kawasaki Medical School, 577 Matsushima, Kurashiki City, Okayama, Japan.

Prokaryotic Development, edited by Y. V. Brun and L. J. Shimkets,
© 2000 American Society for Microbiology, Washington, DC 20005-4171

wide. Roughly 500 million people suffer from trachoma in the world today, of which 5 to 7 million are blind (Schachter and Dawson, 1990). While sexually transmitted conditions were investigated by these early chlamydiologists, it was not until more recently that clinicians have come to appreciate the magnitude of sexually transmitted *Chlamydia trachomatis* infections in human populations (Mardh, 1992). Additionally, we are now aware that serious sequelae can result from a cryptic sexually transmitted *C. trachomatis* infection. The financial costs of sexually transmitted chlamydial diseases are immense. In the United States alone over $4 billion in annual medical expenses are a result of diagnosis and treatment of chlamydial genital infections (Washington et al., 1987). A recent addition to the history of chlamydial disease is the identification of *Chlamydia pneumoniae*, the causative agent of pneumonias and other respiratory conditions (Grayston, 1992). Infection with this species has also been identified as a risk factor for atherosclerosis—a fascinating association between a non-life-threatening acute infection and a subsequent chronic and life-threatening disease (Kuo et al., 1995). *Chlamydia psittaci* and *Chlamydia pecorum* are primarily associated with diseases of animals in many diverse phylogenetic groups, but humans can also serve as incidental hosts of *C. psittaci*, and respiratory psittacosis can be very serious. While there is vast diversity in the diseases caused by the chlamydiae, there are some basic common threads that link all of the causative agents. One of these is the developmental cycle, which is virtually identical in all chlamydial species.

Early Investigations of the Developmental Cycle

Although the labeling of chlamydiae as a virus was incorrect, three compelling aspects of their life cycle led scientists to draw this conclusion. Limited infectivity could pass through filters as small as 0.35 μm, an early hallmark for classification as a virus (Krumwiede et al., 1930). Second, the agent required the presence of living cells to multiply, a trait not unique to viruses

but consistent with that classification. Finally, and probably most convincingly, the infectious agent had an eclipse, or latent phase: a period in a single developmental cycle during which no infectious progeny could be recovered (Collier, 1962). As will be discussed below, these three "virus-like" traits of what we now call the chlamydiae tell much about their growth and development. Eventually, however, many aspects of chlamydial biology led Moulder and others to determine that the agents were of bacterial origin (Moulder, 1964). These later investigators based their evaluation on the presence of both DNA and RNA in the infectious agent, identification of a unit membrane at each step of the developmental cycle, and the sensitivity of the agent to known inhibitors of bacterial growth.

The insight provided by these early investigators tells much about the nature of the developmental cycle. All chlamydiae are obligate intracellular pathogens, requiring host cell biosynthetic machinery for several metabolic functions. The chlamydiae are, for example, auxotrophic for each nucleoside triphosphate except CTP (McClarty, 1994), and analysis of the recently completed *C. trachomatis* genome (Stephens et al., 1998; available through the Chlamydia Genome Project website [University of California, Berkeley]) shows that the chlamydiae also lack much of the machinery required for amino acid synthesis. The entire developmental cycle occurs within a membrane-bound vesicle known as an inclusion, nomenclature that has remained from the early investigations of psittacosis virus. The inclusion is a nonacidified vacuole that interacts with the exocytic arm of the host cell vesicular trafficking pathway (Hackstadt et al., 1997; Moulder, 1991). The determination of the limited filterability of the virus pointed to a relatively small agent, and small particles—termed elementary bodies (EBs)—were described by the first chlamydiologists. We now know that the electron-dense, 0.3- to 0.4-μm-diameter EBs represent the infectious form of the chlamydiae. Light and electron microscopic observations by early scientists also showed other forms associated with the infectious agent. A large, less dense form called the initial body—currently named the reticulate body (RB)—was proposed by early investigators as the source of the infectious particles (Bedson and Bland, 1934). Reticulate bodies are much larger (0.5 to 1 μm in diameter) pleomorphic forms with an electron-lucent cytoplasm. Alternate, less common structures, currently termed intermediate bodies and aberrant RBs, were also observed by these investigators. A discussion of each of these structures will be presented below.

A final trait of the chlamydiae established by these early scientists was the latent phase, a time within the developmental cycle during which no infectivity could be recovered (Girardi et al., 1952). The presence of a latent phase was a key trait that added strength to descriptions of a viral etiology in chlamydial disease. Although the connection was initially tenuous, early investigators correlated this latent phase with the exclusive presence of the RB, a finding that points to another basic trait of the developmental cycle. RBs, while actively metabolic and able to divide, are absolutely noninfectious. Conversely, EBs are infectious but metabolically inactive in the extracellular environment.

Although limited by technical and intellectual constraints of the era, these early investigators laid out the framework for the elucidation of the developmental cycle. During the course of this discussion we present the current state of knowledge regarding each of the developmental forms and describe the known molecular events associated with the completion of the life cycle. But, like these early investigators, we find ourselves still limited in many ways by technical considerations. First, the inability to grow chlamydiae outside of host cells affects the types of experimentation that can be conducted. Almost all studies of chlamydial growth and metabolism must be conducted within the confines of a complicated host-cell background (Moulder, 1991). More importantly however, the lack of a genetic system for introducing genes or creating specific mutations has created difficult barriers to logical experimentation.

For example, the production of auxotrophic chlamydiae and the subsequent analysis of a phenotype, etc., are not currently possible within the chlamydial system. While the sequencing of the *C. trachomatis* and *C. pneumoniae* genomes has greatly enhanced our understanding of chlamydial biology, these limitations still hinder progress.

STEPS IN THE CHLAMYDIAL DEVELOPMENTAL CYCLE

The central core of the developmental cycle is the alternating and complementary nature of the distinct developmental forms. A simplified drawing of the basic developmental cycle is presented in Fig. 1, and electron microscopic images of the inclusion and developmental forms are presented in Fig. 2 and 3. The cycle begins when an elementary body comes in contact with a target cell and is endocytosed

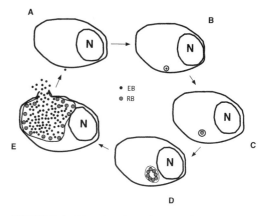

FIGURE 1 Line drawing of a generalized chlamydial developmental cycle. Infection begins when an infectious but metabolically inactive EB comes in contact with a host cell (A) and is endocytosed (B). The phagocytic vacuole (the inclusion) migrates toward the Golgi apparatus, and the EB differentiates into the noninfectious but metabolically active RB (C). RB division ensues, and the inclusion increases in size (D). Reticulate bodies then begin to reorganize back into EBs, and the inclusion grows until it occupies the entire cytoplasm of the infected cell (E). The inclusion lyses, the host cell lyses, and EBs are freed to infect another cell. While there are differences in this cycle among the different chlamydial strains and species, the general process is similar. N, nucleus.

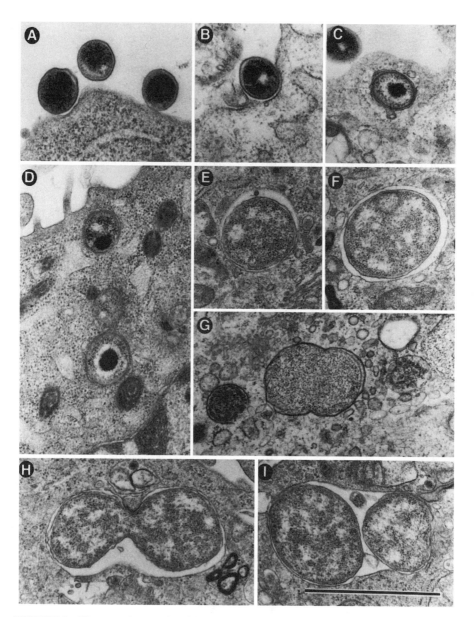

FIGURE 2 Electron micrographs of the sequential changes from attachment of EBs to host cells through the division of RBs. These micrographs show *C. psittaci* Cal 10 infecting L929 cells. (A and B) 30 min p.i. Electrostatic and receptor-mediated adhesions between the EB and the cell can be observed. (C) 60 min p.i. Note the nucleoid structure and the apparent fusion of a tiny vesicle to the phagocyte. (D) 100 min p.i. Two developmental forms can be observed: a typical EB very soon after phagocytosis (top) and another located in the deeper cytoplasm. Note the larger size and apparent reorganization of the nucleoid. The images in panels C and D also represent the sequential change from the EB to intermediate body—a structure that is found both at the beginning and the end of the cycle. (E and F) 6 h p.i. The conversion from EBs to RBs is complete. At this point inclusions will be found in the region of the Golgi apparatus. Note the larger size of the RB in panel F. This may represent growth prior to the initial division process. (G) 8 h p.i. Note the RB constriction prior to binary fission and the fine comb-like structure on the right side of the RB. (H and I) 10 h p.i. Terminal stages of first division. Mitochondria can be seen in the vicinity of the inclusion, but not associated with the inclusion membrane. Magnification of all micrographs, ×60,000; bar in panel I = 1 μm.

(Fig. 1A and 2A to C). In a primary infection these cells are commonly mucosal epithelial cells of the eye, genitourinary tract, or respiratory tract. After bacterial uptake, the phagosomal vacuole—the early inclusion—migrates to the perinuclear area of the cell, and EBs differentiate into RBs (Fig. 1B and 2D to F). RB multiplication by binary fission ensues (Fig. 2G to I), with the inclusion increasing in size in parallel with the increasing numbers of bacteria (Fig. 1C and 3A and B). At a point approximately halfway through a single cycle, RBs begin to condense back into EBs. At each transitional point in the cycle, intermediate forms are also present (Fig. 2C and D and 3D), representing EBs developing into RBs and vice versa. Large numbers of EBs begin to accumulate in the inclusion, and the inclusion expands to occupy essentially the entire cytoplasm of the cell (Fig. 1D and 3E and F). In most strains this is followed by lysis of the inclusion, lysis of the host cell, and release of EBs to infect another cell (Fig. 1E). For most chlamydiae this cycle is complete within 48 to 72 h postinfection, and burst sizes range from approximately 100 to 1,000/infected cell. While there are differences in certain aspects of the cycle among the different strains of chlamydiae, the basic productive developmental cycle is the same: EB to RB to EB.

STRUCTURE OF THE EB

EBs are small coccoid electron-dense structures (Fig. 2A and 4A). The electron-dense center is bound by inner and outer membranes that have lipid compositions more similar to that of the infected host cell mitochondria than to those of other bacteria (Wylie et al., 1997). The EB is quite resistant to physical disruption, as is evidenced by a moderate sonication step incorporated into purification protocols (Caldwell, 1981). This resistance is achieved in the absence of detectable concentrations of peptidoglycan (PG) (Fox et al., 1990). The absence of PG defines a major distinction between the chlamydiae and many other bacteria, particularly those that form true endospores, in which PG is a major constituent of the cell wall. The

inability to detect PG in EBs or RBs is paradoxical (Moulder, 1993). Contemporary methodologies for detection of PG or its precursors within developmental forms have been unsuccessful, and there is no ultrastructural evidence supporting the idea that PG is found in the chlamydiae. However, two alternate lines of evidence suggest strongly that chlamydiae produce PG. First, production of EBs within infected cells is completely blocked by inhibitors of PG synthesis, such as beta-lactam antibiotics and D-cycloserine (Moulder et al., 1963; Weiss, 1950). These drugs inhibit different steps of the PG synthesis pathway and likely also block PG synthesis in the chlamydiae. Additionally, sequencing of the C. trachomatis genome has revealed a virtually complete set of genes encoding proteins associated with PG production. The evidence suggests, therefore, that the chlamydiae produce and utilize PG, but innovative approaches will be required to demonstrate its presence and elucidate its function.

Even if PG is present in EBs the amount produced is not large enough to explain the cell wall structural stability, and therefore this integrity must be supplied by other means. EB cell wall structure and EB infectivity are both resistant to sonication. The cell wall structure is also resistant to treatment with detergents such as N-lauryl sarcosine (Hatch et al., 1984). This treatment removes all of the central electron-dense material within the EB but leaves the cell walls morphologically intact. However, both infectivity and cell wall structural integrity are highly sensitive to a combination of detergent and a reducing agent such as 2-mercaptoethanol or dithiothreitol (Newhall and Jones, 1983; Hatch et al., 1984). Reducing agents are commonly used to break disulfide bonds between and within proteins, and these results suggest that the structural integrity of the EB cell wall is possibly a function of protein arrays linked by disulfide bonds between cysteine molecules. Electrophoresis and autoradiography of ^{35}S-labeled EB lysates showed three dominant labeled protein species—a doublet at 58 to 60 kDa, a protein of approximately 40

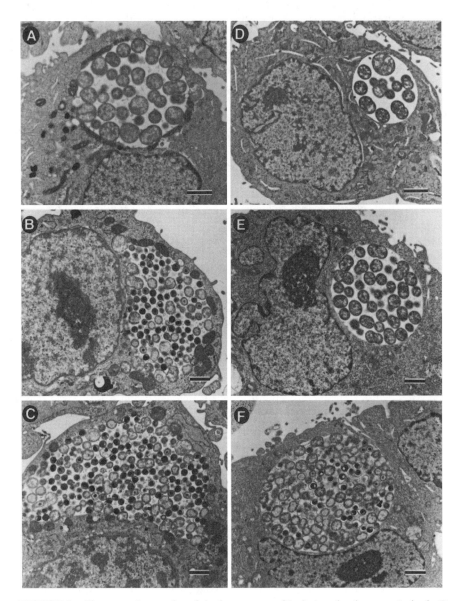

FIGURE 3 Electron micrographs of the later stages of inclusion development in both *C. psittaci* and *C. pneumoniae*. (A to C) *C. psittaci* Cal 10-infected L929 cells cultured at 37°C. (A) 18 h p.i.; (B) 24 h p.i.; (C) 34 h p.i. In each panel, note the relative numbers of RBs and EBs and the association between the inclusion and host cell mitochondria. (D to F) *C. pneumoniae* TW-183 in HEp-2 cells, cultured at 37°C. (D) 18 h p.i.; (E) 24 h p.i.; (F) 34 h p.i. Note the apparent lack of mitochondrial association, the more spherical nature of the inclusion, and the slight compression of the host nuclei. The inclusions in panels A, D, and E contain exclusively RBs, while the remaining panels show the asynchronous nature of the later inclusions. Bars = 1 μm.

kDa that varies in mass among strains and species of chlamydiae, and a protein with an apparent mass of 12 kDa. These same proteins remain associated with cell wall preparations after treatment with N-lauryl sarcosine. However, treatment of detergent-extracted cell wall complexes with 2-mercaptoethanol results in loss of integrity of the cell wall and the solubilization of the three labeled proteins. These three cysteine-rich proteins were also shown to be dominant proteins on polyacrylamide gels of purified EBs and EB cell wall extracts (Fig. 4). The genes encoding these proteins were subsequently cloned and sequenced. The 40-kDa protein, termed Omp1 or MOMP (for major outer membrane protein), has attracted considerable attention as a candidate adhesin and porin and as a prime candidate immunogen for vaccination trials (Baehr et al., 1988; Stephens et al., 1986) (for a summary of all described gene designations, see Bavoil et al., 1996, or the genome project website). The gene encoding MOMP, ompA, is transcribed from early in development through the entire cycle. ompA encodes an integral membrane protein with a classical type II signal peptide. The protein consists of four major variable domains within a conserved backbone structure, and differences within the variable domains define the multiple serovars of C. trachomatis (Yuan et al., 1989; Stephens et al., 1987). MOMP is the most abundant protein in the chlamydial cell wall, is surface exposed, and is likely the dominant adhesin for C. trachomatis and C. psittaci attachment to target cells (Hatch et al., 1981; Su et al., 1990, 1996). Antibodies to conformational domains of native MOMP bind to live EBs and are neutralizing in vitro (Su and Caldwell, 1991; Caldwell and Perry, 1982; Zhang et al., 1989). The possibility that MOMP is a porin is supported by early studies showing that treatment of EBs with reducing agents leads to increased metabolic activity and RNA synthesis. While the resulting synthesis was transient and relatively minor, it suggests that a reduced EB is more porous to nucleotide triphosphates or metabolic precursors or possi-

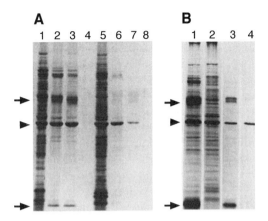

FIGURE 4 Effects of detergents and reducing agents on the integrity of EB and RB developmental forms. In both panels the arrows point to the 60-kDa EnvB (Omp2) and the 12-kDa EnvA (Omp3) proteins, and the arrowhead points to the 40-kDa MOMP band. (A) Sodium dodecyl sulfate-polyacrylamide gel electrophoresis (SDS-PAGE) profiles of insoluble residues of C. psittaci EBs and RBs after extraction with detergents plus or minus 2-mercaptoethanol (2-ME). Purified developmental forms were treated with the extracting agent and subjected to high-speed centrifugation. The material in the pellets was then prepared for standard PAGE. Lanes: 1, EB extracted with phosphate-buffered saline (PBS); 2, EBs extracted with Sarkosyl; 3, EBs extracted with SDS; 4, EBs extracted with SDS plus 2-ME; 5, RBs extracted with PBS; 6, RBs extracted with Sarkosyl; 7, RBs extracted with SDS; 8, RBs extracted with SDS plus 2-ME. Note the absence of detectable EnvA and EnvB in the RB preparations but the abundance of MOMP in both EBs and RBs. (B) A similar experiment with EBs cultured in the presence of either [^{35}S]cysteine or [^{35}S]methionine. Purified EBs were then extracted with either PBS (lanes 1 and 2) or SDS (lanes 3 and 4). Insoluble material was collected by centrifugation and prepared for standard PAGE. The resulting gel was dried and exposed to film. The samples in lanes 1 and 3 represent EB labeled with [^{35}S]cysteine. The samples in lanes 2 and 4 represent EB labeled with [^{35}S]methionine. Note the distinction between detectable SDS-insoluble EB proteins under the two labeling conditions. (Data reproduced from Hatch et al., 1984, with permission of the authors and ASM.)

bly that enzymes responsible for these processes are more active in a reduced environment. Further evidence supporting a role of reducing agents in the enhancement of porosity of the outer membrane is found in the liposome-swelling assays of Bavoil et al. (1984). Liposomes formed with chlamydial outer membrane complexes, which are highly enriched for MOMP proteins, have demonstrable porin activity in a nonreduced environment, but reduction and alkylation of the structure results in 10-fold-higher porin function. While these experiments demonstrate that a reducing environment likely augments porosity of the cell wall, the data support but do not prove the role of MOMP as the major porin. Recent work by Wyllie et al. (1998) strengthened the likelihood that MOMP of *C. psittaci* is a membrane-localized porin by demonstrating that the porosity of lipid bilayers containing solubilized MOMP is eliminated by the addition of conformation-specific anti-MOMP monoclonal antibodies. Collectively, these data present evidence that MOMP functions as a porin and that the conformation of MOMP, and possibly other proteins in the EB cell wall, is sensitive to the redox potential at the bacterial surface.

The additional cysteine-rich proteins, because of their putative association with the outer membrane of detergent-treated EBs, have been labeled Omp2 (a 60-kDa cysteine-rich protein) and Omp3 (a 12-kDa cysteine-rich protein), encoded by the genes *omcB* and *omcA*, respectively. In *C. psittaci* the corresponding proteins are also designated EnvB (60 kDa) and EnvA (12 kDa), encoded by *envB* and *envA*. These genes are transcribed as an operon, with both bicistronic and monocistronic RNA species present in developing RBs (Lambden et al., 1990). While *ompA* transcripts and MOMP protein synthesis can be detected at all but the earliest time points of a developmental cycle, *omcA* and *B* are definite examples of late chlamydial genes (Watson et al., 1995; Hatch et al., 1984). Each of these proteins, MOMP, Omp2, and Omp3, is targeted to the periplasm and outer membrane via amino-terminal signal sequences. *C. trachomatis* MOMP has on average 2 to 3% cysteine, while Omp2 and Omp3 are 4.5 and 14.7% cysteine, respectively. The 60-kDa cysteine-rich proteins are quite conserved among the different chlamydiae, with homologous genes from all species

having high sequence identity. This is not the case, however, with EnvA and Omp3, with approximately 54% identity among the different species. The position of the cysteine residues in the small cysteine-rich protein is, however, largely conserved. Both EnvA and Omp3 have signal sequences characteristic of lipoproteins, and EnvA incorporates labeled palmitate during growth (Everett et al., 1994).

The architecture of the chlamydial cell wall has been investigated using several approaches. Of primary interest is the surface exposure of membrane proteins. This has ramifications for the ability of chlamydiae to contact target host cells, to serve as environmental sensors, and to function as possible targets for antibody-mediated host immune mechanisms. Surface exposure was demonstrated for *C. psittaci* and *C. trachomatis* MOMP by several techniques. First, antibodies specific for conformation-dependent MOMP epitopes prevented binding of EBs to target cells. Secondly, MOMP is sensitive to limited proteolysis by trypsin treatment of whole EBs. Finally, MOMP can also be detected at the surface by iodination techniques and by the use of labeled anti-MOMP antibodies. Surprisingly, MOMP from *C. pneumoniae*, which is otherwise quite similar to MOMP from other species, cannot be accessed at the native EB surface (Knudsen et al., 1998). This suggests that the architecture of the *C. pneumoniae* cell surface may be different than that of *C. trachomatis* or *C. psittaci*.

Assessment of the localization of the other cysteine-rich protein in the cell wall has not been so straightforward. An in vitro ligand binding assay and limited trypsin proteolysis experimentation support the conclusion that EnvB of *C. psittaci* GPIC may be surface exposed (Ting et al., 1995), but this has not been supported by similar studies with other chlamydiae. Recent efforts by Everett and Hatch (1995) provide the most contemporary model for localization of EnvA and EnvB in the cell wall. These authors examined the effects of selected detergents, in the presence and absence of reducing agents, on the integrity of EBs and EB cell wall preparations. They propose that

EnvA is attached to the outer membrane via its lipid tail and resides within the periplasm of the bacterial cell. EnvB is present in the periplasm and is cross-linked to EnvA via disulfide linkages. MOMP is present as cross-linked, surface-exposed multimers that may or may not be cross-linked to EnvA. The structural stability of the EB cell wall is therefore primarily provided by the extensive intra- and intermolecular cross-linking of EnvB and EnvA. MOMP cross-linking may also be important in the structural stability of the EB cell wall.

The outer membrane of the chlamydial developmental forms also contains lipopolysaccharide (LPS), a truncated but otherwise typical molecule that is structurally conserved among all chlamydial species (Nurminen et al., 1983; Caldwell and Hitchcock, 1984). LPS is loosely associated with both the EB and RB cell surfaces, and antibody against LPS does not block infection of host cells. There are reports that LPS is shed into the host cell following infection, but this is controversial (Karimi et al., 1989; Baumann et al., 1992). The enzyme responsible for producing the genus-specific LPS determinant is a saccharide transferase that acts on an LPS precursor (Brade et al., 1987). This protein (KDO transferase) is encoded by gseA, a conserved gene present in each chlamydial species. Expression of this gene in Escherichia coli leads to the production of LPS containing the genus-common determinant and to its localization to the E. coli cell surface (Nano and Caldwell, 1985). The role of chlamydial LPS in the infective process or developmental cycle remains unclear.

Recently an additional and unique collection of outer membrane proteins has been identified in each chlamydial species. The members of this protein family are approximately 90 kDa in size, are recognized by antisera directed at EBs, and are present in both EBs and EB cell wall preparations (Cevenini et al., 1991). Although similarly sized antigens are present in many immunoblots presented in the literature, their detailed examination began with the work of Longbottom et al. (1996, 1998), who used monoclonal and polyclonal antibody reagents to clone linked genes encoding antigenic 90-kDa proteins from an ovine abortion strain of C. psittaci. Subsequent studies showed that at least some members of this family are present in the cell wall and are likely surface exposed (Grimwood et al., 1998; Stephens et al., 1998; Knudsen et al., 1998). Because they are candidate outer membrane proteins, they are labeled Pmps (putative outer membrane proteins), and the corresponding genes (in C. trachomatis) are pmpA to -I. Sequence analyses of the C. trachomatis and C. pneumoniae genomes have identified 9 pmp genes in C. trachomatis and, surprisingly, 20 homologous pmp genes in C. pneumoniae (Stephens et al., 1998). Multiple pmp genes are found in C. psittaci as well (Longbottom et al., 1998). No pmp genes have been reported in C. pecorum, but similarly sized EB proteins are observed in immunoblots (Baghian et al., 1996). The total percentage of the genome occupied by pmp genes underscores their likely importance. In the case of C. pneumoniae, 5 to 6% of the genome encodes a member of this gene family. The nature of the Pmp proteins and their role in the infectious process are just beginning to be investigated.

An additional provocative structure found in chlamydial developmental forms is the surface projections originally described by Matsumoto (Fig. 5B and C), with slightly different views presented subsequently by others (Matsumoto et al., 1976; Matsumoto, 1982a, 1982b; Nichols et al., 1985; Gregory et al., 1979). Analysis of EBs by a variety of electron microscopic techniques has identified unique projections extending out from the surface of the EB. These projections are regularly spaced and appear to localize to one surface of the developmental form. As shown in Fig. 5, it may be that the projections found in EBs are distal to the accumulation of nucleoid. Freeze-etch microscopy has also shown that symmetrical rosette structures are present on the inside surface of the inner membrane, and the spacing and distribution of these rosettes are similar to those of the projections on the outer surface (Matsumoto, 1973; Louis et al., 1980). It is likely that

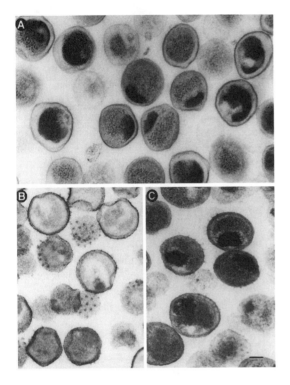

FIGURE 5 Morphologies of highly purified *C. psittaci* EBs fixed differently prior to thin sectioning. (A) EBs were doubly fixed with glutaraldehyde and OsO$_4$ and embedded in Epon. Thin sections were doubly stained with uranylacetate and lead citrate solutions. Note the lack of visible surface projections. (B) EBs were fixed with glutaraldehyde and treated with tannic acid. The sections were examined without subsequent staining with uranylacetate and lead citrate. While the internal structures are not visible, the regular nature of the surface projections is clearly demonstrated. (C) Purified EBs were fixed with glutaraldehyde and treated with ruthenium red. Thin sections were then stained with uranyl acetate and lead citrate. Note that the surface projections are located only on the EB surface opposite the nucleoid structure. Magnifications of all micrographs, ×90,000. Bar (panel C) = 0.1 μm.

these structures serve as the bases of the projections. Surface projections can be found on both RBs and EBs, and the number of projections appears to be highest on growing RBs (Matsumoto, 1982a). In both EBs and RBs these projections are found in patches localized to discrete domains of the developmental form. Electron microscopy of isolated inclusions shows similar projections protruding into or possibly through the surface of the inclusion membrane (Fig. 6). It has recently been proposed that these projections may function in the secretion of proteins from developing RBs, possibly through a type III secretion mechanism (Bavoil and Hsia, 1998). Both *C. psittaci* and *C. trachomatis* encode proteins that likely participate in a type III secretion pathway (Hsia et al., 1997), and the structure of the surface projections resembles the base of a flagellar assembly (secreted and assembled via a type III mechanism), as well as the structures shown by Kubori et al. (1998) to be purified type III secretion assemblies from *Salmonella*. Addi-

tionally, a collection of proteins localized to the inclusion membrane (Inc proteins) are secreted by chlamydiae and have no apparent classical signal sequence at their amino termini (Rockey et al., 1995, 1997; Bannantine et al., 1998b). Therefore, it is likely that the chlamydiae use an alternate secretory pathway, such as a type III pathway, and the surface projections are possible candidates for that function. This is speculative, and it is certainly possible that the projections serve an alternate function. The possibilities include a porin activity to allow material from the cytoplasm to be imported directly into the developmental form and a physical point of attachment for the EB and RB to bind at host cell surfaces either at the plasma membrane or within the inclusion. Although their actual function remains to be elucidated, it is reasonable to propose that the surface projections are involved with some aspect of the interaction between the intracellular environment and the infecting EBs and/or developing RBs.

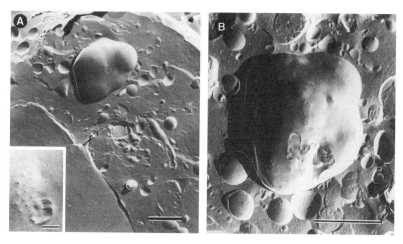

FIGURE 6   Freeze-etch micrographs of the external surfaces of chlamydial inclusions. (A) *C. psittaci* Cal 10 inclusion 16 h p.i. Note the surface projections enlarged in the inset (×76,000). (B) Similar micrograph of Cal 10 at 18 h p.i. Magnification, ×30,000. Bars = 1 μm (full-scale images) and 0.1 μm (inset).

Moving from the surface toward the center, we now explore the nature of the electron-dense cytoplasm of the EB. At first glance the dense nature of the EB, its resistance to physical disruption, and its temporal association with initial and terminal events in chlamydial development suggest analogies between EBs and the endospores of other bacterial species. These analogies, however, are most relevant in comparisons between the EB and endospore cytoplasmic compartments. As we know from other chapters in this text, endospores contain tightly packed arrays of DNA, RNA, and large amounts of DNA binding proteins. That EBs contain DNA and RNA was determined early in the investigation of the chlamydiae (Tamura et al., 1967), and much recent work has focused on the nature of protein–DNA interactions that may cause nucleoid compaction. The identification of mechanisms involved in nucleoid condensation began as several groups identified proteins in EB extracts that strongly bound both proteins and nucleic acids in vitro. The first of these studies demonstrated two DNA binding proteins of 17 and 26 kDa present in EBs but absent in RBs (Wagar and Stephens, 1988). Collectively these studies identified a conserved 18-kDa species and a family of bind-

ing proteins varying in size (26 to 32 kDa) by serovar (Wagar and Stephens, 1988; Hackstadt, 1986; Perara et al., 1992). The initial experimental approaches varied among research groups, and therefore the proposed function varied from the binding of nucleic acids to possibly a candidate adhesin. Subsequent research was directed at cloning the genes encoding each binding protein, and as the sequences of the genes were identified, their roles in chlamydial biology became clearer. These genes encode highly basic proteins, each having sequence identity with eukaryotic histone H1. The 18-kDa protein (Hcl) is 35% identical to sea urchin histone H1 over a span of 106 amino acids (Hackstadt et al., 1991), and the larger protein (Hc2) has 47% identity over a central 66 amino acid stretch (Brickman et al., 1993; Perara et al., 1992). The primary bases for these identities are the high percentages of lysine and alanine in each protein. Hc2, but not Hcl, also has sequence similarity to DNA binding proteins from other prokaryotes, including AlgP of *Pseudomonas aeruginosa* and HrdB of *Streptomyces coelicolor*.

The biology of histone-like proteins is well understood in both eukaryotic and prokaryotic systems. These proteins are responsible for binding DNA, for compacting that DNA into regular arrays, and for affecting transcription and translation. It was proposed that Hc1 and Hc2 were associated with similar functions in the chlamydiae. To explore these possibilities, Hackstadt and colleagues examined the effects of expressing *hctA* (encoding Hc1) in *E. coli*. Induction of *hctA* in this system led to both a concomitant shutdown of overall gene expression and a characteristic condensation of the genomic DNA within the bacterium (Barry et al., 1992). This condensation bore ultrastructural similarity to the nucleoid structure seen in EBs. Analysis of purified nucleoid from these recombinant *E. coli* cells showed chlamydial Hc1 to be tightly bound to the condensed DNA. Although this work was conducted in *E. coli* and not in the native system, these studies strongly suggest that Hc1 is associated with chromosomal condensation and the overall metabolic inactivity of the EB. Subsequent studies by Barry et al. (1993) showed that low-level expression of *hctA* in *E. coli* led to increased transcription of *lacZ* hybrid genes driven by supercoiling-sensitive promoters. These data support the idea that low concentrations of Hc1 may have subtle effects in the maturing developmental form, prior to the nucleoid condensation observed at high Hc1 concentrations. Indeed, Northern blots demonstrate *hctA* RNA present within infected cells several hours prior to the first evidence of nucleoid condensation, while high levels of Hc1 protein accumulate in parallel with the appearance of EBs (Hackstadt et al., 1991). Although HCl has sequence identity with eukaryotic histones, this similarity is limited to amino acids 63 to 125 of Hc1. The amino-terminal half of the molecule does not have sequence identity with other DNA binding proteins or motifs but is highly conserved among the chlamydiae. Experiments with truncated Hc1 proteins localized the DNA binding domain to the region that had identity with H1 and showed that the amino-terminal region of the protein likely had an alternate function (Pedersen et al., 1996b; Remacha et al., 1996). The authors also showed that truncated proteins lacking amino acids 1 to 63 were less efficient at condensing DNA, suggesting a possible role of this domain in protein-protein cooperation. Collectively, these studies suggest that Hc1 is a DNA binding protein with multiple functions, including transcriptional inactivation and nucleoid condensation late in the developmental cycle.

The next question is centered around the role of Hc2, a protein of variable mass that is present in *C. trachomatis* but apparently absent in tested strains of *C. psittaci*. Hc2 has an amino-terminal structure consisting of multiple pentapeptide repeats, each repeat ending with two basic amino acids. The differences in mass of the different proteins were shown by Hackstadt et al. (1993) to result from different amounts of the pentapeptide repeats. Like Hc1, Hc2 may have different functional domains. In this case the pentapeptide repeats have identity with H1 while the carboxy-terminal region lacks such similarity (Brickman et al., 1993; Perara et al., 1992). Expression of *hctB* in *E. coli* resulted in compaction of *E. coli* DNA but to a different degree than that seen with expression of *hctA*. Hc2 also inhibits translation and transcription, possibly at different levels than that seen with Hc1 in vitro (Pedersen et al., 1996a). Therefore, these two DNA binding proteins are involved in the condensation of nucleic acids associated with the formation of EBs and in the inactivation of transcription and translation late in the developmental cycle.

EARLY STEPS IN THE EB-TO-RB TRANSITION

Contact with the host cell and entry of the EB are the first steps in a complicated interaction between the infecting chlamydiae and the invaded host cell. The cell has mechanisms in place to traffic endocytosed material into the lysosomal pathway, leading to the death and digestion of the invader. The chlamydiae must modify their environment, and quickly, to avoid this fate. A major distinction between chlamydiae and other obligate intracellular pathogens lies in their means of addressing the problem of phagolysosomal fusion. One example of a pathogen with a different strategy is the obligately intracellular *Coxiella burnetti*, a species discussed in another chapter of this book. Superficially, *C. burnetii* develops within a parasitophorous vacuole very similar in structure to the chlamydial inclusion. However, careful analysis with labels for different host cell trafficking pathways has shown that *C. burnetii* allows itself to be trafficked to fully functional lysosomes and that only in the acidic environment of the lysosome can the organism develop (Heinzen et al., 1996). Phagosomes containing chlamydiae, on the other hand, quickly leave the lysosomal fusion pathway and reside in the nonacidified exocytic vesicular pathway. Microscopic changes observed following uptake of EBs include increased size of the developmental form, a reduction of the amount of DNA condensation, and a migration of the inclusion to the region of the Golgi apparatus. This morphological transformation is paralleled by a sharp decrease in the infectivity of the intracellular chlamydiae, which commonly approaches zero within 4 h postinfection (Moulder, 1991). Therefore, functionally and morphologically the intracellular chlamydiae are developing into RBs.

A major step in the initiation of the chlamydial developmental cycle is the "conditioning" of the phagosome to become an inclusion. Although the cell biology of inclusion development is poorly understood, the maturation of this vacuole can be monitored by use of the fluorescent lipid N-[7-(4-nitrobenzo-2-oxa-1,3-diazole)]aminocaproylsphingosine (NBD)-ceramide (NBD-cer). NBD-cer is used in many systems to study the process of vesicular trafficking to and from the Golgi apparatus (Pagano et al., 1991). In uninfected host cells incubated at 37°C, exogenously supplied NBD-cer is quickly transferred from the plasma membrane to the Golgi apparatus, enzymatically processed, and then transported back to the surface of the cell via the exocytic vesicle trafficking pathway. In productive chlamydial infections, NBD-cer traffics from the plasma membrane through the Golgi apparatus to the inclusion and then to the intracellular

developmental forms (Hackstadt et al., 1995). Approximately 50% of the NBD-cer is trafficked to the intracellular chlamydiae, where it remains for the rest of the developmental cycle. Accumulation of NBD-cer can be observed in developmental forms as early as 1 to 2 h postinfection (Hackstadt et al., 1996). The transfer of label from the Golgi to the inclusion suggests the intracellular chlamydiae have positioned themselves in an exocytic pathway, deriving lipids and possibly other metabolic precursors from the Golgi vesicular trafficking machinery. Proper trafficking of this label is dependent on early chlamydial protein and RNA synthesis, as inhibitors of these processes block accumulation of NBD-cer in the early inclusion (Scidmore et al., 1996). Infected cells treated with protein synthesis inhibitors show enhanced levels of chlamydial trafficking to lysosomes, leading to their destruction. These inhibitors also block the localization of the inclusion near the Golgi apparatus in the infected cell, as well as the normal EB-to-RB transition.

While EB-to-RB transition and the initial phases of a productive developmental cycle are dependent on bacterial RNA and protein synthesis, very little is known about the initial molecular steps in the intracellular life of the EB or of the signals involved in the EB-to-RB transition. The morphological transition described above initiates within the first 2 h following infection in vitro, and transcriptional and translational events can be documented even more quickly. Protein synthesis can be demonstrated within 15 min of infection, well before there is clear morphological evidence of an EB-to-RB transition (Plaunt and Hatch, 1988). Inhibitors of transcription and translation were used to show that this early protein synthesis involved both preformed transcripts and transcripts generated de novo. The first protein shown to be produced very early in the infectious cycle is that encoded by the Early Unknown Operon of Wichlan and Hatch (1993). This gene (*euo*) and its protein product (EUO) have also been identified in *C. trachomatis* (Kaul et al., 1997). In *C. psittaci*, levels of *euo* transcript peak within 60 min postinfection and are undetectable 8 h postinoculation (p.i.). Curiously, protein levels accumulate throughout the early stages of development and peak 12 h p.i. Protein abundance decreases as EBs begin to accumulate within the inclusion and is undetectable by fluorescence microscopy or immunoblotting late in the developmental cycle (Zhang et al., 1998; Hatch and Rockey, unpublished). Two schools of thought have emerged regarding the possible function of EUO. Kaul et al. (1997) have shown that recombinant EUO has specific proteinase activity directed at chlamydial Hc1. In contrast, Hatch and others have shown EUO to possess specific DNA binding activity (Zhang et al., 1998). While these are quite distinct models, each is consistent with the needs of the changing EBs. Certainly there must be some mechanism to either degrade the chlamydial histones or otherwise separate them from DNA. Also, specific DNA binding proteins are most likely required to selectively activate or repress transcription following infection. It is also conceivable that both models are correct, as EUO may coordinate its proteolytic activity with its DNA binding activity. Work is in progress that hopefully will elucidate the role of EUO in chlamydial development.

Other genes are also transcribed very early in the infectious process. Reverse transcription-PCR has been used by Scidmore-Carlson et al. (1999) to demonstrate transcript encoding the protein IncD as early as 2 h p.i. IncD is a member of a family of proteins that reside in the inclusion membrane of a chlamydia-infected cell. The function of these proteins is unknown, but one member of this family, IncA of *C. psittaci*, is in direct contact with the cytoplasm of the infected cell and can be phosphorylated by host cell protein kinases. Two-dimensional protein electrophoresis of ^{35}S-labeled infected cells has also been used to identify protein species that accumulate early in the developmental cycle. Studies by Lundemose et al. (1990) identified several metabolic and housekeeping proteins that are first produced between 2 and 8 h p.i. In these authors' experiments, the appearance of these proteins preceded the appearance of detectable MOMP. A number of these proteins are apparently specific to the early and middle phases of the growth cycle, as their production and accumulation decreased 24 to 30 h p.i. MOMP, on the other hand, continues to accumulate throughout the cycle. These studies demonstrate that a relatively large collection of proteins that may be essential for early inclusion development are produced very early following infection. It is likely, however, that many of these proteins have very straightforward housekeeping or metabolic functions. Additionally, the production of some of these proteins, such as EUO, may be downregulated as the cycle proceeds, while others are apparently produced throughout development. The continued elucidation of proteins that appear early in the developmental cycle is a major area of research in chlamydial biology.

While transcription and translation can be documented quite early following infection, it is likely that the very initial steps include conformational or secondary-structure changes in the infecting EB. Such changes may represent the initial signaling events responsible for initiating the EB-to-RB transition. The first documented changes in infecting EBs include the reduction of MOMP from a polymeric, disulfide-linked form to a reduced and monomeric form. Recall that MOMP is tightly cross-linked in the EB, resulting in high-molecular-weight matrices. As early as 1 h p.i. monomeric MOMP can be detected within infected cells, presumably resulting from reduction of intermolecular disulfide bonds (Hatch et al., 1986). Monomeric MOMP can then be detected throughout the intracellular phase of the life cycle. In contrast, a similar shift in the degree of cross-linking of Omp2 and Omp3 is not evident, either immediately following infection or as the proteins begin to accumulate late in the cycle. The presence of MOMP monomers is temporally associated with the initiation of EB-to-RB transition, the detection of EUO and *incD* transcript, and the detection of NBD-cer uptake into the developmental forms. A logical model suggests that early conditions in the intracellular-intravacuolar environment, or possibly upon contact with the host cell, trigger the reduction of MOMP. This may augment the porosity of the EB, which could be a first or very early step in the signaling process allowing early transcription and translation and the initiation of the EB-to-RB transition.

Beyond these examples very little is known about the early molecular steps in the chlamydial developmental cycle. This is primarily because we have identified so few early genes and know very little about those identified. Additionally, much will be learned as we develop an understanding of the signals involved in EB-to-RB transition. If such signals can be identified and modeled in vitro, it is possible that the early steps of the EB-to-RB transition could be examined in the absence of a host cell background.

STRUCTURE OF THE RB

With the exception of EUO, very few proteins have been described that are present only in RBs. The first evidence of a protein that is differentially present in RBs and not EBs was the cyclic AMP (cAMP)-binding activity demonstrated by Kaul and Wenman (1986). cAMP-binding activity was found in lysates of RBs but not EBs, but proteins associated with this activity were not purified. A protein with sequence similarity to cAMP-binding protein kinase regulatory proteins is encoded in the genome sequence (open reading frame D235), and it is possible that this is the binding activity identified by these authors. The Inc proteins mentioned earlier can be found in purified RBs but not in EBs, but their final destination is the inclusion membrane of the infected cell. Apparent RB-specific proteins also include p52, shown by Rockey and Rosquist (1994) to be present in RBs but not EBs, and *C. trachomatis* TroA and p242, described by Bannantine et al. (1998a). It is also likely that many metabolic proteins are present in higher amounts in RBs than in EBs, as their activities are needed primarily at the RB stage.

Recall that RBs lack several proteins—Hc1, Hc2, Omp2, and Omp3. The lack of func-

tional histone analogues is important to allow DNA and RNA in the RB to be accessible by transcriptional and translational machinery. The absence in RBs of the heavily cross-linked Omp 2 and -3 leads to a flexible and much more permeable cell wall structure. This leads to exquisite sensitivity of RBs to sonication, in contrast to the insensitivity seen with EBs. MOMP is produced throughout the cycle, but the degree of cross-linking is much reduced in RBs. Because RBs lack the rigidity imposed by the cross-linked outer membrane proteins, they are able to expand during growth and multiplication. The increased porosity of RBs in turn facilitates incorporation of labeled precursors into protein and DNA. These traits of RBs are likely mediated by the changes in cell wall structure that are the hallmarks of the EB-to-RB transition.

MATURATION OF CHLAMYDIAE— THE RB-TO-EB TRANSITION

The activation and subsequent multiplication of RBs leads to an accumulation of these developmental forms within the young and "middle-aged" inclusions (Fig. 3A, D, and E). At time points generally beginning just after the halfway point in the chronology of a cycle, EBs first appear and then begin to rapidly accumulate within the inclusion (Fig. 3B, C, and F). Prior to this point the cycle is relatively synchronous. Very early (less than 1 h p.i.), the intracellular chlamydiae are almost all morphologically EBs, and after about 8 h, RBs are exclusively present in a productive inclusion (Fig. 3A, D, and E). At approximately the midpoint of a single cycle, development becomes asynchronous and both EBs and RBs are found within the inclusion. Late in the cycle EBs dominate, but RBs are still readily detected (Fig. 3C).

Much like the situation with EB-to-RB transition, there is essentially nothing known about the signals responsible for stimulating RB-to-EB conversion. The possibilities include environmental sensing via a two-component regulatory mechanism, by some means through the surface projections, via MOMP-

and LPS-host cell interactions, or via some other completely novel mechanism. One possibility, initially introduced by Hackstadt et al. (1997), involves some means of sensing the level of intimate physical contact between the developmental forms and the host cell membrane. Extracellular EBs have very limited or no contact with host cell membranes or membrane structures. The initial phase of infection involves increasing levels of contact and adherence, either directly or indirectly, between EBs and components of the host cell plasma membrane. Following endocytosis and during the earliest intervals in the phagosomal process, EBs are in intimate contact, on all surfaces, with the vacuolar membrane (Fig. 2). It is possible that this gradation of intimate contact in some way directly stimulates the initiation of the EB-to-RB conversion, possibly leading to production of early proteins such as EUO. This is quickly followed by the transcription of a spectrum of housekeeping genes and the synthesis of proteins involved in anabolism and RB growth. Throughout development one of the most striking traits of growing RBs within the inclusion is the intimate contact retained with the inclusion membrane over at least part of the surface of the RB (Fig. 2 and 3). The groups of projections observed by Matsumoto and others likely represent a domain of the RB surface that is in tight association with the inclusion membrane. Indeed, the number of projections per patch is reportedly higher early in RB growth and diminishes with time. Possibly the level of cross-linking of these projections or other surface moieties or some critical number of contact points per volume of RB is needed to maintain the constitutive phase of RB protein synthesis. After several rounds of RB multiplication there may be crowding at the inclusion membrane, and a critical percentage of the contact with the host membrane might be lost. This may signal the initiation of transcription of late genes, potentially by using the same unidentified signaling mechanisms that are used at the beginning of the developmental cycle. The attractiveness of this model is that it requires only a single signal, the degree of membrane contact, to facilitate both the initial and terminal processes associated with development. Of course this model is purely speculative—the molecular events that surround these transitions remain to be established or even effectively experimented with.

Another question that remains unanswered involves the stoichiometry of RB:progeny-EB production. The number of EBs found in a fully mature *C. trachomatis* inclusion is likely in the high hundreds, but there is no evidence that such high numbers of RBs are ever present in the infected cell. In contrast, while there is evidence of aberrant RBs that contain multiple genomes and possibly multiple centers of chromosome nucleation (Beatty et al., 1995), there is no ultrastructural support for one RB producing many EBs. In fact, ultrastructural data suggest that as histones and the cysteine-rich proteins accumulate, each RB differentiates to form individual EBs. This process involves nucleoid condensation, a significant increase in cell wall rigidity, and an increase in the apparent density of the bacterium. Therefore, the data support the idea that a single RB condenses to a single EB, but this is somewhat in opposition to the rapid production of EBs within terminal infected cells. Many models can be proposed—it is possible that, late in the replicative cycle, one daughter RB continues with logarithmic multiplication while the other differentiates into EBs.

TERMINAL STEPS IN EB RELEASE

With the exception of the analysis of MOMP cross-linking at the end of the cycle, the molecular biology of the end of the developmental cycle is virtually uninvestigated. Detailed structural and ultrastructural analyses of the end stages have been presented. During the last stages of the developmental cycle the inclusion occupies virtually the entire cytoplasm of the infected cell (Figure 1E and 3C and F). In most strains, the fully developed inclusion lyses either just prior to or cotemporally with the host cell plasma membrane. While it is possible that the chlamydiae produce enzymes responsible for inclusion and cell lysis, it is also possible that the membranes simply decay and developmental forms are released. Note that both EBs and RBs are released through these processes, as the terminal stages of the cycle are asynchronous. However, the totally noninfectious RBs have no role in the subsequent infection of a new target cell.

Although inclusion and cell lysis are apparently the most common means to effect the last stages of the developmental cycle, the lytic pathway is not a universal mechanism for freeing EBs and RBs. In HeLa cells infected with *C. trachomatis* serovar D, inclusions appear to exocytose from the cell, releasing the developmental forms but leaving the host cell apparently intact (Todd and Caldwell, 1985). This process was only demonstrated with a single serovar, but the authors propose that it may be a common trait of the less invasive strains of *C. trachomatis*.

ABERRANT RB: AN ALTERATION OF THE DEVELOPMENTAL CYCLE FOR STRESSED CHLAMYDIAE

While the classical chlamydial developmental cycle includes two dominant forms, EBs and RBs, it was apparent to early investigators that additional structures are also present. First, there was the intermediate body, which we have discussed as a structural intermediate in the EB-to-RB or RB-to-EB transition. But investigators also discussed very large initial (reticulate) bodies. These studies were expanded during early experiments testing the antimicrobial sensitivity of the chlamydiae. The early studies of Emilio Weiss (1950) demonstrated that treatment of chlamydia-infected cells with penicillin results in the formation of very large forms that are not infectious and that do not mature to EBs. Subsequent fine-structure analysis by electron microscopy showed the cell walls of these chlamydiae to be ultrastructurally similar to the RB form, with the addition of accumulated membrane within and around the developmental form (Fig. 7A). Otherwise, these large RBs appear much like typical RBs, including the maintenance of surface projection-mediated contacts with the inclusion

membrane (Fig. 7B). While the response to PG synthesis inhibitors initially pointed to an intriguing but potentially artifactual aspect of the growth cycle, subsequent experiments demonstrated that the production of aberrant RBs was not simply a laboratory curiosity. Instead, it was shown that several different treatments had similar effects on the developmental cycle. Culture in the presence of any beta-lactam antibiotic or with D-cycloserine resulted in virtually identical large developmental forms (Moulder et al., 1963). Similar results were also obtained following chlamydial infection of cells cultured in medium lacking cysteine or tryptophan (Allan et al., 1985; Beatty et al., 1994). Examination of the proteins present in these structures demonstrated a profile very similar to that of an RB—lacking the histones and Omp 2 and 3. MOMP production is also downregulated to some extent in the aberrant forms, while certain proteins, such as HSP 60, are not (Beatty et al., 1993). Collectively, these studies demonstrated that application of a bacteriostatic stress upon the chlamydiae seemed to facilitate the interruption of the typical developmental cycle, resulting in the formation of large aberrant RBs. The relevance of this arm of the developmental cycle was further supported by the determination that production of aberrant RBs resulted from culture of infected cells in the presence of gamma interferon and tumor necrosis factor alpha (Shemer and Sarov, 1985; Shemer-Avni et al., 1988). Further, it was shown that this aberrant developmental step leads to the persistence over long periods of viable but nonculturable chlamydiae within infected cells (Beatty et al., 1994). Finally, removal of these stresses several days postinfection results in the condensation of nuclei, the appearance of late proteins, and the production of viable, infectious EBs (Beatty et al., 1995). These are exciting findings, primarily because of their potential relevance to human health. Most of the major sequelae of chlamydial disease are thought to arise from either repeated or persistent chlamydial infection of a host. Many aspects of serious chlamydial disease result from an inappropriate im-

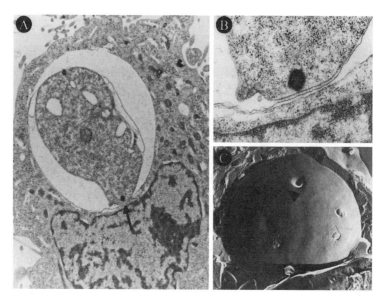

FIGURE 7 Aberrant RB forms produced during culture of *C. trachomatis* L2/434/Bu in the presence of penicillin G (128 µg/ml). (A) Thin section of penicillin-treated, infected cells fixed 24 h p.i. Note the extremely large and vacuolated single RB within a spacious inclusion and the extra folds of membrane adjacent to the RB. (B) A higher magnification of the region of contact between the RB and the inclusion membrane (indicated by an arrow in panel A). Note the connections between the inclusion membrane and RB at the point of contact. (C) Freeze-etch preparation of a similar inclusion. Note the surface projections extending through the surface of the inclusion (arrowhead), as seen in Fig. 6 with normal inclusions. The effect of penicillin on *C. trachomatis* is very similar to that seen with other chlamydiae.

mune response directed at genus-common antigens. Persistence would allow constant presentation to the host immune response of these potentially deleterious immune targets (Beatty et al., 1994). While repeated infection can certainly be documented in many clinical settings, persistence is thought to also have a role. Additionally, this mechanism allows the chlamydiae to find refuge in times of stress, possibly remaining in the aberrant state within a host until conditions permit the production of viable EBs.

SUMMARY

During his early investigations of what was then called psittacosis virus, Bedson (1933) proposed that "the elementary body was not the only form assumed by the virus. . . . Later the virus masses are no longer homogeneous, but are seen to be composed of a number of more or less equal forms 1 µm or more in greatest diameter, and these large forms by repeated division give rise again to the elementary bodies." With this statement the basic chlamydial developmental cycle was described. While there have been some modifications—persistence and aberrant forms, for example—the basic cycle remains EB to RB to EB. Contemporary molecular and cellular approaches are beginning to elucidate the mechanisms involved in the developmental cycle. Research in this difficult field continues to be exciting, and the future likely will see great strides in our understanding of these mechanisms. The recent completion of the *C. trachomatis* genome sequence (and the imminent completion of the *C. pneumoniae* sequence) as well as the anticipated availability of genetic transformation techniques will likely speed up these investigations, leading to a thorough understanding of the developmental cycle. It is hoped that this understanding will ultimately lead to more proactive means for the treatment and prevention of chlamydial disease.

ACKNOWLEDGMENTS

We are grateful to Tom Hatch, Gunna Christiansen, and Allan Herring for discussions of their data. We thank J. Bannantine, C. Sundberg, and D. Grosenbach for editorial comments regarding the manuscript. The assistance of K. Banks in the editorial process is greatly appreciated. We acknowledge the technical assistance of K. Uehira and K. Yamane, Electron Microscope Center, and S. Ohmori, Department of Microbiology, Kawasaki Medical School.

D.D.R. is supported by Public Health Service grant R29AI42869-01. Oregon Agricultural Station Technical Paper 11447.

REFERENCES

Allan, I., T. P. Hatch, and J. H. Pearce. 1985. Influence of cysteine deprivation on chlamydial differentiation from reproductive to infective lifecycle forms. *J. Gen. Microbiol.* **131:**3171–3177.

Baehr, W., Y.-X. Zhang, T. Joseph, H. Su, F. E. Nano, K. D. E. Everett, and H. D. Caldwell. 1988. Mapping antigenic domains expressed by *Chlamydia trachomatis* major outer membrane protein genes. *Proc. Natl. Acad. Sci. USA* **85:** 4000–4004.

Baghian, A., K. Kousoulas, R. Truax, and J. Storz. 1996. Specific antigens of *Chlamydia pecorum* and their homologues in *C. psittaci* and *C. trachomatis. Am. J. Vet. Res.* **57:**1720–1725.

Bannantine, J. P., M. J. Parnell, H. D. Caldwell, and D. D. Rockey. 1998a. Use of a primate model system for identification of *Chlamydia trachomatis* proteins recognized in the context of infection, p. 99–102. *In* R. S. Stephens et al. (ed.), *Chlamydial Infections.* International Chlamydia Symposium, San Francisco, Calif.

Bannantine, J. P., D. D. Rockey, and T. Hackstadt. 1998b. Tandem genes of *Chlamydia psittaci* that encode proteins localized to the inclusion membrane. *Mol. Microbiol.* **28:**1017–1026.

Barry, C. E., III, S. F. Hayes, and T. Hackstadt. 1992. Nucleoid condensation in *Escherichia coli* that express a chlamydial histone homolog. *Science* **256:** 377–379.

Barry, C. E., III, T. J. Brickman, and T. Hackstadt. 1993. Hc1-mediated effects on DNA structure: a potential regulator of chlamydial development. *Mol. Microbiol.* **9:**273–283.

Baumann, M., L. Brade, E. Fasske, and H. Brade. 1992. Staining of surface antigens of *Chlamydia trachomatis* L2 in tissue culture. *Infect. Immun.* **60:** 4433–4438.

Bavoil, P., and R. Hsia. 1998. Type III secretion in *Chlamydia*: a case of deja vu? *Mol. Microbiol.* **28:** 860–862.

Bavoil, P., A. Ohlin, and J. Schachter. 1984. Role of disulfide bonding in outer membrane structure and permeability in *Chlamydia trachomatis. Infect. Immun.* **44:**479–485.

Bavoil, P., R. C. Hsia, and R. G. Rank. 1996. Prospects for a vaccine against chlamydial genital disease. I. Microbiology and pathogenesis. *Bull. Inst. Pasteur* **94:**5–54.

Beatty, W. L., G. I. Byrne, and R. P. Morrison. 1993. Morphologic and antigenic characterization of interferon-gamma mediated persistent *Chlamydia trachomatis* infection in vitro. *Proc. Natl. Acad. Sci. USA* **90:**3998–4002.

Beatty, W. L., G. I. Byrne, and R. P. Morrison. 1994. Repeated and persistent infection with *Chlamydia* and the development of chronic inflammation and disease. *Trends Microbiol.* **2:**94–98.

Beatty, W. L., R. P. Morrison, and G. I. Byrne. 1995. Reactivation of persistent *Chlamydia trachomatis* infection in cell culture. *Infect. Immun.* **63:** 199–205.

Bedson, S. P. 1932. The nature of the elementary bodies in psittacosis. *J. Exp. Pathol.* **13:**65–72.

Bedson, S. P. 1933. Observations of the developmental forms of psittacosis virus. *J. Exp. Pathol.* **14:** 267–277.

Bedson, S. P., and J. O. W. Bland. 1934. The developmental forms of psittacosis virus. *J. Exp. Pathol.* **15:**243–247.

Brade, H., L. Brade, and F. E. Nano. 1987. Chemical and serological investigations on the genus-specific lipopolysaccharide epitope of *Chlamydia. Proc. Natl. Acad. Sci. USA* **84:**2508–2512.

Brickman, T. J., I. Barry, and T. Hackstadt. 1993. Molecular cloning and expression of *hctB* encoding a strain-variant chlamydial histone-like protein with DNA-binding activity. *J. Bacteriol.* **175:** 4274–4281.

Caldwell, H. D., and P. J. Hitchcock. 1984. Monoclonal antibody against a genus-specific antigen of *Chlamydia* species: location of the epitope on chlamydial lipopolysaccharide. *Infect. Immun.* **44:**306–314.

Caldwell, H. D., and L. J. Perry. 1982. Neutralization of *Chlamydia trachomatis* infectivity with antibodies to the major outer membrane protein. *Infect. Immun.* **38:**745–754.

Caldwell, H. D., J. Kromhout, and J. Schachter. 1981. Purification and partial characterization of the major outer membrane protein of *Chlamydia trachomatis*. *Infect. Immun.* **31**:1161–1176.

Cevenini, R., M. Donati, E. Brocchi, F. De Simone, and M. La Placa. 1991. Partial characterization of an 89-kDa highly immunoreactive protein from *Chlamydia psittaci* A/22 causing ovine abortion. *FEMS Microbiol. Lett.* **81**:111–116.

Collier, L. H. 1962. Growth characteristics of inclusion blennorrhea virus in cell cultures. *Ann. N.Y. Acad. Sci.* **98**:42–49.

Everett, K. D. E., and T. P. Hatch. 1995. Architecture of the cell envelope of *Chlamydia psittaci* 6BC. *J. Bacteriol.* **177**:877–882.

Everett, K. D. E., D. M. Desiderio, and T. P. Hatch. 1994. Characterization of lipoprotein EnvA in *Chlamydia psittaci* 6BC. *J. Bacteriol.* **176**:6082–6087.

Fox, A., J. C. Rogers, J. Gilbart, S. Morgan, C. H. Davis, S. Knight, and P. B. Wyrick. 1990. Muramic acid is not detectable in *Chlamydia psittaci* or *Chlamydia trachomatis* by gas chromatography-mass spectrometry. *Infect. Immun.* **58**:835–837.

Girardi, A. J., E. G. Allen, and M. M. Sigel. 1952. Studies on the psittacosis-lymphogranuloma group. II. A non-infectious phase in virus development following adsorption to host tissue. *J. Exp. Med.* **96**:233–246.

Gordon, M. H. 1930. Virus studies concerning the etiology of psittacosis. *Lancet* **218**:1174–1177.

Grayston, J. T. 1992. Infections caused by *Chlamydia pneumoniae* strain TWAR. *Clin. Infect. Dis.* **15**:757–761.

Gregory, W. W., M. Gardner, G. I. Byrne, and J. W. Moulder. 1979. Arrays of hemispheric surface projections on *Chlamydia psittaci* and *Chlamydia trachomatis* observed by scanning electron microscopy. *J. Bacteriol.* **138**:241–244.

Grimwood, J., W. Mitchell, and R. S. Stephens. 1998. Phylogenetic analysis of a multigene family conserved between *Chlamydia trachomatis* and *Chlamydia pneumoniae*, p. 263–266. *In* R. S. Stephens et al. (ed.), *Chlamydial Infections*. International Chlamydia Symposium, San Francisco, Calif.

Hackstadt, T. 1986. Identification and properties of chlamydial polypeptides that bind eucaryotic cell surface components. *J. Bacteriol.* **165**:13–20.

Hackstadt, T., W. Baehr, and Y. Yuan. 1991. *Chlamydia trachomatis* developmentally regulated protein is homologous to eukaryotic histone H1. *Proc. Natl. Acad. Sci. USA* **88**:3937–3941.

Hackstadt, T., T. J. Brickman, C. E. Barry, III, and J. D. Sager. 1993. Diversity in the *Chlamydia trachomatis* histone homolog Hc2. *Gene* **132**:137–141.

Hackstadt, T., M. A. Scidmore, and D. D. Rockey. 1995. Lipid metabolism in *Chlamydia trachomatis* infected cells: directed trafficking of Golgi-derived sphingolipids to the chlamydial inclusion. *Proc. Natl. Acad. Sci. USA* **92**:4877–4881.

Hackstadt, T., D. D. Rockey, R. A. Heinzen, and M. A. Scidmore. 1996. *Chlamydia trachomatis* interrupts an exocytic pathway to acquire endogenously synthesized sphingomyelin in transit from the Golgi apparatus to the plasma membrane. *EMBO J.* **15**:964–977.

Hackstadt, T., E. R. Fischer, M. A. Scidmore, D. D. Rockey, and R. A. Heinzen. 1997. Origins and functions of the chlamydial inclusion. *Trends Microbiol.* **5**:288–293.

Hatch, T., and D. Rockey. Unpublished data.

Hatch, T. P., D. W. Vance, Jr., and E. Al-Hossainy. 1981. Identification of a major envelope protein in *Chlamydia* spp. *J. Bacteriol.* **146**:426–429.

Hatch, T. P., I. Allan, and J. H. Pearce. 1984. Structural and polypeptide differences between envelopes of infective and reproductive life cycle forms of *Chlamydia* spp. *J. Bacteriol.* **157**:13–20.

Hatch, T. P., M. Miceli, and J. E. Sublett. 1986. Synthesis of disulfide-bonded outer membrane proteins during the developmental cycle of *Chlamydia psittaci* and *Chlamydia trachomatis*. *J. Bacteriol.* **165**:379–385.

Heinzen, R. A., M. A. Scidmore, D. D. Rockey, and T. Hackstadt. 1996. Differential interaction with endocytic and exocytic pathways distinguish parasitophorous vacuoles of *Coxiella burnetii* and *Chlamydia trachomatis*. *Infect. Immun.* **64**:796–809.

Hsia, R., Y. Pannekoek, E. Ingerowski, and P. M. Bavoil. 1997. Type III secretion genes identify a putative virulence locus of *Chlamydia*. *Mol. Microbiol.* **25**:351–359.

Karimi, S. T., R. H. Schloemer, and C. E. Wilde III. 1989. Accumulation of chlamydial lipopolysaccharide antigen in the plasma membranes of infected cells. *Infect. Immun.* **57**:1780–1785.

Kaul, R., and W. M. Wenman. 1986. Cyclic AMP inhibits developmental regulation of *Chlamydia trachomatis*. *J. Bacteriol.* **168**:722–727.

Kaul, R., A. Hoang, P. Yau, E. M. Bradbury, and W. M. Wenman. 1997. The chlamydial EUO gene encodes a histone H1-specific protease. *J. Bacteriol.* **179**:5928–5934.

Knudsen, K., A. S. Madsen, P. Mygind, G. Christiansen, and S. Birkelund. 1998. Surface localized proteins of *Chlamydia pneumoniae*, p. 267–270. *In* R. S. Stephens et al. (ed.), *Chlamydial Infections*. International Chlamydia Symposium, San Francisco, Calif.

Krumwiede, C., M. McGrath, and C. Oldenbusch. 1930. The etiology of the disease psittacosis. *Science* **71**:262–263.

Kubori, T., Y. Matsushima, D. Nakamura, J. Uralil, M. Lara-Tejero, A. Sukhan, J. E. Galan, and S. Aizawa. 1998. Supramolecular structure of the *Salmonella typhimurium* type III protein secretion system. *Science* **280**:602–605.

Kuo, C.-C., J. T. Grayston, L. A. Campbell, Y. A. Goo, R. W. Wissler, and E. P. Benditt. 1995. *Chlamydia pneumoniae* (TWAR) in coronary arteries of young adults (15–34 years old). *Proc. Natl. Acad. Sci. USA* **92**:6911–6914.

Lambden, P. R., J. S. Everson, M. E. Ward, and I. N. Clarke. 1990. Sulfur-rich proteins of *Chlamydia trachomatis*: developmentally regulated transcription of polycistronic mRNA from tandem promoters. *Gene* **87**:105–112.

Longbottom, D., M. Russel, G. E. Jones, A. Lainson, and A. J. Herring. 1996. Identification of a multigene family coding for the 90 kDa proteins of the ovine abortion subtype of *Chlamydia psittaci*. *FEMS Microbiol. Lett.* **142**:277–281.

Longbottom, D., M. Russell, S. M. Dunbar, G. E. Jones, and A. J. Herring. 1998. Molecular cloning and characterization of the genes coding for the highly immunogenic cluster of 90-kilodalton envelope proteins from the *Chlamydia psittaci* subtype that causes abortion in sheep. *Infect. Immun.* **66**:1317–1324.

Louis, C., G. Nicolas, F. Eb, J.-F. Lefebvre, and J. Orfila. 1980. Modifications of the envelope of *Chlamydia psittaci* during its developmental cycle: freeze-fracture study of complementary replicas. *J. Bacteriol.* **141**:868–875.

Lundemose, A. G., S. Birkelund, P. M. Larsen, S. J. Fey, and G. Christiansen. 1990. Characterization and identification of early proteins in *Chlamydia trachomatis* serovar L2 by two-dimensional gel electrophoresis. *Infect. Immun.* **58**:2478–2486.

Mardh, P.-A. 1992. Natural history of genital and allied chlamydial infections. *Curr. Opin. Infect. Dis.* **5**:12–17.

Matsumoto, A. 1973. Fine structures of cell envelopes of *Chlamydia* organisms as revealed by freeze-etching and negative staining techniques. *J. Bacteriol.* **116**:1355–1363.

Matsumoto, A. 1982a. Electron microscopic observations of surface projections on *Chlamydia psittaci* reticulate bodies. *J. Bacteriol.* **150**:358–364.

Matsumoto, A. 1982b. Surface projections of *Chlamydia psittaci* elementary bodies as revealed by freeze-deep-etching. *J. Bacteriol.* **151**:1040–1042.

Matsumoto, A., E. Fujiwara, and N. Higashi. 1976. Observations of the surface projections of infectious small cells of *Chlamydia psittaci* in thin sections. *J. Electron Microsc.* **25**:169–170.

McClarty, G. 1994. Chlamydiae and the biochemistry of intracellular parasitism. *Microbiology* **2**:157–164.

Moulder, J. M. 1964. *The Psittacosis Group as Bacteria.* John Wiley and Sons, New York, N.Y.

Moulder, J. W. 1991. Interaction of Chlamydiae and host cells *in vitro*. *Microbiol. Rev.* **55**:143–190.

Moulder, J. W. 1993. Why is *Chlamydia* sensitive to penicillin in the absence of peptidoglycan? *Infect. Agents Dis.* **2**:87–99.

Moulder, J. W., D. L. Novosel, and J. E. Officer. 1963. Inhibition of the growth of agents of the psittacosis group by D-cycloserine and its specific reversal by D-alanine. *J. Bacteriol.* **85**:707–711.

Nano, F. E., and H. D. Caldwell. 1985. Expression of the chlamydial genus-specific lipopolysaccharide epitope in *Escherichia coli*. *Science* **228**:742–744.

Newhall, W. J., and R. B. Jones. 1983. Disulfide-linked oligomers of the major outer membrane protein of chlamydiae. *J. Bacteriol.* **154**:998–1001.

Nichols, B. A., P. Y. Setzer, F. Pang, and C. R. Dawson. 1985. New view of the surface projections of *Chlamydia trachomatis*. *J. Bacteriol.* **164**:344–349.

Nurminen, M., M. Leinonen, P. Saikku, and P. H. Makela. 1983. The genus-specific antigen of *Chlamydia*: resemblance to the lipopolysaccharide of enteric bacteria. *Science* **220**:1279–1281.

Pagano, R. E., O. C. Martin, H. C. Kang, and R. P. Haugland. 1991. A novel fluorescent ceramide analogue for studying membrane traffic in animal cells: accumulation at the Golgi apparatus results in altered spectral properties of the sphingolipid precursor. *J. Cell Biol.* **113**:1267–1279.

Page, L. A. 1966. Revision of the Family Chlamydiaceae Rake (Rickettsiales): unification of the psittacosis-lymphogranuloma venereum-trachoma group of organisms in the genus *Chlamydia* Jones, Rake and Stearns. *Int. J. Syst. Bacteriol.* **16**:223–252.

Pedersen, L. B., S. Birkelund, and G. Christiansen. 1996a. Purification of recombinant *Chlamydia trachomatis* H1-like protein Hc2, and comparative functional analysis of Hc2 and Hc1. *Mol. Microbiol.* **20**:295–311.

Pedersen, L. B., S. Birkelund, A. Holm, S. Ostergaard, and G. Christiansen. 1996b. The 18-kilodalton *Chlamydia trachomatis* histone H1-like protein (Hc1) contains a potential N-terminal dimerization site and a C-terminal nucleic acid-binding domain. *J. Bacteriol.* **178**:994–1002.

Perara, E., D. Ganem, and J. N. Engel. 1992. A developmentally regulated chlamydial gene with apparent homology to eukaryotic histone H1. *Proc. Natl. Acad. Sci. USA* **89**:2125–2129.

Plaunt, M. R., and T. P. Hatch. 1988. Protein synthesis early in the developmental cycle of *Chlamydia psittaci*. *Infect. Immun.* **56**:3021–3025.

Remacha, M., R. Kaul, R. Sherburne, and W. M. Wenman. 1996. Functional domains of chlamydial H1-like protein. *Biochem. J.* **315**:481–486.

Rockey, D. D., and J. L. Rosquist. 1994. Protein

antigens of *Chlamydia psittaci* present in infected cells but not detected in the infectious elementary body. *Infect. Immun.* **62:**106–112.

Rockey, D. D., R. A. Heinzen, and T. Hackstadt. 1995. Cloning and characterization of a *Chlamydia psittaci* gene coding for a protein localized to the inclusion membrane of infected cells. *Mol. Microbiol.* **15:**617–626.

Rockey, D. D., D. Grosenbach, D. E. Hruby, M. G. Peacock, R. A. Heinzen, and T. Hackstadt. 1997. *Chlamydia psittaci* IncA is phosphorylated by the host cell and is exposed on the cytoplasmic face of the developing inclusion. *Mol. Microbiol.* **24:** 217–228.

Schachter, J. 1978. Chlamydial infections. *N. Engl. J. Med.* **298:**428–434.

Schachter, J., and C. R. Dawson. 1990. The epidemiology of trachoma predicts more blindness in the future. *Scand. J. Infect. Dis.* **69:**55–62.

Scidmore, M. A., D. D. Rockey, E. R. Fischer, R. A. Heinzen, and T. Hackstadt. 1996. Vesicular interactions of the *Chlamydia trachomatis* inclusion are determined by chlamydial early protein synthesis rather than route of entry. *Infect. Immun.* **64:**5366–5372.

Scidmore-Carlson, M. A., E. I. Shaw, C. A. Dooley, E. R. Fischer, and T. Hackstadt. 1999. Identification and characterization of a *Chlamydia trachomatis* early operon encoding four novel inclusion membrane proteins. *Mol. Microbiol.* **33:** 753–765.

Shemer, Y., and I. Sarov. 1985. Inhibition of growth of *Chlamydia trachomatis* by human gamma interferon. *Infect. Immun.* **48:**592–596.

Shemer-Avni, Y., D. Wallach, and I. Sarov. 1988. Inhibition of *Chlamydia trachomatis* growth by recombinant tumor necrosis factor. *Infect. Immun.* **56:** 2503–2506.

Stephens, R. S., G. Mullenbach, R. Sanchez-Pescador, and N. Agabian. 1986. Sequence analysis of the major outer membrane protein gene from *Chlamydia trachomatis* serovar L2. *J. Bacteriol.* **168:** 1277–1282.

Stephens, R. S., R. Sanchez-Pescador, E. A. Wagar, C. Inouye, and M. S. Urdea. 1987. Diversity of *Chlamydia trachomatis* major outer membrane protein genes. *J. Bacteriol.* **169:**3879–3885.

Stephens, R. S., S. Kalman, C. Lammel, J. Fan, R. Maratha, L. Aravind, W. Mitchell, L. Olinger, R. L. Tatusov, Q. Zhao, E. U. Kooning, and R. W. Davis. 1998. Genome sequence of an obligate intracellular pathogen of humans: *Chlamydia trachomatis*. *Science* **282:**754–759.

Su, H., and H. D. Caldwell. 1991. In vitro neutralization of *Chlamydia trachomatis* by monovalent Fab antibody specific to the major outer membrane protein. *Infect. Immun.* **59:**2843–2845.

Su, H., N. G. Watkins, Y.-X. Zhang, and H. D. Caldwell. 1990. *Chlamydia trachomatis*-host cell interactions: role of the chlamydial major outer membrane protein as an adhesin. *Infect. Immun.* **58:** 1017–1025.

Su, H., L. Raymond, D. D. Rockey, E. Fischer, T. Hackstadt, and H. D. Caldwell. 1996. A recombinant *Chlamydia trachomatis* major outer membrane protein binds to heparan sulfate receptors on epithelial cells. *Proc. Natl. Acad. Sci. USA* **93:** 11143–11148.

Tamura, A., A. Matsumoto, and N. Higashi. 1967. Purification and chemical composition of reticulate bodies of the meningopneumonitis organisms. *J. Bacteriol.* **93:**2003–2008.

Thygeson, P. 1962. Trachoma virus: historical background and review of isolates. *Ann. N.Y. Acad. Sci.* **98:**6–13.

Ting, L. M., R. C. Hsia, C. G. Haidaris, and P. M. Bavoil. 1995. Interaction of outer envelope proteins of *Chlamydia psittaci* GPIC with the HeLa cell surface. *Infect. Immun.* **63:**3600–3608.

Todd, W. J., and H. D. Caldwell. 1985. The interaction of *Chlamydia trachomatis* with host cells: ultrastructural studies of the mechanism of release of a biovar II strain from HeLa 229 cells. *J. Infect. Dis.* **151:**1037–1044.

University of California, Berkeley. 1999. Chlamydia Genome Project Database. [Online.] http://chlamydia-www.berkeley.edu:4231/.

Wagar, E. A., and R. S. Stephens. 1988. Development-form-specific DNA-binding proteins in *Chlamydia* spp. *Infect. Immun.* **56:**1678–1684.

Washington, A. E., R. E. Johnson, and L. L. Sanders, Jr. 1987. *Chlamydia trachomatis* infections in the United States: what are they costing us? *JAMA* **257:**2070–2072.

Watson, M. W., I. N. Clarke, J. S. Everson, and P. R. Lambden. 1995. The CrP operon of *Chlamydia psittaci* and *Chlamydia pneumoniae*. *Microbiology* **141:** 2489–2497.

Weiss, E. 1950. The effect of antibiotics on agents of the psittacosis-lymphogranuloma group. I. The effect of penicillin. *J. Infect. Dis.* **87:**249–263.

Wichlan, D. G., and T. P. Hatch. 1993. Identification of an early-stage gene of *Chlamydia psittaci*. *J. Bacteriol.* **175:**2936–2942.

Wylie, J. L., G. M. Hatch, and G. McClarty. 1997. Host cell phospholipids are trafficked to and then modified by *Chlamydia trachomatis*. *J. Bacteriol.* **179:** 7233–7242.

Wyllie, S., R. H. Ashley, D. Longbottom, and A. J. Herring. 1998. The major outer membrane protein of *Chlamydia psittaci* functions as a porin-like ion channel. *Infect. Immun.* **66:**5202–5207.

Yuan, Y., Y.-X. Zhang, N. G. Watkins, and H. D. Caldwell. 1989. Nucleotide and deduced amino acid sequences for the four variable domains of the major outer membrane proteins of the 15 *Chlamydia trachomatis* serovars. *Infect. Immun.* **57:** 1040–1049.

Zhang, L., A. L. Douglas, and T. P. Hatch. 1998. Characterization of a *Chlamydia psittaci* DNA binding protein (EUO) synthesized during the early and middle phases of the developmental cycle. *Infect. Immun.* **66:**1167–1173.

Zhang, Y.-X., S. J. Stewart, and H. D. Caldwell. 1989. Protective monoclonal antibodies to *Chlamydia trachomatis* serovar- and serogroup-specific major outer membrane protein determinants. *Infect. Immun.* **57:**636–638.

DEVELOPMENTAL CYCLE OF
COXIELLA BURNETII

James E. Samuel

21

INTRODUCTION

Coxiella and *Chlamydia* are both obligate intra-cellular bacteria that replicate within a parasito-phorous vacuole (PV) in a wide array of eukar-yotic cells. Both organisms assume several morphologically distinct forms. The nature of the developmental cycle for *Chlamydia* has been extensively investigated, and a model to explain these events has become fairly detailed (see chapter 20). In contrast, details about a potential developmental cycle for *Coxiella* are much less clear. Early studies describing mor-phological variants based on electron micros-copy and biochemical analysis provided the basis for a variety of general models of devel-opment but remained largely untested. Re-cent advances in the cloning of variant specific proteins has led to a new appreciation of the *Coxiella* developmental cycle. This review summarizes the current state of research in this area and draws upon several recent reviews (Heinzen, 1997; Heinzen et al., 1999) to pres-ent a more defined and testable model of devel-opment for this unusual pathogen.

Coxiella burnetii is the etiologic agent of human acute Q fever and occasional chronic

manifestations such as endocarditis and hepati-tis. Acute Q fever typically presents as a flu-like illness, with symptoms including a periorbital headache, cyclic fever, and myalgia (Raoult and Marrie, 1995). Clinically the illness falls within the group of FUO (fever of unknown origin) syndromes and is not commonly recog-nized or diagnosed in either developed or de-veloping countries (Fournier et al., 1998). For example, a recent retrospective surveillance study in Japan, a country previously thought not to have Q fever, indicated seropositivity rates of >20% for healthy individuals at risk, which strongly suggests that infection with *C. burnetii* should be considered a reemerging in-fectious disease (Htwe et al., 1993). Humans become infected primarily by inhalation of aerosolized contaminated soil, especially asso-ciated with domestic animal production of cat-tle, goats, and sheep. However, a large variety of transmission cycles have been documented involving such wide-ranging reservoirs as wild animals, arthropods, parturition of domestic cats, and milk production (Marrie, 1990). This is in part due to the wide host range and world-wide distribution of *C. burnetii* (Babudieri, 1959). The wide host range and ubiquitous na-ture of this obligate intracellular organism also reveal clues about its natural resistance to ex-treme stress.

J. E. Samuel, Department of Medical Microbiology and Im-munology, Texas A&M University System Health Science Center, College Station, TX 77843-1114.

Prokaryotic Development, edited by Y. V. Brun and L. J. Shimkets,
© 2000 American Society for Microbiology, Washington, DC 20005-4171

The original phylogenetic placement of *C. burnetii* in the family *Rickettsiaceae*, which included many other obligate intracellular organisms such as *Rickettsia* and *Ehrlichia*, was based upon its fastidious and slow intracellular growth. More recently, 16S rRNA sequence comparison of many of these organisms predicted a more appropriate grouping of most members of the *Rickettsia* group into the α-proteobacteria (Weisburg et al., 1985). However, *Coxiella* was determined to be most closely related to *Legionella* spp. and therefore resides within the γ subdivision of proteobacteria, phylogenetically distant from other classic rickettsia except for *Wolbachia* (Weisburg et al., 1989). The moderate relatedness of *Coxiella* and *Legionella*, revealed by 16S rRNA comparison, has proven consistent when several cloned proteins, such as CbMip and LpMip (Cianciotto et al., 1995; Mo et al., 1995), were compared and may prove valuable in understanding the similarities between *C. burnetii* and *Legionella* interactions with host cells. Genes from *Coxiella* have been expressed well by surrogate bacterial hosts, primarily *Escherichia coli*, and the regulatory regions in their promoters appear similar (Mallavia, 1991). Several studies have analyzed the genetic diversity of these organisms, and several subgroups of isolates have been proposed (divergent at the strain level, with some associated with a particular disease, e.g., chronic versus acute) (Hendrix et al., 1991; Mallavia, 1991; Samuel et al., 1985). Additional biochemical markers were also able to separate isolates into the same groupings, but the hypothesis that specific isolates or strain groups are distinct in their pathogenic potential remains untested. The chromosome of one isolate was recently mapped by infrequent cutting restriction enzymes and estimated to be approximately 2.1 Mbp and potentially linear in organization (Willems et al., 1998). This study speculated that other isolates may contain a chromosome that ranges in size from 1.5 to 2.4 Mbp.

SURVIVAL IN THE EXTRACELLULAR ENVIRONMENT AND UNDER EXTREME CONDITIONS

One of the hallmark properties of *C. burnetii* is its ability to remain infectious in extreme environmental circumstances. The organism has been shown to be highly resistant to elevated temperature and desiccation, osmotic shock, UV light, and chemical disinfectants. The heat resistance was graphically displayed by its survival in milk heated to temperatures of 61°C for 30 min, which allowed numerous cases of milk-transmitted Q fever during the early stages of establishing pasteurization standards and subsequently led to an increase in the standard temperature (Lennette et al., 1952). The organism remains infectious in dried blood for >180 days and in tick feces for up to 586 days (Williams, 1991). In addition, Babudieri (1959) reported resistance to 1% phenol or 0.5% formalin treatment for 24 h. These and numerous other reports of resistance to environmental challenge, when compared with other bacteria, have supported the idea that *Coxiella* has evolved unique mechanisms to remain viable while in a nonreplicating state.

UPTAKE BY AND SURVIVAL WITHIN HOST CELLS

C. burnetii is unique among intracellular bacteria in that it resides within a PV that has characteristics of a secondary lysosome. In vitro, *C. burnetii* passively enters (viable and nonviable organisms are equally internalized) and replicates within a wide variety of epithelial, fibroblast, and macrophage-like cell lines (Baca et al., 1993a). In vivo, the initial target is the alveolar macrophage, although the organism can subsequently disseminate to replicate within a wide variety of tissues. Internalization into host cells occurs by a microfilament-dependent parasite-directed endocytic process (Baca et al., 1993a). The bacterial ligand(s) and host receptor mediating this process are undefined, but entry of virulent (smooth-lipopolysaccharide) organisms via an $\alpha_v\beta_3$ integrin versus avirulent (rough-lipopolysaccharide) organisms via the CR3 complement receptor may provide privilege entry into nonactivated macrophages (Meconi et al., 1998). The nascent parasite-containing phagosomes mature through the endocytic pathway, eventually acquiring the properties of secondary lysosomes. The lysosomal markers 5′-nucleotidase, acid phosphatase, cathepsin D, and lysosome-associated membrane glycoproteins 1 and 2 colocalize with the lumen of the PV or the PV membrane, and the PV acidifies to a pH of ~4.8 (Akporiaye and Baca, 1983; Burton et al., 1978; Heinzen et al., 1996b). *C. burnetii* has an absolute requirement for this moderately acidic pH to activate metabolism and exploits the only intracellular niche capable of establishing such a pH (Hackstadt and Williams, 1981). The organism replicates to high numbers, albeit at a low rate (8- to 12-h doubling time) (Baca and Paretsky, 1983), within this vacuole, despite the presence of toxic host factors, such as acid hydrolases, oxygen and nitrogen radicals, and defensins, which are normally considered bactericidal. Minimal cytopathic effects are noted in infected cells. Consequently, *C. burnetii* is inefficient at forming plaques on most cell monolayers, suggesting the lack of a membranolytic exit system.

Several of the enzyme systems required to survive in the phagolysosomal compartment have been characterized. To survive the oxygen-dependent microbicidal activity within nonprofessional and professional phagocytic cells, *C. burnetii* expresses bacterial catalase and superoxide dismutase (SOD). The gene which encodes catalase has not been characterized, and only a cytoplasmically localized enzyme with a neutral pH optimum has been suggested (Akporiaye and Baca, 1983). A gene encoding SOD was cloned and shown to be similar to genes encoding cytoplasmic iron-containing enzymes (Heizen et al., 1992). A periplasmically localized copper-zinc SOD enzyme, recently shown to be essential for stationary-phase intracellular survival and growth in *Legionella pneumophila* (St. John and Steinman, 1996) and *Salmonella typhimurium* (De Groote et al., 1997), has not been identified. Another enzyme which may help ablate the oxidative burst, as seen with *Legionella micdadei* and *Leishmania*, is acid phosphatase (ACP). Although the gene encoding a *C. burnetii* ACP has not been cloned, ACP distinct from the host enzyme was clearly demonstrated and was able to inhibit the neutrophil oxidative burst associated with phagocytosis (Baca et al., 1993b; Li et al.,

1996). Two additional periplasmic and outer-membrane-localized enzymes have been characterized by cloning and are likely required for correct protein folding in the predicted highly reductive, low-pH environment of the phagolysosome. Com-1 was shown to be a disulfide bond-forming enzyme, containing active site homology with DsbA and -C in *E. coli* and a low-pH-adapted specific activity (Hendrix et al., 1993). *Shigella flexneri* mutants deficient in periplasmic disulfide bond-forming enzymes are attenuated for intracellular survival (Watarai et al., 1995). Cb-Mip, a homologue of the macrophage infectivity potentiator (Mip) of *L. pneumophila*, is a peptidylprolyl isomerase, and *L. pneumophila mip* mutants are reduced in intracellular survival (Cianciotto et al., 1995).

EARLY OBSERVATIONS ABOUT MORPHOLOGICAL VARIANTS

Ultrastructure

The ability to survive in harsh intracellular and extracellular environments is one aspect of *Coxiella* biology that has led investigators to propose a developmental cycle. Equally important have been the observations about different morphological variants. Filterability and pleomorphic nature were noted by the first investigators to describe this agent (Davis and Cox, 1938). Organisms filtered through 0.4-μm-pore-size filters remained infectious for guinea pigs and were antigenically distinct from forms retained by the filter (Kordova, 1960). These filterable forms were subsequently shown to undergo a series of antigenic changes when used to infect tissue culture cells, which provided support for the first model of a *C. burnetii* developmental cycle (Kordova and Kocacova, 1968). Electron microscopic studies in the 1950s and 1960s visualized the pleomorphic nature of organisms inside vacuoles, and staining studies suggested the larger cells were more permeable to negative stain (Anacker et al., 1964; Nermut et al., 1968; Rosenberg and Kordova, 1960; Stocker et al., 1956). An elegant series of electron micrographic studies by McCaul and collaborators examined these forms within infected cells and after purification from host tissue (McCaul, 1991; McCaul and Williams, 1981; McCaul et al., 1991a, 1994). These studies differentiated the forms into several groups, principally large-cell variants (LCV) and small-cell variants (SCV). These studies also identified a phenomenon of an occasional, electron-dense, membrane-bound polar body within LCV. These forms appeared infrequently but seemed to result from an asymmetric septation and led to the idea of a spore-like particle (SLP) (McCaul and Williams, 1981).

Separation of Variants: LCV, SCV, SDC, and SLP

The separation of morphological variants based upon differences in buoyancy during equilibrium density centrifugation in cesium chloride (Wiebe et al., 1972), sucrose (Canonicao et al., 1972), or Renografin (Wachter et al., 1975) has provided a method to compare proteins and antigens within each form (see Fig. 1). Cesium chloride separation has provided the most consistent separation method and is based upon permeability of LCV but not SCV to the salt, resulting in LCV partitioning as a denser form. Although the LCV and SCV forms appear in approximately equal proportions when infected tissue culture cells are visualized by transmission electron microscopy (TEM), this separation yields severalfold more SCV than LCV. This is likely due to lysis of the more osmotically fragile LCV. Other methods have been employed to distinguish variant forms based upon differential sensitivities to physical disruption and osmotic stress. The forms that correspond to LCV have been lysed by hypo-osmotic shock to leave a population of small, resistant cells (McCaul et al., 1981). Osmotic-shock-resistant cells were further separated into pressure-resistant and pressure-sensitive forms by French press breakage (18,000 to 50,000 lb/in^2) (Amano et al., 1984). Small, osmotically resistant cells were composed of similar proportions of pressure-resistant and -sensitive forms, and the resistant forms were designated small dense cells (SDC) (McCaul et al., 1991a).

Metabolic Activity and Infectivity of Forms

To develop a model of the function of these distinct forms of C. burnetii, the metabolic activity and ability to cause infection are important parameters to evaluate. Information relating to these issues has remained scant, however, due to the difficulty associated with experimental design with this obligate intracellular organism. Although C. burnetii only replicates within host cells, organisms separated from host components are metabolically active in axenic medium, able to transport various amino acids and selected sugars and to incorporate them into a variety of macromolecules (Hackstadt and Williams, 1981; Hendrix and Mallavia, 1984). In fact, the acidophilic nature of the organism, which allows it to occupy the phagolysosomal compartment, was established by using this assay system. One early study attempted to combine a strategy to essentially separate LCV from SCV after metabolic activation in axenic medium to assess their relative metabolic activities (McCaul et al., 1981). Purified organisms were metabolically labeled with [^{14}C]-glucose or [^{14}C]glutamate and subsequently treated by osmotic shock and sonication to remove LCV. The starting material contained similar numbers of large and small cells. The resulting resistant forms, containing no LCV-like forms as determined by microscopy, had incorporated ~14- to 43-fold less metabolic label of glutamate or glucose, suggesting that the majority of metabolic activity, as measured by incorporation or CO_2 evolution, in axenic buffer was due to LCV.

Wiebe et al. (1972) also compared the abilities of LCV and SCV, separated by cesium chloride gradients, to cause infection. Both forms remained infectious for tissue culture cells or embryonated hen's eggs, but neither subpopulation remained homogeneous. Both forms led to infected cells in which a mixture of LCV and SCV appeared in the phagolysosomes. Heinzen (1997) reported that when a mixture of SCV and LCV was used to infect Vero cells, both cell types were evident as early as 2 h postinfection. Therefore, these prelimi-

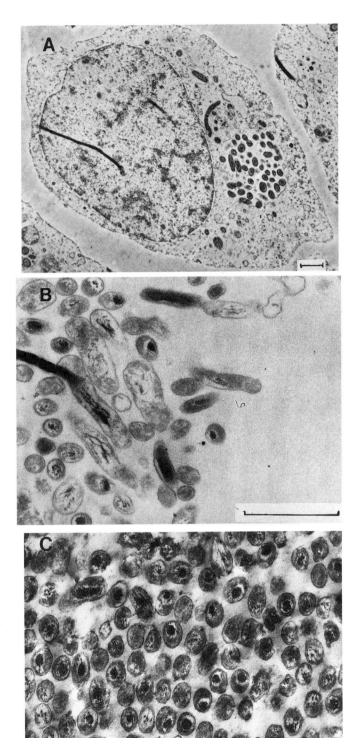

FIGURE 1 Transmission electron micrograph of bacteria in a vacuole and separated into SCV and LCV. (A) Persistently infected L929 cells were prepared for TEM. Note a membrane-bound vacuole with multiple *C. burnetii* cells with pleomorphic morphology, including large cells with diffuse nucleoids (LCV) and small cells with compact nucleoids (SCV). Purified *C. burnetii* cells were separated into LCV (B) and SCV (C) by cesium chloride equilibrium gradient centrifugation. Bars, 1 μm.

nary studies support the contention that both LCV and SCV are infectious forms in a developmental cycle and that the LCV do not represent noninfectious, degenerating SCV (Heinzen, 1997; Wiebe et al., 1972).

A series of studies by Thompson and coworkers supported the idea that the *C. burnetii* within the PV contains cell types with different metabolic activities (Redd and Thompson, 1995; Thompson et al., 1984; Zuerner and Thompson, 1983). These investigators isolated two populations of organisms during infection of tissue culture cells, namely, naturally released cells (NRC), those that accumulate in the medium of infected host cells as a result of natural processes, and mechanically released cells (MRC), those released by sonic disruption of infected cells. MRC transported and incorporated severalfold-higher levels of a radiolabeled nucleoside or amino acid than NRC when incubated in an axenic buffer. The sodium dodecyl sulfate-polyacrylamide gel electrophoresis (SDS-PAGE) profiles of labeled proteins from NRC and MRC also showed distinct differences. Each of these observations suggests that NRC, the prototype extracellular form, and MRC, a mixture of intracellular forms, represent different stages of *Coxiella* development.

Prior Models of the Developmental Cycle

Many of these studies have been used to develop a working model of events within the PV that *C. burnetii* undergoes to survive intracellularly. Particularly, McCaul and coworkers developed several related models to explain events visualized by TEM (McCaul, 1991; McCaul and Williams, 1981). In the most recent model (McCaul, 1991), SCV are the primary infecting particle, entering an endosome that, once fused with lysosomes, activates vegetative growth in the form of "intermediate-cell" forms. Intermediate cells lose resistance properties but maintain SCV morphology. Further bacterial growth leads to SCV becoming LCV, but all "activated" cells undergo binary fission. Unknown environmental signals induce LCV to become "mother" cells, which

then undergo sporogenesis. These spores are released from the mother cell and mature into small "resting" cells, which ultimately are released from lysing host cells and can initiate a new round of infection and replication. Many aspects of this model are highly speculative, including the concept of sporogenesis and synchronicity of an infection cycle.

DIFFERENTIAL EXPRESSION OF STAGE-SPECIFIC PROTEINS

The description of specific molecular mechanisms used by facultative bacteria employing the generation of clones with an altered phenotype has not been effectively approached with obligate intracellular bacteria. Consequently, the approach that has been most illuminating to date has been the cloning of specific genes and the correlation of their functions as understood in other bacterial models. This approach has now begun to define the functional role of *C. burnetii* morphological variants. The following set of proteins have been recently characterized in the context of a developmental model for *Coxiella*.

P1, a Major Outer Membrane Protein

Only a few outer membrane proteins have been characterized by either purification of the native protein or cloning. Surface iodination of intact organisms radiolabeled a major-molecular-mass species of ~29 kDa (Hendrix et al., 1993; Williams and Waag, 1991). The outer membrane was isolated from purified *C. burnetii* cells, and a 29.5-kDa integral major outer membrane protein, designated P1, was partially purified and used to generate a set of monoclonal antibodies (Snyder et al., 1984; Snyder and Williams, 1986; Williams et al., 1984). Partially purified antigen was found to have porin activity in liposome-swelling assays, and anti-P1 monoclonal antibodies inhibited [^{14}C]glucose uptake by *C. burnetii* in axenic medium (Banerjee-Bhatnagar et al., 1996). Attempts to clone the gene that encodes this antigen have proven unsuccessful, but we recently purified P1 to homogeneity and generated N-terminal amino acid sequence for new cloning studies (Varghees et al., 1998). McCaul and co-workers utilized P1-specific antibodies to determine expression by various morphological variants and found a clear differential expression of this major outer membrane putative porin (McCaul et al., 1991). Specifically, LCV expressed an abundant amount of P1, pressure-sensitive SCV expressed significantly less P1, and SDC did not contain a detectable level of P1. In these studies, both Western blot and immunogold TEM analyses were used to detect expression of P1.

Histone-Like Protein Hq-1

Heinzen and coworkers recently compared differential expression of proteins by LCV and SCV separated by the cesium chloride technique (Wiebe et al., 1972). The first class of proteins that was identified consisted of basic DNA binding proteins. Studies with *Chlamydia trachomatis* and growth stage-dependent expression analysis had identified two histone-like proteins expressed only by the extracellularly stable, metabolically less active *Chlamydia* elementary body (EB) (Brickman et al., 1993; Hackstadt et al., 1991). The intracellular, highly metabolic, replicatively active reticulate bodies (RB) do not produce significant histone-like protein, as detected by the DNA-protein interaction assay termed Southwestern blotting. These histone-like proteins were associated with the condensed nucleoids of EBs, which led to the speculation that the proteins inhibited replication and transcriptional activity in the nonreplicating forms. Because a *C. burnetii* developmental cycle may have analogous stages in which SCV would be comparable to EB and LCV comparable to RB, these investigators used Southwestern blotting to detect histone-like protein homologues. Two DNA binding proteins were identified for *C. burnetii*: a ~20-kDa protein (Hq1) was highly enriched in SCV, while a ~14-kDa protein was detected in equal proportions in LCV and SCV. By using heparin-coupled agarose affinity chromatography to purify Hq1, the N-terminal sequence for the protein was deduced.

Degenerate oligonucleotides were developed from this sequence, and the gene encoding Hq1 was cloned. The derived amino acid sequence from the *hq1* open reading frame predicted a protein of ~13 kDa with a very basic pI (13.1). Hq1 was 34% identical to the eukaryotic histone protein H1 and 26% identical to *C. trachomatis* Hc1. Therefore, the relatedness of Hq1 and Hc1 suggests that their developmentally regulated expression is due to a direct role in gene regulation and replication control. Recent completion of the *C. trachomatis* genome sequence led to the speculation that *Chlamydia* acquired several genes from the eukaryotic host, including *hc1*, and a similar horizontal transfer event may have led to *hq1* acquisition by *Coxiella* (Stephens et al., 1998). In fact, the acquisition of eukaryotic histone-like proteins may have permitted each organism to evolve a developmental cycle unique to several obligate intracellular bacteria.

ScvA

Heinzen and coworkers also identified a basic DNA binding peptide of ~4 kDa (Heinzen et al., 1996a). Preparative isoelectric focusing was used to purify this protein, and the N-terminal amino acid sequence was derived. Again, degenerate oligonucleotide probes were used to clone the gene encoding this peptide. The peptide was also apparently developmentally regulated and associated with only SCV chromatin and was designated ScvA (for SCV antigen). Although database comparison did not identify a bacterial homologue for ScvA, it does have unusual and potentially functionally relevant properties. More than 50% of the peptide is composed of arginine, glutamine, and proline (23, 23, and 13%, respectively). It was shown to be exclusively associated with the central core of condensed chromatin by immunogold labeling and electron microscopic visualization. The authors speculated on several intriguing roles that ScvA may play in *C. burnetii* development and survival. The spores of *Bacillus* spp. produce a family of small acid-soluble proteins (SASP) that have low molecular mass (5 to 11 kDa), bind DNA nonspecifically, and protect spore DNA from environmental insult (see chapter 9) (Setlow, 1988). This protein is degraded by SASP-specific proteases during spore germination to generate amino acids that are then available for biosynthetic needs. Heinzen et al. (1996a) presented immunoblot evidence to show that ScvA is degraded upon infection of host cells by SCV. Earlier studies of transport and incorporation of radiolabeled amino acids and sugars demonstrated that in an axenic, acid-activated medium, *C. burnetii* preferred glutamate as an energy source. Therefore, it is logical to speculate that ScvA may be a mechanism to survive extracellular stress and to provide an immediate carbon source upon uptake by host cells.

EF-Ts and EF-Tu

We recently compared the protein profiles of SCV and LCV by Western blotting with a panel of monoclonal antibodies generated against purified *C. burnetii* (Fig. 2) (Seshadri et al., submitted). Two proteins were up-regulated in LCV compared with SCV proteins. A λZapII gene bank of *C. burnetii* chromosomal DNA was screened, and an immunoreactive clone was characterized by sequence analysis. The open reading frame which encoded a 33-kDa immunoreactive protein was found to be located in an *rpsB-tsf* operon and to express the translational element elongation factor (EF)-

Ts. A ~45-kDa LCV-specific protein was detected by Western blotting by the second monoclonal antibody, but immunoreactive clones were not identified by screening several gene banks of *Coxiella* DNA. Serendipitously, reactivity with the monoclonal antibody was observed with a *C. trachomatis* protein of 45 kDa. By using His-tagged recombinant EF-Tu from *Chlamydia*, this antibody was confirmed to be specific for EF-Tu of *Chlamydia* (Cousineau et al., 1992; Zhang et al., 1994). EF-Tu genes have frequently been reported to be unstable in cloning experiments, so PCR cloning of the *C. burnetii* genes encoding EF-Tu was done with conserved upstream and downstream elements. EF-Tu was demonstrated to be detected only in LCV extracts, not SCV extracts. The differential expression by only a subset of organisms was also demonstrated in vitro within infected cells by indirect immunofluorescence. Several implications can be proposed based upon the differential expression of two key translational elements by LCV. This provides genetic evidence to support the hypothesis of earlier studies suggesting increased metabolic activity either for LCV or by a subset of intraphagosomal organisms (McCaul et al., 1981; Thompson et al., 1984). EF-Tu in particular has been shown to be significantly upregulated by both transcriptional and posttranscriptional mechanisms and can account for >5% of the total protein synthesis of log-phase

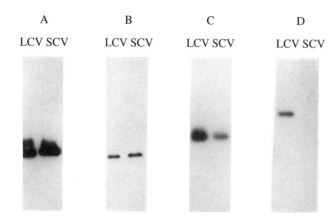

| A | B | C | D |
| LCV SCV | LCV SCV | LCV SCV | LCV SCV |

FIGURE 2 Differential protein expression of *C. burnetii* proteins detected with monoclonal antibodies. LCV and SCV were separated by cesium chloride equilibrium gradient centrifugation, and proteins were subsequently separated by SDS-PAGE. Monoclonal antibodies specific for each antigen were reacted on Western blots. (A) Monoclonal antibody specific for Com-1 (27 kDa); (B) monoclonal antibody specific for CbMip (25 kDa); (C) monoclonal antibody specific for EF-Ts (32 kDa); (D) monoclonal antibody specific for EF-Tu (45 kDa).

E. coli. An LCV-specific protein will also be valuable in combination with SCV-specific proteins like ScvA to further characterize these two populations.

Other Candidate Proteins for Differential Expression

Heinzen and coworkers provide both single- and two-dimensional SDS-PAGE analysis of SCV and LCV proteins, noting several differences beyond P1, Hq1, ScvA, EF-Ts, and EF-Tu (Heinzen, 1997; Heinzen and Hackstadt, 1996). We have confirmed that P1 is up-regulated in LCV compared with SCV by both Western blotting and surface iodination of intact organisms. We also detected at least one ~32-kDa surface protein that appeared SCV specific and which was not EF-Ts (Fig. 3). Finally, the SDC are a major subset of small cells that have remained relatively uncharacterized

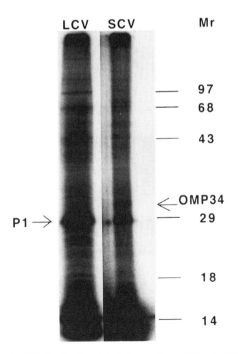

FIGURE 3 Surface-labeled proteins. LCV and SCV were surface iodinated with Iodogen beads, and proteins were separated by SDS-PAGE followed by autoradiographic detection of labeled proteins. The major outer membrane protein, P1, is identified. An apparent SVC upregulated surface protein at ~34 kDa is noted.

antigenically. We have noted that proteins from pressure-resistant cells do not enter SDS-PAGE unless rigorously solubilized with high SDS (>8%) plus boiling for a long time in urea (>8 M). Therefore, the antigenic composition as measured by Western blotting has been difficult to assess.

POTENTIAL ENVIRONMENTAL MODULATORS AND REGULATORY ELEMENTS

One of the observations that appear to distinguish a *C. burnetii* developmental cycle from that of *Chlamydia* is that a mixture of SCV and LCV appears in all vacuoles, whether newly occupied with a few organisms or containing hundreds of organisms. Therefore, predominance of one form does not appear to synchronize with any stage of infection (e.g., early versus late). Attempts to drive *Coxiella* toward one stage by modifying the condition of host cells have generally been unsuccessful, although maintenance of infected cells in culture without a change of medium (6 weeks or longer) was reported to increase the percentage of LCV with SLPs (Heinzen, 1997; Schaal et al., 1987). As noted by Heinzen, depletion of carbon, nitrogen, or phosphorus has been shown to induce morphogenesis in the forms of sporulation in *Bacillus* and in cyst formation in *Azotobacter*. Therefore, a similar mechanism might occur in *Coxiella*, but no data clearly support one of these as a key nutrient regulating development. A second potential regulatory signal for developmental changes could be pH. Activation of metabolism at pHs of ~4.5 to 5.0 and inhibition of replication by lysosomotrophic agents demonstrated the moderate acidophilic nature of the organism (Hackstadt and Williams, 1981). The intracellular pH of the organism is normally maintained at neutrality, but upon depletion of glutamate, the intracellular pH drops below 6.0 (Hackstadt, 1983; Hackstadt and Williams, 1983, 1984). Therefore, a sensing mechanism for nutrient depletion might involve intracellular pH balance mediated by available glutamate. As reported in a review by Heinzen (1997), an experiment to test pH activation of development was performed with bafilomycin A, an inhibitor of vacuolar-type (H⁺) ATPase that is required for effective acidification of endosomes, including those occupied by *C. burnetii*. Bafilomycin A-treated infected cells had no apparent changes in the ratio of LCV to SCV, nor was an increase in the frequency of SLP compared to untreated cells observed by TEM. Therefore, pH did not appear to affect *C. burnetii* development.

DEVELOPMENTAL MODEL OF *C. BURNETII* SURVIVAL

There are superficial similarities between the developmental cycles of *C. burnetii* and *C. trachomatis*, since both are obligate intracellular bacterial pathogens that undergo an infectious cycle inside a membrane-bound vacuole (Moulder, 1991). The most obvious ultrastructural similarity is the condensed chromatin of both chlamydial EB and SCV, the compaction of which is mediated in both cases by lysine-rich histone-like proteins (Hackstadt et al., 1991; Heinzen and Hackstadt, 1996). SCV and EB are both metabolically less active, resistant small cell types that are adapted for extracellular survival. Like LCV, the RB are fragile and less stable outside the host. Notwithstanding these similar biological properties, there are some notable differences. For example, SCV undergo binary fission whereas EB do not divide, and LCV, but not RB, are infectious.

A model for understanding the role of the major morphological forms of *C. burnetii* found in the PV includes aspects of both log phase-to-stationary phase differentiation common to most bacteria and the generation of resistant extracellular forms required for obligate intracellular organisms to survive in the environment (Fig. 4). Evidence suggests that LCV are more metabolically and replicatively active than SCV. As such, they might play a more important role than the SCV in cell-to-cell spread within the host. Conversely, the SCV, especially as SDC, might be able to survive the degradative enzymes and peptides of the phagolysosome for extended periods, and this is possibly the stage responsible for long-term extracellular survival and aerosol transmission of the agent. Because *C. burnetii* does not actively lyse host cells and is frequently transmitted by an aerosol that results from desiccated infected tissues dropped on the soil, a sustained supply of resistant cell forms is critical to its survival. Although the SLP has been hypothesized to be the precursor of the SCV, there is no teleological reason to invoke a "spore-like" stage for the developmental cycle, as the physical properties of the SCV are sufficient to explain *C. burnetii*'s extracellular stability. Indeed, the characteristics of *C. burnetii* resistance to various physical stresses are intermediate between those of vegetative cells and spores (Baca and Paretsky, 1983).

The limited molecular studies of developmentally specific proteins allow some speculation about the mechanism of gene regulation by SCV and LCV forms within the context of this model. SCV and LCV might be analogous to stationary- and log-phase growth forms, respectively. Consistent with this idea is the observation that EF-Tu is detected only in LCV. EF-Tu expression can reach 5% of the total synthesized protein in actively dividing, log-phase *E. coli*, and this expression is dramatically reduced in stationary phase. Moreover, histone-like proteins, similar to the SCV-specific *C. burnetii* protein Hq1, have been shown to be involved in the transition to stationary phase in *E. coli* (Hengge-Aronis, 1996). The histone-like protein HU is expressed as a dimer from two genes, *hupA* and *hupB*. The αβ heterodimer predominates during stationary phase and is required for survival after prolonged starvation (Claret and Rouviere-Yaniv, 1997). The histone-like protein H-NS is involved in establishing stationary phase by posttranscriptional regulation of RpoS (Barth et al., 1995). If SCV-to-LCV differentiation is analogous to stationary phase-to-log phase transition, then a *C. burnetii* RpoS homologue may act as a master regulator of SCV-specific genes.

A model in which different cell forms ex-

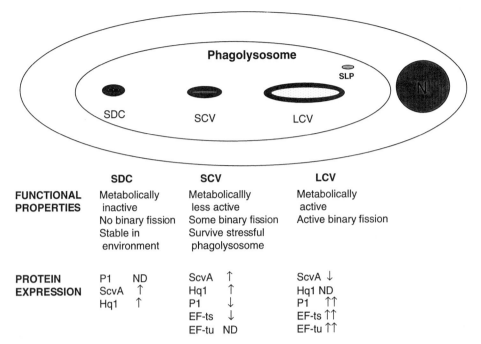

FIGURE 4 Model of LCV, SCV, SDC, and SLP. A working model of *C. burnetii* developmental stages derived from studies summarized in this review. A single infected cell is represented containing a phagolysosome with all four forms. SLP have only been observed infrequently. The ability of each form to develop into one of the alternate forms is unconfirmed. The functional properties assigned to each form and the level of expression of specific proteins is indicated for high (↑↑), medium (↑), relatively low (↓), and not detected (ND).

press different genes while occupying the same vacuolar environment presents an obvious paradox. A similar paradox was noted by Abshire and Neidhardt for *S. typhimurium* replicating in macrophage vacuoles (Abshire and Neidhardt, 1993). Their data support a model in which intracellular *S. typhimurium* populations are a mixture of stationary-phase and actively dividing log-phase organisms.

SUMMARY AND QUESTIONS FOR THE FUTURE

Cellular differentiation in prokaryotes is an adaptive response that involves alterations in gene expression. Morphological differentiation into a resistant, resting form is an obvious advantage for a pathogen such as *C. burnetii* for which the primary mode of disease transmission is inhalation of contaminated aerosols from

natural environments. Studies investigating morphological differentiation and programmed gene expression in *C. burnetii* could be aided by the ability to transiently synchronize the cell cycle. The cloning of genes preferentially transcribed by SCV and LCV and analysis of their products for function are needed to test the morphogenesis model. A major advance in the development of genetic systems for *C. burnetii* was recently reported by Suhan et al. (1996), who successfully transformed the organism. We have been able to repeat transformation experiments with the same shuttle vector and have included an additional marker (green fluorescence protein). These exciting developments will hopefully lead to workable systems, such as gene inactivation, promoter expression, and transdominant complementation, and ultimately to more powerful tools for elucidating critical genetic events in *C. burnetii* morphogenesis.

REFERENCES

Abshire, K. Z., and F. C. Neidhardt. 1993. Growth rate paradox of *Salmonella typhimurium* within host macrophages. *J. Bacteriol.* **175:** 3744–3748.

Akporiaye, E. T., and O. G. Baca. 1983. Superoxide anion production and superoxide dismutase and catalase activities in *Coxiella burnetii. J. Bacteriol.* **154:**520–523.

Amano, K., J. C. Williams, T. E. McCaul, and M. G. Peacock. 1984. Biochemical and immunological properties of *Coxiella burnetii* cell wall and peptidoglycan-protein complex fractions. *J. Bacteriol.* **160:**982–988.

Anacker, R. L., K. Fukushi, E. G. Pickens, and D. B. Lackman. 1964. Electron microscopic observations of the development of *Coxiella burnetii* in the chick yolk sac. *J. Bacteriol.* **88:**1130–1138.

Babudieri, C. 1959. Q fever: a zoonosis. *Adv. Vet. Sci.* **5:**81–84.

Baca, O. G., and D. Paretsky. 1983. Q fever and *Coxiella burnetii*: a model for host-parasite interactions. *Microbiol. Rev.* **47:**127–149.

Baca, O. G., D. A. Klassen, and A. S. Aragon. 1993a. Entry of *Coxiella burnetii* into host cells. *Acta Virol.* **37:**143–155.

Baca, O. G., M. J. Roman, R. H. Glew, R. F. Christner, J. E. Buhler, and A. S. Aragon. 1993b. Acid phosphatase activity in *Coxiella burnetii*: a possible virulence factor. *Infect. Immun.* **61:** 4232–4239.

Banerjee-Bhatnagar, N., C. R. Bolt, and J. C. Williams. 1996. Pore-forming activity of *Coxiella burnetii* outer membrane protein oligomer comprised of 29.5 and 31-kDa polypeptides. *Ann. N.Y. Acad. Sci.* **791:**378–401.

Barth, M., C. Marshall, A. Muffler, S. Fischer, and R. Hengge-Aronis. 1995. Role for the histone-like protein H-NS in growth phase-dependent and osmotic regulation of σ^s and many σ^s-dependent genes in *Escherichia coli. J. Bacteriol.* **177:** 3455–3464.

Brickman, T. J., C. E. Barry, and T. Hackstadt. 1993. Molecular cloning and expression of *kctB* encoding a strain-variant chlamydial histone-like protein with DNA-binding activity. *J. Bacteriol.* **175:** 4274–4281.

Burton, P. R., J. Stueckemann, R. M. Welsh, and D. Paretsky. 1978. Some ultrastructural effects of persistent infections by the rickettsia *Coxiella burnetii* in mouse L cells and green monkey kidney (Vero) cells. *Infect. Immun.* **21:**556–566.

Canonicao, P. G., M. J. Van Zwieten, and W. A. Christmas. 1972. Purification of large quantities of

Coxiella burnetii rickettsia by density gradient zonal centrifugation. *Appl. Microbiol.* **23:**1015–1022.

Cianciotto, N. P., W. O'Connell, G. A. Dasch, and L. P. Mallavia. 1995. Detection of mip-like sequences and Mip-related proteins within the family Rickettsiaceae. *Curr. Microbiol.* **30:**149–153.

Claret, L., and J. Rouviere-Yaniv. 1997. Variation in HU composition during growth of *Escherichia coli*: the heterodimer is required for long term survival. *J. Mol. Biol.* **273:**93–104.

Cousineau, R., C. Cerpa, J. Lefebvre, and R. Cedergren. 1992. The sequence of the gene encoding elongation factor Tu from *Chlamydia trachomatis* compared with those of other organisms. *Gene* **120:**33–41.

Davis, G. E., and H. R. Cox. 1938. A filter-passing infectious agent isolated from ticks. I. Isolation from Dermacentor andersonii, reactions in animals, and filtration. *Public Health Rep.* **53:**2259.

De Groote, M. A., U. A. Ochsner, M. U. Shiloh, C. Nathan, J. M. McCord, M. C. Dinauer, S. J. Libby, A. Vazquez-Torres, Y. Xu, and F. C. Fang. 1997. Periplasmic superoxide dismutase protects *Salmonella* from products of phagocyte NAPDH-oxidase and nitric oxide synthase. *Proc. Natl. Acad. Sci. USA* **94:**13997–14001.

Fournier, P.-E., T. J. Marrie, and D. Raoult. 1998. Diagnosis of Q fever. *J. Clin. Microbiol.* **36:** 1823–1834.

Hackstadt, T. 1983. Estimation of the cytoplasmic pH of *Coxiella burnetii* and effect of substrate oxidation on proton motive force. *J. Bacteriol.* **154:** 591–597.

Hackstadt, T., and J. C. Williams. 1981. Biochemical stratagem for obligate parasitism of eukaryotic cells by *Coxiella burnetii. Proc. Natl. Acad. Sci.* **78:** 3240–3244.

Hackstadt, T., and J. C. Williams. 1983. pH dependence of the *Coxiella burnetii* glutamate transport system. *J. Bacteriol.* **154:**598–603.

Hackstadt, T., and J. C. Williams. 1984. Metabolic adaptations of *Coxiella burnetii* to intraphagolysosomal growth, p. 266–268. *In* L. Lieve and D. Schlessinger (ed.), *Microbiology—1984.* American Society for Microbiology, Washington, D.C.

Hackstadt, T., W. Baehr, and Y. Ying. 1991. *Chlamydia trachomatis* developmentally regulated protein is homologous to eukaryotic histone H1. *Proc. Natl. Acad. Sci. USA* **88:**3937–3941.

Heinzen, R. A. 1997. Intracellular development of *Coxiella burnetii*, p. 99–129. *In* B. Anderson, M. Bendinelli, and H. Friedman (ed.), *Rickettsial Infection and Immunity.* Plenum Press, New York, N.Y.

Heinzen, R. A., and T. Hackstadt. 1996. A developmental-stage-specific histone H1 homolog of *Coxiella burnetii. J. Bacteriol.* **178:**5049–5052.

Heinzen, R. A., M. E. Frazier, and L. P. Mallavia.

1992. *Coxiella burnetii* superoxide dismutase gene: cloning, sequencing, and expression in *Escherichia coli*. *Infect. Immun.* **60:**3814–3823.

Heinzen, R. A., D. Howe, L. P. Mallavia, D. D. Rockey, and T. Hackstadt. 1996a. Developmentally regulated synthesis of an unusually small, basic peptide by *Coxiella burnetii*. *Mol. Microbiol.* **22:** 9–19.

Heinzen, R. A., M. A. Scidmore, D. D. Rockey, and T. Hackstadt. 1996b. Differential interaction with endocytic and exocytic pathways distinguish parasitophorous vacuoles of *Coxiella burnettii* and *Chlamydia trachomatis*. *Infect. Immun.* **64:**796–809.

Heinzen, R. A., T. Hackstadt, and J. E. Samuel. 1999. Developmental biology of *Coxiella burnetii*. *Trends Microbiol.* **7:**149–154.

Hendrix, L., and P. Mallavia. 1984. Active transport of proline by *Coxiella burnetti*. *J. Gen. Microbiol.* **130:**2857–2863.

Hendrix, L. R., J. E. Samuel, and L. P. Mallavia. 1991. Differentiation of *Coxiella burnetii* isolates by analysis of restriction-endonuclease-digested DNA separated by SDS-PAGE. *J. Gen. Microbiol.* **137:** 269–276.

Hendrix, L. R., L. P. Mallavia, and J. E. Samuel. 1993. Cloning and sequencing of *Coxiella burnetii* outer membrane protein gene *com1*. *Infect. Immun.* **61:**470–477.

Hengge-Aronis, R. 1996. Back to log phase: Sigma S as a global regulator in the osmotic control of gene expression in *Escherichia coli*. *Mol. Microbiol.* **21:**887–893.

Htwe, K. K., T. Yoshida, S. Hayashi, T. Miyake, K.-I. Amano, C. Morita, T. Yamaguchi, H. Fukushi, and K. Hirai. 1993. Prevalence of antibodies to *Coxiella burnetii* in Japan. *J. Clin. Microbiol.* **31:**722–723.

Kordova, N. 1960. Study of antigenicity and immunogenicity of filterable particles of *Coxiella burnetii*. *Acta Virol.* **4:**56–62.

Kordova, N., and E. Kocacova. 1968. Appearance of antigens in tissue culture cells inoculated with filterable particles of *Coxiella burnetii* as revealed by fluorescent antibodies. *Acta Virol.* **12:**40–463.

Lennette, E. H., W. H. Clark, M. M. Abinanti, O. Burnetti, and J. M. Covert. 1952. Q fever studies. XIII. The effect of pasteurization of Coxiella burnetii in naturally infected milk. *Am. J. Hyg.* **55:**246.

Li, Y. P., G. Curley, M. Lopez, M. Chavez, R. Glew, A. Aragon, H. Kumar, and O. G. Baca. 1996. Protein-tyrosine phosphatase activity of *Coxiella burnetii* that inhibits human neutrophils. *Acta Virol.* **40:**163–172.

Mallavia, L. P. 1991. Genetics of rickettsiae. *Eur. J. Epidemiol.* **7:**213–221.

Marrie, T. J. 1990. Epidemiology of Q fever, p.

49–70. *In* T. J. Marrie (ed.), *Q Fever*, vol. 1. *The Disease*. CRC Press, Boca Raton, Fla.

McCaul, T. F. 1991. The developmental cycle of *Coxiella burnetii*, p. 223–258. *In* J. C. Williams and H. A. Thompson (ed.), *Q Fever: The Biology of Coxiella burnetii*. CRC Press, Boca Raton, Fla.

McCaul, T. F., and J. C. Williams. 1981. Developmental cycle of *Coxiella burnetii*: structure and morphogenesis of vegetative and sporogenic differentiations. *J. Bacteriol.* **147:**1063–1076.

McCaul, T. F., T. Hackstadt, and J. C. Williams. 1981. Ultrastructural and biological aspects of *Coxiella burnetii* under physical disruptions, p. 267. *In* W. Burgdorfer and R. L. Anacker (ed.), *Rickettsiae and Rickettsial Diseases*. Academic Press, New York, N.Y.

McCaul, T. F., N. Banerjee-Bhatnagar, and J. C. Williams. 1991a. Antigenic differences between *Coxiella burnetii* cells revealed by postembedding immunoelectron microscopy and immunoblotting. *Infect. Immun.* **59:**3243–3253.

McCaul, T. F., J. C. Williams, and H. A. Thompson. 1991b. Electron microscopy of *Coxiella burnetii* in tissue culture induction of cell types as products of developmental cycle. *Acta Virol.* **35:** 545–556.

McCaul, T. F., A. J. Dare, J. P. Gannon, and J. J. Galbraith. 1994. In vivo endogenous spore formation by *Coxiella burnetii* in Q fever endocarditis. *J. Clin. Pathol.* **47:**978–981.

Meconi, S., V. Jacomo, P. Boquet, D. Raoult, J. Mege, and C. Capo. 1998. *Coxiella burnetii* induces reorganization of the actin cytoskeleton in human monocytes. *Infect. Immun.* **66:**5527–5533.

Mo, Y. Y., N. P. Cianciotto, and L. P. Mallavia. 1995. Molecular cloning of a Coxiella burnetii gene encoding a macrophage infectivity potentiator (Mip) analogue. *Microbiology* **141:**2861–2871.

Moulder, J. W. 1991. Interaction of Chlamydiae and host cells in vitro. *Microbiol. Rev.* **55:**143–190.

Nermut, M. V., S. Schramek, and R. Brezina. 1968. Electron microscopy of *Coxiella burnetii* phase I and phase II. *Acta Virol.* **12:**446–452.

Raoult, D., and T. Marrie. 1995. Q fever. *Clin. Infect. Dis.* **20:**489–496.

Redd, T., and H. A. Thompson. 1995. Secretion of proteins by *Coxiella burnetii*. *Microbiology* **141:** 363–369.

Rosenberg, M., and N. Kordova. 1960. Study of intracellular forms of *Coxiella burnetii* in the electron microscope. *Acta Virol.* **4:**52–61.

Samuel, J. E., M. E. Frazier, and L. P. Mallavia. 1985. Correlation of plasmid type and disease caused by *Coxiella burnetii*. *Infect. Immun.* **49:** 775–779.

Schaal, F., H. Krauss, N. Jekov, and L. Rantamaki. 1987. Electron micrographic observations on the morphogenesis of "spore-like particles" of *Coxiella burnetii* in cell culture. *Acta Medit. Patol. Inf. Trop.* **6:**329–338.

Seshadri, R., L. R. Hendrix, and J. E. Samuel. Differential expression of translational elements by lifecycle variants of *Coxiella burnetii*. Submitted for publication.

Setlow, P. 1988. Small, acid soluble spore proteins of *Bacillus* species: structure, synthesis, genetics, and degradation. *J. Bacteriol.* **42:**319–338.

Snyder, C. E. J., and J. C. Williams. 1986. Purification and chemical characterization of a major outer membrane protein from *Coxiella burnetii*, p. 193. *In Abstracts of the Annual Meeting of the American Society for Microbiology*, 1986. American Society for Microbiology, Washington, D.C.

Snyder, C. E., R. F. Wachter, and J. D. White. 1984. Identification of *Coxiella burnetii* antigens and attempts at envelope fractionation, p. 263–265. *In* L. Lieve and D. Schlessinger (eds.), *Microbiology—1984*. American Society for Microbiology, Washington, D.C.

Stephens, R. S., S. Kalman, C. Lammel, J. Fan, R. Marathe, L. Aravind, W. Mitchell, L. Olinger, R. L. Tatusov, Q. Zhao, E. V. Koonin, and R. W. Davis. 1998. Genome sequence of an obligate intracellular pathogen of humans: *Chlamydia trachomatis*. *Science* **282:**754–759.

St. John, G., and H. M. Steinman. 1996. Periplasmic copper-zinc superoxide dismutase of *Legionella pneumophila*: role in stationary-phase survival. *J. Bacteriol.* **178:**1578–1584.

Stocker, M. P., K. M. Smith, and P. Fiset. 1956. Internal structure of *Rickettsia burnetii* as shown by electron microscopy of thin sections. *J. Gen. Microbiol.* **15:**632–635.

Suhan, M. L., C. Shu-Yin, and H. A. Thompson. 1996. Transformation of *Coxiella burnetii* to ampicillin resistance. *J. Bacteriol.* **178:**2701–2708.

Thompson, H. A., R. L. Zuerner, and T. Tedd. 1984. Protein synthesis in *Coxiella burnetii*, p. 288–291. *In* L. Lieve and D. Schlessinger (ed.), *Microbiology—1984*. American Society for Microbiology, Washington, D.C.

Varghees, S., K. Kiss, and J. E. Samuel. 1998. *Characterization of the Major Outer Membrane Protein from Coxiella burnetii*. Texas Branch of the American Society for Microbiology.

Wachter, R. F., G. P. Briggs, J. D. Gangemi, and C. E. Pedersen. 1975. Changes in buoyant density relationships of two cell types of *Coxiella burnetii* phase I. *Infect. Immun.* **12:**433–436.

Watarai, M., T. Tobe, M. Yoshikawa, and C. Sasakawa. 1995. Disulfide oxidoreductase activity of *Shigella flexneri* is required for release of Ipa proteins and invasion of epithelial cells. *Proc. Natl. Acad. Sci. USA* **92:**4927–4931.

Weisburg, W. G., C. R. Woese, M. E. Dobson, and E. Weiss. 1985. A common origin of rickettsiae and certain plant pathogens. *Science* **2230:** 556–558.

Weisburg, W. G., M. E. Dobson, J. E. Samuel, G. A. Dasch, L. P. Mallavia, O. G. Baca, L. Mandelco, J. E. Sechrest, E. Weiss, and C. R. Woese. 1989. Phylogenetic diversity of the rickettsiae. *J. Bacteriol.* **171:**4202–4206.

Wiebe, M. E., P. R. Burton, and D. M. Shankel. 1972. Isolation and characterization of two cell types of *Coxiella burnetii*. *J. Bacteriol.* **110:**368–377.

Willems, H., C. Jager, and G. Bajer. 1998. Physical and genetic map of the obligate intracellular bacterium *Coxiella burnetii*. *J. Bacteriol.* **180:**3816–3822.

Williams, J. C. 1991. Infectivity, virulence, and pathogenicity of *Coxiella burnetii* for various hosts, p. 21–72. *In* J. C. Williams and H. A. Thompson (ed.), *Q Fever: The Biology of Coxiella burnetii*. CRC Press, Boca Raton, Fla.

Williams, J. C., and D. M. Waag. 1991. Antigens, virulence factors and biological response modifiers of *Coxiella burnetii*: strategies for vaccine development, p. 175–222. *In* J. C. Williams and H. A. Thompson (ed.), *Q Fever: The Biology of Coxiella burnetii*. CRC Press, Boca Raton, Fla.

Williams, J. C., M. R. Johnston, M. G. Peacock, L. A. Thomas, S. Stewart, and J. L. Portis. 1984. Monoclonal antibodies distinguish phase variants of *Coxiella burnetii*. *Infect. Immun.* **43:**421–428.

Zhang, Y., Y. Shi, M. Zhou, and G. A. Petsko. 1994. Cloning, sequencing, and expression in *Escherichia coli* of the gene encoding a 45-kDa protein, elongation factor Tu, from *Chlamydia trachomatis* serovar F. *J. Bacteriol.* **176:**1184–1187.

Zuerner, R. L., and H. A. Thompson. 1983. Protein synthesis by intact *Coxiella burnetii* cells. *J. Bacteriol.* **156:**186–191.

DIFFERENTIATION OF FREE-LIVING RHIZOBIA INTO ENDOSYMBIOTIC BACTEROIDS

William Margolin

22

A group of α-proteobacteria, collectively called the rhizobia, are uniquely able to establish a complex and highly evolved mutualistic symbiosis with the roots of leguminous plants. This association results in the fixation of atmospheric nitrogen, which plays an important role in the global nitrogen cycle and also in regeneration of soil nitrogen levels for agriculture.

This group of bacteria includes the genera *Rhizobium*, *Mesorhizobium*, *Azorhizobium*, and *Sinorhizobium*, whose members are considered "fast-growing" strains, and *Bradyrhizobium*, whose members are "slow-growing" strains. The symbiosis between rhizobia and legumes is species specific, so that *Sinorhizobium meliloti* (formerly *Rhizobium meliloti*) associates with alfalfa and *Bradyrhizobium japonicum* associates with soybeans, for example. All rhizobia are obligately aerobic, gram-negative motile rods that inhabit the soil and grow slowly. However, the nearby presence of a legume plant initiates a remarkable transformation in these "free-living" bacteria. An exchange of chemical signals between the plant roots and the rhizobia culminates in the formation of nodules on the roots (Fig. 1) and the controlled invasion of the bacteria into cells within the nodules. Once the bacteria become intracellular, they differentiate into the endosymbiotic form, called bacteroids. Bacteroids are defined here as rhizobia present within plant cells regardless of bacterial morphology or physiology (Sutton, 1981), although often bacteroids are morphologically, ultrastructurally, and physiologically distinct from free-living bacteria (Fig. 2). Finally, both the nodules and the bacteroids senesce and eventually degrade.

Differentiation of rhizobia during symbiosis is complicated. Each stage of the process occurs in a different part of the nodule. Moreover, multiple events are necessary at each step, resulting in many factors that are not absolutely required. Quantitation of the efficiency of the interaction is difficult due to the multiple factors involved, the different plant tissues penetrated, and innate plant-to-plant variability. The plant often determines the physiological characteristics of the symbiosis, influencing everything from gene expression to bacteroid morphology. As a result, factors required for one rhizobial species to form an effective symbiosis may not be required for another, making it more difficult to generalize about what is significant. Finally, whereas other bacterial development systems can be studied in broth cul-

W. Margolin, Department of Microbiology and Molecular Genetics, University of Texas Medical School, Houston, TX 77030.

Prokaryotic Development, edited by Y. V. Brun and L. J. Shimkets,
© 2000 American Society for Microbiology, Washington, DC 20005-4171

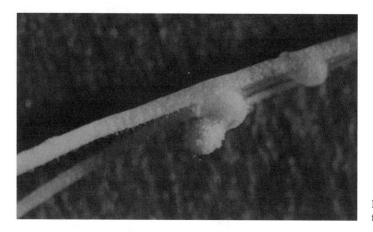

FIGURE 1 Root nodules on alfalfa induced by *S. meliloti*.

tures or on agar plates, the physiologically relevant development of bacteroids must be studied in situ in the plant. It should be noted, however, that bacteroid-like changes in rhizobia can be induced in culture under certain conditions.

There are many fascinating aspects of the symbiosis. One is how the plant recognizes rhizobia as nonpathogens and how it facilitates the invasion of the bacteria into its tissues and ultimately its cytoplasm. Interesting comparisons can be made with animal pathogens such as *Chlamydia*, which like rhizobia have distinct extracellular and intracellular states. Un-

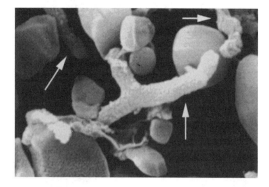

FIGURE 2 Scanning electron micrograph of *S. meliloti* bacteroids in an alfalfa nodule homogenate. The arrows highlight bacteroids, which include a Y-shaped cell in the center of the image and two other elongated cells. The other large structures visible are starch granules.

like pathogens such as *Chlamydia*, however, intracellular rhizobia are sequestered without invoking either a pathogenic response or a damaging host defense response. Another interesting aspect of the symbiosis concerns the selective pressures on the rhizobial population. Whereas the population as a whole benefits nutritionally from the symbiosis, the subset of cells that become bacteroids undergo terminal differentiation and eventually are destroyed. In a sense this partitioning is similar to what happens to mother cells of endospore-forming bacilli, stalk cells of myxobacteria, or heterocysts of cyanobacteria, all of which are also excluded from the germ line. How the population is partitioned to optimize its survival is an interesting and complex ecological problem.

The rhizobia-legume interaction is by far the best-studied symbiosis, in large part because of its agricultural importance and the relative ease of growing and genetically manipulating rhizobia in the absence of the host plant. The last decade has witnessed tremendous strides in understanding the genes and molecules involved in the critical early signal exchange between the plant and microbe, as well as the initial molecular effects on the plant. The subsequent processes of bacterial invasion and differentiation have been described extensively, but the molecular mechanisms behind the later stages are still poorly understood. The advent of genomics and several new experimental strategies is starting to clarify these mechanisms. In this review, the complexities of bacterial invasion and bacteroid differentiation are outlined. Because the early plant-bacterial signal exchange has been the subject of many excellent recent reviews (Kijne, 1992; Spaink, 1995; Dénarié et al., 1996; Long, 1996; Schultze and Kondorosi, 1996, 1998; Bladergroen and Spaink, 1998; Kamst et al., 1998; Albrecht et al., 1999), this review will emphasize the later stages about which less is known. The strategies that both bacteria and plant use to maintain the symbiosis and prevent pathogenesis will also be discussed. The new genetic technologies that hold great promise for uncovering the genes regulating the bacterial contribution to the symbiosis will be covered at the end of the chapter.

GENERAL SCHEME OF DEVELOPMENT

The formation of effective nodules containing differentiated, nitrogen-fixing bacteroids consists of a defined series of stages. These stages, at first glance, appear to be roughly analogous to the steps in the differentiation of cyanobacterial heterocysts, *Bacillus* endospores, *Myxococcus* fruiting bodies, or *Streptomyces* aerial spores in that there is a definite beginning, middle, and culmination and the stages can be recognized both genetically and morphologically. In fact, each stage in bacteroid differentiation can

be observed in specific zones within some types of nodules.

In response to chemical signals secreted by plant roots, rhizobia attach to root hairs, which are cells on the root surface that project outward into the soil. The bacteria in turn synthesize and secrete Nod factors, a specific family of signal molecules which initiate a series of events in the plant leading to formation of the nodule. One of the first effects of Nod factor secretion is the deformation of root hairs in the vicinity of the rhizobia, causing the bacteria to be trapped in a "shepherd's crook" type of structure (Fig. 3). A tube-like invagination of the plant cell wall at the site of the trapped bacterial microcolony allows the bacteria to enter the root hair, although the bacteria are still outside the plant cytoplasm. The bacteria grow and proliferate down this tube, called the infection thread. Meanwhile, the thread invaginates further into the root cortex, where plant cell divisions are simultaneously driving nodule development. The invaginations branch within the growing nodule and terminate in certain nodule cells. At this point, the bacteria are released from the infection thread into the plant cells by an endocytosis-like process. The result is that a subset of nodule cells become full of bacteria, which are enveloped in a membrane derived from the plant, the peribacteroid membrane (PBM). Depending on the species, the bacteria then may or may

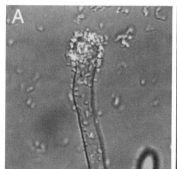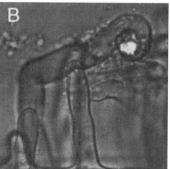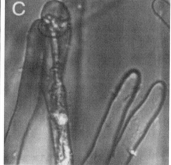

FIGURE 3 Initial interaction between *S. meliloti* and alfalfa root hairs. *S. meliloti* cells expressing GFP (appearing bright on the darker background) were photographed by confocal microscopy at different stages, including initial attachment to the root hair (A), entrapment by the curled root hair (B), and migration down the infection thread (C). (Images courtesy of Daniel J. Gage, University of Connecticut.)

not continue to multiply for a short period so that there may be multiple bacteria per PBM. The bacteria then differentiate into bacteroids and begin fixing nitrogen. The structure consisting of the PBM with the bacteroids inside is termed the symbiosome. It comprises three membranes, the outermost from the plant surrounding the innermost two from the bacteria. The PBM is extremely important for the symbiosis, serving as both a physical barrier between symbiont and host and a selective conduit for import and export of metabolites (Verma and Hong, 1996). In symbioses involving nontropical legumes producing "indeterminate" nodules with a persistent growth zone, such as *S. meliloti*-alfalfa, nodule growth accompanies the stages of symbiosis and an elongated nodule results. In such systems, a cross section of the nodule reveals a developmental gradient from early, near the actively growing part of the nodule, or meristem, to late, at the base of the nodule closest to the root (Fig. 4). Therefore, the developmental stages can be visualized, much as with other bacterial differentiation systems.

As in other bacterial differentiation systems, the morphological stages evident in the symbiosis and its nonessential nature have allowed the isolation of knockout mutants that arrest at various stages. Mutants that exhibit the most obvious phenotypes include those that cannot stimulate the formation of nodules (Nod⁻) and those that form nodules and can be released into symbiosomes but fail to fix nitrogen (Fix⁻). A great deal has been learned about genes required for nodule formation and nitrogen fixation by using these mutants. In addition, mutants defective in some of the intermediate stages, such as induction of infection thread formation and release into the symbiosome, have also been isolated, and information about these stages is being gleaned from the identities of the genes affected. The following sections outline the various stages.

ATTRACTION AND ATTACHMENT TO PLANT ROOTS

The commitment of free-living rhizobia to become differentiated symbionts begins with the secretion of flavonoid compounds by the roots of the host plants. This signaling is in a broad sense equivalent to the initial sensing of nutrient limitation by other bacteria that differentiate, because it results in the expression of new genes that ultimately generate the nitrogen-fixing nodule. However, the presence of the plant root, not nutrient limitation, appears to be what the bacteria are sensing and what stimulates the interaction to begin. Flavonoids act as both chemoattractants and specific inducers of rhizobial gene expression (see below).

Chemotaxis plays an important role in the initial attraction of rhizobia to plant root hairs. Interestingly, *S. meliloti* bacteria swim by changing the rotational speed of their multiple flagella instead of changing their direction of rotation (Gotz and Schmitt, 1987; Armitage and Schmitt, 1997). Rhizobia move by chemotaxis toward sugars, amino acids, organic acids, and a variety of plant flavonoid inducers (Caetano-Anolles et al., 1988; Robinson and Bauer, 1993). Mutants defective in chemotaxis do not compete as well for root colonization but nevertheless are often able to form effective nodules (Ames and Bergman, 1981; Yost et al., 1998).

Once they have been attracted to root hairs, the bacteria then attach to certain portions of the root surface, often via their cell poles (Schmidt, 1979; Dazzo et al., 1984). This polar attachment may be mediated by polarly localized lectins (Bohlool and Schmidt, 1976; Loh et al., 1993) and is an interesting example of bacterial cellular asymmetry (Maddock et al., 1993). Plant lectins play some role in host-specific attachment (Brewin and Kardailsky, 1997; Kijne et al., 1997) but do not seem to be required at early stages (van Rhijn et al., 1998). Attachment has been postulated to occur in two steps (Dazzo et al., 1984). The first step involves rhicadhesin, a calcium-dependent, non-host-specific adhesin (Smit et al., 1987, 1992). Cellulose microfibrils, fimbriae (Vesper and Bauer, 1986; Smit et al., 1987), and plant lectins then appear to mediate subsequent "firm" attachment; mutants that fail to display microfibrils still attach, but not as strongly. The physiological state of the rhizobia, including their ionic environment and nutritional status, plays a significant role in attachment and might

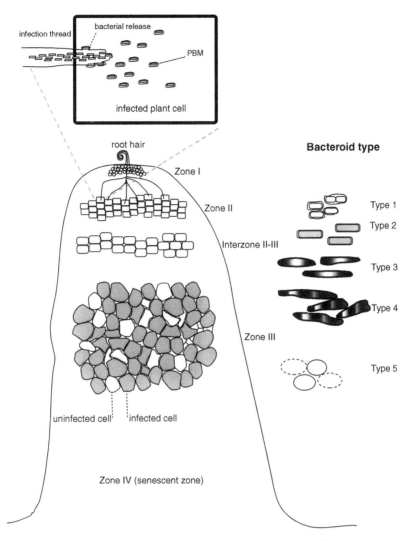

FIGURE 4 Outline of bacteroid development, from release to differentiation. A 4-week-old indeterminate nodule like that formed by the *S. meliloti*-alfalfa symbiosis is shown, as adapted from the results of Vasse et al. (1990). The inset at the top depicts bacterial proliferation within the infection thread and subsequent endocytosis into the plant cell and engulfment by the PBM. The cells within the nodule represent a sample of each zone, denoted at the right of the nodule. Bacteroid types and zones are as described in the text and in Vasse et al., 1990.

explain the high degree of variability in the results from studies of this process (Smit et al., 1992).

EARLY SIGNAL EXCHANGE BETWEEN BACTERIA AND HOST

As mentioned above, a variety of plant flavonoid compounds serve as the first known signals that attract rhizobia to roots in order to begin the nodulation process. In addition to promoting chemotaxis and formation of microcolonies on root hairs of the specific host, flavonoids have a critical role in stimulating gene expression in the bacteria. These expressed genes are termed *nod* genes because they are essential for stimulating nodule formation. They encode a group of enzymes which synthesize a second signal, Nod factor, that is directed toward the plant. The *nod* genes are often clustered with other genes important for symbiosis. These clusters are termed symbiosis islands, by analogy with pathogenicity islands in bacterial pathogens (Freiberg et al., 1997; Mergaert et al., 1997; Sullivan and Ronson, 1998), and suggest that symbiotic functions have been acquired by lateral transfer (Sullivan and Ronson, 1998). In general, symbiosis islands are distributed throughout different replicons. In *Rhizobium* strain NGR234, for example, symbiosis islands are present on the known symbiosis plasmid pNGR234a and the large megaplasmid pNGR234b, as well as the chromosome (Flores et al., 1998).

The induction of *nod* gene expression by flavonoids occurs via activation of a positive transcriptional regulator, NodD, a member of the LysR family of positive regulators (Schell, 1993). In *B. japonicum*, a second regulatory system, NodVW, is also involved in signal transduction between flavonoids and *nod* genes (Gottfert et al., 1990; Loh et al., 1997). In contrast to NodD, NodVW is a two-component sensor-response regulator pair. The mechanism by which flavonoids convert the Nod regulatory proteins into active transcriptional regulators is unknown. However, the NodD proteins appear to be specific transducers of the flavonoid signals, because species-specific activation of NodD proteins promote *nod* gene expression only in the presence of the cognate host plant flavonoid (Spaink et al., 1987). As a result, different NodD proteins, including multiple NodD proteins in a given species (there are three in *S. meliloti*), may recognize different plant inducers. Other important regulators of *nod* genes include NolR, which represses expression of *nodD* and a subset of other *nod* genes (Cren et al., 1995); NolA1 of *B. japonicum*, which encodes a MerR-type transcriptional regulator (Loh et al., 1999); and SyrM, a NodD-like regulator (Barnett and Long, 1990; Demont et al., 1994).

The enzymes encoded by the *nod* genes synthesize a family of lipo-chitooligosaccharides,

called Nod factors, that stimulate nodule morphogenesis (Dénarié et al., 1996; Long, 1996; Schultze and Kondorosi, 1996; Mergaert et al., 1997; Albrecht et al., 1999). This is the second major signal in the signal exchange pathway after the flavonoids and serves as the bacterial response. The Nod factors have in common a chitin oligomer backbone (N-acetyl-D-glucosamine), and their diversity is due to modification of this backbone by a variety of side groups. These modifications, which include sulfate and different acyl chains, help to confer species specificity in these factors (Schulze et al., 1994; Long, 1996). Remarkably, addition of very small amounts of purified Nod factors can stimulate nodules as well as preinfection thread structures (van Brussel et al., 1992) in the absence of bacteria (Finan et al., 1985; Truchet et al., 1991). This implies that the bacterial factor irreversibly and specifically induces a unique plant developmental program. In fact, rhizobia lacking *nod* genes but supplied with purified Nod factor are capable of invading legumes and forming Fix$^+$ (nitrogen-fixing) nodules, suggesting that continuous synthesis of Nod factor is not necessary for bacteroid differentiation or effective symbiosis once bacteria are inside the plant (Relic et al., 1994).

Soon after its addition, Nod factor triggers early molecular developmental responses from root hair cells, such as membrane depolarization, ion fluxes, and calcium oscillations (Ehrhardt et al., 1992, 1996; Felle et al., 1998). These responses occur prior to well-known visible alterations, such as root hair curling (Yao and Vincent, 1969) and cortical cell divisions (Hirsch et al., 1984). It is possible that Nod factors may elicit these and other responses via a G-protein signaling cascade (Pingret et al., 1998).

Other major changes in the root soon follow these initial responses. For example, Nod factor-dependent cytoskeletal disorganization and subsequent reorganization take place in the cortical cells of the root (Dudley et al., 1987; van Brussel et al., 1992; van Spronsen et al., 1994; Yang et al., 1994; Timmers et al., 1998). These events are likely prerequisites for cortical cell divisions, which are essential to form the nodule. If the bacteria do in fact co-opt and stimulate plant signal transduction pathways, it would be reminiscent of the various effects bacterial pathogens exert on their host cells (Finlay and Cossart, 1997; Finlay and Falkow, 1997). This possibility, along with the known presence of gene clusters or islands and a type III secretion system typical of pathogens (Viprey et al., 1998), makes it reasonable to think of rhizobia as sophisticated parasites instead of true symbionts (Vance, 1983; Djordjevic et al., 1987; Long and Staskawicz, 1993).

The signal transduction pathway leading to the formation of the nodule must have unique characteristics, because nodules are completely different from other normal plant structures. Therefore, it is not surprising that a specific set of plant proteins, called nodulins, are induced exclusively in nodules by infecting rhizobia (Legocki and Verma, 1980). One class of nodulins is induced by rhizobia relatively early in nodule development. The roles of these early nodulins are diverse and far from understood (Long, 1996; Verma and Hong, 1996). One potentially important early nodulin is the rip-1 peroxidase, which is expressed strongly but transiently within just 3 h after inoculation with bacteria (Cook et al., 1995). Another recently isolated early nodulin is a chitinase that appears transiently during early stages of invasion. It may serve to limit the influence of Nod factors by degrading them (Goormachtig et al., 1998). Another class of nodulins are preferentially expressed late in the infection process. Some of these include leghemoglobin (see below) and well-known enzymes involved in nitrogen metabolism unique to the nodule environment, but new late nodulins are being identified at an increasingly rapid pace. For example, a screen for additional late nodulins by using expressed sequence tags (Szczyglowski et al., 1998) has uncovered a protein phosphatase 2C homologue that may play a role in signal transduction in mature nodules (Kapranov et al., 1998). Other potentially important late nodulins include ion pumps that localize to the PBM, one of which will be discussed later.

INVASION AND ROLE OF CELL SURFACE POLYSACCHARIDES

Initial invagination of the infection thread is highly localized and dependent on the bacteria (Brewin, 1991). Infection threads may be formed in a two-step process that includes a Nod factor-induced localized breach of the cell wall, followed by complete degradation of the cell wall by the bacteria (Turgeon and Bauer, 1985; van Spronsen et al., 1994). The first step, the degradation of the cell wall by the plant, may result from a special symbiotic polygalacturonase that is specifically induced in the plant by rhizobial invasion (Muñoz et al., 1998). Several pectinolytic and cellulolytic enzymes synthesized by rhizobia are candidates for the second step (Mateos et al., 1992). Infection threads are not always required for nodulation, however. Certain species of rhizobia can penetrate the nodule by a process called "crack entry" (Rana and Krishnan, 1995).

Cell surface polysaccharides play an important role in several *Rhizobium*-host interactions (Kannenberg and Brewin, 1994; Leigh and Walker, 1994; González et al., 1996b; Niehaus and Becker, 1998). There are three major classes of cell surface polysaccharides: extracellular polysaccharide (EPS), lipopolysaccharide (LPS), and capsule, which includes K antigen (KPS). EPS is the major extracellular material, whereas capsule is associated with but not anchored to the outer membrane; LPS is anchored to the outer membrane and thus remains close to the cell surface. Antigens from both EPS and KPS can reach significant distances away from the cell surface.

The structure and function of rhizobial EPS has been studied most extensively in *S. meliloti*. In this species, EPS consists of two known subtypes, EPS I (succinoglycan) and EPS II. Synthesis of succinoglycan and EPS II is dependent on large clusters of *exo-exs* and *exp* genes, respectively, which mostly reside in two large clusters on the second megaplasmid (the first megaplasmid carries the *nod* genes). Genes encoding two regulators of succinoglycan synthesis, *exoR* and *exoS*, map to the main chromosome. Mutants lacking succinoglycan were first isolated by the inability of colonies to fluoresce on plates containing the laundry whitener Calcofluor. Such mutants still complete Nod factor-dependent events, such as root hair deformation, cortical cell divisions, and infection thread initiation. However, the infection threads soon abort, resulting in small, ineffective nodules lacking bacteroids (Leigh et al., 1985). Using rhizobia expressing green fluorescent protein (GFP), it was recently found that succinoglycan is specifically involved in initiating infection threads and allowing their extension within root hairs (Cheng and Walker, 1998). Therefore, *S. meliloti* appears to require both Nod factor and succinoglycan for invasion. Like Nod factor, succinoglycan is able to function in *trans*, because coinfection of a *nod* mutant with an *exo* mutant results in effective nodules carrying a mixed population of bacteroids (Klein et al., 1988).

As long as succinoglycan is present, EPS II is not required for normal nodulation by *S. meliloti*, and it is actually not normally synthesized in *S. meliloti* Rm1021 except when the *exp* genes are overexpressed (Glazebrook and Walker, 1989) or during phosphate limitation (Zhan et al., 1991). However, in many *S. meliloti* strains lacking succinoglycan, EPS II (which does not stain with Calcofluor) is required to form Fix⁺ nodules on alfalfa. Interestingly, this requirement does not hold for closely related host species that normally are nodulated by wild-type *S. meliloti*. Therefore, EPSs affect host range.

The activity of EPS in *trans* is consistent with other recent evidence that certain forms of EPS constitute a specific bacterial signal to the plant (Battisti et al., 1992; González et al., 1996b). Specifically, low concentrations of a low-molecular-weight class of either succinoglycan or EPS II are required for successful invasion in the absence of EPS; it appears that trimer oligosaccharides are the active signal in the case of succinoglycan (González et al., 1998; Wang et al., 1999). Also consistent with EPS being a specific signal are the findings that small changes in the succinoglycan structure have significant effects on invasion, whereas large alterations in succinoglycan quantities have relatively small effects.

If EPS is a specific bacterial signal for invasion, it may be transient, because there is evidence that EPS synthesis is downregulated in bacteroids. Analysis of *exo-phoA* fusions revealed that most expression was in the infection zone and not in bacteroids (Latchford et al., 1991; Reuber et al., 1991). In addition, histochemical examination of the peribacteroid space failed to detect EPS (Vasse et al., 1990). If EPS is in fact downregulated in bacteroids, it may serve to prevent interference with metabolite exchange with the plant.

Rhizobial LPS also plays a role in the symbiosis, although a specific function for LPS in signaling has not been established and species-specific differences are common. For example, most mutants of *S. meliloti* that are defective in LPS form effective nodules, but *Rhizobium leguminosarum* LPS mutants form empty nodules (Clover et al., 1989; Priefer, 1989). Interestingly, however, one *S. meliloti* LPS mutant able to nodulate alfalfa is defective in a later stage of nodulation of an alternate host plant, *Medicago truncatula*, and elicits a host defense response (Niehaus et al., 1998). This evidence suggests that LPS may be involved in suppressing plant defense reactions to the invading bacteria.

K antigen (KPS) is another important cell surface polysaccharide. It was found that the invasion defect exhibited by *exoB* mutants of *S. meliloti*, which fail to synthesize either succinoglycan or EPS II, could be suppressed by the expression of a low-molecular-weight form of *S. meliloti* K antigen (Petrovics et al., 1993; Reuhs et al., 1993, 1995). This surprising result indicates that a symbiotic form of KPS can substitute for both succinoglycan and EPS II to allow *S. meliloti* invasion and subsequent formation of Fix⁺ bacteroids. This function of K antigen, however, is not essential to form Fix⁺ nodules and can be replaced by EPS (Putnoky et al., 1990). These results suggest that at least in *S. meliloti*, the various surface polysaccharides are functionally redundant, with either EPS I, EPS II, or K antigen alone being able to substitute for the others for successful invasion and differentiation. There is evidence that other rhizobial species can nodulate successfully without EPS (Milner et al., 1992), perhaps because of this postulated redundancy. Interestingly, the expression of KPS in *Rhizobium fredii* has been shown to be regulated by flavonoids, suggesting that KPS may also be a specific signaling molecule (Reuhs et al., 1994). Further studies of the structures and roles of polysaccharides in other rhizobial species will determine whether this apparent redundancy is the exception or the rule.

Immunological studies with monoclonal antibodies have documented changes in LPS structure during bacteroid differentiation and when free-living rhizobia are cultured in different media (de Maagd et al., 1988; Sindhu et al., 1990; Kannenberg et al., 1994; Lucas et al., 1996). Therefore, in studies of LPS, it is necessary to distinguish between symbiotic and nonsymbiotic forms of the molecules in order to determine their roles. At present, it is thought that LPS is more important for the later stages of the symbiosis, including the endocytotic release from infection threads and coordination of bacteroid mass increase and division with the PBM (Brewin, 1991). This is a reasonable hypothesis, considering the intimate cell-cell contact during these stages and the importance of LPS in myxobacterial development, which also requires direct cell-cell contact (Bowden and Kaplan, 1998). In addition, the role of LPS in invasion and colonization by rhizobia may turn out to have similarities to the role of LPS synthesized by true plant pathogens that cause damage to the plant (Leigh and Coplin, 1992).

SUPPRESSION OF HOST DEFENSES

Eukaryotic cells normally try to destroy invading bacteria with their lysosome systems. If rhizobia were true pathogens, then the plant would normally mount a defense response and lyse the bacteria. In fact, some mutants of rhizobia exist that fail to produce a functional nodule, and the result is lysis of the bacteria near the host nucleus (Werner et al., 1985).

Both EPS and LPS may play a role in suppression of the host defense response, because mutants in EPS and LPS that fail to successfully nodulate often induce local plant cell necrosis and other signs of a directed plant defense response to the invading rhizobia (Kannenberg and Brewin, 1994; Niehaus and Becker, 1998; Niehaus et al., 1998). The mechanism by which this occurs is not known. One hypothesis is that the plant alters its gene expression so that controlled invasion can occur. For example, it may produce lower levels of defense proteins or different groups of proteins than normally used for defense (Estabrook and Sengupta-Gopalan, 1991). Interestingly, even in normal infections most infection threads are aborted; this could be due to a basal-level plant defense response, so that only a few infection threads remain in order to keep the bacterial burden low. It is unknown what actually causes most infection threads to abort.

Once trapped in the symbiosome by the PBM, the developing bacteroids are separated from the host plant cytoplasm and therefore become controllable by the plant. If there is a problem in forming the PBM or maintaining its integrity, then a pathogenic response ensues, along with the resulting host defense response. As a result, many bacteroids may be degraded, even early in the symbiosis, as a normal part of turnover (Roth and Stacey, 1991). One attractive hypothesis is that the symbiosome is actually a lysosome in disguise. In this model, the continuous release of ammonia from bacteroids neutralizes the pH of the compartment; bacteroids unable to maintain sufficient ammonia export would be degraded (Kannenberg and Brewin, 1989; Mylona et al., 1995). Premature degradation of bacteroids of some *nif* and *fix* mutants does indeed occur (Hirsch and Smith, 1987).

RELEASE FROM INFECTION THREADS AND ENTRY INTO THE PLANT CELL CYTOPLASM

As the infection threads extend and branch into the developing nodule tissue, the rhizobia spread down the growing threads by growth and division. There is no evidence for any bacterial motility directed toward the infection thread tip. In the first use of GFP to monitor patterns of growth and movement of rhizobia in the plant, it was postulated that only bacteria near the growing tip of the thread are actively multiplying (Gage et al., 1996). If this reflects the actual bacterial population, this observation suggests that cells destined to be released from the thread and become bacteroids arise from a small fraction of the total number of cells within the thread.

Release from the infection thread into the plant cell cytoplasm occurs via an endocytosis-like mechanism. It appears that most bacteria are released from lateral segments, near the growing terminus of the infection thread, that are devoid of plant cell wall material (Newcomb, 1981; Verma, 1992). The droplets containing the bacteria are then engulfed by a plant-derived plasma membrane, such that the bacteria are still external to the plant cytoplasm (Fig. 4).

As a result of the endocytosis, the bacteroids face the outside of the plant membrane. This membrane, the PBM, serves as both a physical barrier between partners and a structure through which metabolites are selectively passed (Verma and Hong, 1996; Verma, 1998). Several putative protein channels have been localized to the PBM, and these proteins are probably involved in selective import and export (Miao et al., 1992; Tyerman et al., 1995). Shortly after bacterial release, PBM synthesis and dynamics must be coordinated with bacterial growth and division within the newly formed symbiosome. Otherwise the proliferation of the released bacteria could overrun the plant's ability to contain them. This coordination may be accomplished via close physical contact between the PBM and the bacteroid membrane (Robertson and Lyttleton, 1984). The synthesis of PBM is probably regulated by alterations in Golgi vesicle trafficking (Robertson et al., 1978b). This is significant, because a very large increase of membrane surface area is required to surround all bacteroids in an infected cell.

A mutant, bacA, was isolated that specifically blocks release of bacteria from the infection thread. Because of this block, the resulting nodules are Fix⁻. The bacA gene was found to encode a homologue of sbmA, a microcin transporter of Escherichia coli (Glazebrook et al., 1993). The similarity between BacA and SbmA was confirmed by showing that E. coli sbmA expression could complement a bacA deletion mutant to allow formation of Fix⁺ nodules (Ichige and Walker, 1997). The functional homology with a microcin uptake system suggests that BacA is involved in import of small molecules during bacteroid differentiation. However, the bacA mutant displayed higher resistance to a variety of aminoglycosides, ethanol, and detergents, suggesting that BacA may instead have a general role in membrane integrity during the formation of the symbiosome. Future studies of bacA, including generation of point mutations that may not perturb the membrane but would still be defective in drug uptake, should shed light on its function. Because of its placement in the bacteroid development pathway between nod mutants and fix mutants, the bacA mutant also serves as a useful molecular handle on understanding intermediate events in differentiation, such as release from the infection thread (see below).

BACTEROID DEVELOPMENT

Once they are engulfed by the PBM to form the symbiosome, major physiological and morphological changes in the bacteria take place that culminate in the differentiation into mature bacteroids. In indeterminate nodules formed by species such as S. meliloti, temporal changes can be visualized as a spatial gradient within the developing nodule (Vasse et al., 1990) and are described below. About 20 to 50% of nodule cells remain uninfected. These cells are involved in transporting and assimilating the nitrogen fixed by the infected cells (Verma, 1998).

In a detailed study of the infection process of S. meliloti, five distinct stages of bacteroid development were defined and correlated with distinct nodule zones (Vasse et al., 1990). These stages are outlined in Fig. 4. Type 1 bacteroids, recently released from infection threads, are found in zone II. These bacteroids are capable of growth and cell division, are rod shaped, and are surrounded by an irregular PBM that perhaps is still in the process of developing. Type 2 bacteroids, also found in nodule zone II, no longer divide but still grow and thus are elongated. The PBM is more tightly associated with the bacteroids and is more regularly shaped. Type 3 bacteroids are found in interzone II-III. These bacteroids no longer elongate, fill most of the host cells, and have smooth PBMs that often contact other PBMs. Type 4 bacteroids are found in the part of zone III nearest zone II. These bacteroids appear to be banded, due to the strong staining of polysomes in distinct segments of the cells. Only type 4 bacteroids are active in nitrogen fixation, as judged by their ability to reduce acetylene. Finally, type 5 bacteroids are found in the part of zone III farthest from the nodule tip. These bacteroids are characterized by a loss of the strong banding pattern and display signs of senescence, particularly in zone IV. Fix⁻ mutants are able to develop up to the type 2 and 3 stages but never reach the type 4 stage.

Interestingly, bacteroid morphologies, which sometimes include X and Y shapes (Fig. 2), are largely dictated by the plant host species. For example, infection of different plant hosts by broad-host-range rhizobia yield strikingly different bacteroid morphologies (Kidby and Goodchild, 1966; Sutton, 1981; Kannenberg and Brewin, 1994). This indicates that the physiological and metabolic constraints imposed by the PBM may globally regulate the bacteroid differentiation program.

Other physical properties of bacteroids also often change dramatically. Their cell walls often become visibly thinner, and changes in the outer membrane are also apparent (MacKenzie et al., 1973; Robertson et al., 1978a; Bal et al., 1980) in many, but not all cases (van Brussel et al., 1979). Increased drug resistance and osmotic sensitivity of bacteroids have also been reported, which probably result from significant membrane alterations (Sutton and Pa-

terson, 1979). The presence or absence of polar granules of poly-β-hydroxybutyrate (PHB), which are thought to be energy storage reserves, are determined by the bacteroid state and depend on the species. For example, free-living *B. japonicum* bacteria lack PHB, but bacteroids accumulate it (Ching et al., 1977). This would make sense if PHB were used for energy storage in bacteroids. On the other hand, *S. meliloti* bacteroids are devoid of PHB granules commonly present in free-living *S. meliloti* (Paau et al., 1980). Therefore, the role of PHB during bacteroid development remains unclear.

Changes in macromolecular composition also occur during the transition to the bacteroid state. The DNA content of the bacteroid nucleoids of several species increases severalfold (Paau et al., 1979), and DNA density and supercoiling also change (Wheatcroft et al., 1990). There is a decline in total RNA and the RNA/DNA ratio, not surprisingly considering that the growth rate of bacteroids is close to zero (Dilworth and Williams, 1967; Sutton, 1974; Bisseling et al., 1979). Nevertheless, the rate of protein synthesis increases (Sutton et al., 1977), which correlates with the dramatic increase in ribosome density in type 4 bacteroids. This suggests that there might be uncoupling of normal growth rate control of ribosome assembly in bacteroids.

Much of the increase in protein-synthetic activity comes from the induction of genes with roles in nitrogen fixation and bacteroid metabolism (see below). However, in addition to the induction of expression of many genes during bacteroid differentiation, there also must be some genes that are turned off. Because Nod factors have such a strong and irreversible effect on nodule morphogenesis at early times in the symbiosis, it would seem to be advantageous for the symbiosis to turn off *nod* genes in bacteroids, and this appears to be the case (Sharma and Signer, 1990; Schlaman et al., 1991). This suppression may be due to inhibition of NodD activity (Schlaman et al., 1992), but there is also recent evidence that one of the alternative NodD proteins, NodD2, is directly involved in repression of *nod* expression in strain NGR234 (Fellay et al., 1998). Other genes active early in the symbiosis also may be turned off. For example, expression of some genes involved in EPS synthesis is suppressed (Latchford et al., 1991; Reuber et al., 1991). In addition, synthesis of SyrM, a regulatory protein mentioned earlier that is activated by NodD, is negatively regulated by SyrB (Barnett and Long, 1997). In contrast, another regulator, SyrA, which stimulates EPS synthesis, is only expressed at high levels in intracellular rhizobia (Barnett et al., 1998). However, SyrA negatively regulates *agpA*, which encodes a sugar transporter, suggesting that *agpA* is downregulated specifically during late stages of the infection (Gage and Long, 1998). One hypothesis is that this sugar transport system is important for free-living growth in the rhizosphere, where there is an abundance of α-galactosides, but that potential downregulation of such a transport system in bacteroids by SyrA may promote bacteroid survival by preventing import of substrates that are specifically toxic at that stage of development.

Bacteroids probably also experience osmotic and pH stresses while in the symbiosome. *S. meliloti* appears to be unusual among nonphotosynthetic bacteria in that it can use sucrose as an osmoprotectant (Gouffi et al., 1998). Bacteroids may counteract osmotic stress by importing other osmoprotectants, such as glycine betaines, and inducing synthesis of endogenous compatible solutes, such as glutamate (Fougere and Le Rudulier, 1990; Talibart et al., 1994). Stresses due to alkaline pH in the symbiosome may be counteracted by a recently discovered K^+ efflux system encoded by the *pha* genes. *pha* mutants are K^+ sensitive and also defective in invasion, suggesting that pH stresses may occur before bacterial release from the infection thread (Putnoky et al., 1998).

Despite the complex interaction with plants required for bacteroid differentiation, bacteroid-like properties can be induced in free-living rhizobia under certain conditions. For example, under specialized culture conditions, nitrogenase can be induced in vegetative cells

(Kurz and La Rue, 1975; McComb et al., 1975; Pagan et al., 1975). Moreover, when free-living rhizobia are grown in a layer of soft agar medium, cells in the middle microaerobic layer of this agar synthesize nitrogenase and also assume bacteroid-like morphologies (Pankhurst and Craig, 1978). Rhizobia grown on various organic acids, such as succinate, also assume bacteroid-like forms, characterized by branched, swollen cells (Jordan and Coulter, 1965; Urban and Dazzo, 1982). Not all cultured rhizobia that are bacteroid-like in appearance are capable of fixing nitrogen, indicating that the morphological changes can occur without symbiotic function (Kaneshiro et al., 1983).

The branching and swelling of many bacteroids during their differentiation may be simply analogous to filamentation by *E. coli* and other rod-shaped bacteria, which occurs by default when cell mass increases but cell division stops. In support of this idea, inhibiting DNA replication and cell division of several rhizobial species, even in normal rich growth medium, causes induction of bacteroid-like branching (Kaneshiro et al., 1983; Latch and Margolin, 1997). Interestingly, *Agrobacterium* and *Caulobacter* species, other members of the α-proteobacterial group, also branch in response to cell cycle perturbation (Latch and Margolin, 1997; Wright et al., 1997). How branches are formed in normally rod-shaped cells is not known, but one possible explanation is that this group of bacteria grow by polar extension, similar to fission yeast (Verde, 1998), and that the polar-extension mechanism is regulated by cell cycle cues. The mechanism by which the host plant may exert specific control over the cell cycle of bacteroids is not known. A second, nonessential homologue of the cell division gene *ftsZ* might be such a target in *S. meliloti* (Margolin and Long, 1994).

METABOLITE EXCHANGE WITH THE PLANT

The PBM is a selective barrier, allowing certain metabolites to travel between host and symbiont. These metabolites include factors required for bacteroid respiration and nitrogen fixation, as well as byproducts of nitrogen fixation that are exported to the plant. There are many factors required by bacteroids, such as iron and molybdenum for the nitrogenase enzyme, a carbon source, an energy source, and oxygen for aerobic respiration (Dilworth and Glenn, 1984). Although little is known about how the metals are imported, the transport of carbon sources and oxygen is better understood (see below).

The byproduct of nitrogen fixation, ammonia, is not assimilated by bacteroids (Brown and Dilworth, 1975; Ludwig and Signer, 1977). Recent data suggest that this occurs at least in part because the expression of genes for ammonia assimilation is downregulated in bacteroids by inhibiting the activity of the transcriptional activator NtrC (Tate et al., 1998). Interestingly, *amtB* is a highly conserved putative ammonium transporter that may facilitate movement of ammonia by stimulating the interconversion of NH_3 and NH_4^+ (van Heeswijk et al., 1996; Soupéne et al., 1998).

After ammonia leaves the bacteroid, it presumably moves from the symbiosome space into the plant cells via the PBM. Recently, a PBM ammonia-specific transporter has been cloned and characterized (Tyerman et al., 1995; Kaiser et al., 1998). It is not yet established whether this protein acts passively as an ion channel or actively as a true transporter. It is also not known whether other yet-uncharacterized ammonium transporters or channels exist in the PBM.

It is thought that after ammonia is released into the plant cell, the ammonia is assimilated by the action of glutamine synthetase and glutamate synthetase. However, the hypothesis that ammonia is the primary means of transporting fixed nitrogen out of bacteroids has recently come into question. Experiments on isolated bacteroids by using radiolabeled nitrogen indicate that instead, alanine may be the major exported form of nitrogenase-fixed nitrogen in *B. japonicum* bacteroids (Waters et al., 1998). It was found that excretion of alanine depends on nitrogenase activity and increases linearly with time. Because glutamate is the major nitrogenous compound resulting from nitrogen fixation in plants, alanine might represent a convenient and stable vehicle for the transport of nitrogen to the plant. If it is true that ammonia is not transported in large amounts, then the relevance of the recently discovered PBM ammonia transporter is less clear.

Bacteroids are totally dependent on the plant for nutrients, both for respiration and for nitrogen fixation (Dilworth and Glenn, 1984). Because nitrogen fixation is very energy intensive, large amounts of nutrients must be obtained from stores inside the bacteroids and from nutrients transported from the plant. C_4 dicarboxylic acids such as malate, which is derived from catabolism of sucrose in the leaves of the host plant, are major energy sources for bacteroids. Once inside the bacteroid cells, malate and other dicarboxylic acids are probably shunted through the Krebs cycle via acetylcoenzyme A. As with ammonium export, bacteroids contain transporters for dicarboxylic acid import. In *S. meliloti*, *dctA* is essential for dicarboxylic acid transport and *dctA* mutants are Fix⁻ (Long et al., 1988; Engelke et al., 1989; Yarosh et al., 1989). In free-living rhizobia, *dctA* expression is regulated by the two-component regulators DctB and DctD. However, during symbiosis these regulators appear to act only early in the infection process in the infection thread, whereas another regulator appears to be responsible for *dctA* expression in mature bacteroids (Boesten et al., 1998). The presence of a second regulatory system in later stages is consistent with the increased requirement for dicarboxylic acids in mature bacteroids actively fixing nitrogen.

Other nutrients are also required by bacteroids. For example, the *S. meliloti trpE(G)* locus encodes an enzyme in anthranilate biosynthesis that is expressed in nodules and is necessary for successful symbiosis (Barsomian et al., 1992). Aromatic amino acid biosynthesis mutants that are blocked at or before *trpE(G)* either form normal nodules or form detective nodules lacking bacteroids. In both cases, however, the invasion zones are unusually elongated. Because blocking later steps in tryptophan biosynthesis has no effect on symbiosis, anthranilate itself appears to be important and perhaps acts as an iron scavenger.

REGULATION OF NITROGEN FIXATION

Fix⁻ nodules generated by *nif* mutants can still harbor normally differentiated bacteroids, indicating, as do the bacteroid-like cells in culture mentioned above, that morphological differentiation of bacteroids can occur independently of nitrogen fixation (Dickstein et al., 1988). The *nif* and *fix* genes required for symbiotic nitrogen fixation encode a variety of proteins involved in the late stages of the symbiosis, including nitrogenase and its cofactors (Fischer, 1994). The regulation of the nitrogen fixation genes is the best understood aspect of gene regulation during bacteroid differentiation.

A major regulator of genes and proteins involved in nitrogen fixation is the level of oxygen (Batut and Boistard, 1994). Nitrogenase is sensitive to oxygen, and yet ATP production by the obligately aerobic nitrogen-fixing bacteroids requires oxygen for oxidative phosphorylation. This problem has been solved in three ways: (i) sequestration of the nitrogen fixation zone inside the root cortex, which acts as a diffusion barrier for oxygen; (ii) synthesis of a high-efficiency terminal oxidase; and (iii) the exploitation of an oxygen carrier protein (Fischer, 1996). This carrier, leghemoglobin, is present in the plant cells and is the legume equivalent of animal hemoglobin. Leghemoglobin serves to maintain a very low level of free oxygen in the symbiosome and at the same time efficiently transports this controlled level of oxygen to the bacteroids for respiration.

Leghemoglobin consists of two parts: the leghemoglobin apoprotein and the heme moiety, which must be synthesized separately. The apoprotein is clearly synthesized by the plant host (Sedloi-Lumbroso et al., 1978; Jensen et al., 1981). For many years, it was thought that the heme moiety of leghemoglobin was synthesized by the rhizobial partner, making

for a remarkable case of symbiotic synthesis of an enzyme (O'Brian et al., 1987; Sangwan and O'Brian, 1991). However, recent evidence suggests that this idea is incorrect and that the plant may synthesize most or all of the leghemoglobin needed for the symbiosis (O'Brian, 1996; Santana et al., 1998). The bacterial heme may have a regulatory role in bacteroid development, perhaps in part by influencing the activity of FixL (see below).

Oxygen levels ultimately regulate the expression of nitrogen fixation genes via two regulatory systems. One is the two-component system FixLJ of *S. meliloti*, of which the FixL sensor kinase is a hemoprotein that probably senses oxygen directly (Gilles-González et al., 1991). FixL autophosphorylates and also transfers a phosphate to the FixJ response regulator, which acts to induce transcription of FixK (see below). The other regulator is NifA, a homologue of the σ^{54}-associated response regulator NtrC, which is redox sensitive (Fischer, 1996). It appears that a cascade of regulators is involved in expression of nitrogen fixation genes, analogous to the regulatory cascades found in other prokaryotic developmental systems, such as that of *Bacillus subtilis* (David et al., 1988). However, the complexity of the regulatory network leading to expression of nitrogen fixation genes is becoming more apparent as different rhizobial species are studied. For example, whereas FixL of *S. meliloti* is a hemoprotein, its homologue in *Rhizobium etli* lacks heme and hence is unlikely to sense oxygen directly (D'Hooghe et al., 1998). *R. etli* also harbors two divergent homologues of NifA which appear to have different roles in symbiosis (Michiels et al., 1998).

One member of the rhizobial regulatory cascade dependent on low oxygen concentrations is FixK, a member of the Fnr family of oxygen-dependent regulators (Batut et al., 1989). Expression of *fixK* is induced by low oxygen via the FixLJ system (Batut and Boistard, 1994). FixK, in turn, activates expression of several operons, including *fixGHIS* and *fixNOQP*, which together encode the terminal oxidase essential for rhizobial respiration during the microaerobic conditions of the symbiosis (Preisig et al., 1993). The role of FixK is complex. It exerts feedback control on its own expression via FixT, represses FixLJ, and in *B. japonicum* activates a second *fixK* homologue (Foussard et al., 1997; Nellen-Anthamatten et al., 1998). The second FixK appears to be a bacteroid-specific transcriptional regulator (Durmowicz and Maier, 1998). The tight coupling of oxygen control of gene expression with location in the nodule can be uncoupled by constitutive expression of nitrogen fixation genes or by incubating the nodule under low-oxygen conditions (Soupéne et al., 1995).

Regulation of nitrogen metabolism in rhizobia may also influence the success of nitrogen fixation and bacteroid differentiation. PII protein, encoded by the *glnB* gene, is normally uridylylated under conditions of limiting nitrogen; the uridylylated form of PII is responsible for activating glutamine synthetase I, a key enzyme in the assimilation of ammonium. Bacteroids of *S. meliloti* mutants with *glnB* deleted are able to synthesize nitrogenase but are nevertheless still nitrogen starved and display delays in infection and defects in bacteroid differentiation (Arcondeguy et al., 1997). It is thought that PII affects both rhizobial infection and bacteroid differentiation in addition to its predicted role in nitrogen metabolism, perhaps by regulating ammonium export from bacteroids.

ARE BACTEROIDS TERMINALLY DIFFERENTIATED?

The nodule and the bacteroids within them function at peak capacity for several weeks and then degrade. Bacteroid membranes disappear, and the cell wall undergoes major morphological alterations. The destruction of the membrane may result in part from oxygen radicals transferred from leghemoglobin to the membrane, resulting in lipid peroxidation (Moreau et al., 1996). Other factors are likely to play a role in bacteroid disintegration, but the end result is that most bacteroids probably perish.

Historically, there have been conflicting reports about the fate of bacteroids. Early studies based on micromanipulation indicated clearly that mature bacteroids were nonviable (Almon, 1933). Mature bacteroids from nodules were unable to multiply on various media and were incapable of renodulating a host plant. The conclusions from this and other studies as well as ultrastructural studies that showed physical degradation of the bacteroids were that bacteroids are terminally differentiated. This view persisted until the late 1970s, when two groups claimed that 90% of isolated bacteroids were viable when incubated in media containing osmoprotectants (Gresshoff et al., 1977; Tsien et al., 1977). These results suggested that a possible flaw in previous studies showing no viability was that the bacteroids were cultured on the wrong media for regrowth, too different from the osmotically distinct environment of the symbiosome. However, it was also possible that the "viable bacteroids" were not bacteroids at all but instead were free-living bacteria that still persisted within infection threads or were undifferentiated bacteria residing inside plant cells. In an attempt to circumvent these problems, bacteroids of *B. japonicum* were purified from nodule cell protoplasts, thus separating them from extracellular bacteria. These bacteroids were able to form colonies on conventional media supplemented with an osmoprotectant (Gresshoff and Rolfe, 1978).

Nevertheless, the possibility still remained that the colony formation was due to growth of viable, undifferentiated *B. japonicum* cells present inside the plant cells. This was addressed by a careful electron microscopic examination of *S. meliloti* bacteroids at various stages of maturity and senescence (Paau et al., 1980). This study demonstrated the presence of "vegetative" rhizobia directly adjacent to enlarged, degraded bacteroids in the senescent zone of the nodules. These rhizobia were suggested to be vegetative because, like true free-living forms, they were small (2 μm in length), were not surrounded by a PBM, and contained PHB granules, which are present in free-living *S. meliloti* cells but not in bacteroids. It is difficult to imagine how such small rhizobia with vegetative morphology could have arisen from the large, distorted, degraded bacteroids, because no intermediate morphological stage has ever been documented and because division of the obviously degraded bacteroids is highly unlikely. Instead, it is more likely that these vegetative forms were released from infection threads that extended into disorganized plant cells in the senescent zone IV and that these disorganized cells failed to surround the bacteria with a PBM.

A complication in the study of rhizobial differentiation is the diversity of bacteroid morphologies and presumably differentiation mechanisms among different species. For example, one possible explanation for many conflicting results is that bacteroids in certain species, such as *B. japonicum*, which does not undergo major morphological changes, might be capable of redifferentiation. Moreover, the precise stage at which a vegetative rhizobial cell becomes a differentiated bacteroid remains unclear. Despite the obvious change between a free-living cell in the infection thread and a cell inside the PBM, it is likely that bacteria isolated at this stage would have a greater chance of regrowth than cells which have undergone morphological changes to become mature bacteroids. This question was directly addressed by time-lapse microphotography of the growth of bacteroids and free-living cells isolated from nodules of clover and soybeans. The results clearly indicated that while small rod-shaped bacteria could grow, bacteroids could not (Zhao et al., 1985). Isolation of *S. meliloti* bacteroids by Percoll gradients to mitigate osmotic shock upon isolation (Reibach et al., 1981) also provided strong evidence that mature bacteroids are no longer capable of regrowth (McRae et al., 1989).

In the rhizobia-legume symbiosis, terminal differentiation is not limited to the rhizobia. It also occurs in the cells of the host and results in the degradation of the nodule. This raises interesting and important evolutionary and ecological issues. In the plant host, the senescence and degradation of the nodules appear outwardly to reflect programmed cell death. Whether this process is mechanistically equiva-lent to apoptosis remains to be determined (Hochman, 1997). However, the death of the nodule does not affect the reproductive capacity of the plant because the plant germ line remains intact. Therefore, there should be no selective pressure against nodule degradation. On the other hand, it is logical to assume that the selection pressure on an individual invading rhizobial cell would be to prevent its differentiation into a bacteroid or, failing that, to allow persistence of the bacteroid within the plant cells. Because phylogenetic analysis indicates that some mitochondria are most closely related to α-proteobacteria such as rhizobia, it is reasonable to suppose that such organelles could have arisen from a rhizobial endosymbiont that somehow persisted within the host cell by synchronizing its cell cycle with that of the host (Verma and Long, 1983).

What about selection pressure against entering plant cells? In the infection thread, it appears that bacteroids mainly originate from bacteria near the leading edge of tip growth (Gage et al., 1996). This observation suggests that a complex selection process must be at work. If all rhizobia stayed in the back of the thread to avoid becoming bacteroids, then presumably no bacteroids would be formed. This would harm not only the plant but also the rhizobial population in the soil around the root (rhizosphere) by withholding nutrients derived from fixed nitrogen. Indeed, although many bacteroids die in the act of nitrogen fixation, it is thought that their fixation of nitrogen benefits far greater numbers of rhizobia present in the rhizosphere (Reyes and Schmidt, 1979). For example, it is likely that large numbers of rhizobia arise from the release of nondifferentiated bacteria from infection threads and senescent nodules. Moreover, the presence of more healthy plants and associated rhizobia increases the amount of potentially metabolizable compounds secreted by their roots. One such class of compounds are the opine-like rhizopines, which are synthesized and catabolized by rhizobia. Interestingly, rhizopine synthesis in *S. meliloti* is upregulated late in the symbiosis and may increase the survival chances of nondifferentiated rhizobia (Murphy et al., 1988). If this type of cross-feeding occurs, then rhizobia can be viewed more as parasites that sacrifice some of their own for the benefit of the population. The mechanism by which the rhizobia are partitioned into invading and noninvading and into bacteroid and nonbacteroid is of great ecological significance and awaits further studies. Perhaps a diffusible factor, such as the peptide recently implicated in cyanobacterial heterocyst differentiation (Yoon and Golden, 1998), will be found to be involved in this partitioning process.

NEW STRATEGIES TO ISOLATE GENES INVOLVED IN DIFFERENTIATION

The isolation of *nod* and *fix* mutants has been extremely important in the elucidation of genes important for the early and late stages of symbiosis. As with other bacterial differentiation systems, these mutants were obtained by mutagenesis and then screening for arrest of development at a particular stage. The Nod⁻ phenotype is the absence of root nodules on a susceptible plant, and the Fix⁻ phenotype is the presence of white, often small nodules instead of pink nodules. While these phenotypes are fairly straightforward, screening for mutants blocked in the intermediate stages, such as release of bacteria from the infection thread or bacteroid development independent of nitrogen fixation, is considerably more challenging. In fact, the number of genes required for nitrogen fixation renders a screen for other defects in bacteroid development by visual inspection of nodules potentially very frustrating, because most of the mutations will be in *nif* and *fix* genes already discovered. Nevertheless, several new strategies hold promise for discovering new genes involved in bacteroid development.

The first strategy involves utilizing genome sequences to find potential genes. This approach has been hastened by the completion of the sequence of the symbiotic megaplasmid of strain NGR234 (Freiberg et al., 1997). As the list of known genes and their functions grows, the likelihood of inferring the function of a rhizobial gene from its sequence increases as well. Such potential differentiation genes can also be identified by upstream regulatory sequences that bind bacteroid-specific regulators. For example, a search for σ⁵⁴ promoters or *nod* boxes by genome "mining" may turn up previously unknown genes that are induced upon entry into the plant. These genes can then be knocked out by reverse genetics, and the phenotype can be assessed. In addition, new genomic data might allow better characterization of interesting mutations in bacteroid development that were isolated prior to the advent of genomics (Noel et al., 1982; Putnoky et al., 1988).

Another strategy involves using reporter fusions to find genes whose expression is regulated during a specific developmental stage. Whereas employing this approach in situ would be very laborious, another approach is to use reporters to identify genes that are regulated by stage-specific factors in free-living rhizobia. In one strategy, Tn*phoA* can be used to generate reporter fusions to periplasmic proteins (Manoil and Beckwith, 1985). The first Tn*phoA* system for rhizobia was developed in *S. meliloti* (Long et al., 1988), and a series of Fix⁻ mutants were isolated that were defective in late stages of the symbiosis. A more recent study more fully exploited the properties of Tn*phoA* by searching for random *phoA* fusions specifically induced by several known symbiotic regulators, such as NodD, SyrM, and SyrA (Gage and Long, 1998). This strategy was used to identify *agpA*, a periplasmic galactoside permease that is downregulated in the nodule by SyrA (see above).

A third strategy that holds great promise is promoter trapping or in vivo expression technology, which has been used to identify bacterial genes that are specifically expressed in infected animals (Mahan et al., 1993; Slauch et al., 1994; Heithoff et al., 1997). As adapted to bacteroid differentiation, the approach is to clone a library of *S. meliloti* genomic DNA fragments upstream of a promoterless gene, the expression of which is essential for bacteroid development and Fix⁺ nodules. This gene itself would be fused to a reporter, such as *gusA*, which encodes β-glucuronidase and can be assayed colorimetrically. In an initial screen, fusions that had high constitutive activities would be discarded. The fusions with low or no activities would then be used to infect alfalfa. Such a strategy has recently been used with *bacA* as the promoterless gene (Oke and Long, 1999). Because *bacA* is essential for bacterial release from infection threads, screening for promoter-*bacA* fusions that allow a Δ*bacA* strain to make Fix⁺ nodules should enrich for cloned promoters that are specifically expressed in the nodule. A search downstream from these promoters should uncover genes that are specifically expressed while bacteria are in the infection thread or in the process of differentiating into bacteroids. Although some of the genes isolated by this method have turned out to be *nif* or *fix* genes, other genes with potentially novel functions have been recovered from this screen (Oke and Long, 1999).

Knockout strategies are useful because it is easy to map and clone the affected genes and because many functions in bacteroid development are probably not essential for free-living cells. This approach has some limitations, however. First, gene redundancy is common in rhizobia. For example, multiple homologues of the chaperonin GroEL, the cell division protein FtsZ, and the alternative sigma factor RpoN (encoding σ⁵⁴) have been found in individual species (Margolin and Long, 1994; Ogawa and Long, 1995; Michiels et al., 1998). Therefore, knocking out one gene in an important gene family may not yield a phenotype or may result in a partial phenotype. When considered in the context of the complexity of the process and the variability due to the plant system, partial phenotypes can sometimes be difficult to sort out. Second, some genes involved in bacteroid development are bound to also be important for free-living growth, and knockout mutations of these genes may not be viable. Therefore, another potentially fruitful strategy is to isolate conditional mutants, such as temperature-sensitive mutants.

The advent of these different genetic strate-gies opens a new chapter in the study of bacteroid differentiation. Combining these tools with the ability to visualize fluorescent rhizobia within a living nodule by the use of GFP should yield much new information that was not previously accessible. Whereas previous studies have focused on describing the process, the new genetic and cytological tools that have emerged in the last few years should allow a more detailed dissection of the determinants involved in forming these unique bacterial cell types.

REFERENCES

Albrecht, C., R. Geurts, and T. Bisseling. 1999. Legume nodulation and mycorrhizae formation; two extremes in host specificity meet. *EMBO J.* **18**:281–288.

Almon, L. 1933. Concerning the reproduction of bacteroids. *Zentbl. Bakteriol. Parasitenkol. Infectionskr. Hyg. Abt. II* **87**:289–297.

Ames, P., and K. Bergman. 1981. Competitive advantage provided by bacterial motility in the formation of nodules by *Rhizobium meliloti*. *J. Bacteriol.* **148**:728–729.

Arcondeguy, T., I. Huez, P. Tillard, C. Gangneux, F. de Billy, A. Gojon, G. Truchet, and D. Kahn. 1997. The *Rhizobium meliloti* PII protein, which controls bacterial nitrogen metabolism, affects alfalfa nodule development. *Genes Dev.* **11**: 1194–1206.

Armitage, J. P., and R. Schmitt. 1997. Bacterial chemotaxis: *Rhodobacter sphaeroides* and *Sinorhizobium meliloti*—variations on a theme? *Microbiology* **143**:3671–3682.

Bal, A. K., S. Shantharam, and D. P. Verma. 1980. Changes in the outer cell wall of *Rhizobium* during development of root nodule symbiosis in soybean. *Can. J. Microbiol.* **26**:1096–1103.

Barnett, M. J., and S. R. Long. 1990. DNA sequence and translational product of a new nodulation-regulatory locus: *syrM* has sequence similarity to NodD proteins. *J. Bacteriol.* **172**:3695–3700.

Barnett, M. J., and S. R. Long. 1997. Identification and characterization of a gene on *Rhizobium meliloti* pSyma, *syrB*, that negatively affects *syrM* expression. *Mol. Plant-Microbe Interact.* **10**:550–559.

Barnett, M. J., J. A. Swanson, and S. R. Long. 1998. Multiple genetic controls on *Rhizobium meliloti syrA*, a regulator of exopolysaccharide abundance. *Genetics* **148**:19–32.

Barsomian, G. D., A. Urzainqui, K. Lohman, and G. C. Walker. 1992. *Rhizobium meliloti* mutants unable to synthesize anthranilate display a novel symbiotic phenotype. *J. Bacteriol.* **174:** 4416–4426.

Battisti, L., J. C. Lara, and J. A. Leigh. 1992. Specific oligosaccharide form of the *Rhizobium meliloti* exopolysaccharide promotes nodule invasion in alfalfa. *Proc. Natl. Acad. Sci. USA* **89:**5625–5629.

Batut, J., and P. Boistard. 1994. Oxygen control in *Rhizobium. Antonie Leeuwenhoek* **66:**129–150.

Batut, J., M. L. Daveran-Mingot, M. David, J. Jacobs, A. M. Garnerone, and D. Kahn. 1989. *fixK*, a gene homologous with *fnr* and *crp* from *Escherichia coli*, regulates nitrogen fixation genes both positively and negatively in *Rhizobium meliloti. EMBO J.* **8:**1279–1286.

Bisseling, T., R. C. van den Bos, M. W. Weststrate, M. J. Hakkaart, and A. van Kammen. 1979. Development of the nitrogen-fixing and protein-synthesizing apparatus of bacteroids in pea root nodules. *Biochim. Biophys. Acta* **562:**515–526.

Bladergroen, M. R., and H. P. Spaink. 1998. Genes and signal molecules involved in the rhizobia-Leguminoseae symbiosis. *Curr. Opin. Plant Biol.* **1:**353–359.

Boesten, B., J. Batut, and P. Boistard. 1998. DctBD-dependent and -independent expression of the *Sinorhizobium (Rhizobium) meliloti* C4-dicarboxylate transport gene (*dctA*) during symbiosis. *Mol. Plant-Microbe Interact.* **11:**878–886.

Bohlool, B. B., and E. L. Schmidt. 1976. Immunofluorescent polar tips of *Rhizobium japonicum*: possible site of attachment or lectin binding. *J. Bacteriol.* **125:**1188–1194.

Bowden, M. G., and H. B. Kaplan. 1998. The *Myxococcus xanthus* lipopolysaccharide O-antigen is required for social motility and multicellular development. *Mol. Microbiol.* **30:**275–284.

Brewin, N. J. 1991. Development of the legume root nodule. *Annu. Rev. Cell Biol.* **7:**191–226.

Brewin, N. J., and I. V. Kardailsky. 1997. Legume lectins and nodulation by *Rhizobium. Trends Plant Sci.* **2:**92–98.

Brown, C. M., and M. J. Dilworth. 1975. Ammonia assimilation by rhizobium cultures and bacteroids. *J. Gen. Microbiol.* **86:**39–48.

Caetano-Anolles, G., D. K. Crist-Estes, and W. D. Bauer. 1988. Chemotaxis of *Rhizobium meliloti* to the plant flavone luteolin requires functional nodulation genes. *J. Bacteriol.* **170:**3164–3169.

Cheng, H.-P., and G. C. Walker. 1998. Succinoglycan is required for initiation and elongation of infection threads during nodulation of alfalfa by *Rhizobium meliloti. J. Bacteriol.* **180:**5183–5191.

Ching, T. M., S. Hedtke, and W. Newcomb. 1977. Isolation of bacteria, transforming bacteria, and bacteroids from soybean nodules. *Plant Physiol.* **60:**771–774.

Clover, R. H., J. Kieber, and E. R. Signer. 1989.

Lipopolysaccharide mutants of *Rhizobium meliloti* are not defective in symbiosis. *J. Bacteriol.* **171:** 3961–3967.

Cook, D., D. Dreyer, D. Bonnet, M. Howell, E. Nony, and K. VandenBosch. 1995. Transient induction of a peroxidase gene in *Medicago trunculata* precedes infection by *Rhizobium meliloti. Plant Cell* **7:**43–55.

Cren, M., A. Kondorosi, and E. Kondorosi. 1995. NolR controls expression of the *Rhizobium meliloti* nodulation genes involved in the core Nod factor synthesis. *Mol. Microbiol.* **15:**733–747.

David, M., M. L. Daveran, J. Batut, A. Dedieu, O. Domergue, J. Ghai, C. Hertig, P. Boistard, and D. Kahn. 1988. Cascade regulation of *nif* gene expression in *Rhizobium meliloti. Cell* **54:**671–683.

Dazzo, F. B., G. L. Truchet, J. E. Sherwood, E. M. Hrabak, M. Abe, and S. H. Pankratz. 1984. Specific phases of root hair attachment in the *Rhizobium trifolii*-clover symbiosis. *Appl. Environ. Microbiol.* **48:**1140–1150.

de Maagd, R. A., C. A. Wijffelman, E. Pees, and B. J. Lugtenberg. 1988. Detection and subcellular localization of two Sym plasmid-dependent proteins of *Rhizobium leguminosarum* biovar viciae. *J. Bacteriol.* **170:**4424–4427.

Demont, N., M. Ardourel, F. Maillet, D. Promé, M. Ferro, J. C. Promé, and J. Dénarié. 1994. The *Rhizobium meliloti* regulatory *nodD3* and *syrM* genes control the synthesis of a particular class of nodulation factors N-acylated by (omega-1)-hydroxylated fatty acids. *EMBO J.* **13:**2139–2149.

Dénarié, J., F. Debelle, and J. C. Promé. 1996. *Rhizobium* lipo-chitooligosaccharide nodulation factors: signaling molecules mediating recognition and morphogenesis. *Annu. Rev. Biochem.* **65:** 503–535.

D'Hooghe, I., J. Michiels, and J. Vanderleyden. 1998. The *Rhizobium etli* FixL protein differs in structure from other known FixL proteins. *Mol. Gen. Genet.* **257:**576–580.

Dickstein, R., T. Bisseling, V. N. Reinhold, and F. M. Ausubel. 1988. Expression of nodule-specific genes in alfalfa root nodules blocked at an early stage of development. *Genes Dev.* **2:**677–687.

Dilworth, M., and A. Glenn. 1984. How does a legume nodule work? *Trends Biochem. Sci.* **9:** 519–523.

Dilworth, M., and D. C. Williams. 1967. Nucleic acid changes in bacteroids of *Rhizobium lupini* during nodule development. *J. Gen. Microbiol.* **48:** 31–36.

Djordjevic, M. A., D. W. Gabriel, and B. G. Rolfe. 1987. *Rhizobium*—the refined parasite of legumes. *Annu. Rev. Phytopathol.* **25:**145–168.

Dudley, M. E., T. W. Jacobs, and S. R. Long. 1987. Microscopic studies of cell divisions induced in alfalfa roots by *Rhizobium meliloti*. *Planta* **171:** 289–301.

Durmowicz, M. C., and R. J. Maier. 1998. The FixK2 protein is involved in regulation of symbiotic hydrogenase expression in *Bradyrhizobium japonicum*. *J. Bacteriol.* **180:**3253–3256.

Ehrhardt, D. W., E. M. Atkinson, and S. R. Long. 1992. Depolarization of alfalfa root hair membrane potential by *Rhizobium meliloti* Nod factors. *Science* **256:**998–1000.

Ehrhardt, D. W., R. Wais, and S. R. Long. 1996. Calcium spiking in plant root hairs responding to *Rhizobium* nodulation signals. *Cell* **85:**673–681.

Engelke, T., D. Jording, D. Kapp, and A. Pühler. 1989. Identification and sequence analysis of the *Rhizobium meliloti dctA* gene encoding the C4-dicarboxylate carrier. *J. Bacteriol.* **171:**5551–5560.

Estabrook, E. M., and C. Sengupta-Gopalan. 1991. Differential expression of phenylalanine ammonia-lyase and chalcone synthase during soybean nodule development. *Plant Cell* **3:**299–308.

Fellay, R., M. Hanin, G. Montorzi, J. Frey, C. Freiberg, W. Golinowski, C. Staehelin, W. J. Broughton, and S. Jabbouri. 1998. *nodD2* of *Rhizobium* sp. NGR234 is involved in the repression of the *nodABC* operon. *Mol. Microbiol.* **27:** 1039–1050.

Felle, H. H., E. Kondorosi, A. Kondorosi, and M. Schultze. 1998. The role of ion fluxes in Nod factor signalling in *Medicago sativa*. *Plant J.* **13:** 455–464.

Finan, T. M., A. M. Hirsch, J. A. Leigh, E. Johansen, G. A. Kuldau, S. Deegan, G. C. Walker, and E. R. Signer. 1985. Symbiotic mutants of *Rhizobium meliloti* that uncouple plant from bacterial differentiation. *Cell* **40:**869–877.

Finlay, B. B., and P. Cossart. 1997. Exploitation of mammalian host cell functions by bacterial pathogens. *Science* **276:**718–725.

Finlay, B. B., and S. Falkow. 1997. Common themes in microbial pathogenicity revisited. *Microbiol. Mol. Biol. Rev.* **61:**136–169.

Fischer, H. M. 1994. Genetic regulation of nitrogen fixation in rhizobia. *Microbiol. Rev.* **58:**352–386.

Fischer, H. M. 1996. Environmental regulation of rhizobial symbiotic nitrogen fixation genes. *Trends Microbiol.* **4:**317–320.

Flores, M., P. Mavingui, L. Girard, X. Perret, W. J. Broughton, E. Martinez-Romero, G. Davila, and R. Palacios. 1998. Three replicons of *Rhizobium* sp. strain NGR234 harbor symbiotic gene sequences. *J. Bacteriol.* **180:**6052–6053.

Fougere, F., and D. Le Rudulier. 1990. Uptake of glycine betaine and its analogues by bacteroids of *Rhizobium meliloti*. *J. Gen. Microbiol.* **136:**157–163.

Foussard, M., A.-M. Garnerone, F. Ni, E. Soupéne, P. Boistard, and J. Batut. 1997. Negative autoregulation of the *Rhizobium meliloti fixK* gene is indirect and requires a newly identified regulator, FixT. *Mol. Microbiol.* **25:**27–37.

Freiberg, C., R. Fellay, A. Bairoch, W. J. Broughton, A. Rosenthal, and X. Perret. 1997. Molecular basis of symbiosis between *Rhizobium* and legumes. *Nature* **387:**394–401.

Gage, D. J., and S. R. Long. 1998. α-Galactoside uptake in *Rhizobium meliloti*: isolation and characterization of *agpA*, a gene encoding a periplasmic binding protein required for melibiose and raffinose utilization. *J. Bacteriol.* **180:**5739–5748.

Gage, D. J., T. Bobo, and S. R. Long. 1996. Use of green fluorescent protein to visualize the early events of symbiosis between *Rhizobium meliloti* and alfalfa (*Medicago sativa*). *J. Bacteriol.* **178:**7159–7166.

Gilles-González, M. A., G. S. Ditta, and D. R. Helinski. 1991. A haemoprotein with kinase activity encoded by the oxygen sensor of *Rhizobium meliloti*. *Nature* **350:**170–172.

Glazebrook, J., and G. C. Walker. 1989. A novel exopolysaccharide can function in place of the calcofluor-binding exopolysaccharide in nodulation of alfalfa by *Rhizobium meliloti*. *Cell* **56:**661–672.

Glazebrook, J., A. Ichige, and G. C. Walker. 1993. A *Rhizobium meliloti* homolog of the *Escherichia coli* peptide-antibiotic transport protein SbmA is essential for bacteroid development. *Genes Dev.* **7:**1485–1497.

González, J. E., B. L. Reuhs, and G. C. Walker. 1996a. Low molecular weight EPS II of *Rhizobium meliloti* allows nodule invasion in *Medicago sativa*. *Proc. Natl. Acad. Sci. USA* **93:**8636–8641.

González, J. E., G. M. York, and G. C. Walker. 1996b. *Rhizobium meliloti* exopolysaccharides: synthesis and symbiotic function. *Gene* **179:**141–146.

González, J. E., C. E. Semino, L. X. Wang, L. E. Castellano-Torres, and G. C. Walker. 1998. Biosynthetic control of molecular weight in the polymerization of the octasaccharide subunits of succinoglycan, a symbiotically important exopolysaccharide of *Rhizobium meliloti*. *Proc. Natl. Acad. Sci USA* **95:**13477–13482.

Goormachtig, S., S. Lievens, W. Van de Velde, M. Van Montagu, and M. Holsters. 1998. Srchi13, a novel early nodulin from *Sesbania rostrata*, is related to acidic class III chitinases. *Plant Cell* **10:** 905–915.

Gottfert, M., P. Grob, and H. Hennecke. 1990. Proposed regulatory pathway encoded by the *nodV* and *nodW* genes, determinants of host specificity in *Bradyrhizobium japonicum*. *Proc. Natl. Acad. Sci. USA* **87:**2680–2684.

Gotz, R., and R. Schmitt. 1987. *Rhizobium meliloti* swims by unidirectional, intermittent rotation of right-handed flagellar helices. *J. Bacteriol.* **169:** 3146–3150.

Gouffi, K., V. Pichereau, J. P. Rolland, D. Thomas, T. Bernard, and C. Blanco. 1998. Sucrose is a nonaccumulated osmoprotectant in *Sinorhizobium meliloti*. *J. Bacteriol.* **180:**5044–5051.

Gresshoff, P. M., and B. G. Rolfe. 1978. Viability of *Rhizobium* bacteroids isolated from soybean nodule protoplasts. *Planta* **142:**329–333.

Gresshoff, P. M., M. L. Skotnicki, J. F. Eadie, and B. G. Rolfe. 1977. Viability of *Rhizobium trifolii* bacteroids from clover root nodules. *Plant Sci. Lett.* **10:**299–304.

Heithoff, D. M., C. P. Conner, and M. J. Mahan. 1997. Dissecting the biology of a pathogen during infection. *Trends Microbiol.* **5:**509–513.

Hirsch, A. M., and C. A. Smith. 1987. Effects of *Rhizobium meliloti nif* and *fix* mutants on alfalfa root nodule development. *J. Bacteriol.* **169:**1137–1146.

Hirsch, A. M., K. J. Wilson, J. D. Jones, M. Bang, V. V. Walker, and F. M. Ausubel. 1984. *Rhizobium meliloti* nodulation genes allow *Agrobacterium tumefaciens* and *Escherichia coli* to form pseudonodules on alfalfa. *J. Bacteriol.* **158:**1133–1143.

Hochman, A. 1997. Programmed cell death in prokaryotes. *Crit. Rev. Microbiol.* **23:**207–214.

Ichige, A., and G. C. Walker. 1997. Genetic analysis of the *Rhizobium meliloti bacA* gene: functional interchangeability with the *Escherichia coli sbmA* gene and phenotypes of mutants. *J. Bacteriol.* **179:**209–216.

Jensen, E. O., K. Paludan, J. J. Hyldig-Nielsen, P. Jorgensen, and K. A. Marcker. 1981. The structure of a chromosomal leghaemoglobin gene from soybean. *Nature* **291:**677–679.

Jordan, D. C., and W. H. Coulter. 1965. On the cytology and synthetic capacities of natural and artificially produced bacteroids of *Rhizobium leguminosarum*. *Can. J. Microbiol.* **11:**709–720.

Kaiser, B. N., P. M. Finnegan, S. D. Tyerman, L. F. Whitehead, F. J. Bergersen, D. A. Day, and M. K. Udvardi. 1998. Characterization of an ammonium transport protein from the peribacteroid membrane of soybean nodules. *Science* **281:**1202–1206.

Kamst, E., H. P. Spaink, and D. Kafetzopoulos. 1998. Biosynthesis and secretion of rhizobial lipochitin-oligosaccharide signal molecules. *Subcell. Biochem.* **29:**29–71.

Kaneshiro, T., F. L. Baker, and D. E. Johnson. 1983. Pleomorphism and acetylene-reducing activity of free-living rhizobia. *J. Bacteriol.* **153:**1045–1050.

Kannenberg, E. L., and N. J. Brewin. 1989. Expression of a cell surface antigen from *Rhizobium leguminosarum* 3841 is regulated by oxygen and pH. *J. Bacteriol.* **171:**4543–4548.

Kannenberg, E. L., and N. J. Brewin. 1994. Host-plant invasion by *Rhizobium*: the role of cell-surface components. *Trends Microbiol.* **2:**277–283.

Kannenberg, E. L., S. Perotto, V. Bianciotto, E. A. Rathbun, and N. J. Brewin. 1994. Lipopolysaccharide epitope expression of *Rhizobium* bacteroids as revealed by in situ immunolabelling of pea root nodule sections. *J. Bacteriol.* **176:**2021–2032.

Kapranov, P., T. J. Jensen, C. Poulsen, F. J. de Bruijn, and K. Szczyglowski. 1999. A protein phosphatase 2C gene, LjNPP2C1, from *Lotus japonicus* induced during root nodule development. *Proc. Natl. Acad. Sci. USA* **96:**1738-1743.

Kidby, D. K., and D. J. Goodchild. 1966. Host influence on the ultrastructure of root nodules of *Lupinus luteus* and *Ornithopus sativus*. *J. Gen. Microbiol.* **45:**147–152.

Kijne, J. W. 1992. The *Rhizobium* infection process, p. 349–398. *In* G. Stacey, R. H. Burris, and H. J. Evans (ed.), *Biological Nitrogen Fixation*. Chapman and Hall, New York, N.Y.

Kijne, J. W., M. A. Bauchrowitz, and C. L. Diaz. 1997. Root lectins and rhizobia. *Plant Physiol.* **115:**869–873.

Klein, S., A. M. Hirsch, C. A. Smith, and E. R. Signer. 1988. Interaction of *nod* and *exo Rhizobium meliloti* in alfalfa nodulation. *Mol. Plant-Microbe Interact.* **1:**94–100.

Kurz, W. G. W., and T. A. La Rue. 1975. Nitrogenase activity in rhizobia in the absence of plant host. *Nature* **256:**407–408.

Latch, J. N., and W. Margolin. 1997. Generation of buds, swellings, and branches instead of filaments after blocking the cell cycle of *Rhizobium meliloti*. *J. Bacteriol.* **179:**2373–2381.

Latchford, J. W., D. Borthakur, and A. W. Johnston. 1991. The products of *Rhizobium* genes, *psi* and *pss*, which affect exopolysaccharide production, are associated with the bacterial cell surface. *Mol. Microbiol.* **5:**2107–2114.

Legocki, R. P., and D. P. Verma. 1980. Identification of "nodule-specific" host proteins (nodulins) involved in the development of rhizobium-legume symbiosis. *Cell* **20:**153–163.

Leigh, J. A., and D. L. Coplin. 1992. Exopolysaccharides in plant-bacterial interactions. *Annu. Rev. Microbiol.* **46:**307–346.

Leigh, J. A., and G. C. Walker. 1994. Exopolysaccharides of *Rhizobium*: synthesis, regulation and symbiotic function. *Trends Genet.* **10:**63–67.

Leigh, J. A., E. R. Signer, and G. C. Walker. 1985. Exopolysaccharide-deficient mutants of *Rhizobium meliloti* that form ineffective nodules. *Proc. Natl. Acad. Sci. USA* **82:**6231–6235.

Loh, J., M. Garcia, and G. Stacey. 1997. NodV and NodW, a second flavonoid recognition system regulating *nod* gene expression in *Bradyrhizobium japonicum*. *J. Bacteriol.* **179:**3013–3020.

Loh, J., M. G. Stacey, M. J. Sadowsky, and G. Stacey. 1999. The *Bradyrhizobium japonicum nolA* gene encodes three functionally distinct proteins. *J. Bacteriol.* **181:**1544–1554.

Loh, J. T., S. C. Ho, A. W. de Feijter, J. L. Wang, and M. Schindler. 1993. Carbohydrate binding activities of *Bradyrhizobium japonicum*: unipolar localization of the lectin BJ38 on the bacterial cell surface. *Proc. Natl. Acad. Sci. USA* **90:**3033–3037.

Long, S., S. McCune, and G. C. Walker. 1988. Symbiotic loci of *Rhizobium meliloti* identified by random Tn*phoA* mutagenesis. *J. Bacteriol.* **170:**4257–4265.

Long, S. R. 1996. *Rhizobium* symbiosis: Nod factors in perspective. *Plant Cell* **8:**1885–1898.

Long, S. R., and B. J. Staskawicz. 1993. Prokaryotic plant parasites. *Cell* **73:**921–935.

Lucas, M. M., J. L. Peart, N. J. Brewin, and E. L. Kannenberg. 1996. Isolation of monoclonal antibodies reacting with the core component of lipopolysaccharide from *Rhizobium leguminosarum* strain 3841 and mutant derivatives. *J. Bacteriol.* **178:**2727–2733.

Ludwig, R. A., and E. R. Signer. 1977. Glutamine synthetase and control of nitrogen fixation in *Rhizobium*. *Nature* **267:**245–248.

MacKenzie, C. R., W. J. Vail, and D. C. Jordan. 1973. Ultrastructure of free-living and nitrogen-fixing forms of *Rhizobium meliloti* as revealed by freeze-etching. *J. Bacteriol.* **113:**387–393.

Maddock, J. R., M. R. Alley, and L. Shapiro. 1993. Polarized cells, polar actions. *J. Bacteriol.* **175:**7125–7129.

Mahan, M. J., J. M. Slauch, and J. J. Mekalanos. 1993. Selection of bacterial virulence genes that are specifically induced in host tissues. *Science* **259:**686–688.

Manoil, C., and J. Beckwith. 1985. Tn*phoA*: a transposon probe for protein export. *Proc. Natl. Acad. Sci. USA* **82:**8129–8133.

Margolin, W., and S. R. Long. 1994. *Rhizobium meliloti* contains a novel second homolog of the cell division gene *ftsZ*. *J. Bacteriol.* **176:**2033–2043.

Mateos, P. F., J. I. Jimenez-Zurdo, J. Chen, A. S. Squartini, S. K. Haack, E. Martinez-Molina, D. H. Hubbell, and F. B. Dazzo. 1992. Cell-associated pectinolytic and cellulolytic enzymes in *Rhizobium leguminosarum* biovar trifolii. *Appl. Environ. Microbiol.* **58:**1816–1822.

McComb, J. A., J. Elliott, and M. J. Dilworth. 1975. Acetylene reduction by *Rhizobium* in pure culture. *Nature* **256:**409–410.

McRae, D. G., R. W. Miller, and W. B. Berndt. 1989. Viability of alfalfa nodule bacteroids isolated by density gradient centrifugation. *Symbiosis* **7:**67–80.

Mergaert, P., M. Van Montagu, and M. Holsters. 1997. Molecular mechanisms of Nod factor diversity. *Mol. Microbiol.* **25:**811–817.

Miao, G. H., Z. Hong, and D. P. Verma. 1992. Topology and phosphorylation of soybean nodulin-26, an intrinsic protein of the peribacteroid membrane. *J. Cell Biol.* **118:**481–490.

Michiels, J., M. Moris, B. Dombrecht, C. Verreth, and J. Vanderleyden. 1998. Differential regulation of *Rhizobium etli rpoN2* gene expression during symbiosis and free-living growth. *J. Bacteriol.* **180:**3620–3628.

Milner, J. L., R. S. Araujo, and J. Handelsman. 1992. Molecular and symbiotic characterization of exopolysaccharide-deficient mutants of *Rhizobium tropici* strain CIAT899. *Mol. Microbiol.* **6:**3137–3147.

Moreau, S., M. J. Davies, C. Mathieu, D. Herouart, and A. Puppo. 1996. Leghemoglobin-derived radicals. Evidence for multiple protein-derived radicals and the initiation of peribacteroid membrane damage. *J. Biol. Chem.* **271:**32557–32562.

Muñoz, J. A., C. Coronado, J. Perez-Hormaeche, A. Kondorosi, P. Ratet, and A. J. Palomares. 1998. MsPG3, a medicago sativa polygalacturonase gene expressed during the alfalfa-*Rhizobium meliloti* interaction. *Proc. Natl. Acad. Sci. USA* **95:**9687–9692.

Murphy, P. J., N. Heycke, S. P. Trenz, P. Ratet, F. J. de Bruijn, and J. Schell. 1988. Synthesis of an opine-like compound, a rhizopine, in alfalfa nodules is symbiotically regulated. *Proc. Natl. Acad. Sci. USA* **85:**9133–9137.

Mylona, P., K. Pawlowski, and T. Bisseling. 1995. Symbiotic nitrogen fixation. *Plant Cell* **7:**869–885.

Nellen-Anthamatten, D., P. Rossi, O. Preisig, I. Kullik, M. Babst, H. M. Fischer, and H. Hennecke. 1998. *Bradyrhizobium japonicum* FixK2, a crucial distributor in the FixLJ-dependent regulatory cascade for control of genes inducible by low oxygen levels. *J. Bacteriol.* **180:**5251–5255.

Newcomb, W. 1981. Nodule morphogenesis and differentiation. *Int. Rev. Cytol.* **13**(Suppl.):247–296.

Niehaus, K., and A. Becker. 1998. The role of microbial surface polysaccharides in the *Rhizobium*-legume interaction. *Subcell. Biochem.* **29:**73–116.

Niehaus, K., A. Lagares, and A. Pühler. 1998. A *Sinorhizobium meliloti* lipopolysaccharide mutant induces effective nodules on the host plant *Medicago sativa* (alfalfa) but fails to establish a symbiosis with *Medicago truncatula*. *Mol. Plant-Microbe Interact.* **11:**906–914.

Noel, K. D., G. Stacey, S. R. Tandon, L. E. Silver, and W. J. Brill. 1982. *Rhizobium japonicum* mutants defective in symbiotic nitrogen fixation. *J. Bacteriol.* **152:**485–494.

O'Brian, M. R. 1996. Heme synthesis in the rhizobium-legume symbiosis: a palette for bacterial and eukaryotic pigments. *J. Bacteriol.* **178:**2471–2478.

O'Brian, M. R., P. M. Kirshbom, and R. J. Maier. 1987. Bacterial heme synthesis is required for expression of the leghemoglobin holoprotein but not the apoprotein in soybean root nodules. *Proc. Natl. Acad. Sci. USA* **84:**8390–8393.

Ogawa, J., and S. R. Long. 1995. The *Rhizobium meliloti groELc* locus is required for regulation of early *nod* genes by the transcription activator NodD. *Genes Dev.* **9:**714–729.

Oke, V., and S. R. Long. 1999. Bacterial genes induced within the nodule during the *Rhizobium*-legume symbiosis. *Mol. Microbiol.* **32:**837–849.

Paau, A. S., J. Oro, and J. R. Cowles. 1979. DNA content of free living rhizobia and bacteroids of various *Rhizobium*-legume associations. *Plant Physiol.* **63:**402–405.

Paau, A. S., C. B. Bloch, and W. J. Brill. 1980. Developmental fate of *Rhizobium meliloti* bacteroids in alfalfa nodules. *J. Bacteriol.* **143:**1480–1490.

Pagan, J. D., J. J. Child, W. R. Scowcroft, and A. H. Gibson. 1975. Nitrogen fixation by *Rhizobium* cultured on a defined medium. *Nature* **256:**406–407.

Pankhurst, C. E., and A. S. Craig. 1978. Effect of oxygen concentration, temperature and combined nitrogen on the morphology and nitrogenase activity of *Rhizobium* sp. strain 32H1 in agar culture. *J. Gen. Microbiol.* **106:**207–219.

Petrovics, G., P. Putnoky, B. Reuhs, J. Kim, T. A. Thorp, K. D. Noel, R. W. Carlson, and A. Kondorosi. 1993. The presence of a novel type of surface polysaccharide in *Rhizobium meliloti* requires a new fatty acid synthase-like gene cluster involved in symbiotic nodule development. *Mol. Microbiol.* **8:**1083–1094.

Pingret, J. L., E. P. Journet, and D. G. Barker. 1998. *Rhizobium* Nod factor signaling. Evidence for a G protein-mediated transduction mechanism. *Plant Cell* **10:**659–672.

Preisig, O., D. Anthamatten, and H. Hennecke. 1993. Genes for a microaerobically induced oxidase complex in *Bradyrhizobium japonicum* are essential for a nitrogen-fixing endosymbiosis. *Proc. Natl. Acad. Sci. USA* **90:**3309–3313.

Priefer, U. B. 1989. Genes involved in lipopolysaccharide production and symbiosis are clustered on the chromosome of *Rhizobium leguminosarum* biovar viciae VF39. *J. Bacteriol.* **171:**6161–6168.

Putnoky, P., E. Grosskopf, D. T. Ha, G. B. Kiss, and A. Kondorosi. 1988. *Rhizobium fix* genes mediate at least two communication steps in symbiotic nodule development. *J. Cell Biol.* **106:**597–607.

Putnoky, P., G. Petrovics, A. Kereszt, E. Grosskopf, D. T. Ha, Z. Banfalvi, and A. Kondorosi. 1990. *Rhizobium meliloti* lipopolysaccharide and exopolysaccharide can have the same function in the plant-bacterium interaction. *J. Bacteriol.* **172:**5450–5458.

Putnoky, P., A. Kereszt, T. Nakamura, G. Endre, E. Grosskopf, P. Kiss, and A. Kondorosi. 1998. The *pha* gene cluster of *Rhizobium meliloti* involved in pH adaptation and symbiosis encodes a novel type of $K+$ efflux system. *Mol. Microbiol.* **28:**1091–1101.

Rana, D., and H. B. Krishnan. 1995. A new root-nodulating symbiont of the tropical legume Sesbania, *Rhizobium* sp. SIN-1, is closely related to *R. galegae*, a species that nodulates temperate legumes. *FEMS Microbiol. Lett.* **134:**19–25.

Reibach, P. H., P. L. Mask, and J. G. Streeter. 1981. A rapid one-step method for the isolation of bacteroids from root nodules of soybean plants, utilizing self-generating Percoll gradients. *Can. J. Microbiol.* **27:**491–495.

Relic, B., X. Perret, M. T. Estrada-Garcia, J. Kopcinska, W. Golinowski, H. B. Krishnan, S. G. Pueppke, and W. J. Broughton. 1994. Nod factors of *Rhizobium* are a key to the legume door. *Mol. Microbiol.* **13:**171–178.

Reuber, T. L., S. Long, and G. C. Walker. 1991. Regulation of *Rhizobium meliloti exo* genes in free-living cells and in planta examined by using Tn*phoA* fusions. *J. Bacteriol.* **173:**426–434.

Reuhs, B. L., R. W. Carlson, and J. S. Kim. 1993. *Rhizobium fredii* and *Rhizobium meliloti* produce 3-deoxy-D-manno-2-octulosonic acid-containing polysaccharides that are structurally analogous to group II K antigens (capsular polysaccharides) found in *Escherichia coli. J. Bacteriol.* **175:**3570–3580.

Reuhs, B. L., J. S. Kim, A. Badgett, and R. W. Carlson. 1994. Production of cell-associated polysaccharides of *Rhizobium fredii* USDA205 is modulated by apigenin and host root extract. *Mol. Plant-Microbe Interact.* **7:**240–247.

Reuhs, B. L., M. N. Williams, J. S. Kim, R. W. Carlson, and F. Cote. 1995. Suppression of the Fix⁻ phenotype of *Rhizobium meliloti exoB* mutants by *lpsZ* is correlated to a modified expression of the K polysaccharide. *J. Bacteriol.* **177:**4289–4296.

Reyes, V. G., and E. L. Schmidt. 1979. Population densities of *Rhizobium japonicum* strain 123 estimated directly in soil and rhizospheres. *Appl. Environ. Microbiol.* **37:**854–858.

Robertson, J. G., and P. Lyttleton. 1984. Division of peribacteroid membranes in root nodules of white clover. *J. Cell Sci.* **69:**147–157.

Robertson, J. G., P. Lyttleton, S. Bullivant, and G. F. Grayston. 1978a. Membranes in lupin root nodules. I. The role of Golgi bodies in the biogenesis of infection threads and peribacteroid membranes. *J. Cell Sci.* **30:**129–149.

Robertson, J. G., M. P. Warburton, P. Lyttleton, A. M. Fordyce, and S. Bullivant. 1978b. Membranes in lupin root nodules. II. Preparation and properties of peribacteroid membranes and bacteroid envelope inner membranes from developing lupin nodules. *J. Cell Sci.* **30:**151–174.

Robinson, J. B., and W. D. Bauer. 1993. Relationships between C4 dicarboxylic acid transport and chemotaxis in *Rhizobium meliloti*. *J. Bacteriol.* **175:** 2284–2291.

Roth, L. E., and G. Stacey. 1991. *Rhizobium*-legume symbiosis, p. 255–302. *In* M. Dworkin (ed.), *Microbial Cell-Cell Interactions.* American Society for Microbiology, Washington, D.C.

Sangwan, I., and M. R. O'Brian. 1991. Evidence for an inter-organismic heme biosynthetic pathway in symbiotic soybean root nodules. *Science* **251:** 1220–1222.

Santana, M. A., K. Pihakaski-Maunsbach, N. Sandal, K. A. Marcker, and A. G. Smith. 1998. Evidence that the plant host synthesizes the heme moiety of leghemoglobin in root nodules. *Plant Physiol.* **116:**1259–1269.

Schell, M. A. 1993. Molecular biology of the LysR family of transcriptional regulators. *Annu. Rev. Microbiol.* **47:**597–626.

Schlaman, H. R., B. Horvath, E. Vijgenboom, R. J. Okker, and B. J. Lugtenberg. 1991. Suppression of nodulation gene expression in bacteroids of *Rhizobium leguminosarum* biovar viciae. *J. Bacteriol.* **173:**4277–4287.

Schlaman, H. R., B. J. Lugtenberg, and R. J. Okker. 1992. The NodD protein does not bind to the promoters of inducible nodulation genes in extracts of bacteroids of *Rhizobium leguminosarum* biovar viciae. *J. Bacteriol.* **174:**6109–6116.

Schmidt, E. L. 1979. Initiation of plant root-microbe interactions. *Annu. Rev. Microbiol.* **33:**355–376.

Schultze, M., and A. Kondorosi. 1996. The role of lipochitooligosaccharides in root nodule organogenesis and plant cell growth. *Curr. Opin. Genet. Dev.* **6:**631–638. (Erratum, **6:**773.)

Schultze, M., and A. Kondorosi. 1998. Regulation of symbiotic root nodule development. *Annu. Rev. Genet.* **32:**33–57.

Schultze, M., E. Kondorosi, P. Ratet, M. Buiré, and A. Kondorosi. 1994. Cell and molecular biology of *Rhizobium*-plant interactions. *Int. Rev. Cytol.* **156:**1–75.

Sedloi-Lumbroso, R., L. Kleiman, and H. M. Schulman. 1978. Biochemical evidence that leghaemoglobin genes are present in the soybean but not the *Rhizobium* genome. *Nature* **273:**558–560.

Sharma, S. B., and E. R. Signer. 1990. Temporal and spatial regulation of the symbiotic genes of *Rhizobium meliloti* in planta revealed by transposon Tn5-*gusA*. *Genes Dev.* **4:**344–356.

Sindhu, S. S., N. J. Brewin, and E. L. Kannenberg. 1990. Immunochemical analysis of lipopolysaccharides from free-living and endosymbiotic forms of *Rhizobium leguminosarum*. *J. Bacteriol.* **172:** 1804–1813.

Slauch, J. M., M. J. Mahan, and J. J. Mekalanos. 1994. In vivo expression technology for selection of bacterial genes specifically induced in host tissues. *Methods Enzymol.* **235:**481–492.

Smit, G., J. W. Kijne, and B. J. Lugtenberg. 1987. Involvement of both cellulose fibrils and a Ca_2^+-dependent adhesin in the attachment of *Rhizobium leguminosarum* to pea root hair tips. *J. Bacteriol.* **169:** 4294–4301.

Smit, G., S. Swart, B. Lugtenberg, and J. Kijne. 1992. Molecular mechanisms of attachment of *Rhizobium* bacteria to plant roots. *Mol. Microbiol.* **6:** 2897–2903.

Soupéne, E., M. Foussard, P. Boistard, G. Truchet, and J. Batut. 1995. Oxygen as a key developmental regulator of *Rhizobium meliloti* N_2-fixation gene expression within the alfalfa root nodule. *Proc. Natl. Acad. Sci. USA* **92:**3759–3763.

Soupéne, E., L. He, D. Yan, and S. Kustu. 1998. Ammonia acquisition in enteric bacteria: physiological role of the ammonium/methylammonium transport B (AmtB) protein. *Proc. Natl. Acad. Sci. USA* **95:**7030–7034.

Spaink, H. 1995. The molecular basis of infection and nodulation by Rhizobia: the ins and outs of sympathogenesis. *Annu. Rev. Phytopathol.* **33:** 345–368.

Spaink, H. P., C. A. Wijffelman, W. Pees, R. J. H. Okker, and B. J. J. Lugtenberg. 1987. *Rhizobium* nodulation gene *nodD* as a determinant of host specificity. *Nature* **328:**337–340.

Sullivan, J. T., and C. W. Ronson. 1998. Evolution of rhizobia by acquisition of a 500-kb symbiosis island that integrates into a phe-tRNA gene. *Proc. Natl. Acad. Sci USA* **95:**5145–5149.

Sulton, W. D. 1974. Some features of the DNA of *Rhizobium* bacteroids and bacteria. *Biochim. Biophys. Acta* **366:**1–10.

Sutton, W. D. 1981. The *Rhizobium* bacteroid state. *Int. Rev. Cytol.* **13** (Suppl.):149–177.

Sutton, W. D., and A. D. Paterson. 1979. The detergent sensitivity of *Rhizobium* bacteroids and bacteria. *Plant. Sci. Lett.* **16:**377–385.

Sutton, W. D., N. M. Jepsen, and B. D. Shaw. 1977. Changes in the number, viability, and amino-acid-incorporating activity of *Rhizobium* bacteroids during lupin nodule development. *Plant Physiol.* **59:**741–744.

Szczyglowski, K., P. Kapranov, D. Hamburger,

and F. J. de Bruijn. 1998. The *Lotus japonicus* LjNOD70 nodulin gene encodes a protein with similarities to transporters. *Plant Mol. Biol.* **37:** 651–661.

Talibart, R., M. Jebbar, G. Gouesbet, S. Himdi-Kabbab, H. Wroblewski, C. Blanco, and T. Bernard. 1994. Osmoadaptation in rhizobia: ectoine-induced salt tolerance. *J. Bacteriol.* **176:** 5210–5217.

Tate, R., A. Riccio, M. Merrick, and E. J. Patriarca. 1998. The *Rhizobium etli amtB* gene coding for an NH$_4^+$ transporter is down-regulated early during bacteroid differentiation. *Mol. Plant-Microbe Interact.* **11:**188–198.

Timmers, A. C., M. C. Auriac, F. de Billy, and G. Truchet. 1998. Nod factor internalization and microtubular cytoskeleton changes occur concomitantly during nodule differentiation in alfalfa. *Development* **125:**339–349.

Truchet, G., P. Roche, P. Lerouge, J. Vasse, S. Camut, F. de Billy, J.-C. Promé, and J. Dénarié. 1991. Sulphated lipo-oligosaccharide signals of *Rhizobium meliloti* elicit root nodule organogenesis in alfalfa. *Nature* **351:**670–673.

Tsien, H. C., P. S. Cain, and E. L. Schmidt. 1977. Viability of *Rhizobium* bacteroids. *Appl. Environ. Microbiol.* **34:**854–856.

Turgeon, B. G., and W. D. Bauer. 1985. Ultrastructure of infection-thread development during infection of soybean by *Rhizobium japonicum*. *Planta* **163:**328–349.

Tyerman, S. D., L. F. Whitehead, and D. A. Day. 1995. A channel-like transporter for NH$_4^+$ on the symbiotic interface of N$_2$ fixing plants. *Nature* **378:** 629–632.

Urban, J. E., and F. B. Dazzo. 1982. Succinate-induced morphology of *Rhizobium trifolii* 0403 resembles that of bacteroids in clover nodules. *Appl. Environ. Microbiol.* **44:**219–226.

van Brussel, A. A., J. W. Costerton, and J. J. Child. 1979. Nitrogen fixation by *Rhizobium* sp. 32H1. A morphological and ultrastructural comparison of asymbiotic and symbiotic nitrogen-fixing forms. *Can. J. Microbiol.* **25:**352–361.

van Brussel, A. A. N., R. Bakhuizen, P. van Spronsen, H. P. Spaink, T. Tak, B. J. J. Lugtenberg, and J. Kijne. 1992. Induction of preinfection thread structures in the host plant by lipo-oligosaccharides of *Rhizobium*. *Science* **257:**70–72.

Vance, C. P. 1983. *Rhizobium* infection and nodulation: a beneficial plant disease? *Annu. Rev. Microbiol.* **37:**399–424.

van Heeswijk, W. C., S. Hoving, D. Molenaar, B. Stegeman, D. Kahn, and H. V. Westerhoff. 1996. An alternative PII protein in the regulation of glutamine synthetase in *Escherichia coli*. *Mol. Microbiol.* **21:**133–146.

van Rhijn, P., R. B. Goldberg, and A. M. Hirsch. 1998. *Lotus corniculatus* nodulation specificity is changed by the presence of a soybean lectin gene. *Plant Cell* **10:**1233–1250.

van Spronsen, P. C., R. Bakhuizen, A. A. van Brussel, and J. W. Kijne. 1994. Cell wall degradation during infection thread formation by the root nodule bacterium *Rhizobium leguminosarum* is a two-step process. *Eur. J. Cell Biol.* **64:**88–94.

Vasse, J., F. de Billy, S. Camut, and G. Truchet. 1990. Correlation between ultrastructural differentiation of bacteroids and nitrogen fixation in alfalfa nodules. *J. Bacteriol.* **172:**4295–4306.

Verde, F. 1998. On growth and form: control of cell morphogenesis in fission yeast. *Curr. Opin. Microbiol.* **1:**712–718.

Verma, D. P. 1998. Developmental and metabolic adaptations during symbiosis between legume hosts and rhizobia. *Subcell. Biochem.* **29:**1–28.

Verma, D. P., and Z. Hong. 1996. Biogenesis of the peribacteroid membrane in root nodules. *Trends Microbiol.* **4:**364–368.

Verma, D. P. S. 1992. Signals in root nodule organogenesis and endocytosis of *Rhizobium*. *Plant Cell* **4:** 373–382.

Verma, D. P. S., and S. Long. 1983. The molecular biology of *Rhizobium*-legume symbiosis. *Int. Rev. Cytol.* **14**(Suppl.):211–245.

Vesper, S. J., and W. D. Bauer. 1986. Role of pili in *Rhizobium japonicum* attachment to soybean roots. *Appl. Environ. Microbiol.* **52:**134–141.

Viprey, V., A. Del Greco, W. Golinowski, W. J. Broughton, and X. Perret. 1998. Symbiotic implications of type III protein secretion machinery in *Rhizobium*. *Mol. Microbiol.* **28:**1381–1389.

Wang, L.-X., Y. Wang, B. Pellock, and G. C. Walker. 1999. Structural characterization of the symbiotically important low-molecular-weight succinoglycan of *Sinorhizobium meliloti*. *J. Bacteriol.* **181,** in press.

Waters, J. K., B. L. Hughes II, L. C. Purcell, K. O. Gerhardt, T. P. Mawhinney, and D. W. Emerich. 1998. Alanine, not ammonia, is excreted from N$_2$-fixing soybean nodule bacteroids. *Proc. Natl. Acad. Sci. USA* **95:**12038–12042.

Werner, D., R. B. Mellor, M. G. Hahn, and H. Grisebach. 1985. Glyceollin I accumulation in an ineffective type of soybean nodule with an early loss of peribacteroid membrane. *Z. Naturforsch. Teil C* **40:**179–181.

Wheatcroft, R., D. G. McRae, and R. W. Miller. 1990. Changes in the *Rhizobium meliloti* genome and the ability to detect supercoiled plasmids during bacteroid development. *Mol. Plant-Microbe Interact.* **3:**9–17.

Wright, R., C. Stephens, and L. Shapiro. 1997. The CcrM DNA methyltransferase is widespread in the alpha subdivision of proteobacteria, and its essential functions are conserved in *Rhizobium meliloti* and *Caulobacter crescentus*. *J. Bacteriol.* **179:**5869–5877.

Yang, W. C., C. de Blank, I. Meskiene, H. Hirt, J. Bakker, A. van Kammen, H. Franssen, and T. Bisseling. 1994. *Rhizobium* Nod factors reactivate the cell cycle during infection and nodule primordium formation, but the cycle is only completed in primordium formation. *Plant Cell* **6:**1415–1426.

Yao, P. Y., and J. M. Vincent. 1969. Host specificity in the root hair "curling factor" of *Rhizobium* sp. *Aust. J. Biol. Sci.* **22:**413–422.

Yarosh, O. K., T. C. Charles, and T. M. Finan. 1989. Analysis of C4-dicarboxylate transport genes in *Rhizobium meliloti*. *Mol. Microbiol.* **3:**813–823.

Yoon, H. S., and J. W. Golden. 1998. Heterocyst pattern formation controlled by a diffusible peptide. *Science* **282:**935–938.

Yost, C. K., P. Rochepeau, and M. F. Hynes. 1998. *Rhizobium leguminosarum* contains a group of genes that appear to code for methyl-accepting chemotaxis proteins. *Microbiology* **144:**1945–1956.

Zhan, H. J., C. C. Lee, and J. A. Leigh. 1991. Induction of the second exopolysaccharide (EPSb) in *Rhizobium meliloti* SU47 by low phosphate concentrations. *J. Bacteriol.* **173:**7391–7394.

Zhao, J. C., Y. T. Tchan, and J. M. Vincent. 1985. Reproductive capacity of bacteroids in nodules of *Trifolium repens* L. and *Glycine max* (L.) Merr. *Planta* **163:**473–482.

Index